W0106768

Proteins
Structure and Function

Edited by
James J. L'Italien
Molecular Genetics, Inc,
Minnetonka, Minnesota

PLENUM PRESS • NEW YORK AND LONDON

Library of Congress Cataloging in Publication Data

Proteins: structure and function.

"Selected proceedings of the First Symposium of American Protein Chemists on Modern Methods in Protein Chemistry, held September 30–October 3, 1985, in San Diego, California"—T.p. verso.
    Bibliography: p.
    Includes index.
    1. Proteins—Analysis—Congresses. 2. Amino acid sequence—Congresses. I. L'Italien, James J. II. American Protein Chemistry Society on Modern Methods in Protein Chemistry. Symposium (1st: 1985: San Diego, Calif.)
QP551.P69779 1987   574.19′245   87-22011
ISBN-13: 978-1-4612-9001-8   e-ISBN-13: 978-1-4613-1787-6
DOI: 10.1007/ 978-1-4613-1787-6

Selected proceedings of the first symposium of American Protein Chemists on Modern Methods in Protein Chemistry, held September 30–October 3, 1985, in San Diego, California

© 1987 Plenum Press, New York
Softcover reprint of the hardcover 1st edition 1987
A Division of Plenum Publishing Corporation
233 Spring Street, New York, N.Y. 10013

This volume surveys the current status of many of the important methods and approaches which are central to the study of protein structure and function. Many of the articles in this volume are written to emphasize the general utility of the method or approach which is at its core, and to provide sufficient literature references to enable the reader to adapt the method or approach to other applications. It is hoped that this volume will provide a source from which newcomers as well as experienced scientists may become more familiar with recent developments and future trends in some of the important areas of protein research.

The articles which comprise this book are selected proceedings from the Symposium of American Protein Chemists, which was held in San Diego, California, September 30 to October 3, 1985. The goal of the organizers of this first symposium was to provide a forum for discussion and interaction among scientists whose interests span the broad spectrum of protein structure and function research. The concept and timing of the symposium was well received as evidenced by the approximately 500 delegates to the symposium. The inaugural meeting was marked by a strong scientific program with over 140 papers presented in either a lecture or poster format. A majority of the symposium attendees took part in an open forum on the relative merits and timeliness of forming a scholarly society of scientists interested in protein structure and function. The outcome of this forum was the formation of "The Protein Society."

From an organizational perspective, this book is divided into three major sections. The first of these sections, entitled "Methods of Polypeptide Purification and Characterization," addresses recent development dealing with microanalytical advances in polypeptide isolation and structural elucidation. Topics in this section include applications of microbore HPLC to polypeptide isolation; direct electrotransfer of polypeptide for microsequence analysis; micropreparative isolation of polypeptides by HPLC and immunoaffinity chromatography for microsequence analysis; and microanalytical polypeptide applications of FAB mass spectrometry. The second major section is entitled "Analysis of Protein Structure and Function." This section focuses upon recent developments, as well as new applications of established methods, in determining polypeptide structure/function relationships. Papers in this section highlight the use of molecular biological approaches for site directed mutagenesis and active site studies; biochemical approaches to active site localization; and chemical and immunological approaches to domain and topographical studies. This section also includes papers on the characterization of proteases and the identification of sites of covalent Post-Translational modification. The third section of the book is a summary of microsequencing workshop which was held in conjunction with the symposium. This unprecedented event permitted all conference attendees the opportunity to request an aliquot of an unknown polypeptide which they could characterize by their method(s) of

choice. This "survey" provides and interesting perspective on which
techniques scientists feel are most reliable for characterization of an
unknown "micro" sample and furthermore, permits a comparison of results
obtained by various methods on the same unknown sample.

James J. L'Italien

CONTENTS

A. METHODS OF POLYPEPTIDE PURIFICATION AND CHARACTERIZATION

I. Chromatographic Methods of Polypeptide Purification

II. Electrophortic Methods of Polypeptide Purification

III. Chemical Modification of Polypeptides

IV. Amino Acid Analysis of Polypeptides

V. Mass Spectrometry of Polypeptides

VI.   Microsequence Analysis of Polypeptides by Edman Degraduation

VII.   Analysis of PTH Amino Acids

VIII. Computer Analysis of Protein Sequence Data

IX. Miscellaneous Methods of Polypeptide Characterization

B. ANALYSIS OF POLYPEPTIDE STRUCTURE AND FUNCTION

X. Site-Directected Mutagenesis

XI. Active Site Studies

XII. Domain and Topographical Studies

XIII. Characterization of Proteases

XIV. Identification of Sites of Post-Translateral Modification

XV. New Sequences

A. METHODS OF POLYPEPTIDE PURIFICATION AND CHARACTERIZATION

I. Chromatographic Methods of Polypeptide Purification

USE OF SHORT MICROBORE HPLC COLUMNS FOR THE CONCENTRATION, SEPARATION AND
RECOVERY OF SUBNANOMOLE AMOUNTS OF PROTEIN AND POLYPEPTIDES FOR
MICROSEQUENCE ANALYSIS

Richard J. Simpson*, Boris Grego*, Michael R. Rubira*,
Lindsay G. Sparrow+ and Edouard C. Nice≠

*Joint Protein Structure Laboratory, Ludwig Institute/Walter
and Eliza Hall Institute of Medical Research, Parkville
3050, Australia and ≠Ludwig Institute for Cancer Research
(Melbourne Branch), Parkville 3050, Australia and +CSIRO
(Division of Protein Chemistry), Parkville, Victoria 3052,
Australia

Recent advances in protein sequencing technology which have led to
the development of the gas phase sequenator (1) now permit protein
sequence information to be obtained from as little as 10-50 picomoles of
material(2-4). Although a number of high resolution techniques permit the
isolation of subnanomole amounts of protein or polypeptide in a pure form
(e.g. HPLC(5,6), affinity chromatography(7,8) and polyacrylamide gel
electrophoresis(8-10)), there are serious problems in recovering the
sample in a form suitable for gas phase sequence analysis, i.e. in a small
volume and free of interfering compounds. The ability to purify and to
manipulate minute quantities of protein (e.g. concentrate, reduce and
alkylate, desalt, generate fragments, etc.) is crucial to the success in
obtaining accurate sequence at subnanomole levels.

Unfortunately, conventional concentration procedures (e.g.
lyophilization, evaporation or organic solvent precipitation) and buffer-
exchange methods (e.g. dialysis, gel-permeation chromatography) frequently
result in severe sample loss(4,11-13); this problem is particularly acute
when working at subnanomole levels.

Using conventional HPLC methods (4.6mm ID columns) or affinity
chromatography procedures, pure preparations of protein can be readily
obtained. However the samples are usually recovered in large volumes
(>1ml) which often contain interfering buffer salts. While electrophoretic elution methods have been successfully employed to recover
proteins from gels in small volumes suitable for structural analysis(9,10)
this procedure can be technically exacting and frequently results in the
electroeluate being contaminated with intolerably high levels of SDS and
gel-related artifacts which interfere with the sequence chemistry.
Although these impurities can be selectively removed by organic solvent
precipitation of the protein(7) this procedure is slow and can result in
poor recoveries.

In this paper we summarize the results of recent experiments using
short microbore columns for the manipulation of peptides and proteins
prior to gas phase amino acid sequence analysis (see Fig. 1).

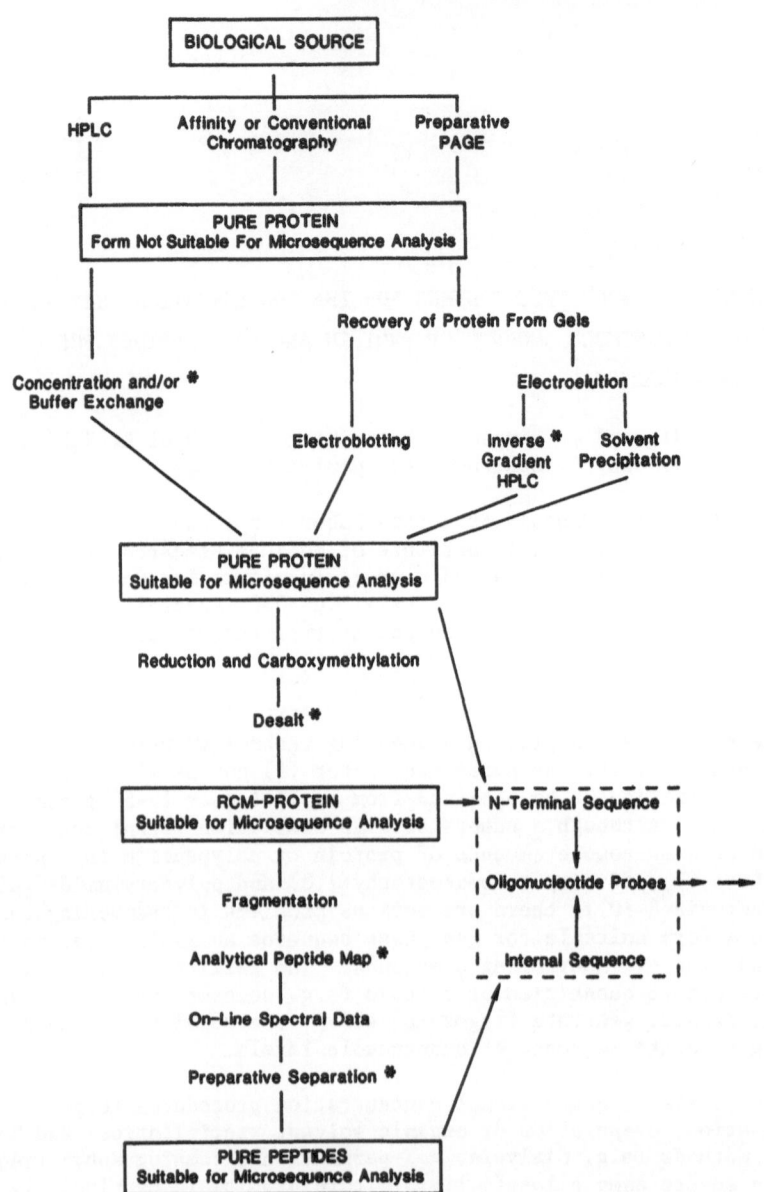

Figure 1: Strategies for the micromanipulation of subnanomole amounts of proteins and polypeptides for microsequence analysis. Asterisk (*) denotes stages where microbore column HPLC has been employed.

INSTRUMENTATION

Chromatography was performed using commercially available HPLC instruments designed for low dispersion chromatography: (i) Perkin Elmer, model LC4 fitted with a variable wavelength spectrophotometer (model LC95) and a fluorimeter (model LS4), and (ii) Hewlett Packard, model 1090 fitted with a diode-array detector (model 1040A). Chromatographic and spectral data were processed as described previously(4). Both instruments were equipped with Rheodyne injection valves (model 7125) and 2ml loops which permitted fractions from conventional HPLC columns (volume *ca.* 1ml) to be

diluted (*ca.* 2-fold) and directly injected onto microbore columns. Microbore separations can also be achieved on other HPLC instruments by
introducing simple modifications designed to minimise problems associated
with large pre-column volumes (typically, 2-5ml)(14-16). Details of
specific columns and chromatographic conditions used are given in the
figure legends.

GENERAL CONSIDERATIONS OF MICROBORE CHROMATOGRAPHY

Column Dimensions

To maintain equivalent linear flow velocities microbore columns are
operated at low flow rates(17) (e.g. 50µl/min for a 1.0mm ID column
compared with 1ml/min for a 4.6mm ID column). Provided column efficiency
is maintained, peptides and proteins are recovered in reduced volumes at
increased concentrations leading to increased sensitivity of detection.
Thus, proteins can be detected at 1-20ng sensitivity using short microbore
columns(6,14). This sensitivity has facilitated the specific activity
determination of radiolabeled proteins(14,18). Furthermore, using
interactive supports (e.g. RP or IEX) short columns (<10cm) offer two
distinct advantages over longer columns (e.g., 25cm): (i) similar
resolution with less sample dilution; and (ii) reduced operating pressures
which enable large sample volumes to be loaded at high flow rates.

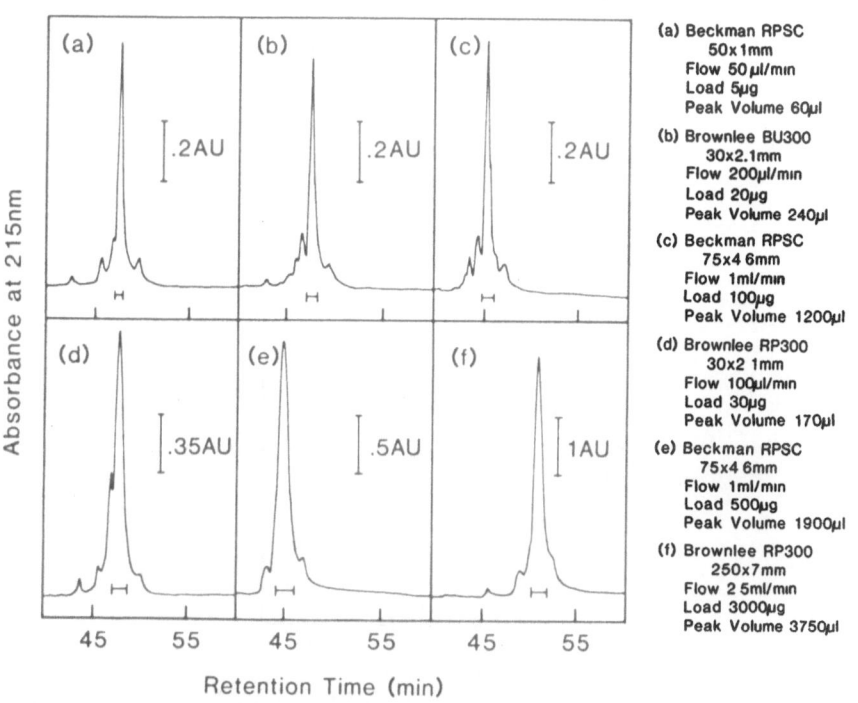

*Figure 2: Chromatographic behaviour of bacterially synthetized
mGM-CSF(19) on columns of varying internal diameter. Chromatographic
conditions: linear 60-min gradient from 0.15% (v/v) aqueous TFA to 60%
acetonitrile/40% water containing 0.1% (v/v) TFA. Column temperature,
45°C. Samples were loaded at 2ml/min (except for panel f); gradient and
operating flow rates were initiated 2 min later. (a,b and c) effect of
column internal diameter on protein resolution and detection. (d,e and f)
effect of mass load on protein resolution and peak recovery volume (⊢—⊣).*

Using similar 30nm pore size, short-alkyl chain column supports (e.g. Beckman RPSC, Brownlee BU-300, Brownlee RP-300) it can be seen (Fig. 2) that a reduction in column internal diameter results in little practical loss in column resolution as evidenced by comparable peak shapes (panels a,b and c). A comparison of column peak widths and flow rates (panels a,b and c, Fig. 2) indicates that the recovered peak volumes are directly proportional to the square of the column radius at equivalent linear flow velocity.

Sample Loading and Recovery

In a recent study(14) where we evaluated the protein capacity of a number of different silica supports packed into columns of varying dimensions, in excess of 4mg of protein could be loaded (by trace enrichment) onto the smallest column used (Hypersil ODS, 75x1.0mm ID) and recovered quantitatively (>90%) with an 80-fold concentration factor. However, to maintain optimal resolution, the protein load capacity is considerably lower (typically, 5-10µg for a 1.0mm ID column or 20-40µg for a 2.1mm ID column; see effect of sample load in Fig. 2 (a),(b),(c) analytical, (d),(e),(f) preparative).

Many of the proteins of current biological interest, e.g. growth factors, hormone receptors are conveniently isolated only in microgram quantities(7,8,18,20-27); such protein levels (20-60µg) can be easily handled on short microbore columns with high chromatographic efficiency(14) as well as recovery in reduced peak volumes with concomitant increased sensitivity of detection (Fig. 2). In addition, the excellent recoveries observed with these columns allow their sequential use in 'multi-dimensional' purification strategies(6).

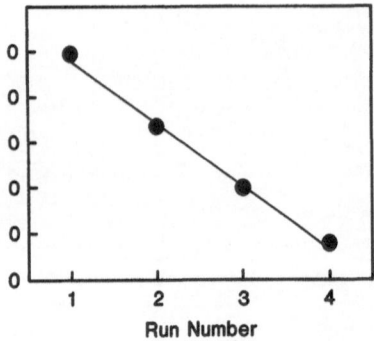

Figure 3: Repetitive yield of mGM-CSF on Brownlee RP-300 (30x2.1mm ID). Bacterially synthesized mGM-CSF (300ng) was chromatographed under identical chromatographic conditions to those given in the legend to Fig. 2(d). Peak fractions of mGM-CSF were recovered manually, diluted 1:1 with primary solvent (0.15% (v/v) TFA and reinjected. Recoveries of mGM-CSF were calculated from peak height measurements and are expressed as a percentage of the material recovered in the first run.

A typical total system recovery for repetitive reinjection of subnanomole levels of protein (300ng mGM-CSF, M_r 14,000) from a microbore column is shown in Fig. 3. The achieved recovery (>80%) includes all handling manipulations. It should be noted that the CSFs are notoriously difficult to micromanipulate at low protein concentrations(18,24,25).

Trace Enrichment

Under gradient elution conditions sample application is not restricted to the small (μl) volumes typically associated with microbore columns, since below critical secondary solvent composition proteins exhibit very large capacity factors (k') on reversed-phase, ion-exchange, or other interactive supports(5). Therefore, under such conditions, large volumes containing nanogram amounts of protein can be trace enriched onto the top of a microbore column without influencing the volumes of the recovered peaks. Samples can be recovered in volumes of 25-80μl(4,14) suitable for direct loading onto the glass fibre sample disk of the gas phase sequenator.

The isolation of low abundancy proteins from large biological samples usually requires a combination of low resolution purification methods to reduce the overall protein level followed by high resolution techniques such as those depicted in Fig. 1. The pure protein, however, is usually recovered in relatively large volumes (800μl) necessitating some form of concentration and buffer exchange to facilitate sequence analysis. Attempts to concentrate samples using conventional concentration methods (e.g. lyophilization) often result in unacceptably high losses(4,11-13). For many proteins this problem can be circumvented by trace-enrichment onto microbore RP-HPLC columns. An example is given in Fig. 4(a) where a preparation of M-CSF (a 70kDa glycoprotein) from a conventional HPLC column (1.6ml volume)(28) was concentrated 14-fold and exchanged into a volatile solvent suitable for N-terminal sequence analysis (Fig. 4(b)).

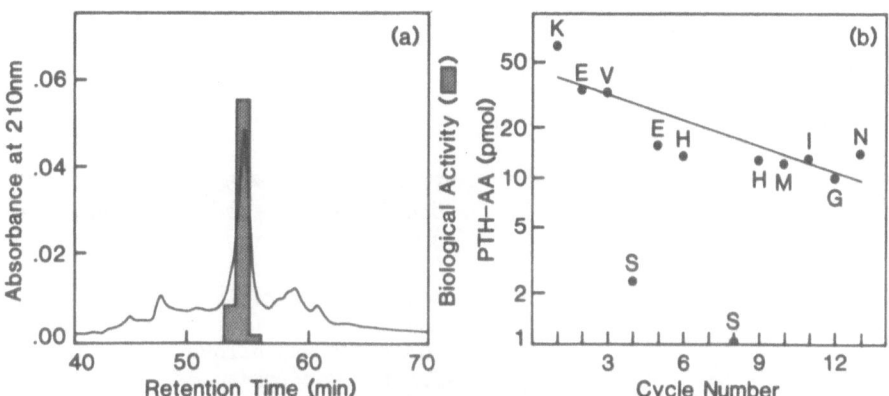

Figure 4(a): Microbore RP-HPLC concentration and buffer exchange of M-CSF for sequence analysis. For chromatographic and sample handling conditions see legend to Fig. 3. Figure 4(b): Amino terminal sequence analysis of M-CSF from Fig. 4(a). Aliquots (80%) from each cycle were analyzed by HPLC(4,29) and the yields were normalized to 100% injection. No Pth-amino acid was identified in cycle 7.

Not all proteins yield useful N-terminal sequence information. For instance, many proteins have blocked N-terminii so that direct sequence analysis gives no data; in many cases meaningful sequence information can be obtained from this refractory proteins following *in situ* CNBr cleavage(30). For large M_r proteins, e.g. murine transferrin receptor(7) and cell surface alloantigen PC-1(8) the sequence data obtained may be uninterpretable due to rapid increase in Pth amino acid background levels. Even if limited N-terminal sequence information is obtained, as was the case with G-CSF (data not given), this information may not be conducive for constructing high-stringency oligonucleotide probes.

One way to circumvent the problems mentioned above is to fragment the intact molecule into peptides, either enzymically (e.g. trypsin) or chemically (e.g. CNBr). Sequence information obtained from peptides allows: (i) the determination of protein structure by rapid DNA methodologies using synthetic oligonucleotide probes and (ii) confirmation of the correct reading frame of putative protein structures deduced from DNA sequence studies.

Many native proteins are resistant to proteolysis. This problem is often overcome by first reducing and alkylating the protein under denaturing conditions in SDS(8), 6M guanidine hydrochloride(31) or 8M urea(32,33). Since the conventional methods for recovery of reduced, carboxymethylated (RCM) proteins from reaction mixtures (e.g. dialysis or conventional gel-permeation chromatography) result in significant loss of sample when working with low amounts of protein, alternative procedures have been introduced, e.g. HPLC (reversed phase, size exclusion (SEC)) and organic solvent precipitation(8). The use of reagents containing radiolabel (e.g. ^{14}C iodoacetic acid) or specific chromophores (e.g. DABIAA(34)) allows the RCM-protein to be clearly identified. An example of desalting by microbore RP-HPLC is shown in Fig. 5. A comparison of the

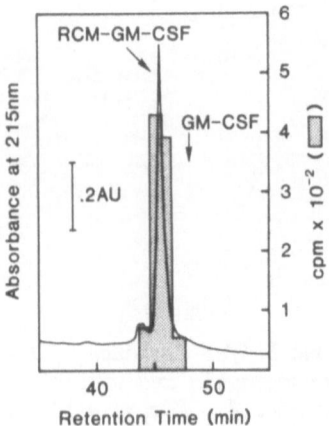

Figure 5: Microbore RP-HPLC desalting of reduced, S-[C^{14}]-carboxymethylated mGM-CSF. Chromatographic column and loading conditions were the same as in Fig. 2(d); 1-min fractions were collected and aliquots were taken for the monitoring of radioactivity.

retention time of native GM-CSF with that of RCM-GM-CSF chromatographed under identical conditions reveals only a slight difference (Fig. 5). It should be noted that with other proteins the change in their retention position on RP-HPLC upon reduction and S-carboxymethylation may be more pronounced(29). This alteration in retention behaviour is due to a combination of protein unfolding and chemical modification, and leads to an alteration in relative hydrophobicity. There is a possibility that the modified protein may become so hydrophobic that it is infinitely retained, thus RP-HPLC is not a panacea for desalting all RCM-proteins. Microbore chromatography with its high sensitivity of detection provides a valuable tool for examining, on an analytical scale (10-50ng protein) the behaviour of such modified proteins using a variety of chromatographic supports and operating conditions. SEC-HPLC can be used for desalting reaction mixtures (Fig. 6), however the peak recovery volumes (500μl) are not

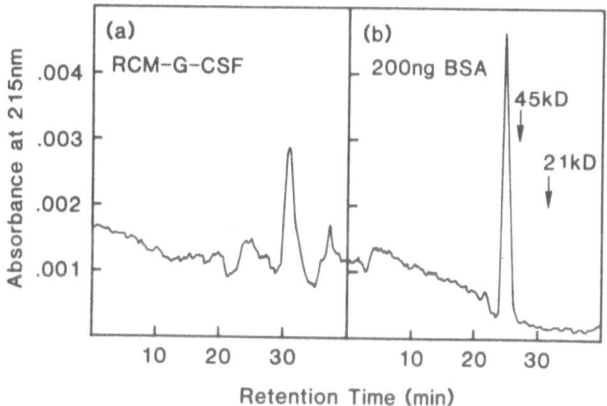

Figure 6: Desalting of RCM-G-CSF by SEC-HPLC. Chromatographic conditions: column, TSK 2000 SW (60x7.0mm ID); solvent, 100mM Na_2HPO_4 buffer, pH7.4, 5% isopropanol, 0.04% Tween 20; Flow rate, 0.5ml/min; column temperature, 25°C. Sample load, approximately 200ng RCM-G-CSF (panel a). A 200ng sample of BSA is shown for comparison (panel b).

ideally suited to subsequent microsequencing. For the purpose of desalting it can be calculated that short SEC-columns (e.g. 10cm) packed with smaller pore size particles (e.g. 6nm) would allow proteins to be recovered in smaller volumes in the column void, well separated from low M_r components (compare with PD-10 columns for conventional desalting). Such SEC-HPLC columns, unfortunately, are not yet commercially available.

Having obtained the protein in a form suitable for digestion one is faced with the dilemma of which fragmentation procedure to adopt. The high sensitivities associated with microbore columns allow them to be used to monitor and optimise peptide mapping strategies(6,22). Analytical amounts of material, similar to that shown in Fig. 7 (40pmol) can be used to monitor the extent of enzymic digestion and to optimize chromatographic separation conditions before the bulk of a sample digest is committed to preparative microbore chromatography. Crucial to the success in obtaining sequence information from peptides using the strategy outlined in Fig. 1 is the ability to purify subnanomole amounts of peptides in high yield in volumes suitable for the gas phase sequencer(1). We demonstrate in Fig. 8 a preparative separation (400pmol) of peptides derived from a tryptic digest of mGM-CSF. Since the purpose of this study was to determine the disulfide assignment of GM-CSF, in this instance the tryptic digestion was performed on the unmodified molecule. Preparative separation of these peptides was achieved using the same column and chromatographic conditions employed for the analytical tryptic peptide map of the RCM-protein shown in Fig. 7; similar peak volumes were obtained in both cases. The major tryptic peptides (T1-T4) were purified to homogeneity using a multi-dimensional HPLC strategy based on the sequential use of alternative mobile phase conditions (e.g. alternative pH or counter ion) on the same column packing(4).

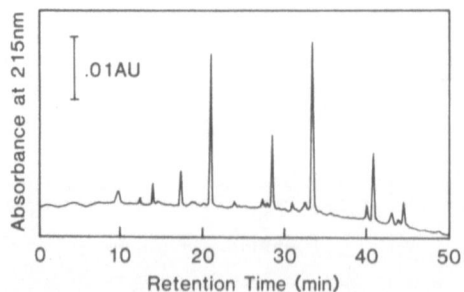

Figure 7: High sensitivity analytical separation of a tryptic digest of RCM-mGM-CSF (40pmol) on microbore RP-HPLC. Chromatographic conditions as for Fig. 2(d).

The diode-array detector is a useful adjunct to this peptide purification strategy. The ability to obtain "on-line" spectral information on submicrogram quantities of peptide(4,6,35), made possible by the increase in solute concentration associated with microbore methods, provides a means for identifying aromatic amino acid-containing peptides. Derivative ultraviolet absorbance spectra (240-320nm) of tryptic peptides T2 and T3 from Fig. 8 are shown in Fig. 9. An inspection of these spectra reveals that peptides T2 and T3 display characteristic second-order-derivative minima(4) for tryptophan (290±2nm) and tyrosine (280±2nm), respectively. The presence of a tryptophan residue in peptide T2 and a tyrosine residue in peptide T3 was confirmed by sequence analysis. From a knowledge of the primary structure of GM-CSF(36), peptide T2 could not be implicated in disulfide bond formation and was not studied further. Another valuable use for spectral analysis is that it provides a means for selecting tryptophan-containing peptides from complex tryptic maps as candidate peptides for obtaining sequence information for oligonucleotide probe construction; these peptides are particularly attractive since they are readily discernible(4) and tryptophan is represented by a unique codon.

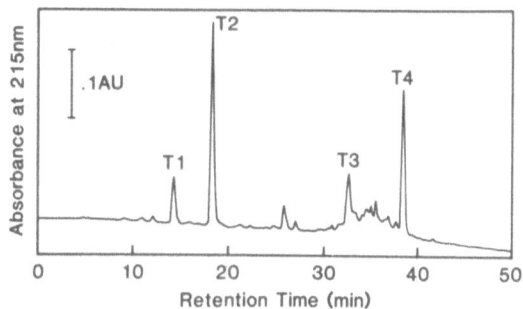

Figure 8: Preparative separation of a tryptic digest of bacterially expressed mGM-CSF (400pmol) on microbore RP-HPLC. Chromatographic conditions as for Fig. 2(d). Peptides T1-T4 were collected manually for further microbore purification by the method described elsewhere(4).

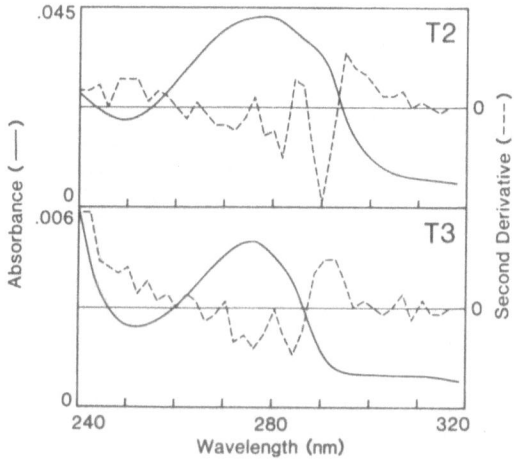

Figure 9: Derivative spectral analysis of tryptic peptides T2 and T3 of mGM-CSF from Fig. 8a obtained with an on-line diode-array detector. Zero order derivative (——), second order derivative (---).

The high resolving capability of SDS–polyacrylamide gels for micro-gram amounts of protein is well known. The most common procedure for recovering proteins from preparative SDS–polyacrylamide slab gels involves localization of the protein by treatment of the gel with 0.25M KCl(37) or staining with Coomassie blue and electroeluting the protein from the gel slice. Although this technique is generally applicable, the electroeluate frequently contains high levels of SDS and gel-artifacts which interfere with sequence analysis. One way of overcoming this problem is to recover the pure protein from the electroeluate by precipitation with 90% (v/v) methanol at –20°C, or alternatively, to directly electroblot the protein from the SDS–polyacrylamide gel onto chemically activated glass fibre sheets which can be applied to the gas phase sequenator without further manipulation(38).

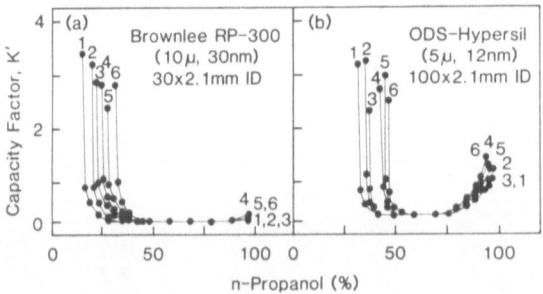

Figure 10: Dependence of protein capacity factor on n-propanol concentration. Data was obtained isocratically using n-propanol/0.1% (v/v) TFA mixtures at a flow rate of 100μl/min; column temperature 25°C. Protein standards (10μg) were: 1, ribonuclease; 2, cytochrome C; 3, α-lactalbumin; 4, serum albumin; 5, trypsin inhibitor; 6, ovalbumin..

We have taken the concept of recovering proteins from electroeluates by organic solvent precipitation and applied it to HPLC columns in the anticipation that, by controlling the sample within the confines of the chromatographic system, sample recovery would be enhanced. This novel HPLC technique relies on an unusual property of certain reversed phase packings which allows considerable retention of proteins by the support at high (80–90%) organic solvent concentrations (see Fig. 10).

As shown in Fig. 10, supports most amenable to this application are not those designed specifically for RP-HPLC of proteins (30nm pore size, short–alkyl chain supports, Fig. 10(a)) but are 6–12nm pore size, 3–5μm particle diameter octyl– or octadecyl bonded silicas (e.g. ODS-Hypersil, Fig. 10(b)). Retained proteins can be recovered by the application of a negative "inverse" organic solvent gradient (i.e. a gradient increasing in water concentration and decreasing in organic solvent concentration). The use of microbore columns results in decreased peak volumes for the same reasons mentioned previously. Optimal mobile phase conditions employ 90% (v/v) n–propanol/10% water as the primary solvent and 40% (v/v) n–propanol/60% water containing 0.1% (v/v) TFA as the secondary solvent combined with fast gradient rates (e.g. 5–10% per min). This procedure allows rapid recovery of protein in yields in excess of 80% (see Table 1).

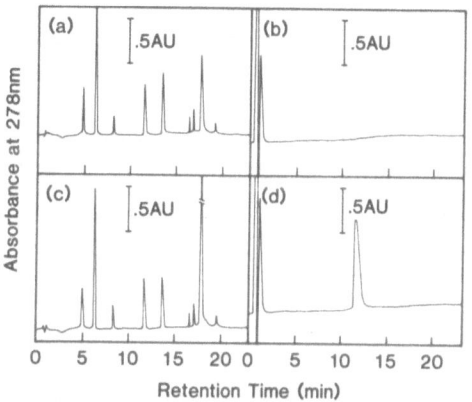

Figure 11: Chromatography of a 32kDa plant glycoprotein recovered from SDS-PAGE by electroelution, on microbore RP-HPLC using conventional and inverse organic solvent gradients. Column: ODS-Hypersil (100x2.1mm ID); flow rate, 50μl/min; column temperature, 25°C. Chromatographic conditions: (a) and (c), linear 20-min gradient between 0.15% (v/v) aqueous TFA and 50% n-propanol/50% water containing 0.1% (v) TFA; (b) and (d), linear inverse 20-min gradient between 90% n-propanol/10% water containing 0.1% (v/v) TFA and 40% n-propanol/60% water containing 0.1% TFA. Sample: (a) and (b), electroelution blank (120μl); (c) and (d); electroelution of a 32kDa glycoprotein.

Inverse-gradient chromatography offers two major advantages over the conventional forward-gradient chromatography for recovering protein from gel-electroeluates. Firstly, by inverse-gradient chromatography SDS Coomassie blue and gel-related artifacts are not retained (compare panels (a) and (b) in Fig. 11) thereby allowing the recovery of pure protein free from these contaminants (panel d, Fig. 11); in conventional forward-gradient chromatography proteins often co-chromatograph with gel-related artifacts (compare panels (a) and (c) in Fig. 11). Secondly, in contrast to inverse-gradient chromatography the retention of proteins on reversed-phase supports used in the conventional mode is highly dependent on the sample SDS concentration (Fig. 12) due to modification of the support by SDS into a dynamic ion-exchanger(39).

The general utility of inverse-gradient chromatography for recovering proteins from SDS-polyacrylamide gel for the purpose of obtaining N-terminal sequence information is demonstrated in Fig. 13 and Table 2.

Table 1. Recovery of ^{125}I-Labeled Protein from SDS-Polyacrylamide Gel Electroeluates Using Inverse-Gradient HPLC

Protein	cpm injected	cpm recovered (%)
Tomato glycoprotein (32kDa)	32,000	94
Tomato glycoprotein (37kDa)	35,000	92
Gastrin-binding protein (80kDa)	109,000	88
Insulin receptor (120kDa)	74,000	85

Table 2. Amino Terminal Sequence Analysis of Proteins Purified from SDS-Polyacrylamide Gel Electroeluates Using Inverse-Gradient HPLC

Protein	Pth-amino acid Yield in 1st cycle (pmol)	Amino amino Residues Identified
S1 allele associated glycoprotein	85	19/20
S3 allele associated glycoprotein	60	20/22
SF11 allele associated glycoprotein	80	28/30
Lyt-2/3 antigen	10	14/14

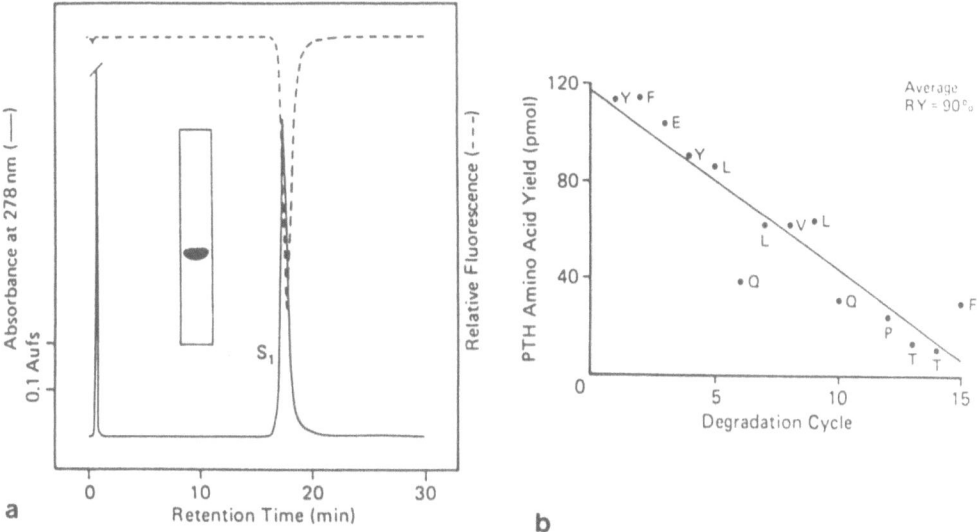

Figure 12: Effect of sample SDS concentration on protein retention using conventional and inverse gradients. Chromatographic conditions: (panels a,c,d,f,g and i), conventional gradients as described in Fig. 11(a); (panels b,e and h) inverse gradients as described in Fig. 11(b). Sample SDS concentration was 0.1% (panels a,b and c), 2.0% (panels d,e and f) and 5.0% (panels g,h and i). Sample, 5μg of serum albumin in 200μl.

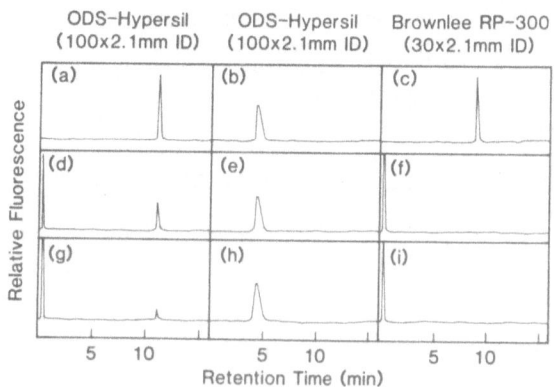

Figure 13: (a) Recovery of S_1 allelic associated glycoprotein from SDS-polyacrylamide gel (inset) electroeluate using inverse gradient HPLC. SDS-PAGE and electroelution conditions are described elsewhere(7). Inverse gradient HPLC conditions are identical to those described in legend to Fig. 11(d). (b) Amino terminal sequence analysis of S_1 allelic associated glycoproteins recovered by inverse-gradient HPLC (Fig. 13(a)). Aliquots (67%) were taken from each degradation cycle. Yields of Pth amino acid have been normalized to 100% injection. No Pth-amino acid was identified in cycle 11.

The recent development of specific column supports designed for the chromatography of proteins (e.g. ion-exchange, hydrophobic interaction, hydroxylapatite) has extended the range of proteins which can be successfully purified by HPLC. These supports are largely based on wide pore silica and non-silica based materials. The protein capacity and trace enrichment potential of these packings renders them suitable for use in microbore columns and should allow multi-dimensional protein purification protocols to be developed at submicrogram levels. We have demonstrated(6) the use of 30nm pore size anion-exchange packings (Brownlee AX-300) in 30x2.1mm ID columns for this purpose.

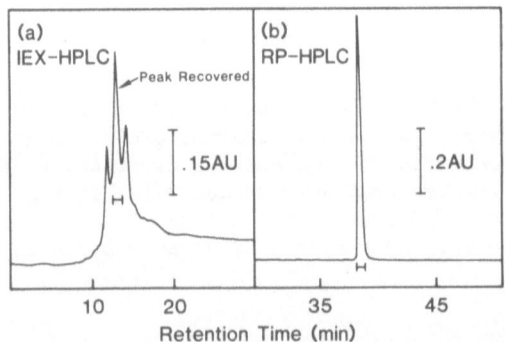

Figure 14: The sequential use of non-porous ion-exchange and microbore reversed phase HPLC for the purification and micromanipulation of h-transferrin. (a) Chromatography of h-transferrin (7.9ug) on Biorad MA7P (7μm spherical non-porous polymethacrylate beads coated with polyethyleneimine, 30x4.6mm ID) flow rate 100μl/min; temperature 45°C. Primary solvent 20mm Tris-HCl buffer, pH6.9. Secondary solvent 20mm Tris-HCl buffer pH6.9 containing 1M NaCl. Linear 60-min gradient. Peak was recovered, as indicated, in a volume of 125μl. (b) Chromatography of peak recovered from Fig. 14(a) on Brownlee BU-300 (30x2.1mm ID) using conditions described in Fig. 2(b).

Currently, we are evaluating a recently developed anion-exchange support based on non-porous polymethacrylate beads(40). The lack of pores in this support results in reduced protein capacity (ca. 300μg for a 30x4.6mm ID column) and a reduction in band broadening due to diffusion into and out of the pores(40). Repetitive yields for multiple injections of human transferrin (determined in a similar manner to that outlined in Fig. 3) were 72%. The application of this column support to the purification of a human transferrin preparation is demonstrated in Fig. 14. It can be seen that considerable purification was achieved. The peak volume recovery (125µl) from this column (30x4.6mm ID) was comparable to that obtained with porous supports packed into microbore columns (30x2.1mm ID). The major peak from this column (Fig. 14(a)) was rechromatographed onto a Brownlee BU-300 (30x2.1mm ID) reversed phase column, enabling the purified protein to be recovered in a volatile solvent (peak volume 80µl) (Fig. 14(b)) suitable for amino acid and sequence analysis. The overall mass recovery (as determined by amino acid analysis) was 4.6µg; this recovery was consistent with the purification observed (*ca.* 60%) from the ion-exchange separation (Fig. 14(a)). The ability to perform separations on microbore columns utilizing a range of packings with alternative selectivities, with the associated advantages of detectability of small amounts of material should further extend the range of proteins which can be purified by HPLC for sequence analysis.

ACKNOWLEDGEMENTS

We thank Dr. D. Metcalf and Dr. N. Nicola (Walter and Eliza Hall Institute, Melbourne, Australia) for providing and assaying CSFs; Dr. J. Delamarter (Biogen, Geneva) for providing bacterially synthesized mGM-CSF; S.-L. Mau (Botany Department, University of Melbourne) for providing electroeluted plant glycoproteins; R.L. Moritz and L. Fabri for skilled technical assistance and S. Blackford for typing the manuscript.

REFERENCES

1. R.M. Hewick, M.W. Hunkapiller, L.E. Hood and W.J. Dreyer, J. Biol. Chem. 256:7990 (1981).
2. M.W. Hunkapiller, R.M. Hewick, W.J. Dreyer and L.E. Hood, Methods Enzymol. 91:399 (1983).
3. L.G. Sparrow, D. Metcalf, M.W. Hunkapiller, L.E. Hood and A.W. Burgess, Proc. Natl. Acad. Sci. USA 82:292 (1985).
4. B. Grego, I.R. van Driel, P.A., Stearne, J.W. Goding, E.C. Nice and R.J. Simpson, Eur. J. Biochem. 148:485 (1985).
5. F.E. Regnier, Methods Enzymol. 91:137 (1983).
6. E.C. Nice, B. Grego and R.J. Simpson, Biochem. Intl. 11:187 (1985).
7. P.A. Stearne, I.R. van Driel, B. Grego, R.J. Simpson and J.W. Goding, J. Immunol. 134:443 (1985).
8. I.R. van Driel, P.A. Stearne, B. Grego, R.J. Simpson and J.W. Goding, J. Immunol. 133:3220 (1984).
9. M.W. Hunkapiller, E. Lujan, F. Ostrander and L.E. Hood, Methods Enzymol. 91:227 (1983).
10. A.S. Bhown and J. Claude-Bennett, Methods Enzymol. 91:450 (1983).
11. F.S. Esch, Anal. Biochem. 136:39 (1984).
12. M.W. Hunkapiller and L.E. Hood, Science 219:650 (1983).
13. J. Rivier, C. Rivier, J. Spiess and W. Vale, Anal. Biochem. 127:258 (1982).
14. E.C. Nice, C.J. Lloyd and A.W. Burgess, J. Chromatogr. 296:153 (1984).
15. R.L. Cunico, R. Simpson, L. Correia and C.T. Wehr, J. Chromatogr. 336:105 (1984).
16. H.E. Schwartz, B.L. Karger and P. Kucera, Anal. Chem. 55:1757 (1983).
17. P. Kucera, J. Chromatogr. 198:93 (1980).
18. A.W. Burgess and E.C. Nice, Methods Enzymol. 116: (in press) (1985).
19. J.F. Delamarter, J.-J. Mermod, C.M. Liang, J.F., Eliason and D. Thatcher, EMBO J., in press.
20. K.C. Zoon, M.E. Smith, P.J. Bridgen, C.B. Anfinsen, M.W. Hunkapiller and L.E. Hood, Science 207:525 (1980).

21. P. Bohlen, A. Baird, F. Esch, N. Ling and D. Gospodarowicz, <u>Proc. Natl. Acad. Sci. USA</u> 81:5364 (1984).

22. M.W. Hunkapiller, J.E. Strickler and K.J. Wilson, <u>Science</u> 226:304 (1984).

23. A.B. Roberts, M.A. Anzano, L.C. Lamb, J.M. Smith and M.B. Sporn, <u>Biochemistry</u> 22:5692 (1983).

24. L.G. Sparrow, D. Metcalf, M.W. Hunkapiller, L.E. Hood and A.W. Burgess, <u>Proc. Natl. Acad. Sci. USA</u> 82:292 (1985).

25. N.A. Nicola, D. Metcalf, M. Matsumoto and G.R. Johnson, <u>J. Biol. Chem.</u> 258:9017 (1983).

26. C.M. Ben-Avram, J.E. Shiveley, R.K. Shadduck, A. Waheed, T. Rajavashisth and A.J. Lusis, <u>Proc. Natl. Acad. Sci. USA</u> 82:4486 (1985).

27. L.M. Petruzzelli, R. Herrera and O.M. Rosen, <u>Proc. Natl. Acad. Sci. USA</u> 81:3327 (1984).

28. A.W. Burgess, D. Metcalf, I.J. Kozka, G. Vairo, J.A. Hamilton, R.J. Simpson and E.C. Nice, <u>J. Biol. Chem.</u> in press.

29. R.J. Simpson, J.A. Smith, R.L. Moritz, M.J. O'Hare, P.S. Rudland, J.R. Morrison, C.J. Lloyd, B. Grego, A.W. Burgess and E.C. Nice, <u>Eur. J. Biochem.</u> (in press).

30. R.J. Simpson and E.C. Nice, <u>Biochem. Intl.</u> 8:787 (1984).

31. R.J. Simpson, G.S. Begg, D.S. Dorow and F.J. Morgan, <u>Biochemistry</u> 19:1814 (1980).

32. A.M. Crestfield, S. Moore and W.H. Stein, <u>J. Biol. Chem.</u> 238:622 (1963).

33. C.H.W. Hirs, <u>Methods Enzymol.</u> 11:199 (1967).

34. J. Dodt, H.P. Muller, U. Seemuller and J-Y. Chang <u>FEBS Lett.</u> 165:180 (1984).

35. B. Grego, I.R. van Driel, J.W. Goding, E.C. Nice and R.J. Simpson <u>Intl. J. Pep. Prot. Res.</u> (in press).

36. N.M. Gough, J. Gough, D. Metcalf, A. Kelso, D. Grail, N.A. Nicola, A.W. Burgess and A.R. Dunn, <u>Nature</u> 309:763 (1984).

37. D.A. Hager and R.R. Burgess, <u>Anal. Biochem.</u> 109:76 (1980).

38. R.H. Aebersold, D.B. Teplow, L.E. Hood and S.B.H. Kent <u>J. Biol. Chem.</u> (in press).

39. J. Knox and J. Jurand, <u>J. Chromatogr.</u> 125:89 (1975).

40. D.J. Burke, J.K. Duncan, L.C. Dunn, L. Cummings, C.J. Siebert and A.S. Ott, <u>J. Chromatogr.</u> submitted.

PRACTICAL AND THEORETICAL ASPECTS OF MICROBORE HPLC

OF PROTEINS, PEPTIDES AND AMINO ACID DERIVATIVES

Kenneth J. Wilson, David R. Dupont, Pau M. Yuan,
Michael W. Hunkapiller and Timothy D. Schlabach

Applied Biosystems, Inc.
850 Lincoln Centre Drive
Foster City, CA 94404 U.S.A.

INTRODUCTION

In the fields of protein chemistry and isolation, there has been a continuing need for procedures that improve sensitivity levels. As isolation protocols became more demanding, the quality of both the instrumentation and chromatographic media improved correspondingly. Similarly, the researcher has had to remain up-to-date on the latest techniques being introduced by a relatively small number of his peers and a larger number of commercial vendors. The directions of these changes have included: improved column packings and design, pumps that are more accurate and have reduced pressure fluctuations, detector designs with improved noise and drift characteristics, and auto-injectors, as well as fraction collectors, that are simpler to use, contain microprocessors and are programmable for certain functions.

Many of these improvements have addressed the needs of the protein chemist for isolating samples at the microgram level or higher. Since the introduction of sequencing techniques at the pmol level, there has been a growing need for continuing not only instrument improvement but also the methods of sample handling. Some of these changes have required that a number of the older, more established protocols be discarded. Examples would be those that incur possible sample losses, ie. lyophilization, dialysis and membrane concentration. Newer methods, such as sample precipitation with an organic solvent, on-column enrichment procedures, or arranging chromatographic separations such that samples can be transferred from one column separation to the next, have been implemented as substitutes.

As the need to move to even higher sensitivities increased there have been a series of specific instrument changes. The reduction of the column internal diameter to either narrowbore (2-3 mm) or microbore (0.5-1.5 mm) dimensions resulted in enhanced mass sensitivity(1). Another improvement has been the introduction(2,3) of a dual syringe solvent delivery system capable of maintaining very accurate volumetric delivery at microliter per min. flowrates, performing gradient elutions and maintaining constant pressures so that pulsations do not interfere with detection limits. There have also been attempts(4) to improve both mass sensitivity and resolution by eliminating connecting tubing between injector, column and detector cell. These particular systems are arranged such that the column fits directly into the injector at one end and the flow cell at the other, a combination that is optimal for isocratic microbore HPLC.

The application of microbore chromatography to protein/peptide isolation and characterization has received only minimal attention, due most probably to the lack of appropriate instrumentation. The investigation by Nice et al.(5) clearly illustrated its utility in the trace enrichment of protein samples. Mass recoveries were in minimal volumes from reverse phase supports, and significant increases in sensitivity were apparent using smaller bore columns. In another publication(6), they effectively used 2mm I.D. reverse phase columns for the high sensitivity isolation of tryptic peptides from murine transferrin receptor.

This communication will illustrate the utility of the reduced diameter columns for the various chromatographic modes of interest to the protein chemist, ie. RPC, IEC and HIC (reverse phase, ion-exchange and hydrophobic interaction chromatographic modes). The observed enhanced sensitivities and reduced elution volumes inherent to these columns are optimal for use in micro-isolations and the characterization of proteins and peptides.

EXPERIMENTAL PROCEDURES

Chromatographic separations were carried out on either of two basic systems. The first was a Model 120B Separation System from Applied Biosystems which was used for all IEC and HIC separations. The single difference between the configuration of this instrument and the PTH Analyzer (Model 120A) is the substitution of a 200 uL dynamic mixer for the static (340 uL)-dynamic mixer (400 uL) combination in the dedicated PTH instrument. The second system used for RPC separations consisted of an essentially identical component combination, ie. a Brownlee Labs MPLC micropump, 200 uL dynamic mixer and either a Kratos 757 or 783 detector. The flow cell dimensions for all experiments (unless otherwise noted) were identical, 12 uL volume and 8 mm path length. Columns were supplied by Brownlee Labs.

Tryptic digestion of apomyoglobin was carried out at room temperature in 0.1 M ammonium bicarbonate for 4 hours using a 4% (w/w) ration of trypsin (Worthington) to substrate. The reaction was stopped by lowering the pH to less than 2 with TFA and the sample recovered by drying (Savant Speed Vac). All proteins used were from Sigma with the exception of apomyoglobin, an Applied Biosystems product (Bioglobin(R)). The water used for buffers was purified through a Milli-Q purification system (Millipore), buffer salts and acetonitrile were from Baker, and TFA from Applied Biosystems. Recombinant interleukin 2 (rIL-2) was kindly supplied by Dr. Carl March of Immunex Corporation, Seattle, WA.

A gas phase sequencer (Model 470A) and on-line PTH amino acid analyzer (Model 120A) were used for sample degradation. In the cases where sample collection for sequencing was done, it was either into a 500 uL Eppendorf tube, or onto a Polybrene containing glass fiber filter. The filters had been 'cleaned' by the TFA procedure and NaCl added to the Polybrene before sequencing was initiated (see Yuan et al., this volume). PTH-residue identifications were carried out as described by Hunkapiller (this volume); data reduction and quantitation using a Nelson 760 interface and a Hewlett Packard 9816 computer.

RESULTS AND DISCUSSION

The introduction(7) of the gas phase sequencer for high sensitivity sequencing has provided a technique suitable for determining partial structures on very small amounts of peptide or protein. In addition to the chromatographic modes that might be used to isolate samples at these levels, other specialized methods have been developed: electroelution/ electrodialysis(8) and electroblotting(9) of proteins separated by either one or two-dimensional PAGE systems. The aforementioned paper by Nice et al.(5) illustrated the utility of short microbore columns in sample trace enrichment and increased levels of detectability for proteins.

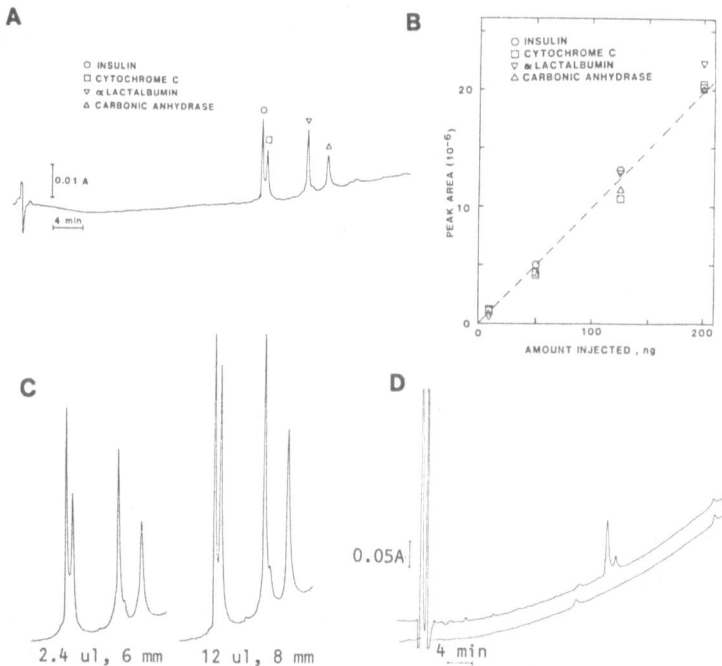

Fig. 1. Microbore RP-HPLC of proteins. A, chromatography was carried out at room temperature on an Aquapore RP-300 (1 x 250 mm) using a linear gradient from 10 to 100% B at 50 uL/min. Buffers were A, 0.1% TFA and B, 60% acetonitrile in A. Detector sensitivity was 0.1 AUFS at 220 nm with the standard size flow cell, chart speed 0.5 cm/min and 50 ng of each protein was injected. B, correlation between integrated peak areas and mass (ng) injected. C, compari- son of detector response as a function of cell dimension; 125 ng/ protein injected. D, injection of rIL-2 (100 ng, about 6 pmol). Elution conditions were identical to Fig. 1A except the gradient was from 50 to 100% B buffer and detection at 0.05 AUFS at 220 nm.

Fig. 1A indicates the sensitivity possible with 1 mm ID RP columns when chromatographing 50 ng amounts of protein. Significant sample loss is not a problem, as measured by peak area of the eluate, over the range of 10 ng to to 200 ng (Fig. 1B). Substituting a microbore cell (2.4 uL, 6 mm) for the standard size (12 uL, 8 mm), indicates minimal band broadening during gradient elution of the same set of proteins (Fig. 1C). It does, however, reduce overall sensitivity. The utility of this methodology for enhanced mass sensitivity is illustrated in Fig. 1D for the chromatography of rIL-2. The higher sensitivity allows for the detection of the main duplex of peaks, and also the background of small peaks eluting throughout the first half of the chromatogram. Note that only 6 pmol of material was required for this analytical characterization.

Through simple reduction in column internal volume, it is possible to significantly increase sensitivity. This is illustrated in Fig. 2 for the tryptic maps of myoglobin. Decreases from 4.6 mm ID (Fig. 2A and B), to 2.1 mm ID (Fig. 2C) and finally to 1 mm ID columns (Fig. 2D), increased sensitivities by factors of approximately 5 and 15 over the standard 4.6 mm ID support. This enhanced detectability obviously allows for mapping to be carried out on smaller amounts of material, as well as the preparative isolation of samples.

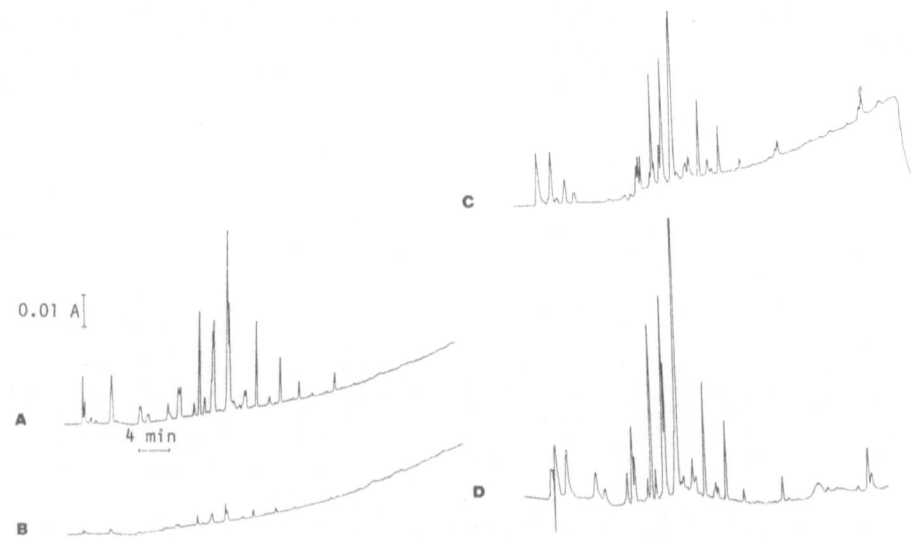

Fig. 2. Peptide mapping with standard, narrow and microbore columns. A tryptic digest of apomyoglobin was chromatographed on the following series of Aquapore RP-300 columns: A, 4.6 x 250 mm at 1 mL/min, 1000 pmol; B, as in A, 100 pmol; C, 2.1 x 220 mm at 200 uL/min, 200 pmol; D, 1.0 x 250 mm at 50 uL/min, 100 pmol. Detector sensitivity (0.1 AUFS at 220 nm) and chart speed (0.5 cm/min) were maintained constant. The linear gradient over 45 minutes was between 0.1% TFA (buffer A) and 60% acetonitrile in 0.1% TFA (buffer B).

The smaller volumes in which the peaks elute from a microbore column (approximately 30-40 uL at 50 uL/min flowrate) allow for collection directly onto the glass fiber filter used in the gas phase sequencer. The results from such an application are given Table I. As estimated by the initial sequencing yields on the collected peaks, peptide recoveries were not quantitative but were acceptable. The ability to collect samples directly onto non-precycled filters has a number of advantages: reduction of sample loss on surfaces, minimizes handling/transfer steps, and provides a simple storage method for numerous fractions, for example those originating from an enzymatic digestion. Additionally, the elimination of precycling increases the available time for sequencing, ie. the instrument is not involved with precycling, and there are reduced chemical costs (see Yuan et al., this volume).

Table 1: Sequencing yields as a function of collection method[a]

| Collection | Peptide, initial yields(%) | | | |
Method	T-3	T-4	T-5	T-6
onto filter	89(E), 21(H)	31(H)	33(V)	38(A)
into tube	111(E), 46(H)	50(H)	43(V)	71(A)
percentages, filter/tube	80 46	62	77	54

a) A tryptic mixture (300 pmol) of apomyoglobin was chromatographed as described in Fig. 4 of the article by Yuan et al. (this volume). The peaks were collected either onto filters containing Polybrene (non- precycled) or into 500 uL Eppendorf tubes. Samples were subsequently sequenced, the initial yields quantitated and compared as a function of collection techniques.

Potential drawbacks to working with such small peak volumes (25-50 uL) are few but extremely important. The most obvious is knowing, or determining the 'dead' volume between the flow cell and the collection point. In order to minimize this volume a 2.4 uL microbore flow cell, rather than the standard 12 uL cell, was employed. A short 8 cm length of 0.004 inch I.D. stainless steel, connected via a 'zero' dead volume female-female union, was employed to direct the column effluent onto the collection surface (either glass filter or Eppendorf tube). Another problem for high sensitivity applications is the slightly higher PTH backgrounds when collecting samples onto non-precycled filters. The somewhat lower yields of those charged PTH residues (Asp, Glu, Arg, His), if appearing within the first couple of cycles of the sample, can also be problematic (see Yuan et al., this volume).

Figures 3 and 4 illustrate the utility of narrow and microbore columns for the chromatography of proteins on IE and HI supports. A standard 300Å pore size was used for higher overall recoveries(10) of the samples chromatographed and 7 u particle sizes for higher efficiencies. Although there are obvious increases in mass sensitivity when decreasing column diameters for these supports, they are not as dramatic as with reverse phase supports. With the exception of the chromatographic results on the anion-exchange support (Fig. 3, right), the separation efficiencies were identical or improved when decreasing the column I.D. to 1 mm.

Detection in the low UV ranges with the high salt-containing buffers used for such separations can present problems. Using a detector sensitivity at 280 nm of 0.02 AUFS for cation- and 0.05 AUFS for anion-exchange HPLC, it was possible to achieve the microgram level of detection. A careful choice of high-purity buffers, as well as source, will allow detection at 220 nm for higher sensitivity IE applications(11). The detection problems with HIC are even more drastic since both refractive index and absorption changes are occurring simultaneously

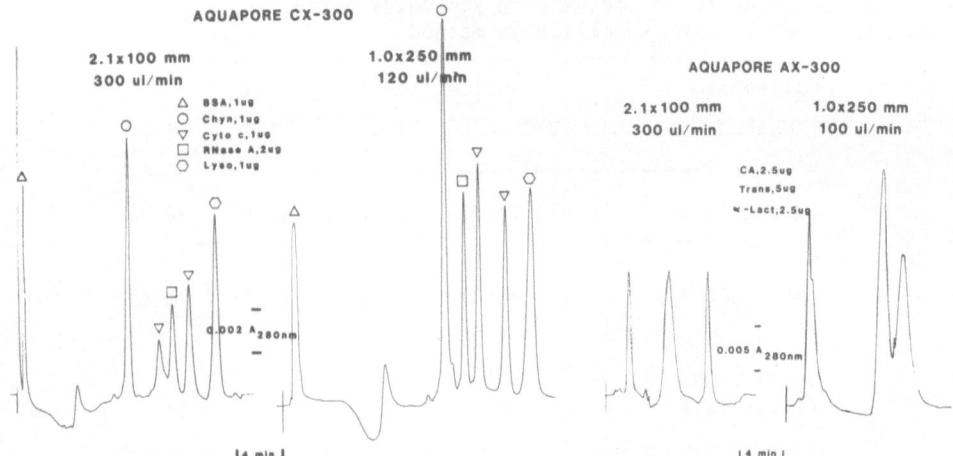

Fig. 3. Narrow and microbore IE of proteins. Cation-exchange separations
(left, CX-300) were carried out using as buffers: A, 25 mM Tris-
acetate, pH 7.0 and B, as in A except containing 0.6 M sodium
acetate. Similarly, anion-exchange chromatography (right, AX-
300) was done using the buffers: A, 25 mM Tris-acetate, pH 8.0
and B, 25 mM Tris-acetate, pH 7.0, 0.5 M sodium acetate. Flow
rates were 300 uL/min for the 2.1 mm and 120 uL/min for the 1.0
mm I.D. columns. Detection was performed at 280 nm employing the
standard 12 uL flow cell.

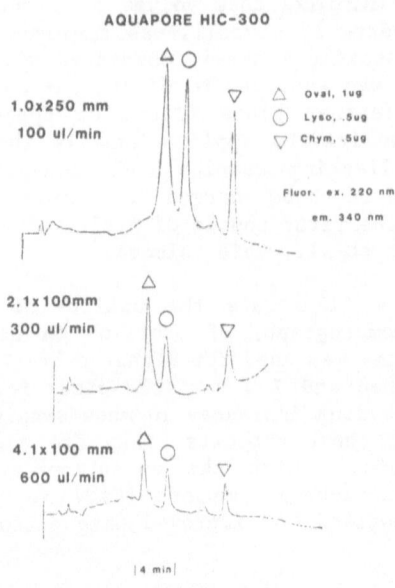

Fig. 4. HIC of proteins. Chromatography was carried out on a propyl-
derivatized 300Å pore size support (7 u particle) using as
buffers: A, 2.0 M ammonium sulfate (BioRad) in 0.25 M ammonium
acetate, pH 7.0 and B, a 1 to 10 dilution of A. Flowrates were
as indicated and detection performed using a Kratos Spectroflow
980 and a standard 5 uL flow cell. Wavelength settings were 220
nm for excitation and 340 nm for emission (filter).

during gradient development. The levels of sensitivity can be increased by using fluorescence detection (Fig. 4). Under these experimental conditions, UV detection levels at 280 nm were approximately 2-4 times less sensitive than fluorescence measurements in high concentrations of ammonium sulfate (1-2 M).

The chromatography of acid hydrolysates of peptides and proteins is an integral part of the characterization of any isolated component. Within the past few years, and coincident with the development of HPLC, there has been a movement away from older post-column derivatization methods of detection and toward pre-column derivatization procedures. Some of these methods, eg. DNS-Cl(12), OPA(13) and PITC(14), allow one to select parameters thought to be essential. Examples would be: sensitivity, ease of use, N-terminal analysis, derivative stability and background limitations. Common to each of the pre-column methods is the need to chromatographically separate not only the amino acid derivatives but also those arising from the hydrolysis and derivatization chemicals.

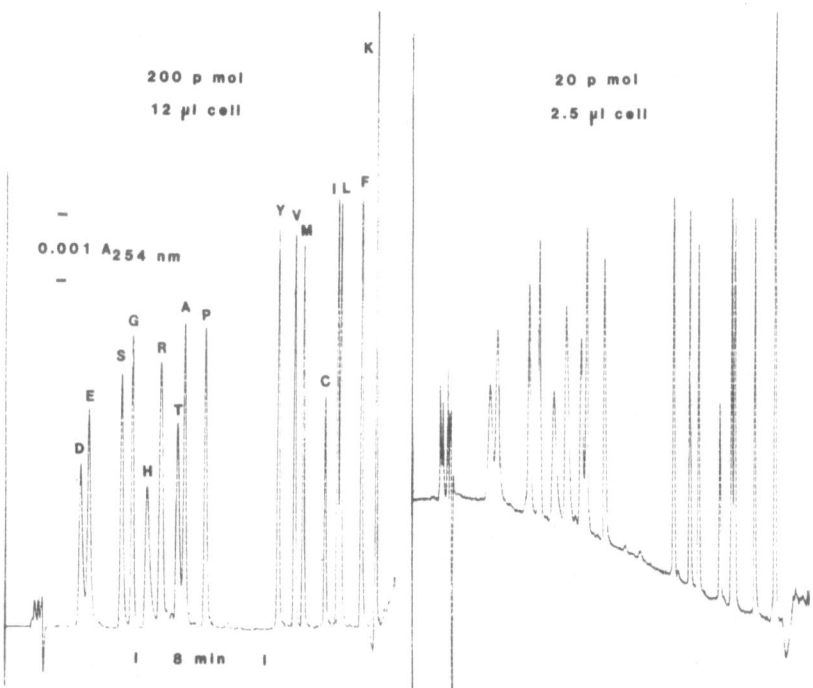

Fig. 5. Narrowbore HPLC of PTC amino acids. Chromatographed were 200 pmol (left) and 20 pmol (right) amounts of the PTC derivatives of most of the amino acids found in 'typical' protein hydrolysates. The buffer system used was: A, 15 mM triethylamine acetate, pH 6.6 and B, acetonitrile. Flowrates were 300 uL/min for the 2.1 x 220 mm RP-18 support at 45°C.

Illustrated in Fig. 5 is the separation of PTC-amino acid mixtures on a narrowbore column under conditions which allow for detection levels in the low pmol range. As was shown above for protein/peptide HPLC separations (Figs. 1-4), an additional increase in sensitivity of approximately 4-fold with a further decrease in column size to 1.0 mm I.D. can also be realized for these amino acid derivatives (results not shown).

The increases in mass sensitivity which occur simply by reducing the *column* internal diameter have been shown for most of the chromatographic supports currently being used in protein and peptide isolation. Integral to such applications is the availability of a chromatographic system with the appropriate pumping and detection capabilities necessary for the gradient elution of narrow and microbore columns. Since the components elute in volumes of 30-50 uL from 1.0 mm I.D. columns, the glass fiber filter of the gas phase sequencer is a suitable sample collection surface. The advantages and disadvantages of such an approach to high sensitivity sequencing have been indicated in both this communication and that from Yuan et al., (this volume).

REFERENCES

1.   Cooke, N.H.C., Olsen, K. and Archer, B.G., LC 2 (1984) 514-524.

2.   Schwartz, H.E. and Brownlee, R.G., Am. Lab. 16 (1984) 43-56.

3.   Schwartz, H.E. and Berry, V.V., LC 3 (1985) 110-124.

4.   Tehrani, A.Y., LC 3 (1985) 42-48.

5.   Nice, E.C., Lloyd, C.J. and Burgess, A.W., J. Chromatogr. 296 (1984) 153-170.

6.   Grego, B., VanDriel, I.R., Stearne, P.A., Goding, J.W., Nice, E.C. and Simpson, R.J., Eur. J. Biochem. 148 (1985) 485-491.

7.   Hewick, R.M., Hunkapiller, M.W., Hood, L.E. and Dreyer, W.J., J. Biol. Chem. 256 (1981) 7990-7997.

8.   Hunkapiller, M.W., Lujan, E., Ostrander, F. and Hood, L.E., Methods Enzymol. 91 (1983) 227-236.

9.   Aebersold, R.H., Teplow, D.B., Hood, L.E. and Kent, S.B.H., J. Biol. Chem., in press.

10.   Wilson, K.J., Van Wieringen, E., Klauser, S. and Berchtold, M.W., J. Chromatogr. 237 (1982) 407-416.

11.   Schlabach, T.D. and Wilson, K.J., unpublished results.

12.   DeJong, C., Hughes, G.J., Van Wieringen, E. and Wilson, K.J., J. Chromatogr. 241 (1982) 345-359.

13.   Jones, B.N. and Gilligan, J.P., J. Chromatogr. 266 (1983) 471-482.

14.   Heinrikson, R.L. and Meredith, S.C., Anal. Biochem. 136 (1984) 65-74.

MICROISOLATION AND SEQUENCE ANALYSIS OF HUMAN EPIDERMAL GROWTH FACTOR

Petro E. Petrides*, Peter Bohlen** and Frederick S. Esch***

* Molecular Oncology Laboratory, Department of Medicine
University of Munich Medical School of Grosshadern
Munich, Germany

** Department of Biochemistry, University of Zurich
Zurich, Switzerland

*** Laboratories for Neuroendocrinology, The Salk
Institute, La Jolla, CA, USA

ABSTRACT

Human epidermal growth factor (EGF) has been characterized by a combination of microisolation, -amino acid analysis and -sequencing. The purification from milk was monitored with a human placental membrane radioreceptor assay using murine EGF as a competitive ligand and was achieved exclusively by the use of reverse phase liquid chromatography (RPLC). The sequential use of preparative, semipreparative and analytical RPLC on octylsilica supports with solvent systems of different solute selectivity such as pyridine formate, triethylammonnium phosphate or perfluorocarboxylic acids in the presence of n-propanol or acetonitrile allowed the purification to homogeneity with five consecutive runs. The molecular weight, amino acid composition and NH_2-terminal sequence of human EGF were determined. Gas phase microsequencing of residues 1 through 17 revealed the following sequence: ASN-SER-ASP-SER-GLU-X-PRO-LEU-SER-HIS-ASP-GLY-TYR-X-LEU-X-ASP which is identical with the NH_2-terminius of urogastrone from human urine. Although human epidermal growth factor has been unequivocally identified in human milk and shown to be identical with urogastrone from human urine, the high resolution techniques employed have also revealed the presence of EGF-related molecules which await further characterization. It is possible that EGF and the EGF-related growth factors possess important regulatory functions in normal growth of the human breast during pregnancy and lactation as well as in abnormal growth during mammary tumor formation and progression.

INTRODUCTION

In 1972, Cohen (1) reported the amino acid sequence of epidermal growth factor, a mitogenic polypeptide, which he had identified in the submaxillary gland of the mouse. Three years later, Gregory (2) described the isolation and characterization of urogastrone, an inhibitory polypeptide of gastric acid secretion present in human urine. Suprisingly, when the primary structure of urogastrone was compared with those of other fully characterized polypeptides a remarkable structural similarity with murine EGF became evident. Of the 53 amino acid residues comprising urogastrone and mEGF, 37 are identical. Since that time it is generally assumed, although not proven, that both biological activities reside in one molecule, i.e. that the human urogastrone and EGF are identical. In order to investigate further the relationship of urogastrone and human EGF, we decided to determine the chemical structure of epidermal growth factor from a human body fluid other than urine for the source of urogastrone. Since EGF activity had also been detected in serum, saliva, amniotic fluid and milk, the latter fluid was chosen as starting material. In our studies we have found that acidified milk – upon fractionation on reverse phase support – contains four polypeptides (A, B, C, and D) capable of competing with murine EGF for human placental membrane receptors (3). Because of their stimulatory effect upon cells grown in semisolid media these polypeptides belong to a group of regulatory polypeptides, the transforming growth factors (TGFs), that induce anchorage independent growth of normal cells the proliferation of which is usually anchorage dependent (4). We have purified the polypeptide present in largest quantity (TGF_D) to homogeneity and determined its molecular weight, amino acid composition and partial NH_2-terminal sequence. Our results strongly support the assumption that TGF_D is EGF and secondly, that human urogastrone and EGF are identical.

METHODS

Polypeptide isolation. The chromatography system has been described (5). Reverse phase columns were either prepared in our laboratory with Lichroprep (RP 8, 2.5 x 6.0 cm, particle size 15 – 25 mμ, pore size 100 A, Merck, Darmstadt) as reported (6) or obtained prepacked from Brownlee Laboratories (RP 300, 0.7 or 0.46 x 25 cm, particle size 10 mμ, pore size 300 A, Santa Clara, CA). Column eluates were monitored with a fluormetric postcolumn derivatization system and/or by UV-absorbance at 280 nm. Mobile phase utilized were 0.36 $\underline{M}$ pyridine formate (pH 3.0), 0.25 $\underline{M}$ triethylammonium phosphate (pH3.0) (7), 0.1 % trifluoracetic (pH 2.10) or 0.05 % heptafluorobutyric acid (pH 2.48) (8). N-propanol or acetonitrile were used as organic modifiers. For isolation batches of 400 ml milk (obtained from the Mother's milk bank in San Jose, CA) were thawed, acidified by the addition of 200 ml glacial acetic acid and centrifuged for 30 minutes at 100.00 x g. After the centrifugation the pellet containing cellular debris and the creamy top layer were discarded and the resulting liquid filtered through nylon gauze. After adjustment to pH 3.0 with pyridine formate the acidified milk was directly pumped onto the preparative RPLC system.

Preparation and iodination of murine EGF. Murine EGF was prepared and iodinated as published (9).

Radioreceptor assay for human TGFs. Preparation of placental membranes and the binding assay were carried out with minor modifications of the procedure described in (10).

SDS-Gel Electrophoresis. Peptides were dissolved in sample buffer and analyzed in a discontinous slab gel system (20% polyacrylamide) in the presence or absence of 50 mM dithiothreitol. After electrophoresis one half of the gel was cut into 1 mm slices and the polypeptides eluted by shaking overnight at 4°C in 1.0 ml of buffer containing 125 mM TrisCl, 1 mM EDTA and 2 mg bovine serum albumin. Two aliquots of 190 μl each were then tested for the presence of active material in the radioreceptor assay. The remainder of the gel was stained with silver.

Amino acid analysis and sequencing. Peptide aliquots of 10 to 20 pmol were subjected to hydrolysis and analyzed with a Liquimat III (Kontron, Zurich, Switzerland) amino acid analyzer (11). Gas phase sequence analyses were performed as previously reported (12) using a gas phase sequencer (Applied Biosystems, Inc., Foster City, Ca).

RESULTS

Purification of TGF$_D$. Preparative RPLC of acidified milk (from one individual donor) revealed two radioreceptor active fractions (pools I and II) eluting at 20.4 and 22.2% n-propanol (fig.1). Both pools were individually rechromatographed in the same solvent system on a semi-preparative RP 300 column using a linear gradient from 10 to 18% n-propanol. Under these conditions the activity present in pool I was separated into bioactive fractions eluting at 14.4 and 14.7% n-propanol which were designated TGF$_A$ and TGF$_B$ (upper part of fig. 2). Chromatography of pool II let to the appearance of two TGFs which eluted at 15.9% (TGF$_C$) and 16.5% (TGF$_D$) n-propanol (lower part of fig. 2). The latter molecule was further purified on an analytical RP 300 column using an acetonitrile gradient from 28 to 32% (over 90 min.) in 0.2% trifluoracetic acid (TFA) as mobile phase (fig. 3A). The bioactive fractions were pooled and rechromatographed on the same column

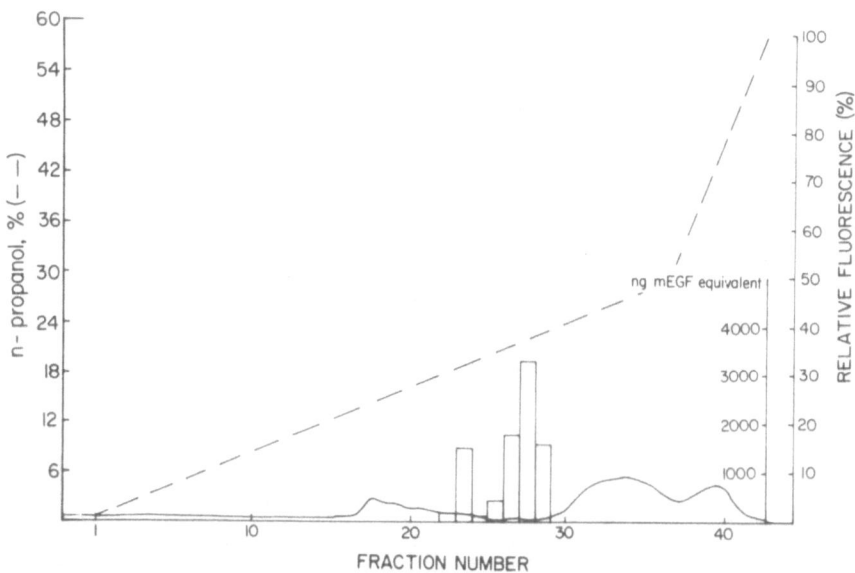

Fig. 1. Preparative RPLC of acidified human milk adjusted with pyridine formate to pH 3.0. The polypeptides were eluted with a linear gradient from 0 to 30% n-propanol over 180 minutes. Flow rate was 2.5 ml/min. 12.5 ml fractions were collected and 10 μl aliquots used for the radioreceptor assay.

using triethylammonnium phosphate as buffer and eluting the polypeptides with an acetonitrile gradient from 21 to 27% over 120 min. In such a system TGF$_D$ is less hydrophobic and elutes at about 24% acetonitrile (results not shown). Final purification was achieved in 0.05% heptafluorobutyric acid where TGF$_D$ is more hydrophobic than in TFA and elutes at about 32% acetonitrile (fig. 3B). The molecule was considered better than 90% pure as judged by the appearance of one symmetrical peak in two solvent systems with different solute selectivity, i.e. in 0.05% heptafluorobutyric acid and rechromatography in 0.1% trifluoroacetic acid.

In addition, the purity was subsequently conformed by analytical gel electrophoresis and gas phase sequencing (see below). From 30 g of protein present in 3000 ml of acidified milk 12 μg (2 nmol) TGF$_D$ was obtained in highly purified form.

Molecular weight of TGF$_D$. Highly purified TGF$_D$ migrated as a single band with an a apparent M$_R$=6000 in SDS-polyacrylamide gel electrophoresis as determined by silver staining of the gel. The same migration pattern was obtained in the absence of reducing agent. The specific association of this band with radioreceptor competing activity was proven by slicing a parallel unstained gel track that contained unreduced sample. One mm slices prepared from this track were

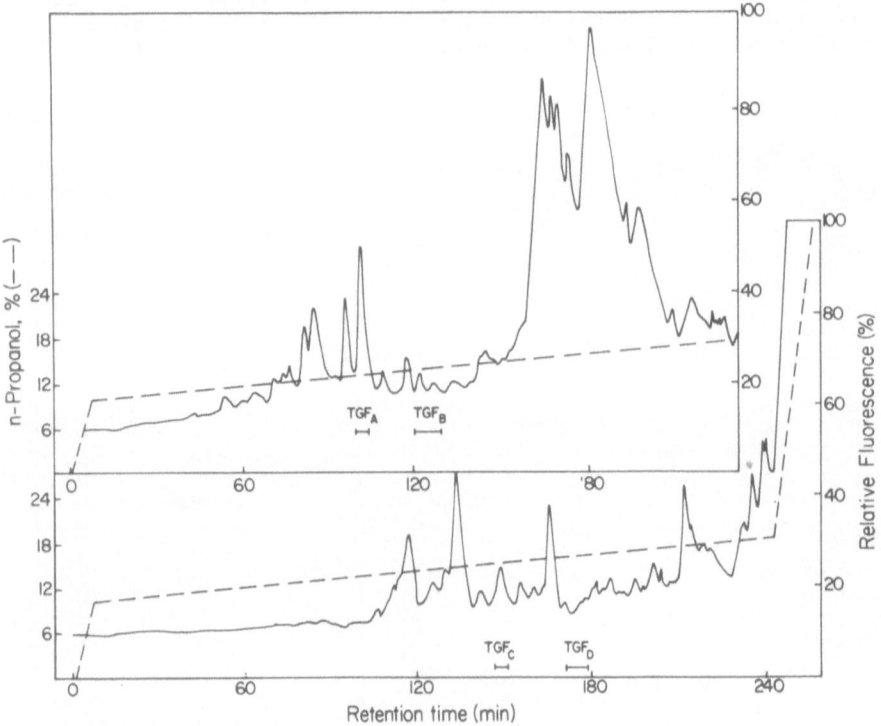

Fig. 2. *Semipreparative RPLC of pools I and II from preparative column. Pool I (upper trace) and pool II (lower trace) were loaded after 1:1 dilution with pyridine formate on a semipreparative RP 300 column in pyridine formate (pH 3.0). The column was eluted with a linear gradient from 10 to 18% n-propanol over 240 min. Flow rate was 1.0 ml/min. 3 ml fractions were collected and 30 μl aliquots used for the radioreceptor assay. The horizontal bars indicate the presence of radioreceptor active material.*

extracted overnight and extracts were assayed for EGF competing activity in the receptor assay. All the extracted radioreceptor activity
was found to migrate in the 6.0 kDa region of the gel (not shown).

Amino acid composition of TGF$_D$. The amino acid composition of TGF$_D$ is
shown in Table 1. Values were calculted assuming a molecular weight
of 6000.

Microsequence analysis of TGF$_D$. Automated Edman degradation of TGF$_D$
was performed with several samples of highly purified TGF$_D$ (amounts

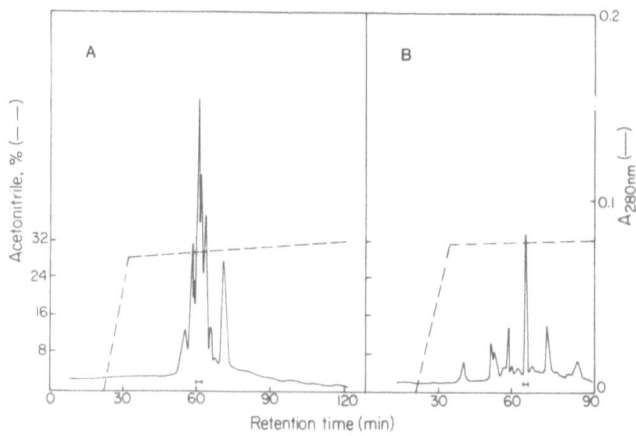

Fig. 3. Further purification of TGF$_D$ on an analytical RP 300 column
in 0.1% trifluoroacetic acid (pH 2.10) (part A) and final purification
in 0.05% heptafluorobutyric acid (pH 2.48) (part B). Flow rate 0.5
ml/min. 1.5 ml fractions were collected and 15 µl aliquots removed
from the radioreceptor assay. The horizontal bars indicate the presence of radioreceptor active material.

ranging from 200 to 400 pmol). The results of one of these runs is
shown in Figure 4: with a 300 pmol sample (based upon amino acid composition) an initial yield of 79.7% and an average repetitive yield of
82.3% were obtained. Unambigious identification of phenylhydantoin
derivatives of amino acids was possible up to residue 17. Cysteine
residues were not positively identified (for instance by preparing the
S-alkylated derivatives) in this analysis, but no other residues were
identified at cycles 6, 14, and 16. Based on this analysis the partial sequences determined of TGF$_D$ corresponds exactly the residues
1-17 of the sequence of human urogastrone.

TABLE 1. Amino Acid Analysis of TGF_D.[a]

	TGF_D	Urogastrone[b]
ASX	6.30 ± 0.32	7
THR	0	0
SER	3.18 ± 0.75	3
GLX	5.85 ± 0.32	5
GLY	4.33 ± 0.43	4
ALA	2.17 ± 0.87	2
VAL	3.56 ± 0.37	3
MET	1.01 ± 0.10	1
ILE	1.91 ± 0.25	2
LEU	5.76 ± 0.91	5
TYR	5.81 ± 0.89	5
PHE	0	0
HIS	1.14 ± 0.37	2
TRP	0.92 ± 0.15	2
LYS	2.14 ± 0.42	2
ARG	2.60 ± 0.24	3
CYS[c]	5.36 ± 0.23	6
PRO	2.31 ± 0.23	1

a   Data are expressed as residues per molecule assuming a molecular
    weight of about 6000. Values are means ± standard deviation of
    three determinations and are not corrected for hydrolysis losses.

b   Values as reported in Ref. 2.

c   Determined as cysteic acid

DISCUSSION

The experimental results described here indicate that human milk contains several polypeptides that able to compete with mEGF for human
placental membrane receptors. One of them, TGF_D, has been purified to
homogeneity with an approach which is novel in that it is entirely
based upon reverse phase liquid chromatography. This obviates the
need for conventional procedures such as ion exchange chromatography,
gel filtration or salt precipitation. The use of solvent systems with
different solute selectivity was allowed the rapid isolation of this
low abundant polypeptide from a complex mixture of biological compounds. The amino acid composition of TGF_D is nearly identical with
that human urogastrone. In fact, the small differences in the amino
acid composition of the two polypeptides may be ascribed to inaccuracies associated with the hydrolysis of very small amounts of polypeptide, i.e. less than 20 pmol. Since NH_2-terminal microsequencing
of TGF_D revealed a structure identical to that of human urogastrone it
is very likely that TGF_D is human EGF and identical with human urogastrone. Our studies also show that human milk contains several
other EgF-like (TGF_A, TGF_B, TGF_C) transforming growth factors (13).

EGF-like transforming growth factors and EGF-receptors have also recently been identified in human breast cancer cells (14). Moreover, urine from normal individuals and cancer patients contains at least five EGF-related TGFs and there are distinct differences with regard to the relative levels of these molecules in cancer patients and normal controls (15). Complete purification and characterization of the TGFs present in human milk, breast tumor tissue as well as urine is necessary in order to establish the molecular nature of these functionally related factors.

Efficient microtechniques such as the one presented here should facilitate the characterization of growth regulating factors and permit the understanding of their relationship in molecular terms as a basis for a better understanding of the (normal) proliferation of human breast cells during pregnancy and lactation and the (abnormal) proliferation during breast tumor formation and progression.

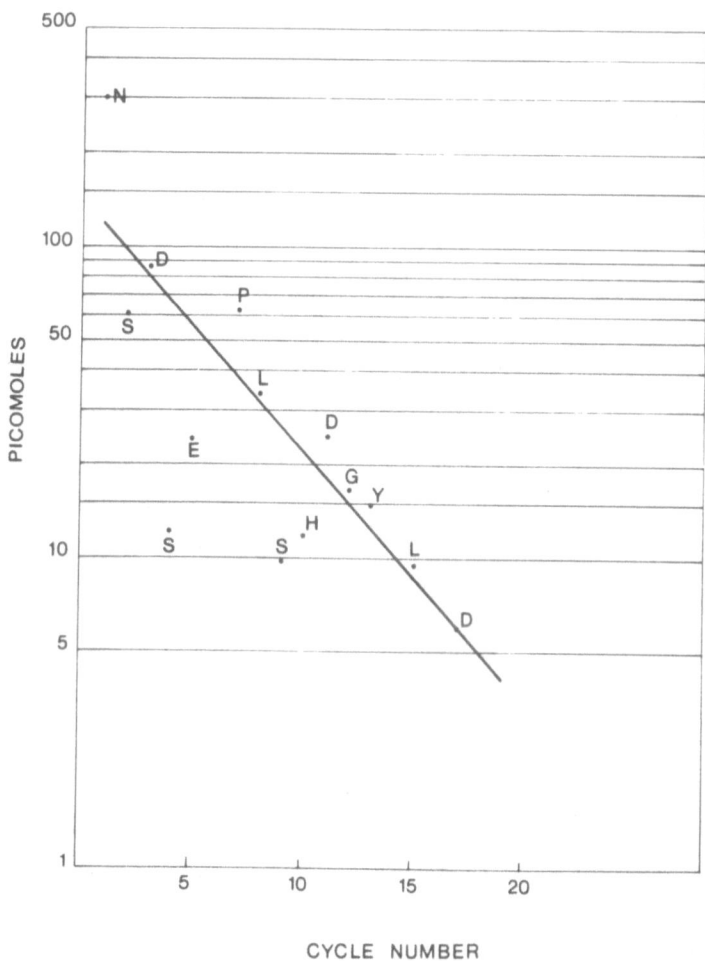

Fig. 4. Gas phase sequence analysis of 300 pmol TGF_D.

REFERENCES

(1) C.R. Savage, et al.: J. Biol. Chem. 247, 7612-7621 (1972)

(2) H. Gregory,: Nature 257, 325-327 (1975)

(3) P.E. Petrides, et al.: FEBS Letters 187, 89-95 (1985)

(4) A.B. Roberts, et al.: Fed. Proc. 42, 2621-2626 (1983)

(5) P.E. Petrides, et al.: Anal. Biochem. 105, 383-388 (1980)

(6) P.E. Petrides, In: Handbook of HPLC for the Separation of Amino Acids, Peptides and Proteins (Hancock, W.S. Ed). Vol. II, pp.327-342, CRC Press, Boca Raton (1984)

(7) J. Rivier,: J. Liq. Chromatogr. 1, 343-346 (1978)

(8) H.P.J. Bennett, et al.: J. Liq. Chromatogr. 3, 1353-1356 (1981)

(9) P.E. Petrides, et al.: Biochem. Biophys. Res. Comm. 125, 218-228 (1984)

(10) Y. Hirata, D.N. Orth: J. Clin. Endocr. Metab. 48, 673-679 (1979)

(11) P. Bohlen, R. Schroeder: Anal. Biochem. 126, 144-152 (1982)

(12) F.S. Esch: Anal. Biochem. 136, 39-47 (1984)

(13) P.E. Petrides, et al: 7th Int. Congr. Endocrinology, Quebec City, Canada, 1159 (1984)

(14) D.S. Salomon, et al.: Cancer Res. 44, 4069-4077 (1984)

(15) E.S. Kimball, et al.: Cancer Res. 44, 3613-3619 (1984)

COMPLETE AMINO ACID SEQUENCE DETERMINATION OF RAT EPIDERMAL GROWTH FACTOR:

CHARACTERIZATION OF A TRUNCATED FORM WITH FULL *IN VITRO* BIOLOGICAL ACTIVITY

Edouard C. Nice≠, John A. Smith*, Robert L. Moritz+, Michael
J. O'Hare*, Philip S. Rudland*, John R. Morrison+,
Christopher J. Lloyd≠, Boris Grego+, Antony W. Burgess≠ and
Richard J. Simpson+

≠Ludwig Institute for Cancer Research (Melbourne Branch)
+Joint Protein Structure Laboratory, Ludwig Institute/The
Walter and Eliza Hall Institute of Medical Research
Parkville 3050, Australia and *Ludwig Institute for Cancer
Research (London Branch), Surrey SM2 5PX, U.K.

Epidermal growth factor (EGF), first isolated by Cohen(1) in 1962 from
adult mouse submaxillary glands, is a single-chain polypeptide of 53
residues with three disulfide bonds(2,3). It is a potent stimulator of
cellular proliferation(4,5) and inhibitor of gastric acid secretion(6).
Although the chemical and biological properties of EGF have been studied
extensively over the past 20 years (for reviews see 7-9) a detailed
understanding of its three-dimensional structure is still not known. The
lack of success with tertiary structure studies (e.g. X-ray crystallography
NMR) may, in part, be attributable to the N-terminal heterogeneity
encountered with many EGF preparations; this heterogeneity has been studied
in detail for two forms of mouse EGF which differ by a lack of one
N-terminal residue(10,11).

Recent studies on rat transforming growth factor-α (rTGF-α)(12-14)
have revealed that this 50-amino-acid-long polypeptide is structurally and
functionally related to EGF. The search for a closer homolog to rTGF-α and
the need for another derivative of EGF for physicochemical studies has led
us to undertake the isolation and characterization of rat EGF (rEGF)(15).

In this communication we report in detail a rapid HPLC methodology for
isolating multiple forms of rEGF (e.g. rEGF1-48, rEGF2-48, rEGF3-48 and
rEGF4-48) from submaxillary glands. All forms appear to be equipotent in
biological activity *in vitro*. The micromanipulative techniques employed in
this study for the purification of protein and derived polypeptides (e.g.
CNBr and chymotryptic fragments) permitted the complete amino acid sequence
determination of rEGF from *ca.* 10μg of protein.

MATERIALS AND METHODS

HPLC. Initial stages of rEGF purification were carried out by HPLC
using an Altex Model 324-40 gradient elution system fitted with a Model 210
injection valve (2ml injection loop). Column eluents were monitored by
absorbance at 215nm or 230nm. Final stages of the purification of rEGF,
and derived peptides, were performed by microbore HPLC using a Hewlett

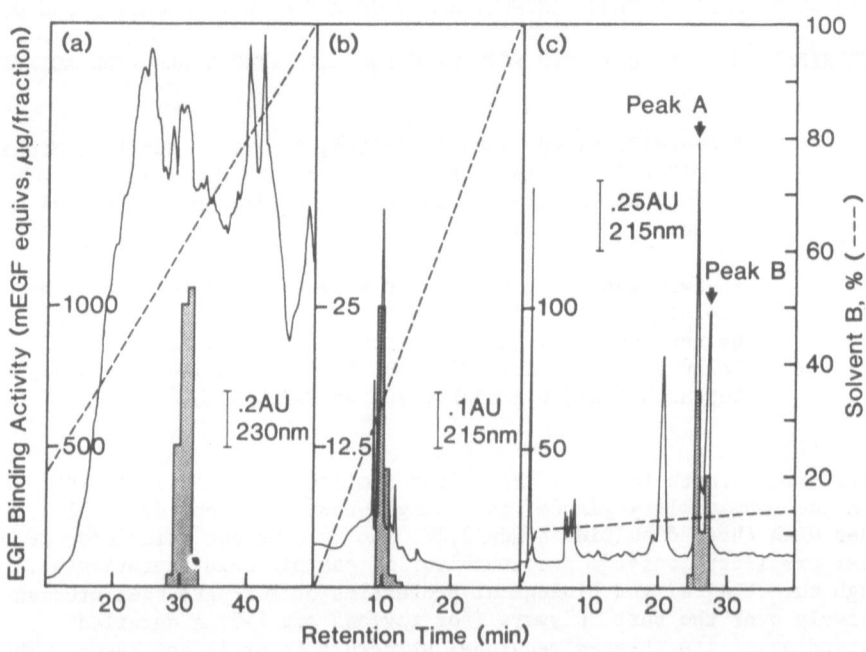

Figure 1. HPLC Purification of Rat EGF.
(a)   Preparative RP-HPLC on ODS-Hypersil (Stage 2). The EGF-containing
*   fraction (5ml) from Stage 1 of the purification was diluted 1:1 with*
*   aqueous 0.15% (v/v) TFA and loaded via the primary solvent pump.*
*   EGF-containing fractions were pooled (horizontal bar).*
(b)   Anion-Exchange HPLC on Mono Q (Stage 3). An aliquot (50μl) from the
*   pool (Fig. 1a) was diluted 10-fold with 20mM Tris-HCl buffer, pH8.5*
*   prior to injection.*
(c)   Analytical RP-HPLC on Ultrapore RPSC (Stage 4). An aliquot (150μl)
*   from the pool (Fig. 1a) was diluted 1:1 with 0.1M sodium phosphate*
*   buffer, pH6.5 for loading.*
All chromatographic conditions are described in Materials and Methods.
Column fractions were monitored for biological activity by competitive
binding(20) to the EGF receptor on A431 cells (▨).

Packard Model 1090 fitted with a model 1040A detector and Rheodyne model 7125 sample injector (2ml injection loop), installed in the column oven compartment.

Preparation of EGF from Rat Submaxillary Glands

Submaxillary glands were removed from adult male Sprague-Dawley rats, flash frozen in liquid nitrogen and stored at -70°C until use. EGF activity was extracted from frozen glands (20g) by homogenization in 1M HCl containing 1% (v/v) TFA, 5% (v/v) formic acid and 1% (w/v) NaCl (7ml/g frozen tissue). The supernatant solution was clarified by centrifugation at 105,000xg and defatted by vacuum filtration through glass wool.

Stage 1: Chromatographic Trace Enrichment of Crude Acid Extract. rEGF activity in the crude acid extract (140ml) was concentrated by trace enrichment onto a "tap-packed" preparative C18 reversed-phase (RP) column with no top frit, (100x4.6mm ID, 55-105µm particle size support, Waters Associates). The acid extract was loaded at 10ml/min by direct pumping (using an Altex Model 110 pump with the frit removed from the top check valve). After loading, the column was washed with 10ml of 5% acetonitrile/95% water containing 0.15% (v/v) TFA. EGF activity was then recovered using 5ml of 40% acetonitrile/60% water containing 0.15% TFA.

Stage 2: Preparative Reversed Phase HPLC. The EGF-containing fraction from Stage 1 was diluted 1:1 with aqueous 0.15% (v/v) TFA and pumped directly onto an ODS-Hypersil column (5µm particle size, 12nm pore size octadecylsilica, 150x4.6mm ID). EGF activity was recovered using a linear 60-min gradient between 0.15% (v/v) TFA and 60% acetonitrile/40% water containing 0.1% (v/v) TFA. The flow rate was 1ml/min and column temperature was 25°C. As shown in Fig. 1a EGF activity was recovered in three 1-ml fractions, eluting between 29-32 min. When these fractions were monitored analytically on a microbore RP-HPLC column under similar gradient conditions (Brownlee RP-300, 30x2.1mm ID; flow rate 100µl/min) *ca.* 70% of the material chromatographed as a single peak (retention time 24.01 min). Spectral analysis across this peak indicated an absence of tryptophan.

Stage 3: Anion-Exchange HPLC. An aliquot (50µl) of the EGF-containing fractions from RP-HPLC (Stage 2) was diluted 10-fold with 20mM Tris-HCl buffer, pH8.5 and loaded onto a Mono Q anion-exchange HPLC column (Pharmacia) equilibrated with the same buffer. The column was developed at 1ml/min and 25°C with a linear 30-min gradient of NaCl (0-0.3M) in the same buffer (Fig. 1b). A single peak of biological activity corresponding to the major UV absorbing peak was recovered (Fig. 1b). Amino-terminal sequence analysis of this material revealed three different Pth-amino acids in the first degradation cycle: asparagine, serine and threonine. The amino acid composition (Table 1) indicates the absence of alanine, phenylalanine, lysine and tryptophan as noted previously(15).

Stage 4: Analytical RP-HPLC. Purified EGF material from either Stage 2 or Stage 3 was further resolved by chromatography on Ultrapore RPSC (5µm particle size, 30nm pore size, dimethylpropylsilica, 75x4.6mm ID, Beckman Inst.) using 0.1M sodium phosphate buffer, pH6.5 with a shallow acetonitrile gradient (Fig. 1c). For loading, samples were diluted 1:1 with starting buffer. For structural studies, samples recovered from either Stage 3 or 4 were concentrated by microbore RP-HPLC(18) (Fig. 2a).

Structural Analysis

Reduction and alkylation of rEGF. Native rEGF from Stage 3 (240pmol/100µl) in 0.2M Tris/HCl buffer, pH8.5, containing 0.002M EDTA and 8.0M urea, was reduced with 0.1M dithiothreitol at 37°C for 3h. Alkylation was

Table 1. *Amino Acid Compositions of rEGF Preparations*[a]

Amino Acid	Residues/mole of rEGF Preparations			
	Stage 3[b]	Stage 4[c]		mEGF[d]
		Peak A	Peak B	
Aspartic acid	6.0 (5–7)	6.1 (6–7)	5.1 (5)	(7)
Threonine	1.1 (1)	1.0 (1)	1.0 (1)	(2)
Serine	2.6 (2–3)	2.7 (2–3)	2.1 (2)	(6)
Glutamic acid	3.2 (3)	3.0 (3)	3.1 (3)	(3)
Proline	2.2 (2)	2.1 (2)	2.0 (2)	(2)
Glycine	5.7 (6)	5.5 (6)	5.9 (6)	(6)
Half–Cystine[e]	4.6 (6)	nd (6)	nd (6)	(6)
Valine	4.3 (5)	4.4 (5)	4.7 (5)	(2)
Methionine	1.0 (1)	0.8 (1)	0.9 (1)	(1)
Isoleucine	1.7 (2)	1.5 (2)	1.6 (2)	(2)
Leucine	2.1 (2)	1.9 (2)	2.0 (2)	(4)
Tyrosine	4.6 (5)	4.5 (5)	4.8 (5)	(5)
Histidine	1.0 (1)	1.2 (1)	1.1 (1)	(1)
Arginine	4.0 (4)	3.7 (4)	3.9 (4)	(4)
Tryptophan[f]	0	–	–	(2)
Total Residues	48	46–48	45	

a. *Values given are for 24h total acid hydrolysis. Values in parenthesis were deduced from the sequence. nd; not determined.*
b. *Mono-Q column, Fig. 1b.*
c. *Ultrapore RPSC column, Fig. 1c.*
d. *Savage et al.(16).*
e. *Determined as RCM-cysteine.*
f. *Tryptophan was determined after hydrolysis with 4M methane-sulfonic acid(17).*

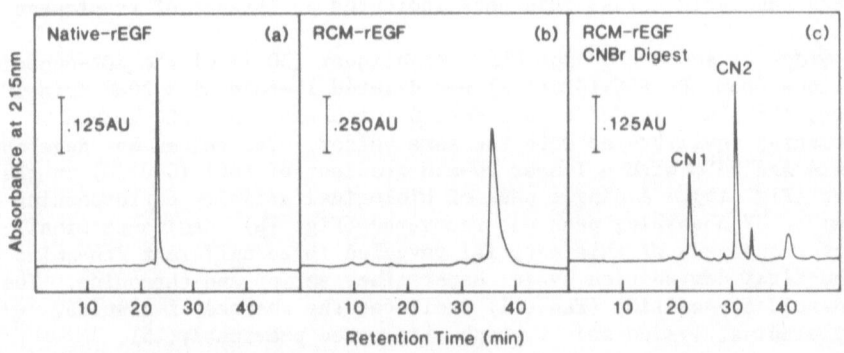

Figure 2. Preparation of Samples for Gas Phase Sequence Analysis. Samples were chromatographed on a Brownlee RP-300 (30x2.1mm ID) column at 45°C. Proteins and peptides were eluted at a flow rate of 100μl/min using a linear 1%/min gradient of acetonitrile in 0.1% (v/v) TFA.
(a) Buffer exchange of an aliquot (75μl) of peak B from Fig. 1c.
(b) Desalting of RCM-rEGF. The sample was loaded at 2ml/min during which time the salt peak elutes in the column void (not shown). The flow rate was adjusted to 100μl/min prior to initiating the gradient.
(c) Separation of CNBr fragments of RCM-rEGF.

achieved by the addition of 10µl iodoacetic acid (15mg iodoacetic acid/100µl 0.1M NaOH) to the mixture and incubation for 15 min at 25°C in the dark. S-Carboxymethyl (RCM)-EGF was desalted by microbore RP-HPLC on a Brownlee RP-300 column (30x2.1mm) (Fig. 2b).

CNBr Treatment. 1µg (240pmol) of RCM-rEGF dissolved in 10µl of 70% (v/v) TFA was cleaved with 1mg of CNBr (1000-fold molar excess over methionine) at 25°C for 18h. The reaction was carried out under nitrogen.

Purification of Peptides by HPLC. Peptide mixtures resulting from enzymatic and chemical cleavage were invariably fractionated on a Brownlee RP-300 column (30x2.1mm ID). Peptides were eluted with a linear 1%/min gradient of acetonitrile in 0.1% (v/v) TFA (first dimension) or 0.1M sodium phosphate buffer, pH6.5 (second dimension). Flow rate for all runs was 100µl/min, column temperature was 45°C. The separation of the peptides derived by CNBr digestion is shown in Fig. 2c. Purified peptides were collected manually after correction for post detector dead volumes.

Automated Edman degradation of peptides was performed using an Applied Biosystems sequencer (model 470A) and Pth-amino acids were identified and quantitated as described elsewhere[19]. Amino acid analysis was performed on a Beckman amino acid analyser (model 6300). Samples were hydrolysed with gaseous HCl generated from 6N HCl containing 0.1% (w/v) phenol.

RESULTS AND DISCUSSION

Isolation of Multiple Forms of rEGF

The HPLC procedures we have developed allow the rapid extraction of two fractions of rEGF activity (peak A and peak B, Fig. 1c) which are equipotent with mEGFα in the EGF competitive binding and [^{3}H]-thymidine incorporation assay (Table 2). Isoelectric points of these rEGFs are in the range of pH5.1 to 5.2. Amino-terminal sequence analysis and amino acid compositions (Table 1) of peaks A and B indicated that peak A contained a mixture of three rEGFs (rEGF -1,-2 and -3) and peak B contained a single polypeptide rEGF-4. These rEGF forms represent different NH_2-terminal deletions: NH_2-Asn-Ser-Asn-Thr-Gly- (rEGF-1), NH_2-Ser-Asn-Thr-Gly- (rEGF-2), NH_2-Asn-Thr-Gly- (rEGF-3) and NH_2-Thr-Gly (rEGF-4).

Table 2: In vitro Biological Activity of rEGF Preparations

EGF Preparation	Binding Affinity[a]	Mitogenic Activity[b]
mEGFα[c]	0.26	0.03
Peak A (rEGFs -1, -2 and -3)	0.20	0.07
Peak B (rEGF-4)	0.21	0.04

a. The EGF concentration (nM) causing 50% [^{125}I]mEGF displacement from human A431 cells[20].
b. The EGF concentration (nM) causing half-maximal stimulation of DNA synthesis in mouse 3T3 cells[20].
c. Purified by the method described elsewhere[20].

Amino Acid Sequence Determination

The sequence of rEGF was determined by automated degradation of the intact molecules from peak A (rEGFs-1, -2, -3 mixture), peak B (pure rEGF-4) and the peptides produced by CNBr cleavage and chymotrypsin digestion of an S-carboxymethylated mixture of rEGFS-1, -2, -3, -4 (Mono Q preparation, Fig. 1b). A summary of the strategy used in determining the sequence is given in Fig. 3. Degradation of the intact molecules established the sequence of the first 26 residues and provided the overlap between the CNBr fragments CN1 and CN2 (Fig. 2c). Sequence analysis of CN2 confirmed the sequence from residues 22 to 26 and established the sequence from residues 27 to 47. The sequence of a chymotryptic peptide (C1)

```
         10        20        30        40
NSNTGCPPSYDGYCLNGGVCMYVESVDRYVCNCVIGYIGERCQHRDLR

    Peak A
NSNTG PPSYDGY                                 lr

               Peak B
   TG PPSYDGY LNGGV MYVES

         CN1                           CN2
 SNTGCPP YDGYCLNGGV  YVESVDRYVCNCVIGYIGERCQHRDLR
                                          C1
                                     IGERCQHRDLR

                P1
            YCLNGGVCM
```

Figure 3. Amino Acid Sequence of rEGF. The amino terminal sequences of rEGF preparations (peaks A and B, Fig. 1c) and specific peptides (italics) are given below the overall sequence (bold type). Prefixes CN, C, P denote peptides generated by cleavage with cyanogen bromide, chymotrypsin and proline-specific endopeptidase(15,31), respectively. Amino acids represented by lower case were identified following digestion of the intact molecule with carboxypeptidase Y(15). The single-letter amino acid notations are used.

provided confirmatory evidence of the sequence from residues 38 to 47 and identified residue 48. The assignment of arginine as the COOH-terminal residue was confirmed by carboxypeptidase digestion of a mixture of the intact molecules (Mono Q preparation, Fig. 1b).

The results presented here permit a proposal for the complete amino acid sequence of rEGF (Fig. 3). There are 48 amino acid residues in the polypeptide chain; the calculated M_r from the sequence analysis is 5377. The amino acid composition, determined from the sequence, is in close agreement with that determined from quantitative amino acid analysis of the protein (Table 1).

A comparison of the sequences of rEGF, mEGF(2,3), hEGF(21), rTGF-α(12), hTGF-α(22) and vaccinia virus polypeptide (VVP)(23) is shown in Fig. 4. There are two particular regions of homology in the EGF family of polypeptides: the highly conserved Asp-Gly-Tyr-Cys-Leu (residues 11-15), also conserved in VVP, and the third disulfide loop region (residues 33-47) where VVP, rTGF-α and hTGF-α also share 8 out of 12 residues. The structural similarity between rEGF and rat and human TGF-α (35%) is not surprising since the two families of polypeptides appear to be functionally related(13,14,24-26). This protein sequence homology can be extended further by a comparison of the cDNAs encoding mEGF(27), hTGF-α(22) and rTGF-α(28) which indicate that these proteins are synthesized from large proteins encoded by mRNAs of 4.5-5.0Kb.

More recently a processed form of the polypeptide encoded by the vaccinia virus genome, which is related to EGF and TGFα, was purified and partially characterized(29). This vaccinia virus growth factor (VVGF), like TGFα, binds to the EGF receptor and stimulates its autophosphorylation. VVGF, unlike EGF and TGFα, is glycosylated(29).

Rat EGF	SNTG[C]PPSYDGY[C]LNGG[G][V][C]MYVESVDRYV[C][N][C]VI[GY]I[G]E[RC]QHRD[L]R
Mouse EGF	NSYPG[C]PSSYDGY[C]LNGG[G][V][C]MHIESLDSYT[C][N][C]VI[GY][S][G]D[RC]QTRD[L]RWWELR
Human EGF	NSDSE[C]PLSHDGY[C]LHD[G][V][C]MYIEALDKYA[C][N][C]VV[GY]I[G]E[RC]QYRD[L]KWWELR
Rat TGF	VVSHFNK[C]PDSHTQY[C]FH-[G][T][C]RFLVQEEKPA[C][V][C]HS[GY][V][G]V[RC]EHAD[L]LA
Human TGF	VVSHFND[C]PDSHTQF[C]FH-[G][T][C]RFLVQEDKPA[C][V][C]HS[GY][V][G]A[RC]EHAD[L]LA
VVP	DIPAIRL[C]GPEGDGY[C]LH-[G][D][C]IHARDIDGMY[C][R][C]SH[GY][T][G]I[RC]QHVV[L]VDYQRS

Figure 4. Comparison of the Amino Acid Sequence of rEGF with mEGF, hEGF, rTGF, hTGF-1 and Vaccinia Virus Polypeptide (VVP). Conserved amino acid residues in all sequences are boxed.

Truncated Carboxyl-Terminus

One striking feature of the rEGF sequence is that it lacks 5 C-terminal residues when compared with mEGF and hEGF. These findings are surprising since it has been reported for both mEGF(3,8) and hEGF(30) that deletion of 5-6 C-terminal residues leads to a 85-90% reduction of *in vitro* potency. The equivalent potency of rEGF (all four forms) and mEGF demonstrated in this study suggests that the conformation of rEGF (which lacks 0-3 N-terminal and 5 C-terminal residues) and mEGF are comparable. Further physicochemical studies need to be undertaken to clarify this point. One of the forms of rEGF purified in this study (rEGF-4 which lacks 3 N-terminal residues) should provide an important model for such structure-function studies.

REFERENCES

1. S. Cohen, J. Biol. Chem. 237:1555 (1962).
2. C.R. Savage, J.H. Hash and S. Cohen, J. Biol. Chem. 248:7669 (1973).
3. C.R. Savage, T. Inagami and S. Cohen, J. Biol. Chem. 247:7612 (1972).
4. S. Cohen, Dev. Biol. 12:394 (1965).
5. C.R. Savage and S. Cohen, Exp. Eye Res. 15:361 (1973).
6. J.M. Bower, R. Camble, H. Gregory, E.L. Gerring and I.R. Willshire, Experientia 31:825 (1975).
7. G. Carpenter and S. Cohen, Ann. Rev. Biochem. 48:193 (1979).
8. S. Cohen, G. Carpenter and K.J. Lembach, Adv. Metab. Disord. 8:265 (1975).
9. J. Schlessinger, A.B. Schreiber, A. Levi, I. Lax, J. Libermann and Y. Yardin, CRC Crit. Rev. Biochem. 14:93 (1983).
10. P.E. Petrides, P. Bohlen and J.E. Shively, Biochem. Biophys. Res. Commun. 125:218 (1984).
11. R.P. Di Augustine, M.P. Walker, D.G. Klapper, R.I. Grove, W.D. Willis, D.U. Harvan and O. Hernandez, J. Biol. Chem. 260:2807 (1985).
12. H. Marquardt, M.W. Hunkapiller, L.E. Hood and G.J. Todaro, Science 223:1079 (1984).
13. G.J. Todaro, C. Fryling and J.E. De Larco (1980) Proc. Natl. Acad. Sci. USA 77:5258 (1980).
14. F.H. Reynolds, G.J. Todaro, C. Fryling and J.R. Stephenson, Nature 292:259 (1981).
15. R.J. Simpson, J.A. Smith, R.L. Moritz, M.J. O'Hare, P.S. Rudland, J.R. Morrison, C.J. Lloyd, B. Grego, A.W. Burgess and E.C. Nice, Eur. J. Biochem., in press.
16. C.R. Savage and S. Cohen, J. Biol. Chem. 247:7609 (1972).
17. R.J. Simpson, M.R. Neuberger and T.-Y. Liu, J. Biol. Chem. 251:1936 (1976).
18. E.C. Nice, G. Grego and R.J. Simpson, Biochem. Intl. 11:187 (1985).
19. B. Grego, I.R. van Driel, P.A. Stearne, J.W. Goding, E.C. Nice and R.J. Simpson, Eur. J. Biochem. 148:485 (1985).
20. A.W. Burgess, J. Knesel, L.G. Sparrow, N.A. Nicola and E.C. Nice, Proc. Natl. Acad. Sci. USA 79:5753 (1982).
21. H. Gregory, Nature 257:325 (1975).
22. R. Derynck, A.B. Roberts, M.E. Winkler, E.Y. Chen and D.V. Goeddel, Cell 38:287 (1984).
23. S. Vankatesan, A. Gershowitz and B. Moss, J. Virol. 44:637 (1982).
24. G. Carpenter, C.M. Stoscheck, Y.A. Preston and J.E. De Larco, Proc. Natl. Acad. Sci. USA 80:5627 (1983).
25. J. Massague, J. Biol. Chem. 258:13614 (1983).
26. A. Ullrich, L. Coussens, J.S. Hayflick, T.J. Dull, A. Gray, A.W. Tam, J. Lee, Y. Yarden, T.A. Libermann, J. Schlessinger, J. Downard, E.L.V. Mayes, N. Whittle, M.D. Waterfield and P.H. Seeburg, Nature 309:418 (1984).
27. J. Scott, M. Urdea, M. Quiroga, R. Sanchez-Pescador, N. Fong, M. Selby, W.J. Rutter and G.I. Bell Science 221:236 (1983).
28. D.C. Lee, T.M. Rose, N.R. Webb and G.J. Todaro, Nature 313:489 (1985).
29. P. Stroobant, A.P. Rice, W.J. Gullick, D.J. Cheng, I.M. Kerr and M.D. Waterfield, Cell 42:383 (1985).
30. M.D. Hollenberg and H. Gregory, Molec. Pharmacol. 17:314 (1980).
31. T. Yoshimoto, R. Walter, and D. Tsura, J. Biol. Chem. 255:4786 (1980).

USE OF HPLC COMPARATIVE PEPTIDE MAPPING IN STRUCTURE/FUNCTION STUDIES

K.R. Williams[*], K.L. Stone[*], M.K. Fritz[*], B.M. Merrill[*],
W.H. Konigsberg[*], M. Pandolfo[+], O. Valentini[+],
S. Riva[+], S. Reddigari[++], G.L. Patel[++], and J.W. Chase[**]

[*]Yale Univ., New Haven, CT. 06510; [+]Istituto di Genetica
Biochimica ed Evoluzionistica, Pavia, Italy; [++]Univ. Georgia
Athens, Georgia 30602; and [**]Albert Einstein College
Bronx, N.Y. 10461

INTRODUCTION

Among the numerous approaches that are frequently used to correlate
a protein's structure with its function are limited proteolysis to
generate "active" fragments; chemical modification with group specific
reagents, in vitro site directed mutagenesis and finally, covalent
crosslinking of proteins to ligands so that individual amino acids near
the active site of the protein can be identified. The ability of peptide
mapping by reverse-phase HPLC to rapidly identify subtle alterations in
the primary structures of proteins enables this technique to play a key
role in these structure/function studies. In this report practical
suggestions will be given for obtaining complete enzymatic digests of
proteins, eliminating HPLC baseline artifacts, maximizing the reproducibility of the resulting retention times, and gas phase sequencing of
the HPLC purified peptides. These methods are generally applicable
and should be of assistance in quickly bringing the full potential of
HPLC comparative peptide mapping to bear on other questions concerning
the interrelationship of protein structure and function. Specific
applications will be given where these techniques have been used to
rapidly verify the location of amino acid substitutions resulting from
in vitro mutagenesis, to identify an active site peptide in a ssDNA
binding protein, and to demonstrate that the most abundant ssDNA binding
protein in higher eucaryotes is actually a proteolytic fragment derived
from a heterogeneous nuclear ribonucleoprotein (hnRNP).

MATERIALS AND METHODS

Providing that the glycerol concentration is below about 20% (v/v)
and the protein concentration is above 0.05 mg/ml then proteins were
prepared for enzymatic digestion by precipitating with a final
concentration of 10% (w/v) trichloroacetic acid. After washing twice
with 0.20 ml cold acetone, the denatured protein was then solubilized in
0.15 ml 8 M urea. While most TCA-precipitated proteins seem to redissolve
readily in 8 M urea, a few of the approximately 15 proteins that we have
tested were not readily solubilized. In these cases a reasonably fine

suspension was made by sonicating the mixture for 5 min in a Branson water-bath sonicator prior to the addition of 0.45 ml of 30 mM NH_4HCO_3. We have so far only found one TCA-precipitated protein that was so insoluble in 2 M urea as to preclude trypsin digestion. In this instance the TCA pellet was redissolved in 6 M GuHCl and the protein was succinylated. Following dialysis and lyophilization, the succinylated protein dissolved readily in 0.15 ml 8 M urea. As previously, 0.45 ml 30 mM NH_4HCO_3 was added prior to trypsin (Cooper Biomedical, lot #34P6977; or Sigma, lot #24F-8011) at a protein/enzyme (w/w) ratio of 25/1. The digest was then continued for 24 hrs at 37°C. If desired, the resulting peptides were then carboxamidomethylated by adding dithiothreitol to a final concentration of 10 mM, followed 10 min later by the addition of 20 mM iodoacetamide. After 20 min at room temperature, the carboxamidomethylation reaction and/or the digest was then stopped by injecting the mixture directly onto a 5 micron Vydac C-4 or C-18 HPLC column (0.46 cm X 25 cm) or a 10 micron Waters Assoc. C-18 μBondapak column (0.39 cm X 30 cm) equilibrated with 0.05% trifluoroacetic acid (TFA) at a flow rate of 0.7 ml/min. Peptides were then eluted by increasing the concentration of solvent B (80% (v/v) acetonitrile in 0.05% TFA) as follows: 0-90 min (0-37.5% B), 90-135 min (37.5% - 75% B) and 135-150 min (75-100% B). Under these conditions both succinylated and non-succinylated peptides separated equally as well.

The Waters Associates HPLC system used for these studies consisted of a model 721 System Controller, a WISP automated sample injector and two model 510 pumps. A 2.5 ml Altex dynamic mixer was placed after the two pumps because a simple, zero dead volume tee was found to give inadequate mixing at pressures below about 2000 psi. This incomplete mixing resulted in less reproducible retention times as well as "wavy" baselines when the ISCO model V^4 absorbance detector that was used was set at 0.1 absorbance units full scale at 210 nm. Two full scale "junk" peaks were seen on blank runs that eluted at about 22% and 29% acetonitrile (v/v) which were subsequently found to result from filtering solvents through Ultipor Nylon 66 (Pall Trinity Micro Corp.) filters. The use of carefully deionized water, a 10 micron pump inlet filter and a 0.5 micron high pressure, in-line filter obviated the need for filtering HPLC solvents prior to use. Maximum reproducibility with respect to retention times was obtained by first briefly degassing solvents by stirring for a few minutes under vacuum and then by keeping the two solvent reservoirs under 2 psi nitrogen pressure in 5 liter Omnifit reagent reservoirs.

Peptides were visualized by their absorbance at 210 nm and were collected manually by peaks. Suitable aliquots were then taken for hydrolysis followed by amino acid analysis on a Beckman 121M analyzer. In some cases the free amino acids were converted to their phenylthiocarbamyl derivatives prior to HPLC analyzsis with the Waters Associates PICO TAG chemistry package. Those peptides selected for sequencing were dried in vacuo prior to redissolving in 0.1 ml 100% trifluoroacetic acid and loading onto an Applied Biosystems Model 470A protein sequencer. The 02NRUN program that came with this instrument was used without modification and the resulting phenylthiohydantoin derivatives were analyzed as previously described[1].

RESULTS AND DISCUSSION

Variations in the Apparent Purity of Commercial Preparations of Trypsin

As shown in Fig. 1 remarkably different chromatograms were obtained when oxidized insulin B chain (Sigma) was digested with L-1-tosylamido-

2-phenylethylchloromethyl ketone (TPCK) treated trypsin obtained from
Cooper Biomedical Corp versus Sigma. In contrast to a previous report[2]
the chromatogram obtained for the Cooper Biomedical Corp. trypsin
indicates that this enzyme can be used without further purification.
Even at a three fold higher ratio of trypsin/protein than what is
recommended (see Materials and Methods) only two major peaks were obtained
corresponding to cleavage after arginine 22 and lysine 29. Use of the
Sigma trypsin results in a 15% decease in the yield of the peptide
corresponding to residues 23-29 and a 90% decrease in the recovery of
the NH_2-terminal peptide (residues 1-22) (Fig. 1). In addition, numerous
other peaks are present which probably result from contamination of the
Sigma trypsin with other proteases. Additional experiments are being
done to determine if the apparent purity of trypsin obtained from these
manufacturers varies appreciably with the particular lot number and to
carry out similar studies on trypsin obtained from other manufacturers.
Irrespective of these results, it is clear that a tryptic peptide map
of oxidized insulin B chain provides an easy, quick means of determining
if a particular preparation of trypsin is suitable for comparative HPLC
peptide mapping. The use of insufficiently pure trypsin decreases the
yield of individual tryptic peptides and increases the complexity of
the resulting maps, making it more difficult to use HPLC peptide mapping
to identify subtle alterations in the primary structures of proteins.

Confirmation of Amino Acid Substitutions Brought About by In Vitro Mutagenesis

HPLC comparative peptide mapping has previously been used to identify

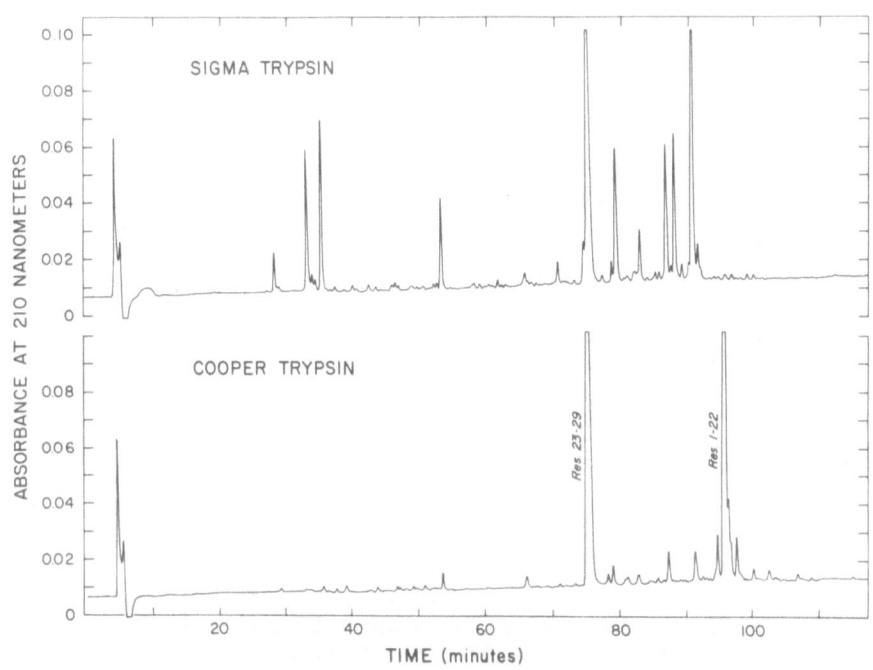

Figure 1: HPLC separation of tryptic peptides from 5 nmol of oxidized
insulin B chain. The insulin was digested for 6 hrs at 37°C in 0.2 ml
of 0.1 M NH_4HCO_3 at a protein/enzyme (w/w) ratio of 7.6 with TPCK treated
trypsin from Sigma (top) or Cooper Biomedical (bottom). The reaction
was stopped by the addition of 5 µl of 5 mM diisopropylfluorophosphate
prior to injecting onto a Vydac C-18 column equilibrated with 0.05%
trifluoroacetic acid. Peptides were eluted with increasing concentrations
of acetonitrile as described in Materials and Methods.

the amino acid replacements that have occurred in two temperature-sensitive mutations in the SSB ssDNA binding protein from E. coli. These studies have demonstrated that the ssb-1 and ssb-113 mutations involve the substitution of tyrosine for histidine 55[3] and serine for proline 176[4] respectively. HPLC gel filtration and sucrose gradient centrifugation have shown that, in contrast to the wild-type SSB, the tetrameric structure of SSB-1 is unstable and gradually dissociates to monomers as the protein concentration is decreased from about 10 to 0.5 μM[3]. It is interesting that an amino acid, histidine 55, that is essential for formation of the SSB tetramer is located so close to an amino acid, phenylalanine 60, that is thought to play a key role in the binding of SSB to ssDNA[1]. Ultraviolet irradiation of the SSB:d(pT)$_8$ complex results in the formation of a single covalent bond between phenylalanine 60 and the oligonucleotide[1]. This result suggests that, like in the case of the binding of other prokaryotic ssDNA binding proteins, aromatic amino acids are also important for the interaction between SSB and ssDNA. Previously, it has been shown that binding of the bacteriophage fd gene 5 protein[5] and the T4 gene 32 protein[6] to ssDNA both involve stacking interactions between aromatic amino acid

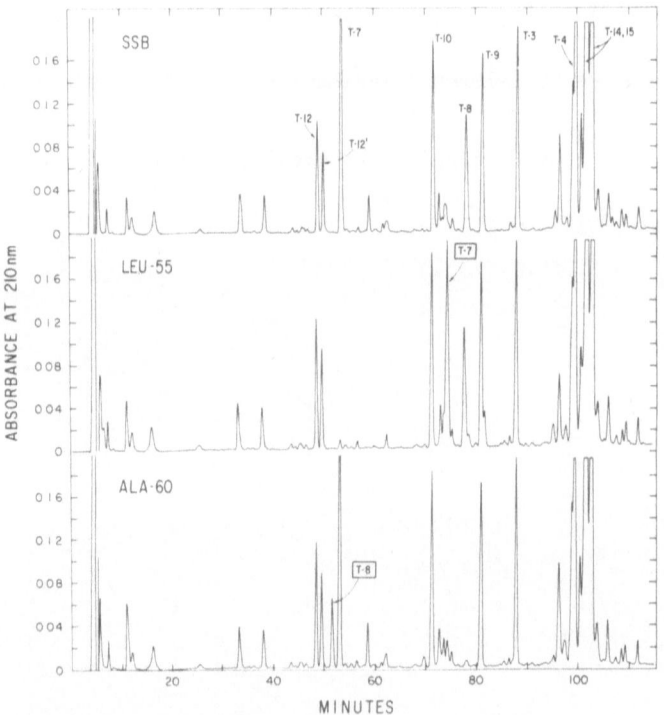

Figure 2: HPLC separation of tryptic peptides from 4 nmol of either the wild-type (top), leucine 55 (middle), or alanine 60 (bottom) SSB protein. The SSB proteins were TCA precipitated and digested with trypsin (Cooper) in 2 M urea at a protein/enzyme (w/w) ratio of 25 as described in Materials and Methods. The reactions were stopped by injecting directly onto a Waters Assoc. C-18 μBondapak colum equilibrated with 0.05% trifluoracetic acid. Peptides were eluted with increasing concentrations of acetonitrile as described in Materials and Methods.

side chains and the nucleotide bases. In order to further explore the role of histidine 55 and phenylalanine 60 in SSB function we have used _in_ _vitro_ mutagenesis to substitute a leucine at position 55 and an alanine at position 60. As shown in Fig. 2 these amino acid substitutions were rapidly verified by comparative HPLC tryptic peptide mapping. The substitution of leucine for histidine 55 increases the elution time of the T-7 tryptic peptide (residues 50-56) from about 54 min to about 74 min. In contrast, substitution of alanine for phenylalanine 60 decreases the elution time for the T-8 tryptic peptide (residues 57-62) from about 78 min to only 52 min. With the exception of these two changes, the chromatograms for the Leu-55 and Ala-60 proteins shown in Fig. 2 are essentially identical with the wild-type profile, thus confirming that only the expected amino acid substitutions have occurred. The importance

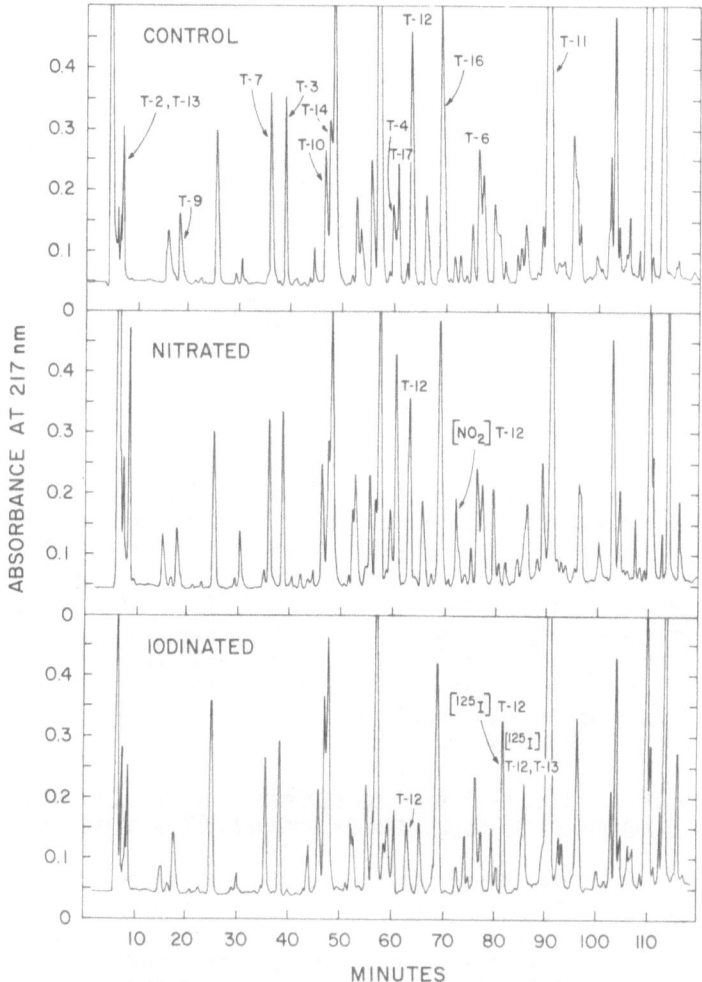

Figure 3: HPLC separation of tryptic peptides from 28.7 nanomole of the native rat liver HDP (top) or from the same amount of this protein that had been nitrated (middle) or iodinated (bottom) to an extent such that 50% of the DNA binding activity had been lost. The proteins were digested and the Waters Assoc. C-18 μBondapak colum was eluted as described in Fig. 2 and in more detail previously[9].

of phenylalanine 60 to SSB binding has been demonstrated by poly[d(A-T)] melting experiments. In 30 mM NaCl SSB decreases the T_M of poly[d(A-T)] by about 33° whereas the Ala-60 protein only decreases the T_M by about 20°. Calculations based on this data suggest that the single leucine for phenylalanine 60 substitution decreases the affinity of SSB for ssDNA by as much as four orders of magnitude.

Identification of an Essential Tyrosine Residue in a Nucleic Acid Helix-destabilizing Protein from Rat Liver

The single-stranded nucleic acid binding proteins that have been reported in rat liver[7] and calf thymus[8] are similar to the E. coli SSB protein in that they also destabilize double-stranded nucleic acids. In order to determine if aromatic amino acids might also play a role in the binding of the rat liver helix-destabilizing protein (HDP) to single stranded nucleic acids, we used chemical reagents to specifically modify tyrosine residues in this protein and then a filter binding assay to determine the effect of this modification on the DNA binding "activity" of the HDP. Our results[9] showed that one of the six tyrosine residues in the rat liver HDP is indeed essential for binding to ssDNA. As shown in Fig. 3 comparative HPLC peptide mapping rapidly identified the location of the essential tyrosine residue. Nitration or iodination of the rat liver HDP to an extent such that 50% of its DNA binding activity is lost decreases the yield of only a single tryptic peptide derived from this protein. As shown in Fig. 3, nitration decreases the yield of the T-12 tryptic peptide, which elutes at about 63 min, by 15% while iodination with Na[^{125}I] decreases the yield by 73%. In both instances, the nitrated or iodinated T-12 peptide elutes later than the unmodified peptide. The only significant [^{125}I] radioactivity detected in the lower chromatogram on Fig. 3 was found in two peaks that corresponded to T-12 and to T-12, T-13, which is an overlapping tryptic peptide that also contained T-12. Both the native and iodinated T-12 were subjected to solid phase sequencing which resulted in the following sequence: Glu-Val-Val-Asp-Ser-Ala-Tyr-Glu-Val-Ile-Lys. When this sequence was searched against the National Biomedical Research Foundations Protein Data Base it was found to exactly match residues 232-242 in porcine M chain lactate dehydrogenase (LDH). That the rat liver HDP actually is LDH was further demonstrated by showing that HPLC tryptic peptide maps for both proteins were identical[9]. While additional experiments are in progress to determine the physiological relevance of the ability of LDH to bind single stranded nucleic acids, it is clear that tyrosine 238 is essential for this "activity" of LDH.

Identification of the UP1 ssDNA Binding Protein from Calf Thymus as a Proteolytic Fragment of a Heterogeneous Nuclear RNA Binding Protein

Immediately following transcription, heterogeneous nuclear RNA (hnRNA) becomes associated with a group of at least 6 proteins which migrate on SDS gel electrophoresis as three closely spaced doublets corresponding to the A_1/A_2, B_1/B_2, and C_1/C_2 proteins that range in size[10] from 32,000 to 42,000 daltons. Because the 22,000 dalton UP1 ssDNA binding protein shares antigenic determinants with all of these hnRNP proteins[11] it seemed possible that UP1 might actually result from proteolytic cleavage of one or more of these proteins. Based on the comparative HPLC peptide maps in Fig. 4, this is indeed the case. Every major peak in the UP1 chromatogram lines up exactly with a peak in the digest of the A_1/A_2 hnRNP proteins. Since UP1 was purified from calf thymus[12] and the A_1/A_2 hnRNP proteins were from HeLa cells it is obvious that the sequence of this fragment of the A_1/A_2 protein is extremely similar if not identical in these two species. Because the NH$_2$-terminus of UP1 is blocked[13], we presume that UP1 represents the first 195 amino

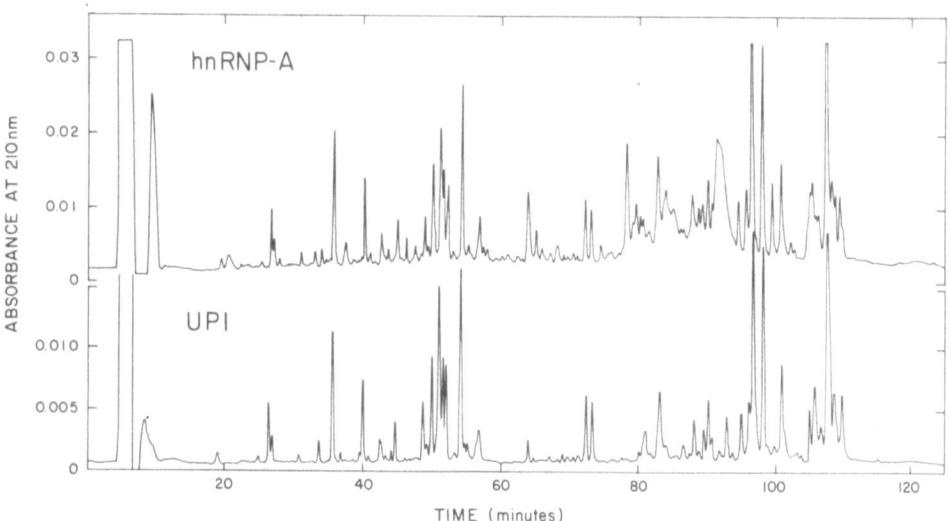

Figure 4: HPLC separation of tryptic peptides from 463 pmol of an approximately equimolar mixture of the A_1 and A_2 hnRNP proteins from HeLa cells (top) or from 500 pmol of the calf thymus UP1 protein (bottom). The proteins were digested in 2 M urea and the Vydac C-18 column was eluted as described in Materials and Methods.

acids in either the A_1 or A_2 proteins. By subtracting the amino acid composition of UP1 from that of the A_1/A_2 proteins it is apparent that the COOH terminal region of at least one of the these hnRNP proteins contains an extremely glycine rich domain (approximately 40%) which is quite different from the NH_2-terminal two-thirds of this protein. Previous studies have demonstrated that UP1 contains a region of internal sequence homology such that when residues 3-93 are aligned with residues 94-194, 33% of the amino acids are identical[12]. Recent partial proteolysis experiments indicate that these two homologous regions correspond to two globular domains, each of which can apparently bind independently to single stranded nucleic acids. Additional studies are in progress to delineate the function of the glycine-rich COOH-terminal domain and, by using ultraviolet light-induced photocrosslinking, to better define which amino acids are actually involved in binding single stranded nucleic acids.

CONCLUSIONS

We have previously used comparative HPLC peptide mapping to rapidly identify several temperature sensitive mutations in ssDNA binding proteins[3,4,14]. These studies demonstrated that even a relatively minor change such as the substitution of a serine for a proline in a 62 residue tryptic peptide resulted in altering the elution position of this peptide on a reverse phase HPLC column by 3 minutes[4,14]. Since similar effects are seen following iodination, nitration, phosphorylation[15], and carboxymethylation[15] of even single amino acids in tryptic peptides, it is apparent that comparative HPLC peptide mapping offers an attractive technique that can be of considerable use in correlating the structure and functions of proteins.

REFERENCES

1. B.M. Merrill, K.R. Williams, J.W. Chase, and W.H. Konigsberg, Photochemical cross-linking of the E. coli single-stranded DNA-binding protein to oligodeoxynucleotides, J. Biol. Chem. 259:10850 (1984).

2. K. Titani, T. Sasagawa, K. Resing, and K. Walsh, A simple and rapid purification of commercial trypsin and chymotrypsin by reverse-phase high performance liquid chromatography, Anal. Biochem. 123:408 (1982).

3. K.R. Williams, J.B., Murphy, and J.W. Chase, Characterization of the structural and functional defect in the E. coli single-stranded DNA binding protein encoded by the ssb-1 mutant gene, J. Biol. Chem. 259:11804 (1984).

4. J.W. Chase, J.J. L'Italien, J.B. Murphy, E.K. Spicer and K.R. Williams, Characterization of the E. coli SSB-113 mutant single-stranded DNA-binding protein, J. Biol. Chem. 259:805 (1984).

5. T.P. O'Connor, and J.E. Coleman, Proton nuclear magnetic resonance (500 MHz) of mono-, di-, tri-, and tetradeoxynucleotide complexes of gene 5 protein, Biochemistry 22:3375 (1983).

6. R.V. Prigodich, J. Casas-Finet, K.R. Williams, W. Konigsberg and J.E. Coleman, ^{1}H NMR (500 MHz) of gene 32 protein-oligonucleotide complexes, Biochemistry 23:522 (1984).

7. G.L. Patel, and P.E. Thompson, Immunoreactive helix-destabilizing protein localized in transcriptionally active regions of Drosophila polytene chromosomes, Proc. Natl. Acad. Sci. USA 77:6749 (1980).

8. G. Herrick, and B. Alberts, Purification and physical characterization of nucleic acid helix-unwinding proteins from calf thymus, J. Biol. Chem. 251:2124 (1976).

9. K.R. Williams, S. Reddigari, and G.L. Patel, Identification of a nucleic acid helix-destabilizing protein from rat liver as lactate dehydrogenase-5, Proc. Natl. Acad. Sci. USA 82:5260 (1985).

10. A.L. Beyer, M.E. Christensen, B.W. Walker, and W.M. LeStourgeon, Identification and characterization of the packaging proteins of core 40S hnRNP particles, Cell 11:127 (1977).

11. O. Valentini, G. Biamonti, M. Pandolfo, C. Morandi, and S. Riva, Mammalian single-stranded DNA binding proteins and heterogeneous nuclear RNA proteins have common antigenic determinants, Nucleic Acids Res. 13:337 (1985).

12. B.M. Merrill, M.B. LoPresti, K.L. Stone, and K.R. Williams, HPLC purification of UP1 and UP2, two related single-stranded nucleic acid binding proteins from calf thymus, J. Biol. Chem. 260: in press (1985).

13. K.R. Williams, K.L. Stone, M.B. LoPresti, B.M. Merrill, and S.R. Planck, Amino acid sequence of the UP1 calf thymus helix-destablizing protein and its homology to an analogous protein from mouse myeloma, Proc. Natl. Acad. Sci. USA 82:5666 (1985).

14. K.R. Williams, J. L'Italien, R. Guggenheimer, L. Sillerud, E. Spicer, J. Chase, and W. Konigsberg, Comparative peptide mapping by HPLC: identification of single amino acid substitutions in temperature sensitive mutants, in: "Methods in Protein Sequence Analysis", M. Elzinga, ed., Humana Press, Clifton, New Jersey, pp 499-507 (1982).

15. K.R. Williams, K.L. Stone, M.B. LoPresti, B.M. Merrill, W.H. Konigsberg, Comparative peptide mapping by HPLC: use of micro-sequencing techniques to identify amino acids that have been altered by mutation, chemical modification, or phosphorylation, Fed. Proc. 44:1617 (1985).

PRECOLUMN DERIVATIZATION OF PEPTIDES WITH PHENYLISOTHIOCYANATE (PITC)

Simon Lemaire and Serge Nolet

Department of Physiology and Pharmacology
Faculté de Médecine, Sherbrooke
Québec, Canada J1H 5N4

SUMMARY

A rapid and ultra sensitive method for the purification and quantitation of peptide hormones is described. The method is based on precolumn derivatization of peptides with phenylisothiocyanate (PITC) to form the phenylthiocarbamoyl derivatives (PTC-peptides). The derivatized peptides are analyzed by reverse-phase high performance liquid chromatography (HPLC) on a Zorbax ODS column and detected at 269 nm at a limit sensitivity of 1 picomole. The eluted material can be either analysed for amino acid composition or structural sequence, using available microanalytic techniques. The HPLC analysis of PTC-dynorphin-(1-13), PTC-dynorphin-(1-11) and PTC-Leu-enkephalin allowed the complete seraration of each peptide with high recovery (>75%). The method can be used at the final purification step of naturally-occurring peptides and for the evaluation of the purity and amount of material than can be submitted either to amino acid analysis or to Edman degradation.

INTRODUCTION

The purification and characterization of peptides at pmolar levels present difficulties in regard to the analysis of their purity and quantity before the determination of their structure. Most purification techniques rely on detection apparatus that necessitate the presence of substantial amounts of material and sometimes may provide signals unrelated to peptide material (salts, impurities). However, some detection methods for peptides such as post-column derivatization with o-phthalaldehyde, ninhydrin or fluorescamine may provide great sensitivity[1] but the products of the reactions are not suitable to Edman degradation[2]. Recently, Chang[3] has introduced a new technique for the isolation and characterization of peptides at pmolar level. His technique is based on the precolumn formation of dimethylaminoazobenzenethiocarbamoyl- (DABTC-) peptides. Such a procedure allows the separation of peptides and their detection at high sensitivity. However, the reaction of peptides with dimethyl-aminobenzene isothiocyanate (DABITC) presents the following difficulties: 1. the reaction is relatively slow and uncomplete; 2. several products can be obtained from a single peptide, depending on the completeness of the reaction. Herein, we present an analytical method which is based on the formation of phenylthiocarbamoyl- (PTC-) peptides followed by their separation by HPLC and their analysis by micro-analytical techniques.

EXPERIMENTAL

Pyridine, dimethyl-N-allylamine (DMAA), triethylamine (TEA), phenylisothiocyanate (PITC), trifluoroacetic acid (TFA), 1-Cl-butane, dimethylformanide (DMF), HCl and o-phthalaldehyde were purchased from Pierce Chemical Co., Rockford, Illinois. N-dymethylaminopropyl-N-ethylcarbodiimide (EDC), aminopolystyrene resin, N-methylmorpholine and N-propylamine were products of Sequemat Inc., Boston, N.J.. All these solvents or reactives were sequenal grade. Acetonitrile and methanol (HPLC grade) were obtained from Fisher, Montreal, Quebec. Polypropylene microcentrifuge tubes (1.5 ml) were products of Walter Sarstedt Inc., Princeton, N.J.. They were extensively washed with HCl (2N), rinced with bidistilled water and air dried. The microcentrifuge (Model 5414, Eppendorf) and the heating block (Model 5320, Eppendorf) were purchased from Mandel Scientific Co. Ltd., Montreal, Quebec. Argon (ultrapure) was obtained from Liquid Carbonic, Montreal. A vortex evaporator (Buchler) and an HPLC system (Waters) including two M-6000 A pumps, one UV detector (Model 410), an automatic injector (WISP 710), a fluorometer (Model 420), a system controller (Model 720) and a data module (Model 730) were also used. Dynorphin-(1-13), dynorphin-(1-11) and Leu-enkephalin were synthetized in our laboratory as previously described[4]

Derivatization of peptides with PITC

The synthetic peptides (0.1 - 5 nmoles) were first added to a polypropylene centrifuge tube in 0.1% TEA. They were dried in the vacuum vortex evaporator and dissolved with 100 µl of DMAA-pyridine-acetonitrile-H_2O (1.18 : 7.5 : 7.5 : 10) adjusted to pH 9.5 with TFA. The tube was filled with a stream of argon and 4 µl of PITC (50% in pyridine) were added on top of the solution. The tube was capped, vortexed and heated at 56°C for 30 min. Excess reactives and by-products were then removed by two washings with 75 µl-portions of 1-Cl-butane. Two ml of methanol were added to the aqueous phase and the sample was dried under vacuum.

Chromatography

Three types of HPLC columns were tested for the separation of PTC-peptides: the µ-Bondapak C18 (10µ, Waters), the ODS-3 (10 µ, Wathman) and the Zorbax ODS (5 µ, Dupont). The latter one was adopted for its greater efficiency. PTC-peptides were separated using either an isocratic system with 44% acetonitrile in 0.08 TFA (pH 2.2; system I) or a gradient system with two eluents; (A) 20% acetonitrile in 0.08% TFA (pH 2.2) and (B) 80% acetonitrile in 0.08% TFA (pH 2.2; system II). The elution was run at 1 ml/min. Linear gradients in system II were effected, starting with 25% B to reach 40%, 70% and 100% B at 5, 30 and 35 min, respectively. The column was washed for 5 min with 100% B and returned to the starting conditions (25% B) so that a new injection could be effected every hour. PTC-peptides were detected by U.V. absorbance at 269 nm. They were collected, lyophilized separately and kept under vacuum at 0°C before further analysis.

Amino acid analysis

Samples of PTC-peptides or underivatized peptides (10 pmoles or more) were poored into a glass tube (50 x 6 mm) in 50 µl of MeOH. The solvent was removed under vacuum. After drying, the vacuum was released and 200 µl of constant boiling HCl (6M) containing 0.1% mercapto ethanol were added to the bottom of the tube. The tube was sealed under high vacuum and the samples were hydrolyzed at 108°C for 24 hours. After hydrolysis, the residual hydrochoric acid was removed under vacuum. Blanks were run by incubating the acid solution in the abscence of peptide. The hydrolysates were dissolved in 100 µl of $Na_2B_4O_2 \cdot 10H_2O$ (0.1 M, pH 9.5) and filtrated through MS1 Cameo filters (0.45µ, MS1 micron Separations Inc., Honcoye Falls, N.Y.). The amino acid analysis was performed by HPLC on a C18/Resolve column (5µ, Waters). The injection was effected in three steps, thus allowing the precolumn derivatization of amino acids with o-phthalaldehyde. One min was allowed for the first two injections with a flow rate of zero. The first injection contained 10 µl of a solution of o-phthalaldehyde (50 mg in methanol: 0.1M $Na_2B_4O_2$, 1.25: 11.2). The second injection was done with 10 µl of a solution of mercaptoethanol (10 µl mercaptoethanol in 6.5 ml of MeOH). The third injection contained the sample in 10 µl of 0.1 M $Na_2B_4O_2$ (pH 9.5). A gradient was then started between the two following eluents (A) MeOH: tetrahydrofuran: 0.05M sodium acetate (pH 7.5), 2 : 2 : 96 and (B) methanol: H_2O, 65 : 35. The reaction was occurring at the beginning of the third injection when the sample and reactives were passing through an enlarged coil (0.5 mm inner diameter, 10 cm long) located at the outlet of the automatic injector. The addition of the reaction coil allowed a great reliability to the system: the fluorescence peaks were proportional to the amounts of amino acids tested and the integrations were quite reproducible from one run to another with the same amount of sample. Derivatized amino acids were eluted separately according to the gradient described in Table 1.

Solid-phase Edman degradation

PTC-peptides (or underivatized peptides) were attached to aminopolystyrene via activation of the C-terminal carboxylgroup by a water soluble carbodiimide (EDC) according to the procedure of L'Italien and Strickler[5]. The coupling yield (20-45%) was generally improved by 5% to 15% when PTC-peptides were used instead of the underivatized compounds. Briefly, the coupling procedure was as follows: the peptide material (0.5 - 1 nmole in 0.1% TEA) was added to a polypropylene centrifuge tube and lyophilized. The resin (15 mg) was

Table 1. Elution gradient for the separation of o-phthalaldehyde-mercaptoethanol derivatives of amino acids by HPLC on C18-Resolve column[a].

Time	Flow	%A	%B	Curve
0	0	100	0	*
2	0.1	100	0	1
2.5	1.0	100	0	6
10	1.0	83	17	6
35	1.0	0	100	6

[a]This gradient was started after the injection of the amino acid sample (third injection) as described under "Experimental".

preincubated with 1 ml of a solution of pyridine:H_2O, 8.1 : 85 (pH 5.0, HCl) for 1 hour at room temperature, washed with three 1.5 ml-portions of H_2O and two 1.5 ml-portions of DMF and added to the peptide sample. The coupling reaction was carried out for 1 hour at 37°C after the addition of 200 µl of DMF, 50 µl of pyridine:H_2O (8.1 : 85, pH 5.0) and 4 mg of EDC (in 100 µl of DMF: H_2O, 80: 20). After the reaction, the resin was washed with two 1 ml-portions of a mixture of the sequencer coupling buffer (pyridine:14% N-methyl-morpholine, 3:2; pH 8.5) and DMF (1:3). The free amino groups were blocked by the addition of 100 µl of PITC (12% in acetonitrile) and a coupling period of 1 hour at 37°C. The resin was washed with four 1 ml-portions of MeOH and two 1 ml-portions of ether and air dried. The peptides were sequenced by solid-phase Edman degradation using unmodified 60 min sequenator program on Sequemat Mini-15 solid-phase sequencer equipped with P-6 autoconvertor[5]. Each cycle from the sequenator (after conversion) was dried twice under the evapo-mix (Buchler). The samples were then dissolved with 50 µl of 50% acetonitrile and applied to a Microsorb C18 column (4.2 x 100 mm, 3 µ; Rainin, Woburn, Mass.). The PTH-amino acids were eluted at room temperature with the following solutions: (A) 20% acetonitrile in 0.1% TFA, (pH 2.2), (B) 60% acetonitrile in 0.1% TFA . A linear gradient between solution A and B was effected starting with 100% of solution A at time 0 to reach 45% of solution B at 25 min.

RESULTS

HPLC of PTC-peptides

   PTC-peptides had a very high retention time on HPLC when analyzed in the same conditions as the underivatized compounds. However, using a Zorbax ODS column (5 µ, Dupont) with a linear gradient of acetonitrile starting at a relatively high concentration of the organic solvent (system II), an excellent separation of PTC-Leu-enkephalin, PTC-Dyn-(1-11) and PTC-Dyn-(1-13) was obtained (Fig. 1). The peaks were collected and lyophilized to yield more than 75% of the starting material (as evaluated by amino acid analysis). An increase in sensitivity was achieved with the isocratic conditions (limit detection: 1 pmole; system I, data not shown); however, some peptides such as Dyn-(1-17) did not separate as well from contaminating by-products under the isocratic conditions. It was useful to use a blank for determining the retention time of excess reactives or by-products (Fig. 1B). However, most by-products were eluted at higher retention times (35 min or more) and their interference could be minimized by increasing the volume of 1-Cl-butane during the washing period of the derivatization procedure.

Amino acid analysis of derivatized peptides

   Samples corresponding to 30 pmoles each of PTC-dynorphin-(1-11) and PTC-dynorphin-(1-13) were submitted to acid digestion followed by amino acid analysis using the precolumn derivatization with o-phthalaldehyde in presence of mercaptoethanol Table 2 shows that the composition of each peptide corresponded to the expected value. Our modified method for precolumn derivatization was quite reliable due to the separate injection of o-phthalaldehyde, mercaptoethanol and the amino acid sample and to the addition of a reaction coil between the automatic injector and the HPLC column. The separation conditions of the derivatized amino acids are indicated in Table 1.

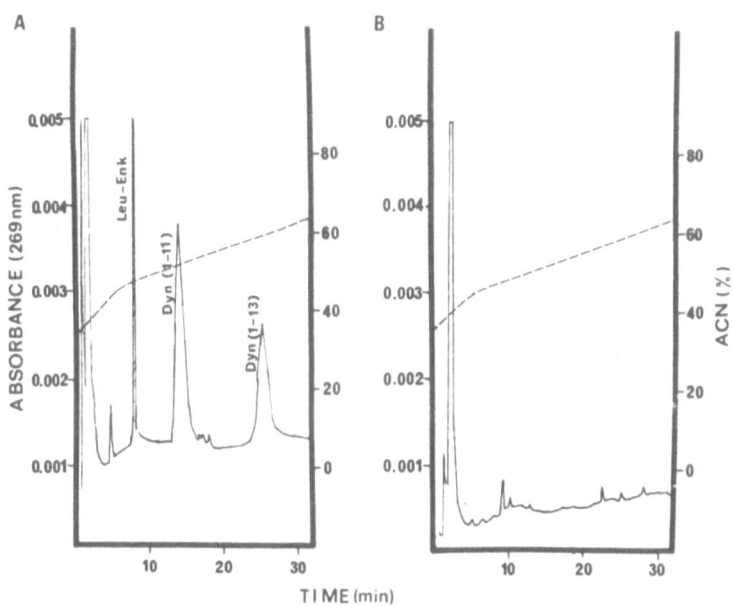

Fig. 1 (A) Separation of a mixture of PTC-Leu-enkephalin, PTC-dynorphin-(1-11) and PTC-dynorphin-(1-13) on HPLC (system II, 100 pmole samples).
(B) Blank.

Table 2. Amino acid composition of acid (HCl) digests of PTC-dynorphin-(1-11) and PTC-dynorphin-(1-13) (30 pmoles samples)[a].

Amino acid	PTC-dynorphin-(1-11) ratio	PTC-dynorphin-(1-13) ratio
Gly	2.12 (2)	2.00 (2)
Arg	2.97 (3)	2.99 (3)
Tyr	0.46 (1)	0.95 (1)
Phe	0.88 (1)	0.76 (1)
Ile	1.00 (1)	1.21 (1)
Leu	1.07 (1)	1.53 (2)
Lys	0.83 (1)	1.99 (2)

[a]PTC-peptides were collected from HPLC (Fig. 1) and 30 pmole samples were hydrolyzed and analyzed by HPLC after derivatization with o-phthalaldehyde as described under "Experimental". Theoretical values are under parentheses; Pro was not detected.

Micro Sequence analysis of PTC-dynorphin-(1-13)

PTC-peptides can be analyzed for their amino acid sequence using any available peptide or protein sequencer provided that their amino group is not blocked by any other group than the PTC function. In addition, for solid-phase sequencing, the C-terminal function must be

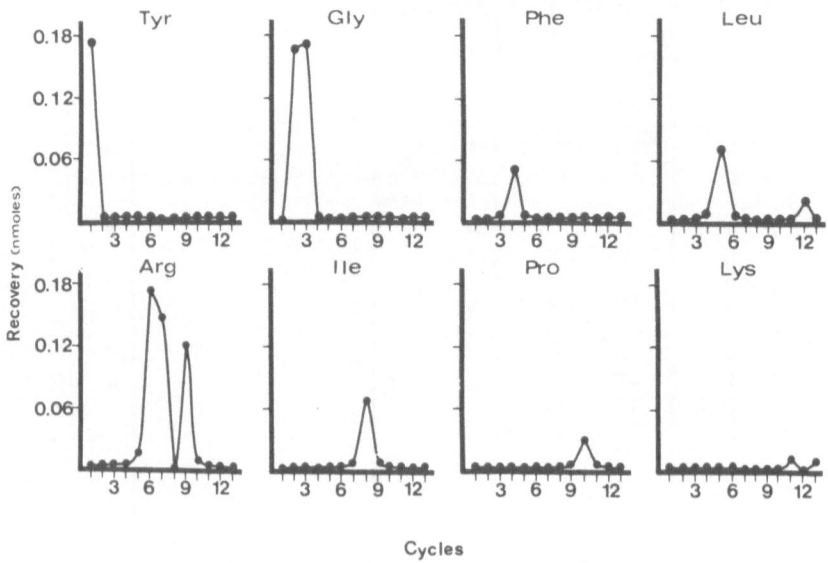

Fig. 2   Solid-phase Edman degradation of PTC-dynorphin-(1-13) (0.5 nmole).   The material was coupled to 15 mg of aminopolystyrene resin and' the degradation was performed on an automatic sequenator (Sequemat).

able to react with a solid-support. As an example, PTC-dynorphin-(1-13) (500 pmoles) was submitted to solid-phase sequencing on a Sequemat sequenator (Fig. 2). The first cycle gave a recovery yield of 36% (Tyr: 0.18 nmole). The recovery of Arg at cycle #9 was 24%. The HPLC chromatograms were all clean and quite unambiguous. The whole structure of PTC-dynorphin-(1-13) was identified (Fig. 2).

DISCUSSION

The advent of HPLC has greatly improved the capacity of separation of peptide mixtures. For peptide detection, HPLC has been combined to various analytical methods including UV absorption at low wavelengts (210-220 nm), post-column derivatization with fluorescamine or ninhydrin and recently to electrochemical detectors for peptides containing Tyr residues[1]. All these methods present great improvements over those used a decade ago. However, while the UV absorption is not selective to peptide material, the derivatization and oxidation techniques give products that are not suitable for amino acid content or sequence analysis. Recently, Chang[3] has developed a detection method (formation of DABTC-derivatives) that can be combined to amino acid content or sequence analyses. However, the derivatization of the side chain of Lys with DABITC is rather slow and often incomplete. On the other hand, PITC reacts quite rapidly and completely with peptides and hopefully gives one single PTC-peptide product (starting with one peptide). PTC-peptides can be separated quite nicely by HPLC, thus providing another separation mean for peptide material (Fig. 1).

Derivatization with PITC has also been used successfully for the analysis of amino acid contents[6]. PTC-amino acids are quite stable and can be separated by HPLC. In the present study, it was found that PTC-peptides are stable as well and slightly acidic chromatographic conditions (0.08% TFA, pH 2.2) do not interfere with subsequent amino acid content or sequence analyses. Our method can be used for the final step of purification of naturally-occurring peptides. The HPLC on Zorbax ODS provides a nice separation of closely related peptides (dynorphin-(1-11) and dynorphin-(1-13)) and still, the eluted material can be recovered and the amount of material can be properly evaluated before any further analysis, with minimal loss of the desired compound. Moreover, the UV detection of PTC-peptides is so much sensitive that only a small percentage of a peptide sample (1 pmole) can be derivatized chromatographed and still provide useful informations on the purity and quantity of the peptide material. However, our method presents some difficulties that have to be worked up: the separation of PTC-peptides requires very special reverse-phase column. For some unknown reasons, very poor separations were obtained on μ-Bondapak (Waters) and ODS C18 columns (Whatman). The method cannot be used in connection with solid-phase sequencing when the peptide has to be attached by the side-chains of the Lys residues. Finally, all the reactives and solvents used for derivatization have to be first grade in order to avoid the appearance of non-peptidic peaks. Most of these problems can be overcome using different methods of separation or sequence analysis. Therefore, the precolumn derivatization of peptides with PITC was found to be a very powerful mean for the characterization of peptides at picomolar levels.

REFERENCES

1. M. W. Hunkapiller, J.E. Strickler and K.J. Wilson. Contemporary methodology for protein structure determination. Sciences 226: 304-311, (1984).

2. P. Edman. Method for determination of the amino acid sequences in peptides. Acta Chem. Scand. 4: 283-293, (1950).

3. J.-Y. Chang. Isolation and characterization of polypeptide at the picomole level. Precolumn fromation of peptide derivatives with dimethylaminoazobenzene isothiocyanate. Biochem. J. 199: 537-545, (1981).

4. A. Turcotte, J.-M. Lalonde, S. St-Pierre and S. Lemaire. Dynorphin-(1-13), Structure-function relationships of Ala-containing analogs. Int. J. Peptide Protein Res. 23: 361-367, (1984).

5. J.J. L'Italien and J.E. Strickler. Application of high-performance liquid chromatographic peptide purification to protein microsequencing by Solid-Phase Edman degradation. Analytical Biochem. 127: 198-212, (1982).

6. B.A. Bidlingmeyer, S.A. Cohen, and T.L. Tarvin. Rapid analysis of amino acids using pre-column derivatization. J. Chromatogr. 336:93-104, (1984).

HIGH-PERFORMANCE LIQUID CHROMATOGRAPHY AS A MEANS OF CHARACTERIZING ISOFORMS OF STEROID HORMONE RECEPTOR PROTEINS

J.L. Wittliff, N.A. Shahabi, S.M. Hyder, L.A. van der Walt, L. Myatt, D.M. Boyle and Y.-J. He

Hormone Receptor Laboratory, Department of Biochemistry, James Graham Brown Cancer Center and University of Louisville School of Medicine, Louisville, KY 40292

INTRODUCTION

Steroid hormone receptors remain one of the most elusive, labile proteins under study in the field of molecular endocrinology. To our knowledge, no one has provided conclusive evidence of the "native state" of either estrogen or progestin receptors in breast and endometrial carcinomas. Most investigators have utilized sucrose gradient centrifugation, conventional column chromatography and/or isoelectric focusing to characterize these proteins in impure preparations[1,2]. Using these techniques, several receptor species have been identified. The origin of the multiple forms of receptors (polymorphism) and their biological significance have been a major focus of our investigations during the past 15 years[1-3]. Clearly, distinct physiologic species exist which we have termed isoforms since each protein binds a particular class of steroid hormone while exhibiting different properties of size, shape and/or surface charge[4]. However, certain components may arise due to processes such as proteolytic cleavage which may occur during homogenization, prolonged incubation and overnight separation. To circumvent the problem of prolonged manipulation in receptor preparations, we developed the use of high-performance liquid chromatography in size exclusion[3,5], ion-exchange[6-12], chromatofocusing[13], and hydrophobic interaction[14] modes for rapid, effective separation of receptor isoforms. Briefly the HPLC columns employed consist of rigid, macro-porous silica-based supports (stationary phase) which give 1) rapid flow rates, 2) low column volume to applied sample ratio and 3) high resolution and recovery. Derivatized groups compose the bonded phase while the solvent represents the mobile phase. These columns require care consisting primarily of the use of guard columns, filtered buffers and stringent wash procedures to give reproducible results. Although most workers in the field utilize radiochemically-labeled ligands, steroid receptors also may be identified by fluorescent ligands[15], radioactive affinity labels[16] or monoclonal antibodies[17].

HPLC Instrumentation

The configuration of the Altex (Beckman) HPLC set-up with in-line technology used in our laboratory is shown in Figure 1. Briefly a two pump system is used for HPIEC, HPCF and HPHIC with UV detector, pH meter (Pharmacia), conductivity meter (Bio-Rad), and gamma radioactivity detector (Beckman) in-line before fraction collection. All chromatography was performed in a Puffer-Hubbard cold box at 3-4°C. HPSEC is conducted isocratically. Pre- and post-column derivatization and fluorescence may also be employed with this arrangement.

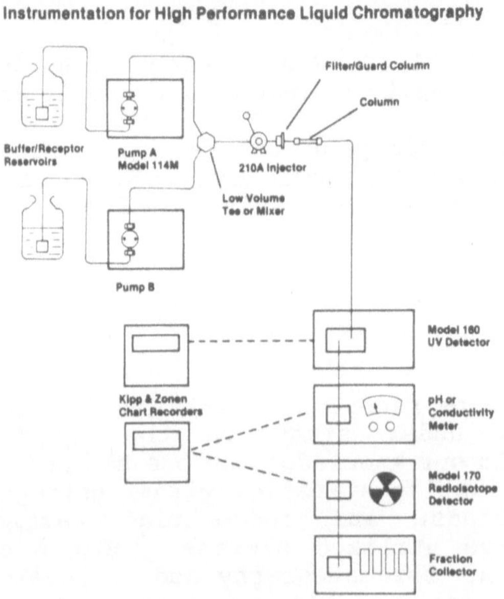

Instrumentation for High Performance Liquid Chromatography

Fig. 1. Diagram of HPLC Set-up with In-line Instrumentation used to Separate and Identify Steroid Receptor Isoforms.

Preparation of cytosolic receptors

All procedures were carried out at 0-4°C. Tissues were homogenized using a Brinkman Polytron (two 10-sec bursts) in 2-4 vol of Tris buffer (10 mM Tris HCl, 1.5 mM EDTA, 10% glycerol, 10 mM monothioglycerol, pH 7.4 at 4°C) or PDEG (10 mM potassium phosphate, 1 mM DTT, 1.5 mM EDTA, and 10% glycerol, pH 7.4) with or without 10 mM sodium molybdate. Cytosols were prepared by centrifugation of the homogenates for 30-60 min at 40,000 rpm in a Beckman Ti 70.1 rotor. Cytosols were incubated at 4°C for 2-24 h with 1-4 nM [^{3}H]R5020, [^{3}H]ORG-2058 or [$16\alpha^{125}$I]iodoestradiol-17ß in the presence (non-specific binding) or absence (total binding) of a 200-fold molar excess of an unlabeled competitor. The incubations were terminated by removing unbound steroid in a pellet derived from an equal volume of a 1% dextran-coated charcoal suspension (1% charcoal, 0.5% dextran). The labeled cytosol was applied to the charcoal pellet, mixed briefly, allowed to stand for 2-10 min and then centrifuged at 600 xg. Cytosol protein concentrations were determined by the methods of Waddell[18] or Bradford[19]. Specific binding capacity was expressed as femtomoles of steroid bound per milligram of cytosol protein.

High-performance size exclusion chromatography (HPSEC)

All chromatography was performed utilizing Spherogel TSK-3000SW size exclusion columns (7.5 x 700 mm) with a Beckman Model 322 HPLC system equipped with an in-line Hitachi Model 100-40 spectrophotometer[5]. The chromatographic column was comprised of two separate units, a short (7.5 x 100 mm) TSK 3000SW guard column and, immediately downstream, the longer (7.5 x 600 mm) TSK 3000SW size exclusion column. Samples were applied in 20-250 µl volumes using a Hamilton syringe and the model 210 sample injection valve. The $P_{50}EDG$ (50 mM NaH_2PO_4/Na_2HPO_4 buffer, pH 7.4 at 4°C, containing 1.5 mM EDTA, 1 mM DTT, 10% (v/v) glycerol). All buffers were filtered with a 0.45-µm filter (Millipore). Elution was carried out at a flow-rate of 0.7 ml/min. Column effluent was collected as 0.5-1 min fractions. Following a day of chromatography, the entire column was washed overnight with filtered, distilled, deionized water. The entire column was washed weekly with a filtered solution of 15% dimethyl sulphoxide in methanol whereas the TSK 3000SW guard column was washed periodically with a solution of 6 M urea. The chromatographic system was stored in filtered, distilled, deionized water.

A representative profile of cytosol proteins separated by HPSEC using a TSK-3000 SW column is shown in Fig. 2. These receptors were extracted in PEDG buffer in the absence and presence of 10 mM molybdate as described in Methods. Under these conditions and in low ionic strength, the specific estrogen binding capacity was distributed primarily as a high molecular weight species (71-76Å) eluting just after the void volume of the column. This component may be analogous to the 8-9S isoform identified by sucrose gradient centrifugation[3]. Non-specific binding was virtually absent whether or not molybdate was included.

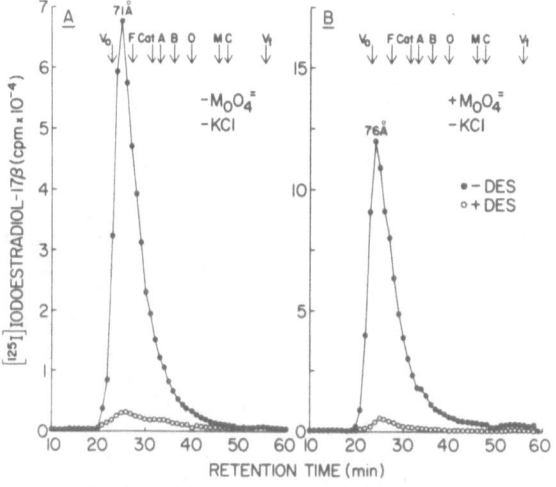

Fig. 2. Separation of Estrogen Receptors by HPSEC. Cytosols were prepared in PEDG buffer either without (A) or with (B) 10 mM sodium molybdate and incubated with [125I]iodoestradiol-17β in the absence (•) or presence (o) of a 200-fold excess of DES. ER were eluted in $P_{50}EDG$ buffer either without (A) or with (B) 10 mM molybdate. Arrows indicate the position of protein markers: ferritin (F), catalase (Cat), human serum albumin (A), bovine serum albumin (B), ovalbumin (O), myoglobin (M), cytochrome C(C); V_O represents the void volume and V_t represents the total volume of the column. Taken from ref. 6.

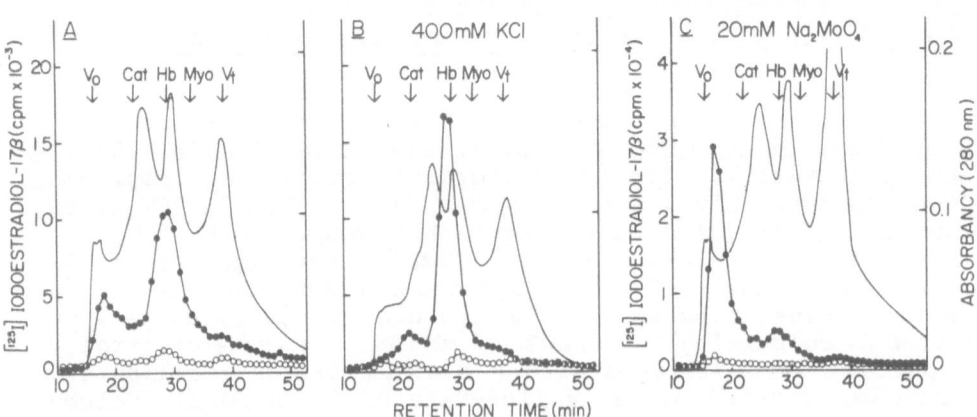

Fig. 3. Influence of KCl and Molybdate on Size Exclusion
Properties of ER in Human Breast Carcinoma. Cytosol preparation
and reaction conditions are described earlier. In each case,
reactions were performed using 4 nM [^{125}I]iodoestradiol-17ß as
ligand in the presence (o) or absence (●) of a 200-fold excess
of DES. Marker proteins were chromatographed individually and as
a group in separate runs and their retention times were determined
from their absorption peaks at 280 nm. (A) The reaction medium
consisted of Tris buffer. (B) The initial reaction medium was
Tris buffer, but KCl was added after 1 hr of incubation to bring
the concentration to 400 mM. The elution buffer also contained
400 mM KCl. (C) The reaction and elution buffers consisted of
Tris buffer with 20 mM molybdate. A total of 0.7 mg of cytosol
protein was applied. V_O = void volume as determined using blue
dextran; V_t = total volume as determined using ^3H_2O. Taken from
ref. 5.

Fig. 3 is a representative HPSEC profile of estrogen
receptors in human breast cancer. It is our experience that 100
mM salt provides the minimum ionic strength required to maintain
a linear relationship in the elution sequence of marker proteins
employed[2,5]. This is due predominantly to the hydrophobic surface
and slightly cationic character of these columns. The column
matrix volume appears to be unaffected by either variable pH or
[salt]. Inclusion of polar organic solvents such as propanol
may improve recovery and sharpen elution profiles.
    Treatment of the cytosol with 400 mM KCl resulted in the
appearance of a 29-32 A form on HPSEC which dominated the profile
(Fig. 3B). There was a small quantity of a component retained at
22 min which appeared to have a Stokes radius of ~48A.
    When 20 mM molybdate was added to the homogenizing buffer,
virtually all of the specific binding was exhibited by a component
eluting after the void volume (Fig. 3C). These data suggest a
component with a Stokes radius of > 70A. Only a small quantity
of the 29-32A species remained.
    We have also utilized HPSEC to study the size isoforms of
progestin receptors in human uterus and breast carcinoma[9,10].
Both 30 and 60 cm TSK3000 SW and columns have been employed. On
the shorter column, receptor-bound ligand appeared as two peaks,
using either [^3H]R5020 or [^3H]ORG-2058 (Fig. 4). The primary
peak associating specifically with R5020 appeared consistently
between fractions 30 and 35 in many uterine cytosols (Fig. 4B).
This receptor isoform represented the majority of (>70%) of
specific binding applied to the column. Recoveries on the 30
cm TSK 3000 SW columns were consistently 87 to 93%. Column

calibration with marker proteins suggested this component to be a very large species of >80A.

A small but distinct and reproducible secondary isoform (Fig. 4B) was demonstrable between fractions 50 and 60 with R5020, (ca. 50A), but only represented 10 to 15% of the specifically bound radioactivity. No free steroid radioactivity appeared in these separations, indicating that dextran coated-charcoal clearance of the unbound ligand was complete and that no discernible column-induced dissociation of steroid occurred. [3H]ORG-2058 also appeared bound specifically by two isoforms with virtually identical ratios as observed for [3H]R5020. However, with ORG-2058 as ligand, the primary receptor isoform was observed between fractions 40 and 47 (Fig. 4A) representative of a protein of ca. 70A. The smaller, secondary receptor peak appeared between fractions 73 and 83 with ORG-2058 when compared with that of R5020, similar to the density gradient sedimentation data reported elsewhere[9].

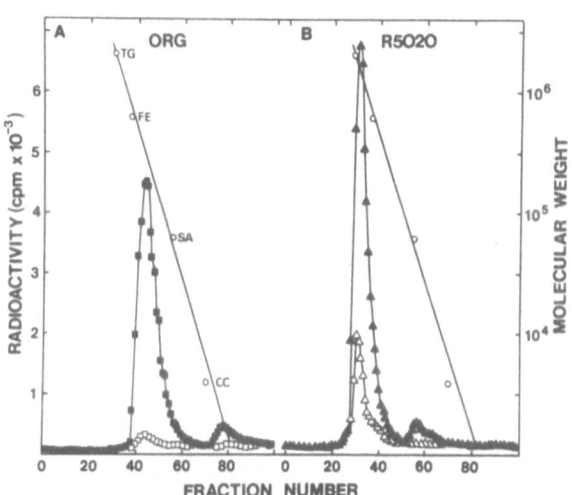

Fig. 4. Separation of Progestin Receptor Isoforms from Human Uterus Using HPSEC. Labeled cytosolic proteins were applied to and eluted from a 30 cm TSK 3000SW column. Progestin receptors labeled with [3H]ORG-2058 and incubated in the presence (□) and absence (■) of excess ORG-2058 are presented in A. The receptor isoform pattern eluting under identical conditions using [3H]R5020 as ligand in the presence (△) and absence (▲) of excess unlabeled steroid is presented in B. The HPSEC system was pre-calibrated with a series of standard proteins, thyroglobulin (TG), ferritin (FE), bovine serum albumin (SA) and cytochrome c(CC). Taken from ref. 9.

High-performance ion-exchange chromatography (HPIEC)

All chromatography was performed in a Puffer-Hubbard cold box at 4°C. A flow-rate of 1.0 ml/min was used for all experiments. Free steroid or the estrogen-labeled cytosols were applied with a Hamilton syringe to the silica-based polyamine-coated SynChropak AX-1000 anion-exchange columns (250 x 4.1 mm I.D.) from SynChrom[7].

HPIEC was performed on a column equilibrated with PDEG (buffer A). The composition of buffer B was identical to that of buffer A except for 500 mM potassium phosphate.

A programmed gradient elution was carried out after injection of sample at time t = 0.5 min in the following manner: 100% buffer

Fig. 5. HPIEC
Separation of Ionic
Isoforms of Estrogen
Receptors from Human
Breast Cancer. Cytosol
was incubated with 5 nM
[^{125}I]iodoestradiol-17ß
in the presence (○) or
absence (●) of 200-fold
excess DES. Elution of
the AX-1000 column was
performed on 200 µl of
cytosol (7.4 mg/ml)
cleared of unbound
ligand at 1.0 ml/min
using a gradient of
potassium phosphate at
pH 7.4 (---). A, 1-ml
fractions were collected
and radio-activity
measured manually with a
gamma counter; B, radio-
activity recorded
continuously using a
Model 170 radioisotope
detector with conduc-
tivity flow cell. Taken
from ref. 8.

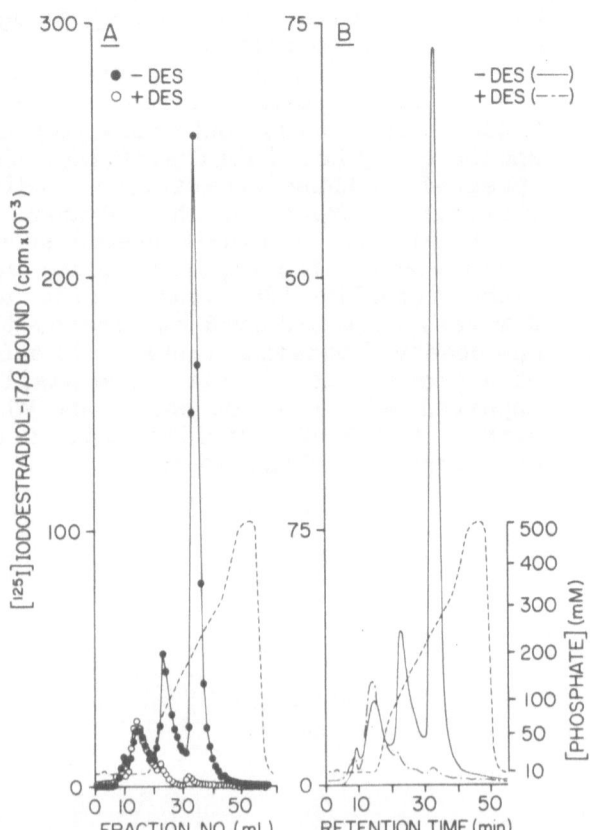

A from 0 to 10 min, 0 to 60% buffer B from 10 to 35 min, 60 to
100% buffer B from 35 to 45 min, 100% to 0% buffer B from 47 to
49 min, 100% buffer A from 49 to 55 min. Potassium phosphate
concentrations were determined by comparison with conductivity
measurements performed with identical buffer solutions containing
varying amounts of phosphate, pH 7.4 at 4°C.

A representative comparison of HPIEC elution profiles of
receptor isoforms determined in-line and manually is shown in
Fig. 5. Fig. 5 compares the radioactivity of profiles obtained
by external manual measurements with a continuous recording of
radioactivity using in-line instrumentation (B). Except for
differences in isotope-counting efficiency observed with the two
methods, the profiles of bound radioactivity were identical with
respect to the types and amounts of receptors isoforms separated.
However, the profile shown in Fig. 5B was generated within 2 h in
a continuous fashion during separation.

The purpose of this study was to develop a rapid, sensitive,
and reproducible means of assessing estrogen receptor
heterogeneity in clinical samples. HPLC is especially appropriate
since this technology is gaining greater acceptance in the
clinical laboratory.

High Performance Chromatofocusing (HPCF)

Free steroid or the estrogen-labeled cytosols were applied
to SynChropak AX-500 (250 x 4.1 mm I.D.) anion-exchange (SynChrom)
with an Altex Model 210 sample injection valve (Beckman)[13].

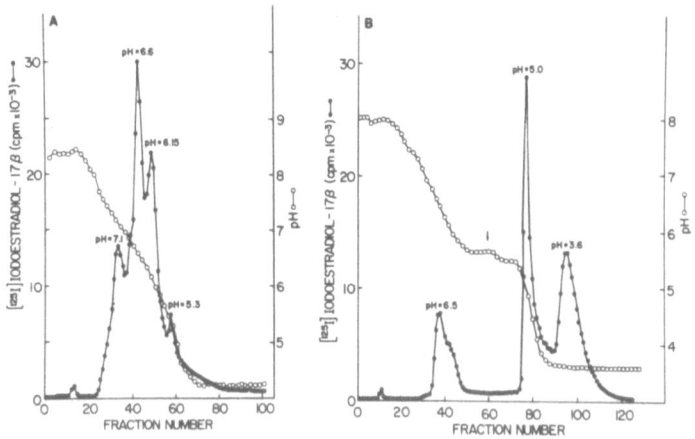

Fig. 6. Separation of isoforms by HPCF. Cytosol was prepared from the uterus of a postmenopausal woman and incubated with [125I]iodoestradiol-17ß as described earlier. Activity was eluted with a 30:70 mixture of Polybuffers 96 and 74 diluted 1:15 with 20% glycerol and adjusted to pH 4.5 (A). The recovery of radioactivity in this representative experiment was 97%. As shown in B, cytosol was prepared and labeled with [125I]iodoestradiol-17ß in the presence of 10 mM sodium molybdate. Receptor preparations (2-4 mg) were eluted with a biphasic pH gradient. The primary eluent was a 30:70 mixture of Polybuffers 96 and 74, diluted 1:10 with 20% glycerol containing 10 mM sodium molybdate and adjusted to pH 5.0. The secondary eluent (initiated at arrow) was Polybuffer 74, diluted 1:10 with 20% glycerol (no molybdate) and adjusted to pH 3.5. The recovery of radioactivity in this representative experiment was 91%. Taken from ref. 13.

Two different column equilibration and elution programs were used depending upon the initial buffer conditions of the receptor preparations. The columns were initially equilibrated to the starting pH (slightly above the desired upper limit) using a common cationic buffer. In the case of HPCF chromatofocusing on AX-500 columns, we have used T25DG (25 mM Tris-HCl containing 1 mM dithiothreitol and 20% (v/v) glycerol) adjusted to pH 8.1-8.3 at 0°C. For chromatofocusing molybdate-stabilized receptor components, 10 mM sodium molybdate was included in the column equilibration buffer. Cytosols were eluted with a 30:70 mixture of Polybuffers 96 and 74. This polyampholyte solution was diluted 10- to 20-fold with 20% glycerol, adjusted to between pH 4.0 and 5.0 at 0-4°C. The cytosols prepared in buffer containing molybdate were eluted sequentially using two separate polyampholyte buffers. The primary eluent was a 30:70 mixture of Polybuffers 96 and 74 diluted 1:10 with 20% glycerol containing 10 mM sodium molybdate, filtered and adjusted to pH 5.0 at 0-4°C. The secondary eluent was Polybuffer 74 diluted 1:8 with 20% glycerol (no molybdate), filtered, and adjusted to pH 3.5 at 0-4°C. For all experiments, 1.0-ml fractions were collected at 1.0 ml/min. pH measurements were generated in-line (Fig. 1) or by external manual determination (Fig. 6).

Fig. 6 illustrates a representative profile of estrogen receptor isoforms separated by HPCF. Note the grossly altered profile in the presence of molybdate. These acidic species appear to be stabilized isoforms which have not been recognized previously[13].

HPLC in the Hydrophobic-Interaction Mode (HPHIC)

All chromatographic procedures were performed at 4°C in a Puffer-Hubbard cold box. Free steroid or estrogen-receptor complexes were applied to the SynChropak 500 propyl (250 x 4.6 mm I.D.) hydrophobic column (SynChrom) with an Altex Model 210 sample injection valve. Elution was carried out with a Beckman 114 solvent delivery module, including a Model 421 system controller.

Several combinations of buffers were tested, and two different column elution programs were used[14]. The column was equilibrated with the high ionic strength buffer and eluted with a reverse salt gradient (500 to 10 mM phosphate). Initially, a long program (105 min) of the reverse phosphate gradient was used with a flow-rate of 0.2 ml/min to allow greater contact time of the receptor with the stationary phase. After 10 min, the flow-rate was increased to 1 ml/min for the next 75 min, during which the reverse salt gradient was developed. The column was washed 20 min with a similar buffer, except that the phosphate concentration was 10 mM. Later in the study, the elution program time was reduced to 60 min, consisting of an initial flow-rate of 1 ml/min for 5 min and then a descending salt gradient elution for 20 min. At the end of the elution program, the column was washed with either water or PEDG buffer for an additional 35 min. This latter step was followed by washing of the column with methanol-PEDG buffer (50:50). Eluted steroid (free and protein-bound) was collected as 1 ml fractions and detected radiometrically.

Moderate concentration of organic solvents, such as methanol and acetonitrile, have been used successfully in reversed-phase HPLC for the purification of various proteins. In this study, we evaluated the use of various organic solvents to retard the interaction of iodoestradiol-17ß with the stationary phase and to isolate labeled receptor as intact isoforms. Our results with acetonitrile-phosphate buffer (20:80) show that [^{125}I]iodoestradiol-17ß was eluted with a retention time of approximately 15 min (data not shown). It was observed that free steroid now eluted within the phosphate gradient in the presence of acetonitrile.

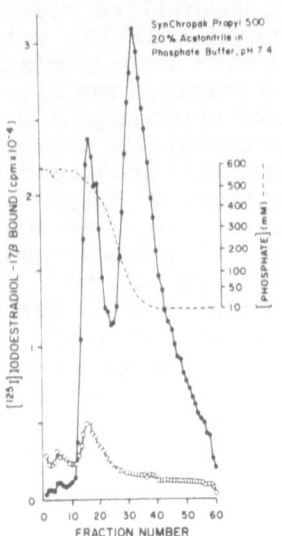

Fig. 7 HPHIC separation of Estrogen REceptor Isoforms from Human Breast Cancer. Labeled estrogen receptors applied to a synChropak 500 propyl column and eluted with a reverse gradient of potassium phosphate buffer containing 20% acetonitrile. Taken from ref. 14.

After establishing the elution position of free iodoestradiol-17ß, we investigated the separation of labeled estrogen receptor by HPHIC. Two [^{125}I]iodoestradiol-binding components were observed in the presence of 20% acetonitrile (Fig. 7). Both of these components appeared to be associated with labeled estrogen in a specific fashion, since unlabeled diethylstilbestrol diminished binding. The first peak of radioactivity was eluted in the same position as unbound steroid (Fig. 7, the inhibition of binding by diethylstilbestrol suggests that peak 1 may consist of both protein-bound and free iodoestradiol-17ß. The peak eluted at approximately 50 mM phosphate appears to consist of specific estrogen-binding components, since ligand association was diminished by inhibitor and the elution position was different from that of free steroid. It should be noted that HPHIC profiles of estrogen receptors in human breast cancer were patient-dependent, suggesting individual tumor variation. Separation profiles should reflect differences in the intrinsic properties (such as amino acid sequence) of estrogen-binding components. The physiological significance of this will have to await progress in the application of HPHIC to steroid hormone receptors.

RECENT APPLICATIONS

Monoclonal Antibody Interactions

Preparations of monoclonal antibodies of the IgG class interact specifically with the estrogen receptor of variouw species[17]. The receptor is not precipitated, but alters the sedimentation behavior of the receptor upon sucrose density gradient centrifugation[2]. In particular, the D547 preparation (provided courtesy of Drs. E.V. Jensen and G.L. Greene) associated with 4S isoforms of the estrogen receptor from human breast cancer, which were extracted, then adjusted to 400 mM KCl. The resulting complex could be detected easily on sucrose density gradients as a large molecular weight species. A similar shift in size was detected by HPSEC on the TSK 4000SW column (Fig. 8). Interestingly, in addition to the expected high molecular weight

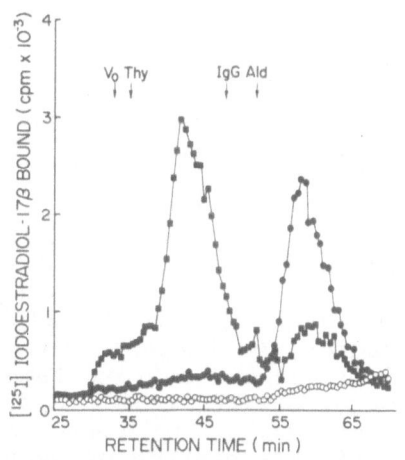

Fig. 8. Influence of D547 monoclonal antibody on the size of the estrogen receptor on HPSEC. The estrogen receptor was extracted from human breast cancer and incubated with [^{125}I]iodoestradiol-17ß in the presence (o) or absence (●) of diethylstilbestrol. The incubate was adjusted to a final concentration of 400 mM KCl. Portions of the incubate were separated on TSK 4000SW column after treatment with non-receptor-reactive antibody (●,o) or D547 monoclonal antibody (■) using Tris Buffer fortified with 400 mM KCl. Various standardproteins were previously separated as indicated: thyroglobulin (Thy), human gammaglobulin (IgG) aldolase (Ald), lysozyme (Ly), cytochrome c(Cc). Blue Dextran was used to estimate the void volume (Vo). Taken from ref. 2.

complex of antibody and receptor, other complexes of greater size were detected. It was also noted a portion of the 4S species did not react with D547 monoclonal antibody.

From studies in our laboratory (cf. 2) we suggest that caution must be observed in using monoclonal antibodies in clinical assays, since certain antigenic determinants may be either unrecognized or inaccessible. This is supported by recent experiments using immobilized monoclonal antibodies[20]. The demonstration that receptors exhibit polymorphism predicts a battery of monoclonal antibodies will be necessary to measure estrogen receptors in a clinically valid fashion.

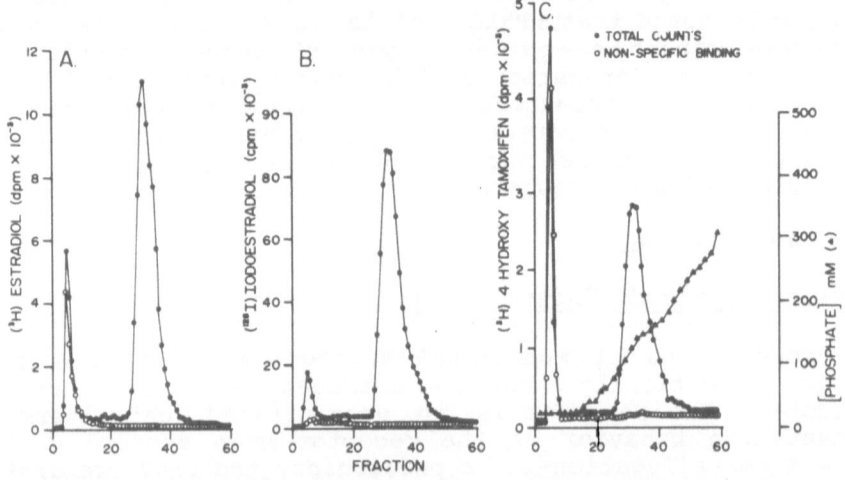

Fig. 9. Influence of Ligand on HPIEC Profiles of Estrogen Receptors. Cytosol prepared from tissue homogenized in the presence of 10 mM sodium molybdate was incubated with either 2 nM [^{125}I]iodoestradiol, 5 nM [^{3}H]estradiol or 10 nM [^{3}H]4-hydroxy tamoxifen each with or without a 200 fold excess of appropriate competitor for 24 hr at 4°C. Labeled cytosols were then fractionated by HPIEC using a 10-500 mM phosphate gradient containing 10 mM molybdate. One ml fractions were collected and radioactivity and conductivity (▲) measured. Total binding (●) and non-specific binding (o) are shown. Taken from ref. 11.

Assessment of Ligand Specificity of Isoforms

HPLC is readily applicable to the determination of the specificity of ligand binding by individual receptor isoforms regardless of the nature of the ligand used. The ligand may be labeled with a fluorescent derivative[15] or with a radioisotope[2].

Fig. 9 provides an example of this application using native estradiol-17ß compared with iodoestradiol-17ß and a potent antiestrogen, 4-hydroxy tamoxifen. When fractionated on the SynChrom AX-1000 column apparently identical ionic species of estrogen receptor were observed regardless of the ligand employed. Specific association of each ligand was found with components eluting at approximately 155 mM phosphate (Fig. 9) in the presence of molybdate. These data suggest the antitumor activity of 4-hydroxy tamoxifen is related to its association with an isoform of the estrogen receptor rather than due to interaction with a separate antiestrogen binding site in cytosol.

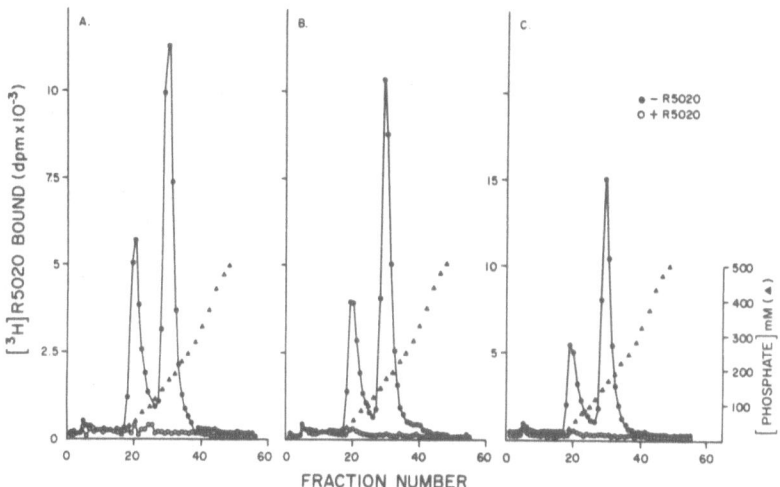

Fig. 10. HPIEC profile from AX-1000 chromatography of progestin receptors of rabbit uteri treated with 0.5 mg tamoxifen/kg. Cytosols were incubated with 10 nM [³H]R5020 in the presence (o) or absence (●) of 200-fold excess of unlabeled R5020. Elution was performed with phosphate gradient (▲) at 1.0 ml/min A,B,C represent uteri from individual animals.

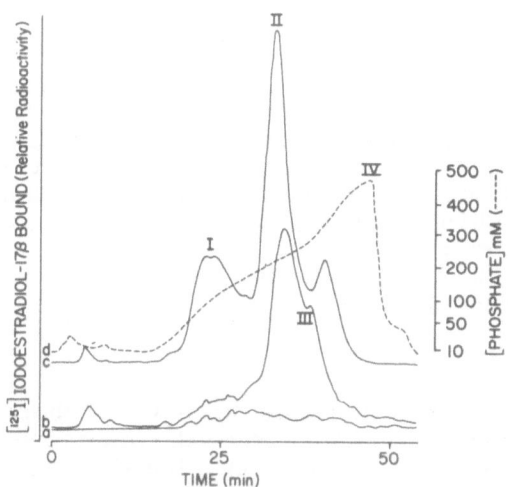

Fig. 11. In-line detection of HPIEC separated ER isoforms from 10 and 19 days lactating mammary glands of the rat. Radioactivity eluted from the AX-1000 ion-exchange column (a,b and c) was monitored with a flow-through Beckman Model-170 radioisotope detector. The phosphate gradient (d) was monitored with a Bio-Rad in-line conductivity meter. The profile shown in (a) represents nonspecific binding. The isoform profile shown in (b) represents estrogen receptors in cytosol of 10 day postpartum mammary gland while that given in (c) represents isoforms form mammary glands of 19 day postpartum rats.

Reproducibility of Receptor Isoform Profiles

The consistency of the molecular heterogeneity of HPLC profiles of receptors has been evaluated with regard to reproducibility of isoforms in the same tissue of different animals. Fig. 10 clearly demonstrates that the HPIEC profiles of progestin receptors in uteri of 3 rabbits were similar with regard to relative elution position and distribution of ligand binding. Furthermore if cytosol is prepared from a single tissue at different times and analyzed by HPIEC, the profiles of both estrogen and progestin receptor isoforms remain relatively consistent. These data suggest isoform distribution is a property of the target organ.

Comparison of Receptor Isoform Profiles During Differentiation and Development

A recent application of HPLC in-line monitoring of receptors has come from studies with differentiation lactating mammary gland of the rat[21]. We have shown for the first time that expression of steroid receptor isoforms in the mammary gland is a function of its stage of differentiation (Fig. 11). In addition, it was demonstrated that this was not due to proteolytic activity since inclusion of a serine protease inhibitor, diisopropylfluorophosphate, did not alter this profile (data not shown). Serine proteases are usually involved in the cleavage of high molecular weight forms of the estrogen receptor to smaller fragments[3].

CONCLUSIONS

In this paper we have briefly discussed our use of HPLC methods in the characterization of labile, regulatory proteins, the steroid hormone receptors. It has been demonstrated that these proteins exhibit molecular heterogeneity although the basis may be physiologic or due to manipulation artifacts or to both. In the least, we now have a means of working with small quantities of these interesting proteins in a rapid mode with retention of biological activity such that a variety of biochemical investigations may be conducted related to their regulation, structure and function.

ACKNOWLEDGEMENTS

Studies from the Hormone Receptor Laboratory have been supported in part by grants from the American Cancer Society (BC-514B), Phi Beta Psi Sorority, USPHS grants CA-19657, CA-34211, CA-32102, and CA-31946 from the National Cancer Institute. The important contributions of numerous research fellows are acknowledged, particularly those of Drs. R.D. Wiehle, G.E. Hoffmann, A. Fuchs, T.W. Hutchens, M. Lonsdorfer, W.B. Mujaji and N. Sato. The authors also express their deepest appreciation to Ms. Dana Gibson for her assistance in the preparation of this type script. N.A.S. and S.M.H. are receipients of research awards from the Graduate School of the University of Louisville. L.M. is a Fulbright Scholar from Hammersmith Hospital, London, England.

REFERENCES

1. J.L. Wittliff, Steroid Binding Proteins in Normal and Neoplastic Mammary Cells, in: "Methods in Cancer Research," H. Busch (ed.), Vol. XI:293-354, New York: Academic Press, 1975.

2. J.L. Wittliff and R.D. Wiehle, Analytical methods for steroid hormone receptors and their quality assurance, in: "Hormonally Sensitive Tumors," V.P. Hollander, ed., Academic Press, Inc. (in press).

3. J.L. Wittliff, P.W. Feldhoff, A. Fuchs, and R.D. Wiehle, Polymorphism of estrogen receptors in human breast cancer, in: "Physiology of Endocrine Disease and Mechanisms of Hormone Action," R.J. Soto, A.F. Denicola, and J.A. Blaquier, eds., Alan R. Liss, Inc., New York (1981).

4. J.L. Wittliff, Separation and characterization of isoforms of steroid hormone receptors using high-performance liquid chromatography, in: "Molecular Mechanisms of Steroid Hormone Action," V.K. Moudgil, ed., Walter de Gruyter and Co., Berlin, Germany, (1985).

5. R.D. Wiehle, G.E. Hofmann, A. Fuchs, and J.L. Wittliff, High-performance size exclusion chromatography as a rapid methods for the separation of steroid hormone receptor. J. Chromatogr., 307:39 (1984).

6. N.A. Shahabi, T.W. Hutchens, J.L. Wittliff, M.E. Kirk, S.D. Halmo, and J.A. Nisker, Physicochemical characterization of estrogen receptors from a rabbit endometrial carcinoma model. Prog. Cancer Res. & Therap., 31:63 (1984).

7. R.D. Wiehle and J.L. Wittliff, Isoforms of estrogen receptors by high-performance ion-exchange chromatography. J. Chromatogr., 297:313 (1984).

8. D.M. Boyle, R.D. Wiehle, N.A. Shahabi, and J.L. Wittliff, A rapid high-resolution procedure for assessment of estrogen receptor heterogeneity in clinical samples. J. Chromatogr. 327:369 (1985).

9. L.A. van der Walt and J.L. Wittliff, High resolution separation of molybdate-stabilized progestin receptors using high performance liquid chromatography. J. Chromatogr. (in press).

10. L.A. van der Walt and J.L. Wittliff, Assessment of progestin receptor polymorphism by various synthetic ligands using HPLC, J. Steroid Biochem. (in press).

11. L. Myatt and J.L. Wittliff, Characterization of non activated and activated estrogen and antiestrogen receptor complex by high performance ion-exchange chromatography. 67th Annual Meeting, The Endocrine Society, 1156 (1985).

12. N.A. Shahabi, R.D. Wiehle, S.M. Hyder, and J.L. Wittliff, Analysis of steroid hormone receptors by multi-dimensional approach utilizing high-performance liquid chromatography (HPLC) methodologies, IV Intl. Symp. HPLC of Proteins, Peptides, and Polynecleotides Abs. (1984).

13. T.W. Hutchens, R.D. Wiehle, N.A. Shahabi, and J.L. Wittliff, Rapid analysis of estrogen receptor heterogeneity by chromatofocusing with high-performance liquid chromatography, J. Chromatogr. 266:115 (1983).

14. S.M. Hyder, R.D. Wiehle, D.W. Brandt, and J.L. Wittliff, High-performance hydrophobic-interaction chromatography of steroid hormone receptors, J. Chromatogr. 32:237 (1985).

15. M. Lonsdorfer, N.C. Clements, Jr., and J.L. Wittliff, Performance Liquid Chromatography in the Elevation of the Synthesis and Binding of Fluorescein-Linked Steroids to Estrogen Receptors., J. Chromatogr. 266:129, 1983.

16. L.S. Dure IV, W.T. Schrader, and B.W. O'Malley, Covalent attachment of a progestational Steroid to chick oviduct progesterone receptor by photoaffinity labelling, Nature 283:784, 1980.

17. G.L. Greene, C. Nolan, J.P. Englar and E.V. Jensen, Monoclonal Antibodies to human estrogen receptor, Proc. Natl. Acad. Sci, USA 77:5115, 1980.

18. W.J. Waddell, A Simple ultraviolet spectrophotmetric method for determining of proteins. J. Lab. Clin. Med., 48:311, 1956.

19. M.M. Bradford, A rapid and sensitve method for the quantitation of microgram quantities of protein utilizing the principle of protein dye binding. Anal. Biochem. 72:248, 1976.

20. N. Sato, S.M. Hyder, L. Chang, A. Thais, and J.L. Wittliff Interaction of Estrogen Receptor Isoforms with Immobilized Monoclonal Antibodies, V Intl. Symp. HPLC of Proteins, Peptides, and Polynecleotides Abs. (1985).

21. S.M. Hyder, R.D. Wiehle, and J.L. Wittliff, Alterations in the estrogen receptor isoforms in the breast and uterus of the rat during differentiation, Fed. Proc. 44:1474 (1985).

RGE SCALE PURIFICATION OF A HUMAN HEPATOMA DERIVED ENDOTHELIAL CELL

JWTH FACTOR

Robert C. Sullivan[+], John A. Smith[*], Ricky Nelson[@],
Yuen W. Shing[+] and Michael Klagsbrun[+#]

Departments of Surgery[+], Childrens Hospital and
Biological Chemistry[#], Harvard Medical School, Boston,
MA 02115; Departments of Molecular Biology and Pathology,
Massachusetts General Hospital, Harvard Medical School,
Boston, MA 02114 and The Monsanto Company[@], St. Louis,
MO.

STRACT

A human hepatoma-derived growth factor (HDGF) was purified to
mogeneity on a large scale by a combination of Biorex-70 cation
change chromatography and heparin-Sepharose affinity chromatography.
GF is a cationic 18,500 molecular weight polypeptide, and is active at
out 1 ng/ml in promoting cell proliferation.

TRODUCTION

Recent studies indicate that many if not all endothelial cell growth
ctors have a strong affinity for heparin (1-7). Heparin affinity
romatography was used by our laboratory to purify a rat chondrosarcoma-
rived endothelial cell growth factor to homogeneity (1). The chondro-
rcoma-derived growth factor (ChDGF) is a cationic 18,000 molecular
ight polypeptide that stimulates capillary endothelial and 3T3 cell
oliferation at about 1 ng/ml. ChDGF is also angiogenic when tested on
e chick chorioallantoic membrane and in the rat cornea (8). In order
carry out structural, immunological and physiological studies, it is
portant to be able to purify tumor-derived endothelial cell growth
ctors on a large scale. Large scale purification was not feasible
ing the transplantable rat chondrosarcoma as a source. In addition,
r potential diagnostic purposes, it would also be important to purify
e endothelial cell growth factor from a human source. In this report
describe the large scale purification of a cationic 18,500 molecular
ight polypeptide, HDGF, from a cell culture line, Sk-Hep-1. The human
patoma cells can be grown in suspension culture in 100 liter bioreact-
s that produce about 5×10^{10} cells per batch. This quantity of cells
ntains about 4,000,000 units of growth factor activity. From this many
lls about 100 ug of HDGF can be purified to homogeneity by a combina-
on of BioRex-70 cation exchange chromatography amd heparin-Sepharose
romatography. Furthermore, HPLC can be used to produce a highly pure
eparation of HDGF free of small molecular weight contaminants such as
lt and amino acids for structural studies such as protein sequence
alysis.

Measurement of Growth Factor Activity

The hepatoma derived-growth factor is mitogenic for both capillary and for 3T3 cells. However for large scale purification it is difficult to grow sufficient amounts of capillary endothelial cells for screening purposes. Accordingly, growth factor activity was determined by measuring the ability of samples to stimulate the incorporation of ^{3}H thymidine into the DNA of confluent quiescent BALB/c 3T3 cells, clone A 31 (9). Briefly, the growth factor assay was carried out in 96-well microtitre plates (Costar). Each well contained approximately 20,000 3T3 cells in 200 ul of Dulbeccos Modified Eagles Medium (Gibco) supplemented with 10% calf serum (Colorado Serum Co., Denver, CO). In the bioassay 0.1-50 ul of sample and 10 ul of tritiated thymidine (ICN, cat.# 24066, 6.7 Ci/mmole, 4 uCi/ml final concentration/well) are added to the 200 ul medium atop the 3T3 cells and incubated for 36-48 hours. Background incorporation is about 1000-2000 cpm and maximal stimulation is about 150,000-200,000 cpm. One unit of growth factor activity is defined as the amount of growth factor required to stimulate half-maximal DNA synthesis in a 36-48 hour incubation period (about 75,000-100,000 cpm). Fractions that were mitogenic for 3T3 cells were also mitogenic for capillary endothelial cells.

Extraction of Hepatoma Cells

The growth factor produced by the human hepatoma cell line, SK-Hep-1, was found to be associated predominantly with the cell rather than the conditioned medium. Sk-Hep-1 (10) cells obtained originally from Dr. J. Fogh of the Sloan-Kettering Institute (New York, N.Y.) were grown in monolayer or suspension with Dulbeccos Modified Eagle's Medium supplemented with 10% calf serum. For large scale culture, SK-Hep-1 cells were grown as follows: SK-Hep-1 cells were grown in monolayer in T-75 cm^2 flasks. At confluence, cells from four flasks were transferred into a 1 liter spinner flask. When the cells reached maximum density (about 10^6 cells/ml) they were transferred and grown first in 3-liter then in twelve-liter spinner flasks. The cells from two to four 12-liter spinner flasks (about 1-5 x 10^5 cells/ml) were introduced into a 100 liter Vibromixer reactor (11). When the cell density reached 5 x 10^5/ml the cells were harvested by centrifugation. A 100 liter reactor produced 5 x 10^{10} SK-Hep-1 cells (150 ml pellet), which were stored at -80°.

For growth factor purification a 150 ml cell pellet was thawed and resuspended in 1500 ml of 1.0M NaCl, 0.01M Tris-HCl pH 7.5. The cells were lysed by homogenization in a Waring Blender for 1 minute at room temperature. The homogenate was stirred overnight at 4° with a magnetic stirring bar and was centrifuged at 25,000 g for 30 minutes. The supernatant (1200 ml) was used for further purification.

Biorex-70 Cation Exchange Chromatography

The clarified extract (approximately 4,000,000 units, 1200 ml) was dialyzed overnight at 4° against 6 volumes of 0.01M Tris-HCl 7.5 to lower the NaCl concentration to 0.15M. The dialyzed extract was centrifuged at 25,000 g for 30 minutes to remove insoluble material. The supernatant was mixed with Biorex-70 (Biorad, 1750 ml) equilibrated with 0.15M NaCl, 0.01M Tris-HCl pH 7.5, overnight at 4° with constant stirring from overhead using a motor driven stirrer (Heller 4 blade propeller, model

GT-21, Thomas Scientific). The Biorex-70 was allowed to settle for 45 minutes and the supernatant was discarded. A slurry of Biorex-70 was poured into a glass column (5 x 100 cm) and was washed with the equilibration buffer (usually 2000 ml) until the absorbance at 280 nm reached baseline (LKB Uvicord-S). The hepatoma-derived growth factor was eluted with 0.6M NaCl 0.01M Tris-HCl pH 7.5 at a flow rate of 60 ml/hour. Fractions (20 ml) were collected and monitored for growth factor activity.

Heparin-Sepharose Affinity Chromatography

The active fractions of Biorex-70 purified hepatoma-derived growth factor were pooled (approximately 2,000,000 units, 700 ml) and applied to a column of heparin-Sepharose (Pharmacia, 2.0 X 13 cm, 40 ml) equilibrated with 0.6M NaCl, 0.01M Tris-HCl pH 7.5 at 4°. After a wash of about 200 ml, hepatoma-derived growth factor was eluted with a 0.6-2.5M NaCl gradient (500 ml) in 0.01M Tris-HCl pH 7.5 at a flow rate of 45 ml/hour. Fractions (7 ml) were collected and monitored for growth factor activity.

Reverse phase chromatography

Active fractions of hepatoma-derived growth factor purified by the heparin-Sepharose chromatography were pooled (1,000,000 units, 100-200 ml), and pumped directly onto an HPLC reverse phase C3 column (Beckman, 4.6mm ID X 7.5 cm) equilibrated with 0.1% TFA (Pierce Chemical Co.), using a Mini Pump (Milton Roy, model 396/2396) at 60 ml/hour at room temperature. The column was rinsed with 0.1% TFA (approximately 10 ml) until the absorbance at 214 nm returned to baseline. The hepatoma-derived growth factor was eluted with a 0-60% gradient of acetonitrile/2-propanol 50/50 v/v (Pierce Chemical Co.) in 0.1% TFA flow rate of 1 ml/minute for 120 minutes. Fractions of 1 ml were collected. The hepatoma-derived growth factor was inactivated by long exposures to TFA. Accordingly, for growth factor assay, aliquots (10 ul) from each fraction of the C3 column were immediately neutralized by dilution with PBS, 0.1% BSA in 96-well plates (200 ul/well). From these dilutions 5-50 ul were tested on 3T3 cells within 2 hours of the HPLC run.

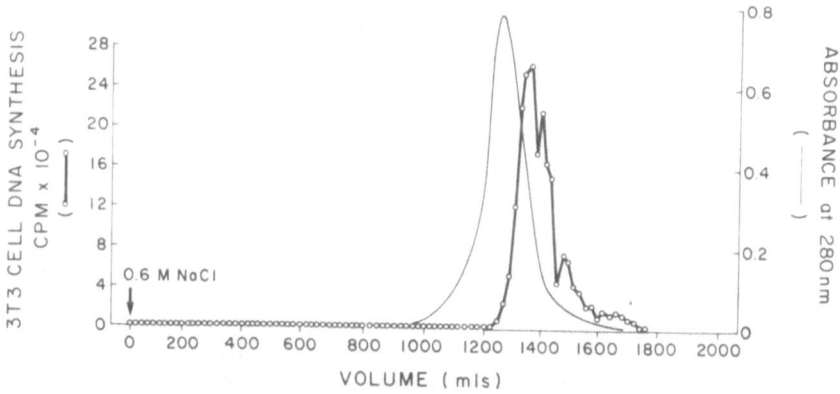

Figure 1: Cationic exchange Chromatography. HDGF (about 4 X 10[6] units, 1250 ml) was clarified by centrifugation and incubated with Biorex-70. The column was washed with 0.15M NaCl, 0.01 M Tris-HCl pH 7.5, and HDGF was eluted with 0.6M NaCl, 0.01M Tris-HCl pH 7.5

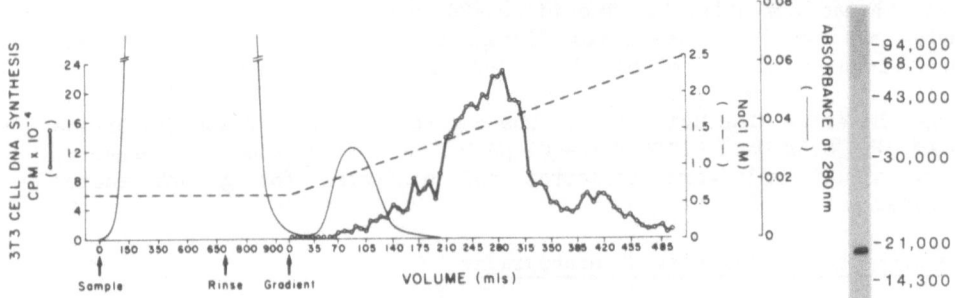

Figure 2: Heparin-Sepharose chromatography. (Left) Biorex-70 purified HDGF (2,380,000 units, 620 ml) was applied directly to a column of heparin-Sepharose (2 X 13 cm, 40 ml) pre-equilibrated with 0.6M NaCl, 0.01M Tris pH 7.5. The column was washed with the equilibration buffer and the HDGF was eluted with a gradient of 0.6-2.5M NaCl in 0.01M Tris-HCl pH 7.5 (500 ml). (Right) About 10,000 units of HDGF was analyzed by SDS PAGE.

SDS Polyacrylamide Gel Electrophoresis

SDS slab gel electrophoresis was performed using the method described by Lamelli (12). A stacking gel of 5% acrylamide and a separating gel of 15% acrylamide with a thickness of 0.75 mm were used. The protein bands were visualized by silver stain (13). The molecular weight markers phosphorylase B (94,000), bovine serum albumin (68,000), ovalbumin (43,000), carbonic anhydrase (30,000), soybean trypsin (21,000), and lysozyme (14,300) were purchased from Biorad. B-lactoglobulin (18,400) was purchased from Sigma.

RESULTS

SK-Hep-1 cells were grown in a suspension culture. Most of the growth factor activity was found to be associated with the cell rather than the conditioned medium (data not shown). Accordingly hepatoma cells (5 X 10^{10}) were homogenized in the presence of 1M NaCl and centrifuged. The clarified extracts contained approximately 4,000,000 units of growth factor activity.

The clarified extract was mixed with the Biorex-70. Greater than 90% of the protein was washed from the Biorex-70 with 0.15M NaCl, whereas all of the hepatoma-derived growth factor remained tightly bound to the resin. The growth factor activity was eluted with 0.6M NaCl (fig.1) as a single peak containing approximately 2,500,000 units with a specific activity of 330 units/ug. The recovery was about 65%.

The active fractions from the Biorex-70 column were pooled and applied to a column of heparin-Sepharose (fig.2, left). The column was washed with 0.6M NaCl removing greater than 90% of the protein, while nearly all of the remaining protein was eluted with 0.6-1.0M NaCl. However, all of the HDGF adhered to the column. When a gradient of NaCl was applied the HDGF eluted at about 1.6M NaCl and appeared as a single band on the SDS PAGE (fig.2, right). The yield of hepatoma-derived growth factor starting with 5 X 10^{10} cells was approximately 800,000

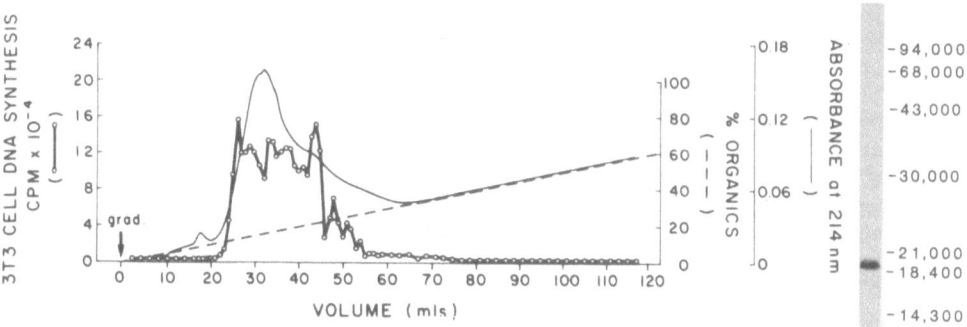

Figure 3: Reverse phase chromatography. (Left) HDGF purified by a combination of Biorex-70 and heparin-Sepharose chromatography (1,250,000 units, 91 ml) was applied directly to a reverse phase C3 column. The column was washed with 0.1% TFA and the HDGF was eluted with a gradient of 0-60% acetonitrile/2-proponal 50/50 v/v (120 minutes, 1 ml/min.). (Right) About 100 ng of reverse phase C3 purified HDGF was analyzed on SDS-PAGE.

units or 144 ug and had a specific activity of about 7,000 units/ug. The overall recovery was 20%.

Structural studies of HDGF require the absence of contaminating species such as amino acids and salts. Therefore, HDGF was de-salted and further purified by chromatography on an HPLC reverse phase C3 column. Heparin-Sepharose purified HDGF (100 ml, 1,000,000 units) was applied to the C3 column using a Milton Roy Mini Pump. The growth factor activity was eluted (fig.3, left) within a range of 12-28% acetonitrile/2-propanol 50/50 v/v as a single band when analyzed on SDS-PAGE (fig.3, right). The recovery of protein was 20%, approximately 50 ug.

DISCUSSION

The availability of human tumor derived endothelial cell growth factors in large amounts is essential in carrying out the physiological, structural and immunological studies that would enhance our understanding of the role of these growth factors in the vascularization of human tumors. A human hepatoma cell line, SK-Hep-1, is an excellent source for the large scale production of the endothelial cell growth factors and provides the following advantages: 1) SK-Hep-1 cells grow in suspension. They can be grown in 100 liter bioreactors (11) that produce approximately 5×10^{10} cells containing 4×10^6 units of growth factor activity. To obtain a comparable number of units from animal tumors would require a much more substantial investment in cost and time. In addition, the relatively high specific activity of hepatoma cell extract compared to tumor tissue extract (8) means that relatively less hepatoma protein must be processed to purify growth factor. 2) Growth factor activity is found to be asociated mostly with the cells rather than with the conditioned medium. Thus it is possible to begin the purification procedure with relatively small cell pellets rather than with large volumes of conditioned medium.

HDGF is a cationic polypeptide with a molecular weight of about 18,500-19,000. It is mitogenic for 3T3 cells with a specific growth factor activity of about 5-7 units/ng and at a concentration of about 0.5 ng/ml in cell culture. Purification on a large scale can be accomplished in just two steps. The first step is batch elution chromatography of cell lysate using the cationic exchanger, Biorex-70. This step is an excellent method for purifying large volumes of extract. The tumor extracts are mixed with the Biorex-70 in a beaker. The resin is poured into a 2 liter column and HDGF is eluted with a batch application of 0.6M NaCl. With this method it is easy to process over 1 liter of tumor extract in a procedure that takes just one day. The recovery of growth factor is over 50%. To complete the purification the active eluate can be applied directly to a heparin-Sepharose column without any further processing. The capacity of the heparin-Sepharose column is so great that over 2,000,000 units can be applied to a small column (40 ml). After application of a gradient of NaCl, HDGF elutes at about 1.6M NaCl as an apparent single band when visualized by silver stain on an SDS polyacrylamide gel. About 500,000-700,000 units/100 ug of purified HDGF can be recovered from 5 X 10^{10} hepatoma cells.

For protein sequence analysis it is important to have preparations that are free of small molecular weight contaminants such as amino acids and salts. It is possible to obtain such highly purified preparations of HDGF by reverse phase chromatography of heparin-Sepharose purified HDGF on HPLC C3 columns. While HPLC on C3 is excellent for preparing HDGF for sequencing, it is not useful for preparing biologically active HDGF. Most of the growth factor activity is inactivated by the exposure of HDGF to 0.1% TFA, pH 2.

The large amounts of purified human HDGF are now being used for sequencing and antibody production. It is hoped that these studies will lead to a better understanding of the role of HDGF in the vascularization of tumors.

REFERENCES

1. Shing, Y., Folkman, J., Sullivan, R., Butterfield, C., Murray, J. & Klagsbrun, M., Science 223, 1296-1298 (1984).

2. Gospodarowicz, D., Cheng, J., Lui, G-M., Baird, A. & Bohlen, P., Proc. Natl. Acad. Sci. U.S.A. 81, 6963-6967 (1984).

3. Klagsbrun, M. & Shing, Y., Proc. Natl. Acad. Sci. U.S.A. 82, 805-809 (1985).

4. Lobb, R.R. & Fett, J.W., Biochemistry 23, 6295-6298 (1984).

5. Maciag, T., Mehlan, T., Friesel, R. & Schreiber, A.B., Science 225, 932-935 (1984).

6. D'Amore, P.A. & Klagsbrun, M., J. Cell Biol. 99, 1545-1549 (1984).

7. Sullivan, R. & Klagsbrun, M., J. Biol. Chem. 260, 2399-2403 (1985).

8. Shing, Y., Folkman, J., Haudenschild, C., Lund, D., Crum, R., & Klagsbrun, M., J. Cell Biochem. (1985) In press.

9. Klagsbrun, M., Langer, R., Levenson, R., Smith, S., & Lillehei, C., Exp. Cell Res. 105, 99-108 (1977).

10. Fogh, J.M. & Orfeo, R., J. Natl. Cancer Inst. 59, 221-225 (1977).

11. Tolbert, W.R., Schoenfeld, R.A., Lewis, C. & Feder, J., Biotech.

    Bioeng. 24, 1671-1679 (1982).

12. Laemmli, U.K., Nature 227, 680-685 (1970).

13. Oakley, B .R., Kirsch, D.R. & Morris, N.R., Anal. Biochem. 105, 361-363 (1980).

VERY RAPID MICROANALYSIS OF IgG IN ASCITES FLUIDS BY HPLC USING A NOVEL

ANION-EXCHANGE COLUMN

David J. Burke* and J. Keith Duncan

Bio-Rad Laboratories,

1414 Harbor Way South, Richmond, CA 94801

ABSTRACT

An HPLC system has been optimized for the rapid resolution and quantitation of IgG in mouse ascites fluids, making use of newly developed HPLC columns (MA7P). Chromatography using MA7P columns is characterized by remarkably narrow band widths and very short retention times. The HPLC system is capable of analyzing the IgG content of a typical ascites fluid with a cycle-to-cycle time of under 5 minutes, regardless of IgG subtype. Separation times under 1 minute are possible. Recoveries of 100 μg injections of IgG and other proteins are quantitative. Using a UV monitor, the system was optimized to generate a linear plot of peak area vs. amount of IgG injected from 100 ng to 3 μg per injection. These properties make the MA7P column very useful for both analytical and micropreparative applications. Ascites fluids from eleven different IgG_1-producing hybridomas were analyzed. The IgG concentrations in those eleven ascites varied considerably. Retention times of the IgG varied slightly but significantly from hybridoma to hybridoma. Some monoclonals produce heterogeneous IgG. This heterogeneity can be detected by chromatography using MA7P columns.

INTRODUCTION

Monoclonal antibodies are being used in an increasingly large number of applications (1,2). Consequently, there is an increasing demand for methods which allow rapid analysis of ascites fluids and purified antibodies. High Performance Liquid Chromatography (HPLC) techniques allow rapid analysis of single samples or small numbers of samples (3-7). However, these procedures are time consuming when large numbers of samples must be analyzed using the same instrument. Therefore, an analytical HPLC technique with very short analysis times is highly desirable.

The recently introduced Microanalyzer™ MA7P cartridge column is a small bed volume column packed with a non-porous support which has a very high selectivity (3). This combination of properties allows for the very rapid and highly sensitive resolution of small aliquots of protein solutions. In this paper, data is presented on the use of the MA7P column for rapid analysis of IgG in ascites fluids.

The MA7P column is significantly faster than other HPLC columns available for this type of analysis. The HPLC system is useful both for quantitating the amount of IgG in the ascites fluid and for monitoring purification of the monoclonal antibody. Previous work with Bio-Gel HPHT columns has demonstrated that monoclonal antibodies are heterogeneous(8). Such heterogeneity is also observed in chromatograms using the MA7P column (3). Analysis of this heterogeneity may be helpful in selecting ascites with a high proportion of "active IgG". Since recovery of proteins from the MA7P column is excellent at very low protein inputs, the column can also be used for micropreparative applications.

MATERIALS AND METHODS

Materials

All buffer solutions were made with distilled, deionized water and reagent grade solutes. The following four ascites fluids (IgG subtype in parenthesis) were obtained from Sigma : MOPC 21 (IgG_1); FLOPC 21 (IgG_3); UPC 10 (IgG_{2a}); and MOPC 141 (IgG_{2b}). Ascites fluid 13H1 was obtained from Dr. Larry Stanker, Lawrence Livermore National Laboratory (Livermore, CA). Ascites fluids BB/A and BBLT-1 through BBLT-11 (Table 3) were obtained from Dr. Barry Bredt, Bio-Rad Clinical Division (Richmond, CA).

Methods

Hydroxylapatite (HPHT) fractionation of ascites BB/A was performed by Ms. Theresa Chow of Bio-Rad Laboratories, using the suggested procedure (8). The Protein-A MAPS kit with buffers was obtained from Bio-Rad, and was used according to the recommended protocol (9). Protein assays were performed as described by Bradford (10), using reagents from Bio-Rad.

Columns and Sample Preparation

Microanalyzer™ MA7P cartridge columns (4.6mm X 30mm), Bio-Gel HPHT hydroxylapatite columns (7.8mm X !00mm), and Bio-Gel TSK DEAE-5-PW columns (7.5mm X 75mm) were obtained from Bio-Rad Laboratories (Richmond, CA). Aquapore AX-300 cartridge columns (4.6mm X 30mm) were obtained from Rainin (Berkeley, CA). Bakerbond MAb material was obtained from J.T. Baker, Phillipsburg, NJ, and was packed into 4.6mm X 30mm cartridge columns. Mono Q columns (5.0mm X 50mm) and Polyanion SI columns (5.0mm X 50mm) were from Pharmacia (Uppsala, Sweden). Cartridge columns were housed in Bio-Rad cartridge holders. Samples were clarified by centrifugation (Eppendorf model 5414 for 2 min) and diluted with low salt buffer. Injections of 20 µl contained 10-50 µg of total protein.

HPLC System

The HPLC system used in these studies was a Bio-Rad Protein Microanalyzer System, consisting of two Model 1330 pumps, a gradient mixer (1.8 ml volume), and either a Model 7125 manual injector or a Model AS-48 Autosampler. Connections between the mixer and the injector, between the injector and the column, and between the column and the detector were kept to a minimum by using short lengths of 0.01 inch (i.d.) tubing. The tubing length from the injector to the detector was 10 cm. Since the detector cell had a volume of 8 µl, the extra-column volume (injector through detector) was about 20 µl. The system was operated by an Apple IIe computer with dual disk drive, ProFile hard disk option, and Bio-Rad Gradient Processor System (Version 3.7) software. Data from the Bio-Rad Model 1305A detector was integrated with a Model 3392A integrator, interfaced with the computer.

RESULTS

The chromatograms in Figure 1 demonstrate that HPLC using the MA7P column can resolve IgG from other components of various mouse ascites fluids in a short time. The four panels show chromatograms of ascites fluids from myeloma tumors producing immunoglobulins with four different subtypes. In order to identify the IgG peak, the IgG was purified from

each ascites by affinity chromatography using Affi-Gel Protein A. After
desalting on a Bio-Gel P6 column, the purified IgG was analyzed using the
MA7P column. Chromatograms of the affinity-purified IgG species are shown
as the lower traces of the four panels of Figure 1. For three of the four
ascites fluids, the IgG peak is readily resolved from the other peaks.
The ascites fluid from the FLOPC 21 cell line contained very little IgG and
that IgG eluted in a very broad band (see Figure 1).

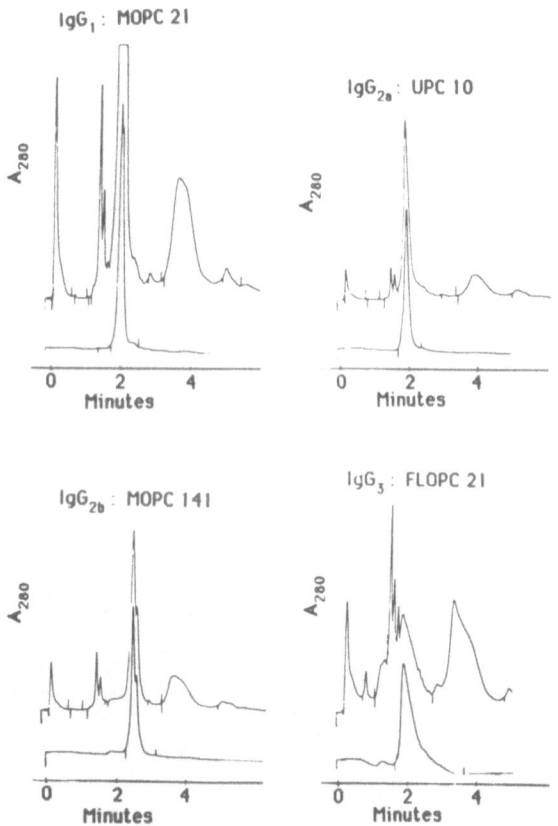

*Figure 1. HPLC chromatograms of four ascites with different IgG subtypes.
In each panel, the upper trace is the ascites and the lower trace is the
IgG purified from that ascites by chromatography on Affi-Gel Protein A.
Column : MA7P. Buffer A : 20 mM tris, pH 8.5; buffer B : 20 mM tris plus
500 mM NaCl, pH 8.5. Gradient : 0-100 % B in 5.0 minutes at 2.0 ml/min.*

In Figure 2, the MA7P column is compared with other HPLC columns for
the ability to rapidly resolve the IgG in ascites fluid from the MOPC 21
myeloma line. The tallest peak in each chromatogram is the IgG peak. In
each case, the columns were optimized in an attempt to resolve the
components in less than five minutes. All columns except the Bakerbond MAb

column were packed by the manufacturer. The "analytical" Bakerbond column is quite large when compared to the other columns in this study. Therefore, a smaller column was packed for comparative purposes. Column dimensions and chromatographic conditions are listed in the Figure.

For the PEI-silica columns (Bakerbond, AX-300, Polyanion SI), buffers similar to those recommended by Baker (11) were selected. Flow rates were selected near the maximum rates recommended by the manufacturers. In the case of AX-300, equally good results were obtained

A. MA7P
(4.6 mm id X 30 mm)

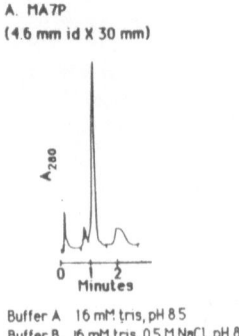

Buffer A 16 mM tris, pH 8.5
Buffer B 16 mM tris, 0.5 M NaCl, pH 8.5
Flow Rate 2.0 ml/min
Gradient 0 to 100 % B from 0.0 to 2.0 min

B. AX-300
(4.6 mm id X 30 mm)

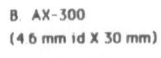
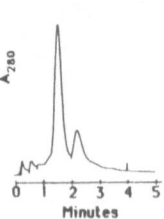

Buffer A 16 mM sodium phosphate, pH 6.8
Buffer B 500 mM sodium phosphate, pH 6.3
Flow Rate 2.0 ml/min
Gradient 0 to 100 % B from 0.0 to 5.0 min

C. Bakerbond MAb
(4.6 mm id X 30 mm)

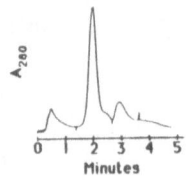

Buffer A 16 mM sodium phosphate, pH 6.8
Buffer B 500 mM sodium phosphate, pH 6.3
Flow Rate 2.0 ml/min
Gradient 0 to 25 % B from 0.0 to 5.0 min

D. Polyanion SI
(5.0 mm id X 50 mm)

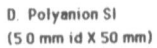
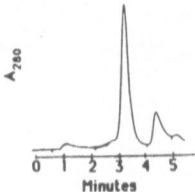

Buffer A 16 mM sodium phosphate, pH 6.8
Buffer B 500 mM sodium phosphate, pH 6.3
Flow Rate 1.0 ml/min
Gradient 0 to 50 % B from 0.0 to 5.0 min

E. Mono Q A_{280}
(5.0 mm id X 50 mm)

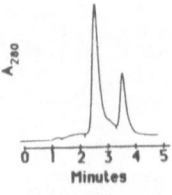

Buffer A 16 mM tris, pH 7.7
Buffer B 16 mM tris, 0.5 M NaCl, pH 7.7
Flow Rate 1.0 ml/min
Gradient 20 to 100 % B from 0.0 to 2.0 min, hold at 100 % for 3 min

F. Bio-Gel TSK-DEAE-5PW
(7.5 mm id X 75 mm)

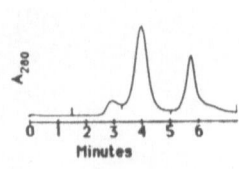

Buffer A 16 mM tris, pH 7.7
Buffer B 16 mM tris, 0.5 M NaCl, pH 7.7
Flow Rate 1.2 ml/min
Gradient 20 to 100 % B from 0.0 to 5.0 min, hold at 100 % for 3 min

Figure 2. HPLC chromatograms of ascites fluid from MOPC 21 using six different anion-exchange columns.

Table 1. Quantitative chromatographic data

Column	Retention time	Area/Height	Resolution*	Area Percent
A. MA7P	1.13 min	0.127	0.97	66 %
B. AX-300	1.58 min	0.349	0.38	65 %
C. Bakerbond	1.92 min	0.388	0.64	73 %
D. Polyanion SI	3.35 min	0.425	0.73	77 %
E. Mono Q	2.69 min	0.336	0.73	64 %
F. DEAE-5PW	4.05 min	0.568	0.79	57 %
G. MA7P (Fig.3)	0.37 min	0.041	0.57	68 %

* Resolution of the IgG/albumin pair using the equation $R = (\Delta t)/2(w_1 + w_2)$ where Δt is the difference in retention times and w_1 and w_2 are the peak widths at half height.

using a gradient similar to that used for the MA7P column (not shown). However, since silica-based columns deteriorate under basic conditions,no further attempts were made to optimize silica columns at higher pH. The gradient used for the Mono Q column was an improved version of a gradient

Table 2 . Recoveries of IgG and other proteins from MA7P columns. For experimental details, see text.

IgG	% Recovered
IgG (from hybridoma 13H1)	91
IgG (from myeloma MOPC 21)	98
IgG (from myeloma UPC 10)	97
IgG (from myeloma MOPC 141)	95
IgG (from myeloma FLOPC 21)	97
Average IgG	96

Other proteins	% Recovered
Hemoglobin A	93
BSA	100
Ovalbumin	95
Conalbumin	93
Carbonic anhydrase	97
Soybean trypsin inhibitor	96
Insulin	91
Glucagon	89
Average other proteins	94

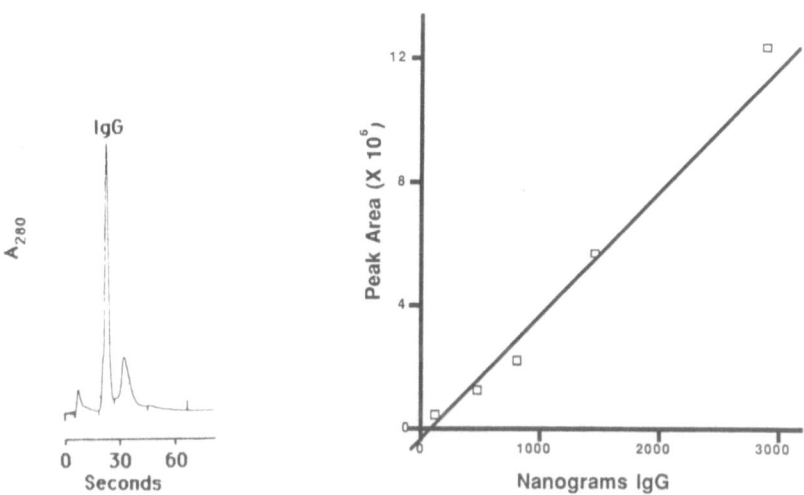

Figure 3 (left). Very rapid separation of ascites fluid from MOPC 21. Flow rate 4.5 ml/min. Buffers as in Figure 1. Step gradient from 0-100 % B at 6 seconds after injection. With the mixer used, this generated a gradient approximately 1 minute long.

Figure 4 (right). Plot showing the linearity of peak area vs. amount of protein injected for affinity-purified IgG from ascites 13H1. Each point represents the average peak area for quadruplicate injections of serially diluted protein solutions. Column : MA7P. Buffers as in Figure 1. Gradient: 15-50 % B at 0.7 ml/min. Detection : absorbance at 225 nm.

published earlier (12). A similar gradient was optimal for the TSK DEAE-5PW column. The performance of the MA7P column was best in pH range from 7.5-8.5. For both the TSK DEAE-5PW and the Mono Q columns, chromatography at higher pH resulted in later elution times for the peaks of interest and required extremely high salt concentrations.

As can be seen in Figure 2, all of these columns can separate the major components of this ascites in a relatively short time. However, the MA7P column is superior when both resolution and speed are important. This

TABLE 3   Retention times and area percents of IgG$_1$ in ascites fluids from 11 different hybridomas. For experimental details, see text.

Ascites	Retention time	Area percent
BBLT-1	1.76 min	55.1
BBLT-2	1.65 min	46.4
BBLT-3	1.73 min	25.3
BBLT-4	1.72 min	58.2
BBLT-5	1.67 min	13.1
BBLT-6	2.00 min	69.8
BBLT-7	1.64 min	27.0
BBLT-8	1.65 min	40.0
BBLT-9	1.81 min	36.7
BBLT-10	1.68 min	46.6
BBLT-11	1.67 min	23.6
Average	1.73 ± 0.1 min	40.2 ± 16.3 %

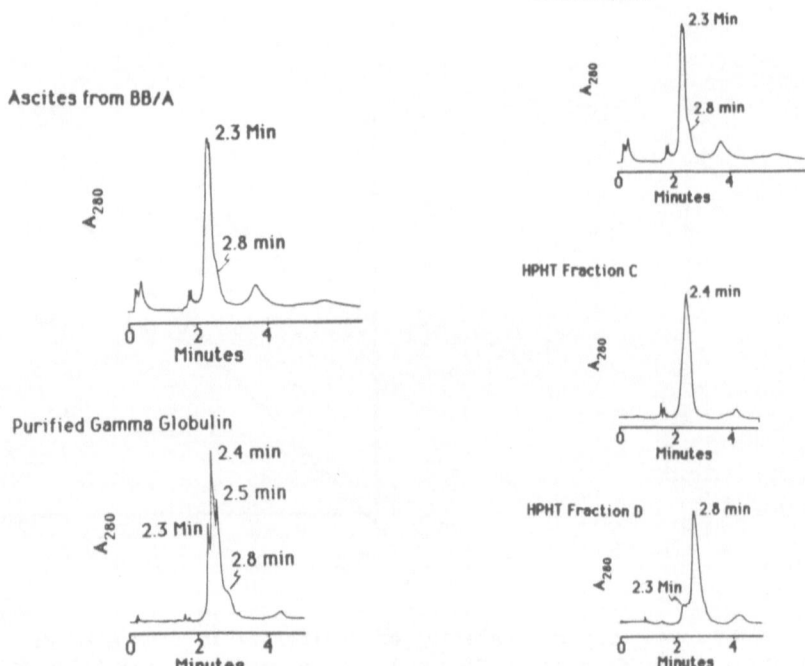

Figure 5 (left). Heterogeneous IgG. The upper panel is a chromatogram of ascites fluid BB/A showing multiple peaks in the IgG region. The lower panel is a chromatogram of the gamma globulin fraction precipitated from that ascites by ammonium sulfate. Chromatography as in Figure 1.

Figure 6 (right). Analysis of ascites BB/A and two hydroxylapatite (HPHT) fractions isolated from that ascites. Chromatography as in Figure 1.

is demonstrated more clearly in Table 1, which contains data on peak widths (expressed as area/height ratios) and elution times for the IgG peaks for the separations in Figure 2, as well as values for resolution of IgG from albumin (the final peak in each of the chromatograms). The MA7P chromatogram is characterized by excellent resolution, a very early elution time and a very narrow peak width for the IgG band. This narrowness of peak width can be largely attributed to the fact that the packing material is non-porous (3). This affects peak width in two ways : 1. there is no band broadening due to diffusion of solutes into and out of pores; 2. a column packed with this material has a smaller total solution volume (since there is little pore volume) than an identical column packed with a porous material having the same average particle diameter. The resolving power of the MA7P column is also shown by the ability to resolve several small peaks eluting just before the IgG peak (Figures 1 and 2). Using other columns, these peaks are not clearly resolved, even in considerably longer chromatograms (see also ref.3).

The MA7P chromatogram in Figure 2 had sufficiently good resolution that still faster separations seemed possible. The chromatogram in Figure 3 shows that extremely fast separations can be accomplished using the MA7P column. In this chromatogram, resolution of the small peaks eluting just before the IgG peak is lost. However, quantitation of the IgG peak was similar between this chromatogram (68 % IgG) and the chromatogram in Figure 2 (66 % IgG). For the separation of IgG from albumin in this chromatogram, R = 0.57 (see Table 1).

An important characteristic for analytical and micropreparative columns is that recoveries of protein are high, even at low protein input. Recoveries of a variety of purified proteins (including IgG) are greater than 90 % when eluted from MA7P columns (Table 2). In this experiment, 100 µg of protein was injected onto the column in low salt (20 mM tris, pH 8.5), then the bound protein was eluted with high salt (20 mM tris plus 500 mM NaCl, pH 8.5). The eluted protein peak was collected and quantitated using the Bradford assay (10) with the protein of interest serving as the standard. Previously, it was shown that the curve of peak area vs. mass of protein injected was linear in the range from 2 to 200 µg protein (3). By modifying the chromatographic conditions to increase the sensitivity, it was possible to show that the curve of peak area vs. amount of IgG injected is linear in the range from 100 ng to 3 µg (Figure 4).

As shown in Figure 1, it was possible to resolve the IgG from other components in ascites fluids with separation times under five minutes and that the same method could be used to analyze the purity of an IgG preparation. The IgG purified from those four ascites by Affi-Gel Protein A chromatography was substantially free of contaminating proteins (Figure 1).

Some myeloma and hybridoma lines produce heterogeneous IgG (8). This heterogeneity may arise from expression of heavy or light chain genes present in the original myeloma line (from which the hybridoma was derived). Such heterogeneity can also be seen using the MA7P column (3). An example of a cell line producing a heterogenous IgG is shown in Figure 1. The IgG from the myeloma line MOPC 141 has 2 partially resolved IgG bands when analyzed using MA7P chromatography. Both bands are retained by Affi-Gel Protein A. A second example is shown with ascites fluid from the line BB/A (Figure 5). Chromatograms of both the ascites and the partially purified IgG fraction from BB/A are characterized by multiple peaks. These peaks were partially resolved by hydroxylapatite (HPHT) chromatography (not shown). Two HPHT fractions were analyzed and shown to consist of IgG species with different retention times (Figure 6). Both fractions were biologically active. The structural differences are under investigation.

The retention times and area percents for the IgG peaks in ascites fluids from eleven different hybridomas are shown in Table 3. Each of the eleven cell lines was shown to produce IgG_1 as the major subclass. In this experiment, the chromatography was as in Figure 1 except that the gradient was completed in three minutes rather than 5 minutes. Under these conditions, the retention times for the IgG peaks varied only slightly. However, the percent IgG in the ascites fluids varied considerably among these cell lines.

CONCLUSIONS

HPLC is a valuable tool for rapid analysis of proteinaceous samples, including ascites fluids. Of the columns tested, the MA7P column most capably achieved the desired combination of speed, resolution, and high recovery. The techniques described here are sufficiently fast that a large number of samples can be handled in a reasonable amount of time with a minimum of sample preparation. Quantitation is possible without the need for the fixing, staining, and densitometric scanning required by electrophoretic methods. In addition, the HPLC system allows the user to collect fractions for micropreparative applications.

REFERENCES

1. D. H. Katz (ed.), "Monoclonal Antibodies and T Cell Products" (CRC Press, Boca Raton, FA, 1982).
2. J. G. R. Hurrell (ed.), "Monoclonal Hybridoma Antibodies : Techniques and Applications" (CRC Press, Boca Raton, FA, 1982).
3. D. J. Burke, J. K. Duncan, L. C. Dunn, L. Cummings, C. J. Siebert, and G. S. Ott, J. Chromatog. (1985), in press.
4. W. Kopaciewicz, M. A. Rounds, J. Fausnaugh, and F. E. Regnier, J. Chromatogr., 266 (1983) 3-21.
5. M. Flaschner, H. Ramsden, and L. J. Crane, Anal. Biochem., 135 (1983) 340-344.
6. Y. Kato, K. Nakamura, and T. Hashimoto, J. Chromatogr., 266 (1983) 385.
7. P. G. Stanton, R. J. Simpson, F. Lambrou, and M. T. W. Hearn, J. Chromatogr., 266 (1983) 273-279.
8. H. Juarez-Salinas, G. S. Ott, J.-C. Chen, T. L. Brooks, and L. H. Stanker, Methods in Enzymology, 121 (1985), in press.
9. H. Juarez-Salinas, W. L. Bigbee, G. B. LaMotte, and G. S. Ott, Methods in Enzymology, 121 (1985), in press.
10. M. Bradford, Anal. Biochem. 72 (1976) 248.
11. _____, "Bakerbond MAb™. A Major Advance in Monoclonal Antibody Purification". Technical Bulletin (J.T. Baker, Phillipsburg, NJ, 1984).
12. S. W. Burchiel, J. R. Billman, and T. R. Alber, J. Immunol. Meth. 69 (1984) 33-42.

APPLICATION OF IMMUNOAFFINITY CHROMATOGRAPHY TO THE PURIFICATION OF POLYPEPTIDES FOR MICROSEQUENCE ANALYSIS

James J. L'Italien

Molecular Genetics, Inc.
Minnetonka, MN 55343

ABSTRACT

Recent advances in the micropurification of polypeptides for structural analysis have centered upon individual improvements in the areas of HPLC (microbore), electrophoresis (electroblot) and bioaffinity chromatography (monoclonal antibody-immunoaffinity). With the advent of monoclonal antibody technology, immunoaffinity chromatography is rapidly becoming the most powerful single separation method for the complete, rapid purification of a protein of interest from complex mixtures of polypeptides. The numerous advantages of this method include its selectivity and the ability to efficiently recover polypeptides, which exist in trace concentrations, from crude cell lysates. Through the judicious selection of buffer and elution conditions, immunoaffinity techniques can often be linked with other micropurification methods such as SDS-PAGE and/or HPLC to eliminate intermediate handling steps and thus streamline overall purification schemes. This report will focus on pertinent aspects of optimization in the preparation and use of immunoaffinity columns for the microscale purification of polypeptides for microstructural analysis. Preferred methods for the selection and purification of monoclonal antibodies will be outlined in addition to an evaluation of immobilization methods. Several successful application of monoclonal immunoaffinity chromatography to the purification of polypeptides for microsequence analysis will be also be included.

INTRODUCTION

Bioaffinity chromatography in its various forms (liquid, lectin and antibody) has been extensively used since the early 1960's (1). In recent years there has been an ever increasing interest in immunoaffinity methods due to the advent of monoclonal antibody technology (2) which has expanded the usefulness and applicability of this technique by permitting investigators to obtain large amounts homogenous antibody with specificity for a single antigenic site. In addition to the specificity, the antibody exhibits homogenous binding and release kinetics for this antigenic site. Thus, the antigen is intereacting with the antibody through only one attachment site and the activity of all antibody molecules in this population is identical for that site. This has removed the limitations of antibody supply and permits the use of milder elution conditions to obtain maximal recovery of bound antigen while maintaining optimal antibody activity. Previously, immunoaffinity chromatography with monospecific polyclonal antibodies would exhibit the potential for attachment at several sites on the antigen with various avidities. Thus, generally creating the need for relatively harsh elution conditions such as chaotropic agents to obtain reasonable antigen recovery. An important correlate in the use of monoclonal antibodies for the micropurification of an antigen of interest is that purified protein is not required to prepare the antibody. In contrast to the preparation of polyclonal antibodies, where the specificity of the preparation is dependent upon the purity of the starting immunogen, monoclonal antibodies are the product of a single lymphocyte which produce only homogenous antibody for a specific antigen. The relevence of this attribute is that an investigator can use biological and/or immunological methods such as ELISA (3), immunoprecipitation (4), and/or western blot (5) to identify a monoclonal antibody specific for the antigen (polypeptide) of interest and then employ that monoclonal directly to purifiy that polypeptide. It should be pointed out that not all monoclonal antibodies to a particular polypeptide may be usable in immunoaffinity chromatography. The inability of certain monoclonal antibodies to be used may result from the inability to immobilize to a solid support, inability to immobilize to the support while retaining activity, and/or the inability of the antibody to release the antigen without irreversible denaturation of the antibody, or the antigen. Thus, even after the specificity of a monoclonal antibody has been ascertained, the usefulness of that antibody for immunoaffinity chromatography needs to be established. This report will focus on methods which have been found useful for the screening and optimization of immunoaffinity chromatography using monoclonal antibodies. This will be followed by selected applications of the methods described.

MATERIAL AND METHODS

Antibody purification by Protein A sepharose (Pharmacia) chromatography was performed in 1 ml columns after rehydration of the support in phosphate buffered saline (PBS). All columns were precycled through a blank elution cycle prior to use. Following requilbration, one ml aliquots of cleared ascites were passed over each column (one column was prepared for each elution method tested). The unbound fraction of the ascites was rinsed from the columns with 5 mls of PBS prior to elution of the antibody fraction. Elution of the bound fraction was accomplished at pH11 with 50 $m\underline{M}$ triethlamine (6), at pH2.5 with 1 $\underline{M}$ acetic acid (7), or at pH6 with 3 $\underline{M}$ potassium isothiocyanate (8). The recovered antibody was dialysed versus water (2 changes) and 100 $m\underline{M}$ HEPES (pH7.5). Recovery and purity of antibody were judged by gel scan versus quantiative markers and unfractionated ascites. The relative activity was determined by an antigen down ELISA.

Antibody purification by hydroxylapatite chromatography was performed with a 7.8 mm X 10 cm hydroxylapatite HPLC column (Bio-Rad) with a guard column (9). In these experiments the murine ascites fluid was diluted 1:1 with buffer A (10 $m\underline{M}$ sodium phosphate, 0.01 $m\underline{M}$ calcium chloride, 0.02% sodium azide, pH6.8) and centrifuged for 1 hour at 35K, 4°C. The diluted ascites was then passed through a 0.2u Gelman filter, with the filter then being rinsed with an equal volume of buffer A to give approximately a 1:3 final volume ratio of ascites to buffer A. For several antibodies tested it was found that optimal column resolution was obtained when 0.5 ml or less of ascities was loaded onto the column. This was accomplished by loading 2 mls of the 1:3 ascities, buffer A solution at a flow rate of 0.5 mls/min. The use of faster flow rates or greater volumes of ascites often led to high pressure problems with the precolumn. Several gradient and flow rate conditions were evaluated. Optimal conditions were determined to be: initial conditions of 20% buffer B at 0.5 ml/min; a linear gradient of 20 to 100% B over 24 minutes at 0.5 ml/min; isocratic at 100% B for 6 minutes at 0.5 ml/min; a linear gradient of 100 to 20% B over 10 minutes at 0.5 ml/min; re-equilabration at 20% B over 40 minutes at 0.5 ml/min (may be shortened at higher flow rates). The B buffer used is 500 $\underline{m}M$ sodium phosphate (pH6.8), 0.01 $\underline{m}M$ calcium chloride, 0.02% sodium azide. The long re-equilabration time was the minimum time necessary to bring the column to starting conditions based upon conductivity measurements. Antibody fractions were dialysed, and subjected to purity and activity assessment as described above.

Antibody purification by cation exchange chromatography was performed with a 7.5 mm X 7.5 cm DEAE 5 PW column (Waters) (10). Ascites was diluted 1:3 with buffer A (20 $\underline{m}M$ Tris, pH8.5), and 2 mls of diluted ascites loaded per injection. Unbound material was eluted from the column with 5 mls of buffer A prior to elution of the bound protein with a 25 minute linear gradient of buffer B (20 $\underline{m}M$ Tris, 0.5 $\underline{M}$ sodium chloride, pH8.5) into buffer A. All column procedures were performed at 1.0 ml/min and ambient temperatures. Dialysis, purity and activity were performed as previously described.

Immunobilization of antibody to preactivated agarose supports (11) was evaluated by looking at several parameters to determine the optimal conditions of antibody binding and activity for those antibodies used in this study. Briefly, purified IgG was dialysed into 100 $\underline{m}M$ HEPES (pH7.5) buffer and adjusted to the concentrations used in the study based upon the BCA protein assay (Pierce Chem Co.). The agarose support (Affi-10, Bio-Rad) was activated by placing the desired volume of support into a test tube, centrifugying it for 15-20 sec in a bench top clinical centrifuge and removing its storage solution (isopropanol) with a pasteur pepette. The process was quickly repeated 2 times after adding a volume of water (equal to the column of support) and repeating the centrifugation and removal steps. The desired volume and concentration of antibody in 100 $\underline{m}M$ HEPES (pH7.5) was then added, the antibody/agarose was agitated and the reaction was allowed to proceed at 4°C for 4-8 hours with shaking. (The process from activation of the support to addition of the antibody was routinely accomplished in less than 2 minutes). The unbound antibody was removed from the support by centrifugation of the reaction mixture, removal of the supernatents and rinsing the support with 100 $\underline{m}M$ HEPES (pH7.5). The excess reaction sites were blocked by addition of 1 $\underline{M}$ ethanolamine (100 μl per ml of support) to the 100 $\underline{m}M$ HEPES. The blocking reaction was carried out at 4°C overnight with shaking. The support was then washed 100 $\underline{m}M$ HEPES or PBS to remove excess blocking reagents prior to evaluation of its activity. The amount of antibody bound was determined by the difference in concentration of the starting antibody concentration (times volume) and the post reaction concentration and volume of the reaction mixture (plus rinse).

Antigen purification by affinity chromatography was accomplished by equilbrating the column in the sample application buffer, applying the sample over the column 2-3 times, removal of unbound material by rinsing the column with 5 column volumes of sample application buffer (6). A wash buffer is then applied to the column (sample application buffer plus 0.5 - 1 M salt) to remove non-specifically bound material. The column effluent from the wash buffer should initially be collected and analysed to ascertain its contents. Excess wash buffer is removed from the column by elution with 5 column of water (or volatile physiological pH buffer such ammonium bicarbonate). Elution of the bound protein was accomplished by elution with 5 column volumes of 50 mM triethylamine (pH11). For the chromatography presented here, virus infected cell lysates were used. The sample application buffer was the same buffer used to lyse the virus infected cells (100 mM Tris, 150 mM sodium chloride, pH8, 1% deoxycholate, 1% nonident 40) and the wash buffer was 100 mM Tris, 500 mM sodium chloride, 1% nonident 40(pH8).

Antigen purification by electrophoresis was performed as previously described (6). Antigen purification by HPLC was performed on a Vydac C4 column using 0.1% TFA in both aqueous phase and acetonitrile. Protein sequence analysis was performed using a gas-phase sequencer (12). Peptide mapping was performed on a Vydac C18 column using 10 mM potassium phosphate as the aqueous phase and acetonitrile as the organic modifier (13). Linear gradients at 1 ml per min were performed as described in the figure legends.

RESULTS AND DISCUSSION

Important considerations in the evaluation and optimixzation of immunoaffinity methods for a particular separation problems are: 1) the selection and purification of the antibody, 2) the preparation of the immunoaffinity support and 3) the development of elution conditions. Antibody selection for the preparation of the immunoaffinity support should include evaluation of the antibody specificity, the reversability of the antibody/antigen interaction under conditions which do not lead to irreversible denaturation of the antibody (or antigen) and the compatability of the antibody with the immobilization method used to prepare the immunoaffinity support. The specificity of the antibody may be established by microimmunological techniques such as immunoprecipitation (4) or western blotting (5). Once specificity of the antibody has been established it is necessary to screen antibodies for their potential usefulness in immunoaffinity chromatography by deter-

Table I. COMPARISON OF RAPID IgG PURIFICATION METHODS

COLUMN	ELUTION CONDITIONS	pH	% IgG RECOVERY	% IgG PURITY	RELATIVE ACTIVITY
Protein A Sepharose	50 mM TEA	11	24%	100%	>95%
Protein A Sepharose	1M HOAc	2	20%	100%	90%
Protein A Sepharose	3M KSCN	6	24%	100%	75%
Hydroxylapatite HPLC	0.3M NaCl	6.8	93%	96%	>95%
DEAE HPLC	0.3M NaCl	8.5	67%	64%	>95%

mining whether they will bind to the support, whether they will retain their activity after immobilized and whether antibody activity is retained following elution of the antigen. These functional properties of the antibody are best determined by preparation and evaluation of mini-immunoaffinity columsn as described below.

Antigen purification is the first step in the evaluation of antibody for the preparation of immunoaffinity supports. While there are numerous ways in which antibodies may be purified, this report summarizes the evaluation and comparison of several methods which were deemed appropriate for the purification of several differnt antibodies in quantaties sufficient for preparation and evaluation of monoclonal mini-immunoaffinity columns. Table I contains a comparison and various methods used to purify antibody from small amounts of murine ascites fluid.

For the monoclonal antibodies used these experiments, the best overall recovery, considering purity and activity was obtained with hydroxylapatite. Some problems were encountered with the precolumn and the column capacity was limited to approximately 20 mg of the total protein thus necessitating 2 runs to purify IgG from 1 ml of ascites. In subsequent work we have employed hydroxylapatite in small open columns using similar gradients with results similar to those reported here. The advantage of these columns is that several can be run at one time and once the antibody is selected these columns exhibit linear scale-up if necessary. The antibody purified by the DEAE columns exhibited excellent activity but its purity was the poorest of the methods tested here and its overall recovery was less satisfactory than hydroxylapatite. The protein A data was interesting from several perspectives. The purity of antibody recovered from the protein A columns was excellent (there was no detectable non-IgG present). The overall recovery of antibody from the protein A columns was significantly lower than expected. These results, however, may be typical of "new" columns (the support was previously unused but had been preconditioned by running 2 blank cycles which included all the steps of the usual procedure except the ascites step). The activity of the protein A columns was confirmed by the fact that there was no antibody in the unretained column fraction, thus ruling out the possibility that the IgG loss was due to a lack of retention to the protein A column.

The antibody eluted from protein A in base (50 mM TEA, pH11) retained its activity within the limits of the experiment. Antibody eluted in acid (1 M acetic acid, pH2) lost approximately 10% of its activity (due to antibody denaturation which occured during pH change, and was evident following dialysis as a precipitate). The use of chaotropic agents to elute the antibody from protein A resulted in a loss of approximately 25% of antibody activity. A heavy precipitate was observed in this fraction following dialysis. The extent of activity loss by irreversible denaturation (through precipitation) by using 1 M acetic acid (7) or 3 M potassium isothiocyante (8) was unexpected, as these are classical eluents used for elution from protein A and which are recommended by the manufacturer for the protein A support (14) and antibody column chromatography (11). This observation has been substantiated with several other monolconal antibodies. Obviously, as evidenced by the literature, these eluents can be used but if one is working with limiting amounts of antibody, one may wish to screen their antibody to observe the effects of these conditions.

The information on antibody activity in the presence of acid, base and chaotropic agents is valuable not only for selection of optimal antibody purification conditions but also for the selection of elution conditions for the immunoaffinity columns. The rational being that conditions which denature the antibody will render it inactive whether it is free or bound to a solid support. Thus, screening the activity of the antibody following treatment with acid, base or chaotropic agents will aid in the selection of elution condition for the immunoaffinity columns.

Preparation of immunoaffinity mini-columns for evaluation of the antibody is the next step. Towards this end various antibody concentrations and several antibody solution/support volume ratios were evaluated to investigate conditions which would permit optimal conditions for antibody immunobilization while retaining maximal antibody activity. Thus, establishing working guidelines for immunoaffinity mini-column preparation. In this study (Table II) 6 aliquots of the same antibody (concentration and/or volume used vary) were immobilixed to 2 ml of hydroxysuccinimide ester agarose (Bio-Rad Affi-10) as described above. The data which was acquired in this study illustrates several interesting points. Comparison of the percent IgG coupled for samples with the same total amount of antibody show that samples with higher concentration couple more effecicently (compare samples 1 and 2, Table II). The percent of antibody bound generally increased with increasing antibody concentration to a maximum of 97% of total available antibody for the concentrations used in this study. The point of maximal antibody % binding was reached when only approximately 60% of the total capacity of the support for this antibody was used. Another important area of comparison was the capacity of the various supports for their antigen. The results, here were that the antigen recovery per ml of support, as might be expected, generally increased as the amount of antibody bound increased to the maximal percent bound than tapered off. There was one exception, however, which occured at low antibody substitution on the column (all of these studies were performed under conditions in which the antigen was in excess). One method of evaluating column efficiency is to calculate the percentage of usable binding sites per column (μ mole of antigen recovered/μ mole of antibody X 1 μ mole antibody bound/2 μ mole binding sites X 100%) when the column is used under conditions of antigen excess. Comparison of this parameter (Table II) illustrates that

Table II. **EVALUATION OF COUPLING METHODS**

	IgG CONC. (mg/ml)	IgG VOL. (ml)	TOTAL IgG (mg)	% IgG COUPLED	IgG BOUND PER ml SUPPORT (nMoles)	nMole OF BINDING SITES PER ml	Ag RECOVERED PER ml SUPPORT (nMole)	% USABLE IgG BINDING SITES
1	1.05	4	4.2	54%	7.6	15.2	0.95	6.3%
2	2.09	2	4.2	73%	10.2	20.4	2.95	14.5%
3	2.09	4	8.4	68%	19.0	38	1.75	4.6%
4	11.0	4	44	87%	128	256	1.90	0.74%
5	25.6	2	51.2	97%	166	332	3.45	1.04%
6	25.6	4	102.4	78%	266	532	2.45	0.46%

colum efficiency decreases with respect to the usability of binding sites as the amount of antibody bound increases. Thus, while the greater percentage of antibody bound (97%) and the highest overall capacity for binding antigen (3.45 μM/ml) were obtained using an antibody concentration of 25.6 mg/ml per ml support, only 1% of the potential antibody binding sites were usable. In contrast, when equal volumes of the same antibody at greater than ten fold lower concentration (2.09 mg/ml) was used for immobilization with the same volume of support, 73% of the antibody was bound yielding an antigen capacity of 2.95 μM/ml of support, making use of 14% of the potential antibody binding capacity. To look at this result in another way, the column with the lower degree of antibody substitution had 86% of the antibody binding capacity using less than 10% of the antibody needed to prepare the more highly substituted column. Thus the optimal use of antibody as judged by percent of useable binding sties in this instance has an inverse relationship with optimization of bound antibody. Therefore, the investigator must determine the conditions which result in a reasonable use of antibody to obtain the desired amount of antigen.

Following preparation of the immunoaffinity support, the next important step is selection of elution conditions. Important considerations in the selection of elution conditions are: maximization of antigen recovery while maintaining activity of the immobilized antibody and the antigen; and secondly, to maintain compatability with subsequent purification characterization steps. The evaluation of elution conditions and antibody activity can be quickly ascertained by treatment of antibody with the elution buffers and evaluation by antigen down ELISA. Alternatively conditions for the efficient disruption of the antibody-antigen interaction and effects of these conditions on antibody activity can be screened by taking small aliquots of the antibody bound support performing batchwise elution using centrifugation to precipitate the support. In this way several sets of conditions can be evaluated using only small amounts of the support should the conditions used lead to inactivity of the immobilized antibody. In screening several sets of conditions it was observed that elution with 50 mM triethylamine (pH11) was consistantly the best elution buffer for maximal antigen recovery and maintance of complete antibody activity. This has been verfied using 5 monoclonal antibodies to 5 different antigens with the majority of these immobilized antibodies retaining significant acitivity for up to 1 year while using the solvent and elution conditions described here.

The procedures described here have been applied to the purification and characterization of 5 viral glycoproteins to the present time. A representative example of the recovery and purity of these glycoproteins following monoclonal antibody immunoaffinity chromatography is given in Table III. In this example approximately 330 μg of the glycoprotein mixture of interest was present in 15 mls of virus infected cell lysate as judged by ELISA. The initial protein concentration was calculated to be 22 μg/ml with a relative purity estimated at less than 0.1% (the presence of 1% deoxycholate in the lysate impeded more accurate measurement). The 15 mls of lysate were passed over a one ml (volume) immunoaffinity column and, following the procedure previously outlined approximately 293 μg of the protein mixture of interest was eluted from the column with 50 mM triethylamine in approximately 2.5 mls. This represents a recovery of approximately 89% of the protein initially present with a purity estimated at 90% by SDS-PAGE. The sample was then further purified by SDS-PAGE and electroelution to yield an estimated 188 μg of protein at 95% purity. The percent recovery for the electroelution step was approximately

Table III.

RAPID TWO-STEP PROTEIN PURIFICATION:
IMMUNOAFFINITY CHROMATOGRAPHY/ELECTROPHORESIS

STEP	PROCEDURE	AMT. (ug)	VOL. (ml)	CONC. (ug/ml)	% RECOVERY	% PURE
1	Cell lysate	330	15	22	–	< 0.1%
2	Immunoaffinity chromatography	293	2.5	117	89%	90%
3	PAGE/ electroelution	188	0.5	376	64%	95%

Overall recovery was 57%

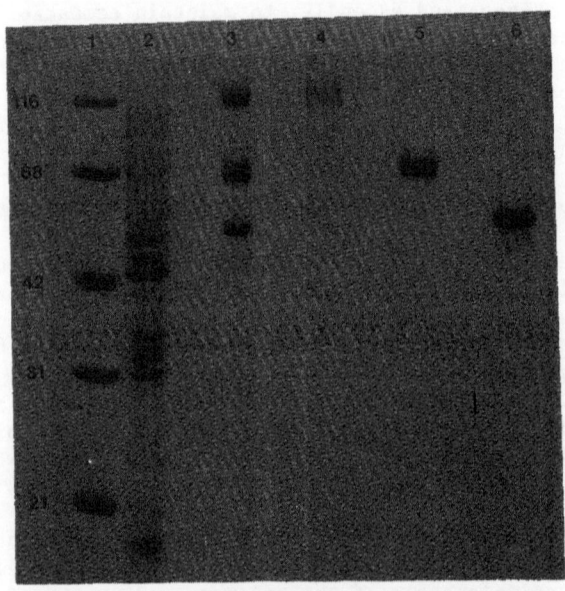

Figure 1.

Sodium dodecyl sulphate - polyacrylamide gel electrophoresis (SDS-PAGE) analysis of the three steps in the purification process referred to in Table III. Briefly, aliquots of samples from each of the three steps in this 2-step purification scheme were heated in the presence of SDS/reducing agent and subjected to SDS-PAGE as described by Laemli (ref. 15). Lane 1 represents molecular weight markers whose molecular weight in Kilodaltons is depicted on the left. Lane 2 represents an aliquot of the cell lysate. Lane 3 represents an aliquot of the protein recovered from immunoaffinity chromatography. The remaining lanes represent aliquots of each of the protein bands recovered following SDS-PAGE purification and electroelution (as described in ref. 6). Lane 4, 5, and 6 were designated GP-1, GP-2 and GP-3 respectively.

AMINO TERMINAL SEQUENCE OF AFFINITY/PAGE PURIFIED GP-2 AND GP-3

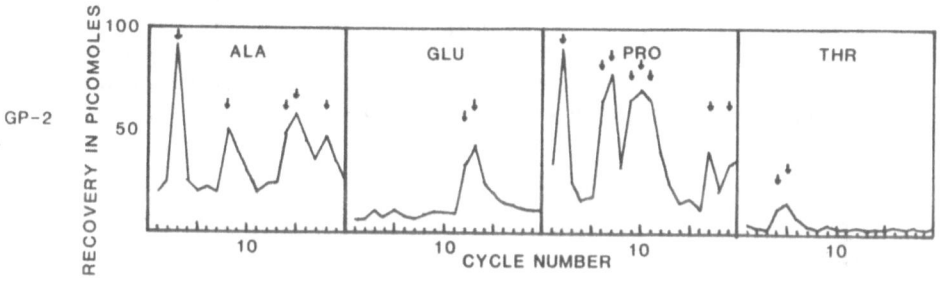

R-P-A-T-T-P-P-A-P-P-P-E-E-A-A-S-P-A-P-P

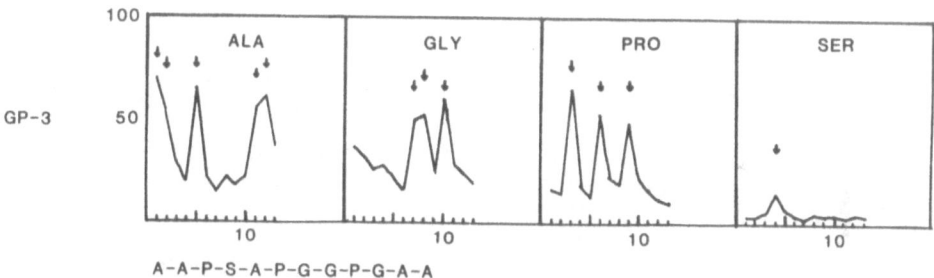

A-A-P-S-A-P-G-G-P-G-A-A

Figure 2.

This figure illustrates the amino terminal sequence obtained by
gas-phase sequence analysis (ref. 12) of each of the proteins purified
as described in the text and illustrates in Fig. 1.

64% and the overall recovery for the two step process was 57% to yield 188 μg of protein at 95% purity. The product from the steps in this purification scheme are visually represented in Figure 1. This figure clearly illustrates the complexity of the initial cell lysate (lane 2). The presence of three polypeptide bands in the affinity purified protein fraction (lane 3) was initially perplexing and lead to the need for further purification by SDS-PAGE electroelution for separation of the 3 constituant protein bands (lanes 4, 5, and 6). Each of the purified protein bands was then subjected to amino terminal sequence analysis in a gas-phase protein sequencer. The results of this analysis for each of the 3 polypeptides bands (designated GP-1, GP-2 and GP-3 for the 124 Kdalton band, 10 Kdalton band and 54 K dalton band respectively) are depicted in Figure 2. As seen in this figure, GP-2 and GP-3 each gave different sequences and each sequence contained only a single signal indicating that they were pure from a sequencing perspective. The subsequent analysis of GP-1 revealed complete identity with the GP-2 sequence through the readible sequence (approximately 15 residues). The presence of a proteolytic cleavage site in the GP-1 polypeptide which results in the formation of the GP-2 and GP-3 polypeptides was confirmed as the DNA sequence for this protein became available. The pertinent data is complied in Figure 3 illustrating the relationship between GP-1 and its proteolytic cleavage fragments.

STRUCTURAL ANALYSIS OF AFFINITY PURIFIED ANTIGENS

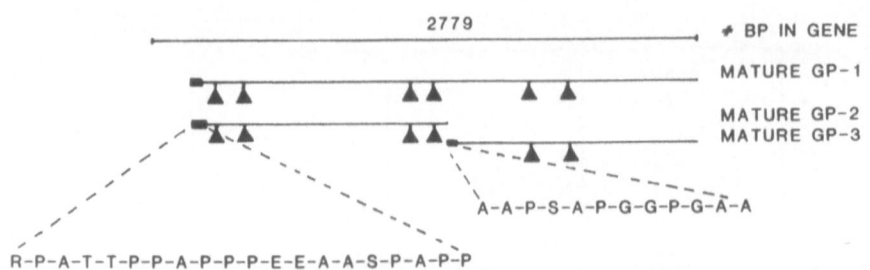

	# AMINO ACIDS	CALCULATED MW	#POTENTIAL GLYCOSILATION SITES	MW(SDS-PAGE)
GP-1	866	99,590	6	124,000
GP-2	438	50,370	4	70,000
GP-3	428	49,220	2	54,000

Figure 3.

This figure summarizes the data available for each of the 3 polypeptides illustrated in Figure 1 and described in the text. The amino terminal protein sequence (Figure 2) was integrated with the data available by SDS-PAGE (Figure 1) and the DNA-sequence (ref. 16) as it became available. The protein sequence of GP-1 and GP-2 established the amino terminus of the mature, processed polypeptide while the protein sequence of GP-3 established a second protolytic site in the mature protein.

REFERENCES

1. P. Cuatrecasas and C.B. Anfinsen (1971) in Meth. in Enzymology 22, 345-378.

2. G. Kohler and C. Milstein (1975) Nature 256, 495-497.

3. E. Engvall and P. Perlmann (1971) Immunochem. 8, 871-874.

4. S.W. Kessler (1975) J. Immunol. 115, 1617-1624.

5. W.N. Burnett (1981) Anal. Biochem. 112, 195-203.

6. J.J. L'Italien, in J. Shively (Editor), Methods of Protein Micro-Characterization, Humana Press, Clifton, N.J., 1986. Ch. 10, pp. 279-314.

7. C.C. Patrick and G. Virella (1978) Immunochem 15, 137-139.

8. M.R. Mackenzie, N.L. Warner, and G.F. Mitchell (1978) J. Immunol. 120, 1493-1496.

9. H. Juarez-Salinas, S. Engelhorn, W.L. Bigbee, M.A. Lowry and L.H. Stanker (1984) Bio Techniques 2, 164-169.

10. M.J. Gemski, B.P. Doctor, M.K. Gentry, M.G. Pluskal and M.P. Strickler (1985) Bio Techniques 3, 378-384.

11. Bio-Rad Bulletin 1085. Activated Affinity Supports: Affi-Gel 10 and 15 (1982) Bio-Rad Laboratories.

12. R.M. Hewick, M.W. Hunkapiller, L.E. Hood, and W.J. Dreyer, (1981), J. Biol. Chem. 256, 7990-7997.

13. J.J. L'Italien and R.A. Laursen (1981) J. Biol. Chem. 256, 8092-8101.

14. In Affinity Chromatography, Principles and Methods Pharmacia Ljungfortagan AB, Orebro, Sweden, 1983, pp 48-57.

15. U.K. Laemmli, (1970) Nature (London), 227, 680-685.

16. T.E. Zamb, (1985) in preparation.

II. Electrophortic Methods of Polypeptide Purification

A NOVEL APPROACH TO ISOLATION OF PROTEINS FOR MICROSEQUENCE ANALYSIS:

ELECTROBLOTTING ONTO ACTIVATED GLASS

Ruedi Aebersold, David B. Teplow, Leroy E. Hood and
Stephen B. H. Kent

Division of Biology, 147-75
California Institute of Technology
Pasadena, California 91125

INTRODUCTION

One of the major goals of modern molecular biology is to understand the structure and function of proteins and their respective genes. To that end recombinant DNA and protein chemical techniques can act in a synergistic manner. Protein sequencing can provide information with which to construct synthetic oligonucleotides for use in the isolation of genes or to confirm that the correct gene has been isolated (1,2). Alternatively protein sequence data can be used to demonstrate the in vivo expression of polypeptides, whose existence can only be predicted based on the presence of open reading frames in cloned DNA (3).

Currently, high sensitivity protein sequencing in a gas phase sequenator (4) with optimized HPLC detection of the resulting phenyl-thiohydantoin (PTH) amino acids (5) can provide limited amino terminal sequence data from 10-20 pmoles of sequencable protein. This corresponds to 0.5-1µg of a 50,000 dalton protein. The preparation of such minute amounts of protein in high yields in a form suitable for automated sequence analysis represents a considerable technical challenge, for a number of reasons: (i) the protein usually has to be purified to homogeneity from a complex mixture of unrelated proteins; (ii) damage to the polypeptide chain, such as amino-terminal blocking or peptide bond cleavage, has to be prevented; (iii) isolated proteins must be in a form which does not interfere with the Edman degradation. In particular, soluble compounds containing primary or secondary amino groups should be avoided; (iv) the number of handling and transfer steps should be minimized since in each transfer significant amounts of protein can be lost.

An optimal isolation technique should be capable of resolving a wide range of proteins, take place in a defined, chemically inert environment and lead to immobilization of the polypeptides free of contaminants as soon as possible after their separation. The most general, highly resolving techniques for separating microgram quantities of proteins use gel electrophoresis. One-dimensional gel electrophoresis methods such as SDS polyacrylamide gel electrophoresis (SDS-PAGE) and isoelectric focusing (IEF) have considerable resolving power and two-dimensional (2-D) gel electrophoresis (IEF, followed by SDS-PAGE) is capable of resolving several thousand polypeptides in a highly reproducible manner (6, 7). Here we describe a novel method which takes advantage of 1-D and 2-D gel electrophoretic techniques to isolate proteins for sequencing. It is based on the direct electrophoretic transfer ("electroblotting") of proteins from these gels onto chemically activated glass fiber sheets and subsequent high sensitivity detection of the blotted proteins. Proteins isolated in this way are suitable for use in the gas phase microsequenator without further manipulation (8). [An oral presentation of results contained in this report was made by one of us (SBHK) at the Methods in Protein Sequence Analysis meeting in Cambridge, U.K., on August 2, 1984].

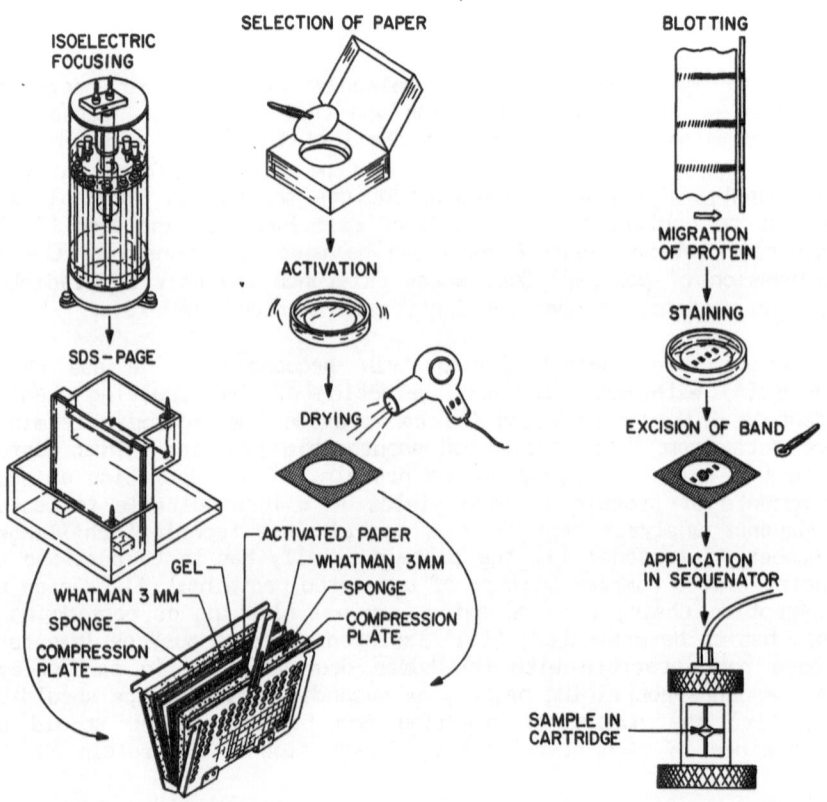

Fig. 1. Principles of blotting. Proteins to be electroblotted may be prepared by one- and/or two-dimensional gel electrophoresis. After electrophoresis, the gel is mounted in a standard blotting apparatus against a sheet of chemically activated glass fiber filter paper. The proteins are electrophoretically transferred onto the paper and then visualized by staining with Coomassie blue, fluorescent dyes, or by autoradiography. After drying, the stained protein bands are excised and loaded directly into a gas phase sequenator.

Principle of the method

Electroblotting of proteins for direct sequence analysis consists of the following steps (Fig. 1): (i) preparation of the glass fiber sheet; (ii) separation of proteins by SDS-PAGE, IEF in immobilized pH gradients (IPG-IEF) or 2D-PAGE; (iii) electrophoretic transfer of the separated proteins onto chemically modified glass fiber paper; (iv) detection of the transferred molecules and excision of the protein-containing bands or spots.

Three variations of the method are currently used, two of which are shown in Fig. 2. Proteins separated by SDS-PAGE or 2-D electrophoresis are transferred at high pH onto glass fiber sheets covalently modified with quaternary ammonium (QA-glass) or with aminopropyl groups (AP-glass), where they are adsorbed by ionic and possibly hydrophobic inter-actions. Proteins isolated by IPG-IEF (i.e., in the absence of SDS) are transferred at low pH onto glass fiber sheets that have been etched with TFA, where they are retained by ionic and possibly hydrophobic inter-actions. The third variant, in which proteins are electroblotted at high pH onto glass fiber paper derivatized with p-phenylenediisothio-cyanate (DITC-glass) for covalent attachment and solid phase sequencing is described in an accompanying paper in this volume (Kent et al.).

Preparation and characteristics of glass fiber supports

A variety of supports have been used for analytical electro-blotting. These include nitrocellulose and charge modified nylon sheets (9, 10). However, neither of these supports is compatible with the Edman chemistry used in a gas phase sequenator. Nitrocellulose dissolves and the charge modified nylon collapses into a solid pellet. Glass fiber discs, however, have been used routinely as the support for proteins and peptides in the gas phase sequenator. They are stable to the sequencing chemistry and are not known to interfere with the chemistry nor introduce background contaminants in the subsequent PTH analysis. Therefore, we decided to develop glass fiber paper as a support for electroblotting and subsequent sequencing.

TFA-etched glass fiber sheets. Circles of Whatman GF/C or GF/F glass fiber paper (11cm to 15cm in diameter) were immersed in anhydrous trifluoroacetic acid (TFA) (Pierce, Chemical Co., sequenal grade, Rockford, IL, 61105) in a covered Petri dish and kept for 1 h at room temperature, after which the TFA was decanted. If more than one sheet was used at a time, extreme care was taken to remove any air bubbles between sheets so that the TFA contacted the glass fiber sheets com-pletely and uniformly. This was best achieved by sliding the sheets into the TFA one at a time. The glass fiber sheets were dried by blowing a stream of cold air over them. It was essential to continue this until no traces of TFA remained. Activated sheets remained usable for at least several weeks when stored at room temperature in a glass container (e.g., a covered Petri dish). CAUTION: Because of the extremely corrosive nature of TFA, all manipulations were performed in an efficient fume hood using gloves, face and eye protection.

Preparation of aminopropyl- and quaternary ammonium glass fiber sheets (Fig. 3). TFA etched GF/C or GF/F glass fiber sheets were immersed in a freshly prepared 2% (v/v) solution of N-trimethoxy-silylpropyl-N,N,N-trimethylammonium chloride (Petrarch Systems Inc., Bristol, PA 19007, Cat. # T-2925) in 95% acetone in water for 2 min in a glass Petri dish. The sheets were then washed at least 10 times in acetone (5 min each wash) on a shaking platform to remove excess

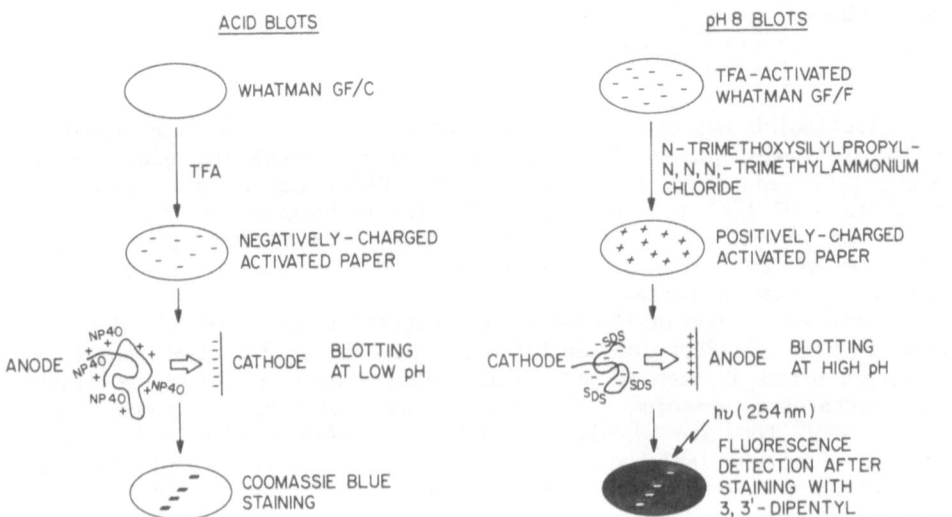

ACID BLOTS

pH 8 BLOTS

Fig. 2.   Mechanisms of protein transfer and detection. Acid blotting is
performed using Whatman glass fiber filter discs activated by
immersion in anhydrous TFA. After complete removal of excess
acid, the activated paper is placed in the blotting apparatus
next to a IPG-IEF gel and electrophoresis performed in dilute
acetic acid, containing NP-40. At this low pH, the proteins
have a net positive charge and migrate toward the cathode onto
the paper. The transferred proteins are visualized by staining
with Coomassie blue. Blotting at high pH (8.3) is performed
similarly, except that the TFA-activated paper is further
derivatized with a quaternary ammonium compound and the
proteins from an SDS-polyacrylamide gel migrate toward the
anode onto the paper, due to the presence of SDS. The proteins
are visualized by treatment with a fluorescent dye and
irradiation with a UV (254 nm) lamp.

reagent.   Curing of the silane linkage was achieved by drying the
sheets, supported on Whatman #1 paper, in an oven for 45 min at 110°C.
The short reaction time, extensive washing and exact curing conditions
were critical to the preparation of derivatized glass suitable for high
efficiency sequencing.

The amount of QA groups deposited onto the glass fiber paper is not
easily determined. Therefore, as a control for these manipulations, the
same modification procedure was used with 3-aminopropyltriethoxysilane
(Pierce Chemical Co., Rockford, IL, 61105, cat. #80370). Deposition of
3-aminopropyl groups on the glass fiber sheets and the completeness of
the washing process was demonstrated by quantitative ninhydrin determinations of free amino groups. This was done by drying down 3 ml of
the acetone washes in 10 x 75 mm glass tubes under a stream of nitrogen,
or by placing a 5 mm diameter disc of the cured aminopropyl-glass fiber

paper in a glass tube of the same dimensions, then carrying out the quantitative ninhydrin determination, as previously described (11). With this analytical control, the modification procedure can be optimized for AP-glass and then similar conditions applied to preparation of QA-glass. Typically, after TFA etching, 10 μmoles of amino-groups could be deposited per g of GF/F glass fiber paper. Modified sheets could be stored for at least several weeks at room temperature. They were chemically and physically stable under conditions used for the Edman degradation and had a capacity of 7-10 μg of protein per cm^2 (2-3 μg/band for a typical analytical gel).

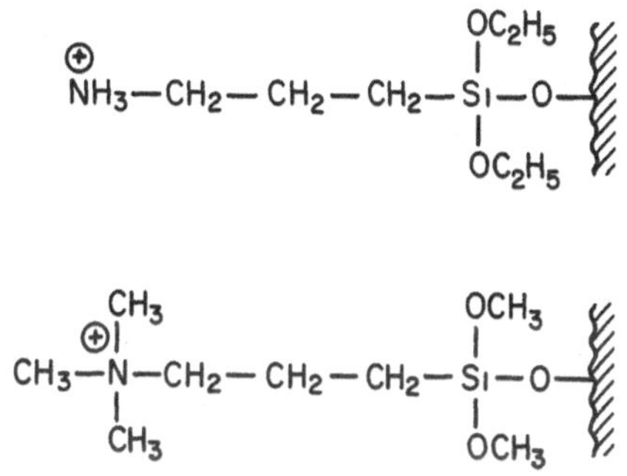

$$\overset{\oplus}{NH_3}-CH_2-CH_2-CH_2-\underset{\underset{OC_2H_5}{|}}{\overset{\overset{OC_2H_5}{|}}{Si}}-O\rightleftharpoons \quad \text{Aminopropyl-glass (AP-glass)}$$

$$CH_3-\underset{\underset{CH_3}{|}}{\overset{\overset{CH_3}{\oplus|}}{N}}-CH_2-CH_2-CH_2-\underset{\underset{OCH_3}{|}}{\overset{\overset{OCH_3}{|}}{Si}}-O\rightleftharpoons \quad \text{Quaternary ammonium-glass (QA-glass)}$$

Fig. 3. Chemistry of activation. 3-Aminopropyltriethoxysilane is used to produce glass fiber supports containing aminopropyl groups (AP-glass) capable of ionic interactions with proteins, or for use as a substrate for further derivatizations. N-trimethoxy-silylpropyl-N,N,N-trimethylammoniuim chloride is used to produce derivatized glass fiber supports containing quaternary ammonium functions (QA-glass) which can participate in ionic interactions with proteins.

Electrophoresis

One-dimensional gel electrophoretic techniques such as SDS-PAGE and IPG-IEF, as well as 2-D gel electrophoresis, are compatible with electroblotting onto activated glass. The separation method used should be selected according to the properties of the protein(s) under study. For IEF, only immobilized pH gradients are compatible with microisolation, since soluble ampholytes strongly bind to proteins and interfere with the Edman chemistry. In classical 2-D electrophoresis, after electrofocusing proteins are electrophoresed through an SDS-polyacrylamide gel where the soluble ampholytes are quantitatively removed and therefore cause no problems during sequencing.

In the experiments described here, SDS polyacrylamide gels were run essentially according to Laemmli (12). Modifications to this procedure and specific precautions to prevent amino-terminal blocking of proteins have been described (13). Two-dimensional gels were run according to the method of O'Farrell (6).

Normally 0.5 mm thick gels were run. However, gels of thickness 1.5 mm or more could also be used successfully by increasing the transfer time proportionately in the electroblotting step. IEF gels containing immobilized pH gradients were prepared using commercially available reagents (Immobilines, LKB-Produkter AB, Bromma, Sweden) and run according to the method of Gianazza (14, 15). Between pH 4 and pH 10, gradients one to six pH units wide could be used. The plastic sheets normally used to support these gels interfered with electroblotting, therefore gels were polymerized between two glass plates that had been extensively silanized (Prosil 28, Speciality Chemicals, Gainesville, Florida, 3260), according to the manufacturer's directions.

Electroblotting

We have used electroblotting from polyacrylamide, IPG-IEF, and 2-D gels to prepare a wide variety of proteins for sequencing. Figure 4 shows an experiment in which six proteins, with molecular weights ranging from 14,000 to 116,000, were electroblotted under high pH conditions from a SDS-polyacrylamide gel. The intense staining of the blot and concomitant lack of staining of the gel demonstrate that the transfer of the proteins has occurred with high efficiency.

Quantitation of blotting from SDS-polyacrylamide gels at high pH and from IPG-IEF gels at low pH is shown in Table 1. Transfer efficiencies were calculated using radiolabeled proteins of known specific activity. Recoveries of the protein standards were uniformly

Fig. 4. Electroblotting at high pH. E. coli β-galactosidase (β-gal, 116 kDa), bovine serum albumin (BSA, 68 kDa), bovine carbonic anhydrase (CA, 29 kDa), soybean trypsin inhibitor (STI, 21 kDa), sperm whale myoglobin (Mb, 17 kDa), and bovine α-lactal-bumin (α-lac, 14 kDa) were each loaded onto two gels and then subjected to SDS-PAGE (12.5% T). One gel was blotted at pH 8.3 onto quaternary ammonium activated GF/F paper (QA-glass) and then this blotted gel (b) and a second reference gel (a) were stained with Coomassie blue. The paper (blot) (c) was stained with 3,3'-dipentyloxacarbocyanine iodide and photographed under 254 nm UV light.

high, ranging from 70 to 99%. Our experience with SDS-PAGE blots using a wide range of proteins (see Table 3 below) and with complex mixtures containing many different protein species has confirmed this observation. Yields between 70-100% are routinely obtained, even with large (100-200 kDa) proteins. We have observed comparable recoveries from gels of low and high percent acrylamide, and from gradient gels.

With IPG-IEF gels, both intact proteins of various sizes, with isoelectric points from 4.5 to 8.5 (Table 1), as well as cyanogen bromide derived peptides have been successfully transferred with high yields.

A detailed description of the blotting experiment is as follows: Electroblotting onto TFA etched, and QA- or AP-glass fiber paper was performed in a BioRad "Trans-Blot" cell with model 250/2.5 constant voltage power supply. Immediately after protein separation the sandwich was assembled in the following way (Fig. 1): sponge; Whatman 3MM paper, cut to the same size as the sponge; activated GF/C or GF/F (back-up sheet); activated GF/C or GF/F sheet; the gel; Whatman 3MM paper, cut to the size of the sponge; sponge. Care was taken to eliminate any bubbles between the gel and the activated glass fiber sheet. This was best achieved by pressing with a glass rod or spatula on the glass fiber sheet placed on top of the gel. The sandwich was then assembled and placed into the blotting chamber, which was equilibrated to 0-5°C using a cooling coil and a refrigerated recirculating water bath. For low pH blotting the activated sheets were oriented closest to the cathode (negative electrode). For high pH blotting the opposite orientation was

Table 1. Blotting efficiency from SDS-polyacrylamide and IPG-IEF gels

Protein	$M_r \times 10^{-3}$	pI	Low pH[a]		High pH[b]	
			Found on blot (%)	Found in gel (%)	Found on blot (%)	Found in gel (%)
Bovine α-lactalbumin	14	4.6	103	<5	94	<5
Sperm whale myoglobin	17[c]	8.35	82	<5	75	<5
	17[d]	8.5	100	<5		
Soybean trypsin inhibitor	21	4.5	70	30	99	<5
Bovine carbonic anhydrase	29[c]	5.8	81	19	74	<5
	29[d]	6.1	72	28		
Bovine serum albumin	68[c]	4.9	96	<5	76	<5
	68[d]	5.1	96	<5		
E. coli β-Galactosidase	116	4.5	n.d.	n.d.	81	<5

[a] 5 μg of each protein loaded onto the IPG-IEF gel.
[b] 3 μg of each protein loaded onto the SDS-PAGE gel.
[c] acidic band
[d] basic band

used. For low pH blotting the transfer solution was 1% acetic acid, containing 0.5% NP-40 (Particle Data Laboratories, Elmhurst, IL, 60126). For high pH blotting the buffer was 25mM Tris-HCl, pH 8.3, containing 0.5 mM DTT. In our initial experiments we included 192 mM glycine, but this was found to be unnecessary and has since been decreased to 10 mM or eliminated completely. Blotting was performed for 2 h at 50-70 V (250-700 mA) at ~2-5°C for 12.5% T (total acrylamide concentration), 0.5 mm thick gels. For thicker gels, or gels with a higher % T, blotting time was increased. The current was dependent on the dimensions of the blotting chamber, the buffer and the temperature. The solution in the blotting chamber was magnetically stirred throughout the procedure.

Detection of electroblotted proteins

Proteins blotted onto the acid-etched glass fiber sheets can be detected by Coomassie blue staining. The detection limit is about 50 ng, 5-10 times more sensitive than Coomassie blue staining of proteins in gels. As has been previously observed (16), the staining and destaining are also much more rapid for the blotted proteins. Note that because the proteins are blotted before staining, the staining procedure cannot interfere with transfer of the protein. Furthermore, provided all excess Coomassie blue is washed from the surrounding glass fiber sheet, the stained protein is sequenced without the dye interfering with the subsequent PTH analysis, as it remains bound to the protein in the sequenator throughout the sequencing process.

The staining procedure consisted of the following steps. After completion of the transfer, TFA-etched glass fiber sheets were placed in staining solution [0.5% (w/v) Coomassie blue R-250, (Serva, Heidelberg, FRG), 30% (v/v) isopropanol, 10% (v/v) acetic acid, in distilled water] for 2 min. Destaining was performed under warm (40°C) running distilled water, by shaking the blot in distilled water, or by using destaining solution (16.5% (v/v) methanol, 5% (v/v) acetic acid, in distilled water). As discussed above thorough destaining was necessary to avoid artefact peaks in phenylthiohydantoin (PTH) analysis after sequencing. Protein bands treated with destaining solution were washed with distilled water to remove all traces of acid before storage.

Coomassie blue, a negatively-charged dye, strongly interacted with quaternary amino-groups. Therefore, this dye was not suitable for detection of proteins on QA-glass. Radioiodinated proteins could easily be detected on that support by exposing Kodak XAR-5 X-ray film to the blot. For non-radiolabeled proteins, however, it was necessary to develop a new staining method that would be compatible with the presence of the positively charged functionalities in the sheets as well as with subsequent sequencing of the proteins. This method used the fluorescent dye 3,3'-dipentyloxacarbocyanine iodide (Molecular Probes, Inc., Junction City, OR, 97448, Cat. #D272), which shows a change in its absorption and emission spectra upon binding to proteins. This dye has previously been used for fluorescence studies of membranes (17). Detection was based on hydrophobic interaction between the proteins and the two pentyl chains attached to the dye.

To perform the staining, the blotted paper was partially dried under a stream of air, at 40°C in an oven for approximately 20 min, or by repeatedly pressing against fresh sheets of Whatman #1 paper. The staining solution was prepared by completely dissolving 3 mg of 3,3'-dipentyloxacarbocyanine iodide in 3 ml of methanol and then adding this to 27 ml of 0.05 M NaHCO$_3$, pH 8.2-8.6. The partially dried blot was added to the staining solution in a Petri dish and shaken for 15 min at

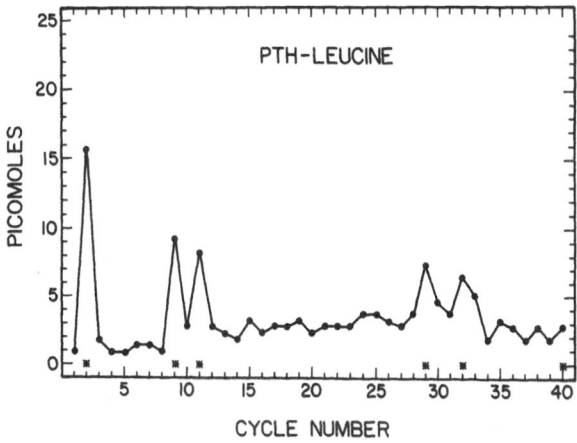

Fig. 5. Sequence analysis of electroblotted sperm whale myoglobin. 41
pmoles (0.74 µg of Mb) were run in SDS-PAGE, electroblotted and
sequenced on a gas phase sequenator. The yields of PTH leucine
in each cycle plotted. Asterisks (*) indicate positions in the
polypeptide where leucine is known to occur.

room temperature. Each blot was stained separately in fresh staining
solution.   Protein quantities of more than 1 µg could be located
immediately as orange bands against a background of the yellow dye
solution. Smaller amounts of protein (50 ng-1 µg) were detected on the
wet blot, using 254 nm UV light, as fluorescent green-yellow bands on a
dark background. Destaining was generally unnecessary, but could be
performed by washing the blot with the same buffer without added dye.
In the sequenator cartridge the dye was completely removed from the
protein with the first ethyl acetate or butyl chloride wash.

<u>Sequencing of electroblotted proteins</u>

    Protein samples prepared by either blotting method sequenced
efficiently.   Using radiolabeled proteins of known specific activity
(determined by amino acid analysis) initial yields of sequencable
material were determined for several proteins (Table 2). Low and high
pH blotting gave an initial PTH signal corresponding to 60-75% of the
protein present. The repetitive yields were measured for the same set
of proteins based on the amount of PTH amino acids present in successive
cycles of the Edman degradation. As can be seen in Table 2 and Fig. 5,
repetitive stepwise yields were typically 93-96% for amounts of protein
in the 50 pmole range.   This is about 2 per cent better than we
routinely obtain sequencing the same proteins on polybrene-coated GF/C
discs.

    The use of radiolabeled proteins enabled us to quantitate protein
losses from the glass fiber support during sequencing. There were
slightly smaller losses from the acid etched support than from the

Table 2. Sequencing yields from electroblotted proteins

Protein	Low pH (IPG–IEF)				High pH (SDS-PAGE)			
	Amount loaded (pmole)	Initial yield (%)	Repetitive yield (%)	Sample loss/cycle (%)	Amount loaded (pmole)	Initial yield (%)	Repetitive yield (%)	Sample loss/cycle (%)
Bovine α-lactalbumin	40	66	94	0.4	45	61	94	1.4
Sperm whale myoglobin	n.d	n.d	n.d	n.d	41	76	93	1.9
Soybean trypsin inhibitor	157	64	96	0.6	33	62	94	0.8
Bovine serum albumin	50	34[a]	96	0.2	29	66	93	0.9

[a] Protein stored in urea for a period of time.

covalently-modified glass fiber support, but both were low, at about 1% per cycle, for a wide variety of proteins (14 - 68 kd).

In addition to the improved repetitive stepwise yields, sequencing of blotted proteins prepared by either methodology yielded PTH analyses with extraordinarily low backgrounds of by-product peaks. This is dramatically illustrated in Fig. 6 where the PTH analyses from the first cycles of sequencing runs of two different proteins (bovine serum albumin and soybean trypsin inhibitor) are shown. Even though high pH blotting onto QA-glass had been carried out in 25 mM Tris, 0.5 mM DTT, 192 mM glycine (which is not routinely used), there was very little contamination of the PTH derivatives with PTH glycine or with phenylthiocarbamyl-dimethylamine (PTC-DMA), diphenylthiourea (DPTU), or diphenylurea (DPU). Levels of PTH glycine and PTC-DMA are typically the equivalent of a PTH amino acid signal of 3 pmole or less, while the level of DPTU plus DPU is usually the equivalent of 10-20 pmoles. Apparently, much of the background normally seen in HPLC analysis of PTH samples from the gas phase sequenator must arise because of the use of polybrene to immobilize the sample.

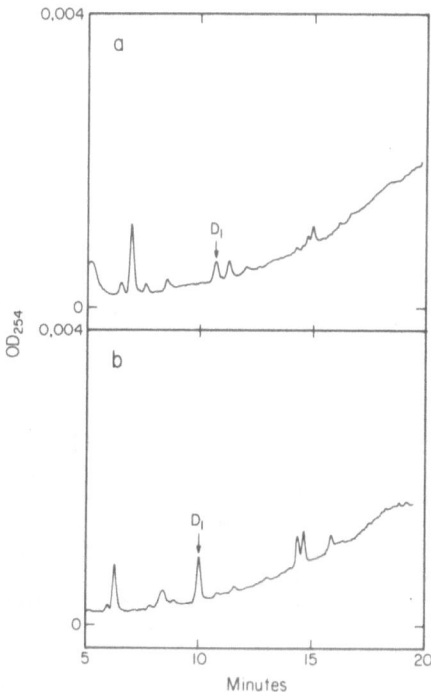

Fig. 6. Analysis of PTH amino acids derived from sequencing electro-blotted proteins. Samples of bovine serum albumin (BSA) and soybean trypsin inhibitor (STI) were subjected to SDS-PAGE, high pH blotted and sequenced. Presented are the HPLC analyses of the amino-terminal PTH amino acids of: (a) STI; (b) BSA. The symbol D_1 indicates the position of PTH Asp, N-terminal amino acid of both proteins.

This combination of high initial yields, high repetitive stepwise yields and very low backgrounds in the PTH analysis has allowed us to undertake sequencing of very small amounts of proteins. The results of some sequencing runs are summarized in Table 3.

Electroblotted and stained proteins could be stored for at least several weeks at -20°C before sequencing, without noticeable deterioration. Transferred proteins were sequenced as follows. Protein bands, detected after Coomassie or fluorescent staining procedures, or by autoradiography, were cut out of the glass fiber sheets and placed in the cartridge of a gas phase sequenator without any further treatment. Well resolved single bands were cut out as discs to fit directly into the upper block of the cartridge. Narrowly spaced bands on the blot were cut out and several pieces (e.g., from different lanes in a gel) were loaded on top of the Zitex support in the cartridge. If small pieces or strips of glass fiber paper were placed in the cartridge, they were covered by a disc of untreated GF/C to avoid channeling of reagents and solvents during sequencing. The insertion of more than two complete layers of glass fiber strips into the cartridge impeded the flow of reagents and gave incomplete reactions and poor sequencing results. Thus, particular attention was paid to solvent delivery rates, especially S2 (ethyl acetate), for the first few cycles in a sequencing run to ensure proper performance.

The protein sequence determination was performed as described (4), with HCl-methanol conversion. The sequencing program was started by a subroutine consisting of a 10 min TFA treatment followed by a 90 s ethyl acetate or butyl chloride wash before the first coupling was performed. These washes completely removed 3,3'-dipentyloxacarbocyanine iodide but not Coomassie blue. The resulting PTH derivatives were analyzed by HPLC on an IBM cyano column essentially as described by Hunkapiller and Hood (5), except that 5-7% (v/v) tetrahydrofuran was added to the "A" buffer and the pH was adjusted to 5.1. PTH amino acids were quantitated by comparison of peak heights with standards.

Troubleshooting

For the preparation of QA-glass, attention should focus on the preparation of the quaternary ammonium solution and the washing process. N-trimethoxysilylpropyl-N,N,N-trimethylammonium chloride should be a clear, slightly yellowish solution in methanol and should be stored under N_2 or Ar in sealed aliquots in a dessicator. The dilute solution used for derivatization should be freshly prepared before use. If the reagent is exposed to moisture or the derivatization solution is stored for an extended period of time, trimethoxysilýl groups are hydrolyzed and and start to crosslink individual molecules, resulting in visible precipitates. It is crucial to thoroughly wash the paper after its derivatization to remove these aggregates and uncoupled monomers as they can interfere with the transfer of proteins from the gel to the paper.

As mentioned earlier there is no easy way to directly detect and quantitate QA-groups and to measure the efficiency of the washing process. Therefore, we have used the following procedure to test both the stability of the silane layers and the efficiency of the washing process. An SDS gel, without a stacking gel, was poured and subjected to high pH electroblotting with QA-glass without any protein applied. Immediately after blotting, the gel was stained for 10 minutes with Coomassie blue and then destained. The QA glass was discarded. If the gel destained completely the conditions for the modification of the

Table 3. Proteins sequenced using the electroblotting technique

Protein	$M_r \times 10^{-3}$	Sample preparation	Sequencable amount (pmoles)	Number of residues determined
Rat heart gap junction autolytic fragment	46	high pH blot	17	22/24
Rat heart gap junction autolytic fragment	31	high pH blot	10	20/25
Mouse liver gap junction	28	high pH blot	12	16/18
Mouse liver membrane protein (gap junction associated)	21	high pH blot	5	10/16
Mouse liver membrane protein (gap junction associated)	16	high pH blot	140	19[a]
IgG heavy chain	58	high pH blot	110	20[a]
IgG light chain	28	high pH blot	12	13/17
Cholinergic differentiation factor	48	high pH blot	10	10/13
HF-treated cholinergic differentiation factor	28	high pH blot	12	9/12
Soluble binding protein	150	high pH blot	33	21/24
Bovine serum albumin	68	high pH blot[b]	12	13/17
Soybean trypsin inhibitor	21	high pH blot[b]	11	12/13
Sperm whale myoglobin	17	high pH blot[b]	12	10/11
Bovine α-lactalbumin	14	high pH blot[b]	12	11/13
Bovine α-lactalbumin	14	low pH blot	40	22/24
Soybean trypsin inhibitor	17	low pH blot	157	18[a]
Bovine serum albumin	68	low pH blot	50	15[a]
Yeast single strand DNA binding protein acidic CNBr fragment	n.d.	low pH blot	15	8/10
Yellow fever virus protein	45	spotted[c]	400	20[a]
Mu phage transposase	70	spotted[d]	5	8/11
Scrapie-associated protein cyanogen bromide fragment	1.8	spotted[c]	50	12/14
[Arg[8]]-Vasopressin	1	spotted[c]	60	8/9
Synthetic peptide	4.5	spotted[c]	200	36/40

(continued)

Table 3. Continued

Notes:
[a]Run terminated at this point.
[b]From two-dimensional gel.
[c]Spotted onto QA-glass disc in water, dilute acid, or 0.1% TFA-aqueous acetonitrile.
[d]Spotted onto QA-glass disc in 300 mM KCl.

---

paper were satisfactory. If the gel did not destain and a strongly stained film was observed on the surface, it demonstrated that QA groups were transferred onto the gel, either due to a lack of stability of the silane layers (polymers) or due to insufficient washing. In an electroblotting experiment, poor transfer would occur if this paper were to be used.

Extensions of the method

In addition to the applications described above we have extended the method. We have found the chemically-modified glass fiber supports to be useful for the direct application of proteins or peptides isolated in solution by a variety of chromatographic methods, including ion exchange and reverse phase HPLC.

Another application where the high pH electroblotting technique is useful involves "multi-dimensional" analysis to obtain internal amino acid sequence information from a protein. A mixture of proteins is separated by SDS-PAGE and a gel slice containing the protein of interest is cut out. The protein is digested in the gel slice either chemically (18, 19) or enzymatically (20) and the resulting fragments applied to a second SDS-PAGE gel of higher %T. After separation, all the fragments are blotted onto an activated glass fiber sheet and individually subjected to sequence analysis.

An alternative to multidimensional analyses is the recovery of electroblotted proteins from the support using a volatile buffer. Recovered proteins can subsequently undergo peptide mapping by HPLC after partial or complete chemical or enzymatic cleavage, chemical or enzymatic deglycosylation, complete hydrolysis for analysis of the amino acid composition, or the biological activity can be analyzed.

SUMMARY

We have described a novel approach to the preparation of proteins for amino acid sequence determination in the gas phase sequenator. Proteins are separated by SDS-PAGE, IPG-IEF, or 2D-electrophoresis, before their electrophoretic transfer from the polyacrylamide gels onto chemically activated glass fiber sheets, subsequent detection, and direct use of the glass fiber-bound protein in sequence determination. This approach offers a number of advantages over previous methods of isolating proteins for sequencing. All three separation techniques are highly resolving and generally applicable to a wide range of proteins. The electroblotting method is straightforward and does not require expensive or complex equipment. The protein is prepared for sequencing without ever being directly handled or removed from solution until it is attached to the support on which it is sequenced. In this form, it can be isolated without the extraordinary precautions often required to minimize sample losses, with high recoveries of even nanogram amounts of most proteins. Because the stained protein is sequenced in situ on the glass fiber blot, there is no doubt as to the presence or identity of the protein being sequenced. Protein samples prepared in this way are inserted directly into the sequenator without further manipulation.

Perhaps the greatest single advantage of the electroblotting procedure is that it involves the concurrent preparation of large numbers of proteins for sequencing. Depending on the samples, it is possible to isolate hundreds of proteins for sequencing in a single operation and to sequence these rapidly without precycling.

ACKNOWLEDGEMENTS

The contributions of Dr. Lloyd Smith, for suggesting the use of fluorescent dyes, and the expert technical contributions of John Kim, Eva Lujan, Wade Hines and Chin Sook Kim are gratefully acknowledged. This work was supported by research grants from the NIH, the NSF (Biol. Inst. Grant #DMB-8500298), the Monsanto Company, and Upjohn Pharmaceuticals. R.A. was the recipient of a EMBO Long-Term Fellowship.

REFERENCES

1. Oesch, B., Westaway, D., Wälchli, M., Mc Kinley, M. P., Kent, S. B. H., Aebersold, R., Barry, R. A., Tempst, P., Teplow, D. B. Hood, L. E., Prusiner, S. B., and Weissman, C., Cell 40:735-746 (1985).
2. Hurley, J. B., Fong, H. K. W., Teplow, D. B., Dreyer, W. J. and Simon, M. I., Proc. Natl. Acad. Sci. USA 81:6948-6952 (1984).
3. Hylka, V. W., Teplow, D. B., Kent, S. B. H. and Straus, D. S., J. Biol. Chem., in press (1985).
4. Hewick, R. M., Hunkapiller, M. W., Hood, L. E., and Dreyer, W. J., J. Biol. Chem. 256:7990-7997 (1981).
5. Hunkapiller, M. W., and Hood, L. E., Meth. Enzymol. 91:486-494 (1983).
6. O'Farrell, P. M., J. Biol. Chem. 250:4007-4021 (1975).
7. Anderson, N. G. and Anderson, N. L., Anal. Biochem. 85:331-340 (1978).
8. Aebersold, R., Teplow, D. B., Hood, L. E., and Kent, S. B. H., J. Biol. Chem. in press (1985).
9. Towbin, H., Staehelin, T., and Gordon, J., Proc. Natl. Acad. Sci. USA 76:4350-4354 (1979).
10 Gershoni, J. M., and Palade, G. E., Anal. Biochem. 124:396-405 (1982).
11. Sarin, V. K., Kent, S. B. H., Tam, J. P., and Merrifield, R. B., Anal. Biochem. 117:147-157 (1981).
12. Laemmli, U. K., Nature 227:680-686 (1970).
13. Hunkapiller, M. W., Lujan, E., Ostrander, F., and Hood, L. E., Meth. Enzymol. 91:227-236 (1983).
14. Gianazza, E., Celentano, F., Dossi, G., Bjellqvist, B. R., and Righetti, P. G., Electrophoresis 5:88-97 (1985).
15. Gianazza, E., Giacon, P., Sahlin, B. and Righetti, P. G., Electrophoresis 6:53-56 (1985).
16. BioRad Technical Bulletin #1080 (1983).
17. Pick, U., and Avron, M., Biochem. Biophys. Acta 440:189-204 (1976).
18. Nikodem, V., and Fresco, J. R., Anal. Biochem. 97:382-386 (1979).
19. Detke, S., and Keller, J. M., J. Biol. Chem. 257:3905-3911 (1982).
20. Cleveland, D. W., Fischer, S. G., Kirschner, M. W., and Laemmli, U. K., J. Biol. Chem. 252:1102-1106 (1977).

BLOTTING OF PROTEIN-DETERGENT COMPLEXES ON GLASS AND GLASS COATED WITH POLYBASES : A COMPARATIVE STUDY. ACID HYDROLYSIS AND GAS-PHASE SEQUENCING OF GEL-SEPARATED PROTEINS IMMOBILIZED ON POLYBRENE GLASS-FIBER

Joel Vanderderckhove

Laboratory of Genetics, State University Ghent
Ledeganckstraat 35, B-9000, Belgium

INTRODUCTION

The transfer onto immobilizing membranes of proteins which have been separated on polyacrylamide gels (popularly referred to as the Western blot) has become an important tool in protein chemistry (for a comprehensive review see[1]). Originally, transferred proteins were covalently bound onto the membrane[2,3], but this method showed low coupling yields and was therefore replaced by the introduction of microporous nitrocellulose membranes to which proteins could be bound by non-covalent interactions[4]. This principle was later improved by using a nylon based, positively-charged membrane[5]. Although proteins immobilized in this way could be easily assayed for their antigenicity or their enzymatic or lectin-binding activity[1], the membranes used for these analyses were sensitive to the chemical treatments necessary for acid hydrolysis or for the Edman-degradation based chemistry. In order to combine the simple, fast, inexpensive, high-resolution purification procedure of polyacrylamide gel electrophoresis with the recently developed technique of protein gas-phase sequencing[6], it was necessary to introduce a blotting support which is both able to bind every kind of protein eluting from a gel, and resistant against chemicals and solvents used in Edman chemistry.

The immobilization procedures discussed in this paper are based on the idea that complexes of proteins with ionic detergents will bind onto hydrophobic surfaces that have a charge opposite to that of the detergent used. Thus, sodium dodecylsulfate (SDS)-protein complexes will bind to apolar, positively charged surfaces, while cetyltrimethylammonium bromide (cetavlon) -protein complexes will bind to apolar, negatively charged supports.

In this paper we describe different protein immobilization conditions and compare them in terms of protein binding-capacity, ease of operation, and possibilities to be combined with currently used gas-phase sequencing strategies. Immobilization of proteins from SDS-containing gels onto glass-fiber sheets coated with a non-covalently bound monolayer of the quaternary ammonium polybase, Polybrene, was found to be the most suitable of all tested procedures. Details for a practical execution of this technique are provided and examples of acid hydrolysis and gas-phase sequencing of proteins immobilized in the 1-20 µg range, are shown. This technique is found to be especially suitable for obtaining sequence information from proteins which are partially enriched by single step procedures (e.g. by affinity-chromatography or immune-precipitation) and fractionated or purified by SDS-polyacrylamide gel electrophoresis. It may therefore serve as a new tool in obtaining partial sequence information that can either serve as guide for synthesis of specific DNA probes, or confirm the protein sequence predicted from DNA sequencing. In addition, this technique promises to become a major tool in mapping the epitopes of monoclonal antibodies, in allocating the sites of post-translational processing, and in assessing protein homologies on SDS-polyacrylmide gel peptide maps of partially digested proteins.

MATERIALS AND METHODS

Protein Immobilization on Unmodified Glass

Proteins are separated on 1.5 mm thick SDS-polyacrylamide slab gels with acrylamide concentrations between 7.5 % and 20 %. Acrylamide concentrations are selected such that most of the proteins to be eluted migrate in the lower half of the gel (more efficient elution). Between 1-70 µg of protein mixture is loaded in a 9 mm broad slot and separated essentially as described by Laemmli[7]. Reference proteins (5 µg each; e.g. the Bio Rad low molecular mass protein mixture) are run in parallel and will serve as a control for all further manipulations. At the end of the separation, proteins are slightly fixed in the gel by washing for 15 min in 50 % methanol-water (by vol). The gel is then further washed for at least 2 hr with several 100 ml changes of a solution containing 0.5 % acetic acid and 0.1 % cetavlon (v/w). In this step, SDS is exchanged by the invert detergent cetavlon. Glass microfiber sheets (GF/C 46 x 57 cm, Whatman Ltd., England) are cut to the size of the gel, briefly washed in water and layered onto the cetavlon-gel. The blotting sandwich, in which nitrocellulose is replaced by the glass-fiber, is constructed as originally described by Towbin et al.[4]. The blotting is carried out for 20 h at 4 V/cm in 0,5 % acetic acid. The protein-cetavlon complex migrates to the cathode side. Immobilized proteins are detected by staining for 2 min with a solution containing 0.25 % Coomassie brilliant blue, 45 % methanol and 9 % acetic acid (w/v/v), and destained with water and finally with 5 % methanol, 7.5 % acetic acid (v/v).

Protein Immobilization on Polybrene Coated Glass

The protein mixture and references are separated as described above. GF/C glass-fiber sheets are cut at the size of the gel and dipped in a 3 mg/ml solution of Polybrene (Janssen Chimica Belgium) in water and allowed to dry in the air. Alternatively, the sheets may be suspended with paper clams and wetted by allowing the Polybrene solution to drip from a pasteur pipette over the suspended sheet. The dried sheets may either be used immediately or stored.

At the end of the electrophoresis, the gel is equilibrated for at least 2 h with 100 ml of transfer buffer to which is added 0.1 % SDS. The transfer buffer consists of 50 mM sodium borate, pH 8.0 and 0.02 % 2-mercaptoethanol. In most experiments we have also included 20 % methanol in the transfer buffer in order to avoid extensive swelling especially of high-percentage polyacrylamide gels during the blotting. Methanol is not necessary for protein immobilization.

The dried Polybrene glass-fiber sheets are washed for 2 min in 100 ml water in order to remove excess Polybrene and mounted in the blotting

sandwich : a piece of scouring pad, 3 layers of wet Whatman 3MM paper, the
SDS-gel, two or three sequentially layered glass-fiber sheets, 3 layers of
wet Whatman 3MM paper and the second piece of scouring pad. The sandwich
is pressed between two metal screens and immersed in the blotting apparatus (e.g. a Gel Destainer GD-4, Pharmacia, or a Bio Rad Transblot Cell).
The blotting is carried out for 20 h at 4 V/cm in the blotting buffer
given above (without SDS). The proteins move towards the anode side.

At the end of the blotting process, the glass-fiber sheets are
separated from the gel and washed immediately with a buffer containing
20 mM NaCl, 10 mM borate, pH 8.0. This washing procedure, intended to
exchange bound glycinate for chloride, is repeated three more times in
100 ml of the exchange buffer and concluded by a short wash (1 min) with
distilled water. The glass-fiber sheets are allowed to dry in the air by
suspending with paper   clams. Immobilized proteins are detected by
dipping the carefully dried blots in a diluted solution of fluorescamine.
Usually concentrations of 1 mg/500 ml in acetone are sufficient for
protein visualization. When less protein is present, a more concentrated
solution may be used (1mg /200 ml). Note however, that higher fluram
concentrations will result in increased blocking of the NH_2-terminal
groups. Fluorescence reaches its highest sensitivity after 5-10 min and
is stable in the dark for more than 4 weeks. Results are recorded by
illumination under UV light using a mineral light lamp (Ultraviolet
Products, San Gabriel USA) and photographed with a Polaroid camera.

Acid Hydrolysis of Proteins Immobilized on Polybrene-Glass

The portion of the glass-fiber sheet containing the protein is cut
out and placed in a Pyrex tube 0.6 x 6 cm to which is added 400 µl 6N HCl
containing 0.05 % 2-mercaptoethanol. The tube is sealed under vacuum and
incubated for 22 h at 110 °C. After hydrolysis, the content is dried
quickly in a vacuum dessicator over NaOH and resuspended in 400 µl 0,2 M
sodium citrate buffer pH 2.2. The glass-fiber is removed by filtration
through a Millex HV4 filter (Millipore) and an aliquot of the amino acid
mixture is applied for amino-acid analysis. Our analyses were carried out
with a Biotronik amino-acid analyser equipped with a fluorescence detector
recording the o-phthaldialdehyde reaction products[8].

Gas-Phase Amino-Acid Sequencing of Polybrene-Glass-Immobilized Proteins

The portion of the glass-fiber sheet carrying the immobilized
fluorescamine stained protein is either punched out with a 12 mm diameter
cork bore, or, when the protein bands are too close to each other, cut out
with scissors. The piece of glass-fiber is then mounted in the cartridge
of the gas-phase sequenator. When rectangular pieces are mounted, care
has to be taken so that they do not slip between the seal of the cartridge
during assembly. Note that it is not essential that an intact 12 mm
diameter disc be mounted for proper performance. The gas-phase sequenator
(Applied Biosystems Inc., USA) assembled and operated essentially as
described by Hewick et al.[6], is then run without changes in its program.
The stepwise liberated phenylthiohydantoin (PTH)-amino-acids are analysed
using a Cyano-HPLC analytical column (IBM Instruments) and the gradient
elution system described by Hunkapiller and Hood[9]. For these analyses we
used a Waters HPLC system, including two M 6000 A pumps, a WISP 710 B
autoinjector, a M 721 system controller and a fixed wave-length detector
M 441 (detecting at 254 nm). The recovery of the PTH-derivatives was
measured by comparing with a reference mixture using an integrative
recorder (Waters Data Module M 730).

Two-Dimensional Polyacrylamide Gel Electrophoresis

Chicken embryonic skeletal muscle tissue (18 days) was dissolved in 10% 2-mercaptoethanol and 3% SDS, centrifuged and aliquots containing appropriate amounts of protein were lyophilized. The dried material was redissolved in lysis buffer[10] and the proteins were separated in the two-dimensional polyacrylamide gel system as described by Garrels[10].

RESULTS AND DISCUSSION

Figure 1 shows the results of blotting experiments carried out with the Bio Rad standard protein mixture separated on a 12.5% polyacrylamide gel. One slot of the gel was stained with Coomassie blue (Fig. 1 A) and the other slots were used for different types of glass-fiber blotting. Lanes B and C represent a Coomassie-stained blot of the proteins immobilized onto the unmodified glass surface as cetavlon-protein complexes. The binding seems to be general since all proteins bind to the first (B) and second (C) layer. In lane D, we still observe the presence of a certain amount of protein left in the gel, indicating that proteins elute less efficiently than as SDS-protein complexes (see Fig. 1, lanes E-G). This low transfer efficiency may be due to incomplete exchange of SDS by

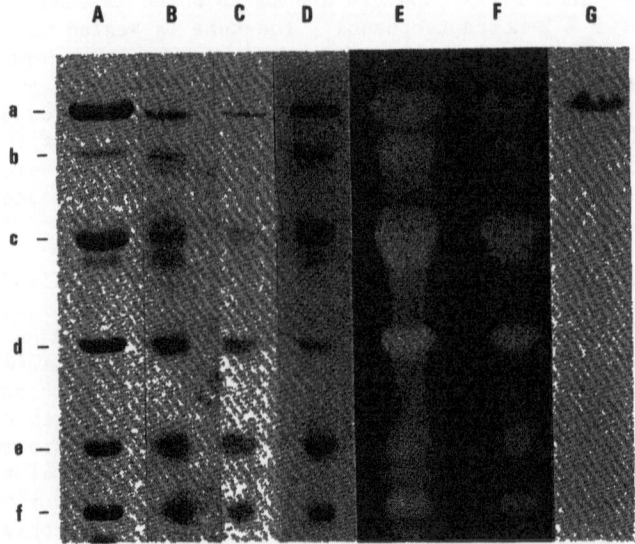

Fig. 1. *Glass-fiber blots of a reference protein mixture. The protein mixture contains phoshorylase b(a), bovine serum albumin (b), ovalbumin (c), carbonic anhydrase (d), soybean trypsin inhibitor (e), and lysozyme (f), separated on a 12.5 % gel. (A) The pattern after staining the gel with Coomassie blue; (B) and (C) first and second layer of the blots stained with Coomassie after immobilization as cetavlon-protein complex; (D) the Coomassie-stained pattern of the proteins remaining in the gel; (E) and (F) the first and second layer of the fluorescamine stained blots after immobilization as SDS-protein complex on Polybrene glass; (G) Coomassie-stained proteins remaining in the gel after the blotting process.*

cetavlon. It may therefore be possible to obtain better yields when proteins are separated in cetavlon-containing gels as described recently[11].

Lanes E and F illustrate the fluorescamine-stained blots on Polybrene-glass-fiber. In G we see the Coomassie-stained pattern of the proteins remaining in the gel. These patterns demonstrate that, with the exception of the largest protein of the mixture, all standard proteins are now quantitatively electro-eluted out of the gel and immobilized on at least two sequentially placed layers of Polybrene-glass-fiber. Phosphorylase b is not completely eluted due to its high molecular mass so that for this protein a longer elution period seems to be necessary. Proteins which are present in undersaturation amounts (e.g. the degradation products of the major proteins) are already fully retained on the first layer. This observation has a practical implication : in cases where the major protein is not fully separated from minor contaminants, it may be advantageous to proceed with the major protein bands immobilized on the second glass fiber layer, since the minor protein band will not have passed through the first layer.

The Capacity of Various Blotting Techniques

The fact that proteins, when applied in sufficient amounts, are also bound on the second and sometimes also on the third layer, indicated that the glass-fiber retains a limited capacity of protein binding. We have measured this capacity for the cetavlon-protein-glass-blot and the SDS-protein-Polybrene-glass-blot. Therefore we determined by amino-acid analysis the amount of protein retained on saturated areas of the sheets (areas in the first layer corresponding with the protein spot on the second sheet). The results were then normalized to record μg protein/cm^2 and are summarized in Table 1. Maximum loading capacities were determined for six different proteins : sperm-whale skeletal muscle myoglobin, chicken egg phosvitin (a highly phosphorylated protein), rabbit skeletal muscle actin, bovine chymotrypsinogen, human serum albumin and bovine pancreatic DNase I. Values were lower for the cetavlon-blots than for the SDS-Polybrene procedure. Since in addition, cetavlon blotting was more time consuming and resulted in less efficient elution of proteins (see Fig. 1), we mainly concentrated on the SDS-Polybrene procedure. Here,

Table 1. Binding Capacity of Different Glass-Blots for Various Proteins from Polyacrylamide Gel.

Protein $\mu g/cm^2$	Cetavlon-protein on Unmodified Glass	SDS-protein on Polybrene-Glass
Phosvitin	3	8
Chymotrypsinogen	10	15
Myoglobin	10	21
DNase I	13	21
Human Serum Albumin	18	23
Actin	15	28

binding capacities varied for most proteins between 15 μg/cm^2 and 28 μg/cm^2 but were low for phosvitin (8 μg/cm^2). The variations in the binding capacities for different proteins may reflect differential binding of SDS to proteins as it may be supposed that proteins bind to Polybrene-glass mainly as their SDS-complexes.

Amino-Acid Analysis

The hydrolysates used for calculation of the capacity, also served to ascertain whether the composition calculated from hydrolysates of Polybrene-glass-immobilized proteins agreed with the real protein compositions. In Table 2 we have summarized the amino-acid compositions obtained from immobilized actin and chymotrypsinogen. The results of these studies, combined with more recent analyses from other proteins allow to emphasize several points concerning this type of hydrolysis. First, all calculated compositions were in good agreement with the values expected from known sequences. Only concentrations of glycine (which may be too high) and methionine (which may be too low) differed sometimes from the expected values. We do not have an obvious explanation for the destruction of methionine, but the high glycine contents may originate from incomplete removal of bound glycinate. This glycine contamination which was sometimes present in molar excess over the protein, has however never been found to interfere with the first step of the gas-phase sequence analysis (see below). This suggests that the non-exchangeable glycinate may be covalently bound to the Polybrene glass and released only during acid hydrolysis but not under the milder conditions of the Edman degradation. Second, neither artefactual peaks, nor abnormally high levels of ammonia were observed during amino acid analysis; this is a major advantage in comparison to hydrolysis of the polyacrylamide gel pieces containing the

Table 2. Amino-Acid Compositions Determined from Polybrene-Glass Blots

| | Actin | | Chymotrypsinogen | |
	Deduced from sequence	Calculated from blot	Deduced from sequence	Calculated from blot
Asx	35	33.9	23	23.4
Thr	27	26.4	23	21.7
Ser	23	23.4	28	24.5
Glx	39	42.2	15	13.9
Gly	28	31.9	23	29.0
Ala	29	30.6	22	24.6
Val	21	19.2	23	25.6
Met	16	13.9	2	1.9
Ile	30	26.6	10	9.9
Leu	26	27.4	19	21.1
Tyr	16	16.6	4	6.5
Phe	12	12.1	6	6.5
Lys	19	17.7	14	15.2
His	9	8.9	2	2.1
Arg	17	18.3	4	5.1

Pro, Trp, Cys = not determined

protein. Third, lysine, which forms acid-stable fluorescamine derivatives is generally recovered in yields up to 90 % of the expected values, indicating that the dilute fluram-stain only consumed a very small amount of amino groups. In conclusion, Polybrene-glass-fiber immobilized proteins can be identified further by their amino-acid composition, provided that the points indicated above are considered.

Gas-Phase Sequencing of Polybrene-Glass-Fiber Immobilized Proteins

The most important application of the glass-blotting technique is that immobilized proteins can now be directly sequenced on the gas-phase sequenator without further manipulations. In this respect, glass-blotting may be considered as a new tool fitting into a general purification strategy for proteins of which partial sequence information is needed.

The combination of glass-blotting with gas-phase sequencing was first tested with immobilized myoglobin. The run involved 18 cycles which allowed a calculation of the initial coupling yield and repetitive yield. We found a repetitive yield (91 % - 92 %) similar to that measured for a "classical" myoglobin run, but we observed a considerably lower initial coupling yield for the Polybrene-immobilized protein (30 % compared to 60 % for the classical run). This lower yield suggests that the NH_2-terminus became partially blocked during the gel separation and/or the blotting process. There may be several possible causes : reaction with impurities in the components of the gel or buffers and the formation of

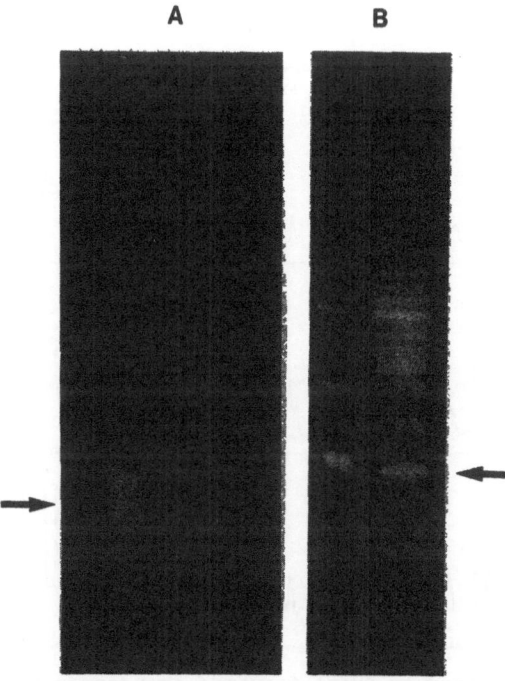

Fig. 2. *Blots on Polybrene-coated glass-fiber of a commercial preparation of Streptavidin (A) and the total lysate of an* E. coli *strain expressing human γ-interferon. The two inner lanes of both figures represent the Bio Rad protein reference mixture. Separation is on 17.5 % polyacrylamide gels. Arrows indicate the bands which were cut out for gas-sequencing (see Figs. 3 and 4).*

Schiff's bases with traces of aldehyde present in the methanol used in the transfer buffer, or in the acetone used to detect the proteins. In order to further increase the overall sequencing efficiency we are currently undertaking experiments in which the protein is reversibly blocked at the NH$_2$-terminus (e. g. by the acid labile exo-cis-3,6-endoxo-Δ^4-tetrahydrophthalic anhydride or by the DABITC/PITC (4-NN-dimethylaminoazobenzene-4'-isothiocyanate/phenylisothiocyanate) double coupling procedure[13]). More simply it may already be advantageous to omit the methanol from the transfer buffer – of which the function is simply to avoid excessive swelling of the gel –, or to add glycinate to the transfer buffer as scavenger for amino-reactive groups.

Two illustrations of the Polybrene glass blot gas-phase sequencing strategy are given. In one experiment we have separated by gel electrophoresis the two subunits of the biotin binding protein streptavidin. The corresponding blot is shown in Figure 2 A. The glass-fiber strip with the small subunit (500 pmol) was used for gas-phase sequencing and the original traces of the PTH-amino-acid analyses are shown in Figure 3. They allow to propose the following partial NH$_2$-terminal sequence : Ala or Val-Glu-Ala-Gly-Ile-Thr-Gly-Thr-Trp-Tyr-Asn-Glu-Leu-Gly-.

In the second experiment, 15 µg of a total lysate of an Escherichia coli strain, harboring a plasmid encoding the human γ-interferon[14] was separated on a one-dimensional 17.5 % polyacrylamide gel and the separated proteins were transferred onto Polybrene glass-fiber (Fig. 2B). Traces of the PTH-amino acid identification runs are shown in Figure 4. They reveal an interesting point : the manipulated E. coli strain expres-

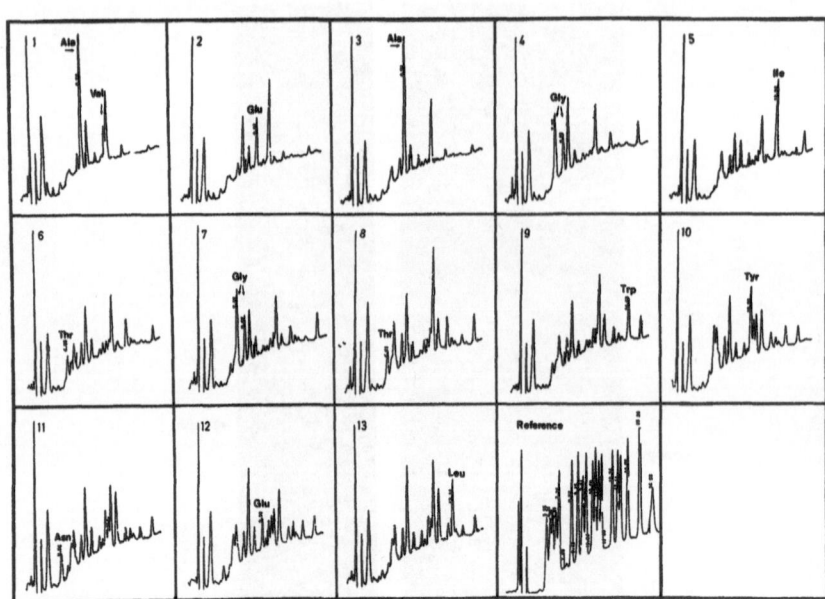

Fig. 3. *HPLC traces from an amino-terminal amino-acid sequence analysis of the small subunit of streptavidin blotted on Polybrene-glass-fiber. The detector was set at 0.01 A$_{254}$ unit full scale. The positions of the PTH-residues, assigned in the traces for cycle 1 through 13 are indicated by the three-letter amino acid designation. The reference mixture contains 100 picomol each; 50 % of each sample was analyzed.*

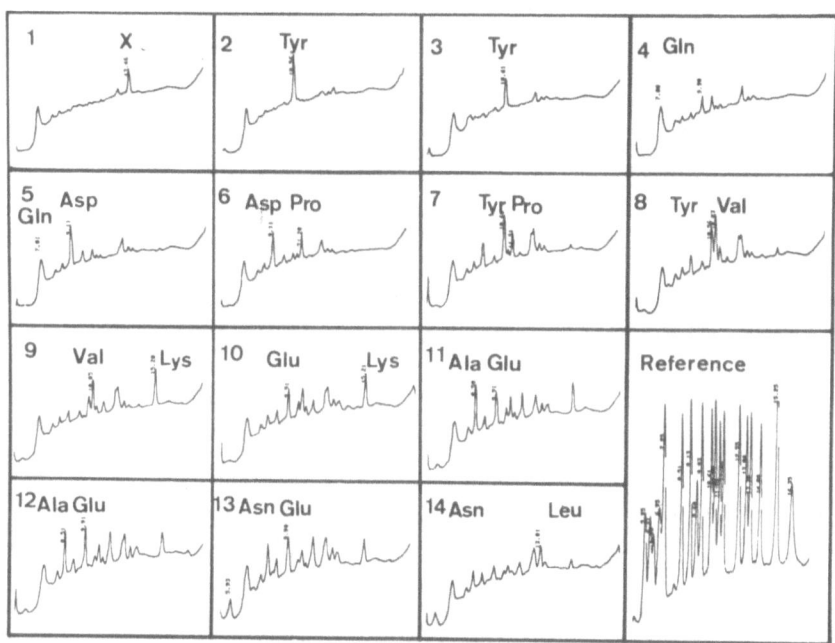

Fig. 4. *HPLC-traces from an amino-terminal amino acid sequence analysis on human γ-interferon. The protein was purified from a total E. coli lysate by one-dimensional gel electrophoresis and transferred on Polybrene-glass-fiber (Fig. 2B). PTH-amino-acids are indicated as described in Fig. 3. X designates a PTH-residue eluting as PTH-Phe although Met was expected from the DNA-sequence. PTH-Cys could not be detected. In each step we identified two residues indicating the presence of two populations of γ-interferon.(see text).*

ses two interferon populations which differ from each other in a deletion of the first NH_2-terminal residue. More than sixty percent shows the NH_2-terminal sequence as expected : Cys-Tyr-Cys-Gln-Asp-Pro-Val-Lys-Glu-Ala-Glu-Asn-Leu- but about forty percent still carries an additional NH_2-terminal amino-acid residue. By HPLC-analysis this additional NH_2-terminal PTH-derivative behaves similar to phenylalanine although such a residue should not be expected from the known DNA sequence of the plasmid construction, where a methionine should be the initiator residue[14]. Further studies are underway to clarify this point.

In conclusion, the technique in which proteins are transferred as SDS-complexes and bound onto glass-fiber coated with a thin layer of non-covalently bound Polybrene, promises to be an extremely powerful new tool in protein chemistry. It is a very simple method, convenient to perform and economical in terms of reagents and time. It can be applied on all kinds of proteins, a fact which may be especially advantageous for highly hydrophobic proteins which can only be purified in the presence of detergents. In its actual form, the glass blotting technique fits very well in a strategy in which proteins are first enriched in a single step procedure (e.g. by affinity chromatography or immune-precipitation) to be then finally purified by a single step gel electrophoretic procedure.

With the future developments of more sensitive PTH-amino acid analysers, it will be possible to take proteins for gas-phase sequencing from a

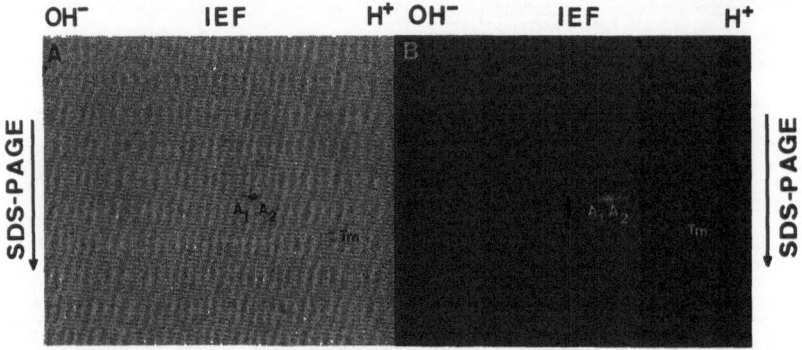

Fig. 5. *Two-dimensional Polyberne-glass blot of the proteins from 18 day chicken embryo muscle tissue. (A) The Coomassie-stained gel; (B) the corresponding Polybrene-glass-blot. A_1, A_2, and Tm refer to β-actin, α-actin and the tropomyosins respectively. H^+ and OH^- indicate the acidic and basic side. Isoelectric focusing (IEF) is carried out in the horizontal direction. SDS/polyacrylamide gel electrophoresis (SDS-PAGE) in the vertical direction.*

2-D-gel. The basis for such an approach is already given here. A protein extract from an 18-day embryonic chicken leg muscle was separated by two-dimensional polyacrylamide gel electrophoresis. Figure 5A shows the protein pattern after staining with Coomassie; figure 5B shows a fluorescamine stained pattern of the corresponding blot on Polybrene glass-fiber. This blot is a nearly exact copy of the proteins visualized in the gel. The major spots are the contractile proteins α- and β-actin and the isoforms of tropomyosin. They represent approximately 0.5 µg of immobilized protein as determined by amino-acid analysis. Such amounts should be sufficient for sequence analysis when PTH-amino acids are detectable at the femtomole level. Thus, 2-D-polyacrylamide gel electrophoresis as a general purification procedure for proteins, combined with glass-fiber blotting and gas-phase sequencing, seems a realistic option for the immediate future. The selection, of the protein of interest in the complete 2-D-protein pattern may be greatly facilitated since it is also possible to use the immuno-peroxidase detection technique for specific antigen location on glass immobilized proteins (results not shown here).

Acknowledgements

J.V. is Research Associate of the Belgian National Fund for Scientific Research (N.F.W.O.). This work was supported by a grant from the N.F.W.O. and in part from the Deutsche Forschungsgemeinschaft. This work was carried out with the valuable assistance of G. Bauw, J. Van Damme, M. Puype and C. Ampe. Appreciation is expressed to Prof. M. Van Montagu for his support and to H. Vanhellemont for preparation of this manuscript.

REFERENCES

1. J. M. Gershoni and G. E. Palade, Protein blotting : principles and applications, Anal. Biochem. 131:1 (1983).

2. J. Renart, J. Reiser, and G. R. Stark, Transfer of proteins from gels to diazobenzyloxymethyl-paper and detection with antisera : A method for studying antibody specificity and antigen structure, Proc. Natl. Acad. Sci. USA 76:3116 (1979).

3. J. Symington, M. Green and K. Brackmann, Immunoautoradiographic detection of proteins after electrophoretic transfer from gels to diazo-paper : Analysis of adenovirus encoded proteins, Proc. Natl. Acad. Sci. USA 78:177 (1981).

4. H. Towbin, T. Staehelin, and J. Gordon, Electrophoretic transfer of proteins from polyacrylamide gels to nitrocellulose sheets : Procedure and some applications, Proc. Natl. Acad. Sci. USA 76:4350 (1079).

5. J. M. Gershoni and G. E. Palade, Electrophoretic transfer of proteins from sodium dodecyl sulfate polyacrylamide gels to a positively charged membrane filter, Anal. Biochem. 124:396 (1982).

6. R. M. Hewick, M. W. Hunkapiller, L. E. Hood, and W. J. Dreyer, A gas-liquid solid phase peptide and protein sequenator, J. Biol. Chem. 256:7990 (1981).

7. U. K. Laemmli, Cleavage of structural proteins during the assembly of the head of bacteriophage T4, Nature (London) 227:680 (1970).

8. J. R. Benson and P. E. Hare, o-Phthalaldehyde : Fluorogenic detection of primary amines in the picomole range. Comparison with fluorescamine and ninhydrin, Proc. Natl. Acad. Sci. USA 72:619 (1975).

9. M. Hunkapiller and L. Hood, Analysis of phenylhthiohydantoins by ultrasensitive gradient high-performance liquid chromatography, Methods Enzymol. 91:486 (1983).

10. J. I. Garrels, Two-dimensional gel electrophoresis and computer analysis of proteins synthesized by clonal lines, J. Biol. Chem. 254:7961 (1979).

11. G. Mócz and M. Bálint, Use of cationic detergents for polyacrylamide gel electrophoresis in multiphasic buffer systems, Anal. Biochem. 143:283 (1984).

12. M. Riley and R. N. Perham, The reversible reaction of protein amino groups with exo-cis-3,6-endoxo-Δ^4-tetrahydrophthalic anhydride. The reaction with lysozyme, Biochem. J. 118:733 (1970).

13. J. Y. Chang, D. Brauer, and B. Wittmann-Liebold, Micro-sequence analysis of peptides and proteins using 4-NN-dimethylaminoazobenzene 4'-isothiocyanate/phenylisothiocyanate double coupling method, FEBS Lett. 93:205 (1978).

14. G. Simons, E. Remaut, B. Allet, and W. Fiers, High-level expression of human interferon gamma in *Escherichia coli* under control of the P_L promoter of bacteriophage lambda, Gene 28:55 (1984).

TWO-DIMENSIONAL GEL ELECTROPHORESIS OF POLYPEPTIDES

Daryll B. DeWald, Lonnie D. Adams, and James D. Pearson

The Upjohn Company, Biotechnology
301 Henrietta St. 7240-209-6
Kalamazoo, MI 49001

INTRODUCTION

We report a method for separating proteins and peptides in the range M_r 2,000 to M_r 200,000 by SDS-polyacrylamide gel electrophoresis without urea or a stacking gel. This system was developed as an alternative to the Swank and Munkres (1) and Kyte and Rodriguez (2) procedures previously used in our laboratory because they were limited to penetration of protein and peptides less than M_r 50,000 and require urea. Other gel systems which cover an extended molecular weight range have been reported (3, 4) but also require urea. The essential features of this gel include gradients of acrylamide, N,N-methylenebisacrylamide (bis), and glycerol in a tris-phosphate buffer system. A two-dimensional separation of an *E. coli* extract first run on a conventional isoelectric focusing tube gel followed by our SDS-PAGE gel is given as an example of its utility in the laboratory.

MATERIALS AND METHODS

Reagents

Acrylamide, N,N-methylenebisacrylamide(bis), SDS, N,N,N',N'-tetra-methylenediamine (TEMED), ammonium persulfate and urea were electrophoresis grade reagents obtained from Bio-Rad (Richmond, California). Phosphoric acid and sodium carbonate were purchased from Mallinckrodt (Paris, Kentucky). Glycerol was obtained from Pierce (Rockford, Illinois). Ampholines (pH range, 3.5-10) were from LKB (Sweden). Tris was purchased from Sigma (St. Louis, Missouri). 2-mercaptoethanol was obtained from Kodak (Rochester, New York).

The molecular weight marker kit of myoglobin fragments was obtained from BDH (Gallard Schlessinger, Carle Place, New York). The "Gel Code" molecular weight markers were prepared in our laboratory but can be purchased from Health Products (South Haven, Michigan). Myosin was purchased from Sigma.

Gel Electrophoresis

We normally prepare twenty-four 180 x 150 x 1.5 mm gradient slab gels at one time so that gels need only be cast once a week. The gels are poured from a gradient maker (Universal Scientific, Atlanta, Georgia) using an ISO-DALT casting box (Health Products, South Haven) as described by Anderson and Anderson (5). The following solutions are required: Solution A, buffer for stock acrylamide, 0.1M phosphoric acid titrated to pH 6.8 with solid tris base; Solution B, stock gel buffer (for 9% acrylamide), 0.22M phosphoric acid titrated to pH 6.8 with solid tris base; Solution C, stock gel buffer (for 26% acrylamide), 0.6M phosphoric acid titrated to pH 6.8 with solid tris base and diluted 3 parts buffer to 1 part glycerol; Solution D, tank buffer, dilute solution A 1:4 (v/v) in water and add solid SDS to a final concentration of 1% (v/v); Solution E, stock acrylamide (for 9%), 400 g of acrylamide and 8g of bis are dissolved in 1 liter of solution A; Solution F, stock acrylamide (for 26%), 400g of acrylamide and 16g of bis are dissolved in 1 liter of solution A.

172 ml Solution E is mixed with 578 ml Solution B and made 0.1% SDS by adding 7.5 ml of 10% SDS (in H_2O). The mixture is degassed, then 120 µl TEMED and 9 ml of 10% ammonium persulfate (in H_2O) are added to initiate polymerization. This final mixture is now 9% acrylamide/0.18% bis.

400 ml Solution F is mixed with 200 ml Solution C and made 0.1% SDS by adding 6 ml of 10% SDS. The mixture is degassed, then 30 µl TEMED and 4 ml of 10% ammonium persulfate are added to initiate polymerization. This final mixture is now 26% acrylamide/1.05% bis.

The acrylamide solutions are quickly loaded into a gradient maker and gels are poured. Water saturated n-butanol layer is misted on top of the gels from a spray bottle, and the gels are allowed to polymerize for 1.5-2 hours. The n-butanol is removed by a distilled water rinse before loading samples.

In conventional nomenclature, this system is 0.56%C to 2.5%C for the bis crosslinker gradient, and 9%T to 26%T for the acrylamide gradient. By definition, %C = g crosslinker x 100 / g acrylamide + g crosslinker, and %T = g acrylamide + g crosslinker/100ml of solution x 100.

The isoelectric focusing tube gel in the first dimension of the *E. coli* cell lysate sample in Figure 2 is not described in detail here, but was prepared and run according to the Anderson method (5, 6) with ISO-DALT equipment (Health Products).

Sample Preparation, Loading, and Staining

An isoelectric focusing (IEF) tube gel is placed on top of the SDS-PAGE second dimension gel in between hot 1% agarose layers dispensed from a pasteur pipet. The agarose solution is prepared in tank buffer. If SDS-PAGE without prior IEF is desired, samples are denatured in an SDS solubilizing buffer (2% NP-40, 1% 2-mercaptoethanol, 1% SDS in 0.125 M tris-HCl, pH 6.8) by heating at 100°C for five minutes. The sample solution is then mixed with an equal volume of hot 1% agarose solution (prepared in tank buffer), and quickly pipeted into a 0.2 ml disposable glass pipet. After agarose solidification, 0.5 cm length gel rods are easily dispensed out of the pipet onto the top of the sizing gel. Thin layers of hot agarose are layered above and below the rods to secure them in place.

After samples are loaded about 10 μl of 0.1% bromophenol blue is layered on top and allowed to diffuse into the agarose. Gels are then electrophoresed at 700-1000 mAmp constant current, depending upon cooling capacity. The electrophoresis process takes 20 to 30 hours, as indicated by the dye front reaching the bottom of the gel.

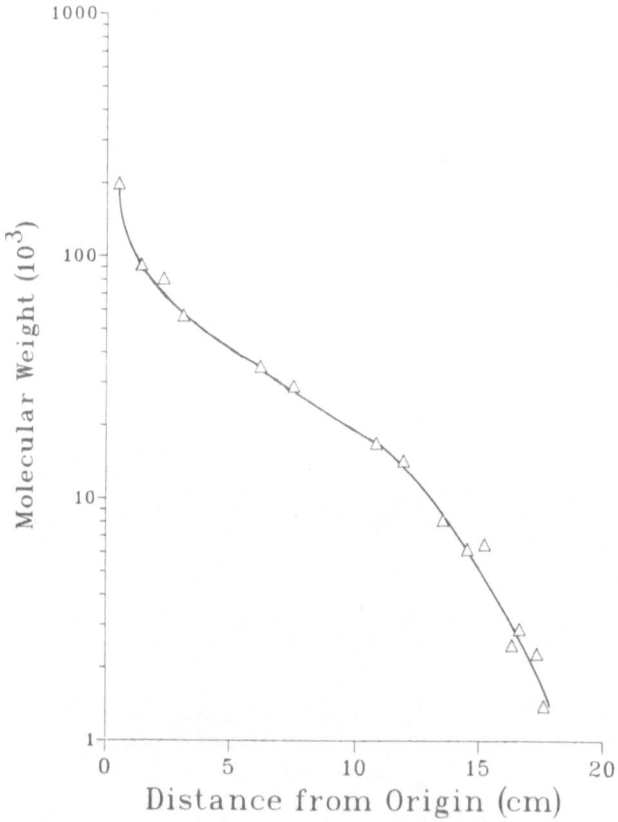

Figure 1: Plot of semi-log molecular weight vs. migration distance of proteins and peptides. Data obtained from a 9-26% acrylamide gradient slab gel as described in Materials and Methods. Mobility standards: 200,00 , myosin; 92,500, 81,000, 57,000, 35,000, 29,000, Gel Code molecular weight markers; 17,000, 14,400, 8,200, 6,200, 2,500, 1,400, BDH myoglobin and CNBr markers; 4,500, 2,900, 2,100, ACTH and ACTH 1-24 and 1-17 fragments. Reprinted from Ref. 14, with permission from the publisher.

Gels can be silver stained by either the Merril *et al.* (7) or Sammons *et al.* (8) method. Fixing was done in 50% ethanol /10% acetic acid for at least 4 hr.

RESULTS AND DISCUSSION

A molecular weight versus migration distance plot is shown in Figure 1 for a collection of fifteen proteins, including myoglobin CNBr fragments and ACTH related peptides. This sizing gel routinely allows an array of separation spanning from M_r 2,000 to M_r 200,000. We have observed that slightly better separation of proteins less than M_r 20,000 can be effected by using a 13-26% rather than a 9-26% acrylamide gradient, but then protein penetration is limited to M_r 90,000 or less. We previously used the 12.5% acrylamide gels described by Swank and Munkres (1), which also gives very good sizing for proteins less than M_r 20,000, but encountered problems maintaining batch to batch reproducibility. The procedure

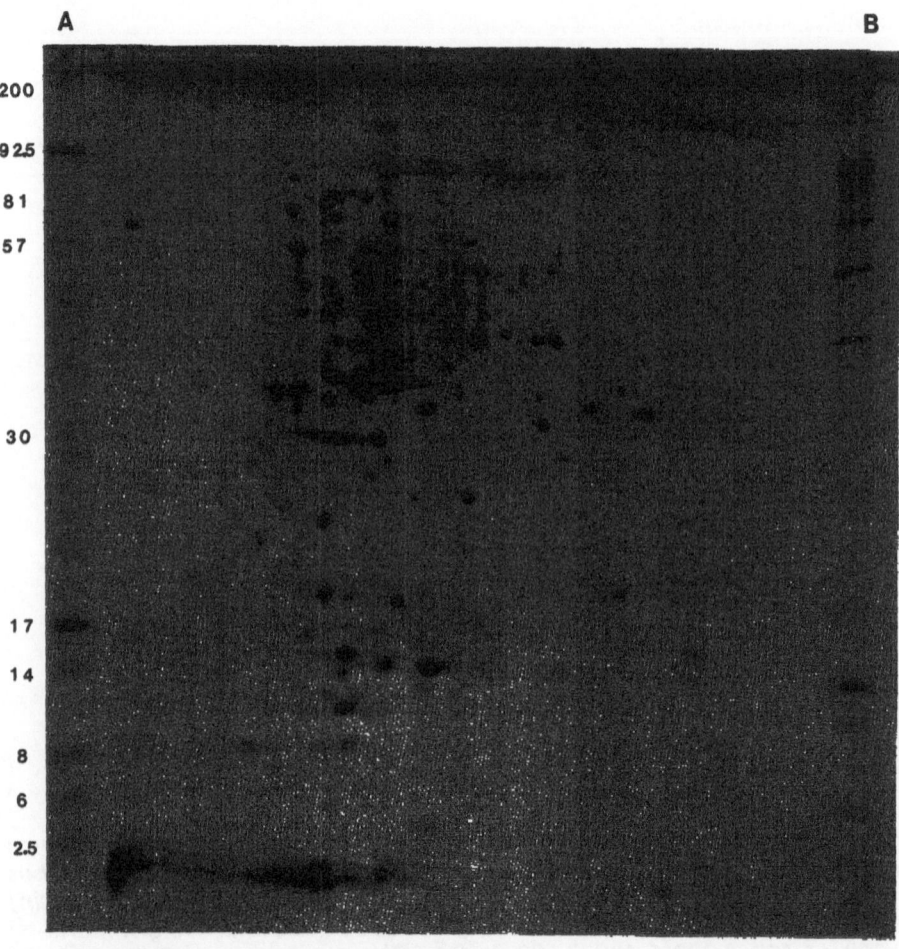

Figure 2: 9-26% acrylamide gradient gel used as the SDS-PAGE second dimension of an isoelectric focusing tube gel (ampholine pH range 3.5-10) with acidic end to the left orientation. The sample is an E. coli cell lysate. Molecular weight markers are identical in lanes A and B, consisting of a mixture of mysoin, Gel code markers prepared in our laboratory, and BDH myoglobin and myoglobin fragments. Silver staining was done by the method of Merril et al. (9).

described in this paper yields very reproducible sizing patterns, probably because the gradients within the matrix mitigate migration variability to a much greater extent than a straight percentage polyacrylamide gel. The use of gradient crosslinkers has been employed in other systems (9, 10) to increase protein resolution, so we incorporated this concept into our matrix. We attempted to optimize our 9-26% acrylamide gradient using fixed percentages of bis within the range 0.2 to 1%, but found those gels inferior to the 0.18 to 1.05% bis crosslinker gradient described in this paper (data not shown).

Our system corroborates the observation of Glyn (11) in that urea is not necessary for polypeptide electrophoresis. Although the use of 7-8M urea has been very popular (1-4, 12), we find that when it is eliminated gels can run at lower temperatures without salting-out effects and that larger proteins can penetrate into the gel matrix. The latter is probably true because elevated urea concentrations reportedly reduces the apparent pore size of the gel (3).

The inclusion of a 0-8% glycerol gradient (6) in these gels may contribute to their pliability during handling, and higher viscosity may impede diffusion of small peptides near the bottom of the gel. A drawback we originally had was drying after staining. We found that by using a modification of Westermann (13), gels could be soaked in 20% glycerol for 2 hr, followed by acetone for 8 hr, and then dried without cracking.

Figure 2 is an example of this broad range sizing method applied to a two-dimensional separation of an *E. coli* cell lysate. The array of proteins resolved extends down to less than M_r 2,000. In this case the gel was silver stained by the method of Merril *et al.* (7). In another study we found that the choice of silver stain has a marked effect on visualizing the array of proteins on a two-dimensional profile of Chinese hamster liver cell homogenate (14). In that study the silver stain methods of Sammons *et al.* (8) and Merril *et al.* were compared. We found that while the Sammons method allowed more sensitivity for detecting spots greater than about M_r 20,000-30,000, the Merril technique resulted in better detection for polypeptides less than about M_r 20,000. Sometimes the degree of difference between the two methods ranged from no stain or a negative stain using one procedure to a distinct dark stain for the other for individual spots. It is therefore advisable to use both staining procedures in initial evaluation of a heterogeneous sample to locate all spots before making a final selection of stain for routine use.

In conclusion, we report a non-urea SDS-PAGE gel system which can be used for separating polypeptides down to M_r 2,000. The technique allows for silver stain detection and may serve as a useful tool for analysis of small polypeptides, by one or two dimensional electrophoresis.

REFERENCES

1. Swank,R.T. and Munkres, K.D. (1971) Anal. Biochem. **39**, 462-477.
2. Kyte, J. and Rodriguez, H. (1983) Anal. Biochem. **133**, 515-522.
3. Anderson, B.L., Berry, R.W., and Telser, A. (1983) Anal. Biochem. **132**, 365-375.
4. Hashimoto, F., Horigome, T., Kanbayashi, M., Yoshida, K., and Sugano, H. (1983) Anal. Biochem. **129**, 192-199.
5. Anderson, N.G., and Anderson, N.L. (1978) Anal. Biochem. **85**, 331-340.
6. Tollaksen,S.L., Anderson,N.L. and Anderson,N.G. (1984) in Operation of the ISO-DALT System (7th edition) ANL-BIM-84-1 Argonne National Laboratory, Argonne, Illinois.

7. Merril, C.R., Goldman, D., Sedman, S.A., and Ebert, M.H. (1981) Science **211**, 1437-1438.

8. Sammons, D.W., Adams, L.D., Vidmar, T.J., Hatfield, C.A., Jones, D.H., Chuba, P.J., and Crooks, S.W. (1984) in Two-Dimensional Gel Electrophoresis of Proteins (Celis, J.E. and Bravo, R., eds.) pp.112-126, Academic Press, New York.

9. Johnson, G. (1979) Biochemical Genetics **17**, pp. 499-516.

10. Moore, D., Sowa, B.A., and Ippen-Ihler, K. (1981) J. Bacteriol. **146**, pp. 251-259.

11. Glyn, M.C.P., Bull, J., and Wright, J. (1982) S. Afr. J. Sci. **78**, 36-37.

12. Burr, F.A. and Burr, B. (1983) in Methods in Enzymology (Fleischer,S. and Fleischer, B.,eds.), vol. **96**, pp. 239-244, Academic Press, New York.

13. Westermann, R. (1985) Electrophoresis **6**, 136-137.

14. DeWald, D.B., Adams,L.D., and Pearson, J.D., Anal. Biochem., in press.

III. Chemical Modification of Polypeptides

CARBOXYMETHYLATED HEMOGLOBIN AS A STRUCTURAL ANALOGUE

FOR CARBAMINO HEMOGLOBIN

Wendy J. Fantl, Alberto Di Donato*, Arthur Arnone**, and
James M. Manning

The Rockefeller University, New York, NY, * Universita' di
Napoli, Naples, Italy, ** The University of Iowa, Iowa City,
IA

INTRODUCTION

The studies in this paper describe the conditions elucidated for the selective reductive carboxymethylation of the α-amino termini of hemoglobin (Hb) and the structural and functional consequences of such a modification. The initial premise for such a modification, $HbNHCH_2COO^-$, was to test its usefulness as a CO_2 or carbamino analogue, $HbNHCOO^-$. The latter compound is formed reversibly and cannot be isolated. The former is irreversibly formed and, prepared in sufficient amounts, could lead to a wealth of information concerning the binding of CO_2 to Hb as well as to the interplay between this effector and other important physiological modulators of Hb, such as anions and protons.

Of the procedures available for the derivatization of the amino groups of a protein, reductive alkylation has proven to be a very useful technique in protein chemistry[1]. In this procedure, the Schiff base produced by nucleophilic attack of a small aliphatic aldehyde on the unprotonated amino function of the protein can be reduced by sodium borohydride or sodium cyanoborohydride (Figure 1). In the present studies the aldehyde used was glyoxylate. The reducing agent was sodium cyanoborohydride, which is specific for protonated Schiff base adducts[2]. At neutral rather than alkaline pH, the reaction is directed primarily toward the predominantly unprotonated terminal amino groups of Hb. The ε-amino groups of lysine residues are mostly protonated at this pH and, therefore, are less reactive. The former sites are considered to be physiologically relevant with respect to reaction and transport of CO_2 [3].

METHODS

Preparation of Hybrids and Identification of the Site of Modification
Hb hybrids, $\alpha_2^{Cm}\beta_2$, $\alpha_2\beta_2^{Cm}$, $\alpha_2^{Cm}\beta_2^{Cm}$, specifically carboxymethylated at the α-amino termini, were prepared as described previously[4].

The site of carboxymethylation was determined by HPLC analysis of the tryptic peptides of a carboxymethylated α-chain and a carboxymethylated β-chain, respectively[4]. The amount of ε-carboxymethyllysine was measured by amino acid analysis after oxidation by performic acid treatment according

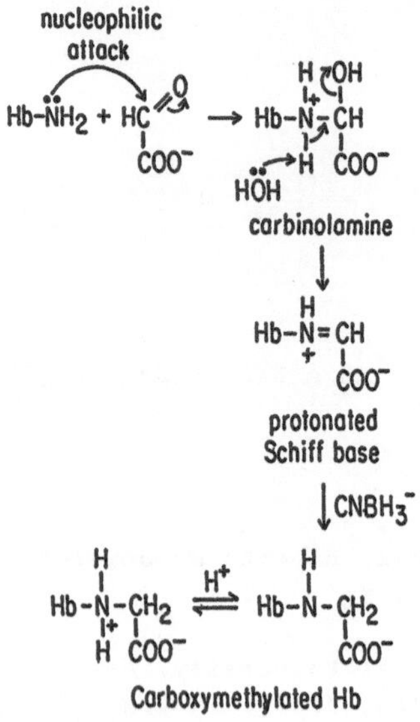

Fig. 1. Scheme for the carboxymethylation of Hb by reductive alkylation with glyoxylate and NaCNBH₃.

to the method of Moore[5]. This treatment is necessary as carboxymethyllysine coelutes with methionine in our chromatographic system. In addition, a Dowex-2 chromatography system was developed to ascertain whether carboxymethylation of the NH_2-terminal valine residues was limited to only the mono derivative at the α-NH_2 group (manuscript in preparation).

<u>Determination of the Oxygen Equilibrium Curves of Carboxymethylated Hb</u> - The hybrid Hb derivatives were prepared for measurement of their oxygen affinity as described previously[4]. In addition to the intrinsic P_{50} values for these hybrids, the effect of chloride and 2,3-DPG upon oxygen affinity was also measured.

<u>Measurement of the Alkaline Bohr and the Acid Effect</u> - These measurements were made by the technique of proton titration[6].

<u>X-Ray Data</u> - An electron density difference map between crystals of $\alpha_2^{Cm}\beta_2^{Cm}$ in the deoxy state compared with deoxy unmodified HbA was constructed, the details of which will be presented in a forthcoming manuscript.

RESULTS

<u>Preparation of Carboxymethylated Hybrids and Identification of the Site of Modification</u> - Carboxymethylation of purified carbon monoxy HbA in the presence of ^{14}C-glyoxylate and sodium cyanoborohydride yielded four derivatives, Hb_0, Hb_1, Hb_2, and Hb_3, which could be separated by DEAE cellulose chromatography. The amount of incorporation of glyoxylate based on the specific activity of each derivative respectively, was 0, 2, 3.5, and 5 to 6 nmols glyoxylate per nmol Hb tetramer[4]. The Hb_1 and Hb_2 fractions were subjected to a chain separation based on a modification of the method described by Bucci and Fronticelli[7]. Equimolar amounts of α^{Cm}-chain (from the chain separation of Hb_1 and Hb_2) were recombined with

unmodified β-chain to produce $\alpha_2{}^{Cm}\beta_2$. Similarly, unmodified α-chain was recombined with β^{Cm}-chain to give $\alpha_2{}^{Cm}\beta_2^{Cm}$ and α^{Cm}-chain was recombined with β^{Cm}-chain to form the hybrid $\alpha_2{}^{Cm}\beta_2$. The hybrids were rechromatographed on CM-52 cellulose, as described previously[4] (Figure 2), for use in the functional studies.

To determine the site of carboxymethylation the chains were again separated[4]. ^{14}C-labeled β^{Cm}-chain from the $\alpha_2\beta_2^{Cm}$ hybrid and ^{14}C-labeled α^{Cm}-chain from the $\alpha_2{}^{Cm}\beta_2$ hybrid were subjected to tryptic digestion and the digests were applied to an HPLC column as described previously[4]. The resultant chromatograms (Figure 3) showed that all the radioactivity was confined to the βT-1 and αT-1 peptides, respectively. Amino acid analysis of the labeled tryptic peptides, αT_1 and $\alpha(T_1 + T_2)$, as well as of βT -1 confirmed that Val-1(α) or Val-1(β) were the sites of carboxymethylation. The presence of ε-$[^{14}C]$-carboxymethyllysine was not detected by amino acid analysis of the fractions from HPLC. In addition, after performic acid oxidation, amino acid analysis of an acid hydrolysate of peak III of $\alpha_2{}^{Cm}\beta_2^{Cm}$ (Figure 2) showed there to be no more than 3% carboxymethyllysine. The Dowex-2 chromatography system (to be described in a manuscript in preparation) indicated that the dicarboxymethylated valine derivative was not present in these samples.

<u>Functional Properties of Carboxymethylated Hybrids</u> - The oxygen equilibrium curves (Figure 4) showed that the P_{50} values increased in the order $\alpha_2\beta_2$ (7 mm Hg), $\alpha_2{}^{Cm}\beta_2$ (12 mm Hg), $\alpha_2\beta_2^{Cm}$ (16 mm Hg), and $\alpha_2{}^{Cm}\beta_2^{Cm}$ (36 mm Hg). The average Hill coefficient of 2.4 for the four hybrids imparts confidence that the cooperative interactions have remained intact

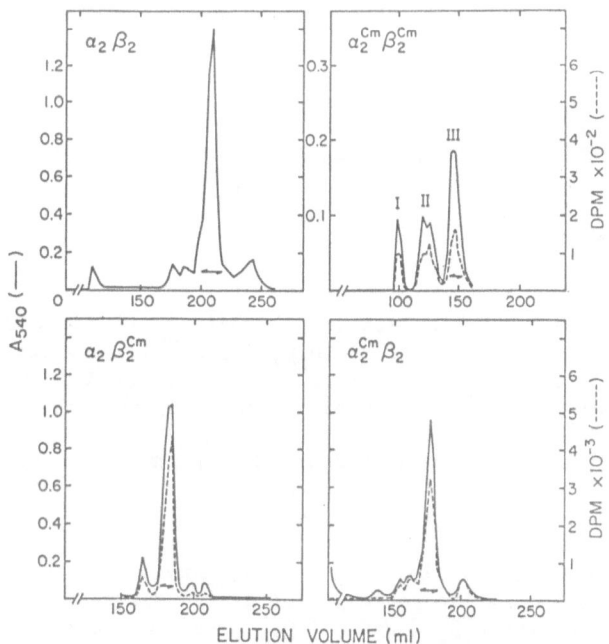

Fig. 2. Rechromatrography of hybrid Hb tetramers on CM-52 exchange resin. The samples were eluted with a gradient of 10 mM potassium phosphate pH 5.85, 1 mM EDTA, to 15 mM potassium phosphate, pH 7.9, 1 mM EDTA. The buffers were saturated with CO.

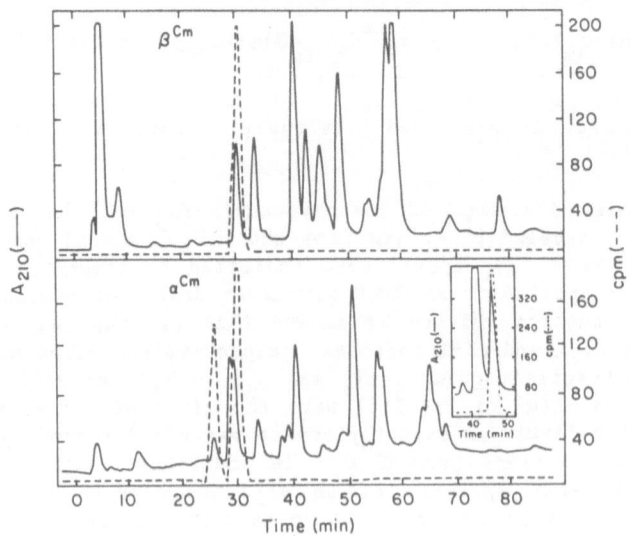

Fig.3. Peptide map of the tryptic digest of the ¹⁴C-carboxymethylated β- and α-chains from HbA. The initial gradient used for each digest was 5 to 70% acetonitrile containing 0.1% trifluoroacetic acid. For rechromatography of the peak eluting at 30 min (of the digest of the α-chain), the fractions comprising this peak were pooled, concentrated, and applied to the HPLC unit with a gradient of 5 to 50% acetontrile containing 0.1% trifluoroacetic acid[4].

after the experimental manipulations. Addition of 2,3-DPG lowers the oxygen affinity about 4-fold both for $\alpha_2\beta_2$ and $\alpha_2^{Cm}\beta_2$, whereas much smaller effects are observed for $\alpha_2\beta_2^{Cm}$ and $\alpha_2^{Cm}\beta_2^{Cm}$[4], where Val-1(β), a residue involved in the binding of this organic phosphate[8], is carboxymethylated.

Chloride binding is diminished in all the modified hybrids, $\alpha_2^{Cm}\beta_2$ (14% reduction), $\alpha_2\beta_2^{Cm}$ (28% reduction), and $\alpha_2^{Cm}\beta_2^{Cm}$ (86% reduction) compared to the binding of this anion in $\alpha_2\beta_2$.

Alkaline Bohr Effect - Under 'stripped' conditions the alkaline Bohr coefficient in 0.1 N chloride for $\alpha_2\beta_2$ and $\alpha_2^{cm}\beta_2$ was very similar, whereas the Bohr coefficient for $\alpha_2\beta_2^{cm}$ and $\alpha_2^{cm}\beta_2^{cm}$ was reduced by about 25% under identical conditions, compared to unmodified HbA (unpublished results).

X-Ray Data - The main conclusions to be drawn from the difference electron density map of deoxy $\alpha_2^{Cm}\beta_2^{Cm}$ are that at the β-chain termini, electron density contiguous with Val-1(β) appeared to interact with Lys-82(β). At the α-chain termini, electron density contiguous with Val-1(α) appeared to interact with Ser-131(α). These latter results are analogous to the previously published data for the electron density map of carbamino Hb[9].

144

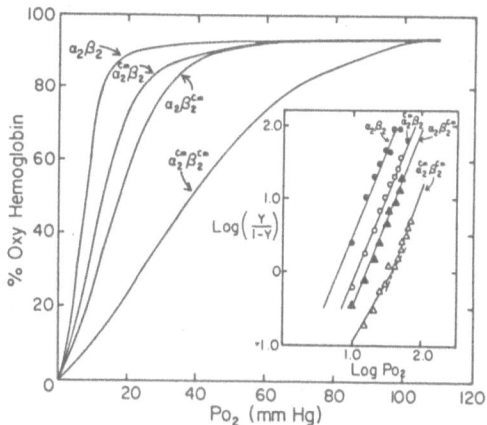

Fig. 4. Oxygen equilibrium curves of unmodified and specifically carboxymethylated hybrids of HbA. The samples were prepared as described previously[4]. 4μl portions were used for determination of the oxygen equilibrium curve in a Hem-O-Scan instrument.

DISCUSSION

The inability of cysteine, methionine, or histidine to participate in Schiff base formation accounts for the absence of carboxymethylation at these sites. Carboxymethylation of the lysine residues of ragweed pollen antigen E by glyoxylate and cyanoborohydride was described by King et al[10].

The purity of the hybrids $\alpha_2\beta_2$, $\alpha_2^{Cm}\beta_2$, $\alpha_2\beta_2^{Cm}$, and $\alpha_2^{Cm}\beta_2^{Cm}$ is shown in Fig 2. Moreover, HPLC analysis, amino acid analysis, and Dowex-2 chromatography ascertained that carboxymethylation was predominantly at the α-amino terminus of each chain and that the monoderivative was the preferred product. This is in contrast to reductive methylation where the dimethyl derivative is the predominant reaction product[1].

The oxygen affinity of the hybrids decreases in the order $\alpha_2\beta_2$, $\alpha_2^{Cm}\beta_2$, and $\alpha_2\beta_2^{Cm}$, and $\alpha_2^{Cm}\beta_2^{Cm}$. This decrease may be compared to a similar decrease observed for unmodified HbA in the presence of increased concentrations of CO_2 (Classical Bohr Effect[11]). In this respect, therefore, carboxymethylated HbA may be considered as a functional analogue of carbamino HbA.

The reduced 2,3-DPG binding observed for $\alpha_2\beta_2^{Cm}$ and $\alpha_2^{Cm}\beta_2^{Cm}$ compared with $\alpha_2\beta_2$ and $\alpha_2^{Cm}\beta_2$ is consistent with the now well-established identification of Val-1(β) as a residue involved in the electrostatic interaction with this organic phosphate[8]. Presumably, *in vivo* there must be some competition between CO_2 and 2,3-DPG for a common binding site.

The reduction in chloride binding for each hybrid may be interpreted with respect to the observations described from the difference electron density maps. In the hybrid $\alpha_2^{Cm}\beta_2$, the covalently attached carboxymethyl group appears to interact with Ser-131(α) and this is also seen for carbamate formation at Val-1(α)[9]. In addition, Val-1(α) has also been identified as an anion binding site[12]. Evidently, two chloride ions can be lodged between Val-1(α_2^{Cm}) and Arg-141(α_2) and between Val-1(α_1) and Ser-131(α_1)[12]. In the $\alpha_2^{Cm}\beta_2$ hybrid, although it appears from the electron density map that the site between Val-1(α_1) and Arg-141(α_2) is vacant, the 14% reduction observed for the binding of this anion might be an indication of a weakened interaction possibly, due to some steric effect.

The 28% reduction for chloride binding to the $\alpha_2\beta_2^{Cm}$ hybrid is possibly a result of the interaction between the carboxymethyl moiety covalently attached to Val-1(β) with Lys-82(β). The latter residue has been demonstrated by a number of studies to be a site for anion binding [13]. The covalently attached carboxymethyl group may preclude anion binding to Lys-82(β).

The 86% reduction in chloride binding to $\alpha_2^{cm}\beta_2^{cm}$ is greater by about 2-fold than the addition of the percentage reduction in chloride binding observed for $\alpha_2^{cm}\beta_2$ and $\alpha_2\beta_2^{cm}$. Possibly this is related to the reduced heme-heme interaction observed at low oxygen tensions (see biphasic Hill plots in Fig. 4).

Under physiological conditions the residues that contribute to the alkaline Bohr effect include Val-1(α), His-122(α), His-146(β), and the extra protons absorbed due to the binding of 2,3-DPG[13]. Under 'stripped' conditions Val-1(α), His-122(α), His-146(β), and chloride binding to Lys-82(β) contribute to the alkaline Bohr effect[13]. It was of interest to evaluate the alkaline Bohr effect for the carboxymethylated hybrids. Under 'stripped' conditions in 0.1 N chloride the alkaline Bohr effect of both $\alpha_2\beta_2$ and $\alpha_2^{Cm}\beta_2$ was comparable. The mechanism by which Val-1(α) contributes to the alkaline Bohr effect is by means of an elevated pk in deoxy Hb due to the oxygen-linked binding of chloride between Val-1(α_1) and Arg-141(α_2)[3]. The similar coefficient for $\alpha_2^{Cm}\beta_2$ compared with $\alpha_2\beta_2$ indicates that the differential pk of Val-1(α) in oxy and deoxy Hb is of the same magnitude as in $\alpha_2\beta_2$ even though chloride binding could be weakened as discussed above.

The decreased Bohr effect (28%) observed for both $\alpha_2\beta_2^{Cm}$ and $\alpha_2^{Cm}\beta_2^{Cm}$ could be a reflection of decreased binding of chloride at Lys-82(β). In view of the data presented above, the lowered Bohr effect of unmodified HbA observed in the presence of CO_2[14] could be due to carbamate formation at Val-1(β) _in vivo_.

Thus, the measurements of oxygen affinity and the alkaline Bohr effect, as well as the difference electron density data support the feasibility of using carboxymethylated Hb as an analogue for carbamino Hb.

REFERENCES

1. Means, G.E. Reductive Alkylation of Proteins. J. Prot. Chem. 3:121 (1984).
2. Jentoft, N., and Dearborn, D.G. Labeling of Proteins by Reductive Methylation Using Sodium Cyanoborohydride. J. Biol. Chem. 254:4359 (1979).
3. Kilmartin, J.V., and Rossi-Bernardi, L. Interaction of Hemoglobin with Hydrogen Ions, Carbon Dioxide, and Organic Phosphates. Physiol. Rev. 53:836 (1973).
4. Di Donato, A., Fantl, W.J., Acharya, A.S., and Manning, J.M. Selective Carboxymethylation of the α-Amino Groups of Hemoglobin. J. Biol. Chem. 258:11890 (1983).

5. Moore, S. On the Determination of Cystine as Cysteic Acid. J. Biol. Chem. 238:235 (1963).

6. Rollema, H.S., de Bruin, S.H., Janssen, L.H.M., and van Os, G.A.J. The Effect of Potassium Chloride on the Bohr Effect of Human Hemoglobin. J. Biol. Chem. 250:1333 (1975).

7. Njikam, N., Jones, W.J., Nigen, A.M., Gillette, P.N., Williams Jr., R.C., and Manning, J.M. Carbamylation of the Chains of Hemoglobin S by Cyanate in vitro and in vivo. J. Biol. Chem. 248:8052 (1973).

8. Arnone, A. X-Ray Diffraction Study of Binding of 2,3-Diphosphoglycerate to Human Deoxyhaemoglobin. Nature 237:146 (1972).

9. Arnone, A., Rogers, P.H., and Briley, P.D. The Binding of CO_2 to Human Deoxygemoglobin: An X-Ray Study Using Low-Salt Crystals, in: Biophysics and Physiology of Carbon Dioxide. C. Bauer, G. Gros, H. Bartels, eds., Springer-Verlag Berlin, Heidelberg, New York.

10. King. T.P., Kochoumian, L., and Lichtenstein, L. Preparation and Immunochemical Properties of Methoxypolyethylene Glycol-Coupled and N-Carboxymethylated Derivatives of Ragweed Pollen Allergen Antigen E. Arch. Biochem. and Biophys. 178:442 (1977).

11. Bohr, C., Hasselbach, K.A., and Krogh, A. Veber Einen in Biologischer Beziehung Wichtigen Einfluss, Den Die Kohlensaurespaunung Des Blutes Auf Dessen Sauerstoffbindung Vbt. Skand. Arch. Physiol. 16:402 (1904).

12. O'Donnell, S., Mandaro, R., Schuster, T.M., and Arnone, A. X-Ray Diffraction and Solution Studies of Specifically Carbamylated Human Hemolgobin A. J. Biol. Chem. 254:12204 (1979).

13. Perutz, M.F., Kilmartin, J.V., Nishikura, K., Fogg, J.H., Butler, P.J.G., and Rollema, H.S. Identification of Residues Contributing to The Bohr Effect of Human Hemoglobin. J. Mol. Biol. 138:649 (1980).

14. Dahms, T., Horvath, S.M., Luzzana, M., Rossi-Bernardi, L., Roughton, F.J.W., and Stella, G. The Regulation of Oxygen Affinity of Human Hemoglobin. J. Physiol. (London) 223:29P (1972).

ACKNOWLEDGEMENTS

The authors wish to thank Dr. A.S. Acharya for help with the HPLC tryptic map. We also would like to thank Judith A. Gallea for her expertise in the preparation of this manuscript. This research was funded by NIH Grant HL-18819.

GOSSYPOL: INTERACTION WITH RIBONUCLEASE A

Hiroshi Ueno[§‡], Samuel S. Koide[†‡], James M. Manning[§] and
Sheldon J. Segal[*‡]

§ The Rockefeller University
Laboratory of Biochemistry
1230 York Avenue
New York, NY 10021

† The Population Council
New York, NY 10021

* The Rockefeller Foundation
New York, NY 10036

‡ The Marine Biological Laboratory
Woods Hole, MA 02543

INTRODUCTION

An important biological effect of gossypol is its anti-fertility action. Clinical studies carried out in China have demonstrated that gossypol is an effective "male pill" because of its low toxicity and reversibility[1]. Inhibition of sperm motility appears to be the main mode of action of gossypol accounting for its anti-fertility activity[2]. The interaction of gossypol with sperm has been studied widely. Although the molecular mechanism of action of gossypol is not clear, an evidence has accululated showing that gossypol interacts with macromolecules to form stable complexes that interferes with the normal function of sperm. In the present investigation, the interaction of gossypol with ribonuclease was studied to clarify the nature of the gossypol-macromolecular linkage. Since gossypol contains two aldehyde groups per molecule (see Figure 1), a formation of a Schiff base between the aldehyde and amino groups of ribonuclease was demonstrated.

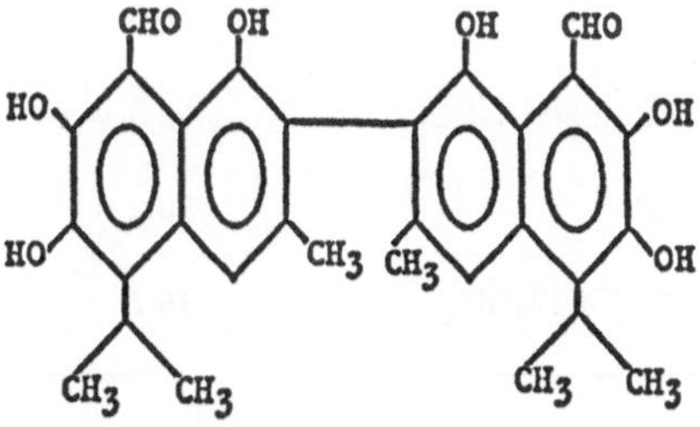

Figure 1: Structure of gossypol.

Recently, the optical isomers of gossypol have been separated[3], and designated as (+)- and (-)-gossypol, respectively based on their optical rotatory activities. We have carried out a comparative study on the reaction of (+)- and (-)-isomers, and (±) racemic mixture of gossypol with ribonuclease.

MATERIALS AND METHODS

Bovine pancreatic ribonuclease A (Type IIA) was purchased from Sigma Chemical Co. Gossypol acetic acid and the crystalline form of (+)- and (-)- gossypol were provided by the Institute of Zoology, Chinese Academy of Sciences, Beijing, China. ^{14}C-Gossypol was prepared by Drs. K. Watanabe and Y. F. Ren of Memorial Sloan-Kettering Cancer Center, New York. Specific activity = 0.76 mCi/mmol.

Preparation of Gossypol Solution

Stock gossypol solutions of 5 mM dissolved in either 100 % ethanol or DMSO were stored at -20 $^{\circ}$C without any detectable decomposition. The Concentration of gossypol solution was determined spectrophotometrically. A small aliquot (about 5-10 μl) was withdrawn and mixed with 1 ml $CHCl_3$ in a quartz cuvette. Absorption spectrum was measured and the concentration of gossypol was estimated based upon absorption at 368 nm (ϵ=14.3 $\times$ 10^3 M^{-1}cm^{-1})[4].

Assays of Ribonuclease A

The modified assay of Anfinsen et al[5] was used[6].

Circular Dichroism Spectrum

Circular dichroism spectra of gossypol-RNase complex were measured with AVIV 60DS, an updated version based Cary 60 with IBM-PC controlled operation system. Typical measurements were carried out by data acquisition of every 1 nm increment from 500 nm to 300 nm. Data from three determinations were accumulated and the average spectrum obtained.

Table 1: Reaction and Stabilization of Gossypol with Ribonuclease.

		TCA precipitation (%)	
		Supernatant + EtOH wash	precipitate
1)	GP[a] alone	98.8	1.2
2)	RNase + GP + NaCNBH$_3$	54.6	45.4
3)	RNase + GP	83.2	16.8
4)	[RNase + GP] + NaBH$_4$	16.7	83.3

[a]   ^{14}C-Gossypol (4.5 mM stock) was used for above experiment.

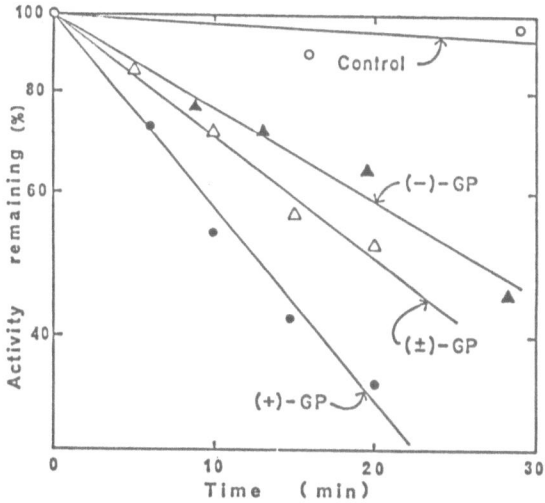

Figure 2: Time-dependent inactivation of RNase by gossypol and its optically active isomers. Fifty μl of gossypol (final concentration = 0.42 mM) was rapidly mixed with 1 ml of RNase (1 mM) in 20 mM Hepes buffer, pH 7.5, at 37 °C. At various times, 10 μl of aliquots was withdrawn, mixed with 5 ml of 0.1 M Tris-HCl buffer, pH 7.5, then RNase activity was measured.

Reaction of Radiolabeled Gossypol-RNase Complex with Reducing Agent

RNase (final concentration = 0.2 mM) was incubated with ^{14}C-gossypol (final concentration = 45 μM) for 2 hr at 25 °C. Reaction 1 shown in Table 1 was gossypol alone; Rx 2 contained RNase and gossypol in the presence of 150 mM of NaCNBH$_3$; Rx 3 contained RNase and gossypol; and Rx 4 was same as Rx 3 except an additional NaBH$_4$ at the end of incubation. Protein was precipitated by adding an equal volume of cold 20 % trichloroacetic acid (TCA). The mixture was stored at -4 °C for overnight and precipitate collected by centrifugation in a microfuge. The radioactivity in each fraction was then measured.

RESULTS

Inhibition of RNase by gossypol

Time-dependent inactivation of RNase with gossypol is shown in Figure 2. Optically inactive gossypol (final concentration = 0.42 mM) gave pseudo-first order inactivation kinetics with $t_{1/2} \sim 20$ min. The kinetic patterns with the optically active isomers, (+)- and (-)-gossypol were similar to that obtained with the racemic mixture. The (+)-gossypol (final concentration = 0.54 mM) showed $t_{1/2} \sim 12.5$ min while the (-)-gossypol (final concentration = 0.47 mM) gave $t_{1/2} \sim 26$ min. It can be concluded that the (+)-isomer is the most potent inhibitor of RNase activity.

Circular dichroism spectrum

Induced circular dichroism at 410 nm was observed when RNase (70 μM) was incubated with gossypol (580 μM) at 25 °C (Figure 3). Optimum intensity was obtained within 1 hr under the experimental condition. No CD peak was initially observed for either gossypol or RNase alone. Addition of NaBH$_4$ significantly reduced the peak height. A red shift of CD peak of (+)-gossypol upon binding with various proteins was observed by Whaley et al[7].

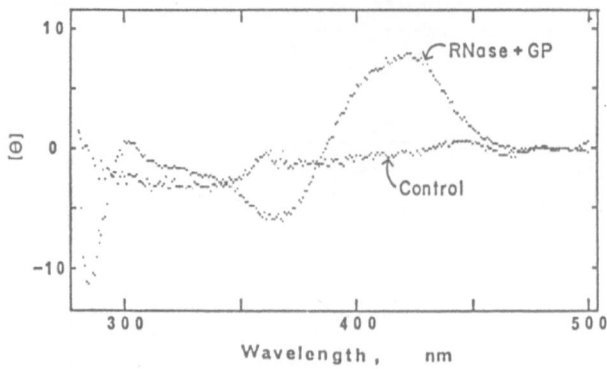

Figure 3. Circular dichroism spectrum of gossypol-RNase complex. Control represents either gossypol alone or RNase alone.

Reaction of gossypol-RNase complex with reducing agent

Kinetic and CD experiments showed formation of a complex between gossypol and RNase. A study was carried out to demonstrate whether or not the complex contains Schiff bases (Table 1). Without reducing agent, over 80 % of the counts were recovered in the supernatant. Because of the poor solubility of gossypol in aqueous solvent, washing was performed with ethanol to avoid non-specific co-precipitation of ^{14}C-gossypol with TCA-RNase. Significant amounts of the label were found associated with the precipitate when treated with the reducing agents, $NaCNBH_3$, which specifically reduce Schiff bases. The results demonstrate the formation of Schiff bases in the reaction mixture. Condition of the experiment used (Rx 2 in Table 1) was similar to the method described by Acharya*et al*[8] to produce reductive alkyl derivatives of hemoglobin.

Separation of RNase-gossypol complex on CM-52 chromatography

RNase-^{14}C-gossypol complex, treated with $NaCNBH_3$, was chromatographed on CM-52 according to Taborsky[9]. Figure 4 shows a typical elution profile. Three major fractions containing radioactivity were pooled (A-C). Each pooled fractions were analyzed for protein content and amino acid composition. The stoichiometry of the radio-label incorporated into the three fractions were determined to be 1.3, 0.7, and 0.4 mol gossypol/mol of RNase, respectively.

DISCUSSION

Gossypol is known to inhibit many enzymes including adenylate cyclase[10], AT-Pase[11], lactate dehydrogenase-X[12], pepsinogen[13], *etc.* The mode of inhibition appears to be due to a modification of the functional amino groups, such as N-terminal NH_2-group or ϵ-amino group of lysine. Finlay*et al*[13] previously demonstrated the formation of Schiff bases between gossypol and Lys-18 and Lys-358 of pepsinogen. They also showed the intra-molecular cross-linkage of gossypol with these two lysine groups. However, Olgiati[10] reported that $NaCNBH_3$ failed to stabilize the gossypol-adenylate cyclase complex. Our present results supports the thesis of Schiff base formation in the gossypol-RNase complex.

A positive circular dichroism seen at 410 nm indicates covalent interaction between gossypol and RNase. Similar CD peak is induced when pyridoxal 5'-phosphate, another yellow chromophore, binds to apo-aspartate aminotransferase or

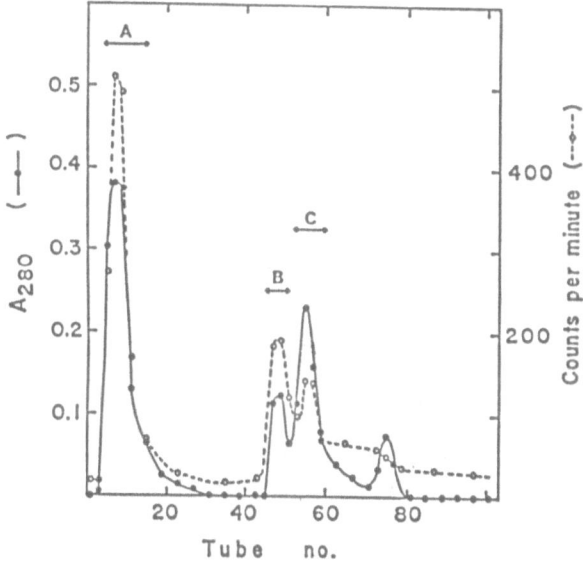

Figure 4. Chromatography of RNase-^{14}C-gossypol complex on CM-52. Each tube contains 2.5 ml.

apo form of other pyridoxal dependent-enzymes[14]. Similar positive CD peak was reported by Whaley *et al*[7] with (+)-gossypol-protein complexes. It is of interest to note that (+)-gossypol alone gives a positive CD peak at 377 nm[3,7]; however, the racemic mixture (optical inactive) shows no CD peak. Thus, when RNase forms a complex with the optical inactive gossypol, either it selectively interacts with the (+)-gossypol in the racemic mixture or forms a complex with both isomers and maintains a stereoconfiguration compatible with (+)-gossypol.

The relationship of optical activity and the anti-fertility action of gossypol has been studied. Matlin *et al*[15], and others reported that (-)-gossypol is responsible for the anti-fertility action *in vivo*. Waller *et al*[16] supports the above idea by demonstrating that the (+)-isomer does not possess this biological activity. However, present results and recent finding by Segal *et al*[17] indicate that (+)-gossypol is more potent than the (-)-isomer in inhibiting the RNase activity and fertilizability of *spisula* sperm under *in vitro* condition.

ACKNOWLEDGMENTS

We wish to thank Maria A. Pospischil for technical assistance. H.U. is a recipient of Rockefeller Foundation Fellowship.

REFERENCES

1. G. -Z. Liu, K. C. Lyle, and J. Cao, 1984, Trial of gossypol as a male contraceptive, *in*: "Gossypol: A potential contraceptive for men", S. J. Segal, ed., Plenum, New York, p.9.
2. M. H. Burgos, C. Y. Chang, L. Nelson, and S. J. Segal, 1980, Gossypol inhibits motility of *Arbacia* sperm *Biol. Bull.* 159:467.
3. D. K. Zheng, Y. K. Si, J. K. Meng, J. Zhou, and L. Huang, 1985, Resolution of racemic gossypol *J. Chem. Soc., Chem. Commun.* 168.
4. C. L. Hoffpauir, J. A. Harris, and J. P. Hughes, 1960, Gossypol acetic acid as a reference standard in determination of gossypol, *J. Ass. Offic. Agr. Chem.* 43:329.

5. C.B. Anfinsen, R. R. Redfield, W. L. Choate, J. Page, and W. R. Carroll, 1954, Studies on the gross structure, cross-linkages, and terminal sequences in ribonuclease, *J. Biol. Chem.* 207:201.

6. P. Blackburn, G. Wilson, and S. Moore, 1977, Ribonuclease Inhibitor from Human Placenta: Purification and properties. *J. Biol. Chem.* 252:5904.

7. K. J. Whaley, D. S. Sampath, and P. Balaram, 1984, A circular dichroism study of (+) gossypol binding to proteins. *Biochem. Biophys. Res. Comm.* 121:953.

8. A. S. Acharya, L. G. Sussman, and J. M. Manning, 1985, Selectivity in modification of the α-amino groups of hemoglobin on reductive alkylation with aliphatic carbonyl compounds *J. Biol. Chem.* 260:6039

9. G. Taborsky, 1959, Chromatography of ribonuclease on carboxymethyl cellulose columns *J. Biol. Chem.* 234:2652.

10. K. L. Olgiati, D. G. Toscano, W. M. Atkins, and W. A. Toscano, Jr., 1984, Gossypol inhibition of adenylate cyclase *Arch. Biochem. Biophys.* 231:411.

11. H. Mohri, K. Matsuda, S. S. Koide, and S. J. Segal, 1982, Effect of gossypol on *Arbacia* sperm ATPase *Biol. Bull.* 163:374

12. C. -Y. G. Lee, Y. S. Moon, J. H. Yuan, and A. F. Chen, 1982, Enzyme inactivation and inhibition by gossypol *Mol. Cell. Biochem.* 47:65

13. T. H. Finlay, E. D. Dharmgrongartama, and G. E. Perlmann, 1973, Mechanism of the gossypol inactivation of pepsinogen *J. Biol. Chem.* 248:4827

14. R. G. Kallen, T. Korpela, A. E. Martell, Y. Matsushima, C. M. Metzler, D. E. Metzler, Yu. V. Moriziv, I. M. Ralston, F. A. Savin, Yu. M. Torchinsky, and H. Ueno, 1985, Chemical and spectroscopic properties, of pyridoxal and pyridoxamine phosphates, *in:* "Transaminases", P. Christen and D. E. Metzler, eds., Wiley-Interscience, New York, p. 37.

15. S. A. Matlin, R. Zhou, G. Bialy, R. P. Blye, R. H. Naqvi, M. C. Lindberg, and S. A. Matlin, 1985, (-)-Gossypol: an active male antifertility agent, *Contraception* 31:141.

16. D. P. Waller, N. Bunyapraphatsara, A. Martin, C. J. Vournazos, M. S. Ahmed, D. D. Soejarto, G. A. Cordell, H. H. S. Fong, L. D. Russell, and J. P. Malone, 1983, Effect of (+)-gossypol on fertility in male hamsters, *J. Androl.* 4:276.

17. S. J. Segal, L. Herlands, and M. K. Sahni, 1985, Effect of optical isomers of gossypol on the ability of *Spisula* sperm to fertilize oocytes, *Biol. Bull.* in press.

STUDIES ON NONENZYMIC GLYCOSYLATION OF PEPTIDES

IN A SIMPLE MODEL SYSTEM

Nobuhiro Mori and James M. Manning

The Rockefeller University

1230 York Avenue, New York, NY

INTRODUCTION

In his review of the chemistry of the Amadori rearrangement, Hodge describes the intermediates and products that were characterized as various osazone and related types of derivatives[1]. Many of these studies were performed during a period in which the major route of isolation was crystallization rather than chromatography. A later study, in which chromatographic approaches were employed, evaluated the reaction of glucose with dipeptides[2]. The conclusion from this latter study was that the major product of such a reaction was the substituted glycosylamine, which was in equilibrium with the starting materials. There was no reported presence of an Amadori rearrrangement product.

In our earlier studies of the reaction of hemoglobin with the three-carbon aldehyde, glyceraldehyde, we found evidence for the product of the Amadori rearrangement at a few of the amino groups of hemoglobin[3],[4]. As shown in Scheme 1 with the model peptide alanylhistidine, the initial reaction with glyceraldehyde involves the formation of the substituted carbinolamine. This intermediate rapidly undergoes dehydration to generate the aldimine or Schiff base intermediate. These intermediates are in equilibrium with the starting materials[5]. The aldimine can further undergo

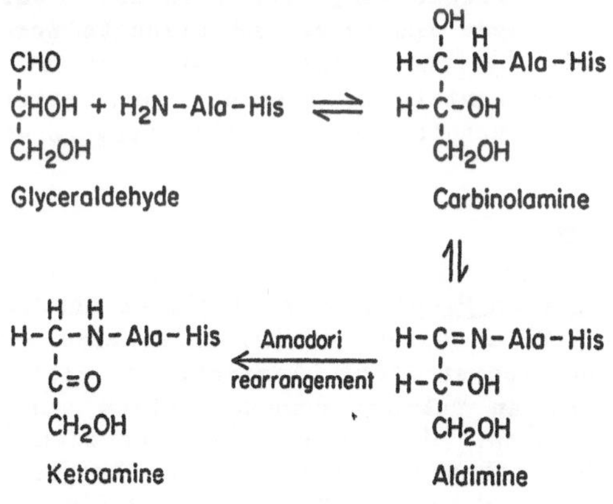

Scheme I. Reaction of L-Ala-L-His with Glyceraldehyde

the Amadori rearrangement under certain conditions to generate the stable ketoamine derivative. It is thought that this latter reaction with hemoglobin is irreversible under most conditions[6].

The present investigation was undertaken to elucidate some of the chemistry of the reaction of glyceraldehyde with small peptides. The ultimate goal is to understand the molecular basis for the fact that only a few of the amino groups of hemoglobin undergo the Amadori rearrangement but most do not. Some of the latter only form the Schiff base intermediate but do not undergo the rearrangement[7]. In addition, we wish to determine the molecular properties and stability of these intermediates and of the ketoamine product.

MATERIALS AND METHODS

The dipeptides, alanylhistidine and valylhistidine, were obtained from Sigma and Bachem, repectively. Crystalline DL-glyceraldehyde was purchased from Sigma. These compounds were judged pure as estimated by elemental analysis, which was kindly performed by Mr. S.T. Bella of this institution.

Reaction of glyceraldehyde with dipeptides - The condensation of glyceraldehyde (50 mM) with either Ala-His or Val-His (1 mM) was carried out as described previously[8]. The products of the reaction were analyzed by amino acid analysis by the system of Spackman, Stein, and Moore[9] on a 0.6 x 20 cm column of the amino acid analyzer with 0.35 N sodium citrate pH 5.28 as the eluent.

For isolation of the intermediates and the product of the reaction, a large scale synthesis was performed as follows: Ala-His (0.2 mmol) was mixed with glyceraldehyde (10 mmol) in 200 ml of 50 mM potassium phosphate, pH 7.0. After 3 hrs at 50°C, the mixture was adjusted to pH 2.2 with conc. HCl and then lyophilized. An aliquot of the reaction mixture (2.0 ml) was applied to an Altex ODS column (10 X 250 mm). Elution was carried out with a linear gradient of 0-50% n-propanol containing 10 mM sodium acetate, pH 5.0. The product was finally desalted on Sephadex G-10. The eluent was 0.34% acetic acid (v/v).

For determination of the amount of carbonyl compounds present in various fractions, aliquots were treated with 2,4 dinitrophenylhydrazine[10]. We found that whereas glyceraldehyde forms a derivative after 30 min with this reagent, the isolated ketoamine required 4 hrs of treatment for full color development. The usable range for this assay was 0.1 to 0.5 mM of carbonyl compound. In some instances the aliquots were also assayed for positive amino groups by the Fluram assay[11] or by ninhydrin[9]. For detection of histidine-containing peptides or derivatives, the Pauly reagent was used[12]. The usable range for this reagent was 0.04 to 0.2 mM.

RESULTS AND DISCUSSION

Analysis of Reaction Products - A sample of the reaction mixture was applied to the 0.6 X 20 cm column of the amino acid analyzer. Unreacted dipeptide Val-His appeared at about 65 minutes of elution time. As shown in Figure 1 panel A, an unknown compound which was ninhydrin-positive eluted at about 30 minutes. Upon reduction of an identical aliquot of the reaction mixture, this unknown material was no longer present (panel B). It seems likely that this material, X, is the aldimine or Schiff base which becomes ninhydrin-negative upon reduction to the secondary amine. Presumably, this aldimine is hydrolyzed to the free peptide in the 100

156

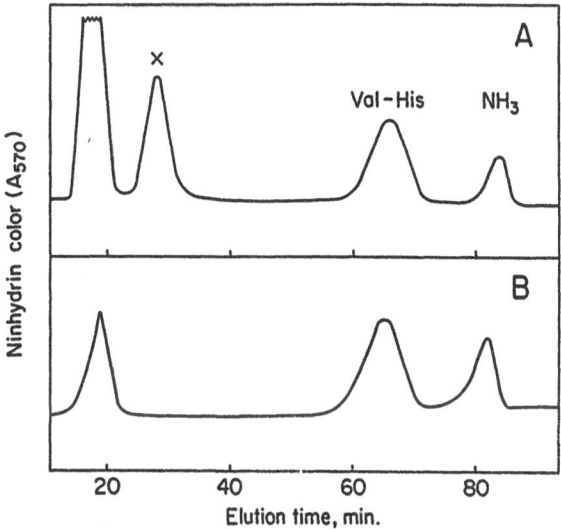

*Figure 1. Amino acid analysis of
the reaction mixture of L-Val-L-His
and glyceraldehyde. The dipeptide
(1 mM) was mixed with DL-glyceraldehyde (50 mM) at 30°C. At
selected times an aliquot was
assayed as described in the text
(panel A). Another aliquot was
reduced with NaBH$_4$ as described
previously[5] and then analyzed
(panel B).*

degree reaction coil in the presence of ninhydrin on the amino acid
analyzer. The ninhydrin-positive material peak eluting near the void
volume of the column (at about 18 minutes) most likely represents free
glyceraldehyde, which gives some type of adduct with ninhydrin prior to
reduction with sodium borohydride. No other ninhydrin-positive compounds
were eluted from the column.

As shown in Figure 2, the disappearance of the starting dipeptide was
measured both by Fluram and by ninhydrin. Although there is a very rapid
disappearance of dipeptide, there is not complete coincidence for the
determinations with these two reagents. It is conceivable that each reacts
with the intermediates to a different extent. The increase in the amount
of aldimine shown in Figure 2 (closed circles) occurs concomitantly with
the disappearance of dipeptide. After a short period, the aldimine reaches
a steady state concentration and then slowly decreases.

It is clear from the data in Figure 2 that there is a substantial
amount of material (75%) which is not accounted for either by the free
dipeptide or by the aldimine after 3 hrs. This result clearly indicates
that the system consisting of glyceraldehyde and the dipeptide is not
simply an equilibrium mixture of the aldimine and a substituted
glycosylamine[5]. Under such conditions, one would expect full recovery
using the assays employed here, if there were complete equilibrium between
the starting material, intermediates, and products.

<u>Isolation of the Ketoamine Product</u> – An HPLC system was developed in
which a large amount of product, which was Pauly-positive but ninhydrin-
negative, could be isolated (Fig. 3). These properties are consistent with

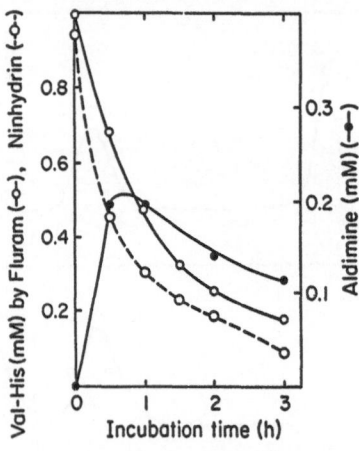

Figure 2. Time course for disappearance of dipeptide and appearance of the aldimine. The reaction conditions were the same as those described in the legend to Figure 1 except that the incubation temperature was 50°C. The analyses were performed as described in the text.

a ketoamine structure. The aldimine intermediate and the unreacted dipeptide are eluted earlier from this column. The yield of ketoamine was calculated to be 60%. The somewhat low yield of the ketoamine may be due to polymerization or other side reactions, which are under study at present. Elemental analysis was consistent with the ketoamine structure as the diacetate salt ($C_{16}H_{25}N_4O_9$):

Calculated	C, 46.04%:	H, 6.00%: N, 13.43%
Found	C, 47.63%:	H, 5.78%: N, 13.38%

The ultraviolet absorption had maxima at 262 nm and 296 nm (Fig. 4). The molar ratio of Pauly-positive to 2,4-dinitrophenylhydrazine-positive material was 1.2/1.

Acid hydrolysis and subsequent amino acid analysis of a weighed amount of the ketoamine yielded the theoretical amount of histidine but the yield of alanine was only about 30%. This result is also consistent with the ketoamine structure since the secondary amine which is formed after Amadori rearrangement would be resistant to acid hydrolysis. The fact that some free alanine is released during acid treatment could be construed as evidence for the partial reversal of the ketoamine reaction (Scheme I). However, this question requires further study. The 2,4-dinitrophenyl-hydrazone derivative of the ketoamine was prepared and washed extensively with HCl and H_2O. Acid hydrolysis generated histidine in 82% yield but less than stoichiometric amount of alanine because of its secondary amine linkage. These results are also consistent with a ketoamine rather than an aldimine type of linkage.

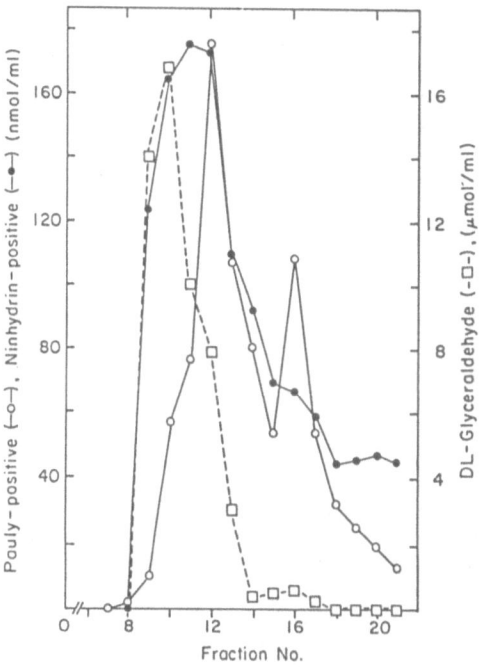

Figure 3. Isolation of the
ketoamine product of the
reaction between glyceraldehyde and L-Ala-L-His. A large
scale synthesis was performed
and an aliquot was applied to
an HPLC column as described in
the text. The column was
eluted at a flow rate of 3
ml/min and 1.5 ml fractions
were collected.

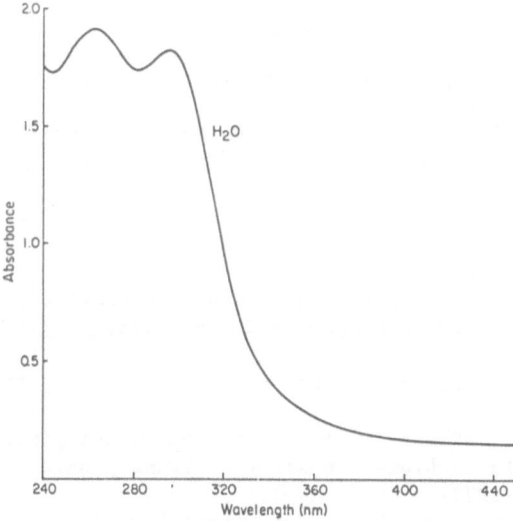

Figure 4. Ultraviolet Spectrum
of the Isolated Dipeptide
Ketoamine.

The chromatographic identification of the intermediates in the reaction should facilitate mechanistic studies on this reaction. The ultimate aim is to extrapolate such information to the behavior of glyceraldehyde with the amino groups of the hemoglobin molecule[3],[4], a more complex system. We were unable to detect any substantial differences in the reactions of either Val-His or Ala-His with glyceraldehyde, thus excluding the possiblity of any special role for the valine residue, which is involved in HbA_{1c} formation. The reason for the rapid Amadori rearrangement could be due to the presence of some amino acid residues close to the site of Schiff base formation. The second amino acid residue of the β-chain of hemoglobin is histidine but whether it is involved in the Amadori rearrangement is not yet clear. Further studies with model peptides of different sequence using the present chromatographic methods may provide answers to these questions.

ACKNOWLEDGEMENTS

We are grateful to Judith A. Gallea for her expert assistance in the preparation of the manuscript. This work was supported in part by NIH Grant HL-18819 and by Grant BRSG-507, RR-07065 from the Biomedical Research Support Grant, Division of Research Resources, National Institutes of Health to Rockefeller University.

REFERENCES

1. J.E. Hodge, The Amadori Rearrangement, Adv. Carbohydrate Chem. 10:169 (1955).

2. H.B.F. Dixon, A Reaction of Glucose with Peptides, Biochem. J. 129:103 (1972).

3. A.S. Acharya, and J.M. Manning, Reactivity of the Amino Groups of Carbonmonoxy hemoglobin S with Glyceraldehyde, J. Biol. Chem. 255:1406 (1980).

4. A.S. Acharya, and J.M. Manning, Amadori Rearrangement of Glyceraldehyde Hemoglobin Schiff Base Adducts: A New Procedure for the Determination of Ketoamine Adducts in Proteins, J. Biol. Chem. 255:7218 (1980).

5. R.E. Feeney, G. Blankenhorn, and H.B.F. Dixon, Carbonyl-Amine Reactions in Protein Chemistry Adv. Protein Chem. 29:135 (1975).

6. R.M. Bookchin, and P.M. Gallop, Structure of Hemoglobin A_{1c}: Nature of the N-Terminal β-Chain Blocking Group, Biochem. Biophys. Res. Commun. 32:86(1968).

7. A.S. Acharya, L.G. Sussman, and J.M. Manning, Schiff Base Adducts of Glyceraldehyde with Hemoglobin Differences in the Amadori Rearrangement at the α-amino Groups J. Biol. Chem. 258:2296 (1983).

8. N. Mori, and J.M. Manning, Studies on the Amadori Rearrangment in a Model System: Chromatographic Isolation of Intermediates and Product, Submitted for publication.

9. D.H. Spackman, W.H. Stein, and S. Moore, Automatic Recording Apparatus for Use in the chromatography of Amino Acids, Anal. Chem. 30:1190 (1958).

10. T.E. Friedemann, Determination of α-Keto Acids, Methods Enzymol., 3:414 (1957).

11. M. Weigele, S. DeBernardo, I. Tengi, and W. Leimgruber, A Novel Reagent for the Fluorometric Assay of Primary Amines, J. Amer. Chem. Soc. 94:5927 (1972).

12. L.A.A. Sluyterman, The Effect of Oxygen Upon the Micro Determination of Histidine with the Aid of the Pauly Reaction, Biochim. Biophys. Acta, 38:218 (1960).

REACTION OF GLYCERALDEHYDE (ALDOTRIOSE) WITH PROTEINS IS A PROTOTYPE OF NONENZYMIC GLYCATION: PROTEIN CROSS-LINKING AS A CONSEQUENCE OF IN VITRO NONENZYMIC GLYCATION

A. Seetharama Acharya

The Rockefeller University
1230 York Avenue, New York, NY

INTRODUCTION

Nonenzymic glycation, a post-translational protein modification reaction, is simply a reflection of the potential of the 'aldehydic' function of aldoses to form a reversible Schiff base adduct with the amino groups of proteins and the subsequent intramolecular rearrangement known as Amadori rearrangement to form a more stable ketoamine adduct[1]. The modification of hemoglobin (Hb) A by glucose, an aldohexose, to form Hb A_{1c} is a continous chemical proccess that occurs in vivo, and is the first known example of nonenzymic glycation[2]. Besides Hb, many other proteins have been shown to undergo nonenzymic glycation in vivo. One of the long-range chemical consequences of in vivo nonenzymic glycation is the covalent cross-linking of proteins[3].

The reaction of glyceraldehyde (2,3-dihydroxypropionaldehyde), an aldotriose, with Hb is mechanistically similar to nonenzymic glycation[4] (Fig. 1). Consistent with this similarity, many of the sites on HbA that are reactive towards glyceraldehyde are the same as those that are

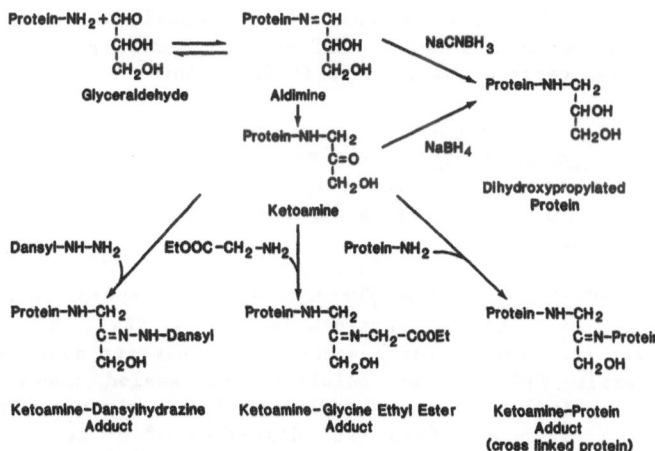

Fig. 1. Schematic representation of reaction of
glyceraldehyde with protein to introduce cross-links.

nonenzymically glycated[5][6]. This study suggested that Amadori rearrangement may be a general property of Schiff base adducts of α-hydroxyaldehydes with the amino groups of proteins[7]. Accordingly, the reaction of glycolaldehyde (aldodiose) with proteins was investigated. The Schiff base adducts of aldodiose with RNase-A were also found to undergo Amadori rearrangement[7]. However, the reaction of aldodiose and aldotriose with proteins occurs at a much faster rate than that of aldohexose. The lower reactivity of aldohexose with proteins is, apparently, a reflection of the fact that only a small fraction of aldohexose is normally present in the 'aldehydic' form, the 'reactive species' for the formation of the initial schiff base adducts[8].

Though the Schiff base adducts of glycolaldehyde undergo the Amadori rearrangement just as the adducts of the higher homologs, there is one significant difference between the Amadori product of aldodiose compared to that of aldotriose and aldohexose. The Amadori product of aldodiose is an aldoamine whereas that of aldotriose and higher homologs is a ketoamine. Since the Amadori rearrangement of aldodiose generates a new 'aldehydic' function _in situ_, the glycolaldehyde is expected to show a latent cross-linking potential. Consistent with this, we have found that glycolaldehyde introduces covalent cross-links into proteins[7].

An aspect of nonenzymic glycation of proteins that has not been investigated in full detail in the past, but may have a significant role in understanding the mechanistic and physiological consequence of nonezymic glycation is the possible reactivity of the carbonyl function of the ketoamine adducts. If the carbonyl function of the ketoamine is reactive towards amino groups of proteins (formation of ketimines), this may provide one of the possible pathways, if not the only one, for the formation of cross-linked proteins as a chemical consequnce of nonenzymic glycation, In principle, the anticipated reactivity of the ketoamine is similar to that of the carbonyl functions dihydroxyacetone or of fructose. This aspect has been investigated in the present study.

MATERIALS AND METHODS

Reaction of Glyceraldehyde with RNase-A - RNase-A (Sigma) (0.5 mM) is phosphate buffered saline, pH 7.4 was incubated at 37°C for indicated period at a given concentration of the aldotriose (Sigma). At the end of the reaction period the sample was desalted by passage through a column (2.2 X 30 cm) of Sephadex G-25 (Pharmacia), equilibrated and eluted with 0.1 M acetic acid. The protein isolated by lyophilization is referred to as glyceraldehyde RNase-A adduct. SDS/polyacrylamide gel electrophoresis of the adducts was carried out as described earlier[7].

Reaction of Glyceraldehyde RNase-A Adduct and Glucose RNase-A Adduct with 2,4-Dinitrophenylhydrazine - The reaction was carried out essentially as described earlier[7]. The derivatized protein was isolated from the excess reagent by gel filtration on Sephadex G-25, equilibrated and eluted with 0.1 M acetic acid.

Reaction of Glucose with RNase-A - The reaction conditions are similar to that used for glyceraldehyde except that the reaction was carried out with 1 M glucose for 3 weeks. The reaction mixture was passed through a sterile filter, the solution was sealed under vacuum and incubated for the desired time. Desalting of the glucose RNase-A adduct was done on a sephadex G-25 as described for the isolation of glyceraldehyde RNase-A adduct.

Reactivity of the Carbonyl Function of the Ketoamine Adduct of Glyceraldehyde - The carbonyl function generated in situ, as the consequence of the Amadori rearrangement of the Schiff base adducts of glyceraldehyde with the amino groups of proteins may have a reactivity towards amino groups comparable, at least, to that of dihydroxyacetone or acetone. The reactivity of the carbonyl function of the 2-oxo-3-hydroxypropyl (OHP) groups on the amino groups of proteins towards substituted hydrazines has been demonstrated previously[4]. However, the reactivity of this carbonyl function towards amino group under physiological conditions has not been investigated so far. Therefore, the OHP RNase-A (the ketoamine adduct) was incubated with ^{14}C-glycine ethyl ester (GEE) for overnight at pH 7.4, 37°C. Gel filtration of the product showed the incorporation of ^{14}C-label into the protein. Quantitation indicated that about 0.6 moles of amino acid is incorporated into the protein. Control experiments with RNase-A as well dihydroxypropylated RNase-A did not incorporate the GEE. The incorporation of GEE clearly reflects the reactivity of the carbonyl functions of OHP groups towards the amino groups.

Cross-Linking Potential of Glyceraldehyde - The reactivity of the carbonyl function of the OHP groups towards the amino groups suggests that glyceraldehyde should also exhibit a latent cross-linking potential[7]. Earlier studies of Nigen and Manning[10] have concluded that glyceraldehdye does not introduce cross-linking into monomeric proteins such as myoglobin, RNase-A, and α-chain of Hb and only a relatively high concentration of glyceraldehyde (50 mM) lead to a 3% of cross-linking between the Hb monomers. However, in that study, the cross-linking introduced on a single 90 minute incubation period was investigated. The cross-linking potential of glycolaldehyde is latent and becomes apparent only on longer incubation periods. The anticipated cross-linking potential of glyceraldehyde will be due to the slow reacting carbonyl group of the 'ketonic' function of ketoamines. This prompted us to re-investigate the cross-linking potential

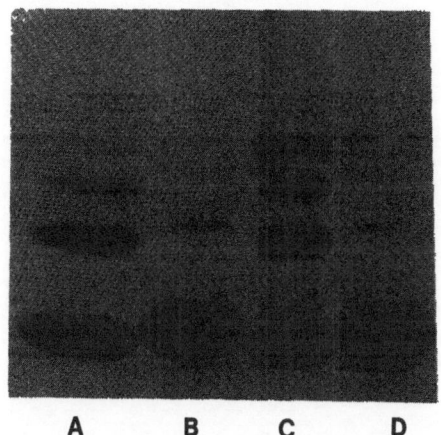

A   B   C   D

Fig. 2. SDS gel electrophoresis of glyceraldehyde RNase-A adduct. Lanes A and B, RNase-A reacted with glyceraldehyde without and with NaCNBH₃. Lanes C and D, RNase-A reacted with glycoladehyde without and with NaCNBH₃.

of this aldotriose. This information is also essential for the eventual use of this aldotriose, as an antisickling agent of therapeutic value.

RNase-A was chosen as the model protein to investigate these aspects. This protein does not contain any prosthetic group and is expected to make the spectral studies of the adducts easier. Glyceraldehyde indeed introduces intermolecular cross-linking on longer incubation periods nearly as effeciently as glycolaldehyde (Fig. 2). Besides, just as in the case of the glycolaldehyde, the presence of NaCNBH$_3$ during the incubation inhibited the cross-linking reaction. In the presence of NaCNBH$_3$, the Schiff base adducts of aldotriose with the amino groups of RNase-A are reduced resulting in the 2,3-dihydroxypropylation of the protein[11] [12]. Thus, inhibiting the formation of ketoamine inhibits the generation of cross-linking.

Fluorescence of RNase-A Glyceraldehyde Adduct – The postulated pathway for the formation of glyceraldehyde cross-linked RNase-A is mechanistically similar to that of glycolaldehyde. Therefore, the spectral and the flourences properties of glyceraldehyde cross-linked RNase-A has been investigated and compared with that generated on reaction with glycolaldehyde (Fig. 3). RNase-A develops a new broad absorption band absorbing the region of 300-350 nm on reaction with glyceraldehyde. The ultraviolet absorption spectrum is qualitatively similar to the spectra of RNase-A glycolaldehyde adduct. This similarity prompted us to investigate the fluorescence properties of RNase-A glyceraldehyde adduct (Fig. 3 inset). The cross-linked material exhibited fluorescence with an emission maximum around 450 nm, and an excitation maximum arount 350 nm. Again, the presence of NaCNBH$_3$ inhibited development of the fluorescence. Thus, the cross-linking, development of absorption, as well as, fluorescence are interrelated. Inhibition of Amadori rearrangement inhibits cross-linking as well as the development of fluorescence. Thus, the cross-linking and the associated spectral changes are the consequence of the presence of reactive carbonyl function of the ketoamine.

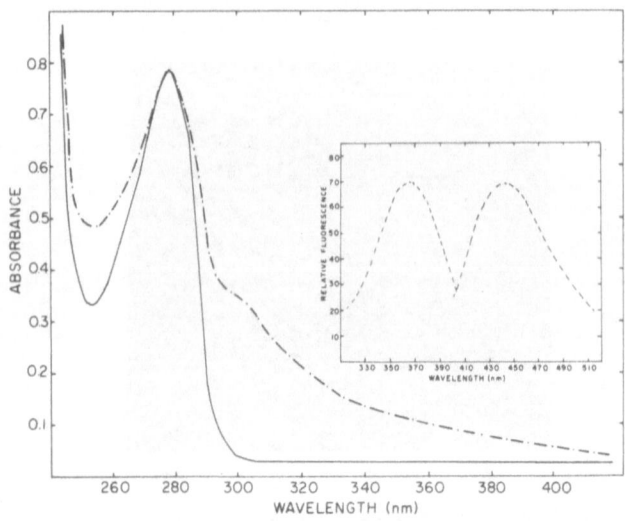

Fig. 3. Absorption and fluorescence spectra
of RNase-A. (---) Absorption spectra of RNase-A,
(- -) RNase-A glyceraldehyde adduct. Inset
shows the fluorescence excitation and emission
spectra of the adduct.

Kinetics of RNase-A Glyceraldehyde Adduct Formation - The progress of cross-linking reaction between glyceraldehyde and RNase-A was followed by monitoring the development of the new absorption band characteristic of the cross-linked product (Fig. 4). Again as in the case of glycolaldehyde reaction, the kinetics of development of new absorption characteristic was biphasic, an initial lag phase during which no new absorption was developed, and a second phase during which the new absorption developed rapidly. The duration of lag period depends on the initial concentration of aldotriose, on increasing the aldotriose concentration the lag period decreased. Pre-incubation of a solution of glyceraldehyde before the addition of RNase-A has no influence on the kinetics of color development. This observation suggests that glyceraldehyde alone and not some condesation product of it that has this potential to introduce covalent cross-links.

In the suggested pathway, the generation of the cross-linked product involves the reaction of the carbonyl group of a 'ketonic function' generated *in situ*. In the case of aldodiose an 'aldehyde group' generated *in situ* is involved in the reaction with the protein amino group. In view of the lower propensity of the carbonyl groups of ketones to form Schiff base adducts with the amino groups, the kinetics of cross-linking of RNase-A by glycolaldehyde and glyceraldehyde has been monitored as reflected in the development of the absorption around 320 nm (Fig. 4). Under identical conditions, the lag period for the development of color is shorter with glycolalehyde (curve A) than that seen with glyceraldehyde (curve B). Besides, the exponential phase for the developemnt of color also proceeds at a slower rate with glyceraldehyde. The propensity of glycolaldehyde and glyceraldehyde to form Schiff base adducts is nearly the same[13]. However, the observed rate of development of color with glyceraldehyde is somewhat slower than that of glycolaldehyde. This may be related to the lower propensity of the carbonyl group of ketone to form Schiff base adducts[13]. The propensity of carbonyl function dihydroxyacetone (ketose) to from Schiff base adduct appears to be at least an order of magnitude slower than that of glyceraldehyde[13]. Dihydroxyacetone, being an α-hydroxycarbonyl compound may be expected to introduce covalent cross-links with proteins like aldoses, however, the reaction sequence leading to cross-linking will

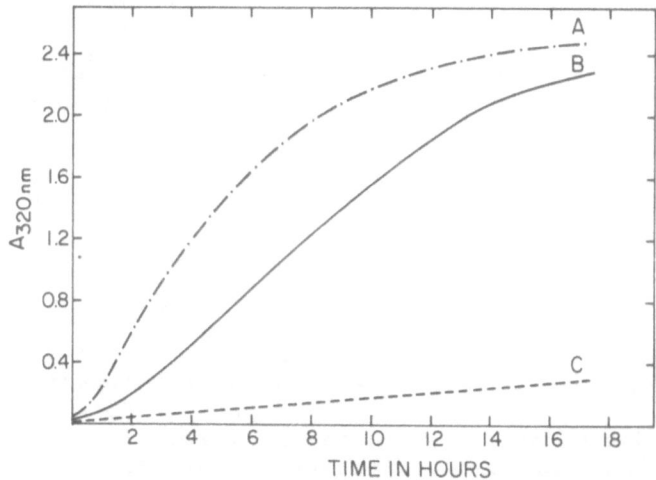

Fig. 4. Kinetics of the development of the new absorption band on reaction of RNase-A with α-hydroxycarbonyl compounds. Glycolaldehyde (A), Glyceraldehyde (B), and Dihydroxyacetone (C).

be reverse to that of glyceraldehyde. Consistent with this the reaction of dihydroxyacetone also generated cross-links. With dihydroxyacetone the reaction as measured by the absorption at 320 nm appears to be linear, though slow. The overall cross-linking reaction with glyceraldehyde proceeds at a much faster rate than that seen with dihydroxyacetone (Fig. 4). A carbonyl function of a ketone is invoved in both these reactions, though in a reverse order. The rate limiting-step in the glyceraldehyde cross-linking reaction (as reflected in the development of 320 nm absorption band) is the second reaction, i.e., that of the carbonyl group of ketoamine (and hence, the lag phase) whereas with dihydroxyacetone it is the first reaction, i.e., formation and Amadori rearrangement of Schiff base adducts of a ketose with the amino groups. The results are also suggestive that the carbonyl function of the ketoamine linkages of aldotriose are more reactive than those of simple aliphatic ketones.

<u>Reactivity of the Carbonyl Function of Ketoamine Linkages of Aldohexose and Consequent In vitro Covalent Cross-Linking</u> –The reactivity of the carbonyl function of the ketoamine of aldotriose suggests that the carbonyl functions of ketoamines of aldohexoses may also exihibit·a similar reactivity. Consistent with this, the nonenzymically glycated RNase-A was found to be reactive towards dinitrophenyl hydrazine (Fig. 5).

With an objective of establishing the mechanistic similarity between the glyceraldehyde induced cross-linking and the glucose mediated cross-linking that occurs as a long-term consequence of nonenzymic glycation, RNase-A was incubated with 1 M glucose at pH 7.4 and 37°C for 21 days under sterile conditions (A high concentration of glucose was used in these studies to offset the low concentration of 'aledhydic' form. The effective concentration of 'reactive' glucose at a given time may be expected to be around 2 to 3 mM.) The incubation resulted in covalent cross-linking of RNase-A with concomitant .development of new ultraviolet absorption band with a maximum around 325 nm, and fluorescence spectra similar to that seen

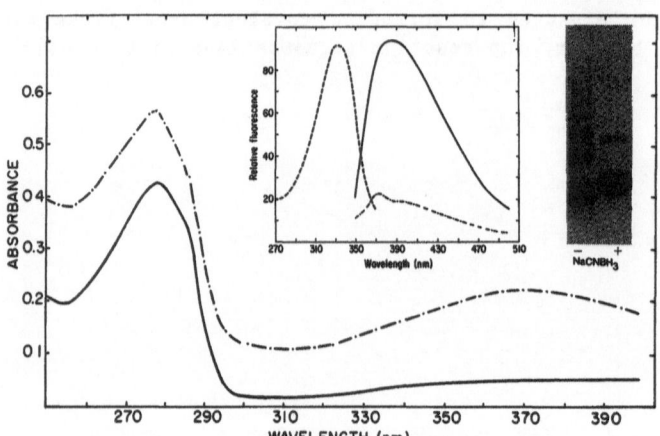

Fig. 5.   Reaction of glucose RNase-A with 2,4-dinitrophenylhydrazine.   (---) Absorption spectra of glucose RNase-A adduct, (- -) and its 2,4-dinitrophenylhydrazine reacted product.   Inset A shows fluorescence emission spectra of RNase-A reacted with glucose (---) in the absence of NaCNBH₃, (- -) of that reacted in the presence of NaCNBH₃, (---) fluorescence excitation spectra.   Inset B shows the SDS gel electrophoresis of RNase-A reacted with glucose in the absence and presence of NaCNBH₃.

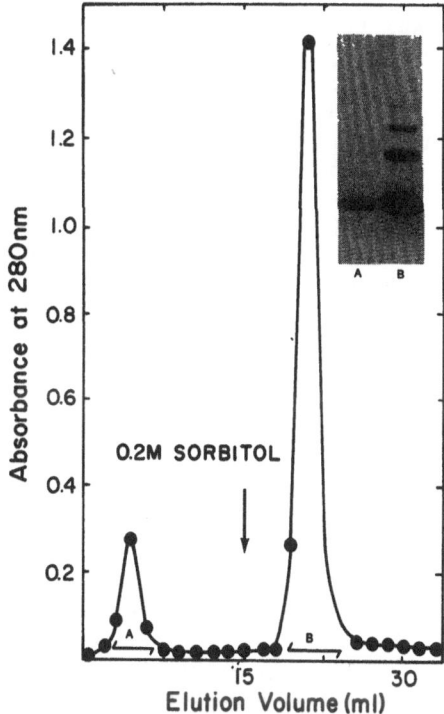

Fig. 6. *Affinity chromatography of glucose RNase-A adduct on Glycogel B. The inset shows the SDS gel electrophoretic pattern of unadsorbed (A) and the sorbitol eluted (B) fraction.*

in the reaction of RNase-A with glyceraldehyde and glycolaldehyde (Fig. 5, inset A). Again as with glycolaldehyde and glyceraldehyde systems, presence of NaCNBH$_3$ inhibited the cross-linking raction, and the development of the absorption, as well as the fluorescence preperties (Fig. 6, inset B). In order to establish further that the cross-linking is a consequence of incorporation of glucose nonenzymically into the protein, the glucose RNase-A adduct prepared above was passed through Glyco-Gel B (affinity chromatography for protein with bound glucose). The glyco-Gel B bound protein was eluted with sorbitol (Fig. 6). Only the glyco gel bound protein had covalent cross-linking (Fig. 6, inset) and the fluorescence demonstrating that the incorporation of glucose, cross-linking and development of fluorescence are interrelated.

DISCUSSION

The results presented here demonstrate the potential of the carbonyl function of the ketoamine of aldotriose and aldohexose to form adducts with the amino functions of GEE and protein. The formation of ketoamine-protein adducts introduces covalent cross-links into proteins. This cross-linking potential of aldotriose and aldohexose is latent in that it becomes apparent only after the completion of the first phase of the reaction, i.e., Amadori rearrangement. Inhibition of the Amadori rearrangement by carrying out the reaction in the presence of NaCNBH$_3$ inhibits the cross-linking reaction as well. The exact structure of the final cross-links introduced by aldotriose is not clear at present. In principle, the ketoamine GEE adduct and ketoamine protein adducts (ketimine structures) should be capable of undergoing further Amadori rearrangement to form aldoamine structures, as well as resonating with an aldimine type of structure wherein the Schiff base at position 1 of aldotriose. Whether the putative rearrangement or the equilibrium between the putative aldimine, aldoamine structure and the ketoamine amino group adduct (ketimine structure) under the physiological conditions provide the stability for the covalent cross-links needs to be investigated. At any rate, the glyceraldehyde-RNase-A reaction discussed here should serve as a good model to understand the mechanistic aspects of the cross-linking reaction mediated by the nonenzymic incorporation of glucose *in vivo*.

REFERENCES

1. F. Wold, *In Vitro* Modification of Proteins, *Annu. Rev. Biochem.* 50:783 (1981).
2. W.R. Holmquist and W.A. Schroeder, A New N-Terminal Blocking Group Involving Schiff Base in Hemoglobin A$_{1c}$, *Biochemistry* 5:2489 (year).
3. V. Monnier and A. Cerami, Nonenzymic Glycosylation and Browning in Diabetes and Aging, *Diabetes* 31(3):57 (1982).
4. A.S. Acharya and J.M. Manning, Amadori Rearrangement of Glyceraldehyde Hemoglobin Schiff Base Adducts, *J. Biol. Chem.* 255:7218 (1980).
5. A.S. Acharya and J.M. Manning, Reactivity of the Amino Groups of Carbonmonoxy Hemoglobin S with Glyceraldehyde, *J. Biol. Chem.* 255:1406 (1980).
6. R. Shapiro, J.M. McManus, C. Zalut, and H.F. Bunn, Sites of Nonenzymic Glycosylation of Human Hemoglobin A, *J. Biol. Chem.* 255:3120 (1981).
7. A.S. Acharya and J.M. Manning, Reaction of Glycolaldehyde with Proteins: Latent Cross-Linking Potential of α-Hydroxyaldehydes, *Proc. Natl. Acad. Sci. USA* 80:3590 (1983).
8. S.J. Angyl, in: 'Assymmetry in Carbohydrates,' E.D. Hamon, ed., Marcel Dekker, Inc., New York, 15–20 (1979).
9. R. Fields and H.B.F. Dixon, Micromethod for Detemination of Reactive Carbonyl Groups in Proteins and Peptides using 2,4-Dinitrophenyl-hydrazine, *Biochem. J.* 121:587 (1972).
10. A.M. Nigen and J.M. Manning, Effects of Glyceraldehyde on the Structural and Functional Properties of Sickle Erythrocytes, *J. Clin. Invest.* 61:11 (1978).
11. A.S. Acharya, L.G. Sussman, and J.M. Manning, Schiff Base Adducts of Glyceraldehyde with Hemoglobin: Differences in the Amadori Rearrangement at the α-Amino Groups, *J. Biol. Chem.* 258:2296 (1983).
12. A.S. Acharya, L.G. Sussman, and B.N. Manjula, Application of Reductive Dihydroxypropylation of Amino Groups of Proteins in Primary Structural Studies: Identification of Phenylthiohydantoin Derivative of ε-Dihydroxypropyl L-Lysine Residues by High Performance Liquid Chromatography, *J. Chromatogr.* 297:37 (1984).
13. A.S. Acharya, L.G. Sussman, and J.M. Manning, Selectivity in the Modification of the α-Amino Groups of Hemoglobin on Reductive Alkylation with Aliphatic Carbonyl Compounds: Influence of Derivatization on the Polymerization of Hemoglobin S, *J. Biol. Chem.* 260:6039 (1985).

ACKNOWLEDGEMENTS

This work has been supported by NIH Grnats HL-27183 and AM-35869 to ASA. ASA is an Established Fellow of New York Heart Association. The assistance of L.G. Sussman, Y.J. Cho, and J.A. Gallea is greatfully acknowledged.

REVERSIBLE DIHYDROXYPROPYLATION OF AMINO GROUPS IN PROTEINS:

APPLICATION IN PRIMARY STRUCTURAL STUDIES OF STREPTOCOCCAL M-PROTEINS

Belur N. Manjula*, Vincent A. Fischetti*,Thomas Fairwell**, and A. Seetharama Acharya*

*The Rockefeller University, 1230 York Avenue, New York, N.Y.
**Molecular Disease Branch
National Heart, Lung, and Blood Institute,NIH, Bethesda, MD

INTRODUCTION

M-protein of group A streptococcus is an immunologically diverse antiphagocytic determinant of the bacteria[1]. In order to better understand the structure-function relationships of the M-proteins, we undertook the determination of their primary structure. The Pep M5 protein, a biologically active, pepsin-derived N-terminal half of the type 5 M protein, was selected for the initial study[2]. Pep M5 protein contains 6 arginines, 35 lysines, and 47 glutamates, with no methionines and tryptophans, thus limiting the choice for obtaining large peptides for sequence studies to cleavage at its arginyl peptide bonds[2]. Though arginine specific clostripain appeared to be satisfactory initially, detailed studies of the peptides generated by clostripain digestion of Pep M5 protein revealed that in addition to the major arginine cleavages, digestion occurred at some of the lysine residues. Furthermore, clostripain failed to cleave some of the arginyl bonds quantitatively[3]. Thus, a better choice for obtaining arginyl peptides appeared to be to take advantage of the high specificity of tryptic cleavage after chemically modifying the ε-amino groups of the Pep M5 protein[4]. The relatively high lysine content of M-protein makes it essential that the reagent used for the modification be very selective to the amino functions and also be capable of derivatizing the lysine residues of the protein completely. Attempts at derivatizing the lysine residues of the Pep M5 protein with citraconic anhydride prior to tryptic digestion were not highly successful. This may be related to the very high lysine content of the Pep M5 protein. Therefore, an alternate procedure for the derivatization lysine residues of the Pep M5 protein appeared to be valuable.

In an unrelated study we have demonstrated that glyceraldehyde (2,3-dihydroxypropionaldehyde) derivatizes the amino groups of proteins by two different modes[5][6] (Fig. 1), the mode of derivatization depending on the presence or absence of the reducing agent, $NaCNBH_3$. In the absence of $NaCNBH_3$, 2-oxo, 3-hydroxypropylation of amino groups takes place (Structure 1, Fig. 1). This reaction is a prototype of nonenzymic glycation, linking the aldose to the protein through a ketoamine linkage. In the presence of $NaCNBH_3$, 2,3-dihydroxypropylation (Structure 2, Fig. 1) of the amino groups of protein takes place, and this reaction (reductive dihydroxypropylation) is analogous to the reductive alkylation. Alternatively, the 2-oxo,3-

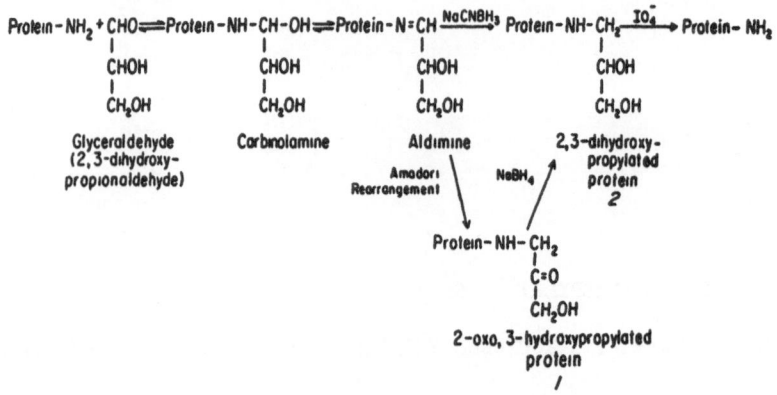

Fig. 1. Schematic representation of reversible chemical modification of amino groups of proteins by glyceraldehyde.

hydroxypropyl groups on the amino groups of proteins could also be reduced to 2,3-dihydroxypropyl (DHP) groups using sodium borohydride. The derivatized lysine, ε-N-2,3-DHP-lysine is stable to acid hydrolysis, elutes slightly ahead of histidine from the amino acid analyzer, and hence can be easily quantitated. The peptide bonds of ε-N-DHP-lysine residues are resistant to tryptic cleavage. This prompted us to investigate the general utility of reductive dihydroxypropylation of amino groups of proteins to limit the tryptic digestion of polypeptide chains to their arginine residues.

MATERIALS AND METHODS

    Reductive Dihydroxypropylation of Proteins - Proteins (0.25 to 0.5 mM) in phosphate bufferred saline, pH 7.4, were incubated with glyceraldehyde in the presence of 10-fold molar excess[4][7] (over the aldehyde) of NaCNBH$_3$ at 37°C for 30 minutes. After the reaction, the derivatized protein samples were dialyzed extensively against 0.1 M ammonium bicarbonate, pH 8.0, and the modified protein was isolated by lyophilization.

    Tryptic Digestion of DHP Protein - The protein samples were taken in 0.1 M ammonium bicarbonate, pH 8.0 and digested with TPCK trypsin at 37°C at an enzyme to protein ratio of 1:100. The digested material was isolated by lyophilization.

    Preparation of Pep M5 and Pep M6 Proteins - The Pep M5 and Pep M6 proteins were isolated by proteolysis of the type 5 and type 6 streptococcal cells with pepsin as previously described[2].

    Sequence Analysis of Protein and DHP Peptides - Amino terminal sequence analyses of the Pep M5 and Pep M6 proteins and their DHP peptides were carried out by automated Edman degradation as previously described[4].

    Periodate Oxidation of DHP protein and DHP Peptides - Routinely, the DHP protein (1 to 2 mg/ml) or DHP peptides (0.1 to 0.5 mg/ml) in 10 mM phosphate buffer, pH 7.4, was incubated with 15 mM sodium metaperiodate for 15 minutes at room temperature. After the oxidation, the protein was isolated either by dialysis or by gel filtration on Sephadex G-25 column equilibrated and eluted with 0.1 M acetic acid. Periodate treated DHP peptides were isolated by RPHPLC.

Reductive Dihydroxypropylation of Amino Groups of RNase-A - The general utility of reductive dihydroxypropylation of ε-amino groups of lysine for limiting the tryptic cleavage to arginine residues of proteins was initially investigated using RNase-A as the model protein. In an attempt to determine whether complete dihydroxypropylation of amino groups of RNase-A could be carried out without using denaturants, RNase-A was modified with 10, 20, 50, and 100 mM glyceraldehyde in the presence of 0.1, 0.2, 0.5, and 1.0 M NaCNBH₃, respectively. Nearly complete (98%) derivatization of the amino groups occurred with 100 mM glyceraldehyde in the presence of 1 M NaCNBH₃. The complete resistance of the peptide bond of ε-N-DHP lysine to tryptic cleavage was investigated by the tryptic digestion of a performic acid oxidized sample of DHP RNase-A. The RPHPLC of a 3 hr tryptic digest showed five components as expected for the four internal arginine residues present in RNase-A. Continuing the tryptic digestion for 24 hrs had little influence on the tryptic peptide maps, thus, demonstrating the complete resistance of the peptide bonds of ε-N-DHP-Lys residues to tryptic digestion.

Application of Dihydroxypropylation in the Sequence Studies of Pep M5 Protein - Dihydroxypropylation of Pep M5 protein using 100 mM glyceraldehyde and 1 M NaCNBH₃ led to the complete modification of the lysine residues of the protein. More than 80% of the modified lysines were present as the mono alkylated derivative. Tryptic digestion of the DHP Pep M5 protein, followed by RPHPLC of the digest yielded the expected arginine peptides of the protein. Sequence analyses of the tryptic peptides of the DHP-Pep M5 protein (Fig. 2), together with the amino terminal sequence of the whole molecule provided the sequence of 173 of the 197 (88%) residues

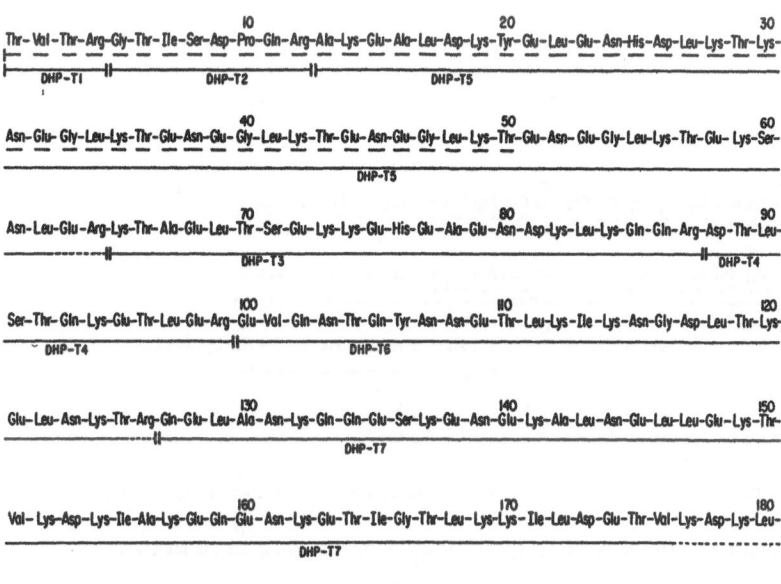

Fig. 2. Complete amino acid sequence of Pep M5 protein. (---) Sequence determined using Pep M5 protein. DHP-T refers to the tryptic peptides of DHP protein. The solid lines represent the extent of sequence established for each peptide. The dashed lines indicate the remainder of that peptide.

of the Pep M5 molecule. This established the framework for the Pep M5 sequence. The complete amino acid sequence of the Pep M5 protein is shown in Fig. 2, along with the data generated from the tryptic peptides of DHP Pep M5 protein[4]. Thus, reductive dihydroxypropylation proved very valuable in the determination of the complete amino acid sequence of the streptococcal Pep M5 protein.

Application of Dihydroxypropylation in the Sequence Studies of Streptococcal Pep M6-Protein - Since the completion of the studies on the primary structure of the Pep M5 protein, studies have been undertaken with another Streptococcal Pep M protein, namely Pep M6 protein, the pepsin-derived N-terminal half of the type 6 M protein. This protein has 6 arginine residues, one of which is at the amino terminus of the molecule. Pep M6 protein was dihydroxypropylated and the peptides generated on tryptic digestion of the DHP protein were fractionated by RPHPLC (Fig. 3). The digest contains 6 major tryptic peptides, designated DHP-T1 through DHP-T6. All but one peptide, DHP-T3 contained arginine. The latter peptide is apparently the carboxy terminal tryptic peptide of the Pep M6 protein. The amino terminal sequences of tryptic peptides of (Fig. 4) the

Fig. 3. HPLC of tryptic peptides of DHP Pep M6 protein.

Pep M6 Arg Val Phe Pro Arg Gly Thr Val Glu Asn(10) Pro Asp Lys Ala Arg Glu Leu Leu Asn Lys(20)

DHP-T2 | DHP-T1 | DHP-T5

(21) Tyr Asp Val Glu Asn Ser Met Leu Gln Ala(30) Asn Asn Asp Lys Leu Thr Thr Glu Asn Asn(40)

DHP-T5

(41) Asn Leu Thr Asp Gln Asn Lys Asn Leu Thr(50) Thr Glu Asn Lys Asn Leu Thr Asp Gln Asn(60) Lys Asn Leu(63) ...

DHP-T5

DHP-T3 Glu(1) Gln Lys Ser Lys Gln Asp Ile Gly Ala(10) Leu Lys Gln Glu Leu Ala(16)

DHP-T4 Leu(1) Thr Thr Glu Asn Lys Gly Leu Thr Lys(10) Lys Leu Ser Glu Ala Glu Glu Glu Ala Ala(20) Asn Lys Glu Arg

DHP-T6 Glu(1) Asn Lys Glu Ala Ile Gly Thr Leu Lys(10) Lys Thr Leu Asp Glu Thr Val Lys Asp Lys(20) Ile Ala Lys Glu Gln

Glu(26) Ser Lys Thr Glu(30) Ile Gly Thr Leu Lys Lys Thr Leu Asp Glu(40) Thr Val Lys Asp Lys Ile Ala Lys Glu Gln(50) Glu Ser ...

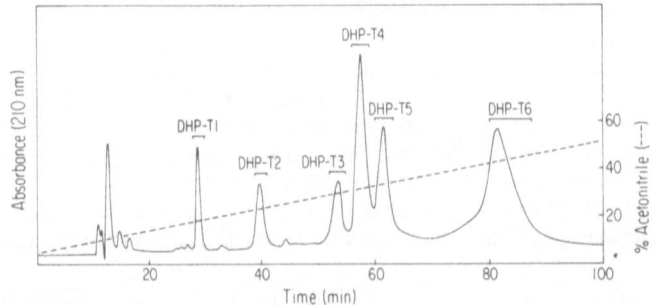

Fig. 4. Amino terminal sequences of Pep M6 protein and its arginyl peptides. The amino terminal sequence of the whole molecule is indicated by dashes. The sequence of DHP-T1, DHP-T2, and DHP-T5 is indicated by the solid line. The sequences of peptides DHP-T3, T4, and T6 are numbered independently with residue 1 corresponding to the amino terminal residue of the peptide.

DHP Pep M6 protein are shown in Fig. 4 along with the amino terminal sequence of the native Pep M6 protein. Knowledge of the amino terminal sequence of the Pep M6 molecule permitted the alignment of DHP-T1, DHP-T2, and DHP-T5. The two large tryptic peptides DHP-T5 and DHP-T6 were sequenced through their amino terminal 48 and 52 residues, respectively. The amino terminal sequence of DHP-T5 provided a 20 residue overlap with the amino terminal sequence of the whole molecule and thus, extended the amino terminal sequence of the Pep M6 protein to 63 residues. The amino acid sequence data thus generated using the tryptic peptides of DHP Pep M6, account for 154 residues (nearly 70%) of the streptococcal Pep M6 protein.

Regeneration of Lysine Residues in DHP Proteins on Periodate Oxidation - The ε-N-2,3-DHP-Lys lysine residues have a substituted α-amino alcoholic function (Fig. 1), and hence, may be expected to be susceptible for periodate oxidation to regenerate the free ε-amino group of these lysine residues. Consistent with this, periodate oxidation of DHP RNase-A (a sample prepared by using 10 mM [^{14}C]-glyceraldehyde and 100 mM NaCNBH$_3$) resulted in the release of more than 95% of the bound [^{14}C]-label, thus demonstrating the reversibility of the dihydroxypropylation reaction (Table I). DHP RNase-A prepared using 20, 50, and 100 mM glyceraldehyde also released significant amounts of the [^{14}C]-label from the protein on periodate oxidation (Table I). The extent of reversibility of dihydroxypropylation decreased as the extent of modification of lysine residues of the protein increased, suggesting that some secondary reactions occur when a large excess of glyceraldehyde (over the total amino groups) is used during the dihydroxypropylation reaction.

The regeneration of lysine residues after the periodate treatment was confirmed by amino acid analysis. ε-N-DHP lysine elutes slightly ahead of histidine (Fig. 5A). Reaction of RNase-A with 50 mM glyceraldehyde in the presence of 500 mM NaCNBH$_3$, resulted in the modification of almost all of the lysine residues of the protein. More than 90% of the modified lysine was present as the mono substituted lysines. On treatment with periodate, the DHP lysine disappeared with a concomitant appearance of lysine (Fig. 5B). The regeneration of lysine from ε-DHP lysine of the DHP RNase-A was calculated to be of the order of 95%.

Regeneration of Trypsin Susceptible Site in a DHP Peptide After Periodate Oxidation - The periodate oxidation of the DHP protein regenerates lysine residues, thus, regenerating trypsin susceptible sites on the protein. This aspect was investigated using a tryptic peptide of DHP Pep M5 protein, namely, DHP-T4. This is a 12 residue peptide corresponding to segment 88-99 of the Pep M5 protein (see Fig. 2) and contains an internal lysine residue at position 94. In the DHP protein, the ε-amino group of this residue is dihydroxypropylated, and hence, the peptide bond of this lysine was not hydrolyzed by trypsin.

Peptide DHP-T4 was isolated from a DHP Pep M5 protein preparation that was derivatized using [^{14}C]-glyceraldehyde in order to make it

Table I. Release of [^{14}C]-Label of DHP RNase-A On Periodate Oxidation

Concentration of [^{14}C]-Glyceraldehyde (mM)	% Lysine Modified	% [^{14}C]-Label of Released on Periodate Oxidation
10	60	94
20	80	90
50	95	85
100	98	75

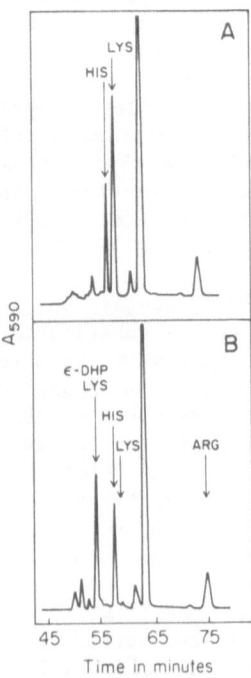

Fig. 5. Amino acid analysis of DHP RNase-A (B) and periodate treated DHP RNase-A (A).

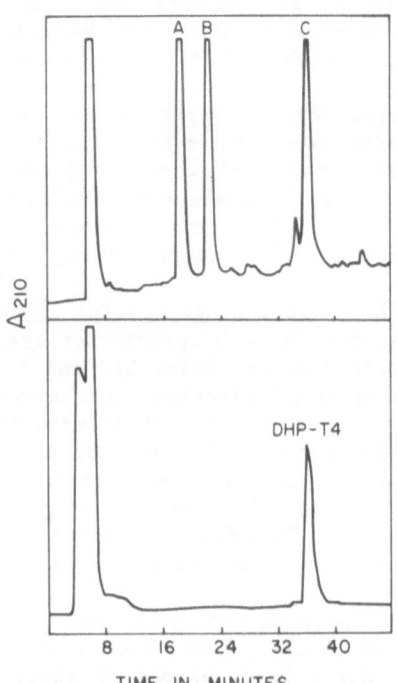

Fig. 6. HPLC analysis of DHP-T4 (bottom) and tryptic digest of periodate treated DHP-T4 (top).

easier to monitor the reversibility of modification. Periodate oxidation of [^{14}C] DHP-T4 resulted in nearly complete release of the [^{14}C]-label of the DHP group from the peptide suggesting the regeneration of the lysine residue. The periodate-treated DHP-T4 eluted on RPHPLC virtually in the same position as the untreated DHP-T4 (Fig. 6, bottom) demonstrating the limited influence of DHP groups on the hydrophobicity of peptides. This periodate treated DHP-T4 was subjected to tryptic digestion (37°C, 1 hr, E:S = 1:100) and the digest analyzed by RPHPLC. Two new peptides (A and B) were formed as a result of this tryptic digestion (Fig. 6, top). Under similar conditions, untreated DHP-T4 was completely resistent to tryptic digestion. This clearly demonstrates the generation of a trypsin susceptible site on periodate oxidation of DHP-T4. Amino acid analyses of peptides A and B revealed that peptide A corresponds to segment 88-94 of the Pep M5 protein while that of peptide B corresponds to segment 95-99 (Table II). The amino acid composition of the material eluting at the original position of DHP-T4 (peak C) was nearly the same as that of DHP-T4 except for the fact that it contained nearly one equivalent of lysine in place of ε-N-DHP-Lys. Redigestion of this material with trypsin yielded peptides A and B again, showing that the digestion was incomplete when the periodate-treated peptide was incubated with trypsin for 1 hr.

It can be seen from Table II, that the content of serine and threonine in peptides A and B is very good showing that the conditions used for periodate oxidation of DHP-T4 is mild, and do not lead to any significant oxidation of the hydroxy amino acids, namely serine and threonine.

174

| | Peptide A | | Peptide B | |
| | | Expected for | | Expected for |
Amino Acid	Found*	Segment 88-94	Found*	Segment 95-99
Asp	1.00	1	–	–
Thr	1.65	2	0.85	1
Ser	0.95	1	–	–
Glu	1.05	1	2.10	2
Leu	1.00	1	1.00	1
Lys	0.90	1	–	–
Arg	–	–	0.90	1

* Calculated assuming the value of Leu as 1.

DISCUSSION

The derivatization of the amino groups of proteins or peptides by dihydroxypropylation and the subsequent removal of the DHP groups, when desired, by mild periodate oxidation complements the other reversible chemical modification procedures for the amino groups[8]. Of the various reversible reagents, maleic, and citraconic anhydrides modify the positively charged amino groups to negatively charged form. The ethylthiotrifluoroacetate converts the amino group to an uncharged form. On the other hand, modification with the alkyl imidates retains the positive charge of the original amino group. Reductive dihydroxypropylation procedure belongs to this last class. The dihydroxypropylation is expected to cause little change in pk_a of the ε-amino groups. The DHP groups on the amino group are stable over a wide range of experimental conditions. The use of tritiated $NaCNBH_3$ will permit the radiolabeling of the alkyl groups introduced. The extent of blocking, and deblocking could also be estimated easily by amino acid analysis.

The selectivity of periodate to cleave α-amino alcohol or substituted α-amino alcohol has been investigated by Feeney and his associates[9] to remove the hydroxyethyl and hydroxyisopropyl groups on the amino groups of proteins. However, in the case of reductive hydroxyethylation a second mole of aldehyde can add relatively easily to the hydroxyethylated amino (a secondary amine) group just as in the case of reductive methylation giving rise to a disubstituted amino group (a tertiary amine). This tertiary amino alcohol present in the disubstituted derivative is resistant to periodate oxidation, thereby preventing (limiting) the reversibility of hydroxyethylation. The extent of reversibility appears to be higher if acetol (α-hydroxy acetone) is used for reductive alkylation (hydroxyisopropylation). Means and Feeney[10] showed that reductive alkylation with acetone (isopropylation) proceeds predominantly to the stage of monoalkylation. The higher reversibility of reductive alkylation by acetol is apparently related to the specificity of this ketone to monoalkylation. However, acetol being a ketone is less reactive with the amino groups, and hence getting a complete modification of the amino groups will be difficult. More recently, the reversibility of reductive glycation on periodate oxidation has also been ivestigated[11]. Reductive glycation, though appears to result in the monoalkylation of amino groups, takes a few days to get a 60-80% modification of the amino groups of proteins. However, the extent of reversibility obtained has been high: 80-95% of amino groups modified by reductive glycation were regenerated with 5 mM periodate in less than 10 minutes.

The reductive dihydroxypropylation of amino groups of proteins appears to proceed predominantly to the stage of mono alkylation. The reductive dihydroxypropylation is complete in about 15 minutes. Since the dialkylation of the type seen with hydroxyethylation does not occur to a significant degree, the reversibility on periodate oxidation is also high (90–95%). Thus, reductive dihydroxypropylation provides the advantages of reductive hydroxyethylation in being a relatively fast reaction, and also that of reductive glycation in terms of high yields of regeneration of lysine residues on periodate oxidation. The results of the present study show that reductive dihydroxypropylation of amino groups of proteins and the regeneration of lysine by mild periodate oxidation is a very useful procedure in the structural studies of proteins and should be a valuable addition to the existing reversible chemical modification procedures.

ACKNOWLEDGEMENTS

This work was supported by NIH Grants HL–36025 and AHA 83–1102 from American Heart Association to BNM, and NIH Grants HL–27183 and AM 35869 to ASA, AI–11822 to VAF, BNM was an Established Investigator of American Heart Association during the tenure of this work, and ASA is an Established Fellow of New York Heart Association. Assistance of Ms. M.L. Schmidt, L.G. Sussman, Y.J. Cho, and J.A. Gallea is very much appreciated.

REFERENCES

1. R.C. Lancefield, Current Knowledge of Type–Specific M–Antigens of Group A Streptococci, J. Immunol 89:307 (1962).
2. B.N. Manjula and V.A. Fischetti, Studies on Group A Streptococcal M Proteins: Purification of Type 5 M Protein and Comparison of its Amino–Terminal Sequence with Two Immunologically Unrelated M Protein Molecules, J. Immunol. 124:261 (1980).
3. B.N. Manjula, S.M. Mische, and V.A. Fischetti, Primary Structure of Streptococcal Pep M Protein: Absence of Extensive Sequence Repeats, Proc. Natl. Acad. Sci. USA 80:5475 (1983).
4. B.N. Manjula, A.S. Acharya, S.M. Mische, T. Fairwell, and V.A. Fischetti, The Complete Amino Acid Sequence of a Biologically Active 197–Residue Fragment of M Protein Isolated from Type 5 Group A Streptococci, J. Biol. Chem. 259:3686 (1984).
5. A.S. Acharya and J.M. Manning, Amadori Rearrangement of Glyceraldehyde–Hemoglobin Schiff Base Adducts: A New Procedure for the Determination of Ketoamine Adducts in Proteins, J. Biol. Chem. 255:7218 (1980).
6. A.S. Acharya, L.G. Sussman, and J.M. Manning, Schiff Base Adducts of Glyceraldehyde with Hemoglobin: Differences in the Amadori Rearrangement at the α–Amino Groups, J. Biol. Chem. 258:2296 (1983).
7. A.S. Acharya, L.G. Sussman, and B.N. Manjula, Application of Reductive Dihydroxypropylation of Amino Groups of Proteins in Primary Structural Studies: Identification of Phenylthiohydantoin Derivative of ε–Dihydroxypropyllysine Residues by High–Performance Liquid Chromatography, J. Chromatogr. 297:37 (1984).
8. L.R. Lundblad and M.C. Noyes, Chemical Reagents for Protein Modification, Vol 1, CRC Press, Inc., Boca Raton, Florida.
9. K.F. Geoghegan, D.M. Ybarra, and R.E. Feeney, Reversibile Reductive Alkylation of Amino Groups in Protein, Biochemistry 18:5392 (1979).
10. G.E. Means and R.E. Feeney, Reductive Alkylation of Amino Groups in Proteins, Biochemistry 7:2192 (1968).
11. W.S.D. Wong, M.M. Kristjansson, D.T. Osuga, and R.E. Feeney, 1–Deoxyglycitolation of Protein Amino Groups and Their Regeneration by Periodate Oxidation, Int. J. Pep. Prot. Res. 26:55 (1985).

ENZYMATIC CARBOXYLMETHYLATION AND NON-ENZYMATIC DEMETHYLATION AT L-ISOASPARTYL RESIDUES: POSSIBLE APPLICATIONS IN PEPTIDE CHEMISTRY

E. David Murray, Jr.[1]

Department of Chemistry and Biochemistry and the
Molecular Biology Institute,
University of California, Los Angeles, CA 90024

INTRODUCTION

Protein carboxyl methyltransferases from human erythrocytes and bovine brain catalyze the methylation of abnormal aspartyl residues in cellular proteins and synthetic peptides (1-2). These sites of methylation appear to include the β-carboxyl group of racemized aspartyl residues in which the α-carbon is in the D-configuration (3) and the α-carboxyl group of isomerized L-aspartyl residues, in which the peptide chain is linked via the side chain β-carboxyl group (4-5). The function of these methyltransferases is not known, but it has been proposed that these enzymes play a surveillance role in the cell to identify abnormal proteins, and may participate in their degradation or rejuvenation.

The development of a synthetic peptide substrate for eucaryotic protein carboxyl methyltransferases has provided an approach to understanding the function of these enzymes. This peptide substrate, L-Val-L-Tyr-L-Pro-L-isoAsp-Gly-L-Ala, is stoichiometrically methylated at the free α-carboxyl group of the isoaspartic acid residue by both the red blood cell and brain methyltransferases (5). Substrates previously identified for these enzymes were proteins and peptides in which only a small subpopulation of the molecules contained altered aspartyl residues capable of acting as substrates (6-10). It was not possible to investigate the metabolism of these methyl esterified proteins, labelled with a radioactive methyl group, because the subpopulation of molecules which contains D-aspartyl and/or L-isoaspartyl residues becomes lost among the much larger population of unmodified proteins following the demethylation reaction. However, the demethylation and subsequent metabolism of stoichiometric methyl-accepting peptides can be tested.

It will be shown that the demethylation reaction of L-isoaspartyl methyl ester residue containing peptide leads to the formation of an intramolecular succinimide ring which, during mild base treatment, opens to yield approximately 25% normal aspartyl residue containing peptide. The potential uses of these methylation and demethylation reactions in peptide and protein chemistry will be addressed.

---

[1] Present address: International Genetic Engineering, Inc.,
1545 17th Street, Santa Monica, CA 90404

Synthetic Peptides. Peptides corresponding to the sequences L-Val-L-Tyr-L-Pro-L-Asp-Gly-L-Ala and L-Val-L-Tyr-L-Pro-D-Asp-Gly-L-Ala in which the aspartyl residues are linked to the glycyl residues either via the α-carboxyl group (normal peptide), β-carboxyl group (isopeptide) or both carboxyl groups (succinimide) were synthesized, purified and characterized previously (5).

High Performance Liquid Chromatography. The system for liquid chromatography consisted of two Waters M-45 pumps, an M660 solvent programmer, a U6K injector, and a model 441 absorbance detector (214 nm). The reverse phase C_{18} columns (4.6 x 250 mm, 5μ m spherical resin) were Econosphere columns from Alltech. Solvent A was 0.1% trifluoroacetic acid in 10% water: 90% acetonitrile. These solvents were prepared using a 10% stock of Pierce Sequenal grade trifluoroacetic acid as described previously (5). The column was equilibrated at room temperature in solvent A. Samples were eluted at 1 ml/min by a linear gradient of 0-40% solvent B in 40 min. This system can resolve the related hexapeptides containing L-isoAsp residues (25 min elution time), L-Asp residues (26.5 min), L-succinimide residues (29 min), and L-isoAsp α-methyl ester residues (30 min)(see below).

Preparation of Erythrocyte Cytosol as a Source or Carboxyl Methyltransferase. Erythrocyte cytosol at a protein concentration of about 40-60 mg/ml was prepared from washed cells by lysis in 4 volumes of 4.5 mM sodium phosphate, 4.5 mM disodium EDTA, 13.5 mM β-mercaptoethanol, and 10% glycerol at pH 8.0 as described (5).

Vapor Diffusion Assays for [3H]Methanol.

Procedure A. Sample (80 μl) is mixed with an equal volume of 0.6 M sodium borate, 1% sodium dodecyl sulfate (pH 10.2) in a 400 μl polypropylene microfuge tube. This tube is placed uncapped in a larger vial containing 2.4 ml of liquid scintillation cocktail as described (5). The amount of [3H]methanol in each sample is determined from the radioactivity which is found in the liquid scintillation fluid after 24 hours at room temperature, after a correction for the efficiency of methanol transfer (about 30%) is made (5).

Procedure B (Filter Paper Assay). A new variation on the methanol vapor diffusion assay for protein carboxyl methyltransferase activity was recently described by Macfarlane (12). Briefly, a base-quenched sample containing protein carboxyl methyl esters is applied to a piece of thick filter paper which is then placed in the neck of a capped scintillation vial containing scintillation fluid. The vial is left undisturbed for several hours at room temperature to allow the radiolabelled methanol to diffuse from the filter paper into the scintillation fluid. After the filter paper is removed, the radioactivity in the vial is determined. Sepcificially, an aliquot (10 μl) of sample is mixed with 30 μl 0.2 M NaOH, 1% SDS, and then immediately applied to a 1x8 cm strip of accordian-pleat-folded filter paper backing for slab gels (Bio-Rad Cat. No. 165-0921). The paper is then inserted into the neck of a 22 ml plastic scintillation vial containing 10 ml of ACS II counting fluid (Amersham). The vial is capped, and after 2 hours, the folded strip of paper is removed and the vial recapped and counted. Control experiments with [3H]methanol and various buffers show that for a volume of 40 μl spotted on the paper, transfer of [3H]methanol is essentially complete (100% yield) after 2 hours (13).

RESULTS

The hexapeptides discussed in this work, corresponding to the sequence L-Val-L-Tyr-L-Pro-L-Asp-Gly-L-Ala in which the aspartyl residues are involved in either normal peptide bonds, isopeptide linkages in which the aspartyl residue is linked to the glycyl residue via the β-carboxyl group, or succinimide derivative were synthesized and characterized previously (5; Fig. 1).

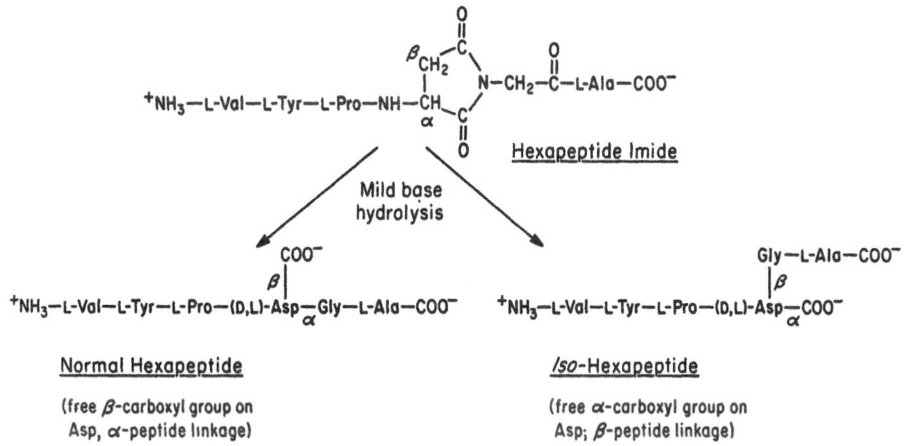

$$
{}^{+}NH_3-L\text{-}Val-L\text{-}Tyr-L\text{-}Pro-NH-\underset{\alpha}{CH}\underset{CH_2}{\overset{\beta}{\big|}}\cdots C(=O)-N-CH_2-C(=O)-L\text{-}Ala-COO^{-}
$$

Hexapeptide Imide

Mild base hydrolysis

$$
{}^{+}NH_3-L\text{-}Val-L\text{-}Tyr-L\text{-}Pro-(D,L)\text{-}\underset{\alpha}{Asp}\overset{\overset{COO^{-}}{\underset{\beta}{|}}}{}-Gly-L\text{-}Ala-COO^{-}
$$

Normal Hexapeptide

(free β-carboxyl group on Asp, α-peptide linkage)

$$
{}^{+}NH_3-L\text{-}Val-L\text{-}Tyr-L\text{-}Pro-(D,L)\text{-}\underset{\alpha}{Asp}\overset{\overset{Gly-L\text{-}Ala-COO^{-}}{\underset{\beta}{|}}}{}-COO^{-}
$$

Iso-Hexapeptide

(free α-carboxyl group on Asp; β-peptide linkage)

Figure 1. Structures of the synthetic hexapeptides used in these studies.

Demethylation of purified L-isohexapeptide methyl ester in buffer occurs at the same rate as demethylation in erythrocyte extracts. Previous studies have shown that the peptide L-Val-L-TyrL-Pro-L-isoAsp-Gly-L-Ala is a high affinity, stoichiometric methyl-accepting substrate of the erythrocyte carboxyl methyltransferase (5). These properties of the peptide make it possible to study the demethylation and subsequent reactions of the methylated peptide. The reaction pathway is shown in Fig. 2. Radiolabelled L-isohexapeptide methyl ester can be purified from an erythrocyte cytosol methylation mixture by HPLC. Incubation of the purified L-isohexapeptide methyl ester in different pH buffers would give a direct measure of k_2 in the absence of any erythrocyte enzymes. A comparison of values, thus obtained, with those from previous demethylation experiments performed in the presence of erythrocyte cytosol would provide information as to whether a methylesterase activity exists in the erythrocyte cytosol. The half-times of demethylation of the HPLC purified L-isohexapeptide methyl ester in buffers of pH 6.0 and 7.4 are 39.7 min and 4.3 min, respectively. These times are essentially identical to those determined by other methods in erythrocyte cytosol (Table I). The rate of demethylation in buffer is very similar to that in cytosol at the same pH, thus providing no evidence for the existence of a methylesterase activity (13).

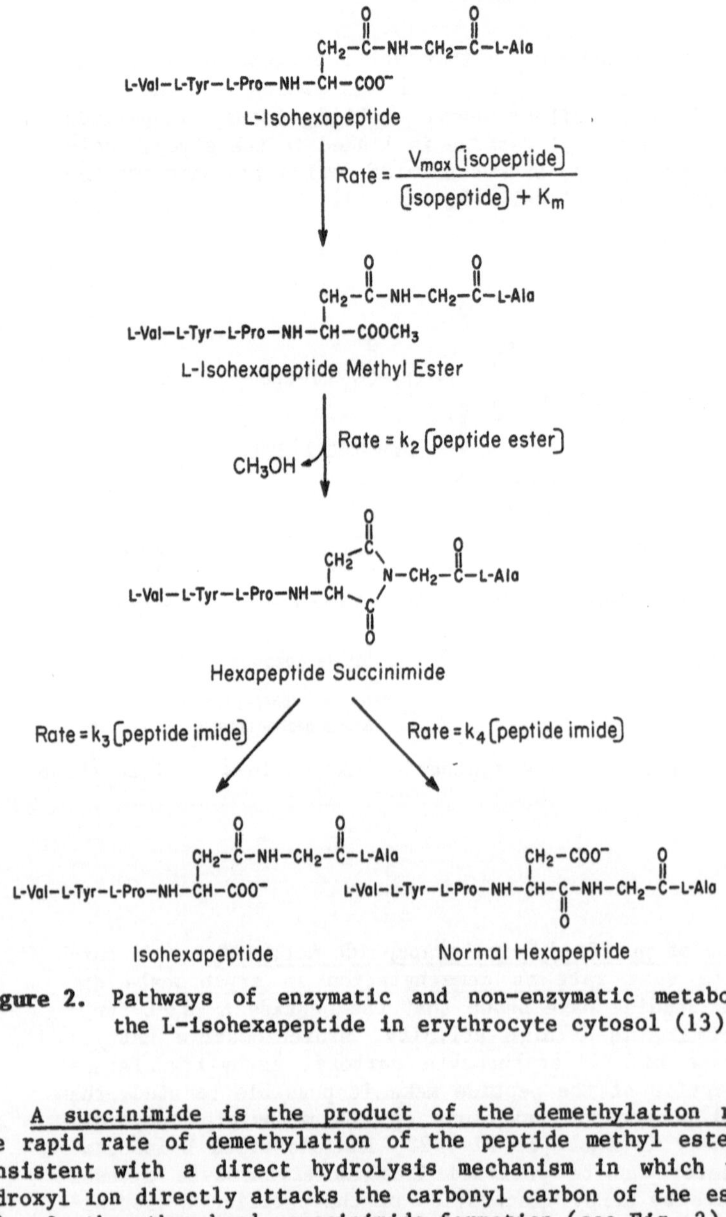

Figure 2. Pathways of enzymatic and non-enzymatic metabolism of the L-isohexapeptide in erythrocyte cytosol (13).

A succinimide is the product of the demethylation reaction. The rapid rate of demethylation of the peptide methyl ester is not consistent with a direct hydrolysis mechanism in which water or hydroxyl ion directly attacks the carbonyl carbon of the ester (14, 15). On the other hand, succinimide formation (see Fig. 2) has been observed with other types of peptide esters and the hexapeptide succinimide appears to be the immediate demethylation product of the hexapeptide methyl ester in these studies. As we have previously characterized the HPLC elution position of the succinimide form of this peptide (5), we performed the experiment shown in Fig. 3 to separate the products of the methylation and demethylation reactions (13). Here, the isohexapeptide was incubated with erythrocyte cytosol and non-isotopically labeled AdoMet. The composition of the incubation mixture was characterized at various times by HPLC, which provides excellent separation of the hexapeptide methyl ester, succinimide, and iso- and normal hexapeptide forms. The data in Fig. 3 shows an initial conversion of the L-isohexapeptide to the methyl ester followed by conversion to a product which elutes in the expected position of the L-succinimide (5). We have confirmed this

Table I

Comparison of Half-Times for the Demethylation and Succinimide Ring-
Opening Reactions of the Hexapeptide Under Various Conditions

pH	Reaction	Medium	t 1/2 min
6.0	Demethylation	Cytosol	39.4[a], 41.8[b]
		Buffer	39.7[c]
	Succinimide ring hydrolysis	Buffer	1660[d]
7.4	Demethylation	Cytosol	3.6[a]
		Buffer	4.3[c]
	Succinimide ring hydrolysis	Buffer	180[d]

[a] Determined from the rate of [3H]methanol production in erythrocyte cytosol in which 1μ M L-isohexapeptide and 13.4 μM [3H] AdoMet were added and free and peptide-bound [3H]methanol (as esters) were distinguished by lyophilization (data not shown).

[b] Determined from the rate of succinimide formation (data of Fig. 3)

[c] Determined by the rate of [3H]methanol production (Macfarlane assay procedure B) from pH 6.0 or 7.4 buffers.

[d] Determined by the production of L-isoaspartyl and L-aspartyl hexapeptides from the L-succinimide as described herein. At pH6, 130 μl of 1.9 mM hexapeptide imide in H_2O was mixed with 370 μl of 0.2 M sodium succinate, pH 6.0 (the final pH was measured at 6.0), and incubated at 37°C. At various time intervals, aliquots (50 μl) were withdrawn and added to 50 μl of a quench solution composed of 4% trifluoroacetic acid. For the pH 7.4 incubation, 400 μl of 0.28 mM hexapeptide imide in H_2O was mixed with 400 μl 0.2 M sodium HEPES , pH 7.4, (final pH was 7.4) and placed in a 37°C bath. At various times, 40 μl aliquots were withdrawn and quenched as described above. The amount of hexapeptide imide remaining was analyzed in each case by HPLC as described above.

identification by isolating the succinimide peak from a 180 min time point of an experiment similar to that shown in Fig. 3. After hydrolysis in 225 mM sodium borate, pH 10.2 for 2 min at 37°C, rechromatography on HPLC revealed two new peaks which eluted in the expected position of the isohexapeptide and the normal hexapeptide. The ratio of absorbance in these peaks of 3.6:1 (isopeptide:normal peptide) is similar to the value of 3.0:1 obtained when the purified hexapeptide succinimide is hydrolyzed (13)(see below).

Hydrolysis of the hexapeptide succinimide. To compare the rate of ring-opening of the L-hexapeptide imide to the demethylation rate of the peptide ester, the hydrolysis of the succinimide to the normal and isopeptide in buffer at pH 6 and pH 7.4 was measured. The relative concentrations of these products were measured by the area of their absorbance peak on HPLC. At pH 6.0, the imide opens with a half-time of 1660 min and at pH 7.4 with a half-time of 180 min (Table I). At both pH values, the products consist of 75% isopeptide and 25% normal peptide.

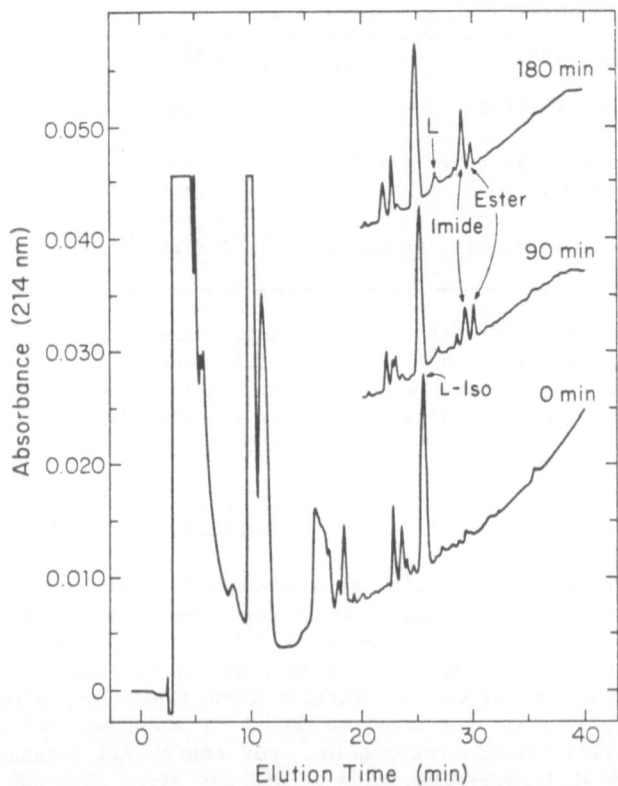

Figure 3. Time course of methylation and demethylation of the L-isohexapeptide in erythrocyte cytosolic extracts as followed by HPLC. L-iso-hexapeptide was methylated using erythrocyte cytosol with non-isotopically labelled AdoMet. The AdoMet stock is prepared as described previously (5) except that the volume of [³H]AdoMet is replaced with an equal volume of 10 mM HCl. In these experiments, one large incubation mixture was prepared and time points were withdrawn from it for quenching. The mixture was made up by mixing in a 1.5 ml microfuge tube 100 μl of 100 μM L-isohexapeptide, 200 μl of 13.3 μM AdoMet, and 400μM sodium citrate, pH 6.0, at 0°C, and the incubation was initiated by addition of 300 μl erythrocyte cytosol and placing the mixture in a 37°C water bath. Aliquots (100 μl) were withdrawn at timed intervals and quenched by addition to a 1.5 ml microfuge tube containing 100 μl of 10% (w/v) trifluoroacetic acid. These samples were analyzed by HPLC as described in "Experimental Procedures", and peaks were identified by co-chromatography with purified peptide standards (13).

DISCUSSION

Non-enzymatic demethylation-succinimide formation at L-isoaspartyl α-methyl ester residues. These results indicate that there is apparently not an esterase activity in erythrocyte cytosol (Table I, 13) and that the L-isoaspartyl α-methyl ester demethylates to form a succinimide derivative of the peptide (Fig. 5, 13). The peptide ester apparently demethylates to the succinimide in both buffer and cytosol as the rates for this reaction are the same in both systems (Table I, 13). The generation of a succinimide from an L-isopartyl residue via methylation/demethylation is potentially useful in that treatment of this succinimide with mild base yields products which are 75% isoaspartyl and 25% aspartyl residues (13). The normal aspartyl product is now restored to its natural configuration and can undergo reactions characteristic of aspartyl residues.

Applications of Methylation and Subsequent Succinimide Formation Involving L-Isoaspartyl Residues. The results described above may be useful to the peptide chemist, especially a sequencer. It is known that isoaspartyl residues are not sequencable in the Edman degradation reaction system used in most sequencers. If, during the sequencing of a peptide, the sequencing stops and it is known from the amino acid composition that one or more aspartyl residues are still to be sequenced, it would be reasonable to check for isoaspartyl residues.

The presence of an isoaspartyl residue in a peptide could be determined analytically by methylation of the peptide with either erythrocyte cytosol (5) or partially purified brain methyltransferase (4) and [^{3}H]AdoMet. If base-labile radioactivity is incorporated into the peptide, this would be indicative of an isoaspartyl residue. This would then provide a possible explanation for the termination of sequencing in the middle of the peptide.

If sufficient quantities of the peptide are available (>10 nmole), it would be reasonable to attempt to repair some of the peptide via a methylation-demethylation-succinimide hydrolysis scheme, followed by purification (via HPLC) of the peptide containing aspartyl residues in the correct peptide bond. There are various ways in which this could be done, but the most direct is to incubate the peptide in the presence of methyltransferase and AdoMet and allow the peptide to methylate and demethylate producing the succinimide derivate of the peptide which could then be isolated. The succinimide-containing peptide could then be treated with base to hydrolyze the imide resulting in approximately 75% isoaspartyl containing peptide. The peptide with the normal aspartyl residue would then be isolated and sequenced or the peptide mixture sequenced directly.

Also, even though solid phase peptide synthesis methods have been developed to reduce or eliminate succinimide formation at aspartyl residues followed by glycine, alanine, or serine (16), the above method provide a way of changing the isoapartyl peptide side products that result from any ring-opening back to normal aspartyl containing peptides.

ACKNOWLEDGEMENTS

During the period of this work, the author was supported in part by U.S. Public Health Service Training Grant GM 07185, while obtaining his doctorate in Dr. Steven Clarke's laboratories at UCLA, which were funded by National Institutes of Health Grants GM26020 and EY 04912 and by a grant-in-aid from the American Heart Association, with funds contributed in part by the Greater Los Angeles Affiliate. The author's attendance at the symposium was funded by International Genetic Engineering, Inc.

REFERENCES

1. Clarke, S. (1985) Ann. Rev. Biochem. 54, 479–506.
2. O'Connor, C.M., Aswad, D.W., and Clarke, S. (1984) Proc. Natl. Acad. Sci. U.S.A. 81, 7557–7564.
3. McFadden, P.N. and Clarke, S. (1983) Proc. Natl. Acad. Sci. USA 81, 7557–7564.
4. Aswad, D.W. (1984) J. Biol. Chem. 259, 10714–10721.
5. Murray, E.D., Jr. and Clarke, S. (1984) J. Biol. Chem. 259, 10722–10732.
8. O'Connor, C.M. and Clarke, S. (1983) J. Biol. Chem. 258, 8485–8492.
9. Kim, S. and Li, C.H. (1979) Int. J. Pept. Protein Res. 13, 281–285.
10. Barber, J.R. and Clarke, S. (1983) J. Biol. Chem. 258, 1189–1196.
11. McFadden, P.N., Horwitz, J. and Clarke, S. (1983) Biochem. Biophys. Res. Commun. 113, 418–424.
12. Macfarlane, D.E. (1984) J. Biol. Chem. 259, 1357–1362.
13. Murray, E.D., Jr. and Clarke, S. (1985) J. Biol. Chem. 261, (in press).
14. Bernhard, S.A. (1983) Anal. N.Y. Acad. Sci. 421, 28–40.
15. Bernhard, S.A., Berger, A., Carter, J.H., Katchalski, E., Sela, M., and Shaltin, Y. (1962) J. Amer. Chem. Soc. 84, 2420–2434. 2586.
16. Stewart, J.M. and Young J.D. (1984) Solid Phase Peptide Synthesis, Second Edition, Pierce Chemical Co., Rockford, Il.

IV. Amino Acid Analysis of Polypeptides

A PRACTICAL GUIDE TO THE GENERAL APPLICATION OF PTC-AMINO ACID ANALYSIS

Robert L. Heinrikson* +, Rene Mora#, and John M. Maraganore*

* The Upjohn Company, Biopolymer Chemistry, 301 Henrietta St., Kalamazoo, MI, 49001
# The University of Chicago, Department of Biochemistry and Molecular Biology, 920 East 58th St., Chicago, IL, 60637
+ Author to whom correspondence should be addressed

The development and widespread application of reverse-phase high-performance liquid chromatography (RP-HPLC) has made a profound impact on the isolation of biological compounds. In protein chemistry alone, literally all phases of the work, from purification of proteins and peptides to analysis of phenylthiohydantoin (PTH) amino acids generated in sequencing, now depend on HPLC procedures. RP-HPLC is posing, as well, an ever more serious challenge to ion-exchange chromatographic procedures for quantitative amino acid analysis (1). The ascendancy of the new methodology follows from the high sensitivities possible from pre-column derivatization, coupled with advantages relative to the lower cost, simplicity, and adaptability of the equipment involved. Moreover, selection of appropriate reagents should provide amino acid derivatives with uniform color constants, thus simplifying systems for their detection and quantitation.

A wide variety of reagents for derivatization of the amino groups of amino acids have been described, some yielding fluorescent (2,3) or color (4,5) properties of the products to facilitate post-column detection with high sensitivity. At the time of this writing, it appears that the most popular pre-column derivatization technique involves the use of the "Edman reagent", phenylisothiocyanate (PITC) to yield the phenylthiocarbamyl (PTC) amino acids, and this will be the subject of the present chapter.

The choice of PITC is obvious; the reagent has been the touchstone of protein sequencing since its introduction by Edman over 30 years ago (6), and there is nothing novel about its application to amino acid analysis. Edman himself described the preparation of PTC-amino acids and showed that all of the amino acids react readily with PITC (7). Moreover, the description of numerous RP-HPLC systems for separating and quantifying the cyclized PTH-amino acids which have appeared continuously starting from the mid-1970's to the present time, held the promise for an even greater success with PTC derivatives, which do not undergo undesirable degradation during cyclization. Thus, in a real sense, the development of the PTC approach to amino acid analysis was a logical extension of developments which took place over several decades.

The first description of the PTC system as a general tool for amino acid analysis of protein and peptide hydrolyzates was published by Heinrikson and Meredith (1) in 1984. This paper gave specific details regarding the chemical derivatization of hydrolyzates, columns, instrumentation, solvent systems, and separation by RP-HPLC of the common and several less common amino acids and derivatives. The thrust of this work demonstrated the flexibility of the method and its general applicability in conjunction with a variety of commercially available HPLC systems and columns. After publication of this paper (1), we learned that Koop *et al.* (8) had earlier applied a similar PTC-amino acid analytical system to identify amino acids released by carboxy-peptidase Y cleavage of an isozyme of cytochrome P-450. Later in 1984, Waters Associates, Milford, MA., began to market a PTC-amino acid analyzer and work station termed the Pico-Tag System (9), and this effort has greatly stimulated interest in the procedure among protein chemists. In fact, several articles in this volume have been devoted to the subject of amino acid analysis by pre-column derivatization with PITC.

The present chapter is written to provide some helpful hints as to the set-up, operation, and trouble-shooting of a PTC-amino acid analysis system and is designed to be of general interest to those who wish to adapt their present HPLC system on a part-time or full-time basis for this application.

HYDROLYSIS OF PEPTIDE AND PROTEIN SAMPLES

Peptide and protein samples for hydrolysis and amino acid analysis using ion-exchange chromatography are required to be salt-free. Similar preparation of samples for RP-HPLC of PTC-amino acids is prefered but not required. In fact, we have found that as much as 4 M salt does not interfere with derivatization of hydrolyzed samples, coupling with PITC, and separation by HPLC. Clearly, salts containing primary or secondary amines must be avoided as such contaminants will react with PITC.

Prepared samples are applied to Corning borosilicate micro-tubes and then dried by evaporation or lyophilization. The micro-tube is then placed in a Pyrex or Kimex thick-walled tube (13 x 100 mm); alternatively, several micro-tubes can be placed in a single Pyrex ignition tube (18 x 150 mm). A small volume of acid (250 - 500 μl) is added to the bottom of the larger tube and, then, the holding tube is sealed *in vacuo*. A vacuum of <25 millitorr should be attained before the tube is sealed. Instrumentation for tube-sealing includes a vacuum pump and gauge, cold trap, and flame torch. One can also obtain a Pico-Tag Work Station, Waters Associates, Milford, MA., wherein all instrumentation for this purpose is provided.

Although liquid phase hydrolysis of protein or peptide samples has been applied with success to PTC-amino acid analysis by RP-HPLC, this approach tends to increase the number and quantity of contaminants. Instead, we suggest the use of vapor phase hydrolysis. Vapor phase hydrolysis of protein or peptide samples can be performed with numerous acids and conditions. Using 6 NHCl (constant boiling), the samples are hydrolyzed for 22 - 24 hrs at 110°C or for 1 hr at 150°C. Alternatively, a mixture of TFA:HCl (2:1, v/v) can be used in a 30 min hydrolysis at 160°C (10). After hydrolysis, the holding tube is opened and the micro-tubes are removed and then rinsed to remove acid from the outside walls, and wiped dry.

The hydrolysis tubes are dried thoroughly *in vacuo* for 20 min with mild heating (40°C) in order to remove any traces of acid. The removal of acid is a crucial step in the derivatization as trace contaminants will lead to cyclization of PTC-amino acids to the anilinothiazolinone- and phenyl-thiohydantoin-amino acids and the obstruction of accurate quantitation.

DERIVATIZATION OF PEPTIDE OR PROTEIN HYDROLYSATES WITH PITC

Instrumentation involved in the derivatization steps can be shared, in part, with that used for hydrolysis. In order to facilitate simultaneous derivatization of many samples at a time, it is suggested that the user obtain a Speed-Vac Concentrator, Savant Instruments. This unit can be attached by high-vacuum tubing to 1 - 2 cold traps containing ethanol and dry ice and these, in turn, are connected to a vacuum pump capable of pulling <50 millitorr. In order to avoid routine contamination of pump oil with organic solvents, we recommend use of a J. Lesker Molecular-Sieve trap, between cold trap and pump. Even with this precautionary measure, pump oil should be changed every 6 - 8 weeks.

After acid has been thoroughly removed from hydrolysates as described above, an aliqout (100 µl) of chasing buffer is added to the sample. This chasing buffer may consist of acetonitrile:pyridine:triethylamine:water (10:5:2:3, v/v) as we have previously described (1). Alternatively, mixtures of methanol:water:triethylamine (2:2:2, v/v) or ethanol:water:triethylamine (2:2:1, v/v) (9) can be used. In every case, purity of solvents is critical. Acetonitrile, ethanol, and methanol should be of HPLC grade. Triethylamine and pyridine, if used, should be of the highest grade commercially available (Sequanal grade, Pierce Chemical Co., Rockford, IL., or Gold Label, Aldrich Chemical Co., Milwaukee, WI.) and then redistilled. Water should be glass-distilled. Applications of the chasing buffer should be followed by vortexing and sonication in a sonicator bath, and, then, the sample can be dried *in vacuo* for 15 minutes.

Subsequent to drying, the samples are redissolved in coupling buffer (20-100 µl), vortexed, and sonicated. The coupling buffer may vary considerably depending upon choice; these include mixtures of: 1) acetonitrile:pyridine:triethylamine:water (10:5:2:3, v/v) with 5 µl of PITC (1); 2) methanol:water:triethylamine:PITC (7:1:1:1, v/v); or 3) ethanol: water:triethylamine: PITC (7:1:1:1, v/v) (9). PITC should be obtained through Pierce Chemical Co., Rockford, IL., (catalog # 26922) in 1 ml ampules. Reaction of hydrolysates with PITC is allowed to proceed for 15 min and, subsequently, coupled samples are dried *in vacuo* for 15 min. The **use of heat** in drying coupled samples is **not recommended** and could account for degradation of PTC-amino acids.

The dried, coupled samples are next treated with acetonitrile, a chasing solvent, in order to help remove organic contaminants generated prior to chasing and coupling steps. Acetonitrile (200 µl) is added to the sample and, then, following vortexing and sonication, the sample is dried *in vacuo* for 15 min; the same treatment is then repeated. Samples thus prepared should be stored, if necessary, at -20°C in the dark. Prior to injection of samples into the HPLC, they should be centrifuged to remove any insoluble materials. PTC-amino acids show some degradation at room temperature in 24 hours, but are relatively stable for this interval at 4°C. Therefore, refrigerated auto-injectors are recommended for programmed analyses.

Problems which may arise in derivatization of samples (Table II) can result from: 1) poor quality of solvents, especially pyridine and triethylamine; 2) inadequate removal of acid; 3) inadequate chasing of sample(s) with both chasing buffer and solvent; and 4) excessive exposure to heat. Coupling efficiency can be measured by comparing yields of PTC-Lys to other PTC-amino acids in standard hydrolysates using extinction coefficients for PTC-amino acids (1).

Several liquid chromatographic systems can be used successfully for the separation of PTC-amino acids but it is recommended that whatever unit is chosen, it should have a ternary solvent system capability. A binary solvent system can be applied, but may prove especially cumbersome and time-consuming. We have had good results with chromatographic systems from IBM (LC 9533), Varian (5060, with ternary option) and Hewlett-Packard (HP 1090). In all cases, reproducibility appears to depend on the quality of pre-column mixing of solvents. Of course, the Waters Pico-Tag System is available for dedicated applications.

We have tested several HPLC columns for separation of PTC-amino acids and have found best results both with respect to reproducibility and resolution with an IBM C-18 column, 5 micron particle size, 25 x 0.46 cm. Such a column is good for 400 - 500 analyses and, after such use, can be regenerated by removal of the top 1 mm of stationary phase and application and in-line packing of the column with additional beads supplied by the manufacturer. Regenerated columns are indistinguishable from new ones and may be used for another 400 - 500 analyses. Column life can also be lengthened by use of a Rheodyne pre-column frit, minimal volume, supplied by several companies. Several other HPLC columns have been applied to PTC-amino acid analysis, of which some are discussed in other chapters in this volume.

Ultraviolet detection systems are, generally, suitable to this application. We have had success with both variable wavelength detectors, at 250 nm, and fixed wavelength detectors, at 254 nm. Identification and quantitative estimation of PTC-amino acids can be achieved with a number of commercially-available recorder-integrators. A sophisticated system we have employed includes a personal computer compatible with the Nelson Analytical Software System. Such instrumentation allows, ultimately, for identification of retention times, peak areas, and nanomolar or picomolar quantitation of PTC-amino acids. Moreover, the same software can be applied throughout the laboratory for data collection from numerous chromatographic applications. The separation of PTC-amino acids requires a heating system (water-jacketed or oven) for the HPLC column. Such instrumentation should be capable of attaining a temperature of 52°C. Finally, it should be noted that laboratories with a large demand for amino acid analysis would benefit greatly from an automatic injection system. As mentioned above, this system should have a refrigeration capacity in order to avoid degradation of PTC-amino acids.

Several solvent systems have been applied with success in the separation of PTC-amino acids (ref. 1; see also other chapters in this volume). We would like to suggest the use of two separation systems, requiring 19 min and 35 min, the programs of which are described in Table I. In both solvent systems, solvent A consists of 0.05 M ammonium acetate, pH 6.0, containing 0.02% sodium azide; solvent B consists of 0.1 M ammonium acetate, pH 6.0, in 50:50 (v/v) acetonitrile:water, containing 0.02 % sodium azide; and solvent C is a mixture of acetonitrile and water (70:30, v/v). Solvents A and B should be titrated to pH 6.0 with glacial acetic acid, and, in the case of solvent B, this should be done at a volume near the final mixture of water and acetonitrile. Solvents A and B should be filtered with a Nucleopore Filtration Apparatus and a 0.2 μm polycarbonate filter. In the case of solvent B, the aqueous solution should be filtered prior to mixing with acetonitrile. We recommend that a stock solution of sodium azide be made and that concentrations of sodium azide in solvents A and B be adjusted by volume. During separation of PTC-amino acids, solvents should be under constant purge with helium.[1] The quality of ammonium acetate is critical both to the quality of separation achieved and to column life. We suggest the use of hydrated ammonium actetate from the Chemical Dynamic Co., Southplainfield, NJ., Ultralog grade, or the Gold Label product from Aldrich Chemical Co., Milwaukee, WI.. Similarly, quality solvents, (HPLC grade, *e.g.* , from Burdick-Jackson, Muskegon, MI), and glass-distilled water should be employed at every step.

Application of the instrumentation and solvent systems mentioned above will allow for the successful separation, identification, and quantitation of PTC-amino acids at the nanomolar to picomolar level. Shown in Fig. 1A is a standard separation of 500 picomoles of coupled amino acid calibration standard (Amino Acid Calibration Standard H, Pierce Chemical

[1]The Varian 5060 chromatographic system requires purging with helium only prior to use in separation of PTC-amino acids.

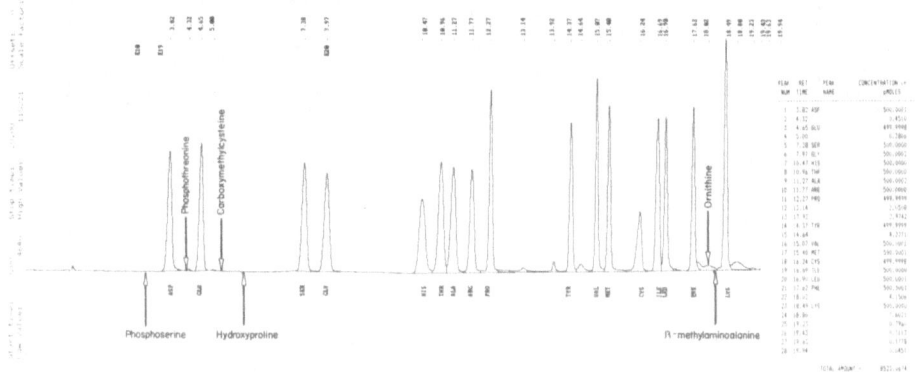

Figure 1A. Separation of PTC-amino acids (500 pmoles) by reverse-phase HPLC using the 19 minute solvent program and designation of retention times for several rare or pre-derivatized PTC-amino acids.

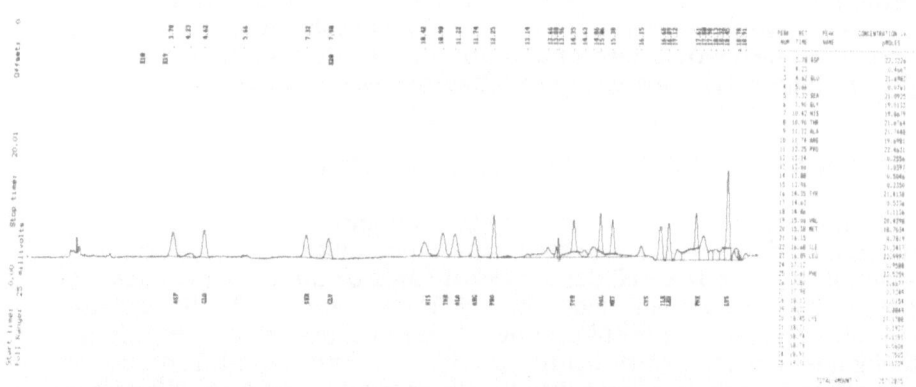

Figure 1B. Separation of PTC-amino acids (20 pmole) by reverse-phase HPLC using the 19 minute solvent program.

Co., Rockford, IL., catalog # 20089) using the 19 min solvent program. A separation at the 20 picomole level is shown in Fig. 1B. Samples dissolved in solvent A may be injected in a wide range of volumes from 1 to 200 µl. The column is pre-heated to 52°C and equilibrated for at least 30 min with solvent A prior to injection of the first sample. Subsequent reequilibration time is only 15 min. The flow rate during separation is kept constant at 2 ml/min. Application of either 19 or 35 min solvent programs and post-run reequilibration of 15 min allows for an amino acid analysis every 34 or 50 min. This represents a considerable improvement on the time involved in separation of amino acids by ion-exchange chromatography, while affording, as well, sensitivity capabilities in the low picomole range.

Table I - PTC-amino acid analysis: Solvent Programs

19 minute analysis				35 minute analysis			
TIME	%A	%B	%C	TIME	%A	%B	%C
0	100	0	0	0	100	0	0
8	85	15	0	15	85	15	0
17	34	66	0	30	50	50	0
19	34	66	0	35	50	50	0
19.1	0	0	100	35.1	0	0	100
21	0	0	100	40	0	0	100
21.1	100	0	0	40.1	100	0	0
35	NEXT INJECTION			55	NEXT INJECTION		

The analysis of a hydrolysate of pepsin is shown in Fig. 2. While PTC-amino acid analysis of peptide or protein hydrolysates involves the introduction of several contaminant peaks not encountered in analyses of PTC-standards, these do not interfere with the identification and quantitation of PTC-amino acids. It should be noted that PTC-amino acid analysis of pepsin shown in Fig. 2 corresponded well with known compositional data. Should contaminant peaks prove to interfere with PTC-amino acids, we would suggest the use of the 35 min solvent program. More likely, however, such would be an indication of solvent or reagent impurity, or faulty technique in chasing and derivatization steps (Table II).

IDENTIFICATION OF MODIFIED AND UNUSUAL AMINO ACIDS

PTC-amino acid analysis by RP-HPLC can also be applied to the identification and quantitation of several modified and rare amino acids (1). The retention times of the PTC-derivatives of some of these amino acids are designated by arrows in Fig. 1A. These amino acids include carboxymethylcysteine, phosphoserine, phosphothreonine, ornithine, β-methylaminoalanine, and hydroxyproline. The identification of hydroxyproline by ion-exchange procedures presents severe complications not present in PTC-amino acid analysis. We have used a modified solvent program in the identification and quantitation of a rare, tri-amino acid, hypuscine. Finally, while not shown and applied to the present solvent programs described herein, RP-HPLC of PTC-amino acids can be used in identification and quantitation of homoserine and homoserine lactone and tryptophan (when methanesulfonic acid hydrolysis is used). Moreover, yields of homoserine and its lactone are improved in the current system by comparison to ion-exchange chromatography as the pH is maintained constant.

CONCLUSIONS

Perhaps the greatest advantage of PTC-amino acid analysis by RP-HPLC, aside from sensitivity to the picomolar range, is the flexibility of the system and its use in general applications. Indeed, amino acid analysis need not involve expensive and dedicated instrumentation. Liquid chromatographic systems are commonplace in today's laboratory and these same general tools for purification and identification can now be applied, when necessary, to amino acid analysis.

We entitle this chapter a **practical guide** in order to allow interested investigators ease in the set-up of PTC-amino acid analysis. Accordingly, we have prepared a trouble-shooting table (Table II) which serves to summarize precautionary measures and to suggest causes of and solutions to problems which may be encountered.

The original description of amino acid analysis by pre-column derivatization with PITC and use of RP-HPLC (1) followed from more than 30 years experience in the use of PITC together with the advent of HPLC. The clear advantages of this method over alternative pre-column and post-column derivatization methods has been discussed previously at length (1). PTC-amino acid analysis was described but a year ago, and already there are several modifications of the original method with respect to: 1) use of different columns and solvent programs to increase speed, reproducibility, and resolution; 2) use of different systems for hydrolysis, coupling, and chasing in order to eliminate contaminants, increase efficiency of derivatization, and avoid side reactions; and, finally, 3) use of PITC-analogs to increase sensitivity. It is clear, then, that current methodology represents a step toward a better means of achieving amino acid analysis; improvements will follow from advances in the chemistry of phenylthiocarbamylation and the application of microbore HPLC technology. Future research in this area

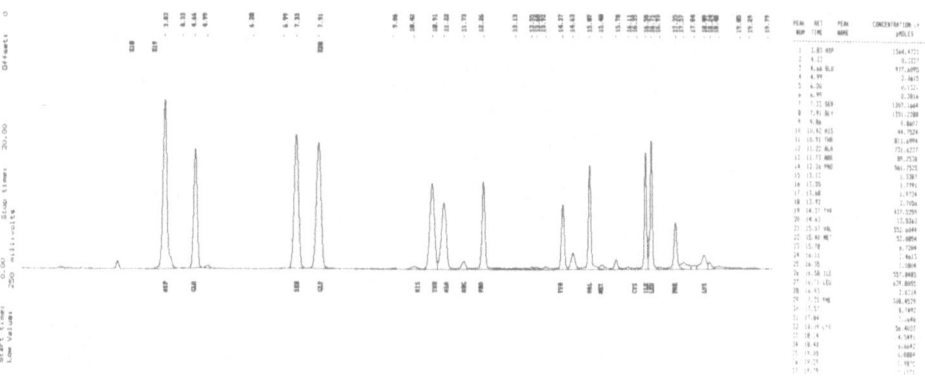

Figure 2 - PTC-Amino acid analysis of a pepsin hydrolysate at the picomolar level using the 19 minute solvent program.

193

will, no doubt, lead to the discovery of new reagents for modification of amino acids, the derivatives of which will allow for greater sensitivity and ease in separation.

TABLE II - Trouble-shooting with PTC-amino acid analysis

Symptom	Probable Cause(s)	Solution(s)
1. Doublet peaks; poor resolution	a. Cyclization to PTH due to heating, aging, or presence of acid during coupling b. Dead space or contamination at top of column	a. Avoid heat or chase acid more extensively b. Clean and top-off column
2. Baseline shifts	a. Solvent impurity b. Imbalance of NaN_3 in solvents A & B	a. Use high-quality solvents b. Use a NaN_3 stock soln. to adjust concentrations
3. Poor reproducibility	a. Improper reequilibration b. Pump problem-air bubbles, piston seals, etc.	a. Equilibrate longer b. Purge line, replace seals
4. Low yields of Lys and other di-PTC amino acids	a. Aged or impure coupling buffer	a. Prepare new buffer or redistill buffer solvents
5. Low yields of Asp & Glu; contaminant near His is large	a. Inadequate drying and chasing of acid b. Excessive heating during evaporation	a. Dry down & chase acid more b. Avoid heat after coupling
6. Contaminant peaks after Met and postrun are large	a. Inadequate chasing & drying of coupling buffer	a. Dry longer or chase more
7. Large peaks after Pro and before Tyr	a Sugar or lipid contaminant	
8. Large peak at Ile	a High salt or other contaminants	a. Prepare a salt-free or clean sample

9. Degradation of vacuum	a. Contamination of pump oil:
	  i. not protected
	  ii. oil needs change
		i. Use filter
		ii. Change oil

REFERENCES

1. Heinrikson, R.L. and Meredith, S.C. (1984) Anal. Biochem., **136**, 65-74.
2. Benson, J.R., and Hare, P.E. (1975) Proc. Nat. Acad. Sci. USA, **72**, 619-622.
3. Cronin, J.R., and Hare, P.E. (1977) Anal. Biochem., **81**, 151-156.

4. Tapuhi, Y., Schmidt, D.E., Lindner, W., and Karger, B.L. (1981) <u>Anal. Biochem.</u>, **115**, 123-129.
5. Chang, J.Y., Martin, P., Bernasconi, R., and Braun, D.G. (1981) <u>FEBS Lett.</u>, **132**, 117-120.
6. Edman, P. (1950) <u>Acta. Chem. Scand.</u>, **4**, 277-283.
7. Isle, D. and Edman, P. (1962) <u>Aust. J. Chem.</u>, **16**, 411-41.
8. Koop, D.R., Morgan, E.T., Tarr, G.E., and Coon, M.J. (1982) <u>J. Biol. Chem.</u>, **257**, 8472-8480.
9. Bidlingmeyer, B. A., Cohen, S. A., and Tarvin, T. L. (1984) <u>J. Chromat.</u>, **336**, 93-104.
10. Tsugita, A., and Scheffler, J.J. (1982) <u>Eur. J. Biochem.</u>, **124**, 585-588.

HIGH PERFORMANCE LIQUID CHROMATOGRAPHY OF PHENYLTHIOHYDANTOIN AND PHENYLTHIOCARBAMYL AMINO ACIDS

Barry N. Jones, Angelo P. Consalvo, Lisa LeSueur, Susan Lovato, Stanley D. Young, James A. Koehn and James P. Gilligan

Unigene Laboratories, Inc.
110 Little Falls Road
Fairfield, NJ 07006

INTRODUCTION

Microsequence analysis of polypeptides by the Edman procedure has become a standard and routine method for the protein chemist (1,2). This cyclic stepwise process is initiated by the reaction of phenylisothiocyanate (PITC) with the amino terminus of a polypeptide. Following this coupling step, the derivatized amino-terminal residue is selectively cleaved from the polypeptide and converted to a phenylthiohydantoin (PTH) amino acid. Various reverse-phase high performance liquid chromatography (HPLC) systems have been reported for the identification and quantitation of the resulting PTH amino acids (3-8). Recently, methods have been described which also use PITC derivatization for amino acid analysis (9,10). This procedure involves the reaction of PITC with amino acids to form phenylthiocarbamyl (PTC) derivatives and their subsequent identification and quantitation by reverse-phase HPLC.

Since both of the above procedures require HPLC analysis, a rapid and sensitive chromatography system was developed which utilizes the same reverse-phase column, flow rate, column temperature, and elution buffers for either type of analysis. This report presents a description of this system as well as its application to the structural analysis of polypeptides.

EXPERIMENTAL

Materials

Phenylisothiocyanate, triethylamine, constant boiling hydrochloric acid, PTH-amino acid standards, and standard mixtures of amino acids (2.5 nmol/μL each) were obtained from Pierce. High purity water was obtained with a system from Hydro Service and Supplies. Acetonitrile was HPLC-grade (EM Science) and used without further treatment. HPLC-grade glacial acetic acid and sodium acetate were purchased from J.T. Baker. All other chemicals were reagent grade.

Microsequence Analysis of Polypeptides

Amino-terminal microsequence analyses were performed with an Applied Biosystems Model 470A protein sequencer (1,2). Dried PTH-amino acid samples were reconstituted with a 30-µL solution of water: acetonitrile, 2:1. Dansyl amide (20 pmol/µL) was incorporated into the reconstitution solution as an internal standard.

Hydrolysis of Polypeptides

For acid hydrolysis, polypeptides were hydrolyzed with HCl/water vapor from constant boiling HCl for 1 hr. at 150°C. This procedure was performed in a Waters Pico-Tag Workstation as described by Bidingmeyer et al. (10). The HCl solution was not modified by the addition of phenol, mercaptoethanol, or thioglycolic acid. For each set of hydrolysis samples, a mixture of amino acids (25 nmol each) was subjected to the identical hydrolysis conditions for use as a calibration standard.

For total enzymatic hydrolysis, peptides were dissolved in 200 mM N-methylmorpholine acetate, pH 7.0, and digested with aminopeptidase M according to the method of Jones and Gilligan (11). This procedure is routinely employed to determine the acid/amide and tryptophan content of small peptides.

Derivatization of Hydrolysates with PITC

Derivatizations were performed in a Waters Pico-Tag Workstation according to the procedure of Bidingmeyer et al. (10). Following derivatization, samples were redissolved in 20-250 µL of 150 mM sodium acetate, pH 6.5. This buffer solution contained benzoic acid (250 pmol/µL) as an internal standard to correct for variations in injection volumes.

Chromatography System

The HPLC system was a Hewlett-Packard 1090 liquid chromatograph equipped with a binary solvent delivery system (DR5), an autosampler, a heated column compartment, and a programmable filter-photometric detector. The detector was fitted with interference filters for monitoring column effluents at either 250, 269, or 313 nm. Gradients were generated by a Hewlett-Packard 85 controller. Samples were injected in volumes ranging from 1 to 25 µL using a Hewlett-Packard variable-volume auto-injector. Chromatographic peaks were recorded and integrated by a Hewlett-Packard 3392A integrator system. The chromatographic unit consisted of a Shandon Hypersil ODS column (5 µm particle size, 25 cm x 4.6 mm I.D.) operated at 50°C and at a flow rate of 1.2 mL/min. Gradients were formed between two degassed solvents. Solvent A was 150 mM sodium acetate, pH 6.5, and Solvent B was acetonitrile. PTC-amino acids were detected at 250 nm. A combination of 269 and 313 nm detection was employed for PTH-amino acid analysis. Two separate gradient programs were used to resolve the PTC-amino acids. For amino acid mixtures obtained from acid hydrolysates, the column was equilibrated with 7%B. After injection, a linear step to 18%B in 8 min. followed by a linear step to 36%B in 5.5 min. was used. For amino acid mixtures obtained from enzymatic digests, the column was equilibrated with 4%B. After injection, a linear step to 16%B in 8 min. followed by a linear step to 38%B in 6.5 min. was employed for separation. A single gradient program was used for all PTH-amino acid analyses. The column was equilibrated at 17%B, then a linear gradient to 49%B over 15 min. was used following sample injection. At the end of each gradient program, a washing step with 70-90%B was employed to remove any residual sample components. The analysis time (sample to sample) for all three programs is approximately 20 min.

RESULTS AND DISCUSSION

Resolution of PTC- and PTH-Amino Acid Mixtures

Typical chromatograms demonstrating the resolution of PITC-derivatized amino acid mixtures are presented in Figure 1. As can be seen from this figure, all of the PTC derivatives are well-resolved. However, as the number of analyses increases, PTC-Arg and PTC-Thr begin to co-elute. Resolution of these two derivatives can be restored by increasing the ionic strength of Solvent A. Thus, for a new column the sodium acetate concentration in Solvent A is 150 mM. After 700-800 analyses, the sodium acetate concentration needed to maintain resolution is 200 mM. The reverse-phase column is routinely discarded after 1000 analyses. It should also be noted from Figure 1 that most of the individual amino acids are stable to the hydrolysis conditions of 150°C for 1 hr. with the exception of serine, threonine, tyrosine, and cystine. These amino acids are partially destroyed during hydrolysis with typical yields of 84%, 89%, 85%, and 56%, respectively. This approximates the destruction observed for these amino acids during acid hydrolysis of polypeptides. Thus, the hydrolyzed amino acid standard can be routinely used for calibration. Alternatively, if the unhydrolyzed standard is used for calibration, correction factors to compensate for destruction can be used; however, the hydrolysis time and temperature must be kept constant (12).

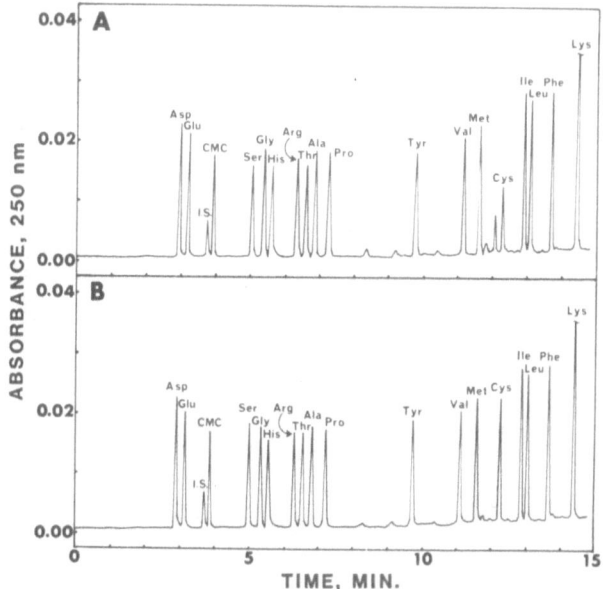

Figure 1: Separation of PTC-amino acids on a Hypersil ODS column (5 μm, 4.6 x 250 mm) using the gradient for acid hydrolysates (see section on "Chromatography System"). A, Standard mixture of amino acids (10 μL, 25 nmol each) dried, subjected to acid hydrolysis for 1 hr. at 150°C, derivatized with PITC, and dissolved in 250 μL Solvent A containing 250 pmol/μL benzoic acid as internal standard (I.S.). Chromatogram obtained following 2 μL injection (200 pmol each). B, Amino acid standard (25 nmol each) derivatized directly with PITC without being exposed to acid hydrolysis conditions. Chromatogram obtained following 2 μL injection (200 pmol each).

The resolution of a mixture of PTH-amino acids is shown in Figure 2A. Sufficient separation is obtained to permit accurate identification and quantitation of the PTH derivatives. With column use, the retention time of PTH-Arg approaches that of PTH-Gly. This is corrected by increasing the ionic strength of Solvent A. Fortuitously, the increase in ionic strength needed to maintain the separation of PTH-Arg and PTH-Gly is sufficient to maintain the resolution of arginine and threonine during PTC-amino acid analysis. For comparison, the separation of a PTC-amino acid mixture is presented in Figure 2B. It should be noted that the two chromatograms in Figure 2 represent consecutive analyses and were obtained without operator intervention. Thus, by the use of an auto-sampler system and a microprocessor-controlled chromatograph, a series of amino acid analysis and protein microsequence analysis samples can be run unattended.

Applications

One of the major uses of a PTH-/PTC-amino acid analyzer system is in the determination of the amino acid composition of purified proteins following acid hydrolysis. This is illustrated in Figure 3 by the PTC-amino acid analysis of a purified recombinant protein with growth hormone activity. Analyses were performed on both the unmodified and carboxy-methylated proteins. The calculated amino acid composition from both analyses is tabulated in Table 1. As can be seen from these results, the calculated compositions are in good agreement with the composition predicted from the amino acid sequence of growth hormone. Calculations

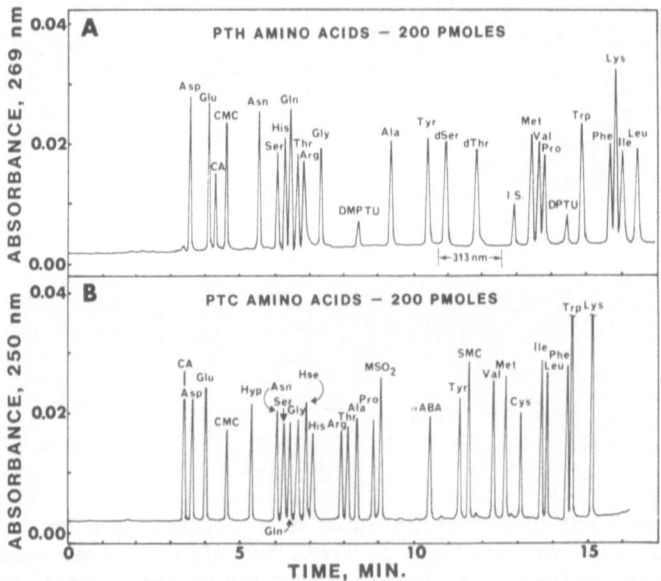

Figure 2: Separation of PTH- and PTC-amino acids on a Hypersil ODS column (5 μm, 4.6 x 250 mm). A, Standard mixture of PTH-amino acids. Detection wavelength was changed to 313 nm between 10.6 and 12.4 min. in the analysis to detect dehydroserine (dSer) and dehydrothreonine (dThr) derivatives. Internal standard (I.S.) is dansyl amide. B, PTC-amino acid standard for enzymatic digests. Non-standard abbreviations used: CA, cysteic acid; CMC, carboxymethyl cysteine; DMPTU, dimethylphenylthiourea; DPTU, diphenylthiourea; Hse, homoserine; MSO_2, methionine sulfone; αABA, 2-aminobutyric acid; SMC, S-methyl cysteine.

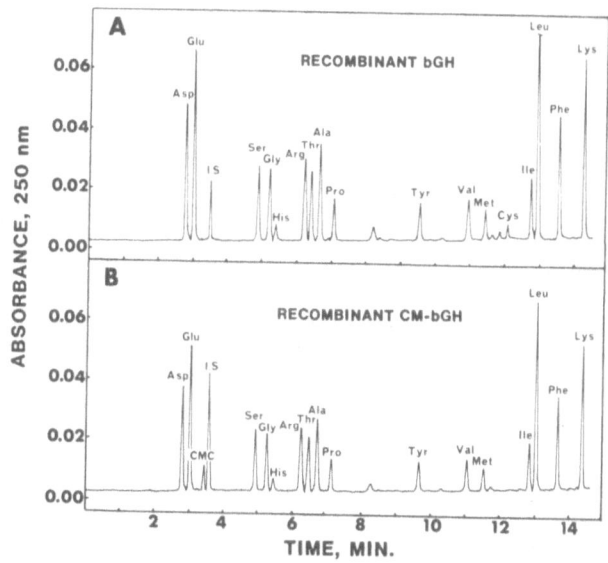

Figure 3: PTC-amino acid analysis of recombinant bovine growth hormone (bGH). A, Unmodified protein (450 pmol) was hydrolyzed, derivatized with PITC, and then dissolved in 50 μL Solvent A. Chromatogram obtained following 3 μL injection. B, Carboxymethylated (CM) protein (150 pmol) was hydrolyzed, derivatized with PITC, and then dissolved in 50 μL Solvent A. Chromatogram obtained following 6 μL injection.

Table 1: Amino acid composition of recombinant bovine growth hormone (bGH)

AMINO ACID	RESIDUES bGH	RESIDUES CM-bGH	RESIDUES FROM SEQUENCE
ASX	15.9	15.3	16
GLX	24.5	24.5	24
CMC	----	4.2	--
SER	13.4	13.6	13
GLY	10.8	11.3	10
HIS	3.0	2.8	3
ARG	13.6	13.7	13
THR	12.1	11.7	12
ALA	15.0	14.7	15
PRO	6.5	6.5	6
TYR	5.9	5.9	6
VAL	6.3	6.1	6
MET	4.2	3.8	4
½CYS	4.0	---	4
ILE	6.2	6.3	7
LEU	25.9	26.4	27
PHE	12.6	12.5	13
LYS	10.3	10.6	11
TRP	N.D.	N.D.	1
			191

SAMPLES WERE HYDROLYZED WITH CONSTANT BOILING HCl AT 150°C FOR 1 HR. THE NUMBER OF RESIDUES WAS CALCULATED ASSUMING 190 RESIDUES PER MOLECULE. TRYPTOPHAN CONTENT WAS NOT DETERMINED (N.D.).

CM-bGH, CARBOXYMETHYLATED BOVINE GROWTH HORMONE.

CMC, CARBOXYMETHYLCYSTEINE.

were based on a hydrolyzed calibration mixture of amino acids and no correction factors were used. It should be noted that both the unmodified and carboxymethylated proteins yield the correct number of cysteinyl (½Cys) residues. Furthermore, based on the amino acid sequence of bovine growth hormone, the molecule contains a total of seven –Leu-Leu– and –Leu-Ile-peptide bonds. These linkages are usually somewhat resistant to acid cleavage. However, from the calculated number of residues for both leucine and isoleucine,.the hydrolysis conditions of 150°C for 1 hr. appear to be sufficient for complete hydrolysis of the protein. Finally, the amount of protein hydrolyzed for the analysis depicted in Figure 3B was 150 pmol. This was more than sufficient to obtain reliable compositional data. Experience has shown that at least 25-50 pmol. of protein should be hydrolyzed for the PTC-amino acid analysis of proteins.

Another routine use of the PTH-/PTC-amino acid analyzer system is providing structural data for peptides obtained from enzymatic digests. To illustrate this application, carboxymethylated growth hormone was digested with trypsin and the resulting tryptic map is shown in Figure 4. Aliquots from each of the 46 peptide fractions along with a calibration mixture of amino acids and a hydrolysis blank were subjected to acid hydrolysis followed by derivatization with PITC. Each hydrolysate was then subjected to PTC-amino acid analysis. The total time required for hydrolysis, derivatization, and HPLC analysis of these samples was only 18 hrs. A representative example of one of these PTC analyses is depicted in Figure 5. From this analysis, the amino acid composition was determined for the peptide in Fraction 5. This peptide was also subjected to microsequence analysis and the resulting PTH analyses are shown in Figure 6. As can be seen from these figures, the results from the PTC-amino acid analysis and the microsequence analysis are in complete agreement. For the purposes of comparison, the PTC analysis for one of the minor peptide

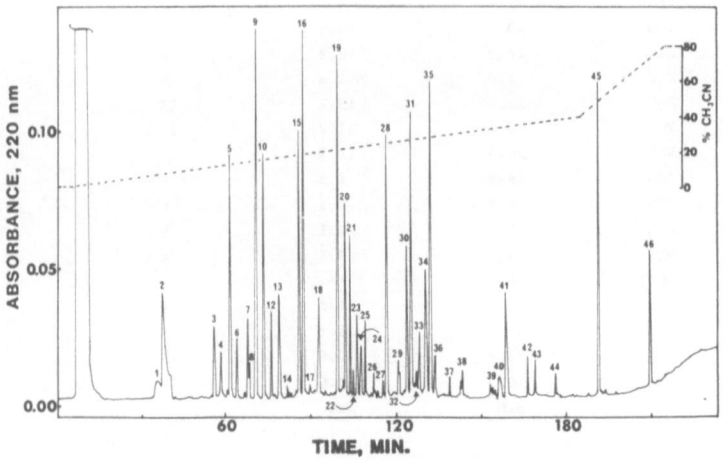

Figure 4: HPLC analysis of a tryptic digest of carboxymethylated bovine growth hormone (3 nmol). Digest was applied to a Vydac C18 column (5 μm, 4.6 x 250 mm) equilibrated with 0.1% trifluoroacetic acid and then eluted with a gradient of acetonitrile. Column flow rate was 0.5 mL/min. All fractions collected for analysis are numbered consecutively.

202

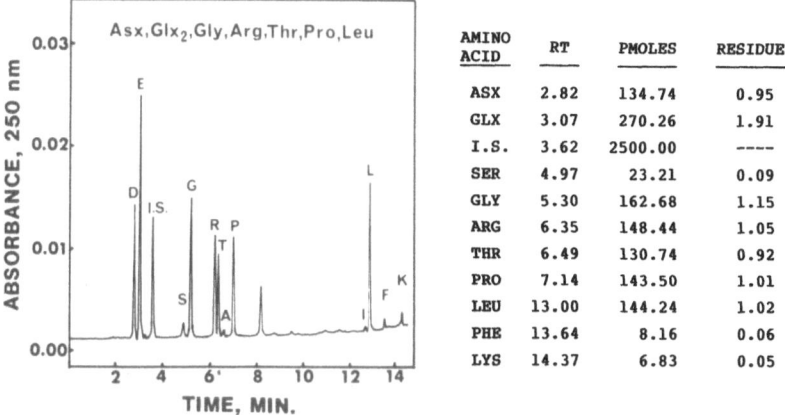

AMINO ACID	RT	PMOLES	RESIDUES
ASX	2.82	134.74	0.95
GLX	3.07	270.26	1.91
I.S.	3.62	2500.00	----
SER	4.97	23.21	0.09
GLY	5.30	162.68	1.15
ARG	6.35	148.44	1.05
THR	6.49	130.74	0.92
PRO	7.14	143.50	1.01
LEU	13.00	144.24	1.02
PHE	13.64	8.16	0.06
LYS	14.37	6.83	0.05

Figure 5: PTC analysis of the tryptic peptide from Fraction 5 (see Figure 4). An aliquot (250 μL) was removed, dried, hydrolyzed, derivatized with PITC, and redissolved in 50 μL Solvent A. Chromatogram was obtained following 5 μL injection. The retention time (RT), pmoles, and number of calculated residues for each amino acid are also shown. Amino acid peaks are labeled by their single-letter code.

peaks (Fraction 22) obtained from the tryptic map is presented in Figure 7. This peptide appears to have been generated by the action of chymotrypsin which should have been an insignificant contaminant in the trypsin preparation.

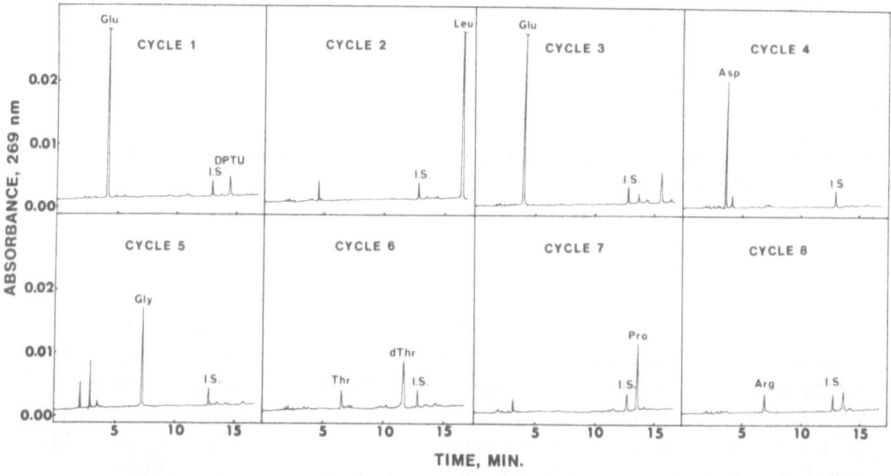

Figure 6: PTH-amino acid analyses obtained for the microsequence analysis of the tryptic peptide from Fraction 5 (see Figure 4). Approximately one nmol. of peptide was subjected to sequence analysis. Dried residues from each cycle were dissolved in 30 μL of aqueous acetonitrile. Chromatograms were obtained following 15 μL injections.

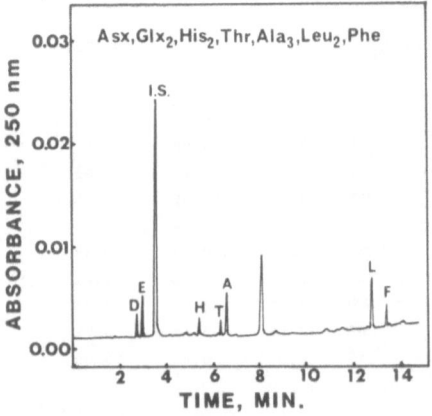

AMINO ACID	RT	PMOLES	RESIDUES
ASX	2.86	23.04	1.10
GLX	3.09	43.25	2.06
I.S.	3.66	5000.00	----
HIS	3.54	40.30	1.92
THR	6.50	19.75	0.94
ALA	6.75	59.82	2.85
LEU	13.01	43.04	2.05
PHE	13.65	22.80	1.09

Figure 7: PTC analysis of the tryptic peptide from Fraction 22 (see Figure 4). An aliquot (250 µL) was removed, dried, hydrolyzed, derivatized with PITC, and redissolved in 50 µL Solvent A. Chromatogram was obtained following 10 µL injection. The retention time (RT), pmoles, and number of calculated residues for each amino acid are also shown. Amino acid peaks are labeled by their single-letter code.

Other routine uses for the PTH-/PTC-amino acid analyzer system include the following: 1) amino acid analysis of peptide hydrolysates generated by total enzymatic digestion, 2) identification and quantitation of free amino acids released during tryptic digestion of polypeptides, 3) sequence analysis using time course hydrolysis with exopeptidases, 4) monitoring of the reaction steps for peptide synthesis, and 5) rapid screening of pharmaceutically-important proteins as a means of quality control.

CONCLUSIONS

The PTH-/PTC-amino acid analytical system described in this report provides a rapid, sensitive, and convenient method for the structural characterization of polypeptides. This evaluation is based upon observations made over the last 8 months during which time a total of 5000 PTC and PTH analyses have been performed.

ACKNOWLEDGMENT

The authors wish to thank Ms. Hazel Deegan for assistance in the preparation of this manuscript.

REFERENCES

1. M.W. Hunkapiller, R.M. Hewick, W.J. Dreyer, and L.E. Hood, High-sensitivity sequencing with a gas-phase sequenator, in: "Methods in Enzymology," C.H.W. Hirs and S.N. Timasheff, eds., Academic Press, New York (1983).

2. F.S. Esch, Polypeptide microsequence analysis with the commercially available gas-phase sequencer, Anal. Biochem. 136:39 (1984).

3. C.L. Zimmerman, E. Appella, and J.J. Pisano, Rapid analysis of amino acid phenylthiohydantoins by high performance liquid chromatography, Anal. Biochem. 77:569 (1977).

4. S.D. Black and M.J. Coon, Simple, rapid, and highly efficient separation of amino acid phenylthiohydantoins by reversed-phase high-performance liquid chromatography, Anal. Biochem. 121:281 (1982).

5. D. Hawke, P. Yuan, and J.E. Shively, Microsequence analysis of peptides and proteins, Anal. Biochem. 120:302 (1982).

6. J.J. L'Italien and J.E. Stricker, Application of high-performance liquid chromatographic peptide purification to protein microsequencing by solid-phase Edman degradation, Anal. Biochem. 127:198 (1982).

7. R. Knecht, U. Seemuller, M. Liersch, H. Fritz, D.G. Braun, and J.Y. Chang, Sequence determination of Eglin C using combined microtechniques of amino acid analysis, peptide isolation, and automatic Edman degradation, Anal. Biochem. 130:65 (1983).

8. H.V.J. Kolbe, R.C. Lu, and H. Wohlrab, Reversed-phase high-performance liquid chromatographic separation and quantitation of phenylthiohydantoin derivatives of 25 amino acids, J. Chromatogr. 327:1 (1985).

9. R.L. Heinrikson and S.C. Meredith, Amino acid analysis by reverse-phase high-performance liquid chromatography: precolumn derivatization with phenylisothiocyanate, Anal. Biochem. 136:65 (1984).

10. B.A. Bidlingmeyer, S.A. Cohen, and T.L. Tarvin, Rapid analysis of amino acids using pre-column derivatization, J. Chromatogr. 266:471 (1983).

11. B.N. Jones and J.P. Gilligan, o-Phthaldialdehyde precolumn derivatization and reversed-phase high-performance liquid chromatography of polypeptide hydrolysates and physiological fluids, J. Chromatogr. 266:471 (1983).

12. Operator's Manual, Pico-Tag Workstation, Waters Associates, Milford, MA.

AMINO ACID ANALYSIS USING PRE-COLUMN DERIVATIZATION WITH

PHENYLISOTHIOCYANATE: MATRIX EFFECTS AND TRYPTOPHAN ANALYSIS

Steven A. Cohen, Thomas L. Tarvin and Brian A. Bidlingmeyer

Waters Chromatography Division of Millipore Corp.
34 Maple Street
Milford, MA 01757

We have recently reported[1-3] on a new methodology for amino acid analysis (AAA) that relies on the reaction of free amino acids with Edman's reagent, phenylisothiocyanate (PITC), to form the phenylthiocarbamyl (PTC) derivatives, which are subsequently separated and quantitated using a reverse phase gradient liquid chromatography (LC) system for analysis. This work extended the methodology for AAA using PITC introduced several years ago by Tarr and co-workers[4,5], and appeared concurrently with a report by Heinrikson and Meredith[6] describing similar methodology for AAA. The method was shown to be reliable, accurate, and extremely rapid (12 minute) with a maximum one picomole detection limit for each amino acid. This new method provides improved capabilities in comparison with conventional ion-exchange, post-column ninhydrin reaction systems. In addition, linear response in the 10-1000 pmol range was demonstrated as well as a high degree of reproducibility[1]. This system has been recently commercialized under the name PICO•TAG™ Amino Acid Analysis System[3].

The current work described here expands the capabilities of this new system to those samples containing nonvolatile matrix contaminants, the removal of which can be difficult, time-consuming, and/or result in sample loss and a decrease in sensitivity. Modifications of the original procedure demonstrate that increasing the polarity of the derivatization reagent mixture improves derivatization efficiency in the presence of many potential interferents.

METHODS

Salt and Detergent Studies

Stock protein and peptide samples were dissolved in water or 0.01 - 0.10 $\underline{M}$ HCl at concentrations of 1-5 mg/ml. Aliquots containing 5 μg were hydrolyzed in constant boiling HCl (~ 5.7$\underline{M}$) containing 0.5% liquified phenol (v/v) at 110°C for 22h or at 150°C for 1h using a vapor phase technique[3]. After hydrolysis and drying under vacuum, a 50 μl aliquot of salt or detergent was added to the sample, water removed under vacuum, and derivatization proceeded in a manner analagous to that previously described[2].

Tryptic peptides from sulfopropylated human placental lysozyme were isolated by LC as previously described[7] using a two-solvent linear gradient system with the weak solvent being 0.1% H_3PO_4 and 0.1$\underline{M}$ NaClO$_4$, and the strong solvent being 75% CH_3CN, 25% H_2O. The peptides were kindly provided by Dr. Daniel Strydom of Harvard Medical School.

Rabbit muscle aldolase was isolated in a single chromatographic step using Accell™ CM ion-exchange medium and a linear gradient running from 50mM sodium acetate pH 6.0 to the same buffer containing 1.0 M NaCl over 20 min. Muscle tissue was minced and homogenized in the starting buffer, centrifuged, and filtered through a Millex® HV filter prior to chromatography. Activity recovery and protein content were measured by standard techniques[8,9].

The two most frequently used reagents for derivatization were Reagent I = ethanol:water:triethylamine (TEA):PITC (7:1:1:1) and Reagent II = methanol:water:TEA:PITC (7:1:1:1).

Hydrolysis of Peptides and Proteins with Methanesulfonic Acid (MSA)

Samples were dried under vacuum in 6 X 50 mm Pyrex® tubes, and 20 μl of 4M MSA containing 0.1% tryptamine HCl (w/v) added to each sample. The tubes were placed in a vacuum vial[3], and 0.1ml of 1% phenol in water (v/v) added to the vial. Hydrolysis was then carried out at 110°C for 22h using a method similar to that described above. After hydrolysis, samples were neutralized with either 4N KOH or TEA.

LC Analysis

Separation and quantitation of PTC amino acids was carried out on a Waters gradient LC system as previously described[2].

RESULTS AND DISCUSSION

Salts and Detergents

The data in Table 1 summarize the results of the AAA of bovine serum albumin using the derivatizing reagent mixtures described above. The samples were salt-free and gave equivalent results; these values were considered our standard composition for BSA, and were used as a benchmark for comparing data obtained in the presence of salts or detergents.

Using Reagent I for derivatization for samples with nonvolatile matrix components resulted in the deleterious effect of salt or detergent on compositional analysis illustrated by the data in Table 2. However, use of Reagent II with moderate levels (200-500 micrograms) of these species present produced accurate amino acid compositions.

Table 1. Comparison of Derivatizing Solutions Used for PITC Amino Acid Analysis[a]

Amino Acid	From Sequence	Reagent I	Reagent II	Average Value
Asp	53	53.3	53.7	53.4
Glu	78	75.2	74.7	75.0
Ser	28	19.7	19.5	19.6
Gly	15	17.9	17.5	17.7
His	17	16.1	16.2	16.2
Arg	23	23.4	23.4	23.4
Thr	34	26.2	26.7	26.5
Ala[b]	46	46	46	46
Pro	28	31.8	32.7	32.3
Tyr	19	19.3	18.6	19.0
Val	36	34.1	33.4	33.8
Met	4	4.1	4.0	4.1
Ile	14	13.0	12.8	12.9
Leu	61	62.7	62.0	62.4
Phe	26	26.7	25.7	26.2
Lys	59	58.7	58.1	58.4

a) The sample, 5 micrograms of bovine serum albumin, was hydrolyzed at 150°C for 1h. One-tenth of the hydrolyzate was used for analysis.
b) Values are normalized to Ala = 46

Table 2. Influence of Matrix Contaminants on Amino Acid Analysis Using PITC Derivatization[a]

Amino Acid	From Sequence	Reagent I[b]	Reagent II[b]	Reagent I[c]	Reagent II[c]
Asp	53	37.4	54.4	35.8	55.1
Glu	78	69.8	83.3	65.9	77.2
Ser	28	24.0	28.1	21.7	24.0
Gly	15	18.0	21.3	15.2	16.7
His	17	14.8	16.2	11.5	13.4
Arg	23	24.0	25.1	23.2	22.3
Thr	34	30.4	32.8	28.3	29.9
Ala[d]	46	46	46	46	46
Pro	28	36.5	40.4	28.2	28.7
Tyr	19	19.8	19.9	17.8	18.5
Val	36	32.5	31.6	34.0	34.1
Met	4	4.0	4.0	2.4	3.0
Ile	14	12.5	12.4	12.6	13.1
Leu	61	62.4	62.9	62.5	62.8
Phe	26	26.5	25.9	27.5	28.9
Lys	59	58.3	54.3	54.0	55.3

a) The samples were the same as described in Table 1 except hydrolysis was at 110°C for 22h.
b) Samples were spiked with 0.05 ml of 0.25M NaCl prior to hydrolysis.
c) Samples were spiked with 0.5 mg of SDS prior to hydrolysis.
d) All values were normalized to Ala according to the actual content of BSA.

Table 3. Analysis of Derivatizing Solution Efficacy with LC Purified Samples Containing Salt[a]

Amino Acid	Peptide 9, Lysozyme[b]		Rabbit Aldolase[c]	
	Reagent I	Reagent II	Sample	Control
Asp	1.6 (3)	2.9	26.3	26.1
Glu	0.1 (0)	0.1	41.6	40.8
Ser	0 (0)	0	16.6	16.5
Gly	0.9 (1)	0.9	33.0	32.0
His	0 (0)	0	11.1	9.4
Arg	1.2 (1)	1.2	18.8	16.6
Thr	1.0 (1)	1.0	19.2	20.5
Ala[d]	2.0 (2)	2.0	42	42
Pro	0 (0)	0	22.5	26.2
Tyr	0.5 (1)	0.6	11.2	11.7
Val	0.1 (0)	0.1	21.4	21.0
Met	0 (0)	0	2.7	3.8
Ile	0 (0)	0	19.1	20.2
Leu	0.2 (0)	0.2	38.1	39.4
Phe	0 (0)	0	8.2	7.7
Lys	0.2 (0)	0.1	30.3	29.3

a) Aliquots of LC fractions were hydrolyzed for 22h with HCl. Values in parentheses are known values for the peptide. Peptide numbering is described in reference 7.
b) Lysozyme peptides were isolated as described in reference 7 and 5% of each fraction taken for AAA.
c) Aldolase samples (1% of the active fraction, 5μg of standard) were derivatized with Reagent II.
d) Values for aldolase were normalized to Ala = 42.

Further evidence for the improvement in amino acid analysis using the more polar derivatizing mixture is demonstrated in Table 3. Since the lysozyme peptide was in a phosphate/perchlorate buffer solution, the use of Reagent I gave poor yields of the acidic amino acids while use of Reagent II gave an accurate composition. In general, this trend towards lower yields of polar amino acids with increasing matrix contamination can be observed even at moderate levels if ethanol is employed in derivatization, while the methanol-based solution gives good results at much higher levels of contamination. This improvement using Reagent II in most cases (data not shown) is similar to that observed using a reagent consisting of ethanol:water:TEA:PITC (7:2:1:1). However, for some lysozyme tryptic peptides, notably Peptide 9, values for the acidic amino acids were low with this modified ethanol reagent, whereas Reagent II afforded nearly quantitative recoveries for Asp and Glu. These observations are consistent with the hypothesis that reagent polarity has a significant effect on polar amnio acid yield in the presence of nonvolatile matrix components. Practical limits with Reagent II are 50 µl of 0.5$\underline{M}$ NaCl, and 0.5 mg of sodium dodecyl sulfate, above which yields of polar amino acids can be less than quantitative.

The aldolase compositions shown in Table 3 compare results of ion-exchange purified protein with a salt-free control. The NaCl concentration of the fractions was approximately 0.8$\underline{M}$, and it can be seen that this level of salt had no effect on the calculated composition.

One problem that has been noted with more than 0.5 mg of salts present is that the solubility of the matrix in any of the reagents is not sufficient to allow all of the sample to dissolve. It is very possible that this solubility effect is the cause of poor derivatization yields. Nonetheless, in samples containing up to 5 mg of sodium acetate, good compositional analyses were obtained even though there was a significant amount of insoluble material observed during the derivatization procedure (data not shown).

Tryptophan Analysis

Measurement of tryptophan content in peptides and proteins is most commonly carried out on samples hydrolyzed with 4$\underline{M}$ MSA[10]. Tables 4 and 5 give results of samples analyzed using the PITC methodology after hydrolysis with MSA using the modification of this earlier work described in the Methods Section. Because of the acids' lack of volatility it was necessary to include a neutralization step prior to derivatization. However, the convenience of batch-wise vacuum sealing, hydrolysis and derivatization can still be readily achieved.

One notable advantage of the PITC method for tryptophan quantitation analysis is the excellent resolution and peak shape obtained (Figure 1), and the consequent ease of quantitation with symmetrical, sharp, well-resolved peaks. This is in marked contrast to the broad top peak usually observed in conventional ion-exchange analysis.

Using Reagent II, good results for Trp were obtained with most yields being above 70%. Lower yields seen with some lysozyme tryptic peptides could be caused by oxidation of tryptophan prior to hydrolysis in the perchlorate buffer. All of the other amino acids gave excellent values, including aspartic and glutamic acids, even though low values for polar residues might have been expected due to the high salt content remaining following neutralization.

210

Table 4. Amino Acid Analysis Using PITC Derivatization of Standard
Samples[a] Hydrolyzed with MSA.

| | Hexapeptide | | ACTH | | Lysozyme | |
	TEA[b]	KOH[c]	KOH [c]	HCl[d]	KOH[c]	HCl[d]
Asp			2.19	2.24 (2)	22.1	23.1 (21)
Glu			4.86	4.99 (5)	5.6	6.2 (5)
Ser			1.52	1.63 (2)	8.6	10.0 (10)
Gly			3.32	3.33 (3)	12.5	12.3 (12)
His			0.97	0.97 (1)	1.1	1.1 (1)
Arg	1.3 (1)	0.8	3.51	3.47 (3)	7.2	12.6 (11)
Thr			0	0.13 (0)	7.3	7.1 (7)
Ala	1.0 (1)	1.0	2.97	2.98 (3)	12[e]	12[e] (12)
Pro			4.02	4.11 (4)	2.4	2.5 (2)
Tyr			1.77	1.67 (2)	3.7	3.1 (3)
Val			2.86	2.88 (3)	6.5	5.7 (6)
Met	0.9 (1)	0.8	0.54	0.85 (1)	1.4	2.1 (2)
Ile			0	0.05 (0)	5.6	5.7 (6)
Leu	1.0 (1)	1.0	2.18	2.21 (2)	8.3	8.6 (8)
Phe	1.0 (1)	1.0	2.76	2.98 (3)	3.3	3.1 (3)
Trp	1.0 (1)	0.9	0.76	0.0 (1)	5.1	0 (6)
Lys			3.96	4.85 (4)	6.6	6.0 (6)

a) The samples (5μg) were hen egg white lysozyme, porcine ACTH and the
hexapeptide Leu-Trp-Met-Arg-Phe-Ala. Values in parentheses are the
number of residue known from sequence analysis.
b,c) Samples were neutralized as described in the Methods Section.
d) Control samples were hydrolyzed with HCl.
e) Values for lysozyme were normalized to the known Ala content.

Table 5. Analysis of MSA Hydrolyzed Lysozyme Peptides Using PITC
Derivatization[a]

Peptide Number[b]	3-4-5	8	7-8
Asp	0.9 (1)	0.8 (1)	2.0 (2)
Glu	0.9 (1)	0.8 (1)	1.0 (1)
Ser	0.9 (1)	0.7 (1)	1.8 (2)
Gly	1.0 (1)	0.9 (1)	2.0 (2)
His	0.0 (0)	0.0 (0)	0.0 (0)
Arg	0.2 (0)	1.1 (1)	1.1 (1)
Thr	0.3 (0)	0.8 (1)	0.9 (1)
Ala	2.1 (2)	0.0 (0)	2.2 (2)
Pro	0.1 (0)	0.0 (0)	0.0 (0)
Tyr	0.0 (0)	0.8 (1)	1.0 (1)
Val	0.1 (0)	0.3 (0)	0.1 (0)
Met	0.1 (0)	0.5 (1)	0.7 (1)
Ile	1.0 (1)	0.0 (0)	0.9 (1)
Leu	0.3 (0)	1.9 (1)	2.0 (2)
Phe	0.1 (0)	0.2 (0)	0.0 (0)
Trp	0.6 (1)	0.6 (1)	1.5 (2)
Lys	0.8 (1)	0.2 (0)	0.8 (1)
Spcys	0.7 (1)	0.0 (0)	0.7 (1)

a) Peptides were isolated as previously described(7). KOH was used for
neutralization after MSA hydrolysis. Values in parenthesis are the
actual numbers from the published sequence.
b) Peptide numbering is described in Reference 7.

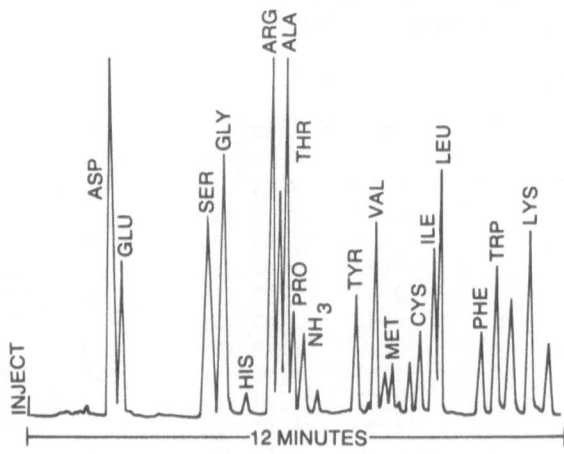

Fig. 1. Chromatogram of PITC-derivatized MSA
hydrolyzate of hen egg white lysozyme.
Sample preparation and the chromatographic
conditions are described in the text. The
data for this analysis is shown in Table 4.

Two problems associated with MSA hydrolysis were detected in some
analyses. Occasionally values for arginine were low; the exact cause is
not known, and is the subject of further study. Unusually high levels of
background due to reagent are occasionally observed in chromatograms near
the tyrosine and valine peaks, and can interfere with their quantitation.
The source is not known, but since this contamination is not seen with
HCl hydrolyzates, one possible source is the tryptamine present in the
commercially available MSA. As arginine, tyrosine and valine are easily
quantitated in HCl hydrolyzates, best compositional analyses will be
obtained if both HCl and MSA hydrolyzed samples are prepared.

CONCLUSIONS

Pre-column derivatization of amino acids using PITC has previously
been demonstrated to be a viable alternative to ion-exchange methodology
for peptide and protein AAA[1-6]. This report has shown that PITC-based
AAA can be flexible enough to analyze a variety of samples regardless of
the presence of involatile matrix components. In particular, samples can
be analyzed if they are in solutions containing levels of phosphate,
perchlorate or chloride salts normally employed in peptide or protein
purifications.

PITC derivatization has also been extended to the analyis of Trp in
MSA hydrolyzates of peptides and proteins. Typical Trp yields are
comparable to those observed in ion-exchange analysis, and the superior
chromatography simplifies and improves Trp analysis.

In summary, the flexibility of PITC use for AAA reinforces our
earlier claim[1-3] that this new technique is the first viable alternative
to ion-exchange analysis. On-going research focussing on methods for
phospho-amino acids, amino sugars, and AAA of physiological fluids will
provide further support for this contention.

ACKNOWLEDGEMENTS

We are extremely greatful to Deborah B. Anderson and Raymond Hanna
for technical assistance, to Katherine Bergeron for manuscript
preparation and to Kevin Monaghan for the artwork.

REFERENCES

1. B. A. Bidlingmeyer, S. A. Cohen and T. L. Tarvin, paper presented at
    the International Symposium on HPLC in the Biological Sciences
    Conference in Melbourne, Australia, February 20-22, 1984.
2. B. A. Bidlingmeyer, S. A. Cohen and T. L. Tarvin, Rapid Analysis of
    Amino Acids Using Pre-Column Derivatization, J. Chromatogr.
    336:93 (1984).
3. S. A. Cohen, T. L. Tarvin and B. A. Bidlingmeyer, Analysis of Amino
    Acids Using Pre-Column Derivatization with Phenylisothiocyanate,
    Am.Lab, August:49 (1984).
4. D. R. Koop, E. T. Morgan, G. E. Tarr and M. J. Coon, Purification
    and Characterization of a Unique Isozyme of Cytochrome P-450
    from Liver Microsomes of Ethanol-Treated Rabbits,J.Biol.Chem.,
    257:8472 (1982).
5. G. E. Tarr, S. D. Black, V. S. Fujita and M. J. Coon, Complete Amino
    Acid Sequence and Predicted Membrane Topology of
    Phenobarbital-Induced Cytochrome P-450 (Isozyme 2) from Rabbit
    Liver Microsomes, Proc. Natl. Acad. Sci. USA, 80:6552 (1983).
6. R. L. Heinrikson and C. S. Meredith, Amino Acid Analysis by Reverse
    Phase HPLC: Precolumn Derivatization with Phenylisothiocyanate,
    Anal. Biochem. 136:65 (1984).
7. J. W. Fett, D. J. Strydom, R. R. Lobb, E. M. Alderman and B. L.
    Vallee, Lysozyme: A Major Secretory Product of a Human Colon
    Carcinoma Cell Line, Biochem. 24:965-975 (1985).
8. Boehringer Mannheim Biochemicals Handbook, p37 (1985).
9   M. Bradford, Anal. Biochem. 72:248 (1976).
10. T.-Y. Lin and Y. H. Chang, J. Biol. Chem. 246:2842 (1971).

AN IMPROVED REVERSE PHASE HPLC SEPARATION AND POSTCOLUMN DETECTION

OF AMINO ACIDS

Stephen Gruber, Noel M. Meltzer, Stanley Stein, and
Guillermo I. Tous

Schering Corporation
60 Orange Street
Bloomfield, N.J.

INTRODUCTION

The original liquid chromatographic method for amino acid analysis was
based on ion-exchange separation of the amino acids and postcolumn reaction
with ninhydrin (1). For over 2 decades this methodology, with minor
improvements, has remained as the most popular approach for amino acid
analysis. For higher sensitivity analysis (picomole level), ninhydrin
may be replaced by fluorescamine (2,3) or o-phthaldialdehyde (OPA) (4,5).
More recently, precolumn derivatization procedures with OPA (6,7,8) and
phenylisothiocyanate (9) have attracted much attention, because they are
based on reverse-phase HPLC rather than ion-exchange chromatography. In
the present article, improvements are reported on the procedure of Radjai
and Hatch (10), which is based on the unique combination of reverse-phase
HPLC and postcolumn reaction with OPA. Each approach offers certain
advantages and disadvantages, which are discussed below.

MATERIALS AND METHODS

Two separate chromatography systems were employed in the present investigation. One consisted (Figure 1) of two high performance liquid chromatography pumps (Waters Associates Models 6K and M-45), one constant flow
pump for reagent delivery (Milton Minipump or Eldex Model A-30-S), a
Farrand Ratio Fluorometer-2 equipped with a Corning 7-60 primary filter
and a Wratten No. 2A secondary filter, and a Waters Associates Model 720
System Controller for generation of elution gradients. The other chromatography system consisted of a LDC/Milton Roy ConstaMetric III HPLC pump,
a Gilson Model 121 fluorometer equipped with the standard OPA filters, an
Eldex Model A-30-S constant flow pump for reagent delivery, and an Eldex
Chromatotrol Model III microprocessor to control an Eldex six-port, low-
pressure solvent selector valve. The latter system was used to characterize
the retention of the first seven amino acids as a function of mobile phase
pH and temperature.

Sample injections were performed using either a Waters Associates WISP automatic injector or a Rheodyne injector valve (Model 7120) equipped with a 50 µl sample loop. An ES Industries C-18 column (4.6 x 100 mm; particle size, 3 µ) was used for chromatographic separations. Chromatographic peaks were integrated by a Perkin-Elmer LIMS, and chromatograms recorded on a two-channel recorder. In the case of gradient elution, gradients were formed between two degassed solvent mixtures. Mobile phase A was 0.07% decyl sodium sulfate prepared in .025% phosphoric acid and mobile phase B was .12% heptane sodium sulfonate in acetonitrile:water (30:70) containing .05% phosphoric acid. The column and reaction coil were maintained at the indicated temperature using a Dupont Instruments Column Heater Model 851101-9801.

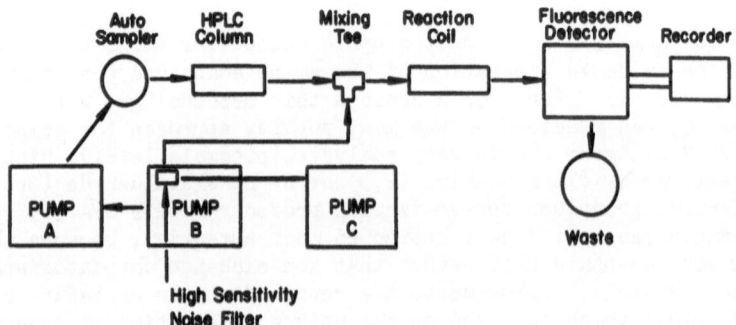

Figure 1. Schematic diagram of HPLC and reagent delivery system for postcolumn OPA detection of amino acids.

Reagents and Standards

All HPLC solvents were distilled-in-glass (Burdick and Jackson or J.T. Baker) and used without further treatment. HPLC quality water was obtained with a Milli Q system from Millipore Corporation. Solutions of the amino acid standards (2.5 µmole/ml), the individual amino acid standards, 2-mercaptoethanol, o-phthaldialdehyde, and Brij-35 were purchased from Sigma Chemical Company. Sequencing grade 6N HCl was obtained from Pierce Chemical Co. Decyl sodium sulfate and heptane sodium sulfonate were purchased from Eastman Kodak and used as received. All other chemicals were reagent quality.

Preparation of the o-Phthaldialdehyde Reagent Solution

A modification of the reagent described by Radjai and Hatch (10) was used. In a 1 L volumetric flask is placed 2 ml of 2-mercaptoethanol. The solution is made to volume with 0.4 M borate buffer (pH 10.4). To this solution is added 10 ml of ethanol containing 800 mg of o-phthaldialdehyde followed by 3 ml of Brij-35. The mixture is filtered through a .45 μ filter, transferred to a solvent delivery bottle, and kept under a helium atmosphere. Under these conditions, the reagent can be used for up to three days without a loss in fluorescent product yield.

Acid Hydrolysis of Proteins

For HCl hydrolysis, proteins (.25-1 nmol) were dried in glass limited volume inserts typically used with the WISP autoinjector (Waters Associates part number 72704) in a Savant Speed-Vac Model RT-100A. The vials were then transferred to a vacuum dessicator using the ceramic plate as a holder. A beaker containing 2 ml of 6N HCl with 1% thioglycolic acid was placed in the bottom of the dessicator. The dessicator was evacuated to approximately 50 milliTorr and then placed in an oven maintained at 115 degrees. Samples are removed at the appropriate times. After hydrolysis, 150 μl of mobile phase A was added to each vial and aliquots were injected onto the reverse-phase column.

Method Validation

The gas-phase hydrolysis and HPLC methods were validated using human serum albumin (HSA, Sigma Chemical Co.) A stock solution of HSA was prepared in 0.1N HCl and aliquots were transferred to glass limited volume insert vials. To each vial was added 15 μl of an internal standard solution (β-alanine, 2.50 μmole/ml). The HSA was hydrolyzed according to the procedure above. The data collected following HPLC was then used to calculate the amino acid composition of HSA and the protein concentration of the solution in order to determine the hydrolysis yield.

RESULTS AND DISCUSSION

Preliminary separations of the standard amino acid mixture including cysteic acid (CA), S-carboxymethylcysteine (CMC), and tryptophan (W), were carried out using mobile phase A adjusted to pH 2.85 and mobile phase B at pH 3.0 at a column temperature of 35 degrees. All the amino acids could be satisfactorily resolved with the exception of threonine (T) and CMC. The affect of temperature on the separation of the first seven amino acids to elute from the column indicated that column temperature could be optimized to obtain separation of T and CMC (Figure 2). When the column temperature effects were first investigated, the pH of mobile phase A was maintained at 2.85. Subsequent studies of the effect of the pH of the column eluant indicated that the pH of mobile phase A had a profound effect on the separation of T and CMC with the separation improving at lower pH values (Figure 3). The pH of mobile phase B was found to have little effect on the resolution of the later eluting amino acids.

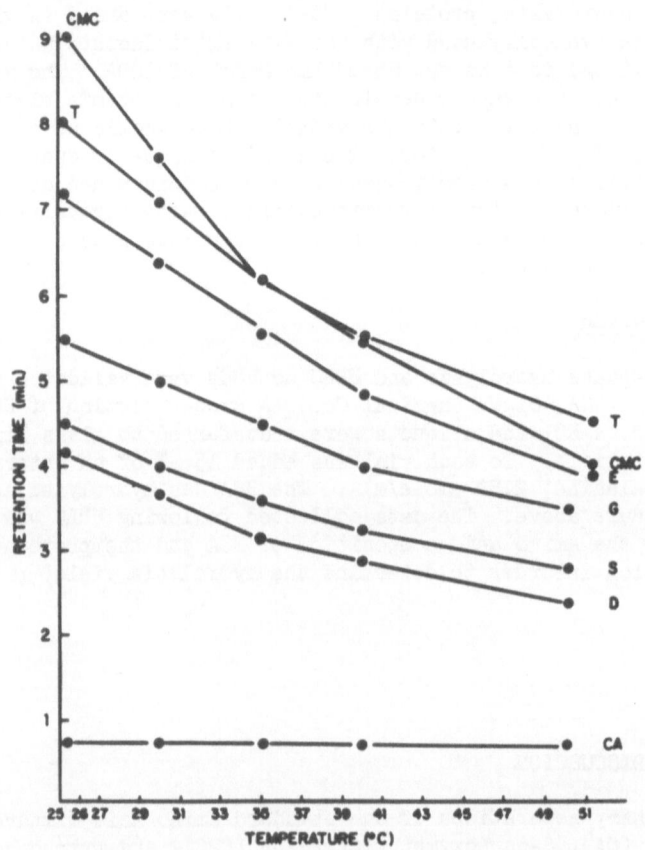

Figure 2. Plot of retention time as a function of column temperature for the elution of CA, D, S, G, E, T and CMC.

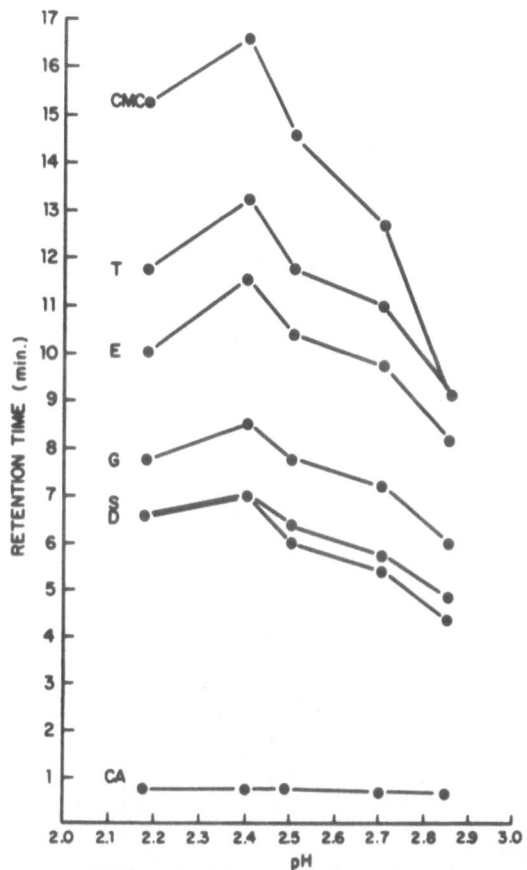

Figure 3. Plot of retention time as a function of mobile phase pH for the elution of CA, D, S, G, E, T and CMC.

Using the mobile phases described in the methods and materials section (mobile phase A, pH 2.5) and a column temperature of 25 degrees, chromatograms of protein hydrolystates and amino acid standards were obtained (Figures 4 and 5, respectively). Excellent separation between all the amino acids is obtained in a run time under one hour. The acid hydrolysis method provides a simple, no-transfer, one container procedure for the hydrolysis of proteins, which is reasonably accurate as indicated by the composition obtained for human serum albumin (Table I). Under optimal conditions, it has been possible to detect as little as 500 fmole of amino acids. The generally used range is 50 pmol to 50 nmol amino acid injected and the response is linear with concentration within this range.

Table I

Calculated Composition of Human Serum Albumin Using Gas-Phase
Hydrolysis and Reverse-Phase Ion-pair HPLC with Postcolumn
OPA Detection

Amino Acid	Calculated Composition	Theoretical Composition[a]
Asd	56.3 ± 4.9[b]	53
Ser	22.3 ± 2.6	24
Gly	13.7 ± 2.2	12
Glv	84.6 ± 3.4	81
Thr	28.4 ± 0.9	28
Ala	62.5 ± 3.1	62
Val	38.0 ± 2.7	41
Met	5.7 ± 0.7	6
Tyr	17.0 ± 0.5	18
Ile	6.3 ± 1.1	8
Leu	58.9 ± 2.6	61
Phe	29.4 ± 0.4	31
His	15.3 ± 0.7	16
Lys	59.5 ± 2.8	60
Trp	0.6 ± 0.1	1
Arg	20.8 ± 3.3	24

[a]From the gene sequence
[b]Mean ± standard deviation of 6 replicate hydrolyses

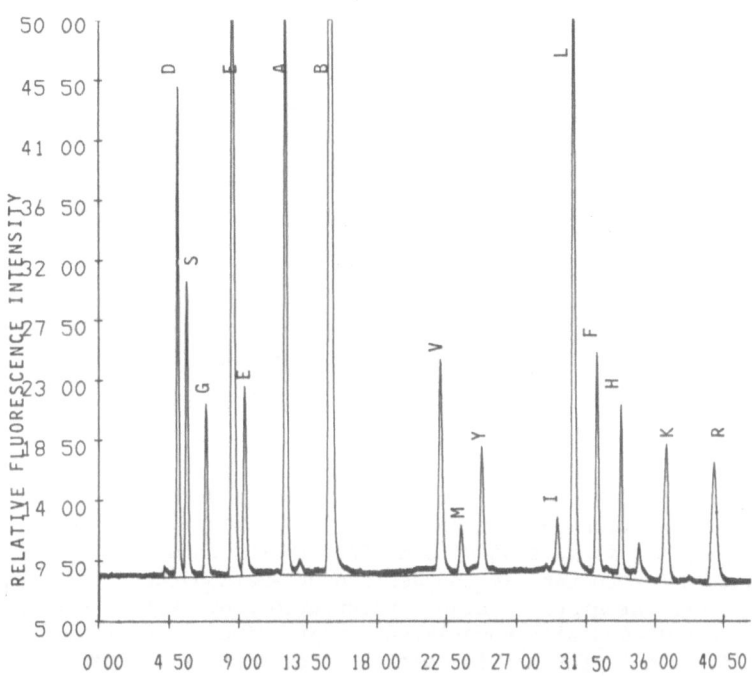

Figure 4. Chromatogram obtained for hydrolysate of 200 pmole of human serum albumin.

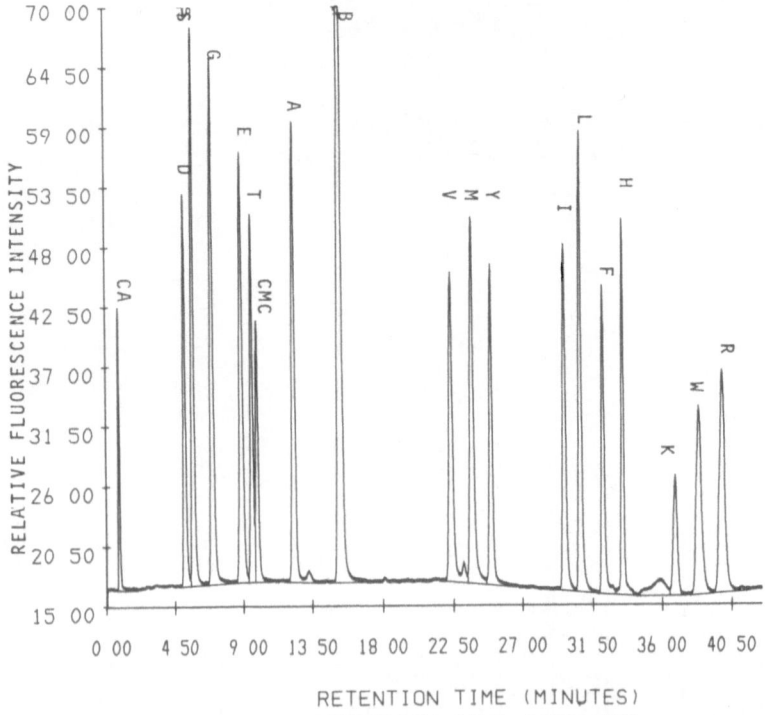

Figure 5. Chromatogram of 2 nmole amino acid standards.

The chromatographic method described presents certain advantages over ion-exchange separations and precolumn derivatization procedures. With reverse-phase HPLC, it is easier to manipulate the chromatographic parameters in order to achieve the desired separation than is the case with ion-exchange chromatography. High concentrations of salts, which are typically contaminated with ammonia and other amines, are avoided. In addition, little interference is observed from ammonia in the reverse-phase separation. An injection of an ammonium acetate solution indicated that ammonia eluted between alanine (A) and β-alanine (B-ala) and the response for ammonia was one-hundredth that of the other amino acids.

Precolumn derivatization techniques with reagents, such as dansyl chloride and phenylisothiocyanate require that the hydrolysates be free from contaminants such as ammonia, primary amines, or other nucleophiles in order to obtain reasonable chromatographic results. Postcolumn derivatization also avoids problems encountered with derivative instability, as is the case with OPA precolumn derivatization.

222

Gas-phase acid hydrolysis of proteins has been described previously in a specialized apparatus (11). The hydrolysis method described above is simple to do and uses common laboratory equipment. It allows for minimal handling of the protein and the absence of physical dilution of the sample with HCl reduces the chances of contamination.

Two drawbacks of the current postcolumn derivatization method are that it does not react with the secondary amino acids, proline and hydroxy-proline, and it has a poor response to cystine and cysteine. Efforts are underway in our laboratory to combine in-line oxidation prior to reaction of the amino acids with the OPA reagent. This has been described by several laboratories as a feasible means of assaying all amino acids (12,13). The ability to separate CMC allows the quantitation of cysteine (or cystine) residues following (reduction and) reaction of the protein with iodoacetic acid. Alternatively, determination of cysteic acid after performic acid oxidation has been found to give satisfactory results. It is, therefore, possible to analyze all amino acids found in protein hydrolysates.

An improved reverse-phase HPLC postcolumn OPA detection of primary amino acids is described. It provides a sensitive, reliable and rapid method for the detection and quantitation of amino acids in protein hydrolysates. It provides a viable alternative to ion-exchange chromatography with postcolumn reaction or precolumn derivatization with reverse-phase HPLC for analysis of amino acids.

REFERENCES

1. D.H. Spackman, W. H. Stein and S. Moore, Anal Chem. 30:1190-1206 (1958).
2. S. Udenfriend, S. Stein, P. Bohlen, W. Dairman, W. Leimgruber and M. Weigele, Science, 178 (1972).
3. S. Stein and L. Brink, 1981, in: "Method in Enzymology," S. Pestka, ed., Vol. 79:20-25, Academic Press, New York.
4. N. Roth, Anal Chem. 43:880-882 (1971).
5. P. Bohlen and M. Mellet, Anal. Biochem. 94:313-321 (1979).
6. J. C. Hodgin, J. Liq. Chromatogr. 2:1047 (1979).
7. P. Lindroth and K. Mopper, Anal. Chem. 51:1667 (1979).
8. B. N. Jones, S. Paabo and S. Stein, J. Liq. Chromatogr. 4:565 (1981).
9. R. L. Heinrikson and S. C. Meredith, Anal. Biochem. 136:65-74 (1984).
10. M. K. Radjai and R. T. Hatch, J. Chromatogr. 196:319-322 (1980).
11. B. A. Bidlingmeyer, S. A. Cohen and T. L. Tarvin, J. Chromatogr. 336:93-104 (1984).
12. Y. Ishida, T. Fujita and K. Asai, J. Chromatogr. 204:143-148 (1981).
13. M. W. Dong and J. R. Gant, J. Chromatogr. 327:17-26 (1985).

HPLC ASSAY OF PHOSPHOAMINO ACIDS BY FLUORESCENCE DETECTION

Michael J. Watson, Joan R. Kanter, and Laurence L. Brunton

Divisions of Pharmacology and Cardiology, M-013H
University of California, San Diego
La Jolla, CA 92093

INTRODUCTION

Phosphorylation of proteins at serine residues and, less frequently, at threonine residues is an active area of research that has grown steadily for almost two decades as a consequence of the discovery that second messengers such as cyclic AMP, cyclic GMP and Ca^{++} can stimulate specific protein kinases (for a recent review, see reference 1). Interest in phosphorylation as a means of regulating protein function has increased recently with the findings that receptors for several polypeptide hormones such as insulin and epidermal growth factor are substrates for and apparently possess hormone-sensitive protein kinase activities with specificity for tyrosine residues of substrates[2-4]. The data of Hunter and Sefton[5] and others have provided additional stimulus: several viral gene products are tyrosine-directed protein kinases, implicating tyrosine phosphorylation as a key reaction in the malignant transformation of mammalian cells by viruses.

In contradistinction to the impression that one gets from textbook examples of protein phosphorylation such as the regulation of glycogen metabolism, the transfer of phosphate from ATP to a protein often occurs with no obvious or immediate functional consequence. Although protein kinase activities may be readily demonstrated by ^{32}P transfer, specific substrates of known importance cannot be easily identified. As a result, there is a great deal of current research on identifying substrates of protein kinases and on determining what amino acids are phosphorylated to what extent and with what physiologic consequence.

The standard technics for characterizing amino acyl phosphate involve transfer of ^{32}P to protein either by *in vivo* labelling of cellular ATP with [32]P_i or by the addition of exogenous activators and/or protein kinases along with [^{32}P]ATP to cell homogenates or subcellular fractions. Under these conditions, 10 to 100 ^{32}P-labelled proteins may be formed, as evidenced by autoradiographs of SDS-polyacrylamide gels separating proteins in the molecular weight range of 10,000 to 200,000 daltons. Specific proteins may be selected immunologically or by criteria such as molecular weight, but in any event, one or many protein substrates, undoubtedly along with many non-substrates, will then be hydrolyzed (usually in acid, since serine phosphate and threonine phosphate are not base stable). Hydrolysates are subjected to one or two dimensional electrophoresis/chromatography[6,7] or occasionally to anion exchange HPLC[8] and then the ^{32}P chromatographing with excess standards (detected by ninhydrin or by absorbance at 206 nm) is quantified. The exact method employed depends on the complexity of the sample; most workers rely on two dimensional electrophoresis.

In our experience, this arduous electrophoretic separation has drawbacks and could benefit from improvements such as:
1. greater speed and ease of use
2. more reliable and greater separation of phosphoamino acids from a variety of other [^{32}P]labelled compounds formed during incubation of whole cells with ^{32}P$_i$
3. the option for quantitation in chemical (absolute) terms in addition to radiochemical (relative) terms. At present, for instance, it is generally not possible to determine the quantity of phosphoamino acids not radiolabelled in a sample, an amount that could be large in the case of phosphate groups that turn over very slowly.

With these improvements as goals, we have developed one HPLC separation of phosphoamino acids and two HPLC separations of the o-phthaldialdehyde (OPA) derivatives of phosphoamino acids. The OPA derivatives can be sensitively detected in the column effluent by fluorescence, providing chemical quantitation in the sub-picomole range.

MATERIALS AND METHODS

We employed a Gilson high pressure liquid chromatograph consisting of two high pressure pumps and a mixer controlled by an Apple IIe computer using Gilson software. For anion exchange procedures we used a Whatman 10 μm SAX column, 4.6x250 mm; for the reverse phase separation we used a column of the same dimensions packed with 10 μm Lichrosorb RP18 from Merck. All solvents and buffers were Fisher HPLC grade, and both these, the OPA reagent, standards and samples were routinely filtererd with 0.45 μm Nuclepore polycarbonate membranes (buffers) or Millipore Millex GV or HV4 syringe filters (standards, samples).

Phosphoamino acid standards were made up as 10 mM stocks in water and were stored frozen (-20°C). Amino acids were made into fluorophores by derivatization with o-phthaldialdehyde (OPA)[9]. OPA reagent for pre-column derivitization was made by dissolving 40 mg OPA (Calbiochem) in 0.5 ml HPLC grade methanol, then adding 9.5 ml of 0.4 M Na borate (pH 9.5) and 10 μl 2-mercaptoethanol. This reagent is stable in the refrigerator for several weeks. All quantitations of sample phosphoamino acids were based on fluorescence standard curves generated using OPA reagent and phosphoamino acid standards made fresh that day.

For the detection of fluorescent derivatives, the column effluent was passed through a Kratos FS 970 fluorometer. This fluorometer collects emitted photons with a hemispherical reflector, thereby gathering a fraction of the emitted fluorescence that is substantially greater than that in instruments with the standard "right angle" configuration[10]. Column effluent was illuminated at 230 nm. Emitted light was monitored at 460 nm after passage through a Kratos 418 filter that effectively removed light of short wavelengths (<410 nm) and passed >99% of light with λ>440 nm (the emission maximum of OPA derivatives is 455-460 nm. The OPA derivatives actually have two usable absorption maxima: λ_1=230 nm, λ_2=340 nm. The longer wavelength would be appropriate for fluorometers utilizing mercury or xenon sources with weak radiant energy in the range of 230 nm. With a deuterium source (as in the Kratos FS 970), there is strong radiant emission in the low ultraviolet so that the 230 nm absorbance is preferred. In our experience, the lower excitation wavelength leads to lower baseline noise and a 3-fold enhancement of sensitivity compared to excitation at 340 nm.

Fluorescence data from the Kratos detector were integrated and quantified using the Gilson Datamaster and Apple computer. UV absorbtion of phosphoamino acids was monitored at 206 nm with a Kratos Spectraflow 757 detector.

RESULTS

Rapid Separation of Phosphoamino Acids by HPLC; Post-column Formation of OPA Derivatives

With the initial aim of rapidly separating phosphoamino acids, we applied them to an anion exchange column, detecting them in the effluent by their absorbance at 206 nm. As anticipated, manipulation of pH and organic phase gave an excellent separation in a 40 min run (Fig. 1). The elution differed from existing methods[8,11] in that it put a wide separation between phosphotyrosine (usually the smallest signal in biological samples) and phosphoserine (usually the largest signal). With the phosphotyrosine eluting first we reasoned that the signal to background ratio for phosphotyrosine should be substantially improved. This separation (Fig. 1) can be used in place of one and two dimensional electrophoretic separations and can be performed relatively quickly. Even the need for collecting fractions for scintillation spectrometry to detect ^{32}P could be circumvented by continuous monitoring of radioactivity in the effluent. If extensive separation of phosphotyrosine and phosphoserine is not required, by changing the mobile phase to 10% methanol in 20 mM KPO_4, pH 3.0, an adequate separation may be obtained in 14 min: phosphothreonine, 8.8 min; phosphotyrosine, 10.8 min; phosphoserine, 12.8 min.

Building on the separation summarized by Fig. 1, we attempted to add a post-column step to provide chemical rather than radiometric quantitation. We chose OPA derivatization, essentially following the procedure of Kucera[12]. Reacting equal portions of column effluent with OPA/borate reagent at room temperature for 2 min gave a good fluorescent signal with a useful sensitivity of about 40 pmoles of phosphoamino acid. However, the noise in the signal was considerable, due in part to the mixing artifacts from the OPA reagents and buffer. Thus, despite the extensive history of post-column derivatization for amino acid analysis[13], we reasoned that removal of excess reagents would lower background and improve sensitivity, thus we turned to pre-column derivatizing.

Formation and Separation of OPA Derivatives of Phosphoamino Acids

Roth[9] described the reaction of OPA with primary amines to make a fluorophore that could be used as a sensitive assay. In this reaction OPA, an SH donor and a primary amine interact at basic pH to form a 1,2 substituted isoindole[14] (Fig. 2). Although ethane thiol gives a more stable product[15], we found its stench so objectionable that we substituted 2-mercaptoethanol (dithiothreitol also works although it yields a somewhat less fluorescent adduct). The reagents are added in great excess so that the fluorescent indole is produced from amines over a large range. At room temperature, the reaction occurs rapidly (Fig. 3) and is stable for at least 90 min. In our buffer systems, the signal from phosphotyrosine exceeds those of equimolar phosphoserine and phosphothreonine, so that the method is conducive to detecting the most scarce phosphoamino acid. For routine use, we mix equal portions (15-60 μl) of sample and OPA reagent

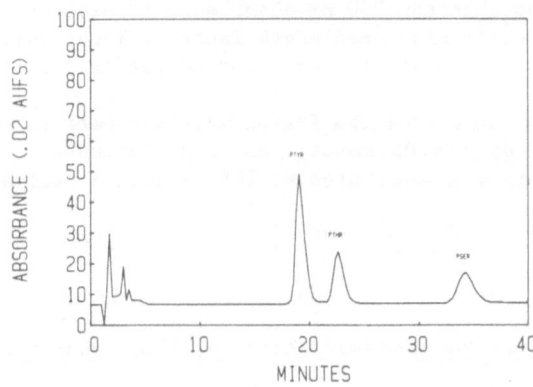

Fig. 1. Separation of phosphoamino acids by anion exchange HPLC. Phospho-amino acids (150 nmoles each of phosphoserine and phosphothreonine, 3 nmoles of phosphotyrosine) were added to a Whatman SAX column and eluted with 20 mM KPO_4, pH 3.0:acetonitrile::1:1 at 2 ml/min. Retention times are: 20.2 min for phosphotyrosine, 23.9 min for phosphothreonine, and 36.3 min for phosphoserine. Full scale absorbance=0.02 absorbance units; λ=206 nm.

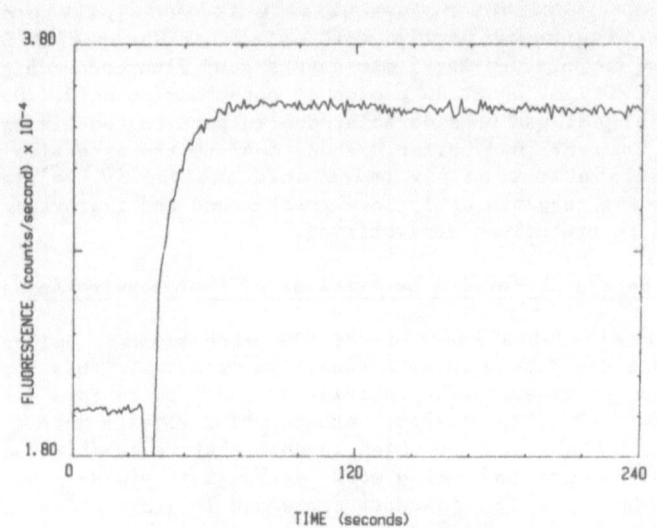

Fig. 2. Fluorogenic reaction of o-phthaldialdehyde and β-mercaptoethanol with a primary amino acid to yield a 1-alkylthio-2-alkyl isoindole as proposed by Simons and Johnson[14].

Fig. 3. Time course of reaction of phosphotyrosine and OPA reagent. At t=30 sec, 760 pmoles phosphotyrosine were added to stirring OPA reagent. Fluorescence (excitation=340 nm; emission=460nm) was monitored with a SPEX 111C Fluorolog spectrofluorometer in the photon counting mode.

228

(see Materials and Methods), and allow the reaction to proceed for exactly 2 min at room temperature before injecting a portion of the mixture onto the HPLC column.

OPA derivatives of phosphoamino acids can be separated by both anion exchange and reverse phase HPLC. From an anion exchange column (Fig. 4), we elute the OPA derivatives with 75 mM potassium phosphate (pH 6, 2 ml/min). Phosphothreonine elutes first (6.6 min) followed by phosphoserine (19.2 min) and phosphotyrosine (33.4 min). Increasing the salt concentration and lowering the pH could be used to speed the elution of the later compounds. In our hands, the isocratic elution works well and provides detectable fluorescence in the range of 2 pmol (see standard curve in inset to Fig. 4). Nonphosphorylated amino acids elute ahead of phosphothreonine, as do nucleotides, NAD and NADP (which form poor fluorescent adducts with OPA).

We obtain sharper peaks with reverse phase HPLC (Fig. 5), for which method we have found a gradient necessary for elution of the OPA derivatives of phosphotyrosine and phosphothreonine. Thus, using 50 mM potassium phosphate, pH 6.5 at 2 ml/min, with a linear methanol gradient (0 to 12% between 0 and 9 min), the average retention times are 11.5 min for phosphoserine, 25.8 min for phosphothreonine, and 29.9 min for phosphotyrosine and may decrease slightly with column age. The sharpness of the peaks makes the method sensitive to as little as 250 fmoles of phosphoamino acid (see insert in Fig. 5). In this method, the frequent contaminants aspartate and glutamate give noticeable peaks at 12 min and 19 min, respectively, thus we have avoided a gradient that would elute OPA-phosphoamino acids more quickly but that would superpose the signals with those of glutamate and aspartate. No other amino acids give signals in this elution; indeed, others are retained by the column and are eluted at day's end with a 100% methanol wash (following a water wash to remove phosphate buffer). As with the elution from the anion exchange column, nucleotides do not form fluorescent adducts that interfere.

Employing the Method for Biologic Samples

We believe that the pre-column derivatizing (Figs. 4 and 5) with fluorescence monitoring of column effluents offers sensitive assays for phosphoamino acids that are faster than current electrophoretic assays and provide a chemical estimate of phosphoamino acid content as well as the opportunity to count fractions for radioactivity. Thus, questions of stoichiometry and of pre-existing, metabolically stable phosphate can now be addressed simultaneously with questions about metabolically active phosphate (^{32}P incorporation) such as that transferred by protein kinases in response to hormonal stimuli.

The application of the method to biologic samples is not as easy as the process appears for phosphoamino acid standards. The preparation of samples for analysis has been well described by Cooper et al.[7] and by Martensen[16] and Martensen and Levine[17]. Two major problems are loss of covalently attached phosphate during hydrolysis and incomplete hydrolysis. One can recover only about 20% of amino acyl phosphate following preparative acid hydrolysis. In base, peptide bonds are quickly and quantitatively broken but serine and threonine phosphates are lost quantitatively. Tyrosine phosphate is largely retained following base treatment, so that this stability is a good criterion of phosphotyrosine. Problems that we are currently addressing are those of incomplete hydrolysis of sample proteins and removal of excess nonphosphoamino acids when large amounts of sample protein are employed.

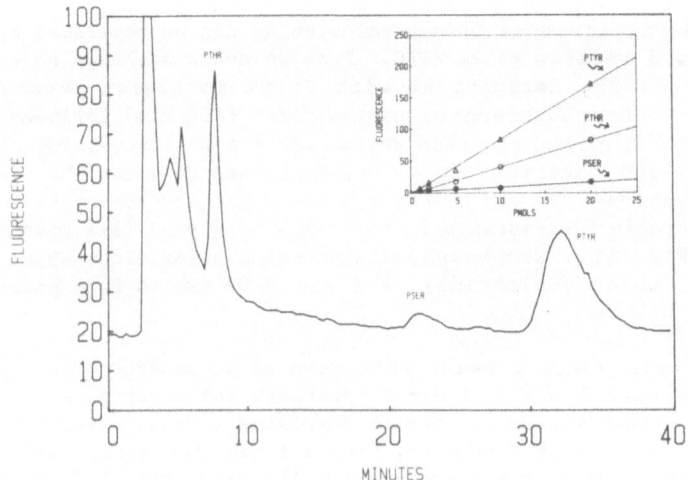

Fig. 4. Isocratic elution of OPA-phosphoamino acid adducts. Sample: 10 pmoles of each phosphoamino acid, derivatized with OPA. Buffer: 75 mM KPO₄, pH 6.0; flow 2 ml/min. Retention times: phosphothreonine, 6.6 min; phosphoserine, 19.2 min; phosphotyrosine, 33.4 min. Insert shows standard curve data. Column: Whatman SAX (250x4.6 mm). Detection: Kratos FS 970 fluorometer, 0.05 µA full scale.

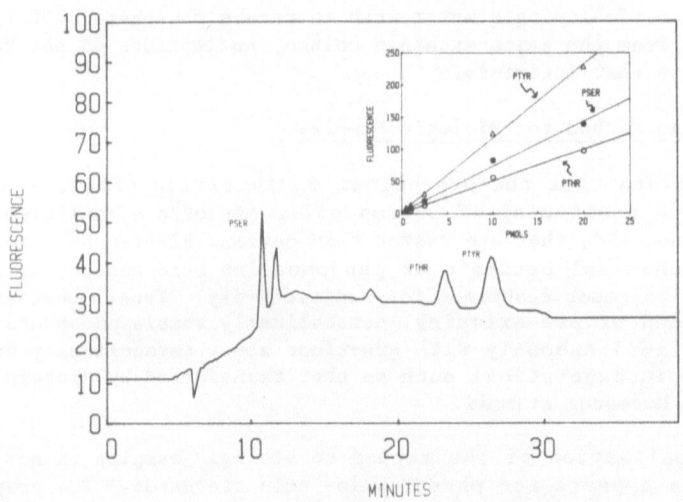

Fig. 5. Reverse phase gradient elution of OPA derivatives of phosphoamino acids. Sample: 5 pmoles of each phosphoamino acid, derivatized with OPA. Mobile phase: 50 mM KPO₄, pH 6.5, with a linear gradient of methanol, 0 to 12%, between 0-9 min, and continuing at 12% methanol to end of run; flow=2 ml/min. The retention times are: phosphoserine, 11.0 min; phosphothreonine, 23.2 min; and phosphotyrosine, 26.5 min. Common contaminants aspartate and glutamate give fluorescence peaks at 12 and 19 min. Insert shows standard curve. Column: Merck 10 µm Lichrosorb (250x4.6 mm). Detection: Kratos FS 970 fluorometer, 0.1 µA full scale.

In our experience, the agreement between ^{32}P incorporation and fluorescent adduct formation is good when model substrates are employed[18]. It would seem likely that use of both ^{32}P incorporation and the OPA method would provide an optimal procedure, whereby ^{32}P could monitor recovery of phosphoamino acid during sample preparation, so that the quantities of amino acyl phosphate detected by the OPA method could be corrected for losses.

ACKNOWLEDGMENT: We appreciate the support of NSF, AHA and Kratos Analytical.

REFERENCES

1. A.C. Nairn, H.C. Hemmings, Jr., and P. Greenguard, Protein kinases in the brain, Ann. Rev. Biochem. 54:931 (1985).
2. M. Kasuga, Y. Zick, B.L. Blith, F.A. Karlsson, H.U. Haring, and C.R. Kahn, Insulin stimulation of phosphorylation of the β subunit of the insulin receptor, J. Biol. Chem. 257:9891 (1982).
3. S. Cohen, G. Carpenter, and L. King, Jr., Epidermal growth factor-receptor-protein kinase interactions: co-purification of receptor and epidermal growth factor enhanced phosphorylation activity, J. Biol. Chem. 255:4834 (1980).
4. T. Hunter, Phosphotyrosine-A new protein modification, Trends Biochem. Sci. 7:246 (1982).
5. T. Hunter and B.M. Sefton, Transforming gene product of Rous sarcoma virus phosphorylates tyrosine. Proc. Natl. Acad. Sci. 77:1311 (1980).
6. L.J. Pike, B. Gallis, J.E. Casnellie, P. Bornstein, and E.G. Krebs, Edpidermal growth factor stimulates the phosphorylation of synthetic tyrosine-containing peptides by A431 cell membranes, Proc. Natl. Acad. Sci. 79:1443 (1982).
7. J.A. Cooper, B.M. Sefton, and T. Hunter, Detection and quantification of phosphotyrosine in proteins, Methods Enzymol. 99:387 (1983).
8. G. Swarup, S. Cohen, D.L. Garbers, Selective dephosphorylation of proteins containing phosphotyrosine by alkaline phosphatases, J. Biol. Chem. 256:8197 (1981).
9. M. Roth, Fluorescence reaction for amino acids, Analytical Chem. 43:880 (1971).
10. R. Weinberger and E. Sapp, Fluorescence detection in liquid chromatography, Am. Lab. 16:121 (1984).
11. J.C. Yang, J.M. Fujitaki, and R.A. Smith, Separation of phosphohydroxyamino acids by high performance liquid chromatography, Analytical Biochem. 122:360 (1982).
12. P. Kucera, and H. Umagot, Design of a post-column fluorescence derivatization system for use with microbore columns. J. Chromatography 255:563 (1983).
13. R. Pfiefer, R. Korol, J. Korpi, R. Bergoyne, and D. McCourt, Practical application of HPLC to amino acid analysis. Am. Lab. 15:78 (1983).
14. S.S. Simons, and D.F. Johnson, Reaction of o-phthalaldehyde and thiols with primary amines: formation of 1-alkyl (and aryl) thio-2-alkylisoindoles. J. Org. Chem. 43:2886 (1978).
15. S.S. Simons, and D.F. Johnson, Ethanethiol: a thiol conveying improved properties to the fluorescent product of o-phthalaldehyde and thiols with amines, Analytical Biochem. 82:250 (1977).
16. T.M. Martensen, Phosphotyrosine in proteins: stability and quantification. J. Biol. Chem. 257:9648 (1982).
17. T.M. Martensen, and R.L. Levine, Base hydrolysis and amino acid analysis for phosphotyrosine in proteins, Meth. Enzymol. 99:402 (1983).
18. Brunton LL, Kanter JR and Watson MJ, in preparation.

AMINO ACID ANALYSIS BY REVERSE-PHASE HIGH PERFORMANCE
LIQUID CHROMATOGRAPHY: SEPARATION OF PHENYLTHIOCARBAMYL
AMINO ACIDS BY SPHERISORB OCTADECYLSILANE COLUMNS

Chao-yuh Yang*+, Felix I. Sepulveda*,
Tseming Yang*, and Wei-yong Huang+

Departments of Medicine* and Biochemistry+
Baylor College of Medicine and The Methodist Hospital
Houston, TX 77030

INTRODUCTION

A liquid chromatographic approach to automatic amino acid analysis first became available to analytical laboratories in 1958[1]. Automated ion-exchange chromatography as developed by Moore, Stein, and Spackman has undergone refinements in the design of narrower bore columns, more sensitive flow cells, smaller resin particle size, and greater electronic amplification of spectroscopic signals. Improvements in the instrumentation design have made it possible to increase the sensitivity in the sub-nanomole range and decrease the analysis time from days to minutes[2].

The basic method remained essentially unchanged until 1983 and is still the most widely employed method for quantitative amino acid analysis. The facility and ease of operation of high performance liquid chromatography (HPLC) has been extended recently to the application of this instrument for the automated analysis of amino acids. The currently available approaches for HPLC analysis of amino acids involve either precolumn modification or postcolumn analytical methods. Precolumn modification methods for amino acid analysis are usually sensitive since they employ fluorescent or chromophore-coupled amino acid derivatives[3-7]. The use of o-phthaldehyde[8-11] also offers high-sensitivity postcolumn detection, but lacks proline detection and, without addition reaction, produces unstable fluorescent products. Finally, it is difficult to quantitate due to the sensitivity to quenching. Quantitation of proline and hydroxyproline requires special oxidative procedures, thus greatly complicating the design of o-phthaldehyde based analyzers. The use of dansyl chloride, a precolumn modifier, is hampered by slow derivative formation, and thus the excess of reagent needed to insure complete reaction of the amino acid interferes in the separation process[6]. An alternative precolumn derivatization system, employing dimethylaminoazobenzene-4'-sulfonyl chloride, also presents difficulties in quantitative reaction with all amino acids, thus limiting the application of this approach[7].

The role of the reagent phenylisothiocyanate [PITC] in amino terminal degradation, discovered by Edman in 1956[12], has become increasingly important. Edman realized early on the advantages of quantitative reaction of PITC with the N-terminal of amino acids. Thus this method survives as the best quantitative N-terminal analytical procedure. The application of PITC for precolumn derivatization of amino acids and the quantitative analysis of PTC derivatives by RP-HPLC has been developed by Heinrikson and Meredith[13]. A commercial system, Pico-Tag, is available from WATERS Associates. The present study describes the use of the Edman reagent PITC as a means for quantitative precolumn derivatization of amino acids with a slightly modified procedure, and the separation of 22 PTC amino acids[14] or 17 PTC amino acids by a RP-HPLC system. Results thus obtained from analysis of HCl hydrolysates of several peptides compare favorably both in sensitivity and precision with those from state of the art ion-exchange analyzers.

MATERIALS AND METHODS

Materials

PITC, triethylamine [TEA], a standard mixture of amino acids, individual amino acids, and 6 N HCl ampules were obtained from Pierce Chemical Co.; acetonitrile (HPLC grade) was obtained from Burdick and Jackson Laboratories Inc.; deionized water was further purified by a Milli-Q water system; oxidized insulin α-chain and glucagon were purchased from Sigma. T19, T21, and T25 peptides were purification products of a tryptic digest of rabbit apolipoproteins A-I[15].

Coupling of Amino Acids with PITC

PITC amino acids were prepared according to the procedure as described by WATERS Associate Pico-Tag procedure[16] with minor modifications. The procedure involves two steps; 1) hydrolysis of the samples, 2) derivatization of the hydrolysate. An aliquot of the peptide to be hydrolyzed (approximately half of the amount of the sample with peak height OD = 0.1), was dried in small pyrex tubes (6 x 50 mm); these tubes then were placed in the reaction vial (WATERS Associates) together with 200 μl of 6 N HCl and 20 μl of 0.1% phenol. The vial was sealed under vacuum for hydrolysis at 110°C for 24 hrs or 150°C for 1 hr. The hydrolyzed samples were dried in a Speedvac (Savant Instruments Inc.) and redried by adding 20 μl of an ethanol solution (1:1:1=EtOH:water:TEA v/v) to remove trace amounts of ammonia.

For derivatization of the hydrolysates, the samples were coupled with 20 ul of PITC solution (EtOH:water:TEA:PITC=7:1:2:1, v/v) for 10 min and dried again in a Speedvac. The sample diluent, 30 μl of a 0.5 M sodium phosphate buffer, pH 7.4, with 5% acetonitrile, was used to reconstitute the samples; 15 μl of the sample were applied on the column and analyzed as described below.

Preparation and Analysis of Amino Acid Standards

The 22 amino acid standard was prepared by mixing 10 μl containing 1 nmol each of amino acid (CMC, Trp, Gln, Cys-SO₃H, hydroxy-Pro, and Asn) with 10 μl of 1 nmol of the Pierce amino acid mixture (Asp, Glu, Ser, Gly, His, Thr, Ala, Arg, Pro, Tyr, Val, Met, Cys, Ile, Leu, Phe, Lys), and then diluted to 1 nmol/100 μl. After evaporation of the amino acid mixture, the 22 PTC amino acids were prepared directly through derivatization, whereas the Pierce amino acid mixture (17 different residues) was hydrolyzed and derivatized along with other samples using the procedure as described before.

Reverse Phase HPLC Separation of PTC Amino Acids

The HPLC instrument used in this study was a Hewlett Packard HP 1090 Liquid Chromatograph. A column packed with Spherisorb ODS II 3 μm (4.6 x 150 mm) was used for the separation of PTC amino acids. The PITC derivatives were monitored at 254 nm. The conditions for the elution were as follows; column temperature - 47°C; flow rate - 0.8 ml/min; eluent A - 0.153 M sodium acetate, 0.05% TEA and 6% acetonitrile (pH 6.4); eluent B - 40% water in acetonitrile; gradient - linear gradient from 0% to 12% B in 10 min, then from 12% to 48% B in 10 min, and finally from 48% B to 100% B in 0.1 min; then isocratic flow at 100% B for 5 min and back to 0% B in 5 min.

RESULTS AND DISCUSSION

The elution profile of 30 pmol of the 17 PTC amino acids on Spherisorb ODS column (4.6 x 150 mm) is shown in Fig. 1. All 17 PTC amino acids were separated completely in 21 min.

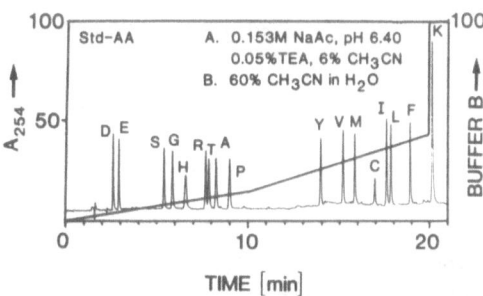

Fig. 1. Gradient elution of 17 PTC-amino acids from a 150 x 4.6 mm Spherisorb ODS-II 3 μm column. Flow rate: 0.8 ml/min. Temperature: 47°C. Sample amount: 30 pmol of each PTC-amino acid.

The characteristics of PTC-amino acid separation on an ODS column with sodium acetate-CH₃CN solvent system are similar to those of dimethylaminoazobenzenethiohydantoin-amino acids and PTH-amino acids. Namely, 1) the retention times of acidic and basic PTC amino acids can be affected by pH changes of the buffer, 2) the retention times of PTC-His and PTC-Arg can be altered by variation of the salt concentration[17-20], and 3) the retention times of PTC-His and PTC-Arg gradually changes during the lifetime of the column[21-24]. An elevated column temperature improves the resolution of the PTC-amino acids and reduces column pressure.

The retention time of PTC-Arg can be shortened by increasing the salt concentration of buffer A. Hence, a better separation of the PTC-Arg peak can be achieved. Using the experimental conditions described, the chromatographic position of PTC-Arg's can be shifted from a position between PTC-His and PTC-Thr to a position just after elution of PTC-Pro. This can be accomplished by changing the sodium acetate concentration from 0.153 M to 0.03 M. Standard analyses were achieved using 50 pmol of each of the 22 PTC amino acids on Spherisorb ODS column, including 0.03 M sodium acetate in buffer A[14]. All amino acid derivatives, including the ammonia and an unknown peak, were well resolved in 22 min. Better resolution of PTC-CysSO₃H and PTC-Asp was possible using an isocratic elution of buffer A for 1.5 min.

Oxidized insulin α-chain, glucagon, and the peptides obtained from the digestion of apolipoprotein A-I[15] were used to compare the experimentally obtained material composition to the known compositions. The result of the amino acid analysis data is shown in Table 1 and 2. It indicates that the optimum level of amino acids detection is at 30 pmol. Our results, as shown in Fig. 1, indicate a reasonable reliability to about 10 pmol. Amino acid analysis by RP-HPLC method has been used extensively in our laboratory with satisfactory results. In this report we have shown the results of only a few examples, which reflect our cumulative experiences using this methodology.

Table 1. Amino acid analysis of glucagon and oxidized insulin A chain by RP-HPLC of PTC-amino acids

Amino Acid	Glucagon		ox. Insulin α-chain	
Cys-SO₃	--		4.12	(4)
Asp	4.10	(4)	1.88	(2)
Thr	2.92	(3)	--	
Ser	3.82	(4)	1.99	(2)
Glu	3.00	(3)	4.22	(4)
Pro	--		--	
Gly	1.08	(1)	1.09	(1)
Ala	1.18	(1)	1.00	(1)
Val	1.10	(1)	1.74	(2)
Met	1.10	(1)	--	
Ile	--		0.76	(1)
Leu	1.86	(2)	2.26	(2)
Tyr	1.92	(2)	1.94	(2)
Phe	1.89	(2)	--	
His	0.82	(1)	--	
Lys	1.11	(1)	--	
Arg	2.10	(2)	--	
Trp	n.d.[a]	(1)	--	
Total	29		21	

[a] n.d. not determined

The main advantages of the RP-HPLC method are the following: 1) excellent resolution of all 22 PTC-amino acids, 2) shortened analysis time in comparison to the original method of PTC-amino acid analysis, 3) increased sensitivity of the detection to the 10 pmol level, and 4) substantially decreased solvent consumption due to the low flow rate.

Modern HPLC instrumentation is capable of improving the sensitivity of analysis through reduction of diameter and length of the column, and through improvements of the packing material. The application of microbore columns for PTC-amino acids analysis has great potential for further increasing the sensitivity of amino acid analysis.

Table 2. Amino acid analysis of 3 tryptic peptides of 2 rabbit apo lipoprotein A-I (T19, T21, and T25)[b]

Peptide	T19	T21	T25
Residue positions	172 - 175	181 - 193	225 - 236
Amino Acid			
Asp			1.7 (2)
Thr			0.9 (1)
Ser		0.8 (1)	0.8 (1)
Glu		3.1 (3)	2.2 (2)
Pro			
Gly		3.2 (3)	
Ala	1.9 (2)	3.0 (3)	1.8 (2)
Val			2.1 (2)
Met			
Ile			
Leu	1.1 (1)	1.0 (1)	1.1 (1)
Tyr		0.9 (1)	
Phe			
His			
Lys		0.9 (1)	1.3 (1)
Arg	1.0 (1)		
Total	4	13	12

[b] Amino acid sequence of peptides: T19 = LAAR;
                                                             T21 = EGGGASLAEYQAK;
                                                             T25 = ASVQNLVDEATK

Acknowledgments

We thank Susan Kelly for the artwork and Marjorie Sampel for her assistance in the preparation of this manuscript. This work was supported in part by the Specialized Center of Research on Atherosclerosis grant (HL-27341) and AHA-TX 85G-202.

References

1. Spackman, D.H, Stein, W.H., and Moore, S. Automatic Recording Apparatus for Use in the Chromatography of Amino Acids, Anal. Chem. 30:1190 (1958).
2. Hare, P.E. Subnanomole-Range Amino Acid Analysis in: "Methods in Enzymology, XLVII Enzyme Structure Part E", C.H.W. Hirs and Serge N. Timasheff, eds. Academic Press, New York (1973).
3. Voelter, W. and Zech, K. High-Performance Liquid Chromatographic Analysis of Amino Acids and Peptide-Hormone Hydrolysates in the Picomole Range, J. Chromatogr. 112:643 (1975).
4. Svedas, K.V.J., Galaev, J.L., Borisov, I.L., and Berezin, I.V. The Interaction of Amino Acids with O-phthaldehyde: A Kinetic Study and Spectrophotometric Assay of the Reaction Product Anal. Biochem. 101:188 (1980).

5. Hodgin. J.C.   The Separation of Pre-Column O-phthaldehyde Derivatized Amino Acids by High Performance Liquid Chromatograpy   J. Liquid Chromatogr. 2:1047 (1979).

6. Yamabe, T., Takai, N., and Nakamura, H.   Analysis of Dns-Amino Acids by Liquid Chromatography.   I, Selection of Optimum Mobile Phase Composition for Separation of Dns-Amino Acids on Polyvinyl Acetate Gel J. Chromatogr. 104:359 (1975).

7. Chang, J.Y., Knecht, R., and Braun, D.G.   Amino Acid Analysis at the Picomole Level, Application to the C-terminal Sequence Analysis of Polypeptides Biochem. J. 199:547-555 (1981).

8. Benson, J.R. and Hare, P.E.   O-Phthaldehyde: Fluorogenic Detection of Primary Amines in the Picomole Range, Comparison with Fluorescamine and Ninhydrin Proc. Nat. Acad. Sci. USA 72:619-622 (1975).

9. Cronin, J.R. and Hare, P.E.   Chromatographic Analysis of Amino Acids and Primary Amines with O-Phthaldehyde Detection   Anal. Biochem. 81:151-156 (1977).

10. Pfeifer, R., Karol, R., Korpi, J., Burgoyne, R., and McCourt, D.   Practical Application of HPLC to Amino Acid Analysis   Amer. Lab. March, 78-87 (1983).

11. Bohlen, P. and Schroeder, R.   High-Sensitivity Amino Acid Analysis; Methodology for the Determination of Amino Acid Compositions with less than 100 Picomoles of Peptides Anal. Biochem. 126:144-152 (1982).

12. Edman, P.   On the Mechanism of the Phenyl Isothiocyanate Degradation of Peptides Acta Chem. Scand. 10:761 (1956).

13. Heinrikson, R.L. and Meredith., S.C.   Amino Acid Analysis by Reversed-Phase High-Performance Liquid Chromatography: Precolumn Derivatization with Phenylthiocyanate Anal. Biochem. 136:65-74 (1984).

14. Yang, C.Y. and Sepulveda, F.I.   "Separation of Phenylthiocarbamyl amino acids by high-performance liquid chromatography on Spherisorb octadecylsilane columns" J. of Chromatography 346:413-416 (1985).

15. Yang, C.Y., Yang, T.M., Pownall, H.J., and Gotto, A.M. Jr.   "The complete primary structure of Apolipoprotein A-I from rabbit high density lipoprotein", submitted to Biochem. Biophys. Res. Commun. (1985).

16. WATERS Associates, Pico-Tag amino acid Analysis System Operator's Manual, manual number 88140.

17. Yang, C.Y. and Wakil, S.J.   Separation of Dimethylaminoazobenzenethiohydantoin Amino Acids by High-Performance Liquid Chromatography at Low Picomole Concentrations Anal. Biochem. 137:54-57 (1984).

18. Sugden, J., Cox, G.B., and Loscombe, C.R.   Chromatographic Behavior of Basic Amino Compounds on Silica and ODS-Silica Using Aqueous Methanol Mobile Phases J. Chromatogr. 149:377-390 (1978).

19. Lundanes, E., and Greibrokk, T.   Reversed-Phase Chromatography of Peptides J. Chromatogr. 149:241-254 (1978).

20. Aunan, W.D.   Separation of Phenylthiohydantoin-Amino Acids by High Performance Liquid Chromatography J. Chromatogr. 173:194-197 (1979).

21. Johnson, N.D., Hunkapillar, M.W., and Hood, L.E.   Analysis of Phenylthiohydantoin Amino Acids by High-Performance Liquid Chromatography on Dupon Zorbax Cyanopropylsilane Columns   Anal. Biochem. 100:335-338 (1979).

22. Scottrup-Jensin. L., Petersen, T.E., and Magnusson, S.   Analysis of Amino Acid Phenylthiohydantoins by High-Performance Liquid Chromatography Using Gradient Elution with Ethanol   Anal. Biochem. 107:456-460 (1980).

23. Dimasi, S.J., Robinson, J.P., and Hash, J.H.   Use of a Ternary Gradient for the Separation of Phenylthiohydantoin-Amino Acids Including the Methyl Esters of Aspartic and Glutamic Acids, by High Performance Liquid Chromatography J. Chromatogr. 213:91-97 (1981).

24. Lottspeich, F.   Identification of the Phenylthiohydantoin Derivatives of Amino Acids by High Pressure Liquid Chromatography Using a Ternary, Isocratic System Hoppe-Seyler's Physiol. Chem. 361:1829-1834 (1980).

V. Mass Spectrometry of Polypeptides

FAST ATOM BOMBARDMENT MASS SPECTROMETRY: APPLICATION

TO PEPTIDE STRUCTURAL ANALYSIS

Blair A. Fraser

Division of Biochemistry and Biophysics
CDB, Food and Drug Administration
Bethesda, MD

Study of structure-function relationships for biological molecules
is a central theme of contemporary biomedical research. Once isolated
and purified, these molecules require structural characterization before
they can be used effectively to study their function in the living
organism. One important subset, polypeptides, are often only available
in picomole quantities. To structurally analyze these small quantities
has been one goal of protein chemistry. Several methods have been
developed to determine the amino acid composition, amino acid sequence,
and covalent chemical structure of polypeptides. Amino acid analysis
requires only picomoles of polypeptide [1]. The amino acid sequence for a
polypeptide can be determined from picomole amounts by automated
sequential Edman degradation followed by high performance liquid
chromatographic identification of the resulting phenylthiohydantoin amino
acids [2]. The molecular weight and other covalent structures can be
inferred from the mass spectrum of picomoles of the polypeptide.
Integration of these technologies is useful in providing structural
information about these small quantities.

Entry of mass spectrometry into protein structural analysis is not
a recent occurrence. For over 20 years, mass spectrometry has
supplemented traditional methods of amino acid sequence analysis [3,4].
Over the last decade, developments in ionization methods, analyzer
design, and computer technology have now made possible the recording of
mass spectra for underivatized polypeptides as large as trypsin
(MW = 23,643) [5,6].

In the mass spectrometer, molecules are ionized, a beam of ions is
produced, and this beam is focused through a combination of electric and
magnetic fields to give a mass spectrum. The ionized molecule, or
molecular ion, is a high energy species which can subsequently decompose
yielding daughter ions and neutral fragments. The mass spectrometer
measures the mass and abundance of each ion and, aided by computer, these
can be displayed as either a histogram or a table. The mass spectrum is
a pattern that can be interpreted and then used, along with other
chemical information, to postulate a structure for the molecule. The
interpretation of the spectrum is not by any means unique but needs to
be used along with other chemical information.

In the past, peptides presented a problem for mass spectrometry because they are polar, labile, involatile polymers. To be analyzed by mass spectrometry, peptides needed to be derivatized. To avoid derivatizing peptides, new approaches to sample introduction and ionization were investigated. Fast atom bombardment [7] and other methods of desorption ionization [8] have emerged over the past several years as solutions to this problem. Combining fast atom bombardment ionization with high mass range analyzers [9], a variety of peptides have now been analyzed [10,11].

Analysis of peptides by fast atom bombardment mass spectrometry requires attention to sample preparation and to operation of the instrumentation. Gained from the mass spectrum can be additional information that may be used in a total approach to structurally analyze a polypeptide, be it from biosynthetic or chemical synthetic origin.

Preparation of the peptide for analysis by fast atom bombardment mass spectrometry [11] is not unlike that required for analysis by sequential Edman degradation. The peptide needs to be a single, homogeneous chemical species, free of contaminants such as salts, other peptides, or other organic molecules that may contribute to increased noise in the mass spectrum. This chemical noise will serve to complicate and confuse the spectral interpretation. High performance liquid chromatography, employing volatile solvents and chemically-bonded, silica-based, chromatographic supports, should always be used as the final step of sample purification. A one microliter volume containing the peptide, dissolved in its appropriate acidified, aqueous, organic solution, is then delivered to one microliter of matrix, glycerol or other low vapor pressure liquid [12], on the sample probe tip, preferably made of gold. The sample probe is introduced into the mass spectrometer source where its tip is bombarded by a beam of noble gas atoms, usually xenon. The matrix and sample molecules are desorbed, ionized during this process, and accelerated from the probe tip. Under computer control, mass spectra are sequentially recorded, usually once every 30 seconds or less. The file of mass spectra is accumulated by the computer for real-time storage on magnetic media and for subsequent display as softcopy and hardcopy.

MOLECULAR WEIGHT DETERMINATION

Once a peptide has been purified, it can then be characterized by amino acid analysis [1] and fast atom bombardment mass spectrometry (Figure 1). It is prudent to obtain both an amino acid composition and fast atom bombardment mass spectra on the smallest quantity of peptide, reserving the bulk of the peptide for additional structural and biological experiments. In most cases, committing only 50 to 100 picomoles of the peptide sample is necessary for analysis by fast atom bombardment mass spectrometry. Since the mass limit, at highest sensitivity, of new magnetic sector instruments has reached 10,000, most peptides purified on octadecylsilanyl-silica high performance liquid chromatography columns are suitable candidates for fast atom bombardment mass spectral analysis.

Once the fast atom bombardment mass spectrum and amino acid analysis have been obtained on about 200 picomoles of peptide, these data can be examined and the peptide can be initially characterized. Comparison of the protonated molecular weight, $(M+H)^+_{AAA}$, calculated from the amino acid composition, with the mass for the protonated molecular ion, $(M+H)^+_{FAB}$, determined by fast atom bombardment mass spectrometry, usually results in one of three conditional relationships (Figure 1).

242

STEP ONE

- Calculate $(M+H)^+$ from Amino Acid Analysis

- Determine $(M+H)^+$ by FAB-MS

STEP TWO

- If $(M+H)^+_{FAB}$ Less Than $(M+H)^+_{AAA}$
    - Presence of ASN, GLN, PCA, α-CONH$_2$
    - Loss of Neutral Species
    - Cyclic Peptide

- If $(M+H)^+_{FAB}$ Equals $(M+H)^+_{AAA}$
    - Absence of ASN, GLN, PCA, α-CONH$_2$
    - Free Amino Terminus
    - Absence of TRP

- If $(M+H)^+_{FAB}$ Greater Than $(M+H)^+_{AAA}$
    - Prosthetic Groups
    - Protecting Groups
    - Oxidized TRP, MET, CYS
    - Undetected or Unusual Amino Acids
    - Reduced Disulfide

ADDITIONAL CONSIDERATIONS

- If $(M+H)^+_{FAB}$ Appears as Multiplet
    - Polyisotopic Elements

- If $(M+H)^+_{FAB}$ Changes with Time
    - Reducible Functionalities
    - Chemical Reactions

- If $(M+H)^+$ Displays Satellite Intensities
    - Cationated Molecular Ions
    - Adduct Ions
    - Multiply-Charged Ions
    - Polymeric Species

Figure 1: Strategy for Peptide Characterization

If the $(M+H)^+_{FAB}$ is less than the $(M+H)^+_{AAA}$, several features of the peptide can be inferred. Since the mass of a carboxamide is lower than a carboxyl group, asparagine, glutamine, or other carboxamides may be present. Illustrated in Figure 2, α-endorphin displays a protonated molecular ion at 1746 mass units while the protonated molecular weight calculated from the amino acid analysis is 1747. This difference is due to the glutamine, a γ-carboxamide, detected as glutamic acid, a γ-carboxylic amino acid, by amino acid analysis. Loss of water or ammonia due to either cyclization of an amino acid residue, e.g., glutamine to pyroglutamic acid, asparagine to aspartimide, cyclization of the peptide, or facile elimination from an amino acid residue or a known prosthetic group may account for a lower than predicted mass for the protonated molecular ion. The protonated molecular ion will also appear at lower mass if an intramolecular disulfide bond has formed and remained intact. Rapid loss of a neutral species from the protonated molecular ion may result in a higher intensity for a fragment ion than for the protonated molecular ion.

If the $(M+H)^+_{FAB}$ equals the $(M+H)^+_{AAA}$, usually little needs to be inferred from this relationship. In most cases, the peptide will have a free amino terminus, no carboxamides, and no undetected amino acids. It is important to be aware of the peptide's lineage prior to analysis because unusual amino acids and cyclic peptides will affect this conditional generalization.

If the $(M+H)^+_{FAB}$ is greater than the $(M+H)^+_{AAA}$, exciting investigation awaits. Prosthetic groups, introduced either as post-translational modifications or by isolation procedures, add mass to the peptide. For synthetic peptides, this condition indicates that protecting groups may not have been fully removed or covalent adducts may have formed during deprotection [13]. Oxidized, uncommon, or undetected amino acids may also account for this increased peptide mass. Illustrated in Figure 2, luteinizing hormone-releasing hormone displays a protonated molecular ion at 1183 mass units (1182.57mu) while the protonated molecular weight calculated from the amino acid analysis is 1015 (1015.48). Addition of tryptophan (186.08mu), presence of the α-carboxamide (16.03mu) instead of the α-carboxyl group (17.01mu), and loss of water (18.01mu) due to formation of the pyroglutamyl amino terminus explains the mass difference of 167.07 mass units.

Careful inspection of the molecular ion region in the fast atom bombardment mass spectrum can reveal additional details (Figure 1). If the protonated molecular ion appears as a multiplet, the peptide may either have amino acids containing characteristic polyisotopic elements, such as sulfur or selenium, or have been modified by a reagent containing a polyisotopic element, such as bromine. Certain prosthetic groups, covalently-attached to the peptide, or certain peptides may sequester polyisotopic ions, e.g., heme-containing peptides, valinomycin.

Examining several sequentially accumulated fast atom bombardment mass spectra provides additional information about any changes in the peptide during the experiment. Reduction of intact disulfide bonds in the peptide will result in a time-dependent appearance of a multipartite protonated molecular ion envelope [10,19]. Illustrated in Figure 2, Lys_8-vasopressin displays a protonated molecular ion at 1056 mass units. The intensity at 1058 mass units is more intense than expected for the isotopic contribution by sulfur suggesting that partial reduction of the disulfide is occurring. Chemical reactions that may occur, such as

• If (M + H)⁺_FAB Less Than (M + H)⁺_AAA

β-ENDORPHIN   Y G G F M T S E K S Q T P L V T

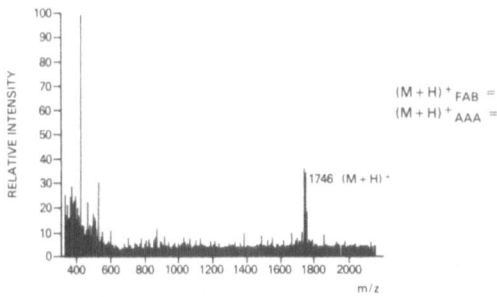

$(M + H)^+_{FAB} = 1746$
$(M + H)^+_{AAA} = 1747$

• If (M + H)⁺_FAB Greater Than (M + H)⁺_AAA

LUTEINIZING HORMONE –   < E H W S Y G L R P G NH₂
RELEASING HORMONE

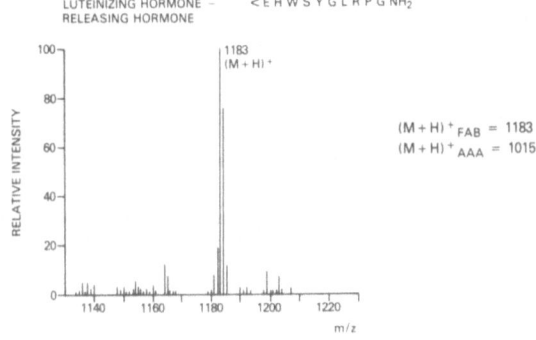

$(M + H)^+_{FAB} = 1183$
$(M + H)^+_{AAA} = 1015$

• If (M + H)⁺_FAB Changes with Time

Lys₈ VASOPRESSIN   H - C Y F Q N C P K G - NH₂

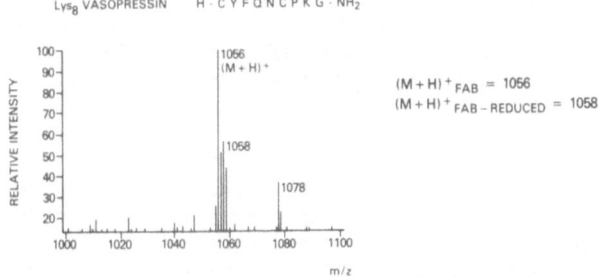

$(M + H)^+_{FAB} = 1056$
$(M + H)^+_{FAB - REDUCED} = 1058$

• If (M + H)⁺ Displays Satellite Intensities

ANGIOTENSIN Ⅲ   R V Y I H P F

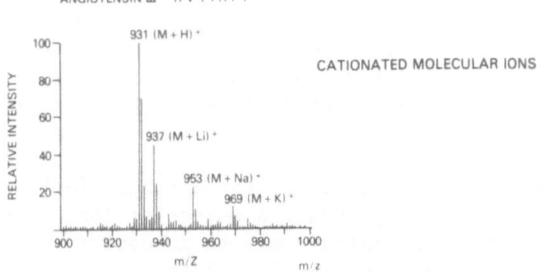

CATIONATED MOLECULAR IONS

Figure 2: Examples of Peptide Characterization

hydrogenation [10], dehydrogenation, dehydration, or dehydrohaloge-
nation [15] yield protonated molecular ion envelopes that change in
appearance over the course of the experiment.

Satellite intensities may arise during the desorption which can
help confirm and verify the identity of the protonated molecular ion. In
general, no matter how pure the water used during the isolation, the
peptide usually displays cationated molecular ions. Most commonly, the
natriated molecular ion, $(M+Na)^+$, appears 22 mass units higher than the
protonated molecular ion. In some instances the kaliated molecular ion,
$(M+K)^+$, is also present. Cationated molecular ions are evident in the
spectrum for angiotensin III (Figure 2). A five molar excess of alkali
hydroxides was added to the peptide prior to analysis. In addition to
the protonated molecular ion at 931 mass units, the lithiated, natriated,
and kaliated molecular ions appear at 937, 953, and 969 mass units,
respectively. Other inorganic and organic cations present with the
peptide, such as ammonia and magnesium, may also appear as cationated
molecular ions.

Adduct ions, those formed by non-covalent attachment of a neutral
species to the protonated molecular ion, appear in the spectra of some
peptides. Materials mixed with the peptide, such as silicone oils,
detergents, and matrix components, contribute to formation of these
adduct ions. Frequently, glycerol from the matrix will attach to the
protonated molecular ion, displaying an intensity 92 mass units higher.

Not frequently observed when analyzing small molecules by other mass
spectrometric methods, peptides analyzed by desorption ionization mass
spectrometry frequently display multiply-protonated molecular ions.
Stabilization of two positive charges on the peptide molecule results in
formation of the doubly-protonated molecular ion, $(M+H_2)^{++}$, which appears
at half the mass that observed for the protonated molecular ion, $(M+H)^+$.
Likewise, the triply-protonated molecular ion, $(M+H_3)^{+++}$, will appear at
one-third the mass that observed for the $(M+H)^+$. Observation of
multiply-protonated molecular ions for a particular peptide helps to
verify the identity of the protonated molecular ion. Multiply-cationated
molecular ions of the peptide are sometimes observed.

Additionally, if several molecules of the peptide attach to each
other, molecular clusters of the peptide may be observed. These
clusters, arising as $(2M+H)^+$, $(3M+H)^+$, etc., also help to confirm the
identity of the protonated molecular ion. In some instances, these
ionized molecular clusters may appear as a multipartite ion intensity
envelope due to both covalent and non-covalent attachment of the peptide
molecules. For example, oxytocin, which contains a single intramolecular
disulfide bond, displays a $(2M+H)^+$ ion envelope composed of three major
intensities; an intensity for the protonated dimer with intact disulfide
bonds, an intensity for the protonated dimer with one intact disulfide
bond, and an intensity for the protonated dimer with no intact disulfide
bonds [14]. Protonated molecular clusters may also display cationated
molecular clusters, implying cluster formation prior to analysis or
during bombardment.

Not every peptide sample analyzed by fast atom bombardment mass
spectrometry will display a protonated molecular ion. Increasing the
amount of peptide used for analysis may be necessary to increase the
signal for the protonated molecular ion above noise. Those peptides
containing covalently bound prosthetic groups usually require more
material for analysis. Peptide mixtures present special problems since

the protonated molecular ions for each may not be distinguishable [11]. Increasing the quantity of the peptide mixture or varying matrix conditions may help. Excess organic and inorganic compounds co-isolated with the desired peptide may increase spectral background and thereby decrease signal-to-noise. The intrinsic charge and polarity of the peptide may influence the analytical outcome. Generally, basic peptides give more intense protonated molecular ions [11]; acidic peptides give more intense proton-abstracted molecular ions [16].

AMINO ACID SEQUENCE INFORMATION

Once the peptide's molecular weight and composition have been determined, amino acid sequence analysis can be undertaken. If the peptide has a free amino terminus, sequential Edman degradation of a portion of the peptide would be most expeditious. If the amino terminus is blocked, the carboxy terminus is free, and the peptide is less than twenty amino acid residues, the peptide can be digested by carboxypeptidase and, at various times, the mixture of peptide homologs can be analyzed by fast atom bombardment mass spectrometry [17,18]. The several spectra can then be used to infer the amino acid sequence. If both termini of the peptide are blocked, enzymatic digestion, separation of the peptide fragments, and characterization by amino acid analysis, fast atom bombardment mass spectrometry, and sequential Edman degradation will provide complementary data that can be used to propose a structure for the peptide. If the peptide exceeds the instrumental molecular weight limits for these technologies, peptide fragments will still need to be used for a rigorous proof of structure.

The entire fast atom bombardment mass spectrum for the peptide can be examined to discern a pattern of ion intensities that may indicate the peptide's structure. Along with other chemical or enzymatic sequence data, this fragmentation pattern should be consistent with the proposed structure for the peptide. In addition to the protonated molecular ion, fast atom bombardment mass spectra display fragment ions resulting from decomposition of the protonated molecular ion. These fragment ion intensities are usually less than 10% the intensity of the protonated molecular ion.

The two most common pathways of decomposition [10], important for discerning the amino acid sequence, are: 1) peptide bond fission with retention of the positive charge on the amino terminal fragment ion and loss of the carboxy terminus as a neutral, yielding predominantly the acylium and aldiminium ion series; 2) peptide bond fission with retention of the positive charge on the carboxy terminal fragment ion and loss of the amino terminus as a neutral fragment, yielding predominantly the ammonium ion series. Each peptide bond is a likely site for fission, depending upon the particular peptide bond strength. The two patterns of fragment ion intensities in the mass spectrum are then interpreted and used along with the other chemical data, to propose a structure for the peptide.

Certain amino acids create ambiguities in the mass spectrum that can usually be resolved by amino acid analysis and sequential Edman degradation. For example, isoleucine, leucine, and hydroxyproline have identical nominal masses; lysine and glutamine have identical nominal masses. Mass spectrometric distinction of hydroxyproline and lysine is possible after chemical modification but leucine and isoleucine need to be distinguished by sequential Edman degradation. Certain peptide structures display characteristic fragment ion intensities, e.g., losses of water from serine or threonine, preferred fragmentation adjacent to a disulfide bond [14].

APPLICATION TO STRUCTURAL ANALYSIS

Several technologies, including high performance liquid chromatography, amino acid analysis, sequential Edman degradation, solid phase peptide synthesis, and mass spectrometry, can be integrated into an overall strategy for the covalent structural analysis of peptides and proteins. Obtaining an accurate molecular weight and an amino acid composition for a polypeptide is a useful first step in structural analysis. New methods of desorption ionization mass spectrometry allow molecular weight determination for polypeptides of molecular weight 23,000. The limit of this technology may not yet have been reached.

Once the peptide's molecular weight and amino acid composition are determined, the peptide may be fragmented and these fragments characterized by a repeat of the same methodology, this time using sequential Edman degradation and fast atom bombardment mass spectrometry. After the structure for the peptide has been determined, the peptide can be synthesized by solid phase chemical methods [13] or biosynthesized employing an appropriate expression system.

The vast majority of polypeptide amino acid sequences have and will continue to be determined by sequential Edman degradation and deduction from the DNA sequence. But throughout the entire process of structurally characterizing a polypeptide, the mass spectrometer can play a useful role. Obtaining an accurate molecular weight, consistent with amino acid composition, for each peptide can help to confirm and verify the amino acid sequence obtained by sequential Edman degradation and deduced from the DNA sequence [19]. Before chromatography, examining a crude synthetic peptide mixture will indicate if the peptide is present. Surveying chromatographic fractions for either the desired synthetic peptide or deletion peptide will help in designing synthetic and chromatographic strategies. Identifying peptides with prosthetic groups, carboxamidated amino acids, uncommon amino acids, or modified amino acids will assist sequential Edman degradation efforts. Evaluation of an enzymatic digest of a polypeptide may help to determine the extent of digestion or its peptide map [19]. Monitoring enzymatic modification of a peptide will aid understanding post-translational modifications. Just the appearance of the desired protonated molecular ion will help to determine if the suspected prosthetic group has been covalently or non-covalently attached to the peptide.

These three technologies, amino acid analysis, sequential Edman degradation, and fast atom bombardment mass spectrometry, can be integrated into a powerful tool for structural analysis of polypeptides. The gene contains the template for polypeptide biosynthesis but the phenotypic expression of that polypeptide's function lies beyond the ribosome. Its function as an effector or receptor lies, in part, in its covalent chemical structure. Determining that covalent structure can now be made easier.

REFERENCES

1. R.L. Henrikson and S.C. Meredith, Amino acid analysis by reverse-phase high performance liquid chromatography: precolumn derivatization with phenylisothiocyanate, Anal. Biochem. 136:65 (1984).

2. M. W. Hunkapiller and L. E. Hood, New protein sequenator with increased sensitivity, Science 207:523 (1980).

3. K. Biemann, Sequencing of proteins, Int. Jour. of Mass Spectrom. and Ion Physics 45:183 (1982).

4. H. R. Morris, Biomolecular mass spectrometry, Int. Jour. of Mass Spectrom. and Ion Physics 45:331 (1982).

5. R. D. MacFarlane, Particle-induced desorption mass spectrometry of large involatile biomolecules: surface chemistry in the high-energy short-time domain, Acc. Chem. Res. 15:268 (1982).

6. B. Sundqvist, I. Kamensky, P. Hakansson, J. Kjellberg, M. Salehpour, S. Widdiyasekera, J. Fohlman, P.A. Peterson and P. Roepstorff, Californium-252 plasma desorption time of flight mass spectroscopy of proteins, Biomed. Mass Spectrom. 11:242 (1984).

7. M. Barber, R. S. Bordoli, R. D. Sedgwick, A. N. Tyler, and B. N. Green, Fast atom bombardment mass spectrometry, in: Recent Developments in Mass Spectrometry in Biochemistry, Medicine, and Environmental Research, Volume 8," A. Frigerio, ed., Elsevier, Amsterdam (1983).

8. K. L. Busch and R. G. Cooks, Mass spectrometry of large, fragile, and involatile molecules, Science 218:247 (1982).

9. A. Dell and G. W. Taylor, High-field-magnet mass spectrometry of biological molecules, Mass. Spectrom. Revs. 3:357 (1984).

10. A. M. Buko, L. R. Phillips, and B. A. Fraser, Peptide studies using a fast atom bombardment high field mass spectrometer and data system. 2. Characteristics of positive ionization of peptides, m/z 858 to m/z 5729, Biomed. Mass Spectrom. 10:408 (1983).

11. A. M. Buko, L. R. Phillips, and B. A. Fraser, Peptide studies using a fast atom bombardment high field mass spectrometer and data system. 1. Sample introduction, date acquisition, and mass calibration. Biomed. Mass Spectrom. 10:324 (1983).

12. J. L. Gower, Matrix compounds for fast atom bombardment mass spectrometry, Biomed. Mass Spectrom. 12:191 (1985).

13. G. Barany and B. Merrifield, Solid-phase peptide synthesis, in: "The Peptides, Volume 2," E. Gross and J. Meienhofer, eds., Academic Press, New York (1980).

14. A. M. Buko and B. A. Fraser, Peptide studies using a fast atom bombardment high field mass spectrometer and data system. 4. Disulfide-containing peptides, Biomed. Mass Spectrom. (in press).

15. S. K. Sethi, C. C. Nelson, and J. A. McCloskey, Dehalogenation reactions in fast atom bombardment mass specrometry, Analyt. Chem. 56:1975 (1984).

16. A. M. Buko, L. R. Phillips, and B. A. Fraser, Peptide studies using a fast atom bombardment high field mass spectrometer and data system. 3. Negative ionization: Mass calibration, data acquisition and structural characterization. Biomed. Mass Spectrom. 10:387 1983.

17. C. V. Bradley, D. H. Williams, and M. R. Hanly, Peptide sequencing using the combination of Edman degradation, carboxypeptidase digestion, and fast atom bombardment mass spectrometry, Biochem. Biophys. Res. Commun. 104:1223 (1982).

18. R. Self and A. Parente, The combined use of enzymatic hydrolysis and fast atom bombardment mass spectrometry for peptide sequencing, Biomed. Mass Spectrom. 10:78 (1983).

19. B. W. Gibson and K. Biemann, Strategy for the mass spectrometric verification and correction of the primary structures of proteins deduced from their DNA sequences, Proc. Natl. Acad. Sci. U.S.A. 81:1956 (1984).

PROTEIN SEQUENCE DETERMINATION AND FAB MASS SPECTROMETRY

Terry D. Lee, Kassu Legesse, and Vickie Spayth

Beckman Research Institute of the City of Hope

Division of Immunology, Duarte, CA 91010

Introduction

From the beginning, FAB (fast atom bombardment) mass spectrometry has held great promise for the structural analysis of peptides and proteins[1,2]. As the technique matures, it is evident that it will play an increasingly important role in primary structure determinations[3]. The purpose of this report is to define areas where FAB can be used to advantage for protein sequence determination with emphasis on the complimentary nature of this mass spectral technique with respect to more classical methods such as amino acid analysis and Edman microsequencing. The ability of FAB to analyze peptides directly without prior derivatization has facilitated the incorporation of this technique into protein sequencing strategies.

Strategic Overview

The primary structure determination of an unknown peptide or protein begins with what is hopefully a pure sample. At this stage, a number of operations can be performed including N- and C-terminal sequence analysis, amino acid composition analysis and mass spectral analysis (Figure 1). These techniques are generally sufficient to sequence small peptides unless sample amounts are severely limited, or unusual structural features are encountered.

For higher molecular weight samples, it is necessary to divide the molecule into more managable sized pieces using enzymes or some other chemical means. There is little that can be done directly with the mixture.

Components are usually separated by HPLC chromatography to yield individual peptide fractions. This separation also yields a chromatogram often referred to as a "map" which is characteristic of the mixture. The HPLC map of a digest can be used to compare proteins that have a large degree of homology, or to compare different preparations of the same protein. Another map known as a "FAB map" results from mass spectral analysis of the digest mixture. Unlike an HPLC map, the components of the FAB map can be largely predicted from the sequence, and molecular weight information is generally more useful that HPLC retention times.

Once the individual peptide fragments have been isolated, the structural analysis problem and methods available to solve it are the same as before, except that now parts of the molecule are found in many separate sample tubes. Not only must each piece be sequenced; but each piece must be properly positioned with respect to the rest. FAB analysis is ideally suited for dealing with multiple samples since only a few minutes is required for each one. A single mass spectrometer can readily handle a sample load that would keep several automated microsequencers busy.

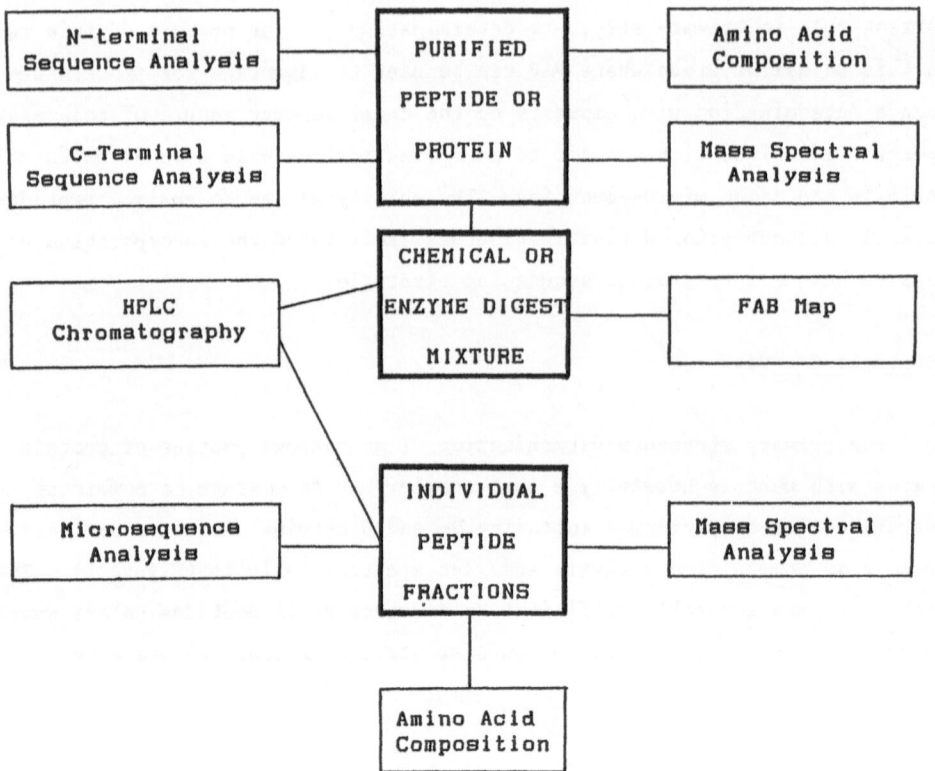

Fig. 1. Diagram of analytical methods available for protein primary structure determination.

FAB peptide spectra are characterized by relatively intense protonated molecular ions and weaker fragment ions. Thus, little difficulty is encountered in determining the molecular weight. For peptides of 100 residues or less, FAB analysis provides a more accurate determination of the molecular weight than is available using any other method. The technique has yet to be applied to any peptide greater than 100 residues. A related technique known as plasma desorption mass spectrometry has successfully been used to analyze proteins as large as 25000 amu [4]. However, it requires special instrumentation not generally available to protein chemists.

The ability to analyze higher molecular weight compounds has necessitated certain adjustments in the way mass values are expressed. For lower molecular weight compounds such as leucine enkephalin there is little difference between nominal mass (calculated using the nearest integral mass value of the most abundant isotope of each element), monoisotopic mass (calculated using the exact mass of the most abundant isotope), or average mass (calculated using the weighted average of the naturally occurring isotopes) (Fig. 2A). As peptide molecular weights increase, mass defects (deviations from nominal mass values) increase and ion clusters become more complex due to the increased probability of having one or more of the less abundant isotopes in a given molecule.

Additionally, there is a significant difference between calculated protonated molecular ion clusters and those observed in a FAB spectrum. Ion clusters are superimposed on a background of peaks at every mass (Fig. 3) which can alter relative intensities of the various peaks in the cluster. Generally, for peptides less than 4000 amu, it is possible to assign with confidence the first peak in the protonated molecular ion envelope as the monoisotopic mass. At higher masses, it becomes increasingly difficult to distinguish the first peak of the cluster from the spectral background. Certainly for an unknown peptide as large as human proinsulin (Figure 2G) or even bovine insulin (Figure 2F) it would be impossible to select the monoisotopic mass. Consequently, there is little advantage in obtaining spectra at unit mass resolution[5].

At lower mass spectrometer resolutions, the individual peaks in the cluster are not resolved. The center of the peak containing all the contributions from different isotopes is close to the calculated average mass value for the ion. Taking spectra at lower instrument resolution settings greatly

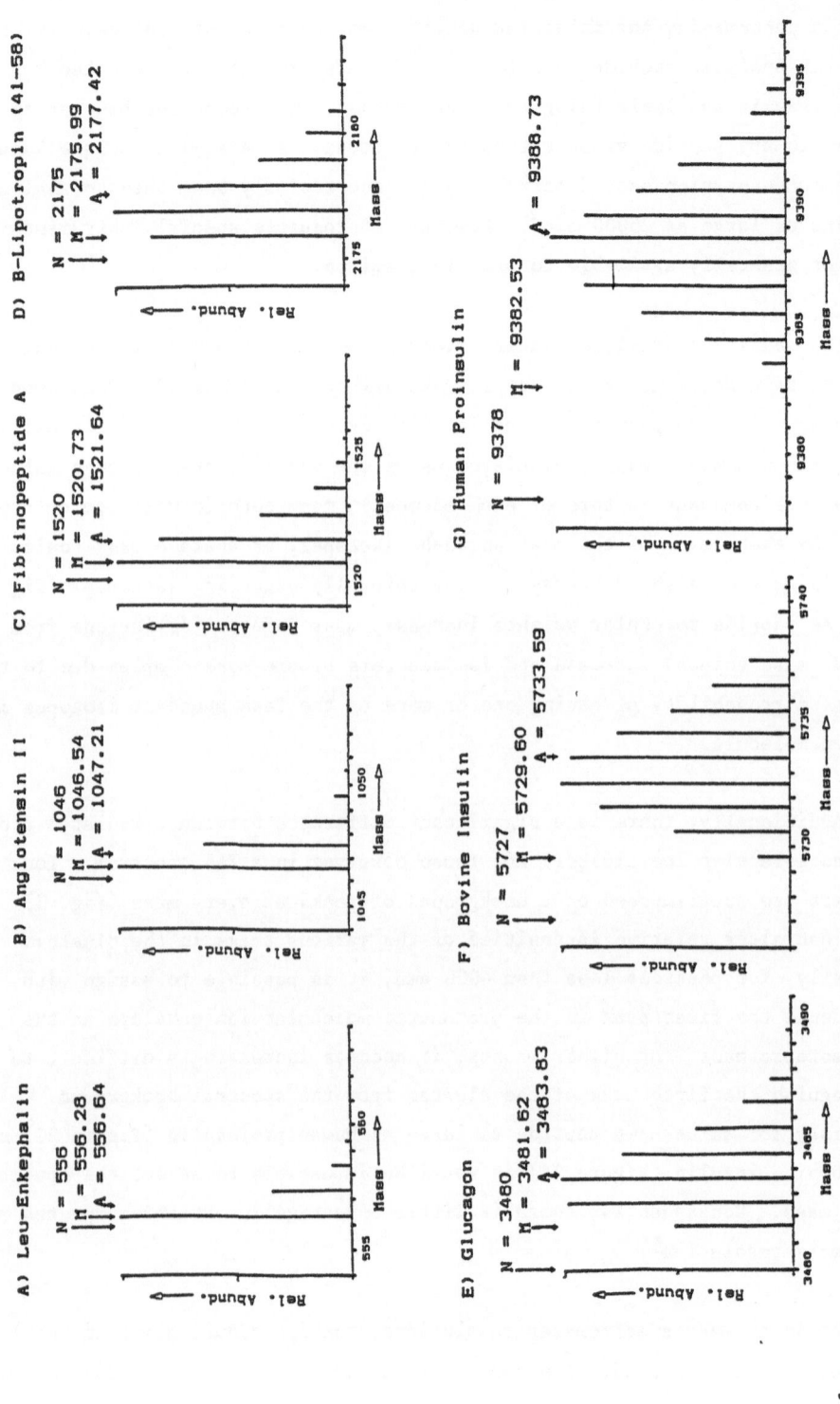

Fig. 2. Nominal (N), Monoisotopic (M), and Average (A) mass values for the protonated molecular ion clusters of selected polypeptides.

enhances the sensitivity of the measurement, and thus reduces the amount of material needed. This is particularly important for high molecular weight ions for which sensitivities can be rather low due to inefficient ionization, detection and transmission through the instrument.

The utility of low resolution mass measurements is illustrated in the case of three peptide hormones isolated from the atrial gland of <u>Aplysia</u> <u>californica</u>[6]. The hormones have molecular weights in excess of 6000 amu and consist of two chains joined by a single disulfide bond. Single scans at 500 resolution (A representative spectrum for one of the hormones is given in Figure 4) were sufficient to confirm that the small chain (peak A) was common to all three peptides, and that differences among the peptides were due to variability in the large chain (peak B)[7]. Also evident in the spectra were doubly protonated molecular ions (MH_2^{++}) which are often observed in the FAB spectra of larger peptides.

Once the region of the molecular ion was located, a more accurate determination of the molecular weight was made by accumulating several scans over a narrow mass range. Using sample amounts of less than 30 pmole (as determined by amino acid composition) molecular weight determinations were made on all three aplysia peptides that were accurate to within 0.4 amu, sufficient to confirm the expected amidation of the C-terminus.

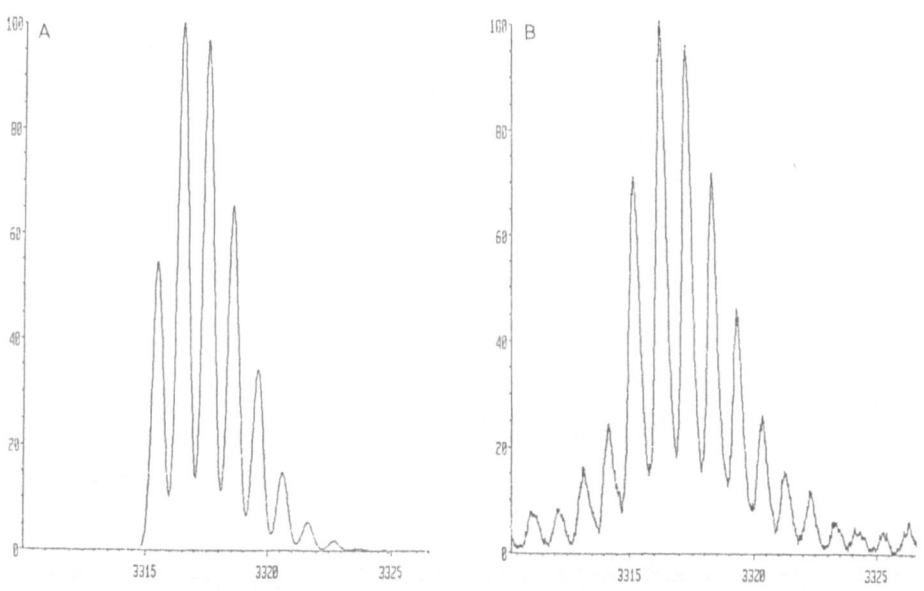

Fig. 3. Calculated (A) and observed (B) ion distribution for the protonated molecular ion cluster of a peptide with the elemental composition $C_{146} H_{220} O_{48} N_{41}$.

For lower molecular weight peptides, sensitivity is limited by interference from the liquid matrix, a necessary part of the FAB measurement. The choice of which matrix to use is rarely obvious. Factors to be considered can be illustrated by considering the analysis of 50 pmole of neurotensin 1-8 in three different matrixes; glycerol (Figure 5A), dithiothreitol: dithioerythitol (5:1) (DTs) (Figure 5B)[8], and thioglycerol (Figure 5C). In glycerol, the molecular ion is less intense than with the DTs or thioglycerol. However, the pattern of the background peaks is very regular and reproducible. Even for lower molecular weight peptides it is often very easy to pick our the molecular ion because it does not fit the pattern. The reproducibility of its spectrum makes glycerol an ideal matrix for background subtraction. A second advantage is that the sample lifetime in the ion source can be as long as 30 min.

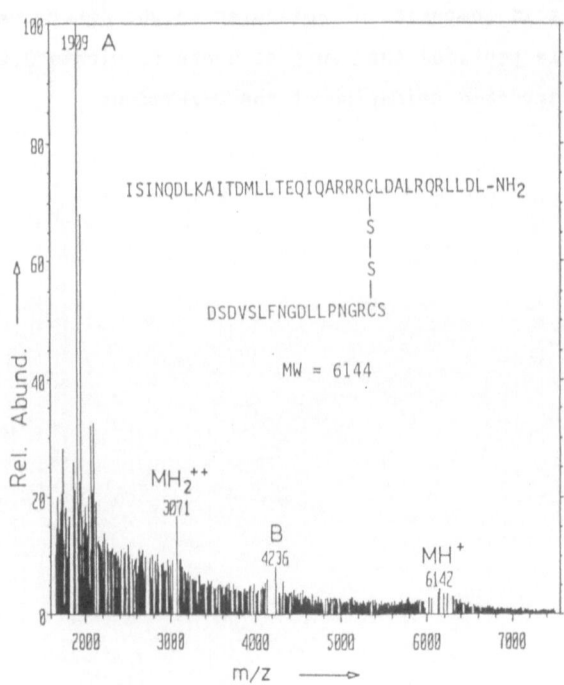

Fig. 4. Structure and low resolution survey scan for Califin A. Mass spectrum was obtained using a JEOL HX100HF mass spectrometer operating at 3KV accelerating potential and a resolution setting of 500. The peptide (12 pmole) was dissolved in approx. 1 μl of a 5:1 mixture of dithiothreitol and dithioerythritol. Total scan time for the mass range m/z 1650 to 7500 was 165 s.

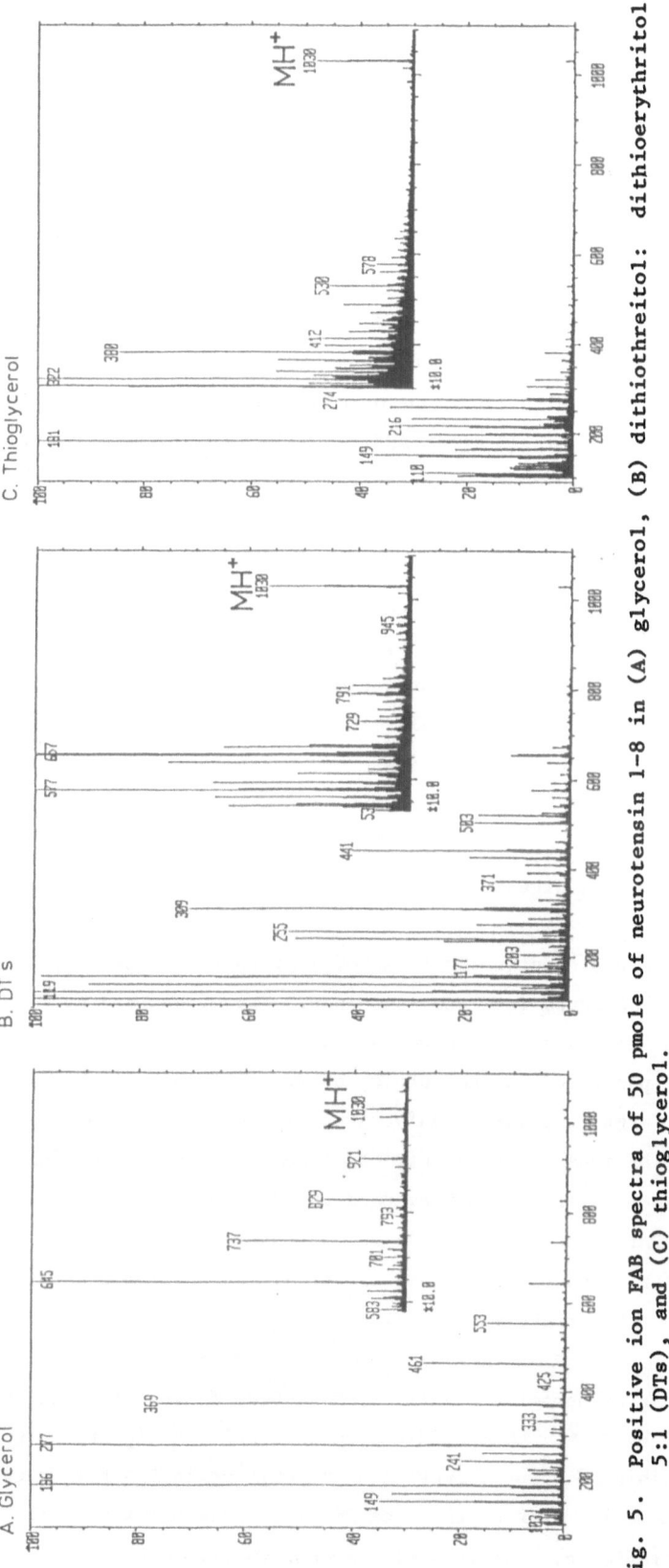

Fig. 5. Positive ion FAB spectra of 50 pmole of neurotensin 1-8 in (A) glycerol, (B) dithiothreitol: dithioerythritol 5:1 (DTs), and (C) thioglycerol.

Thioglycerol nearly always enhances protonated molecular ion intensities and matrix interference is minimal beyond mass 600. Below mass 600, the background is complicated and changes with time. Hence, it is very difficult to obtain good background subtractions. Sample lifetime in the source is only a few minutes due to the volatility of the matrix. The relative intensity of the protonated molecular ion to the matrix ion can change significantly from scan to scan.

The DTs are something of a middle ground between glycerol and thioglycerol. Molecular ions are intense and long lasting. Although interference from the matrix is significant below mass 700, it is generally more reproducible than that of thioglycerol Additionally, adducts with sodium and potassium are more likely to be observed in this matrix.

For a molecular weight determination, all that is necessary is to detect the protonated molecular ion. Minimum sample requirements depend on the properties and purity of the peptide, but generally lie in the range of 1 to 100 pmole. By itself, the molecular weight is of little value. It does not give sequence or even composition information. However, when used in conjunction with other data it provides a firm numerical value with which all other experimental conclusions must agree with or the structure is not solved.

Analysis of Enzyme Digest Mixtures

When the FAB spectrum is taken of a mixture of peptides, a molecular ion is generally observed for each component. Exceptions are usually low molecular weight peptides which have low ionization efficiencies, compete poorly for surface positions in the matrix, or are obscured by matrix ions or fragments from larger peptides. Unfortunately, relative intensities of protonated molecular ions are a function of other factors in addition to concentration; so FAB analysis can not be used to measure the amount of each peptide. Even so, a FAB map can be used to rapidly check large portions of protein structure for which the structure is largely known from cDNA sequencing[9] or homology to other proteins.

A number of hemoglobin variants have been determined using the FAB mapping technique[10-12]. When oxidized whole globin is digested with trypsin, 20 of the possible 29 tryptic fragments are evident in the FAB spectrum of the mixture (Table 1). Six of the nine missing fragments are small, having 1-2 amino acid residues. The remaining 3 contain tryptophan and are destroyed

Table 1. Protonated Molecular Ion m/z Values and Sequences for the
Tryptic Fragments of Oxidized Human Hemoglobin. Those Observed
in the Positive Ion FAB Spectra of the Tryptic Digest Mixture
are Indicated with an Asterisk.

		ALPHA CHAIN			BETA CHAIN
No.	MH+ m/z	Sequence	No.	MH+ m/z	Sequence
1 *	729.4	VLSPADK	1 *	952.5	VHLTLPEEK
2 *	461.3	TNVK	2	932.5	SAVTALWGK
3	532.2	AAWGK	3 *	1314.7	VNVDEVGGEALGR
4 *	1529.8	VGAHAGEYGAEALER	4	1274.8	LLVVYPWTQR
5 *	1103.5	ZFLSFPTTK	5 *	2090.9	FFESFGDLSTPDAVZGNP
6 *	1834.0	TYFPHFDLSHGSAQVK	6	246.2	VK
7 *	398.2	GHGK	7 *	412.2	AHGK
8	147.1	K	8	147.1	K
9 *	3028.5	VADALTNAVAHVDDZPNALSALSDLHAHK	9 *	1669.9	VLGAFSDGLAHLDNLK
10	288.2	LR	10 *	1469.7	GTFALSELHODK
11 *	818.4	VDAPVNFK	11 *	1126.6	LHVDPENFR
12 *	3015.6	LLSHOLLVTLAAHLPAEFTPAVHASLDK	12 *	1768.0	LLGNVLVOVLAHHFGK
13 *	1252.7	FLASVSTVLTSK	13 *	1378.8	EFTPPVQAAYQK
14	338.3	YR	14 *	1239.7	VVAFVANALAHK
			15	319.3	YH

O = Cysteic Acid
Z = Methionine Sulfone

during the oxidation. Peptides containing tryptophan are prominant in the
FAB spectrum of the tryptic digest of unoxidized hemoglobin. In this
instance oxidation was used to obtain complete digestion of the core region.

In the best of cases, all that is needed to determine the location the
compare the FAB map of the normal protein tryptic digest to that of the
abnormal protein. In the case of the Setif variant[13], a new peak is evident
at m/z 866[12] (Fig. 6). Only a substitution of Tyr for Asp or Val for Phe in
the αT11 fragment can account for the observed mass difference. The substitution of Val for Phe can be ruled out based on the behavior of the alpha
chain during electrophoresis. Since only one of the four genes coding for
the protein is altered, the normal αT11 fragment is observed at m/z 818.

Often it is necessary to isolate the abnormal peptide in order to establish the exact position of the substitution. In the case of hemoglobin Tarrant[12,14], the FAB map indicates that the mass of αT12 has shifted from m/z
3015.5 to 3014.5. This one mass unit difference is consistant with amidation
of either the Asp or Glu residue. In this particular instance, the location
of the substitution could be determined based on fragment ions in the FAB
spectrum of the isolated peptide. However, as an added check, the αT12

peptide was subjected to chymotryptic digestion. Peptides in the resulting mixture were consistant with the substitution of Asn for Asp and not with substitution of Gln for Glu (Fig. 7).

FAB Analysis and Automated Edman Microsequencing

The contribution that FAB can make to automated Edman microsequencing can be illustrated by considering the recently completed sequence determination of Apolipoprotein AII from murine strain BALB/c[15]. The complete peptide is 78 residues long and blocked at the amino terminus with a pyro-

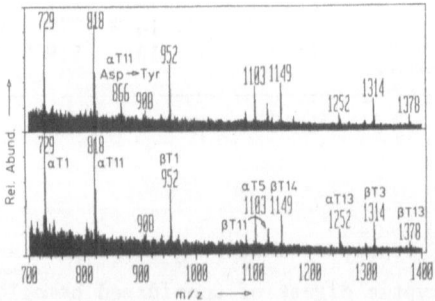

Fig. 6. FAB map of the tryptic digest mixture for oxidized Setif variant hemoglobin (upper) and normal oxidized hemoglobin. Spectra taken with a JEOL HX100HF mass spectrometer operating at 5 KV accelerating potential and a resolution setting of 3,000. Approx. 10 nmole of the peptide mixture was disolved in approx. 1.5 µl of glycerol on a 1mm X 6mm stainless steel sample stage.

glutamic acid residue (Fig. 8). Enzymatic removal of the blocking group permitted an amino terminal microsequecing run complete to the 39th residue. The tryptic digest of the blocked peptide yielded 14 fragments. Complete sequences were obtained by automated Edman microsequencing for only four of the tryptic peptides. The others were either blocked at the amino terminus, had questionable calls at certain cycles, or failed to sequence to the end of the peptide. Amino terminal peptides which could not be sequenced could be assigned based on the mass value of the protonated molecular ions and

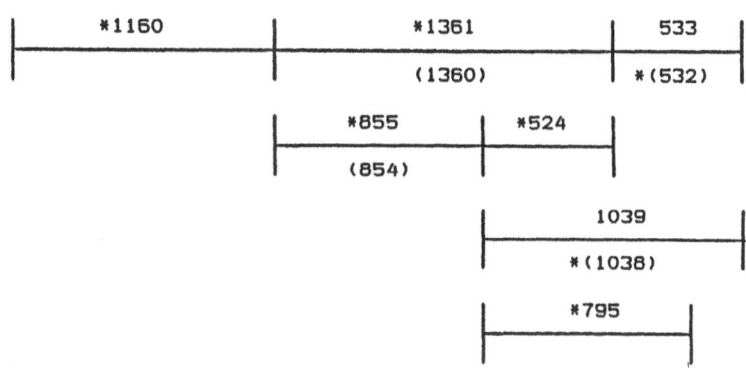

MW = 3016   L L S H C'L L V T L A A H L P A E F T P A V H A S L D K
  (3015)

Fig. 7. Sequence for oxidized alpha hemoglobin tryptic fragment 12 and
single amino acid substitutions (in parentheses) consistant with
the protonated molecular ion observed for Tarrant variant. Ions
observed in the FAB spectrum for the chymotryptic digest of the
Tarrant variant are indicated with an asterisk. (C' = cysteic
acid).

information from the N-terminal microsequencing run. Those which gave
partial sequences (Fig. 8, mass values 1218, 1830, 1483, and 1171) were
similarly assigned.

The C-terminal tryptic peptide was sequenced through to the first proline. Amino acid composition analysis indicated a proline and three alanine
residues still to be accounted for, in agreement with the molecular weight
determined from the FAB spectrum. C-terminal sequence analysis using carboxypeptidase Y yielded only alanine. The mass spectrum was sufficiently
intense that sequence fragment ions were evident sufficient to position the

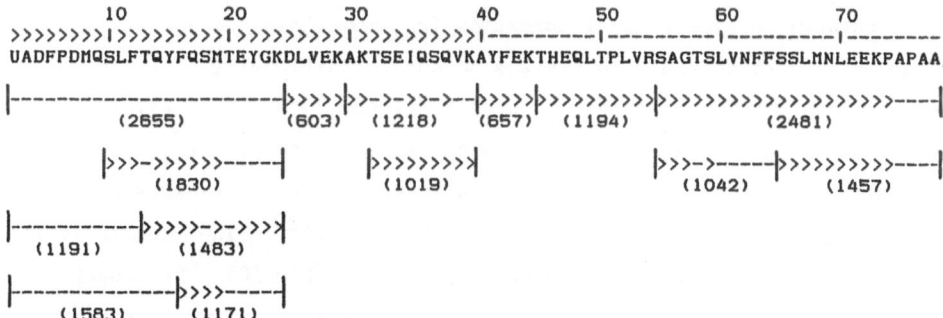

Fig. 8. Sequence and tryptic fragments for apolipoprotein AII isolated
from murine BALB/c. Values for protonated molecular ions are
given in parentheses. Residues sequenced by automated Edman
degradation are indicated by >.

final four residues. Peptides which did not overlap with the amino terminal sequence were aligned by homology to apolipoproteins from other species.

While microsequencing can be done without the aid of mass spectral analysis, the efficiency of the operation is greatly improved by utilizing information from FAB spectra. The mass of a peptide fragment is a hard piece of evidence which can often be used to resolve any ambiguities in a microsequencing run. Additionally, partially sequenced peptides can often be completed using information from mass spectral fragment ions. By helping to account for and complete the sequence of all tryptic fragments, FAB analysis can assist greatly in determining just when enough work has been done and the structure is solved.

Sequence Information from FAB Spectra

Sequence information in FAB spectra is contained in a number of fragment ion series[3]. Basically, there are three different bond types in the peptide backbone: between the alkyl carbon and the carbonyl carbon, between the carbonyl carbon and the amide nitrogen, and between the amide nitrogen and the alkyl carbon. Cleavage of any of these bonds can occur either homolytically to produce a radical ion or heterolytically with or without hydrogen transfer. The charge can be located on either the N-terminal or C-terminal portion. Thus a total of 18 ion series are possible in positive ion spectra. Fortunately, certain of these occur much more frequently than the others although any of them may be found in a given spectrum.

It is relatively straight forward to assign structures to fragment ions in the spectrum of a known peptide simply by looking for the mass values predicted by the sequence. It is much more difficult to interpret the spectrum of an unknown peptide since information from more than one ion series must often be used to obtain a complete sequence. Additionally, more than one sequence can usually be found which will fit the data. Some efforts have been made to computerize the analysis of FAB peptide spectra[16]. A program currently under development systematically constructs possible sequences for six of the more common fragment ion series (Fig. 9)[17]. The process is illustrated for the spectrum of the hexapeptide RE2 (Fig.10). Prominent ions in the spectrum are manually selected and entered into the interpretation program along with the molecular weight and amino acid composition. In this instance, the computer was able to generate 70 unique C-terminal sequences and 33 unique N-terminal sequences (Table 2). N-terminal and C-terminal

Table 2. Number of Possible Sequences Derived by Computer, Interpretation of the FAB Spectrum of Peptide RE2.

Fragment Ion Series	Number of Unique Sequences	Fragment Ion Series	Number of Unique Sequences
A	19	Z	26
B	50	Z"	0
C"	20	Y"	12

Total unique C-terminal sequences = 70.
Total unique N-terminal sequences = 33.
Total unique and complete combined sequences = 37.

sequences were combined to obtain 37 complete sequences, 13 of which had total scores greater that 5 (arbitrary cutoff) (Table 3). The score is a function of the number of ions contributing to the N-terminal and C-terminal parts of a sequence and extent that the two overlap with each other. The sequence with the highest score is the sequence of the peptide.

This computerized approach works well for peptide spectra obtained using 1-10 nmole of sample. At lower sample amounts, weaker fragment ions begin to disappear into the noise of the background. One use of the computer program is to evaluate different techniques for obtaining spectra. Computer analysis of the spectrum of 5 nmoles of a neurotensin fragment (Ac-RRPYIL) (Fig. 11A)

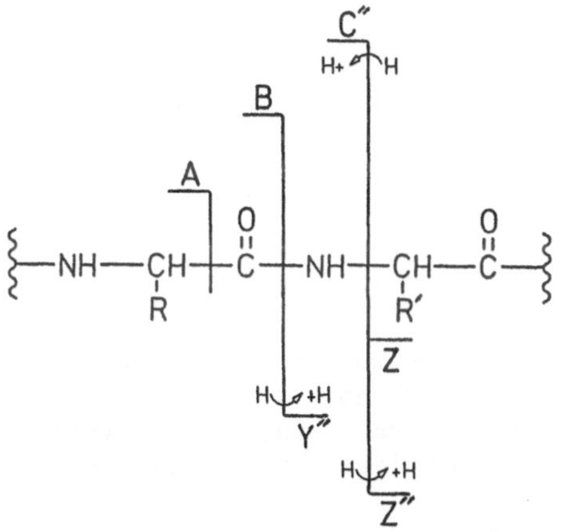

Fig. 9. Nomenclature for the N-terminal (A, B, and C") and C-terminal (Y", Z, and Z") fragment ion series used for computer assisted interpretation of peptide FAB spectra[18].

Table 3. Possible Complete Sequences with Scores >5 Derived by Computer
Interpretation of the FAB Spectrum of Peptide RE2. An asterisk
indicates the correct sequence.

Rank		Sequence	Score	Rank	Sequence	Score
1	*	ELKSEC	24	8	LEKSEC	9
2		ELSKEC	20	9	LESKEC	7
3		ELKSCE	20	10	LEKESC	7
4		ELKESC	20	11	LEKSCE	7
5		ELSKCE	16	12	SCLKEE	6
6		LKCSEE	10	13	SCKLEE	6
7		CLKSEE	10			

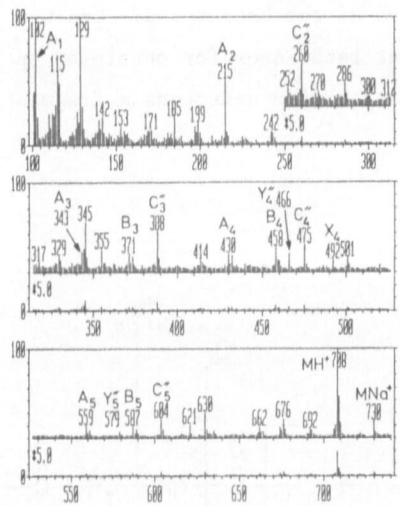

Fig. 10. Positive ion Fab spectrum for the hexapeptide RE2 (ELKSEC).
Spectrum obtained with approx. 10 nmole of peptide dissolved in
approx. 1 µl of glycerol using a JEOL HX100HF mass spectrometer
operating at 5KV accelerating potential and a resolution setting
of 3,000.

Table 4. Scores for complete sequences derived by computer interpretation of FAB Spectra of N-Ac Neurotensin Fragment 8-13. An asterisk indicates the correct sequence.

Sequence	5 nmole Glycerol	5 nmole DTs CSA [a]	0.5 nmole Glycerol Bkgd. Sub.	0.5 nmole DTs CSA B/E Scan[b]
1. *RRPYLL	109	82	44	11
2. RPRYLL	76	36	17	5
3. RRYPLL	57	72	19	9
4. RRPLYL	47	50	34	9
5. RPRLYL	23	23	11	9
6. RYRPLL	6	21		5
7. RRYLLP		43		
8. RRYLPL		43	16	
9. RYRLLP		11		
10. RYRLPL				
11. PRRYLL			9	
12. PRRLYL			4	
13. RPYLLR				7
14. RPLLYP				7
15. RPYLRL				7
16. RPLYRL				7

[a] A mixture of dithiothreitol: dithioerythritol (5:1) and 6 mM camphor sulfonic acid used for sample matrix.

[b] Daughter ion spectrum for MH+ (m/z 859) obtained with a B/E link scan.

yields only 6 possible complete sequences, with the correct one having by far the highest score (Table 4). The same amount of peptide in the DTs matrix to which camphor sulfonic acid has been added to enhance the sample ion intensities results in a more complex spectrum (Fig. 11B)[19]. Although the information content has been altered, the interpretation program has little difficulty finding the correct sequence. At a level of 500 pmoles, interference from the glycerol matrix precludes sequence determination. However, the background subtracted spectrum (Fig. 11C) contains the necessary information. Finally, 500 pmoles of the peptide could be sequenced if a daughter ion spectrum is obtained (Fig. 11D). In this mode of operation, the electric and magnetic sectors of a double focussing mass spectrometer are scanned in such a way that only ions resulting from fragmentation of the protonated molecular ion are recorded. This effectively removes matrix ion interference.

Preliminary data would indicate that sequencing of peptides by computer interpretation of FAB spectra is possible at a level of approximately 500 pmoles provided good amino acid composition data is available. It is

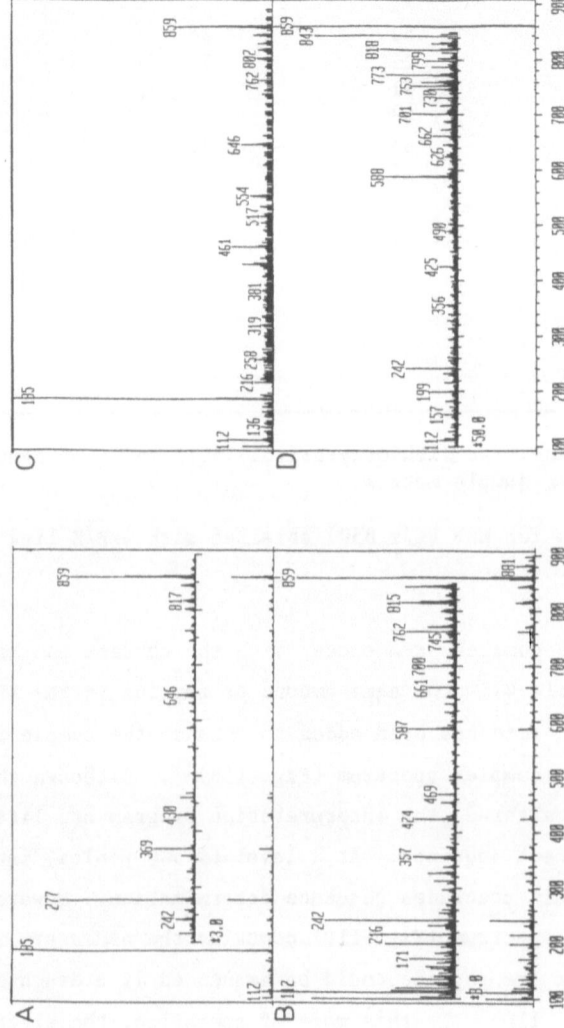

Fig. 11. Postive ion mass spectra for N-Ac neurotensin fragment 8-13: (A) 5 nmole in glycerol, (B) 5 nmole in dithiothreitol : dithioerythritol 5:1 (DTs), (C) 500 pmole in glycerol with background subtraction, and (D) 500 pmole in DTs B/E linked scan daughter ion spectum of the protonated molecular ion.

unlikely that mass spectral methods will ever fully replace chemical
approaches. For one thing, it is not possible to distinquish between leucine
and isoleucine. In the neurotensin peptide just discussed, the sequence
contains an isoleucine and a leucine rather than the two leucines indicated
in Table 4. The exact position of the Ile would have to be determined by
some other means. Even so, mass spectral sequence analysis has the potential
to greatly reduce the time required to sequence a peptide or protein.

References

1. M. Barber, R.S. Bordoli, D. Segwick, A.N. Tyler, and E.T. Whalley,
   Fast Atom Bombardment Mass Spectrometry of Bradykinin and Related
   Oligopeptides, Biomed. Mass Spectrom. 8:337 (1981).
2. D.H. Williams, C.V. Bradley, S. Santikarn, and G. Bojesen,
   Fast-atom-bombardment mass spectrometry. A New technique for the
   determination of molecular weights and amino acid sequences of
   peptides, Biochem. J. 201:105 (1982).
3. T.D. Lee, Fast atom bombardment and secondary ion mass spectrometry
   of peptides and proteins, in: "Microcharacterization of Peptides
   and Proteins," J.E. Shively, ed., The Humana Press, New Jersey,
   in press.
4. B. Sundqvist, P. Roepstorff, J. Fohlman, A. Hedin, P. Hakansson, I.
   Kamensky, M. Lindberg, M. Salehpour and G. Sawe, Molecular weight
   determinations of proteins by californium plasma desorption mass
   spectrometry, Science 226:696 (1984).
5. J.A. Yergey, R.J. Cotter, D. Heller and C. Fenselau, Resolution
   requirements for middle-molecule mass spectrometry, Anal. Chem.
   56:2262 (1984).
6. B.S. Rothman, D.H. Hawke, R.O. Brown, T.D. Lee, A.A. Dehghan, J.E.
   Shively and E. Mayeri, Isolation and primary structure of the
   califins, three biologically active, ELH-like peptides from the
   atrial gland of Aplysia californica, J. Biol. Chem., in press.
7. T.D. Lee, K. Legesse, D.H. Hawke, J.E. Shively, B.S. Rothman and E.
   Mayeri, Routine fast atom bombardment mass spectral analysis of
   high molecular weight peptides - atrial gland peptides from
   Aplysia californica, Bioc. Biop. Res. Commun., in press.
8. J.L. Witten, M.H. Schaffer, M. O'Shea, J.C. Cook, M.E. Hemling and
   K.L. Rinehart, Structures of two cockroach neuropeptides assigned
   by fast atom bombardment mass spectrometry, Bioc. Biop. Res.
   Commun. 124:350 (1984).
9. B.W. Gibson and K. Biemann, Strategy for the mass spectrometric
   verification and correction of the primary structures of proteins
   deduced from their DNA sequences, Proc. Natl. Acad. Sci. USA,
   81:1956 (1984).
10. Y. Wada, A. Hayashi, F. Masanori, I. Katakuse, T. Ichihara, H.
    Nakabushi, T. Matsuo, T. Sakurai and H. Matsuda, Characterization
    of a new fetal hemoglobin variant, Hb F Izumi, by molecular
    secondary ion mass spectrometry, Biochim. Biophys. Acta, 749:244
    (1983).
11. S. Rahbar, J. Louis, T.D. Lee and Y. Asmerom, Hemoglobin North
    Chicago: a new high affinity hemoglobin, Hemoglobin, in press.

12. S. Rahbar, T.D. Lee, J.A. Baker, L.T. Rabinowitz, Y. Asmerom and H.M. Ranney, Reverse phase high-performance liquid chromatography and secondary ion mass spectrometry. A strategy for identification of ten human hemoglobin variants, <u>Hemoglobin</u>, submitted.

13. H. Wacjman, O. Belkohdja and D. Labie, Hb Setif: A new αchain hemoglobin variant with substitution of the residue involved in a hydrogen bond between the subunits, <u>FEBS Lett</u>. 27:298 (1972).

14. W.F. Moo-Penn, M.H. Johnson, S.M. Wilson, B.J. Thierell and R.M. Schmidt, Hemoglobin Tarrant: A new hemoglobin variant in α1β1 contact region showing high oxygen affinity and reduced cooperativity, <u>Biochim. Biophys. Acta</u> 490:443 (1977).

15. C.M. Ben-Avram, T.D. Lee, R.C. LeBoeuf and J.E. Shively, The primary structure of murine apolipoprotein A-II from inbred mouse strain BALB/c, <u>J. Biol. Chem</u>., submitted.

16. T. Sakurai, T. Matsuo, H. Matsuda, and I. Katakuse, PAAS 3: a computer program to determine probable sequence of peptides from mass spectrometric data, <u>Biomed. Mass Spectrom</u>. 11:396 (1984).

17. T. D. Lee and V. Spayth, Computer assisted interpretation of fast atom bombardment mass spectra of peptides, <u>33rd Annual Conf. Mass Spectrom. Allied Topics</u>, 266 (1985).

18. P. Roepstorff and J. Fohlman, Proposal for a common nomenclature for sequence ions in mass spectra of peptides, <u>Biomed. Mass Spectrom</u>. 11:601 (1984).

19. E. DePauw, G. Pelzer, D.V. Dang and J. Marien, On the influence of hydrophobicity in the SIMS spectra of amino acids in glycerol matrix, <u>Bioc. Biop. Res. Commun</u>. 123:27 (1984).

THE USE OF HPLC AND MASS SPECTROMETRIC TECHNIQUES

TO QUANTIFY ENDOGENOUS OPIOID PEPTIDES

Dominic M. Desiderio[1,2], Genevieve H. Fridland[2],
F. S. Tanzer[3], Chhabil Dass[2], Peter Tinsley[2], and
John Killmar[2]

Dept. of Neurology[1] and The Charles B. Stout Neuroscience
Mass Spectrometry Laboratory[2] - College of Medicine, and
Dept. of Biologic and Diagnostic Sciences[3], College of
Dentistry, Univ. Tenn. Center for Health Sciences, 800
Madison Avenue, Memphis, TN 38163

INTRODUCTION

State-of-the-art mass spectrometry (MS) analytical methods have
been developed to quantify endogenous peptides in the peptide-rich
fraction derived from brain, pituitary, tooth pulp, and CSF samples.
Current target peptides include enkephalins with a further goal to
include other neuropeptides such as endorphins, dynorphins, and
substance P. These MS methods are used in conjunction with reversed
phase high performance liquid chromatography (RP-HPLC), radioimmunoassay
(RIA), and radioreceptor assay (RRA) in experiments aimed towards taking
experimental advantage of the unique features of each analytical system
and to determine qualitative and/or quantitative differences of
endogenous peptides in the normal versus stressed physiological states.
MS offers the highest level of molecular specificity currently available
for analytical measurement of important endogenous peptides, and RIA the
highest level of detection sensitivity. In the MS quantitative
analytical mode, fast atom bombardment (FAB) MS produces the protonated
molecular ion - $(M+H)^+$ - of an HPLC-purified peptide, mass spectrometry-
mass spectrometry (MS-MS) selects a unique amino acid sequence-
determining fragment ion, and a stable-isotope incorporated peptide
internal standard added to the biological extract provides optimal
quantification accuracy. RRA uses a receptor-rich preparation derived
from the canine limbic system to effectively screen for the presence of
opioid receptoractive (ra-) peptides in each HPLC fraction and to

provide a semi-quantitation of a peptide. Furthermore, RIA is done on selected fractions.

The objective of this research program is to develop and utilize state-of-the-art MS techniques to analyze endogenous peptides with an aim towards understanding the molecular basis involved in pain, stress, and addiction (1). The reason for undertaking research in these areas is to understand the molecular processes that occur in these physiological events and to provide a more rational basis for effective treatment of pathophysiological disorders.

This research focuses on the neuropeptides that derive from four precursor proteins (2): preproenkephalins A and B, proopiomelanocortin (POMC), and preprotachykinin. An endogenous mixture or constellation composed of these proteins, intermediate precursor molecules, working peptides, and inactive metabolites normally exists in a dynamic homeostatic metabolic relationship in a cell. When one or several of these pathways becomes metabolically deranged, it is crucial to define analytically which peptide pathway is defective.

The analytical needs that currently exist in this research program to effectively address these problems include the ability to quantify a specific peptide with maximum molecular specificity; to screen for a wide range of peptide hydrophobicities in a gradient RP-HPLC chromatographic separation for the presence of specific receptoractivities; to observe any inter-relationships that may exist within and between peptide families and individual peptides; and to study these relationships in the control versus stressed physiological states. We define here the term "molecular specificity" to mean that we know unambiguously the structure of the peptide being measured.

There are several available analytical methodologies that serve effectively to observe the neuropeptide homeostatic system described above. These analytical techniques and their associated basic principles and/or detected species include:

a. radioreceptor assay (RRA) - competitive displacement
    from receptors (3);
b. radioimmunoassay (RIA) - immunoreactivity (4);

c. HPLC - UV absorbance, fluorescence, electroactive
   species;

d. bioassay (BA) - tissue contractions; and

e. mass spectrometry (MS) - protonated molecular ions
   and/or specific amino acid sequence-determining
   fragment ions by MS-MS (1).

It is trivial to state, but important to remember, that each analytical method except MS monitors only a selected and/or unique aspect of the peptide and not necessarily an entire molecular structure. It is the aim of this paper to show that MS and wet chemical sequence determination are the only analytical techniques that convey during measurement crucial structural (amino acid sequence) information. This concept is most important because it must be realized for example that, if one very simply considers the case of a small range of peptides (dipeptides through tridecapeptides), 6.7 billion different peptides are theoretically possible, and that a significant fraction of that number of peptides may occur naturally (5). If one were to depend upon only a 90-minute HPLC gradient separation to resolve completely all of these potential peptides, a chromatographic resolution corresponding to a 500 nanoseconds window is required to separate each peptide from every other possible peptide and other compounds. Conversely, millions of peptides could co-elute within each second of an HPLC gradient. From this so-called "n! problem", which is a simple arithmetic consideration, it is hoped to be proven in this manuscript that a detector with a high level of molecular specificity must be used for measurement of endogenous peptides in a biological extract in contradistinction to only increasing the resolution of the RP-HPLC. The latter line of reasoning and experimentation currently has a nearly impossible goal; however, even if that goal of total chromatographic separation were realized, the molecular specificity of the detector still remains open to question. Put in other words, as the molecular specificity of the detector increases, the required chromatographic resolution decreases. On the other hand, if the chromatographic detector (such as RRA, RIA, HPLC, UV, or BA) does not have a sufficiently high level of molecular specificity, then the resolution of the chromatography must increase commensurately.

Tissue acquisition.   The important experimental step of
acquisition of a variety of tissues has been studied.   Human and canine
pituitary (6), canine brain (7), human CSF (8), human and canine tooth
pulp (9) all have been the targets for the analytical methodology
described in this paper.

Receptorassay.   While RRA has been utilized for many years (10,
11), we will briefly describe an improvement made in our laboratories
(9) wherein we use the canine limbic system as the source for opioid
receptors because of the larger weight of available tissue and the
emotional similarity of the dog to the human in distinction to other
laboratory rodents.   Many workers use whole rat brain, minus cerebellum.
A receptor-rich preparation is derived either from a canine limbic
system synaptosomal (12) or P2 fraction.   A variety of radioligands
(dihydromorphine, etorphine, diketoazocine, substance P, etc.)   can be
utilized to screen for the presence of opioid peptides.   The P2 fraction
can be demonstrated to contain opioid receptors because of the data
shown below, where several unlabeled peptides react with the opioid
receptor preparation to displace the tritiated dihydromorphine.

Radioimmunoassay.   Commercially available (Immunonuclear,
Stillwater, MN) RIA kits are used for this study.

Mass Spectrometry.   Finnigan MAT 731 and VG 7070E-HF/11-250 DS mass
spectrometers are used in this study.   FAB-MS effectively ionizes
peptides (13).   The $(M+H)^+$ ion of a peptide contains molecular weight
information; however, it must be realized that amino acid sequence-
determining information is not readily derived from $(M+H)^+$.   For
example, five different amino acids can be combined in 120 (equals 5!)
unique combinations to yield 120 different peptides.   Therefore, to
extract the required structural information and concomitantly increase
the molecular specificity of this mode of detection, the $(M+H)^+$ ion is
subjected to MS-MS, or linked-field techniques (14) scanning MS, wherein
amino acid sequence-determining fragment ions are detected.   In a B/E (B
= magnetic field, E = electric field) linked-field scan, all of the
product ions that derive only from a selected precursor ion are
detected.   In the case of peptides, the most appropriate precursor ion

is $(M+H)^+$ and the product ions are amino acid sequence-determining (and other) fragment ions. A specific and unique amino acid sequence-determining fragment ion is selected for selected ion monitoring and quantification of the selected peptide (15-19). This process is known as tandem MS, or MS/MS (20, 21). For example, for leucine enkephalin (LE = YGGFL), the fragment ion at m/z 336 corresponds to the C-terminal tripeptide fragment -GFL (plus two hydrogens), which in the linked-field study has been shown to arise only from the $(M+H)^+$ ion at 556 (22). That $(M+H)^+$ ion corresponds to LE, which in turn only comes from the collected HPLC fraction. This set of experimental data firmly establishes the linkage: HPLC fraction $\longrightarrow (M+H)^+ \longrightarrow$ specific fragment ion $\longrightarrow {}^{18}O$ IS, and provides the basis for the optimal molecular specificity readily derivable from MS and MS-MS methods.

Peptide internal standards (IS) have been synthesized that incorporate two ${}^{18}O$ atoms in each free carboxy group (23).

Reversed phase HPLC. RP-HPLC is utilized because of the fast, efficient, and high recovery of peptides (24). A gradient of the organic modifier is utilized with triethylamine:formic acid as the volatile buffer (25, 26) to analyze for a wide range of hydrophobicities.

RESULTS AND DISCUSSION

Fig. 1 contains a gradient RP-HPLC (200 nm) chromatogram of the peptide-rich fraction from human pituitary tumor tissue (6). The left-hand axis relates to receptor activity, which is calculated as picomoles of methionine enkephalin (ME = YGGFM) equivalents mg^{-1} protein, because ME is used for the calibration curve. The bottom axis ranges from 0-90 and relates to the fraction number and elution time (minutes). The organic modifier (acetonitrile) profile ranges from 10-100%. Along the top of the chromatogram are indicated the retention times of known peptide synthetic standards, which are representatives of the peptide families described above. It is crucial for the purposes of this paper to realize that coelution of an HPLC peak with any one of the indicated retention times of the standards can not constitute structural proof of the putative neuropeptides.

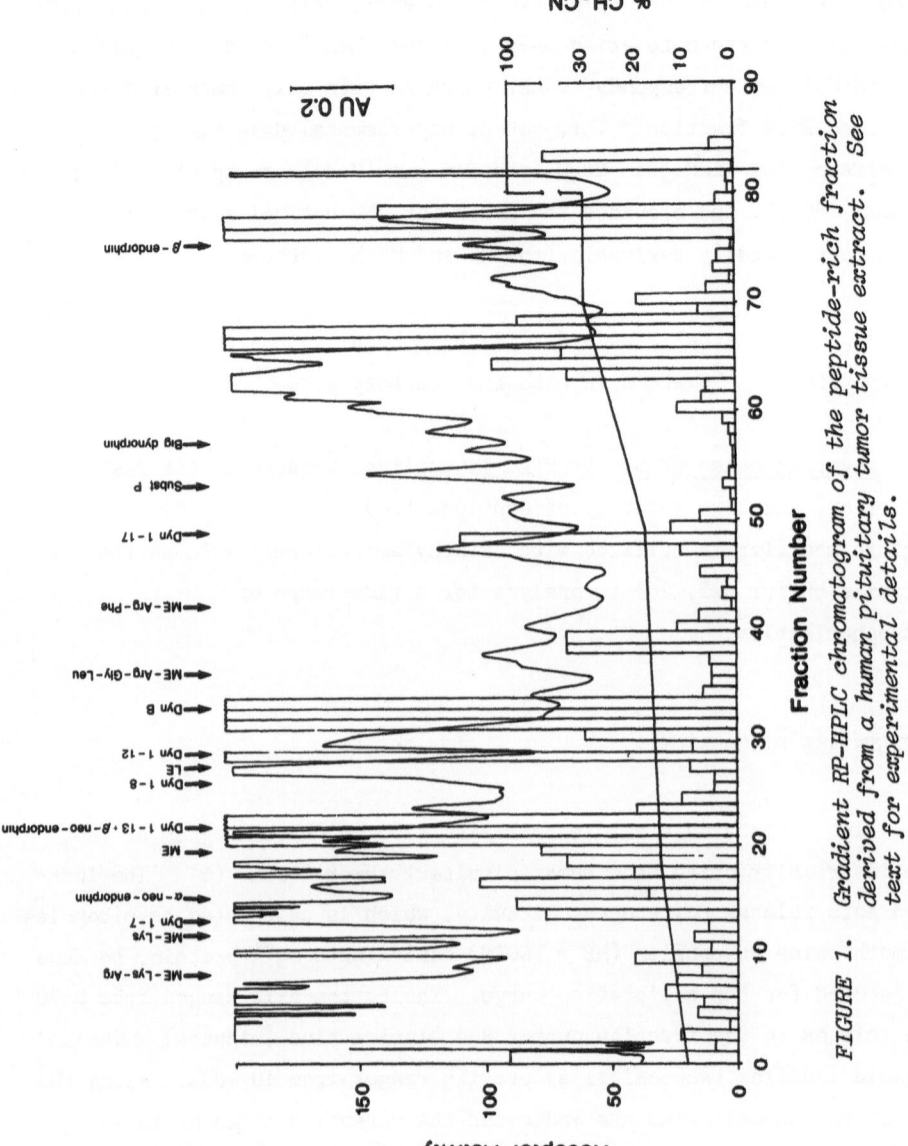

FIGURE 1. Gradient RP-HPLC chromatogram of the peptide-rich fraction derived from a human pituitary tumor tissue extract. See text for experimental details.

In a study (27), we indicated that RP-HPLC fractions can be readily analyzed by a combination of analytical techniques: FAB-MS to define the $(M+H)^+$ ion of compounds in each HPLC fraction, linked-field scanning to determine a unique amino acid sequence-determining fragment ion, RIA, and RRA. In those cases, we have defined the appropriate composite analytical data for LE, ME, and substance P (see Table 1). Furthermore, accurate mass determinations indicate that several HPLC fractions correspond to a "bleed" from the LC column and relate to octadecylsiloxane-related compounds.

In another study on human tooth pulp tissue (9), we utilized RIA to show that the HPLC-purified ME fraction contains, for controls 34.3 $\pm$ 53.0 ng mg^{-1} tooth pulp of ir-ME (n=23). The experimentals are those patients who have their teeth stressed before removal; they have 17.0 + 26.0 ng mg^{-1} of tooth pulp of ir-ME (n = 40). Even though the coefficients of variation of these measurements are relatively high, the two sets of data are statistically significantly different and indicate that orthodontic stress effectively mobilizes at least the methionine enkephalinergic pathway.

The RRA data on the same two sets of human tooth pulp patients, show, for control, receptoractivity of 93 $\pm$ 123.4 (n=21), and for the experimentals, 51.8 $\pm$ 69.2 ng (n=38) of receptoractive ME mg^{-1} of tooth pulp. Furthermore, the RIA level of ME for the first extracted tooth in the experimental patients can be plotted against the amount of orthodontic force that was applied to that tooth before its removal. It

TABLE I

MS, RIA,and RRA Analytical data obtained for three
selected RP-HPLC fractions (27)

	ME	LE	SP
Fraction No.	19	23	54
FAB-MS $(M+H)^+$	574	556	1,348*
Linked-field MS (B/E)	340	336	-
RIA	Yes	Yes	Yes
RRA	Yes	Yes	**

*very weak   **preliminary

was noticed that the concentration of ME tended to decrease as the amount of orthodontic force increased. Regression analysis indicates a significant relationship that can be best represented by the linear equation: RIA = 194.9 - 38.2 (log F).

The measurements on these human teeth were done with RIA and RRA. One of the samples was also tested by FAB-MS. An appropriate $(M+H)^+$ ion was observed at 574 and is consistent with the presence of ME in that tooth.

The data shown in Fig. 2 demonstrate that the canine limbic system receptor-rich preparation does indeed contain intact opioid receptors. The concentration of several unlabeled peptides is plotted versus the displacement of the tritiated dihydromorphine (represented as percentage of control) from the receptor. For example, it can be shown that dihydromorphine can be very effectively displaced from that group of receptors by the enkephalin peptide group, a less efficient displacement by the endorphin group (see insert), inefficient displacement by the dynorphin group, and virtually no displacement by other peptides such as substance P and the pentapeptide metabolites.

MS data are provided in Table 2 (1). Measurements have been obtained for a variety of canine tissue extracts and demonstrate that this technique is sufficiently sensitive at the level of ng peptide g^{-1} biologic tissue.

CONCLUSIONS

Several conclusions derive directly from the research described in this paper:

(1) MS and MS-MS are demonstrated to have sufficient sensitivity for the detection and analytical measurement of selected endogenous opioid peptides in several biologic tissue extracts and fluids;

(2) RP-HPLC is an effective method to separate and collect endogenous peptides;

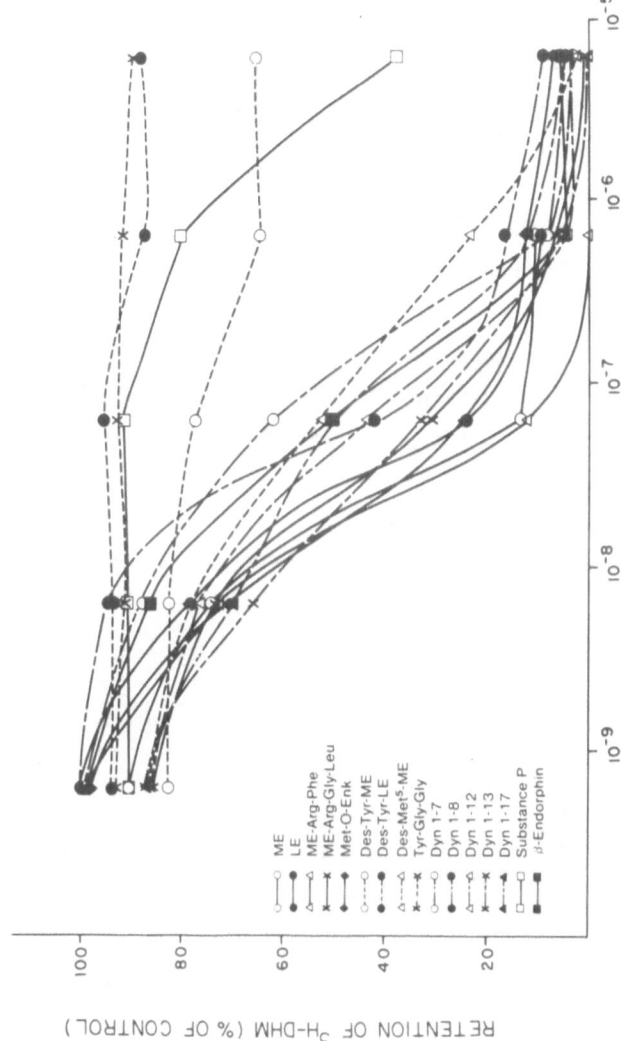

CONCENTRATION OF UNLABELLED PEPTIDES (M)

RETENTION OF ³H-DHM (% OF CONTROL)

ME
LE
ME-Arg-Phe
ME-Arg-Gly-Leu
Met-O-Enk
Des-Tyr-ME
Des-Tyr-LE
Des-Met⁵-ME
Tyr-Gly-Gly
Dyn 1-7
Dyn 1-8
Dyn 1-12
Dyn 1-13
Dyn 1-17
Substance P
β-Endorphin

FIGURE 2.   RRA calibration curves. Canine limbic system synaptosome receptor
preparation; ³H-dihydromorphine ligand; unlabeled peptides listed
in insert. Left axis: percentage retention of ³H-DHM by receptor
preparation. Bottom axis: concentration of peptide.

277

(3) MS-MS has the maximum level of molecular specificity that can be currently obtained for measurement of peptides. This statement is supported by several experimental considerations - a specific RP-HPLC fraction is collected that elutes at a known retention time of a given peptide; for that fraction, an $(M+H)^+$ ion is produced by FAB-MS; a unique amino acid sequence-determining fragment ion is obtained by MS-MS; a synthetic ^{18}O-incorporated peptide internal standard added to the biological extract in the beginning of the separation coelutes with the HPLC-purified endogenous peptide; the appropriate ion currents corresponding to these two fragment ions are accumulated by a microcomputer; and, because the amount of exogenous internal standard added is known, the amount of endogenous peptide is readily calculated. Again, it must be repeated for emphasis that this high level of molecular specificity for an HPLC detector is required because of what we described above as the "n! problem";

TABLE II

Mass spectrometric measurement (1) of enkephalins in canine tissue extracts (ng g^{-1}) and CSF (ng ml^{-1})

SOURCE	ENKEPHALIN	AMOUNT	METHOD (a)
Hypothalamus	LE	170	1
Cerebrospinal fluid	LE	44	1
Pituitary			
Anterior	LE	70	2
	ME	2,950	2
Posterior	LE	2	2
	ME	3,760	2
Caudate Nucleus	LE	1,500	2
Tooth Pulp (b)	LE	30	2
	ME	179	2
Tooth Pulp			
Control	LE	20	2
	ME	487	2
Electrostimulated	LE	45	2
	ME	390	2

(a) Method 1: FDMS-SIM-COM

         2: FAB-CAD-B/E-B'/E'-SIM-COM

(b) Unstimulated pooled tissue from five dogs

(4) MS has sufficient sensitivity to detect endogenous peptides at the nanogram level, which corresponds to the picomole level. This detection sensitivity allows measurements of peptides in several biological tissue extracts; but not yet routinely plasma or CSF. Current developments are tending towards the femtomole level;

(5) The RIA and the RRA data show that we have effectively initiated the metabolism of the ME-containing neuropeptidergic pathway following application of an orthodontic force to human teeth. These data show that we can now begin to study effectively the molecular basis of that stressful condition;

(6) It can be shown for the first time that the RRA screen is extraordinarily efficient for screening for the presence of ra-peptides that cover a wide range of hydrophobicity values. Hydrophobicity correlates only approximately to the molecular weight of the peptide (28). For the RRA, a variety of tritiated ligands can be used to uncover the presence of several peptide receptors including the mu, kappa, sigma, delta, substance P, etc. The canine opioid peptide-rich preparation (limbic system or P2 fraction) is utilized because of the similarity of a dog to human in terms of emotion; and

(7) Finally, we are becoming convinced that, rather than using only one analytical technique, the combination of several techniques such as HPLC/RIA/RRA/MS is the most powerful analytical system available to measure endogenous peptides in the peptide-rich fraction derived from biological extracts by taking advantage of the multiple structural aspects of peptides.

ACKNOWLEDGEMENTS

The authors gratefully acknowledge the financial assistance from the National Institutes of Health (NIH GM-26666) and the typing assistance of Dianne Cubbins.

REFERENCES

1. D. M. Desiderio, "Analysis of Neuropeptides by Liquid Chromatography and Mass Spectrometry", Elsevier, Amsterdam (1984).

2. V. Hollt, Multiple endogenous opioid peptides, Trends in Neurosci., 6:24 (1983).

3. S. R. Childers, Receptors for opiates and opioid peptides, in: "Principles of Receptorology", M.K. Agarwal, ed., Walter de Gruyter, New York (1983).

4. R. S. Yalow, Radioimmunoassay, Ann. Rev. Biophys. Bioeng. 9:327 (1980).

5. G. Fridland and D. M. Desiderio, Profiling of Neuropeptides Using Gradient RP-HPLC with Novel Detection Methodologies, in: "Profiling of Body Fluids and Tissues, J. Chromatogr. Biomed. Applic. Spec. Vol.," C.C. Sweeley and Z. Deyl, eds., Elsevier, Amsterdam, in press.

6. D. M. Desiderio, R.C. Cezayirli, G. Fridland, J.T. Robertson, and H. Sacks, Measurement of RRA opioid peptides in a human pituitary ACTH-secreting tumor, Life Sci., 37:1823 (1985).

7. D. M. Desiderio and H. Takeshita, RRA of opioid peptides in selected canine brain regions, Anal. Lett., in press.

8. D. M. Desiderio, H. Onishi, G. Fridland, G. Wood, and D. Pagidipati, Metabolic profiling of opioid peptides in human CSF with HPLC and RRA, Peptides, submitted.

9. J. Walker, Jr., F.S. Tanzer, E.F. Harris, C. Wakelyn, and D.M. Desiderio, The enkephalin response in human tooth pulp to orthodontic force, Amer. J. Ortho., submitted.

10. R. Simantov and S.H. Snyder, Brain pituitary opiate mechanisms: pituitary opiate receptor binding, radioimmunoassays for methionine enkephalin and leucine enkephalin, and ^{3}H-enkephalin interactions with the opiate receptor, in: "Opiates and Endogenous Opioid Peptides," H. Kosterlitz, ed., Elsevier, Amsterdam (1976).

11. A. Wahlstrom, L. Johansson, and L. Terenius, Characterization of endorphins (endogenous morphine-like factors) in human CSF and brain extracts, in: Opiates and Endogenous Opioid Peptides," H. Kosterlitz, ed., Elsevier, Amsterdam (1976).

12. E. G. Gray and V.P. Whittaker, The isolation of nerve endings from brain: an electron-microscope study of cell fragments derived by homogenization and centrifugation, J. Anat. 96:79 (1962).

13. M. Barber, R.S. Bordoli, R.D. Sedgwick, and A.N. Tyler, Fast atom bombardment mass spectrometry of the angiotensin peptides, Biomed. Mass Spectrom. 9:208 (1982).

14. K.R. Jennings, Observation of metastable transitions in a high performance mass spectrometer, in: "High performance mass spectrometry: chemical applications," M.L. Gross, ed., Amer. Chem. Soc., Washington, D.C. (1978).

15. D.M. Desiderio and S. Yamada, Measurement of endogenous leucine enkephalin in canine thalamus by HPLC and FD-MS, J. Chromatogr. 239:87 (1982).

16. D.M. Desiderio and S. Yamada, FD-MS measurement of picomole amounts of leucine enkephalin in canine spinal cord tissue, Biomed. Mass Spectrom. 10:358 (1983).

17. D.M. Desiderio, I. Katakuse, and M. Kai, Measurement of leucine enkephalin in caudate nucleus tissue with fast atom bombardment-collision activated dissociation-linked field scanning mass spectrometry, Biomed. Mass Spectrom. 10:426 (1983).

18. D.M. Desiderio, High performance liquid chromatography and field desorption mass spectrometry measurement of endogenous amounts of neuropeptides in biologic tissue, in: CRC Handbook of HPLC for the Separation of Amino Acids, Peptides, and Proteins, W.S. Hancock, ed., CRC Press, New Zealand (1984).

19. D.M. Desiderio, High-performance liquid chromatography and mass spectrometry of biologically important peptides, in: "Adv. in Chromatography, Vol. 22," J.C. Giddings, E. Grushka, J. Cazes, and P.R. Brown, eds., Marcel-Dekker, New York (1983).

20. K.L. Busch and R.G. Cooks, Analytical applications of tandem mass spectrometry, in: "Tandem Mass Spectrometry," F.W. McLafferty, ed., Wiley, New York (1983).

21. P.J. Todd and F.W. McLafferty, Collisionally activated dissociation of high kinetic energy ions, in: "Tandem Mass Spectrometry", F. W. McLafferty, ed., Wiley, New York (1983).

22. I. Katakuse and D.M. Desiderio, Positive and negative fast atom bombardment-collision activated dissociation-linked field scanned mass spectra of leucine enkephalin, Int. J. Mass Spectrom. Ion Phys. 54:15 (1983).

23. D.M. Desiderio and M. Kai, Preparation of stable isotope-incorporated peptide internal standard for field desorption mass spectrometry quantification of peptide in biologic tissue, Biomed. Mass Spectrom. 10:471 (1983).

24. J. E. Rivier, Use of trialkyl ammoniun phosphate (TAAP) buffers in RP-HPLC for high resolution and high recovery of peptides and proteins, J. Liq. Chromatogr. 1:343 (1978).

25. D.M. Desiderio and M. L. Cunningham, Triethylamine formate buffer for HPLC-FDMS of oligopeptides, J. Liq. Chromatogr. 4:721 (1981).

26. D. M. Desiderio, M.D. Cunningham, and J. Trimble, High performance (pressure) liquid chromatography separation and quantification of picomole amounts of prostaglandins utilizing a novel triethylamine formate buffer, J. Liq. Chromatogr. 4:1261 (1981).

27. D.M. Desiderio, H. Takeshita, H. Onishi, G. Fridland, and C. Dass, Metabolic profiling of pituitary peptides by a combination of RP-HPLC, RRA, RIA and mass spectrometry, in: Proceedings of the Ninth American Peptide Symposium, V.J. Hruby, C.M. Deber, and K.D. Kopple, eds., Pierce Chemical Co., Rockford, IL, in press.

STRUCTURAL ANALYSIS OF PROTEINS OF THE NERVOUS SYSTEM

Daniel R. Marshak

Laboratory of Cell Biology
National Institute of Mental Health
Bethesda, Maryland

INTRODUCTION

Proteins and peptides are essential to normal development, growth, and function of the nervous system[1]. The structural analysis of these proteins and peptides is complicated by the difficulty in obtaining sufficient material for complete, chemical characterization. It is difficult to obtain a homogeneous population of cells to serve as the starting material for protein isolation because the tissues of the nervous system, including brain, spinal cord, peripheral nerves, and sensory systems, are composed of multiple cell types[2]. Primary cultures of specific populations of cells[3,4] have been useful as models of neural development and function but are limited in their general application. Systems of cultured cells[5,6], such as embryonic neurons, glia, neuroblastomas, or gliomas, that proliferate in culture also have limited roles as models of normal nervous system function. Therefore, upon purification and characterization of a protein of the nervous system, it is imperative to return to the whole organism to demonstrate the physiological function, anatomical distribution, and pharmacological regulation of the protein. In this way, the structural analysis of proteins of the nervous system is one part of a multidisciplinary approach to neuroscience.

PROTEINS IN PROBLEMS OF NEUROSCIENCE

Because of the diversity and complexity of functions of the nervous system, it is useful to define several broad problems of function. These include: (i) signal transduction, (ii) cellular differentiation, and (iii) cell recognition. Within each of these areas, the characterization of proteins can contribute immensely to our understanding of the molecular basis of neurobiological function.

Signal Transduction

Signal transduction refers to the transmission of an electrical or chemical stimulus between cells or within a single cell. Neurons have connections at synapses[7] where chemical mediators transmit a signal from the presynaptic neuron that results in a change in the electrical properties of the postsynaptic neuron. Glial cells may use ionic gradients or protein factors to modify neuronal function[8]. The nervous system also interacts with the endocrine, immune, muscular, and gastrointestinal systems by a variety of chemical signals[9]. Proteins and peptides are the primary messengers for some of these signalling events and also serve as cell surface receptors in the affected tissues[10]. Peptides act as neurotransmitters between cells at synapses and other junctions and may also act as neuromodulators that amplify or attenuate the effects of a primary signal[1]. Membrane proteins can mediate various cell surface phenomena, such as ion conductance and transport processes[11]. Inside cells, a primary signal is transduced by a second messenger, such as calcium ions[12], cyclic adenosine 3'-5' monophosphate (cAMP)[13], and phosphatidyl inositol turnover[14]. Proteins act as enzymes for the production of these messengers, modulators of messenger synthesis, messenger-binding proteins, and targets of the message. Long term regulation of cellular signals might involve the biosynthesis of proteins that regulate gene expression.

Cellular Differentiation

Cellular differentiation involves protein factors that are transferred between cells as well as those produced within a single cell[15,16,17]. For example, differentiation factors appear to be produced by glial cells that influence neuronal development[18,19]. Mitogenic growth factors, such as insulin, are found in the nervous system and are vital to cell survival[20,21]. Neuropeptides might also act as mitogens[22], and proteins as mitogen receptors[10,21]. Intracellular proteins also affect the normal and abnormal differentiation of a cell. Cellular oncogene products[23] appear to influence cellular development at particular stages in the cell cycle and may be involved in tumorigenesis[24]. Other developmental proteins include trans-actng factors that regulate gene expression by modulating chromatin structure or transcriptional activity at specific genes[25]. The search for such factors in the brain may include the identification of brain-specific protein[26] or brain-specific DNA sequences[27]. Thus, the morphology and physiology of neurons and other cells of the nervous system may be controlled, in part, by proteins, many of which have yet to be identified.

Cell Recognition

During development, cells of the nervous system establish specific contacts by means of cell recognition events. Mediating some of these events are large, membrane-associated glycoproteins that are responsible for cell adhesion[28]. Neural cell adhesion molecules comprise a family of glycoproteins whose structures change during development of the embryonic and adult vertebrate brain. Proteins are also involved in events secondary to the primary cell adhesion or other recognition phenomenon. These secondary responses may involve the regulation of gene expression, modulation of cytoskeletal elements, and metabolism. In addition, peptides of the nervous system might act as chemotactic factors, just as such factors act in slime molds[29,30], yeast[31,32], and bacteria[33]. In highly specialized tissues, such as retina, multiple layers of cells carry out specific functions in transmitting visual signals to the brain[34]. Various peptides have been localized in the retina[35], and the organization of these retinal cells might involve peptides, glycoproteins, or other, novel, cell recognition molecules.

STRATEGIES FOR PROTEIN ANALYSIS

In the discussion above, I have suggested that proteins of the nervous system may be classified on the basis of function. However, in the structural analysis of proteins, it is more useful to segregate them by size, while continuing to keep function in mind. Such a size classification is helpful in designing a strategy for the purification, characterization, and application of the protein in question.

284

Proteins of the nervous system may be divided into three general classes according to mass, as shown below in Figure 1. Each class represents a
category of proteins for which a particular strategy for structural analysis
may be useful, and the mass ranges are guides only and not strict limits.
Class I consists of peptides of mass 200-10,000 including neuropeptides,
hormones, and growth factors. The mass range of this class is defined by
the present limits of fast atom bombardment mass spectrometric analysis of
peptides[36]. Class II consists of proteins of mass 10,000 to around 60,000,
including enzymes, regulatory proteins, and certain growth factors. The
upper limit of the mass range in this category is defined by the practical
limits of chemical, amino acid sequence analysis[37,38], although theoretically, there is no limit to this analysis. Class III consists of proteins
of mass 60,000 and greater including large enzymes and enzyme complexes,
cell surface proteins, structural proteins, and certain receptor molecules.
Strategies for this class of proteins demand innovative, non-classical approaches, including special solubilization conditions, affinity chromatography, and partial protein sequence analysis followed by DNA cloning.

In the remaining portions of this article, I will examine examples of
strategies that one might use for the structural analysis of proteins and
peptides. For each class, I will discuss specific examples for which that
general strategy has been successful. These classification systems and strategies are meant as guidelines for the neuroscientist-protein chemist and
should not be treated as dogma. Particularly in laboratories involved in
many different sorts of projects, these strategies may be helpful in decid-
ing on an initial approach to a problem.

Class I Proteins

The peptides and small proteins of Class I correspond to polymers of
two to about ninety amino acid residues. A large fraction of these have
masses between 1000 and 5000, including peptides that have been identified
as chemical mediators[1,36]. These mediators are part of the overall question
of signal transduction in the nervous system, and such peptides have been
detected using both physiological and chemical approaches.

Among those peptides identified physiologically as chemical mediators
are neurotransmitters, neuromodulators, releasing factors, and release-in-
hibitory factors[1]. Two or more of these peptides may be secreted by the
same cell; in some cases a peptide that is a primary mediator may be accompanied by another peptide that modulates the primary response. For example,

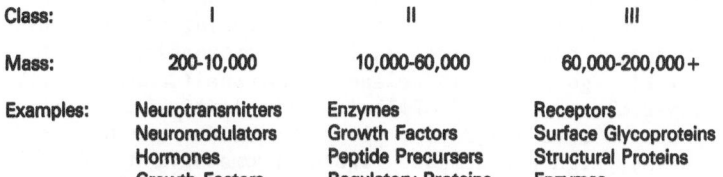

Class:	I	II	III
Mass:	200-10,000	10,000-60,000	60,000-200,000+
Examples:	Neurotransmitters	Enzymes	Receptors
	Neuromodulators	Growth Factors	Surface Glycoproteins
	Hormones	Peptide Precursers	Structural Proteins
	Growth Factors	Regulatory Proteins	Enzymes

Figure 1: Classification of proteins of the nervous system on the basis
of their size and corresponding strategy for structural analysis.

in mammals, corticotrophin releasing factor is secreted from cells of the paraventricular nucleus of the hypothalamus to stimulate adrenal cortico-trophic hormone (ACTH) release from the anterior pituitary lobe[39]. Under certain physiological conditions, the same cells of the hypothalamus secrete vasopressin, which amplifies the corticotrophic response[40,41]. Two landmark discoveries in this field were: (i) the isolation and characterization of corticotrophin releasing factor[42], and (ii) the anatomical and physiological demonstration of two peptides in the same cells of the hypothalamus[40,41].

Other peptides in Class I have been identified chemically as α-carboxy-amidated peptides of the gastrointestinal tract and the brain[43]. Several peptides with α-carboxyamides are localized to and are active in the gut and the brain[1,43,44,45]. These peptides are derived from larger precursors by a processing enzymes: endopeptidases, carboxy-exopeptidases, and acetyl-transferases[45]. The functional significance of this processing system is still an emerging field of study.

A variety of peptides and small proteins in Class I contribute to questions of cellular differentiation. In normal development of the nervous system, neuronal death occurs in an ordered fashion[46], and recently, in a model system, experimental results suggest that morphine, and perhaps opioid peptides can prolong the survival of certain neurons, while leaving others unaffected[47]. In cell culture systems of neuronal or glial cells in serum-free, defined media, cell survival usually requires insulin and transferrin.[5] These small proteins may be involved with survival, differentiation, or pro-liferation of the cells[5,17]. These examples illustrate that the structural analysis of these Class I proteins encompasses fundamental problems of cell growth, differentiation, and death.

Small proteins and peptides might also be active in cell recognition, and in the development of specific cellular connection in the brain. In various microorganisms, peptides play roles as chemotactic factors to stim-ulate cell recognition[29,32,33]. By analogy to these peptides, those of the brains of vertebrates might also function in cell recognition. In yeast, a tridecapeptide promotes the recognition and subsequent reproductive process of the a and α cell types[31,32]. In a slime mold, a novel peptide acts as a chemotactic attractant to promote the formation of a reproductive fruiting body[29]. Bacterial chemotactic factors might also include peptides or other amino acid containing compounds[33]. Thus, it is conceivable that small peptides act as chemotactic factors in the vertebrate nervous system. To this end, it is interesting that the yeast mating factor has significant structural homology with a vertebrate neuropeptide, gonadotropin releasing factor[39,48]. This homology has been reinforced recently by the elucidation of the struc-ture of gonadotropin releasing hormone from lamprey[48]. The role of such peptides with structural homology to chemotactic factors in the cell recog-nition events of the developing vertebrate nervous system are still unclear.

A strategy for the structural analysis of peptides from Class I is shown in Figure 2. The purification of relatively small peptides and pro-teins can take advantage of the differential solubility of the peptide in organic solvents. Many larger proteins precipitate from solvents such as acetone, ethanol, and ether, but small peptides can often be solubilized at a particular pH in an organic solvent. Neurotensin, for example, was iso-lated by a procedure that employed an initial extraction in acidified ace-tone[44]. Similarly, gonadotropin releasing hormone (GnRH) molecules may be extracted in petroleum ether[48]. Following the initial extraction and clas-sical, low pressure chromatography, Class I peptides can be purified further by reversed-phase, high performance liquid chromatography (HPLC)[49]. The most common medium for this fractionation is 5-10 μm silica particles that are bonded with octadecylsilanyl moieties. The mobile phase is commonly an aqueous buffer or dilute acid mixed with increasing proportions of a miscible organic solvent, such as acetonitrile or 2-propanol. Perfluoroalkyl acids are useful as ion pairing reagents at low pH, and trialkylamine-organic acid buffers are useful at neutral pH. The introduction of microbore HPLC has greatly increased the sensitivity of reversed-phase HPLC separations[38,49].

Initial characterization of Class I protein should include an amino acid analysis and elemental analysis. Amino acid analysis can be accomplished rapidly and at high sensitivity using pre-column derivatization and reversed-phase HPLC methodologies[38,49]. Elemental analysis is important to identify potential modification of the peptide, such as sulfation, phosphorylation, or metal ion chelation[12,50,51]. Following this chemical analysis, an accurate molecular weight of the peptide should be determined, usually by fast atom bombardment mass spectrometry[36]. Using a magnetic sector instrument with virtual field technology, one can measure protonated molecular ions of peptides with mass up to 10,000 amu. Effective ionization requires beam desorption methods, such as fast xenon atom bombardment of the peptide dissolved in an appropriate matrix. Molecular spectra can be recorded for picomole quantities of peptide, or less, but the absolute sensitivity of the instrument depends on the peptide structure and ionization properties[36].

Once the chemical composition and mass of a peptide are known, sequence analysis can reveal the covalent structure. Frequently, useful information on the amino acid sequence can be gathered from the pattern of fragment ions from the mass spectrometric analysis of the peptide[36]. However, automated

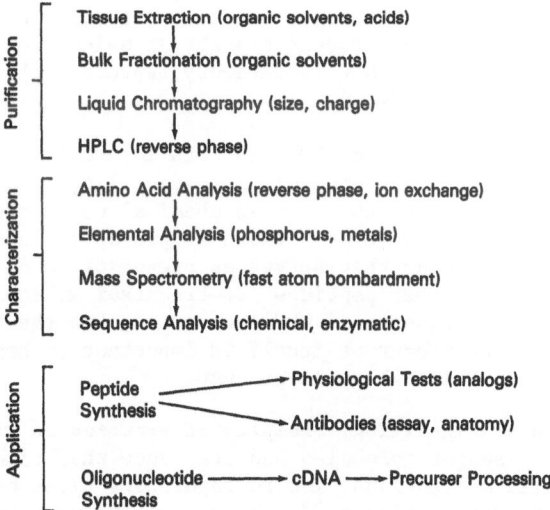

Figure 2: A general strategy for the purification, characterization, and application of proteins and peptides in Class I.

Edman degradations remain the mainstay of sequence analysis[37,38]. In a complementary technique, the mass spectrometer can be used to analyze exopeptidase digestion products of peptides[36]. Such a method provides an additional weapon in the arsenal of peptide structural analysis.

The elucidation of the covalent structure of a Class I peptide is the starting point for the application of this knowledge to problems in neuroscience. Complete chemical synthesis of a peptide can be accomplished using solid-phase techniques, introduced by Merrifield[52]. The synthetic peptide and structural analogs can be used for physiological testing in the whole organism, isolated tissue, and cell culture system. In addition, peptides can serve as antigens for producing polyclonal or monoclonal antibodies. Such antibodies can be useful in designing assays for the natural products and in mapping the anatomical distribution of the peptide. In parallel with synthetic peptides, an oligonucleotide[53,54] can be synthesized based on the predicted DNA sequence that would be complementary to an mRNA species for the peptide. Such an oligonucleotide may then be used to select a particular cloned cDNA molecule from a library of such moecules[53,55]. Cloning and sequence analysis of the cDNA will permit the assignment of a proposed precursor to the peptide, if any, and will provide a reagent to use in probing the gene structure and expression[55,56]. These procedures thus allow one to study the regulation of the peptide at three levels: (i) precursor protein processing, (ii) RNA transcript processing, and (iii) gene expression.[56]

The strategy outlined above and in Figure 2 has been used to isolate and characterize the GnRH molecules of lamprey brain[48]. In this endeavor, an integrated approach to structural analysis employed HPLC, amino acid analysis, chemical sequencing, and fast atom bombardment mass spectrometry. The measured mass of the natural product and the calculated mass of the peptide differed by only 0.05 amu. Subsequently, the synthetic peptide corresponding to the peptide structure was shown to have biological activity in the whole organism. In this fashion, Sherwood and colleagues[48] made use of state-of-the-art protein chemistry technology in all three aspects of peptide analysis: purification, characterization, and application.

Class II Proteins

The proteins in Class II have molecular weights that are generally in the 10,000 to 60,000 range. This mass range comprises molecules that are too large to perform fast atom bombardment mass spectrometry with high resolution and accuracy with current technology, but that are still of reasonable size to perform complete amino acid sequence analysis using chemical methods[37,38]. Several examples of proteins in this category represent the general problems of neuroscience discussed above.

Proteins in Class II are involved in both intercellular and intracellular signal transduction. In intercellular signalling, Class II proteins may be precursors to the peptides that act as chemical mediators. One of the best characterized of the these precursors is pro-opiomelanocortin (POMC)[57], a protein of 241 amino acids that serves as precursor to ACTH, lipotropins, endorphins, and several other peptides[57]. Precursor molecules like POMC can be identified from the sequence of cDNA molecules, but the isolation and characterization of the precursor itself is important in tracing the post-translational fate of this ribosomal product.

In intracellular signalling, examples of proteins in Class II are those that bind second messenger molecules and transduce this signal to metabolic enzymes. The cyclic nucleotides bind to regulatory sites on the cyclic nucleotide-dependent protein kinases[58]. This event removes an inhibition by the regulatory portion of the enzyme, and thus allows phosphorylation of target substrates. Several members of this family have been characterized using traditional biochemical procedures[59,60]. In parallel with the cyclic nucleotide-dependent protein kinases, the calcium modulated proteins transduce intracellular signals through binding calcium ions and conferring calcium ion sensitivity on target enzymes[12,61]. Vertebrate brain calmodulin is a calcium modulated protein of 148 amino acids that has multiple functions and is found in nearly all tissues[12,61]. Troponin-C is a calcium modulated protein, slightly larger than calmodulin, that is found in striated muscle and confers calcium sensitivity on acto-myosin ATPase activity, and thereby muscle contraction, in response to the calcium signal.

Class II proteins have distinct roles in neuronal differentiation. Various neurotrophic proteins have been identified that are freely soluble in aqueous media and stimulate neurite outgrowth and neural development in general[15,16,17]. Nerve growth factor[16] is a dimeric protein that was identified by its ability to maintain the survival of embryonic peripheral neurons in culture and of various neuronal elements in vivo. Beyond neuronal survival, however, a variety of proteins with mass about 20,000 to 45,000 appear to act in promoting neurite extension in cell culture systems of embryonic neurons[17,18,19]. The amino acid sequences of these factors might yield a coherent pattern of structural or functional homology among these differentiation factors. An interesting new aspect to this area is the relationship between neurotrophic growth factors and cellular oncogene products[23,24]. Nerve growth factor appears to stimulate the levels of expression of at least one oncogene in a pheochromacytoma cell line[62]. However, other oncogene products might amplify or attenuate the response to a neurotrophic factor under normal or pathophysiological conditions[23,24]. This field re-

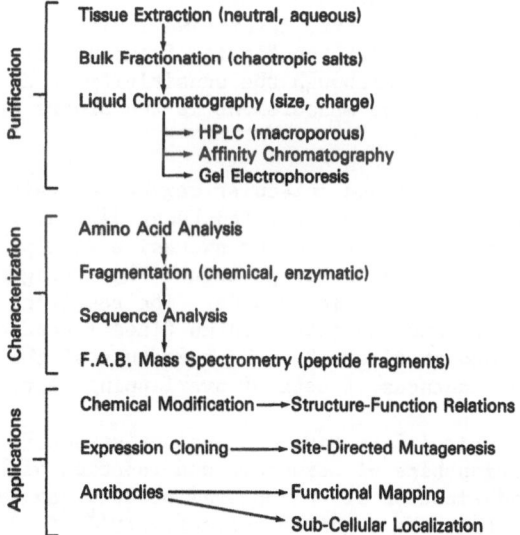

Figure 3: A general strategy for the purification, characterization, and application of proteins in Class II.

mains a fertile line of research, as the structural analysis of neurotrophic factors and oncogene products will form a basis for studies of the differentiation of cells at the phenotypic and genotypic levels.

Events of cell recognition in the developing nervous system may have mediators that are Class II proteins. Intracellular events that are secondary to cell adhesion phenomena might involve soluble enzymes and regulatory proteins such as kinases or phosphatases[58]. In addition, cells of the nervous system may employ cell recognition molecules that are related to the immunoglobulins[63,64] or lymphokines[65] of the immune system. In the brain, such molecules might function as specific cellular markers for cell death, cell-cell recognition, or terminal differentiation. However, such functions are still the subject of conjecture.

A strategy for the structural analysis of Class II proteins is shown in Figure 3. The purification of Class II proteins often begins with an extraction of the tissue with aqueous buffers at neutral pH, in order to prevent precipitation and to retain biological activity. Bulk fractionation using chaotropic salts, such as ammonium sulfate, or using isoelectric point precipitation may be used for the rapid removal of large amounts of contaminating proteins. Following these initial steps, chromatography remains a widely used and successful method of further fractionation. Size exclusion and ion exchange chromatography can be performed with traditional low-pressure apparatus. However, HPLC of proteins can be performed with reversed-phase, ion exchange, or gel exclusion media. For high recovery of proteins, silica matrices with 30-100 nm pore sizes are recommended[36,49]. Affinity chromatography[66] can be an efficient, high yield, purification step employing ligands that specifically bind to the protein. For example, a subset of calcium modulated proteins can be isolated by affinity-based chromatography on matrices of immobilized phenothiazines or naphthalene sulfonamide drugs[67]. Finally, preparative gel electrophoresis may also be used to efficiently isolate Class II proteins[68].

The first step in the characterization of Class II proteins is determining the amino acid composition and an accurate molecular weight. Determinations of molecular weight[69] can be accomplished by centrifugation, gel electrophoesis, and size-exclusion chromatography, but these methods are subject to inaccuracies due to unusual hydrodynamic properties of the protein. Mass spectrometry may be performed on these moderately large proteins using a time-of-flight instrument[36]. Soft, desorptive ionization of the proteins is accomplished by bombardment with the fission fragments from the decay of the isotope Californium-252[36]. Although the sensitivity of this technique is variable, the accuracy of mass measurement is two orders of magnitude above that of gel electrophoresis.

Following compositional and molecular weight analysis, sequence determinations should be performed by automated Edman degradation. In many Class II proteins, a blocked amino terminus or excessive size prevents the complete sequence determination of the intact protein. Thus, fragmentation is required using chemical or enzymatic methods. The resulting peptide fragments can be isolated by HPLC and characterized as Class I peptides, using the strategy outlined above and in Figure 2. The structure of the whole protein can be deduced from the sequences of sets of overlapping peptide fragments[37,38].

Application of this information on covalent structure ought to include studies of the relationships of structure and function of the Class II proteins. Chemical modification of the natural product can serve to identify amino acid residues that are required for activity[70,71]. Alternatively, site-directed mutagenesis[71,72,73] and the production of protein from cloned DNA can yield a library of mutant proteins with abnormal function. The cloning of cDNA molecules in mammalian expression systems[74] permits the study of potential post-translational modifications. Finally, scale-up of the purification of the natural product or the cloned, expressed product can aid in the process of functional mapping of the protein. The cloned product might be expressed in mammalian[74], yeast[75], or bacterial[55] cell systems. This process should be accompanied by antibody production and the use of these antibodies for the inhibition of function in vivo, measurement of the protein in immunoassays, and the mapping of the anatomical and subcellular distribution of the protein.

A successful application of this strategy has been the isolation and characterization of a neurite extension factor from adult bovine brain[19]. This protein appears to have a monomer molecular weight around 11,000, but it forms disulfide-bonded dimers or higher oligomers in the active state. In primary cultures of embryonic chicken neurons, this protein promotes the extension of neural processes in a serum-free, defined medium. Thus, under well-defined conditions, a factor that is directly involved in promoting neurite extension has been isolated and characterized using a Class II protein strategy.

Class III Proteins

Class III generally comprises large proteins of molecular weight 60,000 and higher that demand special methods for solubilization and fractionation due to size, solubility, or localization. In addition to such proteins, Class III includes proteins of somewhat lower molecular weight that require the same special purification methods or that have such extensive modification that normal amino acid sequence analysis is unsuitable. Examination of the general problems of neuroscience leads relentlessly to Class III proteins which are some of the greatest challanges to the modern protein chemist.

Large proteins and protein complexes play vital roles in signal transduction in the nervous system. In intercellular signalling, plasma membrane associated proteins act as receptors for various chemical mediators[1,10]. The acetyl choline receptor from eel electric organ[76] and mammalian muscle[77] have been the subjects of intense study for several years. These multi-subunit proteins have been analyzed by techniques ranging from DNA sequence analysis to functional mapping with antibodies. The receptors for other chemical mediators, such as neuropeptides[10], have not yet been characterized in the detail of the acetyl choline receptor. However, peptide recptors may fall into structural or functional families based on homologies in molecular domains. Class III also includes ion channels[11,12] that mediate signals in cells via ion fluxes across membranes.

Situated in membrane compartments in conjunction with receptors are the enzyme complexes that regulate the production of second messengers for intracellular signalling. Adenylate cyclase[13,78] is an enzyme complex whose catalytic subunit carries out the synthesis of cAMP from ATP, while the regulation of this activity is accomplished by hormone receptors coupled to a family of guanyl-nucleotide binding proteins. One of the common targets of cAMP and calcium ions as messengers is an enzyme complex, phosphorylase kinase, that is crucial for metabolic function. This enzyme catalyzes the phosphorylation of phosphorylase, which catalyzes glycogen breakdown to monosaccharides. Homology of the catalytic subunit of phosphorylase kinase to the cAMP-dependent protein kinases[79,80,81] illustrates the structural, functional, and evolutionary relationship among families of proteins. Other protein kinases may be found among Class III proteins that regulate metabolic events in the nervous sytem, and the identification of the substrates of these phosphorylation events may reveal sets of Class III proteins that are involved in a cellular signal transduction[82] event.

Several different kinds of Class III proteins are involved in neuronal differentiation. Substrate attachment factors, such as laminin and fibronectin, are laid down by cells as a matrix for morphological changes during differentiation[83,84]. These large, insoluble protein polymers have been characterized by means of chemical and molecular biological approaches, appear to have some repetitive sequence elements[84], and may be required for neural crest cell migration and neurite extension[83]. During differentiation of neurons, cytoskeletal proteins must alter their distribution to account for gross morphological changes and changes in the mitotic apparatus. The study of nonmuscle contractile proteins[12,85] has led to the characterization of important Class III proteins in brain and other tissues. Finally, mitogenic growth factors and soluble differentiation factos appear to have membrane associated receptor molecules that fall into Class III. An interesting development has been the recognition that oncogene products may have homologies with the guanyl-nucleotide regulatory proteins of adenylate cyclase[86]. Thus, extracellular, cell surface, and intracellular proteins all are examples of Class III proteins that are important in the differentiation of neurons and the elaboration of neural processes.

A large number of cell recognition events employ proteins of Class III as mediators. One of the central issues in cell recognition is cell adhesion, and specific protein molecules appear to cause these cell-cell interactions. Neural cell adhesion molecules (N-CAMs) [28] are a family of large, membrane associated glycoproteins whose carbohydrate composition changes during neural development. Structural analysis of the protein moieties of N-CAMs can add to our understanding of how neurons establish specific contact. The size (120,000 to 180,000), carbohydrate content, and insolubility of N-CAMs pre-

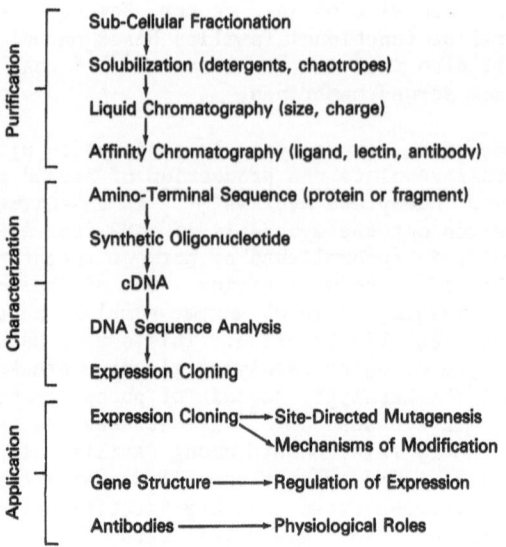

Figure 4: A general strategy for the purification, characterization, and application of proteins in Class III.

sent problems to the protein chemist that are typical of Class III proteins and that require innovative approaches to analysis.

A strategy for the structural analysis of Class III proteins is shown in Figure 4. The purification of these proteins can be optimized by an initial subcellular fractionation of a tissue to enrich for an organellar fraction that is rich in the protein of interest. Organellar fractions such as membrane, secretory vesicle, nuclear, or ribosomal fractions are often a more desirable starting material for protein pruification than whole tissue. Even from a soluble protein fraction, ultracentrifugation can be useful in the partial purification of a Class III protein of very large mass. Solubilization of these proteins from the appropriate fraction might involve a nonionic detergent such as deoxycholate, NP-40, Lubrol, or Triton X-100. In some cases, chaotropic salts can solubilize Class III proteins, and ionic strength and pH may be manipulated for differential solubility of proteins.

Following the initial fractionation, chromatographic methods are often used for final purification. Although size exclusion, ion exchange, and reversed-phase chromatography can be used, the sacrifice in yield is generally unacceptable. Affinity chromatography[66] should be employed whenever possible to maximize yield and purification. Small ligands such as drugs, dyes, and substrate analogs may be useful as immobilized matrices for affinity-based adsorption chromatography[67]. In addition, larger ligands such as lectins[87], binding proteins[12], and antibodies can serve as the basis for affinity supports. Lectins select for specific carbohydrate moieties in glycoproteins, while binding proteins, such as calmodulin[12,61], select for specific target enzymes. Monoclonal antibodies[88] often provide ligands for affinity chromatogrphy with exquisite selectivity for specific brain antigens[89] An ideal strategy for the purification of Class III proteins might combine two or more affinity chromatography steps using different kinds of ligands or antibodies with different specificities.

Because of the size, complexity, and often paucity of Class III proteins, the complete amino acid sequence obtained by chemical means may not be a feasible or cost-effective pursuit. In these cases, partial sequence analysis of the amino terminus of the protein may be sufficient to allow DNA cloning procedures to be fruitful. If the protein has a blocked amino terminus, then enzymatic or chemical cleavage of the molecule should yield fragments that have suitable sequences for cloning efforts. To this end, it is often useful to perform cyanogen bromide cleavage at methionyl residues, and to separate the peptide products by HPLC using diode array detectors of ultraviolet absorbance. This method permits the collection of on-line absorption spectra and the identification of those fragments containing tryptophanyl residues. Sequence analysis of these peptides may yield an amino acid sequence that is most amenable to low redundancy oligonucleotide synthesis[53,54] because tryptophan and methionine have unique codons in the genetic code. These oligonucleotide probes can help select cDNA molecules that contain all or only part of the information on the primary structure of the protein. A probe based on the amino terminal sequence of the protein can help select for a full length cDNA clone. Once the appropriate cDNA clone has been isolated, the DNA can be excised and inserted into expression vectors for production in bacterial[55], yeast[75], or mammalian[74] cells. Mammalian expression systems are preferable since the proper glycosylation or other modification may occur normally. In large scale culture, the cloned protein can be isolated in gram quantities or more, thus facilitating more complete chemical analysis.

Once the primary structure of the protein has been deduced from the peptides and cDNA molecules, a variety of applications of this information are available. Antibodies can be prepared from the natural product or from the expressed, cloned product. Site-directed antibodies can be prepared using synthetic peptides that correspond to specific regions of the protein molecule. These antibodies can be used for functional mapping of the natural product as probes of physiological function, and as tools for anatomical localization. In addition, antibodies can be used to select clones of genes for the protein from expression systems[90]. Expression cloning of cDNA molecules may allow a study of the mechanism of modification of the protein or alternative RNA processing events. The availability of cDNA molecules and antibodies allows the initiation of studies of the structure and expression of the gene for the protein. Thus, these applications lead to a detailed description of the structure, function, and regulation of Class III proteins.

Recently, this strategy has been applied to the determination of the structure of the amino terminal domain of mammalian N-CAMs[91]. As described above, N-CAMs are large, cell surface glycoproteins that mediate cell adhesion during development of the nervous system. Milligram quantities of embryonic and adult N-CAM were purified using a series of monclonal antibody affinity columns. The structural analysis of the amino termini of these molecules allowed the synthesis of peptides and the preparation of site-directed antibodies. The reactivity of the antibodies with native N-CAM confirmed the amino terminal sequence and provided a tool for studying the orientation of the molecules in the plasma membrane. Concomitantly, synthetic oligonucleotides based on the amino terminal sequence are being used to select DNA clones for N-CAMs. In this fashion, an integrated approach to protein structure has yielded the beginning of a complete description of a large, complex, glycoprotein that plays an essential role in the development of the mammalian nervous system.

CONCLUSIONS

In this article I have presented three general strategies for the structural analysis of proteins of the nervous system. Accompanying each strategy are suggestions of the sorts of proteins that might be amenable to the strategy. In each case, I have referred to a specific example of a problem in protein structure for which the strategy has been successful for me and my colleagues. However, these strategies are meant only as guidelines for an approach to a question of protein structure. To this end, the strategy for analysis should be based on the size and chemical properties of the protein, as well as the type of biological problem being investigated.

Each strategy should include three elements: (i) purification, (ii) characterization, and (iii) application. Purification of a protein to chemical homogeneity is essential for using modern tools such as microsequencers or mass spectrometers. Mixtures of peptides can lead to misinterpretation of the data and result in the loss of time and resources. Characterization should make use of all methodologies available, including such non-traditional technologies as DNA cloning and mass spectrometry. In application, the goal of protein chemistry of the nervous system should be the structure, function, and regulation of proteins. This regulation may be at many levels, such as transcription, RNA processing, and protein modification, and in various developmental, normal physiological, or pathophysiological states. An integrated strategy that contains all three elements provides the basis for a complete description of an interesting event in neuroscience.

The discipline of neuroscience and the many problems of development and behavior that are included in this field are fertile grounds for the application of a multidisciplinary approach including modern, protein structural analysis.

REFERENCES

1. D. T. Krieger, M. J. Brownstein, and J. B. Martin, "Brain Peptides," John Wiley & Sons, New York (1983).
2. M. B. Carpenter, "Human Neuroanatomy," The Williams & Wilkins Company, Baltimore (1976).
3. G. Lynch and P. Schubert, The use of in vitro brain slices for multidisciplinary studies of synaptic function, Ann. Rev. Neurosci. 3:1 (1980).
4. H. K. Kimelberg, Primary astrocyte culture- a key to astrocyte function, Cell. Molec. Neurobiol. 3:1 (1983).
5. G. H. Sato, A. B. Pardee, and D. A. Sirbasku, "Growth of cells in hormonally defined media," volume IX, parts A and B, Cold Spring Harbor Press, New York (1982).
6. B. Hamprecht, Cell cultures as models for studying neural functions, Neuropsychopharmacol. Biol. Psychiat. 8:481 (1984).
7. J. C. Eccles, The synapse: from electrical to chemical transmission, Ann. Rev. Neurosci. 5:325 (1982).
8. S. Varon and G. G. Somjen, Neuron-glia interactions, Neurosci. Res. Prog. Bull. 17:1 (1979).
9. E. R. Kandel and J. H. Schwartz, "Principles of Neural Science," Elsevier/North Holland, New York (1981).
10. K.-J. Chang and P. Cuatrecasas, Receptors and second messengers, in: "Brain Peptides," D. T. Krieger, M. J. Brownstein, and J. B. Martin, eds., John Wiley & Sons, New York (1983).
11. S. G. Waxman and J. M. Ritchie, Organization of ion channels in the myelinated nerve fiber, Science 228:1502 (1985).
12. L. J. Van Eldik, J.G. Zendegui, D. R. Marshak, and D. M. Watterson, Calcium-binding proteins and the molecular basis of calcium action, Intl. Rev. Cytol. 77:1 (1982).
13. E. M. Ross and A. G. Gilman, Biochemical properties of hormone-sensitive adenylate cyclase, Ann. Rev. Biochem. 49:533 (1980).
14. L. E. Hokin, Receptors and phosphoinositide-generated second messengers, Ann. Rev. Biochem. 54:205 (1985).
15. H. Theonen and D. Edgar, Neurotrophic Factors, Science 229:238 (1985).
16. B. A. Yanker and E. M. Shooter, The biology and mechanism of action of nerve growth factor, Ann. Rev. Biochem. 51:845 (1982).
17. D. K. Berg, New neuronal growth factors, Ann. Rev. Neurosci. 7:149 (1984).
18. J. Guenther, H. Nick, and D. Monard, A glia-derived neurite-promoting factor with protease inhibitory activity, EMBO J. 4:1963 (1985).
19. D. Kligman and D. R. Marshak, Purification and characterization of a neurite extension factor from bovine brain, Proc. Natl. Acad. Sci. USA, in press (1985).
20. R. James and R. A. Bradshaw, Polypeptide growth factors, Ann. Rev. Biochem. 53:259 (1984).
21. S. A. Hendricks, J. Roth, S. Rishi, and K. L. Becker, Insulin in the nervous system, in: "Brain Peptides," D. T. Krieger, M. J. Brownstein, and J. B. Martin, eds., John Wiley & Sons, New York (1983).
22. M. R. Hanley, Neuropeptides as mitogens, Nature 315:14 (1985).
23. J. M. Bishop, Cellular oncogenes and retroviruses, Ann. Rev. Biochem. 52:301 (1983).
24. P. H. Duesberg, Activated proto-onc genes: sufficient or necessary for cancer?, Science 228:669 (1985).
25. M. Rosenberg and D. Cpurt, Regulatory sequences involved in the promotion and termination of RNA transcription, Ann. Rev. Genetics 13:319 (1979).
26. B. W. Moore, Brain-specific proteins: S100 protein, 14-3-2 protein, and glial fibrillary protein, in: "Advances in Neurochemistry," B. W. Agranoff and M. H. Aperison, eds., Plenum Press, New York (1975).

27. R. J. Milner, F. E. Bloom, C. Lai, R. A. Lerner, and J. G. Sutcliffe, Brain-specific genes have identifier sequences in their introns, Proc. Natl. Acad. Sci. USA 81:713 (1984).

28. G. M. Edelman, Cell adhesion and the molecular processes of morphogenesis, Ann. Rev. Biochem. 54:135 (1985).

29. J. T. Bonner, Chemical signals of social amoebea, Sci. Am. 248:114 (1983).

30. P. Cappuccinelli and J. Ashworth, "Developments and differentiation in cellular slime molds," Elsevier/North Holland, New York (1977).

31. E. P. Sena, D. Radin, J. Welch, and S. Fogel, Synchronous mating in yeasts, Meth. Cell Biol. 11:71 (1975).

32. E. Loumaye, J. Thorner, and K. J. Catt, Yeast mating pheromone activates mammalian gonadotrophs:evolutionary conservation of a reproductive hormone?, Science 218:1323 (1982).

33. D. E. Koshland, Jr., Biochemistry of sensing and adaptation in a simple bacterial system, Ann. Rev. Biochem. 50:765 (1981).

34. P. Sterling, Microcircuitry of the cat retina, Ann. Rev. Neurosci. 6:149 (1983).

35. N. C. Brecha and H. J. Karten, Identification and localization of neuropeptides in the vertebrate retina, in: "Brain Peptides," D. T. Krieger, M. J. Brownstein, and J. B. Martin, eds., John Wiley & Sons, New York (1983).

36. D. R. Marshak and B. A. Fraser, Structural analysis of brain peptides, in: "Brain Peptides," D. T. Krieger, M. J. Brownstein, and J. B. Martin, eds., 2nd edition, John Wiley & Sons, New York, in press (1985).

37. K. A. Walsh, L. H. Ericsson, D. C. Parmelee, and K. Titani, Advances in protein sequencing, Ann. Rev. Biochem. 50:261 (1981).

38. M. W. Hunkapiller, J. E. Strickler, and K. J. Wilson, Contemporary methodology for protein structure determination, Science 226:304 (1984).

39. W. W. Vale, C. Rivier, J. Spiess, and J. Rivier, Corticotrophin releasing factor, in: "Brain Peptides," D. T. Krieger, M. J. Brownstein, and J. B. Martin, eds., John Wiley & Sons, New York (1983).

40. J. Z. Kiss, E. Mezey, and L. Skirboll, Corticotropin releasing factor-immunoreactive neurons of the paraventricular nucleus become vasopressin positive after adrenalectomy, Proc. Natl. Acad. Sci. USA 81:1854 (1984).

41. P. E. Sawchenko, L. W. Swanson, and W. W. Vale, Co-expression of corticotropin releasing factor and vasopressin immunoreactivity in parvocellular neurosecretory neurons of the adrenalectomized rat, Proc. Natl. Acad. Sci. USA 81:1883 (1984).

42. W. W. Vale, J. Spiess, C. Rivier, and J. Rivier, Characterizatio of a 41-residue ovine hypothalamic peptide that stimulates secretion of corticotropin and β-endorphin, Science 213:1394 (1981).

43. K. Tatemoto and V. Mutt, Chemical determination of polypeptide hormones, Proc. Natl. Acad. Sci. USA 75:4115 (1978).

44. S. E. Leeman and R. E. Carraway, Neurotensin discovery, isolation, characterization, synthesis, and possible physiological roles, Ann. N.Y. Acad. Sci. 400:1 (1982).

45. Y. P. Loh, M. J. Brownstein, and H. Gainer, Proteolysis in neuropeptide processing, Ann. Rev. Neurosci. 7:189 (1984).

46. J. W. Truman, Cell death in invertebrate nervous system, Ann. Rev. Neurosci. 7:171 (1984).

47. S. D. Meriney, D. B. Gray, and G. Pilar, Morphine-induced delay of normal cell death in the avian ciliary ganglion, Science 228:1451 (1985).

48. N. M. Sherwood, S. A. Sower, D. R. Marshak, B. A. Fraser, and M. J. Brownstein, Primary structure of gonadotropin-releasing hormone from lamprey brain, J. Biol. Chem., submitted (1985).

49. W. S. Hancock, "Handbook of HPLC Separation of Amino Acids, Peptides, and Proteins," volumes 1 and 2, CRC Press, Boca Raton, Florida (1984).

50. L. E. Eiden, The cell biology of peptidergic neurons: an overview, in: "Neuropeptides in Psychiatric and Neurological Disease," C. B. Nemeroff, ed., The Johns Hopkins University Press, Baltimore (1986).

51. F. Wold, In vivo chemical modification of proteins (post-translational modification), Ann. Rev.Biochem. 50:783 (1981).

52. G. Barany and R. B. Merrifield, Solid-phase peptide synthesis, in: "The Peptides: Analysis, Synthesis, Biology," volume 2, E. Gross and J. Meienhofer, eds., Academic Press, New York (1979).

53. K. Itakura, J. J. Rossi, and R. B. Wallace, Synthesis and use of synthetic oligonucleotides, Ann. Rev. Biochem. 53:323 (1984).

54. M. J. Gait, "Oligonucleotide Synthesis: A Practical Approach," IRL Press, Oxford (1984).

55. T. Maniatis, E. F. Fritsch, amd J. Sambrook, "Molecular Cloning: A Laboratory Manual," Cold Spring Harbor Press, New York (1982).

56. J. Douglass, O. Civelli, and E. Herbert, Polyprotein gene expression: generation of diversity of neuroendocrine peptides, Ann. Rev. Biochem. 53:665 (1984).

57. A. S. Liotta and D. T. Kriger, Pro-opiomelanocortin-related and other pituitary hormones in the central nervous system, in: "Brain Peptides," D. T. Krieger, M. J. Brownstein, and J. B. Martin, eds., John Wiley & Sons, New York, (1983).

58. A. C. Nairn, H. C. Hemming, Jr., and P. Greengard, Protein kinases in the brain, Ann. Rev. Biochem. 54:931 (1985).

59. K. Titani, T. Sasagawa, L. H. Ericsson, S. Kumar, S. B. Smith, E. G. Krebs, and K. A. Walsh, Amino acid sequence of the regulatory subunit of bovine type I adenosine cyclic 3', 5'-phosphate-dependent protein linase, Biochemistry 23:4193 (1984).

60. K. Takio, S. B. Smith, E. G. Krebs, K. A. Walsh, and K. Titani, Amino acid sequence of the regulatory subunit of bovine type II adenosine cyclic 3',5'-phosphate-dependent protein kinase, Biochemistry 23:4200 (1984).

61. C. B. Klee, T.H. Crouch, and P. G. Richman, Calmodulin, Ann. Rev. Biochem. 49:489 (1980).

62. T. Curran and J. I. Morgan, Superinduction of c-fos by nerve growth factor in the presence of peripherally active benzodiazepines, Science 229:1265 (1985).

63. H. N. Eisen, "Immunology," Harper & Row, Hagerstown, Maryland (1974).

64. A. F. Williams, Immunoglobulin-related domains for cell surface recognition, Nature 314:579 (1985).

65. A. Fontana and P. J. Grob, Lymphokines and the brain, Springer Semin. Immunopathol. 7:375 (1984).

66. P. Cuatrecasas, Affinity chromatography, Ann. Rev.Biochem. 43:169 (1974).

67. D. R. Marshak, T. J. Lukas, C. M. Cohen, and D. M. Watterson, Immobilized drugs in protein and peptide isolation," in: "Calmodulin Antagonists and Cellular Physiology," D. J. Hartshorne and H. Hidaka, eds., Academic Press, New York (1985).

68. M. W. Hunkapiller, E. Lujan, F. Ostrander, and L. E. Hood, Isolation of microgram quantities of proteins from polyacrylamide gels for amino acid sequence analysis, Meth. Enzymol. 91:227 (1983).

69. K. E. Van Holde, "Physical Biochemistry," Prentice-Hall, Englewood Cliffs, New Jersey (1971).

70. T. Kaiser, D. S. Lawrence, and S. E. Rokita, The chemical modification of enzymatic specificity, Ann. Rev. Biochem. 54:565 (1985).

71. G. K. Ackers and F. R. Smith, Effects of site-specific amino acid modification on protein interactions and biological function, Ann. Rev. Biochem. 54:597 (1985).

72. R. M. Myers, L. S. Lerman, and T. Maniatis, A general method for saturation mutagenesis of cloned DNA fragments, Science 229:242 (1985).

73. D. Botstein and D. Shortle, Strategies and applications of in vitro mutagenesis, Science 229:1193 (1985).

74. H. Okayama and P. Berg, Bacteriophage lambda vector for transducing a cDNA clone library into mammalian cells, Molec. Cell Biol. 5:1136 (1985).

75. R. A. Smith, M. J. Duncan, and D. T. Moir, Heterologous protein secretion from yeast, Science 229:1219 (1985).

76. B. M. Conti-Tronconi, M. W. Hunkapiller, J. M. Lindstrom, and M. A. Raftery, Subunit structure of the acetyl choline receptor from *Electrophorus electricus*, Proc. Natl. Acad. Sci. USA 79:6489 (1982).

77. B. M. Conti-Tronconi, C. M. Gotti, M. W. Hunkapiller, and M.A. Raftery, Mammalian muscle acetyl choline receptor: a supramolecular structure formed by four related proteins, Science 218:1227 (1982).

78. R. J. Lefkowitz, J. M. Stadel, and M. G. Caron, Adenylate cyclase-coupled beta-adrenergic receptors: structure and mechanisms of activation and desensitization, Ann. Rev. Biochem. 52:159 (1983).

79. P. Cohen, A. Burchell, J. G. Foulkes, P. T. W. Cohen, T. C. Vanaman, and A. C. Nairn, Identification of the Ca^{++} dependent modulator protein as the fourth subunit of rabbit skeletal muscle phosphorylase kinase, FEBS Lett. 92:287 (1978).

80. C. Picton, C. B. Klee, and P. Cohen, Phosphorylase kinase from rabbit skeletal muscle: identification of the calmodulin-binding subunits, Eur. J. Biochem. 111:553 (1980).

81. E. M. Reimann, K. Titani, L. H. Ericsson, R. D. Wade, E. H. Fischer, and K. A. Walsh, Homology of the γ-subunit of phosphorylase b kinase with cAMP-dependent protein kinase, Biochemistry 23:4185 (1984).

82. E. J. Nestler and P. Greengard, "Protein Phosphorylation in the Nervous System," John Wiley & Sons, New York (1984).

83. P. C. Letourneau, Cell-to-substratum adhesion and guidance of axonal elongation, Dev. Biol. 44:92 (1975).

84. K. M. Yamada, Cell surface interactions with extracellular materials, Ann. Rev. Biochem. 52:761 (1983).

85. M. Clarke and J. A. Spudich, Nonmuscle contractile proteins: the role of actin and myosin in cell motility and shape determination, Ann. Rev. Biochem. 46:797 (1977).

86. J. B. Hurley, M. I. Simon, D. B. Teplow, J. D. Robishaw, and A. G. Gilman, Homologies between signal transducing G proteins and ras gene products, Science 226:860 (1984).

87. S. H. Barondes, Lectins: their multiple endogenous cellular functions, Ann. Rev. Biochem. 50:207 (1981).

88. K. L. Valentino, J. Winter, and L. F. Reichardt, Applications of monoclonal antibodies to neuroscience research, Ann. Rev. Neurosci. 8:199 (1985).

89. M. Adinolfi and S. Brown, Monoclonal antibodies against brain antigens, Dev. Med. Child Neurol. 25:651 (1983).

90. R. A. Young and R. W. Davis, Efficient isolation of genes by using antibody probes, Proc. Natl. Acad. Sci. USA 80:1194 (1983).

91. G. Rougon and D. R. Marshak, Structural and immunological characterization of the amino-terminal domain of mammalian neural cell adhesion molecules, J. Biol. Chem., submitted (1985).

PANCREATIC PREPROSOMATOSTATIN PROCESSING: ISOLATION AND STRUCTURE OF

INTERMEDIATES AND FINAL PRODUCTS OF PROCESSING

P. C. Andrews and J. E. Dixon

Department of Biochemistry
Purdue University
West Lafayette, IN 47907

INTRODUCTION

The advent of recombinant DNA methodology has introduced a
situation in which protein structures are being determined indirectly
(via deduction from DNA) at a much faster rate than directly by Edman
degradation and other protein methods. Deduction of a protein sequence
from the DNA is an important first step in structure determination.
However, proteins may undergo a wide range of post-translational
modifications (1), most of which cannot yet be accurately predicted from
the primary structure. These processing events are often crucial for
expression of biological activity. Detailed structure determination
still requires laborious and time consuming methods.

Peptide hormones and neuropeptides represent a class of compounds
which undergo extensive post-translational processing which may include
acetylation, sulfation, glycosylation, amidation, and proteolytic
cleavage among many others. Specific proteolytic cleavage represents
one of the most common processing events for biologically active peptides.
Although proteolytic processing events occur most commonly at single
arginine residues or at adjacent pairs of basic residues, the parameters
that determine which basic sites are actually cleaved are not understood. Knowledge of the precursor sequence is insufficient to predict
cleavage sites. For this reason, while the sequences for many
precursors to small peptides of physiological significance have been
deduced from their cDNA, the steps leading to formation of the final,
physiologically active product are rarely known.

In an attempt to alleviate some of the problems arising in peptide
structure determination, we have utilized fast atom bombardment mass
spectrometry (FAB MS) to examine the post-translational processing of
preprosomatostatin. Particular emphasis has been on proteolytic
processing, although other types of post-translational processing
were also examined. Mammalian preprosomatostatin is processed to
either a 14-residue somatostatin (SST-14) or to a 28-residue somatostatin (SST-28), depending on the tissue in which the single preprosomatostatin gene is expressed. Production of SST-14 and SST-28 from
the same precursor proceeds through the use of alternative processing
sites. Cleavage of preprosomatostatin at a site of two adjacent basic
residues results in production of SST-14. Somatostatin-28 is produced

when the two adjacent basic residues are not utilized but cleavage occurs at a single arginine, twelve residues before the processing site cleaved during SST-14 production.

Alternative processing also occurs in the case of the two somatostatins produced in anglerfish endocrine pancreas (2) although in this case SST-14 and SST-28 are derived from different precursors. The structure of the two anglerfish precursors to somatostatin have been deduced from the cDNA (2), but the processing events leading to the two mature somatostatins remain unknown. In a homologous manner to mammalian preprosomatostatin, preprosomatostatin-I (aPPSST-I) has a cleavage site occurring at two adjacent basic residues resulting in a SST-14 identical to SST-14 isolated from other species. Preprosomatostatin from the second gene (aPPSST-II) appears to have an exclusive cleavage site at a single arginine (3) resulting in a SST-28 containing hydroxylysine (4,5,6). Cleavage at the two adjacent basic residues which would result in a 14-residue somatostatin does not occur during processing of aPPSST-II. Other proteolytic processing sites in anglerfish preprosomatostatin have been unknown. This study examines the final products and intermediates of prosomatostatin processing in anglerfish endocrine pancreas and verifies their structures. The peptides were purified by classical chromatographic methods and by reversed-phase HPLC. Extensive use of tryptic analysis by fast atom bombardment mass spectrometry (FAB MS) was made in order to provide initial verification of structure.[1] Further characterization of the peptides was provided by gas-phase sequencing[1] and amino acid analysis when necessary.

RESULTS AND DISCUSSION

All the peptides described in this manuscript were purified from acid/ethanol extracts of anglerfish endocrine pancreas (7), Gel filtration chromatography of the crude extract resolved a number of absorbance peaks at 280 nm (data not shown). Fractions corresponding to specific molecular weight size ranges were combined and further purified by ion exchange chromatography and/or reversed-phase HPLC (data not shown). In addition to SST-14 and SST-28 which had previously been reported (3,4,5,6,8), six other fragments of the two preprosomatostatins were isolated and then identified via a combination of tryptic analysis by FAB MS, amino acid composition and Edman degradation. The peptides derived from aPPSST-I were found to correspond to aPPSST-I, 26-92, 26-52, and 94-105. Those peptides having aPPSST-II as a precursor were found to correspond to aPPSST-II, 25-96, 25-60, and 63-96.

The use of FAB MS for the analysis of unresolved tryptic digests is illustrated in Fig. 1 for the partially hydroxylated anglerfish SST-28. The major peaks represent the peptide molecular ions [M+H]$^{+}$ (peptide mass + 1 dalton) and are accompanied by their sodium and potassium adducts at M+23 and M+39 daltons respectively. A few interfering ions from the dithioerythritol/dithiothreitol matrix (8) are observed and are marked in Figure 1. No ions resulting from

[1]The authors wish to acknowledge David Heller and Dr. C. Fenselau of the Mid-Atlantic Mass Spectrometry Facility, Baltimore, MD for the mass spectrum reproduced in Fig. 2. All other mass spectra were obtained at the Purdue University facility by Razieh Yazdanparast and Dr. D. Smith. We also wish to acknowledge Dr. M. Hermodson for providing access to a gas-phase sequenator and for helpful discussion. This work was supported by grants from the National Institutes of Health.

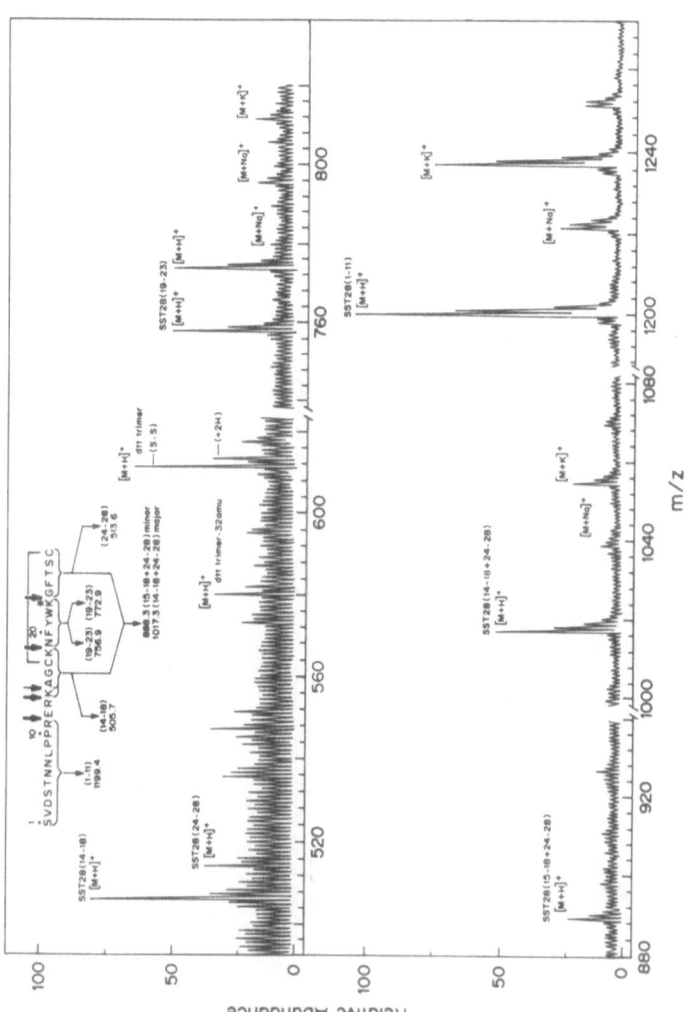

Fig. 1. Fast atom bombardment mass spectrum of trypsinized anglerfish somatostatin-28. The identity of tryptic fragment is indicated just above the appropriate molecular ion [M+H]⁺. The corresponding sodium [M+Na]⁺ and potassium [M+K]⁺ adducts are also indicated. The position of the hydroxylysyl residue is indicated by an asterisk (Insert). Major tryptic cleavage sites are indicated by heavy arrows and the minor cleavage site by a small arrow. Theoretical average masses of the non-charged tryptic fragments are indicated below the sequence. The intact disulfide bond occurring between Cys 17 and Cys 28 is indicated by the line connecting those residues above the peptide sequence.

fragmentation of the peptides by the beam are observed. The site of
the hydroxylated lysyl residue is readily identified from the molecular
ion at 774 daltons, exactly 16 daltons higher than the observed
molecular ion at 758 for non-hydroxylated SST-28, 19-23. The mass
difference between lysine and hydroxylysine is equivalent to one
oxygen atom (16 daltons). This information was previously obtained
using much more laborious, classical methods (4,5). All the other
expected molecular ions were observed including two fragments resulting
from major and minor trypsin cleavages at the two adjacent basic residues,
Arg-13 and Lys-14 (889 and 1018 daltons). One interesting observation
is that the disulfide bond remains intact and a molecular ion may be
observed at 1018 daltons for the two fragments 14-18 and 24-28 connected
via the disulfide bond. This observation suggests that FAB MS of
proteolytic digests could be used to determine disulfide connectivity
in larger proteins. A significant amount of reduced peptide is also
observed (507 and 515 daltons). The presence of this reduced material
may be attributed to <u>in situ</u> reduction by the atomic beam (10).
Any deviation from the sequence deduced from the DNA will result in
a mass change which is potentially detectable using FAB MS. The
advantages which this method has over more classical techniques are:
sensitivity ($\leq$ 200 pmol), simplicity (no need to resolve fragments
chromatographically), speed (15 to 30 min. per analysis), and information
(any deviation from theoretical mass may be quantitated). The presence
of a hydroxylysyl residue in SST-28 was unexpected and could not have
been predicted from the protein sequence.

The mass spectrum of another post-translationally modified somatostatin shown in Figure 2 represents a mixture of the two intact, major
forms of the 22-residue somatostatin from catfish which are both
O-glycosylated (11). This O-glycosylation could also not be predicted
from the peptide sequence. The higher mass peptide differs from the
lower mass peptide by one N-acetyl neuraminic acid residue (291 daltons).
This observation provides substantiating evidence that heterogeneity
in the carbohydrate moiety is responsible for the observed chromatographic heterogeneity of SST-22. Note the skew in the isotopic envelope
due to reduction of the peptide in the atomic beam.

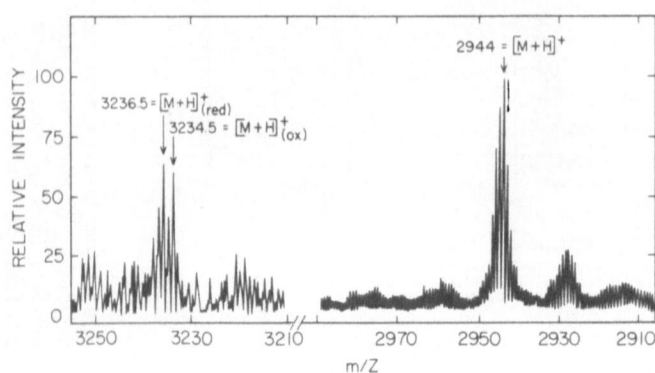

*Fig. 2. Molecular ion regions of the FAB MS of O-glycosylated catfish
somatostatin-22. The theoretical average masses are indicated
next to the mass peaks. The masses for both the peptide
with an intact disulfide bond, [M+H]$^+$$_{(ox)}$, and the reduced
peptide [M+H]$^+$$_{(red)}$, are shown.*

Tryptic analysis by FAB MS of the previously uncharacterized fragments of aPPSST-I isolated in this study, indicate that the structure of aPPSST-I, 26-52 could be identified for most of its length (data not shown). These data indicate that the site of signal cleavage occurs between Cys-25 and Ser-26 of aPPSST-I. This conclusion is confirmed by gas-phase Edman sequencing of both aPPSST-I, 26-52 and 26-92. Tryptic fragments from aPPSST-I, 26-92 could also be identified but the mass of the fragment corresponding to aPPSST-I, 54-92 (4191.1 daltons) waa beyond the capability of the mass spectrometer in its present configuration (Kratos MS50, 23 kgauss magnet, no post-acceleration detector). Subsequent gas-phase Edman sequencing of the aPPSST-I, 54-92 tryptic fragment indicated a Gly for Glu substitution at position 83. This observation is consistent with a corrected cDNA sequence for aPPSST-I (12) and the gene sequence for aPPSST-I (13).

The major products of aPPSST-I processing in this tissue are aPPSST-I, 26-92, 94-105, and 108-121 (SST-14). These results suggest that aPPSST-I 94-105 and 108-121 are rapidly released from the prohormone and that cleavage at the single Arg-53 (Fig. 3) occurs at a much slower rate.

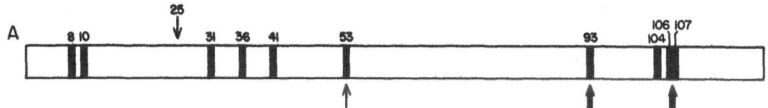

Fig. 3. Diagrammatic representation of anglerfish preprosomatostatin-I. All basic residues are numbered. Narrow bands indicate single Arg sites. Broad bands indicate two adjacent basic residues. Arrows indicate major (large arrows) and minor (small arrow) sites of proteolytic processing. The signal cleavage site occurring after Cys-25 is indicated by a small arrow.

Fragments of aPPSST-I resulting from processing at sites other than those listed above were not observed. However, while this observation suggests that the processing events associated with the non-hormone portion of prosomatostatin-I are specific, alternate, minor cleavage sites cannot be ruled out.

This study accentuates the present inability to assign post-translational modifications solely on the basis of deduced sequence information. It also emphasizes the utility of FAB MS in verifying structures, assigning modified residues, and providing basic information on the nature of the modifications.

REFERENCES

1. Wold, F. (1981) in Ann. Rev. Biochem. (E. E. Snell, ed.) 50:783.
2. Hobart, P., Crawford, R., Shen, L.-P., Pictet, R., and Rutter, W. J. (1980) Nature 288:137.
3. Noe, B. D. and Spiess, J. (1983) Program of the 65th Annual Meeting of the Endocrine Society, San Antonio, TX, June 8-10, p. 85 (abstract).
4. Spiess, J. and Noe, B. D. (1985) Proc. Natl. Acad. Sci. U.S.A. 82:279.
5. Andrews, P. C., Hawke, D., Shively, J. E., and Dixon, J. E. (1985) Endocrinology 116:2677.

6. Spiess, J. and Noe, B. D. (1985) in _Adv_. _Exp_. _Med_. _Biol_.,
    Somatostatin, editors Y. C. Patel, G. S. Tannenbaum,
    Plenum, N.Y.

7. Andrews, P. C. and Ronner, P. (1985) _J_. _Biol_. _Chem_. 260:3910.

8. Noe, B. D., Spiess, J., Rivier, J. E., and Vale, W. (1979)
    Endocrinology 105:1410.

9. Witten, S. L. Scheffer, M. H., O'Shea, M., Cook, J. C., Hemling,
    M. E., and Rinehart, K. L. (1984) _Biochem_. _Biophys_. _Res_. _Commun_.
    124:350.

10. Yazdanparast, R., Andrews, P. C., Dixon, J. E., and Smith, D.,
    submitted for publication.

11. Andrews, P. C., Pubols, M. H., Hermodson, M. A., Sheares, B. T.,
    and Dixon, J. E. (1984) _J_. _Biol_. _Chem_. 259:13267.

12. Goodman, R. H., Jacobs, J. W., Chin, W. W., Lund, P. K., Dee, P. C.,
    and Habener, J. F. (1980) _Proc_. _Natl_. _Acad_. _Sci_. _U.S.A_. 77, 5869.

13. Crawford, R., Hobart, P., and Rutter, W. J., personal communication.

VI. Microsequence Analysis of Polypeptides by Edman Degradation

PROGRESS TOWARD POLYBRENE PURIFICATION

AND UTILITY IN MICRO-PEPTIDE/PROTEIN SEQUENCING

Pau M. Yuan, Sylvia Yuen, John Bergot, Michael W. Hunkapiller and Kenneth J. Wilson

Applied Biosystems, Inc.
850 Lincoln Centre Drive
Foster City, CA 94404 U.S.A.

INTRODUCTION

Since the introduction of automated Edman degradation(1) there have been continuing attempts to minimize sample loss during the organic extraction steps inherent to the chemistry. These are required to remove excess reagent, buffer, reaction by-products and, of course, the ATZ-amino acid prior to its conversion to the PTH derivative for identification. Various attempts have been made to reduce sample loss: addition of blocked proteins(2), synthetic amino acid polymers(3), and chemically introducing groups which render the sample more hydrophilic(4). Another modification was the covalent attachment of the sample onto an appropriately derivatized matrix, ie. solid phase sequencing(5).

The introduction(6,7) of Polybrene, a polymeric quaternary amine as a carrier seemed to solve some of the problems, ie. sample loss due to extraction, especially as the length of the peptide decreased. In both the gas and liquid phase instruments, the addition of 2-5 mg Polybrene needs to be followed by a series of precycling steps prior to sample application. If precycling is not performed, then background peaks are often of such size (amounts) as to obscure some of the PTH-residues as detected by either GLC, TLC or HPLC. Since these precycling steps waste both chemicals and available instrument time, there have been a number of reports describing methods for Polybrene purification prior to sequencing(8,9). These methods have proven useful for sequencing samples at the nanomole level but have yet to be shown applicable on low picomole amounts.

A drawback to using Polybrene has been the reduced yield noted with some of the charged PTH derivatives, most notably PTH-His and PTH-Arg. In the gas phase sequencer PTH-His yields can vary from 5% to 40% depending on the sample and instrument used. Such variations might well arise from inefficient ATZ-His extraction with butylchloride (S3), which in turn could be related to the physical and chemical conditions of the Polybrene film and glass fiber filter during extraction. In addition to reduced yields of these derivatives, lower than expected initial yields as well as lower PTH-Asp and PTH-Glu yields are often encountered.

The utility of the gas phase sequencer at levels below 25 pmol has been demonstrated on numerous occasions for both peptides(10,11) and proteins(12,13). When handling such small amounts of sample there are serious losses encountered at either the isolation or sample recovery stages. The use of microbore HPLC offers a number of distinct improvements, eg. high sensitivities, reduced volumes, easy sample concentration. Using the glass fiber filter of the gas phase instrument as a "fraction collector" can eliminate sample handling steps. Since the presence of Polybrene is a necessity when sequencing low picomole quantities, the manner in which the filter is treated prior to sample collection is important.

The purpose of this communication is three-fold:
1. To illustrate results from Polybrene clean-up and the necessity of precycling;
2. Show how the use of TFA-treated glass fiber filters and NaCl improve recoveries of the charged PTH amino acids; and
3. Indicate how the filters can be employed for "fraction collection" and the optimized conditions for doing so.

EXPERIMENT PROCEDURES

All sequencing experiments were carried out on a Model 470A Gas Phase Sequencer from Applied Biosystems, using chemicals from the same source. PTH amino acid identifications were performed by HPLC using either chromatography on an IBM cyano column as described by Hunkapiller et al.,(14) or with an on-line Model 120A PTH Analyzer and chromatography on a narrow-bore C-18 column (see Hunkapiller, this volume). Data reduction and quantitation was performed using a Nelson 760 interface and a Hewlett Packard 9816 computer. The Polybrene (Biobrene(R)) and apomyoglobin (Bioglobin(R)) are products of Applied Biosystems; angiotensin II was from Bachem.

Tryptic digestion of apomyoglobin was carried out at room temperature in 0.1 M ammonium bicarbonate for 4 hours using a 4% (w/w) ratio of trypsin (Worthington) to substrate. The reaction was stopped by lowering the pH to less than 2 with TFA and the sample recovered by drying (Savant Speed Vac). Reverse phase separations of the tryptic fragments were carried out using a microbore column from Brownlee Labs and a Kratos Model 783 variable wavelength unit with the standard 12 uL, 8 mm flow cell as the detector.

Two different methods were employed in an attempt to pre-purify the Polybrene. The first (method 1) involved liquid-liquid extraction of a 10% (w/v) aqueous solution of crude Polybrene with either chloroform or ethyl acetate for ca. 96 hours, separation of the extracted aqueous phase, and subsequent lyophilization to afford a white solid.

Method 2 for Polybrene purification involved dissolving the crude material in 50% aqueous acetonitrile, adding PITC and TMA under N_2, and reacting for 30 minutes at 44°. The solution was extracted with 3 volumes of heptane, followed by one of ethylacetate, and the aqueous fraction lyophilized. After dissolving in 25% TFA, it was extracted once with ethylacetate, and twice with butylchloride. The aqueous fraction was lyophilized and the Polybrene dissolved at a concentration of 1.8 mg/30 uL H_2O for use.

The glass fiber filters were 'cleaned' by soaking in TFA (R3) for 60 minutes, dried for 2-3 hours under a hood with air, and finally under vacuum overnight.

RESULTS AND DISCUSSION

The necessity of using Polybrene in automated sequencing has been accepted for some years. The degree of purity of this material, however, has been a question of the source and its lot number, the number of pre-cycling steps carried out, and the actual percentage of the degradation cycle utilized. In other words, the amount of background is proportional to the percentage injected at any given cycle for identification. When microsequencing is being carried out, it is often necessary to inject up to 100% of the product from each cycle in order to adequately interpret the results. This implies that the inherent background from the se-quencing chemicals and the instrument, as well as reaction chamber or cartridge, must be at a minimum.

The results illustrated in Fig. 1 indicate the levels of reduced background that are possible. Note that optimal results are from Poly-brene that has simply been precycled 3 times in the instrument, rather than attempting to prepurify it prior to sequencing. The levels of DPTU associated with the pre-cleaned preparations are significantly higher and DMPTU is elevated in the Polybrene that had been carried through an Edman cycle in an extraction flask (Fig. 1C). The presence of the smaller peaks in Fig. 1B and C indicate that additional contamination was also introduced. Lower initial yields for the PTH-Asp in Cycle 1 are obvious (compare Fig. 1A with 1B and 1C) and suggests that either N-terminal blockage might be occurring or reduced ATZ-derivative extraction, or both.

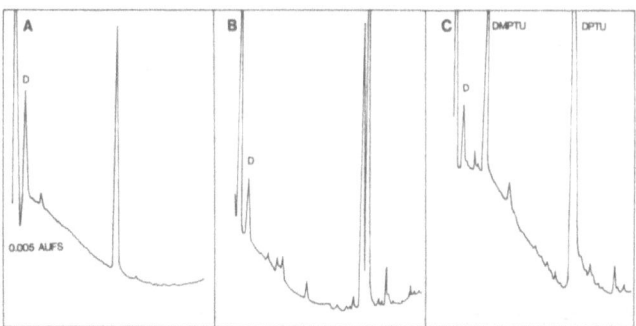

Fig. 1. Comparison of variously prepared Polybrenes for sequencing. A, Polybrene (1.8 mg) was precycled three times on the glass fiber filter prior to addition of 100 pmol angiotensin II. Shown is the HPLC chromatogram of the first cycle of Edman degradation. B, as in A above except that the Polybrene had been pre-purified by method 1 (see Experimental) and applied together with the sample. C, as in B using Polybrene pre-purified by method 2 (see Experimental). Abbreviations: D, aspartic acid; DMPTU, dimethyl-phenylthiourea; DPTU, diphenylthiourea.

This latter possibility, ie. ATZ extraction, has been found to be a limitation with Polybrene (Fig. 2). A simple treatment, viz., etching or cleaning of the filter disc with TFA results in an approximately 50% increase in the initial yield of Asp-1, slightly higher yield of Arg-2 and minimal changes in recovery of the other residues. However, the addition of 3 umol NaCl to the filter drastically improved, to quantita-tive levels, the yield of Asp-1 and significantly increased recoveries of both Arg-2 and His-6. In other words, the overall yields of charged PTH residues can be additively improved by both TFA treatment of the filter

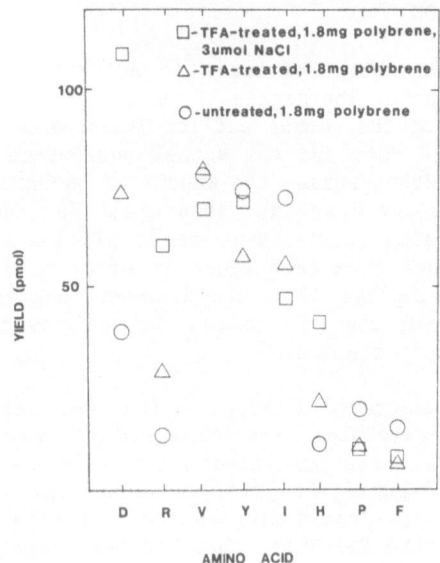

Fig. 2. Effects of TFA treatment of the glass fiber filter and addition of NaCl during sequencing. □ , TFA-treated filter plus 1.8 mg Polybrene plus 3 umol NaCl. △ , TFA-treated filter plus 1.8 mg Polybrene;○, untreated filter plus 1.8 mg Polybrene. Filters were precycled prior to sample application (100 pmol angiotensin II).

and adding NaCl. The recoveries as a function of NaCl concentration are approximately constant over the range of 0.6 to 15 umol. These increased recoveries, ranging from 60% to nearly quantitative, have also been noted when a wide variety of other proteins and peptides were tested in our laboratory.

The chromatograms in Fig. 3 clearly illustrate why precycling is necessary when sequencing at the 100 pmol level or less. Note that not only are both the DMPTU and DPTU peaks reduced by three precycle steps,

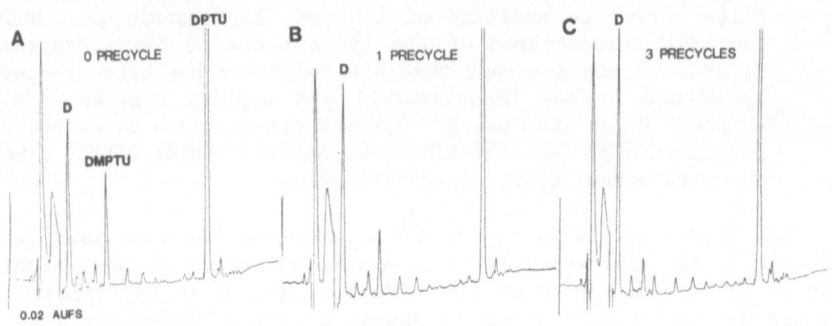

Fig. 3. Precycling effects on background and yield of charged PTH derivatives. TFA-treated filters were loaded with 1.8 mg Polybrene and 3 umol NaCl and precycled 0, 1 or 3 times prior to addition of 100 pmol angiotensin II and sequencing started. Shown are the HPLC chromatograms from the first cycle from either 0 (A), 1 (B), or 3 (C) precycles.

but that there are dramatic increases in Asp-1 recoveries from 53% without precycling to 68% with a single precycle to 100% with three. The peak background present in each of these degradations originates from either contamination in the sample or the instrument itself (chemicals, surfaces, etc.). The presence of this background is a concern with samples collected from microbore HPLC separations (see following results).

As indicated in another paper (see Wilson et al., this volume) smaller diameter columns of either 2.1 mm or 1.0 mm ID, ie. narrow and microbore columns, are convenient substitutions for the standard 4.6 mm ID column when higher sensitivities are required. The chromatogram in Fig. 4 illustrates the level of sensitivity and the preparative separation of tryptic peptides from apomyoglobin. Since maps can be generated at < 25 pmol by microbore HPLC and the gas phase instrument can function well at this range, we have investigated the utility of the fiber filter as a "fraction collector". To this end, repetitive isolations were performed. In the first separation, the indicated peaks (Fig. 4) were collected onto TFA-treated filters containing Polybrene (supported in a 1.5-mL Eppendorf tube) and for the second they were collected into 500-uL Eppendorf tubes. These were subsequently sequenced in the same order, ie. those collected onto the filters were degraded first, followed by application of the corresponding collected peak onto the same filter and sequencing repeated. This then compares the filter-collected sample on non-precycled Polybrene with what would be considered a precycled filter.

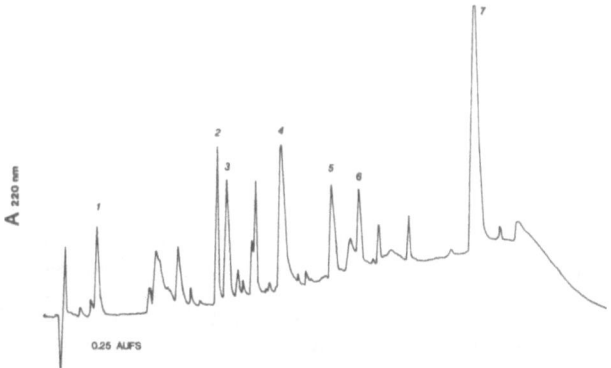

Fig. 4. Tryptic map of apomyoglobin by microbore HPLC. The equivalent of 300 pmol of the digest was loaded onto a 1 x 250 mm Aquapore RP-300 column. Elution was carried out at 50 uL/min and room temperature using a linear gradient from 10 to 100% of buffers A, 0.1% TFA, and B, 60% acetonitrile in 0.1% TFA. The peaks were collected as indicated in the first separation onto glass fiber filters containing 1.8 mg Polybrene or in the second into 500 uL Eppendorf tubes.

Fig. 5 illustrates the type of results found from such an experiment. The peptide, peak 5 from Fig. 4, represents residues 17-31 of myoglobin. Only the first six degradation cycles are shown and represent an initial yield of 40 pmol. These analyses, as well as each of those from the other numbered peaks (results not given), are characterized by:

A. An increased background of free PTH amino acids in the first cycle for the samples collected directly onto the filter;
B. Elevated levels of DMPTU and DPTU present in the initial cycle on non-precycled filters;

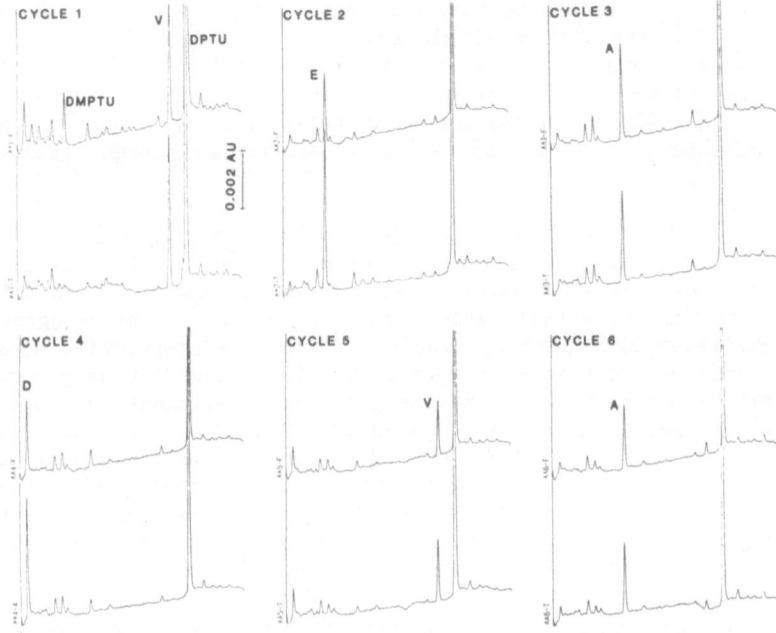

Fig. 5. Comparison of first six degradation cycles of peptide 5 from
Figure 4. The samples were collected either onto the filter
(upper series of chromatograms) or into tubes prior to spotting
onto precycled filters before initiating the degradation (lower
series of chromatograms). PTH analyses were carried out on-line
using a Model 120A Analyzer and injection 40% of the PTH residue
arising from each degradation cycle.

C. Somewhat lower initial yields on the filter-collected peptides,
representing extraction problem rather than α-amino group
blockage; and

D. Reduced recoveries of charged PTH residues; note PTH-Glu and
PTH-Asp yields in the second and fourth cycles are significantly lower on filter-collected samples.

E. Possible destruction of Trp residues in samples which subsequently give rise to multiple peak formation during PTH
analysis (example not given).

These results illustrate both the utility of filter collection and
the potential drawbacks. The limits which we presently set are 25-50
pmol (initial yield) for samples collected onto non-precycled and 10
pmol for precycled filters. Obviously there are a series of important
factors that allow one to carry out successful PTH identification when
working at these levels, most notably the HPLC detection system. The
introduction of the online PTH analyzer has solved a number of the
problems which have limited the ease and reproducibility of PTH analysis
over the past years.

In summary, we have indicated which steps are necessary to implement
for filter precycling and those modifications to the filter (TFA cleaning
and NaCl addition) which will increase initial yields and recoveries of
the charged PTH amino acids. The utility of microbore HPLC and filter
collection of samples has been demonstrated and the present quantity
limitations given. Not until a method has been found to purify Polybrene
and the glass filter sufficiently, will the real utility of microbore
chromatography in conjunction with microlevel gas phase sequencing be
realized.

REFERENCES

1. Edman, P. and Begg, G., *Eur. J. Biochem.* 1 (1967) 80-91.

2. Rochat, H., Bechis, G., Kopeyan, C., Gregoire, J. and Van Rietschoten, J., *FEBS Lett.* 64 (1976) 404-408.

3. Niall, H.D., Jacobs, J.W., Van Rietschoten, J. and Tregear, G.W., *FEBS Lett.* 41 (1974) 62-64.

4. Braunitzer, G., Schrank, B. and Ruhfus, A., *Hoppe-Seyler's Z. Physiol. Chem.* 351 (1970) 1589-1590.

5. Laursen, R.A., *Eur. J. Biochem.* 20 (1971) 89-102.

6. Tarr, G.E., Beecher, J.F., Bell, M. and McKean, D.J., *Anal. Biochem.* 84 (1978) 622-627.

7. Klapper, D.A., Wilde, C.E. and Capra, J.D., *Anal. Biochem.* 85 (1978) 126-131.

8. Wittmann-Liebold, B. (1981) *in* Chemical Synthesis and Sequencing of Peptides and Proteins, (Liu, T.Y., Schechter, A.M., Heinrikson, R.L. and Condliffe, eds.), Elsevier North Holland, Inc., Amsterdam – New York, pp. 76-110.

9. Lai, P.-H., *Anal. Chimica Acta* 163 (1984) 243-248.

10. Downward, J., Parker, P. and Waterfield, M.D., *Nature* 311 (1984) 483-485.

11. Grego, B., VanDriel, I.R., Stearne, P.A., Goding, J.W., Nice, E.C. and Simpson, R.J., *Eur. J. Biochem.* 148 (1985) 485-491.

12. Jong, A.Y.S., Kuo, C. and Campbell, J.L., *J. Biol. Chem.* 259 (1984) 11052-11059.

13. Sparrow, J.G., Metcalf, D., Hunkapiller, M.W., Hood, L.E. and Burgess, A.W., *Proc. Natl. Acad. Sci. U.S.A.* 82 (1985) 292-296.

14. Hunkapiller, M.W. and Hood, L.E., *Methods Enzymol.* 91 (1983) 486-493.

APPROACHES FOR MICROSEQUENCING:

EXTRACTION FROM SDS GEL, COMPOSITION, TERMINAL ANALYSIS

Akira Tsugita and Tatsukai Ataka

Department Chemistry, Faculty of Science
Science University of Tokio
Yamazaki, Noda, 278 Japan

Protein sequencing techniques are ever becoming more accurate, sensitive, quick and simple. This presentation is limited to those techniques recently developed in our laboratory. These are: (1) a simple extraction method from SDS polyacrylamide gels (2) a rapid gas hydrolysis method for protein and (3) a sensitive detection method of Edman degradation.

1) Extraction from the conventional polyacrylamide gel (1)

There are many established extraction procedures from polyacrylamide gels including electroelution. Some of our experiences with electroelution have shown that it allows protease digestion even in the presence of detergent, e.g. sodium dodecyl sulfate (SDS), especially when the sample is crude such as cell extracts. Detergents or ampholine, which is the medium of electrofocussing, are difficult to remove completely and if removed the protein loses its solubility in many cases. We found that 70%-80% formic acid is effective to dissociate proteins from SDS as well as to extract proteins from the polyacrylamide gel.

After electrophoresis the gel is stained by the conventional method with Coomassie blue and the spot (in two-dimensional gels) or band (in one-dimensional gels) is cut out and placed in more than 4 times its volume of 70% formic acid. The gel usually swells to 2.5 times its original volume after equilibration in 70% formic acid. Extraction is carried out by gently shaking (as with DNA extraction) at room temperature for 6-16 hrs, depending on the molecular size of the protein, until equilibrium. The dye may be used as a convenient but minimal indicator of equilibrium because formic acid dissociates the protein-dye complex and in the extraction, equilibrium is quicker for small molecules, such as the dye, than for the protein. This extraction is repeated 2 to 3 times. The extracted

---

* Seiko Electronics and Instruments
  6-31-1, Kameido, Koto-ku, Tokyo 136, JAPAN

solution is evaporated using a rotary evaporator after first adding a drop of octyl alcohol to avoid foaming. The residue is dissolved in 100-300 ul of 70% formic acid and applied to a molecular sieving column of Biogel P10 (100-200 mesh 0.9 X 80 cm) which has been equilibrated with the same acid. The protein is eluted first from the column separated from other small molecules including the detergent, electrophoresis buffers, glycine and the blue dye. The eluted protein in 70% formic acid is ready to use for subsequent protein characterizations such as amino acid composition and N- and C-terminal sequencing. The extracted protein is usually tested for homogeneity by small scale gel electrophoresis and silver staining.

There is some suspicion that the 70% formic acid cleaves peptide bonds in proteins. However, it should be noted that formic acid (66-80%) has been widely used as a solvent in the cyanogen bromide reaction for specific cleavage of the methionine peptide bond. In many cases, no cleavage of peptide bonds by formic acid has been observed. However it is known that the aspartyl - prolyl bond is partially cleaved by 70% formic acid at elevated temperatures. It is wise, therefore, to test for this by preliminarily incubating an aliquot of protein in 70% formic acid at 37 °C for 24-48 hrs, followed by SDS-gel electrophoresis.

Generally, the recovery has been observed to be above 60% and in some cases as high as 95% using 1-100 ug of protein. It should be noted that this procedure does not favour the extraction of extremely high molecular weight proteins because of the time taken to achieve equilibrium. In order to achieve high recovery, cutting out the gel without crushing it or further dividing it into small pieces is most important; the small gel pieces reabsorb a large portion of the protein in the limited volume of formic acid which is needed for the application for column chromatography. Therefore after each extraction, filtration of the solution with a small glass filter is recommendable. In addition, gels more than 1 mm thick need a long time (more than 24 hrs) for extraction.

From our experience we believe that this extraction method has the following advantages: virtually all proteolysis is avoided, exact bands can be cut out, small undesired molecules removed, and after extraction the protein is ready for further sequencing experiments. The above procedure has successfully been employed for a variety of proteins(2,3).

2) Rapid and micro methods for hydrolysis of proteins

One of the classic but still important characterizations of a protein is its amino acid composition. The recent needs are for both rapid and micro-scale protein hydrolysis and a method for the complete hydrolysis of hydrophobic proteins.

Progress of automatic amino acid analysers equipped with post column ninhydrin reaction systems enable a sensitivity of 500 pmol to 2.5 nmol amino acid in full scale to be obtained. HPLC and pre- or post-column derivatization with phenylisothiocyanate or with various fluorescent reagents have achieved pmol level analysis.   Hydrolysis of proteins into amino acids has been carried out by treatment with acid, alkali or enzymes.   Among them the most popular method has been acid hydrolysis with constant boiling HCl (5.7 M) at 106°C for more than 24 hrs.   Elevation of the temperature accelerated the rate of peptide bond hydrolysis as shown in curve a, Fig.1. A temperature elevation of 60°C may accelerate hydrolysis 2^6 =64 times by simple calculation, thus reducing hydrolysis times from hours to only minutes. However, curve a shows considerably less acceleration effect than expected.   An addition of trifluoroacetic acid(TFA) and elevation of the molarity of HCl is needed to approach the theoretical accelerated hydrolysis times(curve b).   TFA is a strong organic acid having a pKa 0.23, a high vapour pressure and low boiling point (72.5°C); the use of such an organic acid may allow the acid to become accessible to the hydrophobic region of the peptide.   A new rapid hydrolysis method has been proposed based on the above observations (4-6).

   In the attempts being made to significantly reduce the amount of protein required to determine the amino acid composition, we have met serious problems of contamination which occur during the process of hydrolysis and/or the following procedure of amino acid analysis.   This contamination is observed both with the conventional 6 M HCl method and our new methods.   The hydrolysis acid contains amino acids or peptides which upon hydrolysis produce amino acids. (Contamination may also come from dust in the air or on the equipment such as pipettes,tubes, etc.). To determine

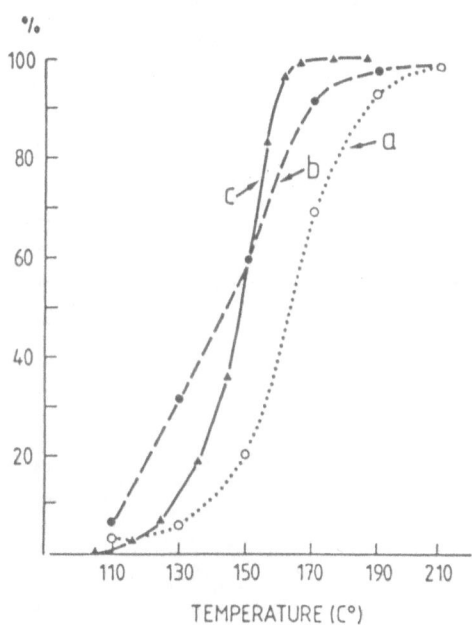

Fig. 1. Hydrolysis of Val-Glu with 6 M HCl (curve a), with TFA:HCl
        (1:2) (curve b) and TFA:HCl:H$_2$O (1:6:3, gas) curve c) at
        various temperatures.

the minimum volume of hydrolysis reagent required, 5 ug of protein were hydrolyzed with various volumes of acid in 0.8 mm X 8 mm test tubes. Experiments showed that 5 ul of acid was sufficient to hydrolyze 5 ug of protein.

When vapour of the acid was used, hydrolysis was still achieved and contamination was found to be considerably reduced. The extent of hydrolysis is more efficient than that by the liquid method (Fig. 3, curve c). Vapour phase hydrolysis of protein at elevated temperatures has been developed.

Protein (100-500ng) is placed in a small test tube (0.4 X 4 cm) and evaporated. The small test tube is placed in a larger test tube (1.2 X 12 cm) which contains 200 ul of a mixture of TFA: HCl: H_2O (1:6:3 v/v). The mixing ratio's were varied and tested for the efficiency of hydrolysis and decomposition of amino acids. Fig. 1, curve c shows the efficiency of hydrolysis at various temperatures. The tube is cooled in an ice bath evacuated for 1 min at 0 °C and vacuum sealed with a flame. Hydrolysis is carried out between 156 °C and 166 °C for 15 and 30 min. After hydrolysis the small test tube is taken out and placed in a vacuum desiccator to take out traces of acid. The residue is dissolved in 0.01 M HCl containing 5% thiodiglycol and applied to the amino acid analyzer. The addition of phenol is not required. Instead of the large test tube, more convenient reaction apparatus such as the Pierce reaction tube has been tested without success. It has been observed that the combination of both high temperature and the aggressiveness of the acid mixture occasionally break down the vacuum and cause oxidative decomposition of serine,

Table 1. Recovery of Standard Amino Acids

	(a)gas phase		(b)liquid phase	
	15 min	30 min	25 min	50 min
Asp	94	97	99	103
Thr	89	82	92	84
Ser	75	61	85	78
Glu	100	100	100	100
Pro	84	85	96	98
Gly	93	91	104	108
Ala	113	110	115	101
Val	102	102	98	98
Met	70	65	87	76
Ile	97	99	93	95
Leu	96	100	92	94
Tyr	92	94	89	81
Phe	99	101	90	87
His	106	101	96	99
Lys	101	101	95	98
Arg	108	106	100	100

Recovery of amino acids after incubation (a) at 158°C by gas phase TFA : HCl (2:3) and (b) at 166°C by liquid phase TFA:HCl:H_2O(1:6:3). Recoveries are expressed as a percentage of the value of glutamic acid taken as 100%.

Table 2. Amino Acid Composition using Gas Phase Hyrolysis

	glucagon			cytochrome C		
	22.5 min	45 min	T	22.5 min	50 min	T
Asp	4.3	4.3	(4)	8.1	8.3	(8)
Thr	2.7	2.6 (2.9)	(3)	9.0	8.1 (9.9)	(10)
Ser	3.3	2.7 (3.8)	(4)	0	0	(10)
Glu	3.0	3.0	(3)	11.9	12.0	(12)
Pro	0	0	(0)	3.8	3.8	(4)
Gly	1.1	1.0	(1)	11.8	12.0	(12)
Ala	1.1	1.1	(1)	6.0	6.0	(6)
Val	1.0	1.0	(1)	2.6	3.0	(3)
Met	0.8	0.6	(1)	1.9	1.8	(2)
Ile	0	0	(0)	4.7	5.5	(6)
Leu	2.0	2.0	(2)	5.4	5.6	(6)
Tyr	1.9	1.9	(2)	3.4	2.6	(4)
Phe	1.8	1.9	(2)	3.6	3.5	(4)
His	0.9	0.9	(1)	3.1	3.0	(3)
Lys	1.0	1.0	(1)	18.8	18.9	(19)
Arg	2.1	2.1	(2)	2.1	2.1	(2)

Gas phase hydrolysis of Protein (500ng) using TFA:HCl:H$_2$O (1:6:3 v/v) at 158°C. The figures in parentheses are extrapolated values at 0 time hydrolysis. Column T gives the amino acid composition of the proteins.

threonine and other amino acids. This method reduces the contamination of the amino acids to less than 1 pmole. Development of a method to avoid contamination such as the one mentioned above is required for the sensitive amino acid analysis at the level of less than 100 pmole full scale (corresponding to 100 ng of protein).

Recovery of standard amino acids and analysis of amino acid composition of proteins are listed in Tables 1 and 2 using the above hydrolysis method followed by analysis with a Durrum D500 amino acid analyzer equipped to a sensitivity of 500 pmole of amino acid full scale (ninhydrin detection system). The method provides sufficient hydrolysis but decomposition of threonine and serine are found to be a little more than that of liquid phase hydrolysis. This method was found to be especially powerful when the protein is extremely hydrophobic such as the proteolipid protein of myelin (3).

3) A sensitive detection method of Edman degradation

We have been recently developing a sensitive detection method for Edman degradation. In the well established Edman degradation, the classical method of coupling with phenylisothiocyanate and the cyclization reaction are conserved however the resulting anilinothiazolinone (ATZ) derivative is subjected to a sensitizing modification. Instead of the conversion to the phenylthiohydantoin derivative, the reaction of the ATZ derivative and amino compounds such as 125-Iodohistamine or 9-aminofluorine results in the phenylthiocarbamyl (PTC) derivative as shown in Fig. 2. Fig. 3. shows one of the typical reactions

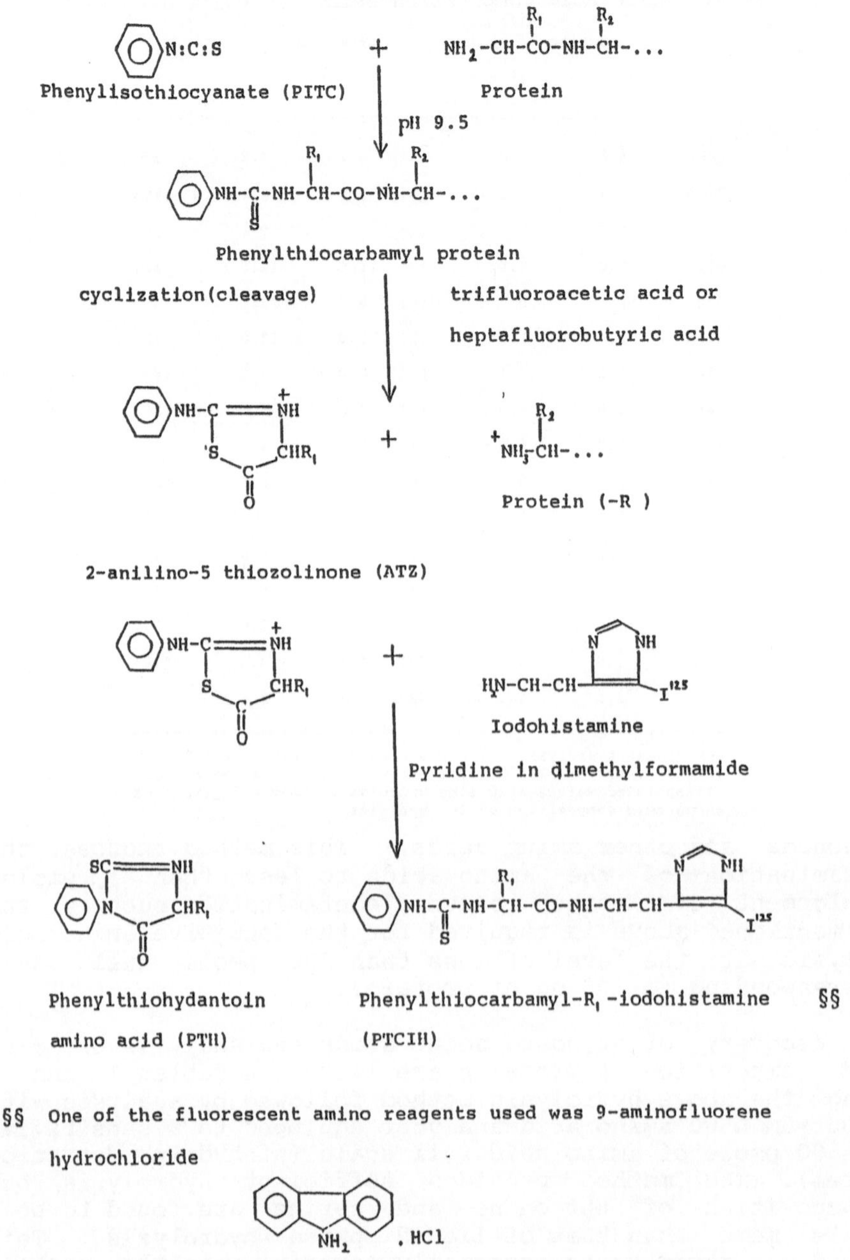

Fig. 2. Reaction Scheme for Sensitizing Edman Degradation.

between the ATZ derivative of leucine (peak a) and iodohistamine on HPLC. The reaction was followed by an increase of the PTC derivative (peak c) and was completed after one hour. The PTC derivative product has been confirmed by NMR and elementary analysis. The yield of recrystallized product was 80% for leucine. These derivatives of amino acids are stable and can be analysed on silica gel thin layer chromatography and by reverse phase HPLC. Fig. 4. shows the reaction products of several amino acids seperated by thin layer chromatography. The theoretical sensitivity of this method with iodohistamine is 0.1-1 fmole and 10-100 fmole with fluorescent NH_2 compounds.

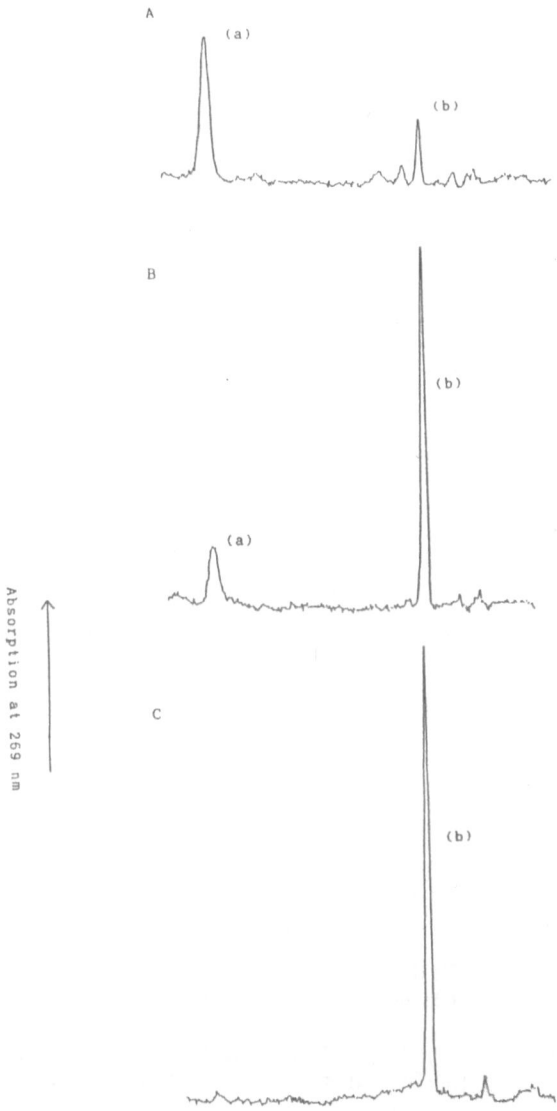

Fig. 3.   Reaction between ATZ-leucine and Iodohistamine:

   A) Just after start of reaction.
   B) Incubation for 30 minutes at 50°C
   C)   Incubation for 1 hour at 50°C

ATZ-leucine (1mg) and iodohistamine (2mg) were mixed in 30 %
pyridine-dimethylformamide.
HPLC column: Servachrom Si 100 Poly RP18, 5μm, 4.6 x 250 mm
Solvent system: a gradient betweeen 0.01 M sodium acetate buffer
pH 5.0, 1 volume and acetonitril-methanol (4:1 v/v) 5 volume.
Peak a: ATZ-leucine
Peak b: PTC-leucine iodohistamine

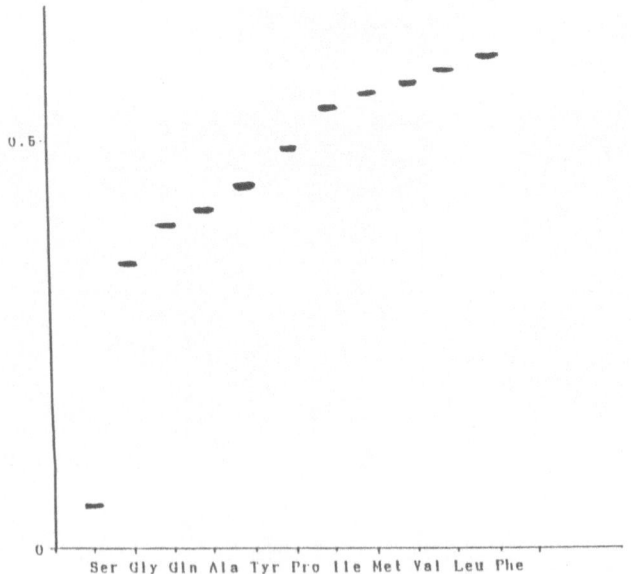

Fig. 4. Thin layer chromatography of iodohistamine derivaties of
PTC-amino acids

Thin layer plate LHP-K Whatman
Solvent Benzene: t-butanol: methanol (8:1:1 v/v)

References

(1) Dianoux, A.-C., Vignais, P.V., Tsugita, A. (1981) FEBS
Lett. 130 119-123

(2) Schenkman, S., Tsugita, A., Schwarts, M., Rosenbusch,
J.P. (1984) J. Biol. Chem. 259 7570-7576

(3) Riccio, P., Tsugita, A., Bobba, A., Vilbois, F.,
Quagliariello, E. (1985) Biochem. Biophys. Res. Comm. 127
489-492

(4) Tsugita, A., Scheffler, J.J. (1982) Proc. Japan Acad.,
58 B 1-4

(5) Tsugita, A., Scheffler, J.-J. (1982) Eur. J. Biochem.
124 585-588

(6) Tsugita, A., Vilbois, F., Jone, C., vd Broek, R. (1984)
Hoppe-Seyler's Z. Physiol. Chem. 365 343-345

MANUAL GAS PHASE MICROSEQUENCING ON FILTERCHIPS FOR SIMULTANEOUS EDMAN DEGRADATION OF MULTIPLE PEPTIDE SAMPLES

E. L. Mehl[*], M. Haniu[&] and J. E. Shively[&].

[*]Max Planck Institute for Psychiatry, Munich, Fed. Rep. of Germany.

[&]Division of Immunology, Beckman Research Institute of the City of Hope, Duarte, CA 91010.

INTRODUCTION

Manual Edman chemistry has been in wide use in the protein chemistry laboratory because of its significantly lower operating costs and ability to handle multiple samples compared to the automated instruments. On the other hand, automated sequencing has the advantages of unattended operation, more reproducible cycles, and less exposure of the sample to air. The need to immobilize peptides in a polybrene film presents handling problems for manual liquid phase samples; ie, the sample polybrene film must be dissolved and redistributed at two points in each Edman cycle, namely with base (usually triethylamine) and with acid (usually heptafluorobutyric acid or trifluoroacetic acid). If the films are not reproducible or droplets form, solvent extractions will not be reproducible, often leading to large variations in background peaks. Film problems will also lead to poor repetitive yields or extraction of the anilinothiazolinone (ATZ) derivatives.

The introduction of automated gas phase sequencing by Hewick et al.[1] suggests a possible solution to handling the sample-polybrene film in manual Edman chemistry; namely that the film could be retained on a glass fiber disk instead of on the inside of a glass tube [2]. Furthermore, the two reagents which normally dissolve the film in liquid phase chemistry could be delivered as gasses, thus avoiding the possibility of physically displacing the film. In this report we describe the design of filterchips which are conveniently adapted to manual gas-phase Edman chemistry. The size of the filters (1 mm) and the ease of handling should allow routine sequencing of up to 32 samples simulatenously.

Apparatus

An example of a filter chip is shown in Figure 1. The filter chip comprises an array of filter capillaries, each containing a 1 mm diameter glass fiber disk. Since the array was designed to fit into a Beckman Microfuge centrifuge, the number of filter capillaries in an array would be five for a "D" holder and 8 for a "B" holder. The number in an array can be modified to suit the experimemter and centrifuge. The microtest tubes

are 0.4 mL Eppendorf tubes which receive liquid from the filter chips
during the centrifugation steps. A detailed drawing of an individual
filter capillary is shown in Figure 2. A 1.2 mm glass fiber disk (Whatman
GF/C) is inserted into the bottom of a conical pipette tip. The orifice
of the piette tip is about 1.0 mm, thus retaining the flexible glass fiber
disk by a compression fitting. The glass fiber disk is retained even when
centrifuged at 10,000 rpm. The length of the pipette tip may range from
0.5 cm or greater depending on the desired volume. A second version of the
filter capillary is also shown in Figure 1, the sandwich or supported disk.

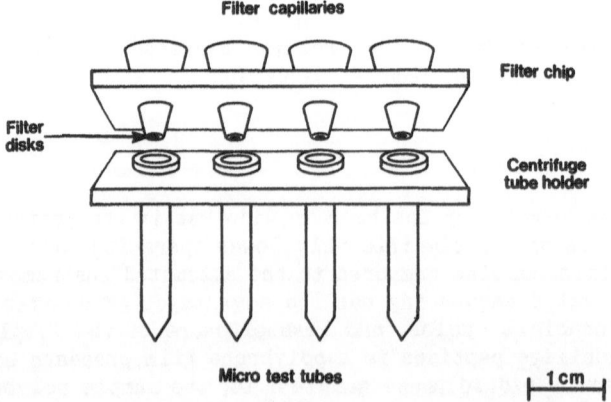

Figure 1. Filter chip and complimentary centrifuge tubes. The example
          shown here consists of four filter capillaries in a common
          holding strip (a filter chip). The filter chip may be lowered
          onto the complimentary microtest tube rack and centrifuged to
          remove liquid from the filter chip. A detail of a single filter
          capillary is shown in Figure 2.

In this case, the disk is supported by a Teflon screen which is welded to
the bottom of the conical tube. This version is suitable for larger
conical tubes (orifice > 1 mm) in which the glass fiber disk would not be
retained during the centrifugation steps. The teflon screen (mesh 0.004
in.) is welded to a polypropylene conical tube by a heating plate set at
140 deg. If desired the disk may be sandwiched between two supporting
teflon screens, a configuration more suitable for retaining small particles
(mesh > 0.004 in.) such as polystyrene beads used in solid phase Edman
chemistry or solid phase peptide synthesis. In this article, only the
filter capillary version will be described.

324

The filter capillary may be filled with liquid in one of three ways: a) the array may be touched to blotting paper, thus drawing liquid into the pipette tip by capillary action, b) liquid can be pipetted into the conical tube from the top, and c) an Eppendorf pipetter may be attached to the conical tube by an adapter (made by cutting down an Eppendorf tube) and used to draw liquid up through the filter disk. The filter capillary may be emptied by centrifugation at 10,000 rpm for 5-10 min in a Microfuge. Sample concentration in the filter capillary may be achieved by filling the conical tube, sealing the open end with a substance such as Parafilm, puncturing the seal with a fine wire or needle, inverting the tube or array disk side up, and allowing the liquid to evaporate from the disk. The conical tube can be a full length Eppendorf tip with a glass fiber disk inserted. The disk is first spotted with 20 ug of precycled polybrene and dried in a gentle stream of nitrogen. Once the sample is applied and concentrated onto the disk, the Eppendorf tip can be cut to fit the array comprising the filter chip. Alternatively, this process can be sped up by placing the filter chips in a vacuum chamber under reduced pressure. In this way several hundred microliters of sample may be concentrated on a single filter disk. The vacuum chamber can be used to maintain an inert atmosphere if desired. A third option is to directly apply the sample to the disk. This requires that the sample is in a volume of a few micro-liters. A single application of 0.5-1.0 uL can be made with a Microman pipetter (Gilson, France; range 0.5-25 uL). The polyethylene tip of the piston is shortened 1 mm in order to make the piston flush with the pipette tip during fluid ejection. If this modification is not made, the piston tip may dislodge the glass fiber disk from the filter chip.

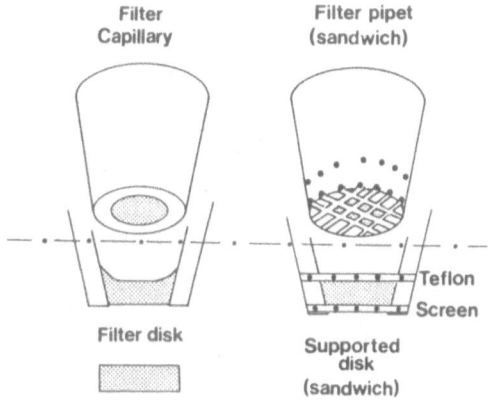

Figure 2. Detail of a conical filter capillary. Left: a glass fiber disk (Whatman GF/C, 0.4 x 1.2 mm) is inserted into a conical pipette tip (orifice 1.0 mm, length 6.5 mm). The disk is retained in the pipette tip without a supporting screen. Right: a filter pipette with a supporting screen. The pipette orifice may range from 1.2-20 mm and length from 7-100 mm. The disk may be supported with one teflon screen at the bottom, sealed to the pipette by heating, or with two teflon screens in a sandwich arrangement as shown. The disk may be replaced with particulates such as polystyrene beads in the sandwich version.

In order to perform gas phase reactions on the filter chips, a vapor reaction vial and vapor flask are required (Figure 3). The filter chip is inserted into the vapor reaction vial which is saturated with the vapor by adding liquid to the inner lining of glass fiber paper. The filter chip is prevented from touching the saturated liner by means of a teflon screen (mesh 0.004 in.) and a centering cone made from polyethylene. The atmosphere in the vapor reaction vial is maintained saturated with vapor by gently blowing nitrogen through a vapor flask to the reaction vial. This system was used for gas phase delivery of triethylamine (TEA)-water vapor or trifluoroacetic acid (TFA) vapor to the filter chips.

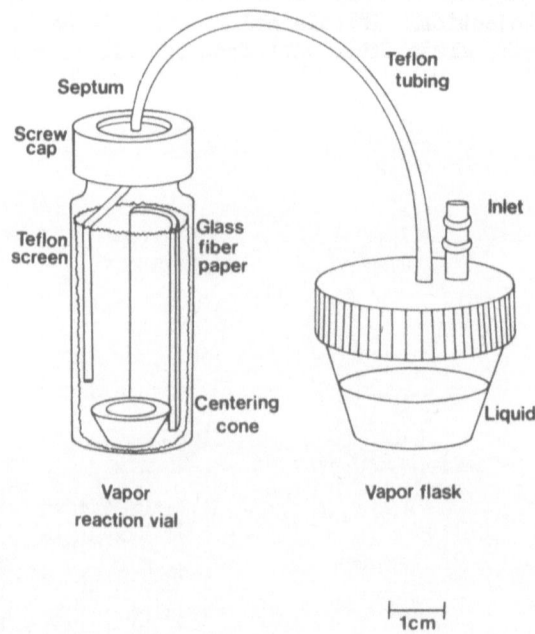

Figure 3. Vapor reaction vial and vapor flask. A screw cap septum vial (14 mL, 13 mm upper i.d., 18 mm lower i.d.) with a septum cap (Pierce, Rockford, IL.) was used. A reagent wetted glass fiber paper (Whatman GF/C, 42 x 25 mm) was retained on the sides of the vial with a Teflon screen (0.004 in mesh, Savillex, Minnetonka, MN.). A conical centering ring was used to prevent contact of the filter chip to the sides of the vial. The vapor flask was constructed from a 7 mL conical flask (PFA, Savillex). The Teflon tubing connecting the flask to the vial was sealed with pull through fittings [3].

Sequencing Protocol

The general principles of manual gas-phase sequencing compared to automated gas-phase chemistry is shown in Table I. The most striking difference is that the size of the disk is much smaller in the filter chip method. The diameter of the disk in automated chemistry is 10 times greater than that used in the filter chip, and the weight of the disk 40 times greater. The reduction in size results in the need for 40 times less polybrene on the disk to retain the sample. The corresponding reductions in the amount of phenylisothiocyanate (PITC) and solvents for extraction used make the manual method more economical and should result in a theoretical lowering of background in sample analysis. The peptide washout problem may be lowered somewhat by the ability to wash the disk at lower temperatures than possible with the automated instrument. A major advantage of the manual method is the ability to process multiple samples simultaneously. Advantages of the automated chemistry over the manual are the better exclusion of oxygen from the system, and the higher repetitive yields (probably due to the total exclusion of air from the system).

The composition of the reagents and solvents in the manual filter chip method is shown in Table II. The cyclic steps using the reagents and solvents are shown in Table III. R1 is blotted onto the filter chip to initiate coupling, and R2 is introduced in the vapor phase to complete the coupling reaction. Excess PITC and reaction by-products are removed by centrifugal wash steps with S1. Cleavage is achieved by first applying 1 uL of R3 to each of the filter disks on the filter chip, followed by the vapor phase introduction of R3. Extraction of the ATZ derivatives of the amino acids is achieved with a centrifugal wash with S3. The ATZ derivatives collected into the 0.4 mL polyethylene tubes are converted to their respective PTH derivatives with R4. During this step, the Edman cycle staring from step 1 is repeated. Alternatively, the ATZ derivative is stored until an appropriate number of cycles has been run, at which point all of the ATZs are converted and then analyzed by HPLC.

Sequencing Results

We routinely test our automated gas-phase sequencer with the hexapeptide Phe-Asn-Gly-Leu-Arg-Gly (3). The hexapeptide ends with a carboxyamido group which usually causes washout of the last residue, thus we identify only the first five residues. A comparison of the automated versus the manual gas phase methods is shown in Table IV. The manual method shows a low initial yield and decreasing yields through four cycles. By cycle five the yield of Arg is only 1% of the starting material. In contrast the automated method gives higher initial yields and little decrease on cycles 2-5. These results suggest that the manual method is far from optimal, but demonstrate that it is indeed possible to perform multiple manual gas phase Edman degradations. The chromatograms for the sequence analysis of the standard hexapeptide are shown in Figure 4. The background peaks are almost entirely eliminated except for a peak eluting before PTH-Arg. Although this background peak does not interfere with sequence analysis, we have observed it before and its identity is unknown. A blank cycle is shown for purposes of comparison to other cycles. The blank was obtained from a filter disk with polybrene but no sample on the same filter chip.

Sequence analysis of 1 nmole of the tetrapeptide Gly-Gly-Trp-Ala is shown in Figure 5. The Ala at cycle 4 is not observed for this peptide in either manual or gas phase sequence analysis. The yield of Gly at cycle 1 is 255 pmoles (102 pmoles at cycle 1, 40% injected), corresponding to an initial yield of 26%. As observed for the synthetic hexapeptide, the

Table I. Comparison of automated and manual gas phase sequencing.

Parameter	Automated	Manual
1. Weight of glass fiber disk	3000 ug	75 ug
2. Weight of polybrene	2000 ug	50 ug
3. PITC usage/cycle	1000 ug	25 ug
4. Solvent usage/cycle	2500 uL	100 uL
5. Wash step temp.	50-55 deg	25 deg
6. Samples per cycle	1	8, 16, or 32
7. Cartridge flow	one direction	back wash poss.
8. Exclusion of oxygen	near perfect	partial only
9. Repetitive yield	90-95%	<90%
10. Scope of application	gas phase	gas, liquid, solid phases

Table II. Composition and usage of reagents and solvents.

Reagent/Solvent	Volume per cycle (uL)
R1 PITC/heptane/1-PrOH/TEA/water (1/9/8/1/1, by vol)	<1
R2 TEA/water (2/98, by vol)	100
R3 TFA (liquid)	<1
TFA (gas)	100
R4 TFA/water (1/3, by vol)	10
S1 Ethyl acetate/1-PrOH (400/1, by vol)	150
S2 Butyl chloride/TFA (400/1, by vol)	20
S3 Butyl chloride	20
S4 Acetonitrile	40

R1 is freshly mixed for each cycle. S1 and S4 have 0.1 g/L of DTT.

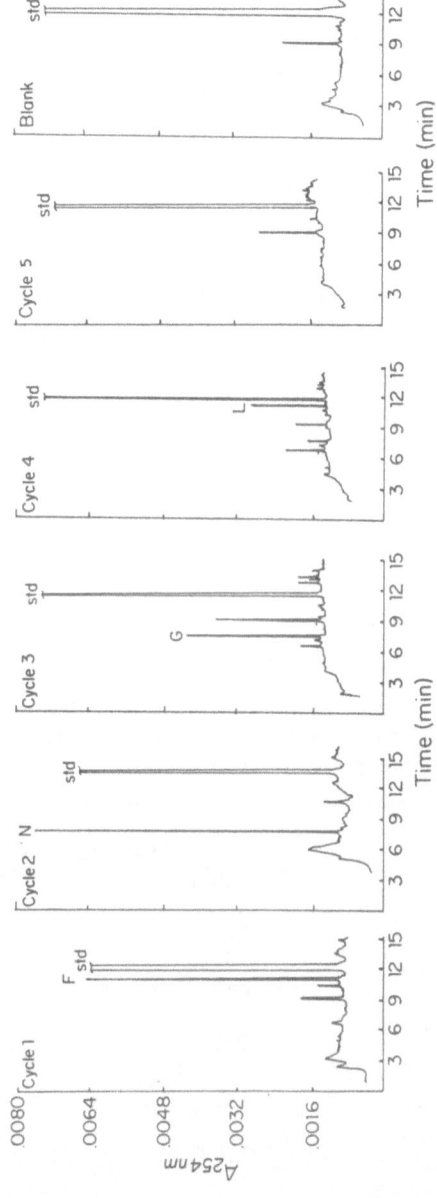

Figure 4. Sequence analysis of a synthetic peptide by manual gas phase chemistry. The peptide Phe-Asn-Gly-Leu-Arg-Gly-amide (500 pmoles) was sequenced. Five cycles plus one blank is shown. The yields are shown in Table IV. Forty percent of each cycle was analysed.

Table III Steps in one cycle of manual gas phase sequencing

Step	Liquid	Gas	Temp.	Duration
1. Coupling	R1	R2	50	10 min
2. Wash	S1		25	3X 50 uL
3. Wash	S2		25	2X 10 uL
4. Cleavage	R3	R3	50	10 min
5. Extraction	S3		25	2X 10 uL
6. Conversion	R4		50	30 min
	S4		25	2X 20 uL

Coupling is initiated with the application of 0.5 uL of R1 and continued
with gas phase delivery of R2 at 50 deg. Allow 1-2 min per wash step.
Cleavage is initiated with the application of 0.5 uL of R3 followed by gas
phase delivery of R3 at 50 deg.

Table IV. Comparison of sequence analysis of a synthetic peptide with the
automated versus the manual gas phase methods

Cycle	Residue	Yield (pmoles)	
		Automated	Manual
1	Phe	170	140
2	Asn	230	127
3	Gly	165	62
4	Leu	235	32
5	Arg	150	5
6	Gly	0	0

The peptide Phe-Asn-Gly-Leu-Arg-Gly-amide (500 pmoles) was sequenced by
both the automated (3) and manual gas phase methods. The yields at each
cycle are given in pmoles.

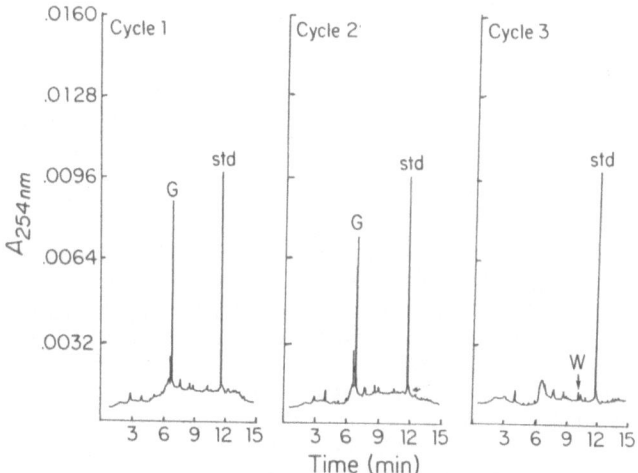

Figure 5. Sequence analysis of the synthetic peptide Gly-Gly-Trp-Ala. The sample (1 nmole) was analyzed as described in Figure 4. The yields for cycles 1-3, normalized to 100% injection were 254, 230, and 16 pmoles.

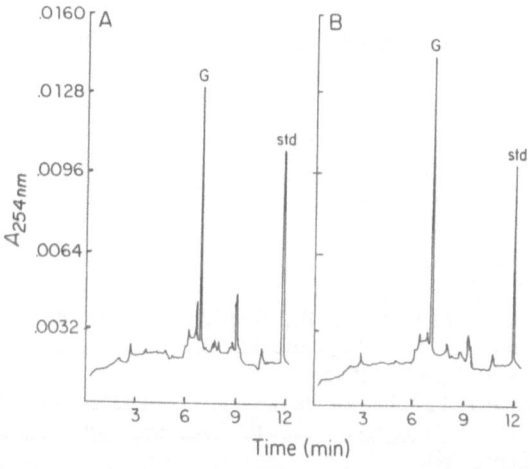

Figure 6. Reduction of background with an 0.2% TFA in butyl chloride wash. Cycle 1 of the analysis of 1 nmole of the peptide Gly-Gly-Trp-Ala. Left: no prewash before cleavage. Right: prewash before cleavage with 0.2% TFA in butyl chloride.

inital yield is low, but acceptable for an as yet unoptimized system. The corresponding yields (corrected for 40% injected) for cycles 2 and 3 are 230 and 16 pmoles, respectively. Identical samples were sequenced on four positions of the filter chip and compared to the example shown in Figure 5. The reproducibility of the initial yield at cycle 1 was about 20%, demonstrating some intersample variation. In a separate experiment the effect of the composition of the solvent wash on background was studied. It was found that if the sample was washed with butyl chloride containing 0.2% TFA after the ethyl acetate wash and prior to cleavage, the background was reduced. These results are shown in Figure 6.

CONCLUSIONS

A great deal of effort was spent on developing an apparatus for manual gas phase Edman chemistry. This report describes the apparatus and its use. Advantages include the ease of handling multiple samples, filling and washing by capillarity and centrifugation, and the small consumption of reagents and solvents. We have not as yet optimized the gas phase chemistry as applied to this apparatus, but even at this stage we are able to obtain reasonably good sequences from short peptides. This has been the strong point of manual Edman chemistry from its inception. We believe that with further optimization of the chemistry and better exclusion of air from the system, we can improve the initial and repetitive yields. The smaller size of the glass fiber disk compared to the traditional gas phase instrument, should lead to lower backgrounds, and therefore greater sensitivity. To date we have realized the lower backgrounds, but not the greater sensitivity. The backgrounds were further reduced by the novel use of a prewash with butyl chloride containing TFA prior to cleavage. It is our hope that this report will stimulate other groups to experiment with manual gas phase chemistry, and that ultimately it will provide an attractive and economical alternative to automated gas phase chemistry in microsequence analysis of peptides and proteins.

ACKNOWLEDGEMENT

This work was partly supported by NIH grants HD 14900 and CA 37808.

REFERENCES

1. Hewick, R.M., Hunkapiller, M.W., Hood, L.E., and Dreyer, W.J., (1981) A gas-liquid solid phase peptide and protein sequenator, J. Biol. Chem. 256:7990-7997 (1981).
2. Tarr, G.E., Manual batchwise sequencing methods, (1982) in Methods in Protein Sequence Analysis, ed. M. Elzinga, pp. 223-232, Humana Press, Clifton, NJ.
3. Hawke, D.H., Harris, D.C., and Shively. J.E., (1985) Microsequence analysis of peptides and proteins. V. Design and performance of a novel gas-liquid-solid phase instrument, Anal. Biochem. 147: 315-330, (1985).

STRATEGIES FOR SEQUENCING PEPTIDE MIXTURES BY SELECTIVE BLOCKING

Michael N. Margolies, Andrew W. Brauer, Rou-Fun Kwong, and
Gary R. Matsueda

Departments of Surgery, Medicine and Pathology
Massachusetts General Hospital and Harvard Medical
School, Boston, MA 02114

A year following the initial description of the automated protein sequencer (1) William Gray proposed that the amino acid sequence of a polypeptide could be deduced from sequence analysis of complex peptide mixtures
produced by several different specific cleavages of that polypeptide, thus
avoiding the necessity for fractionation of multiple peptides (2). While
this approach is theoretically feasible, it remains impractical because the
simultaneous degradation of multiple peptides is required to be uniformly
efficient and the yields of all phenylthiohydantoin (PTH) - amino acids
quantitative. Although advances in the Edman chemistry itself since that
time have been relatively modest, technological advances in instrumentation
have fulfilled Gray's other prediction that automation will be utilized in
sequence analysis (2). Although the deduction of sequences from mixture
analysis remains risky, approaches which permit the unequivocal sequence of
a single peptide in a mixture by specific blocking are desirable in avoiding
sometimes arduous purification procedures. Such strategies are particularly
useful for the examination of sets of homologous proteins or peptides and
for peptides available in limited amounts. Examples of selective blocking
include: 1) succinylation to block the amino-terminus of a parent peptide,
followed by specific internal cleavage permitting the unique sequence of the
deblocked daughter peptide to be determined, 2) selective blocking of contaminating peptides containing amino-terminal glutamine by cyclization in
acid. This report reviews two other selective blocking methods: one which
spares prolyl-peptides and a second which blocks serine-containing peptides.

Bhown and coworkers (3) demonstrated that fluorescamine introduced into
the spinning cup sequencer reduced background arising from acidolytic peptide bond cleavage by blocking primary amino groups at cycles where a proline residue is present at the amino-terminus. The resultant decrease in
background thus permits longer degradations by increasing the ratio of
signal to noise, not only by reducing background but by eliminating the
cumulative overlap ("lag", or "out-of-phase" degradations) at the treatment
cycle. As their method requires manual introduction of fluorescamine at
lower temperature, alternative blocking reagents suitable for automation
were sought. Machleidt and Hofner (4) utilized o-phthalaldehyde (OPA) in
solid phase sequence analysis, as OPA is stable in aqueous solution.
Thereafter we reported the use of OPA in the spinning cup instrument to
prolong degradations on polypeptides (5) and to aid in peptide mixture
analysis (6). Spiess and coworkers (7) utilized OPA for the selective
sequence of prolyl-peptides at the nanomole level. We recently reported

the methodology for OPA blocking using the gas phase sequencer (8,9).

As the details for OPA blockade using the spinning cup sequencer were previously reported (10) we highlight here the advantages and one cautionary note. In Figure 1 is illustrated an extended degradation in a Beckman 890C sequencer, using methanolic HCl conversion (11) on a 50,000 dalton immunoglobulin heavy chain. The rate of increase of background following OPA treatment at proline 41 roughly parallels the initial accrual of background, indicating that the block of primary amines is relatively stable. Use of OPA permitted identification in this degradation for 70 continuous cycles, without alteration of the repetitive yield, provided residual OPA is removed before phenylisothiocyanate (PITC) coupling. The applicability of OPA for selective sequence of a member of a peptide mixture is illustrated in Figure 2. o-Iodosobenzoic acid cleavage of a monoclonal antibody light chain resulted in multiple products with different yields at each of four tryptophan residues (6,10). As an invariant proline residue was known to be present at position 40 (cycle 5 in the second peptide shown relative to a tryptophan at position 35) this fragment was selectively sequenced for 46 additional cycles. A second OPA block was used at a second predicted proline position (cycle 24). In this instance OPA blocking permitted the sequence of a useful large overlap fragment.

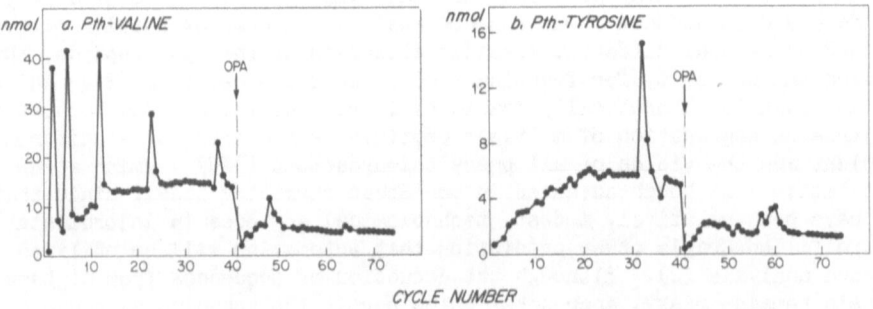

Figure 1. Yields of Pth-Val and Pth-Tyr at each cycle of Edman degradation of a antibody heavy chain. OPA treatment was instituted at cycle 41 (proline) permitting extension of a readable sequence to more than 70 cycles.

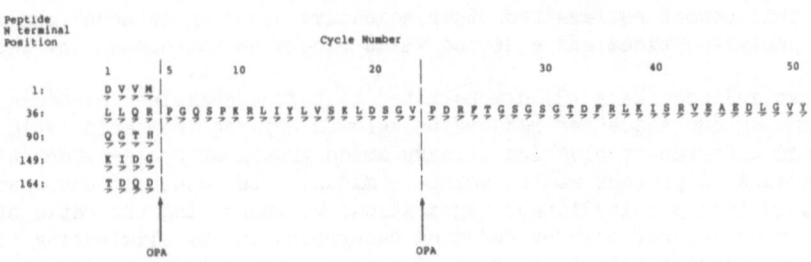

Figure 2: Use of OPA blocking to sequence peptide mixtures where the target peptide contains proline near the amino-terminus (10). A monoclonal antibody light chain (23,000 daltons) was cleaved with o-iodosobenzoic acid and treated with OPA prior to cycles 5 and 24. PTH-amino acids recovered in the first 4 cycles are consistent with cleavage at tryptophan 35, 89, 148 and 163. Following OPA treatment, a single sequence was interpretable for an additional 47 cycles.

Based on the procedure of Bhown et al (3) which employed fluorescamine, in our preliminary experiments following OPA coupling and solvent washing we exposed the protein to cleavage acid. Using the model sequence as shown in Figure 3, when OPA blocking was thus attempted, extensive "preview" was experienced at proline residues 9 and 14 which was abrogated when the hepta-fluorobutyric acid (HFBA) is omitted prior to subsequent PITC coupling of the prolyl-peptide (Figure 3). These results are consistent with the formation of an OPA-prolyl adduct with subsequent cleavage of the prolyl-peptide bond (10). Therefore following OPA treatment, it is essential that PITC coupling precede acid exposure. It is not known whether the preview seen in this experiment was sequence-specific or not.

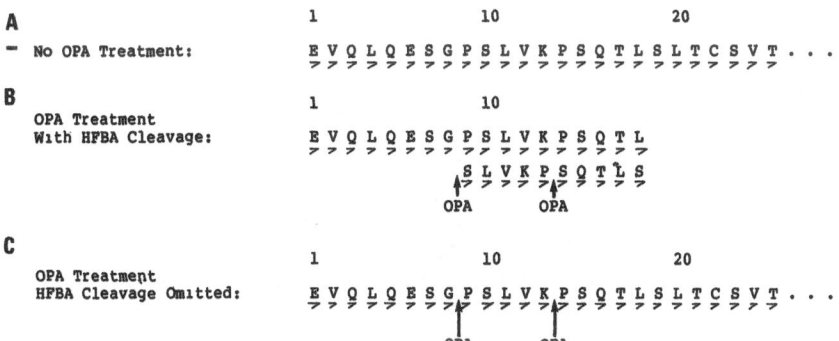

Figure 3. Effect of cleavage acid (HFBA) following OPA treatment at cycles 9 and 14 of Edman degradation on a monoclonal antibody heavy chain (5). When OPA treatment is followed by exposure to cleavage acid (3) two PTH-amino acids were obtained at each cycle, consistent with preview of the expected sequence. Omission of HFBA following OPA coupling prevents preview; the repetitive yield following OPA treatment is thus unchanged.

Recently, OPA blocking in the spinning cup sequencer was used advantageously to elucidate the sequence specificity of an anti-fibrin monoclonal antibody (12). Fibrin-specific antibodies have been sought as potentially useful clinical reagents to detect thrombi *in vivo* and to evaluate thrombolytic disorders. Fibrinogen, with molecular weight 340000 consists of three pairs of non-identical polypeptide chains, designated α, β and γ. Thrombin enzymatically cleaves two pair of small polar peptides from fibrinogen α and β chains (fibrinopeptides A and B) to yield fibrin monomers, which spontaneously polymerize into an insoluble gel or clot. Matsueda and coworkers (13) produced fibrin-specific monoclonal antibodies which recognize the fibrin β chain amino-terminus. These were elicited by immunization with a synthetic heptapeptide-protein conjugate with sequence G-H-R-P-L-D-K corresponding to the human fibrin β chain amino-terminus. These antibodies bind to fibrin but do not bind to fibrinogen. One of these monoclonal antibodies (59D8), in a radioimmunometric assay, binds human and canine fibrin, but fails to bind to porcine, bovine, or ovine fibrins (Figure 4A). In order to correlate this immunoreactivity with the fibrin β chain structure, OPA blocking was used at cycle 4 during Edman degradation based on the fact that human β chains contain a proline at position 4 (Figure 5). Edman degradation using OPA treatment was used to selectively sequence β chains from unfractionated fibrin monomers of each of five different species. In the first three cycles, a mixture of amino acids was found, consistent with the known human and canine sequence; in cycles following OPA blockade only the β chain sequence is detected. For human and canine fibrin β chains which bind antibody 59D8, a leucine occurs at position 5, while among the three species with fibrins which are antibody non-reactive, a tyrosine residue was found at position 5, suggesting that the species selectivity was related to the identity of this residue. To confirm this hypothesis and directly

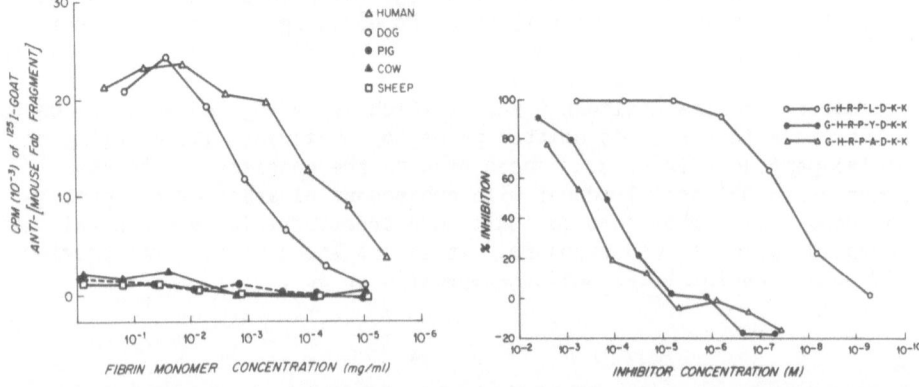

Figure 4A (left). Species specificity of a monoclonal anti-human fibrin antibody (59D8)(12,13) as determined by radioimmunoassay. Fibrin monomer of each species bound to wells of plastic microtiter plates were exposed to the monoclonal antifibrin antibody 59D8. Specifically bound antibody was detected using radioiodinated goat anti-(mouse Fab) antibody.
Figure 4B (right): Inhibition of binding of antifibrin antibody 59D8 to human fibrin monomer by synthetic fibrin-like peptides. The Leu-5 peptide contains the human β chain fibrin N-terminal sequence. The Tyr-5 peptide is homologous to ungulate fibrin chains.

assess the contribution of position 5, fibrin-like peptides were synthesized and used as competitive inhibitors of the binding of fibrin to antibody 59D8 (Figure 4B). Consistent with the results of sequence analysis the leucyl-5 peptide was an effective inhibitor, while the tyrosyl-5 and alanyl-5 inhibitors were weak inhibitors. Thus OPA blockade was useful in this instance to rapidly screen homologous sequences, without the necessity of fractionation of subunits. In addition, since the α chain has a proline at position 2 (Figure 5), OPA treatment was used to elucidate fibrin α-chain amino-terminal sequences of several species (13).

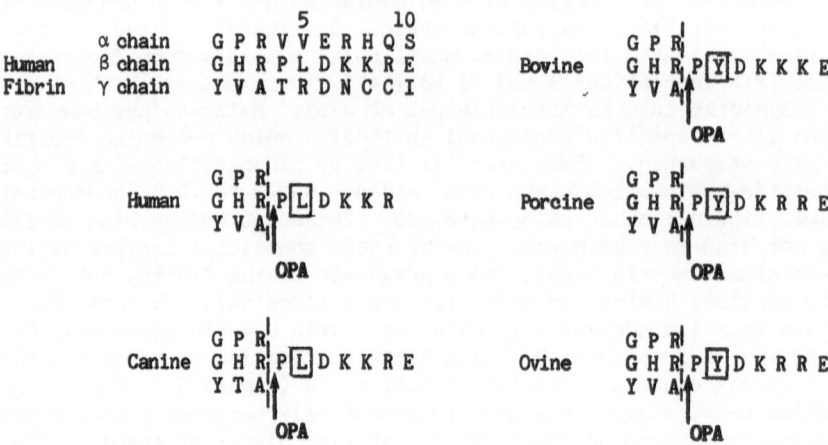

Figure 5. Comparison of fibrin β chain sequences from various species utilizing Edman degradation with OPA treatment (13). Using OPA blocking it was possible to obtain a single sequence for the β chain beginning at cycle 4 (proline) on unfractionated fibrin monomers. Human and canine β chains contain a leucyl residue at position 5 and bind the antifibrin antibody 59D8 (Figure 4A), while bovine, porcine and ovine β chain contain a tyrosyl residue at position 5 and do not bind significantly to the antibody.

OPA blocking may be applied readily using the gas phase sequencer for both extended degradations on polypeptides (8), and for sequence of peptides in the picomolar range. Minor contaminants are seen on HPLC which are easily separable from PTH-amino acid peaks and disappear in 2-3 cycles. OPA (0.1 mg/ml) is delivered in butyl chloride from the S1 reagent reservoir. All sequencer programs used do not include an S1 delivery except at the OPA cycle. The heptane ordinarily present in S1 is mixed (1:2 v/v) with the ethyl acetate in the S2 solvent reservoir without perceptible effect on sequence efficiency. As in the spinning cup program, double coupling with PITC and extended cleavage of the PTC-prolyl peptide is used. An example of the application of OPA to an extended degradation using the gas phase instrument is shown in Figure 6. A 17,800 dalton fibrillar protein from pancreas with unusual solubility properties had been originally thought to contain a single subunit. However, sequence analysis (8) revealed a mixture of two PTH amino acids in equimolar amounts at most cycles, consistent with the presence of two subunits (Figure 6). As proline residues were detected at cycles 3 and 4, OPA blocking at cycle 4 permitted the elucidation of a single sequence of the larger subunit through cycle 48. Thereafter OPA blocking at cycle 3 confirmed the sequence of the second subunit of approximately 3,000 daltons. In Figure 7 is illustrated the use of OPA in sequence of peptides present in limited amounts. A mixture of two tryptic peptides (15 and 17 residues) not resolved by HPLC were obtained from a monoclonal antibody heavy chain. Initial sequence analysis of the mixture (Figure 7, upper panel) showed them to be present in a ratio of 3:2 with a proline at cycle 3 in the peptide of interest. In a second degradation on a repurified sample, the target peptide proved to be only a minor component in the mixture, perhaps owing to cyclization of the amino-terminal glutamine. Nonetheless the entire sequence was determined using OPA blocking on 60 picomoles of peptide. Thus, unlike OPA blocking during degradation on large polypeptides, the background increase following selective OPA block is minor with short and intermediate length peptides. It is thus possible to use OPA in microsequence analysis even where the target is a minor component in the mixture (7).

The occurrence of an undesirable side reaction during automated Edman degradation led us to explore another blocking method which may be used for serine-containing peptides. Prior to the time that Henschen-Edman and coworkers demonstrated that background accumulation during Edman degradation was due to acidolysis at aspartic acid, threonine, and serine (14), several laboratories (15,16) tested the proposal of Begg (17) that hydrolysis was the culprit, and therefore added trifluoroacetic (TFA) anhydride to the cleavage acid to reduce trace water, measured spectrophotometrically in the near infrared. Not only did such a maneuver not reduce the background, but when we invertently overtitrated, the small excess of TFA anhydride resulted in the total block of an immunoglobulin light chain sequence at a serine residue at position 7. The yields in the first 6 cycles were normal. Using the same preparation of HFBA containing TFA anhydride, sequence analysis of myoglobin demonstrated complete blocking at serine cycle 3. In order to determine whether such treatment permitted efficient degradation of peptides in a mixture that did not contain serine at the cycle of interest, we tested the behavior of equimolar mixtures of myoglobin and an immunoglobulin light chain of known sequence using the spinning cup instrument (Figure 8). Arrows indicate the cycles where the sequencer was programmed to deliver HFBA containing 1-3% TFA anhydride (representing a 15-45 fold molar excess over hydroxyl groups, not accounting for water initially present in the HFBA). The HFBA containing the anhydride was delivered from R3, while at all other cycles HFBA (without anhydride) was delivered from R4. In these experiments, myoglobin was blocked to an extent of 90-95% at the serine in cycle 3, while the light chain sequence continued thereafter with a normal repetitive yield (95-96%), indicating that at least under certain conditions the side-reaction of trifluoroacetylation of amino-termini can be minimized,

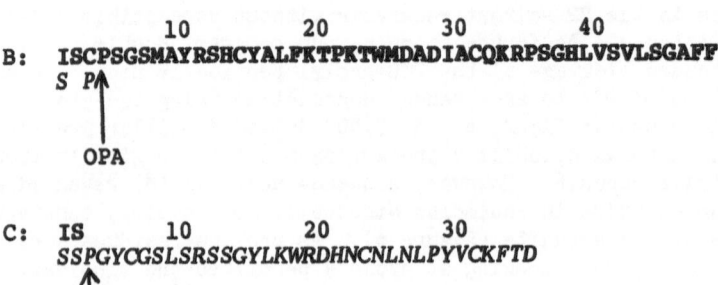

Figure 6. Sequence of the subunits of pancreatic thread protein (8) obtained by the OPA blocking strategy using the gas phase sequencer. A: An initial degradation indicated two PTH amino ac ds at most cycles in equal yields. At cycles where only one PTH amino acid was identified, the increased yield suggested that identical residues were present in both chains. Proline residues were present at cycles 3 and 4. B: OPA treatment at cycle 4 blocked the sequence of the smaller subunit (approximately 3,000 daltons) resulting in a single sequence for the larger subunit (14,500 daltons) for 48 cycles. C: OPA treatment at cycle 3 resulted in a single sequence of the smaller subunit for 35 cycles.

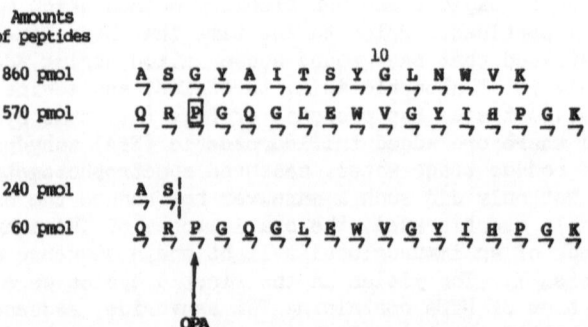

Figure 7. Selective sequence of proline-containing tryptic peptides from an immunoglobulin heavy chain using OPA blocking in the gas phase sequencer. Sequence analysis of a fraction isolated by HPLC (upper panel) was consistent with the presence of two tryptic peptides in the ratio shown. As the sequence of the pentadecapeptide was already known, the sequence of the heptadecapeptide was tentatively deduced by difference. The sequence of the first 4 residues of the heptadecapeptide was known based on a previous degradation of the intact H chain. In a second degradation employing another sample, OPA treatment at cycle 3 permitted determination of the complete sequence of the peptide despite partial N-terminal blocking due to glutamine cyclization (lower panel).

338

presumably because the amino-group is protonated under acidic conditions. Of interest also is that PTH-serine recovery in subsequent cycles (cycles 7, 9, and 10 in Figure 8) were identical to those in the control degradation, suggesting that the TFA groups are removed from side chain hydroxyls during exposure to Quadrol. The HPLC chromatograms from the first 4 cycles of this experiment are shown in Figure 9. Note that at cycle 3 in the experiment using anhydride, serine is absent and in cycle 4 PTH- glutamic acid is absent owing to blocking of myoglobin, while residues from the light chain are present in expected yields as compared to the control. In this initial experiment, anhydride treatment was carried out at the first three cycles. In order to determine at which cycles anhydride treatment was necessary to produce specific blocking, and to avoid unnecessary delivery of anhydride at cycles in target peptides which might contain serine, we examined relative blocking by anhydride treatment during cleavage at cycle 3 only (Figure 8).

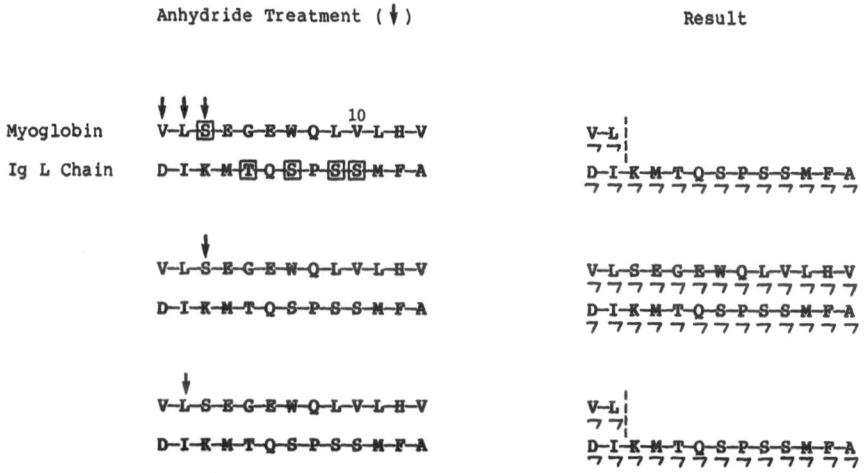

Figure 8. *An equimolar mixture of sperm whale apomyoglobin and an immuno-globulin light chain of known sequence (20) was used to assess blocking at serine residues by TFA anhydride in a Beckman 890C spinning cup sequencer. When TFA anhydride was added to the cleavage acid (HFBA) at each of the first 3 cycles, the sequence of myoglobin was blocked at serine 3; the light chain sequence continued unaltered (upper panel). When anhydride treatment was used at cycle 3 only, the degree of blocking of myoglobin varied between 0 and 40% (middle panel). When anhydride was introduced only during cleavage at cycle 2, selective blocking of the myoglobin sequence resulted (bottom panel).*

When anhydride treatment was limited to this cycle, the sequence of myoglobin was unaffected or blocked partially (up to 40%). However when the anhydride-containing HFBA was delivered only at the preceding cycle (cycle 2), where the seryl-peptide is exposed, complete blocking of myoglobin at cycle 3 serine was seen. These observations suggest two possible mechanisms for blocking: 1) the serine hydroxyl groups are trifluoroacetylated in acid, but when the amino terminus is deprotonated during PITC coupling at basic pH, O-- N acyl migration results in blocking of the amino-terminus (18). 2) Serine hydroxyls are trifluoroacetylated in acid, following by PITC coupling and attack of the sulphur on the methylene group under alkaline conditions, forming a cyclized product as suggested by Laursen (19).

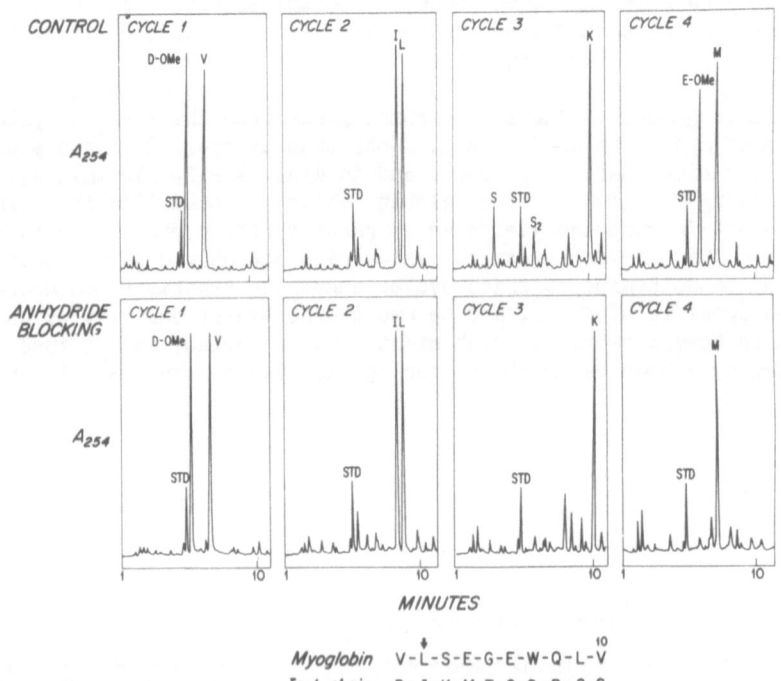

Figure 9. *PTH-amino acids recovered at each cycle of Edman degradation of a mixture of myoglobin and an immunoglobulin light chain where TFA anhydride was included in the cleavage acid at cycle 2 (compare Figure 8, bottom panel). Anhydride treatment resulted in blocking of the sequence of myo-Globin at serine 3. PTH-aspartic acid and PTH-glutamic acid were recovered as the methyl esters (OMe)(11).*

A potential advantage of selective blocking using TFA anhydride is that serine residues are ubiquitous in proteins. However, it should be noted that unlike OPA blocking in which primary amines are blocked and the prolyl containing peptide of interest remains unblocked, in anhydride blocking only the contaminants which have serine at the appropriate cycle are blocked. We did not observe specific blocking at threonine, likely due to steric hindrance to cyclization. It remains to be determined whether the use of TFA anhydride will have practical application for a number of reasons. When a larger excess of anhydride was used, significant non-selective amino-terminal blocking occurred. In these later experiments, we did not use spectrophotometric titration of water in the acid; even with such measurements it is difficult to precisely control the stoichiometry with respect to the amount of water present in the system over time. Unlike the OPA method, where a wide range of ratios of OPA to peptide may be used, the effective range of anhydride concentration providing selective serine blocking, without nonspecific amino-terminal blocking, appears narrow. The problem is likely to be further complicated if attempts are made using the gas phase sequencer, as the anhydride has a lower boiling point than TFA. Further investigation of the reaction mechanisms involved and the reaction conditions are necessary, in order to prove whether such an approach may be utilized in practice or rejected.

Acknowledgements: Supported by NIH grants HL 19259, HL 28015 and CA 24432.

References

1. P. Edman and G. Begg, Eur J Biochem 1:80-91, (1967).
2. W. R. Gray, Nature 220:1300-1304, (1968).
3. A. S. Bhown, J.C. Bennett, P.H. Morgan, and J.E. Mole, Anal Biochem 112:158-162, (1981).
4. W. Machleidt and H. Hofner, in: Methods in Protein Sequence Analysis, Elzinga, M., ed., Humana Press, Clifton, N.J., (1982).
5. E. C. Juszczak and M. N. Margolies, Biochemistry 22:4291-4296, (1983).
6. J. Novotny and M. N. Margolies, Biochemistry 22:1155-1158, (1983).
7. J. Spiess, J. Rivier, and W. Vale, Biochemistry 22:4341-4346, (1983).
8. J. Gross, A. Brauer, R. Bringhurst, C. Corbett, and M. N. Margolies, Proc Natl Acad Sci 82:5627-5631, (1985).
9. R. M. Hewick, M. W. Hunkapillar, L. E. Hood, W. J. Dreyer, J Biol Chem 256:7990, (1981).
10. A. W. Brauer, C. L. Oman, and M. N. Margolies, Anal Biochem 137:134-149, (1984).
11. M. N. Margolies, Brauer, A., Oman, C., Klapper, D.G., and M. J. Horn, in: "Methods in Protein Sequence Analysis", M. Elzinga, ed., Humana Press, Clifton, N.J., (1982).
12. G. R. Matsueda and M. N. Margolies, Biochemistry (in press).
13. K-Y. Kwan, E. Haber and G. R. Matsueda, Science 222:1129-1132, (1983).
14. W. F. Brandt, A. Henschen, and C. v. Holt, Hoppe-Seyler's Z. Physiol. Chem 361:943-952, (1980).
15. H. D. Niall, in: "Methods in Peptide and Protein Sequence Analysis", Chr. Birr., ed., Elsevier, Amsterdam, (1980).
16. A. Henschen-Edman and F. Lottspeich, in: "Methods in Peptide and Protein Sequence Analysis", Chr. Birr. ed., Elsevier, Amsterdam, (1980).
17. G. S. Begg, D. S. Pepper, C. N. Chesterman and F. L. Morgan, Biochemistry 17:1739-1744, (1978).
18. P. Edman and A. Henschen, in: "Protein Sequence Determination", S. B. Needleman, ed., Springer-Verlag, Berlin/New York, (1975).
19. R. A. Laursen, in: "Recent developments in the Chemical Study of Protein Structures," A. Previero and M. A. Coletti-Previero, eds., INSERM, Paris (1971).
20. M. Mudgett-Hunter, M. Anderson, E. Haber, and M. N. Margolies, Mol Immunol 22:477-488, (1985).

AMIDATION OF PROTEIN CARBOXYL GROUPS

George E. Tarr

UM Protein Sequencing Facility
Department of Biological Chemistry
University of Michigan
Ann Arbor, MI 48109

INTRODUCTION

One of the delights of working with proteins, as opposed to nucleic
acids, is their great variation in properties. But for the protein sequencer
concerned only with covalent structure, this is also a problem. For example,
a protein which is acidic or easily precipitated by organic solvents may not
elute from a reversed-phase HPLC column under the standard conditions of
dilute acid and acetonitrile (MeCN) so effective in a majority of cases. Use
of a neutral buffer (a UV-transparent, volatile one is available)[1]
alleviates problems due to acidity but aggrevates those due to precipitation. And many proteins that do behave well on HPLC may balk when treated
with a protease: they are either insoluble in suitable buffers or present a
refractory surface with susceptible cleavage sites buried. The usual
stratagem is to include a denaturant, but the level of urea that trypsin and
most other enzymes will tolerate is often insufficient to promote complete
and repoducible digestion, and detergents -- which also are not always
effective -- interfer with separations employing RP-HPLC. An alternative to
adjusting conditions to try to accomodate each ill-behaved protein is to
modify these proteins at the start in such a way that they are all alike in
their properties and common problems are overcome. Ideally this modification should be specific, complete, stable, universally produced under gentle
conditions, unrestrictive with regard to other manipulations, and advantageous during sequencing. Amidation of carboxyl groups with N,N-dimethyl-
ethylenediamine (DMED) in hexafluoroacetone hydrate (HFA) comes close to
this ideal: 1) The product is more soluble and more polar in dilute acid
(and often at neutrality) than the starting material. 2) The modified
protein is partially unfolded because of its polycationic structure and
this, along with enhanced solubility, gives proteases access to cleavage
sites. 3) Acidic cleavage at Asp-Pro bonds under the conditions of CNBr
fragmentation at Met is completely suppressed. 4) With one exception, each
peptide purified from a digest of modified protein has a carboxyl group only
at the C-terminus: this permits several excellent methods for covalent
attachment for solid-phase sequencing. 5) The C-terminal fragment in a
digest is identifiable -- and potentially purifiable -- as the only peptide
with a C-terminal DMED instead of a carboxyl group. 6) The number of free
acids can be estimated from the DMED recovered on amino acid analysis.
7) Because acidic residues are marked from the beginning there can be no
confusion later during sequencing in the assignment of acids and amides.

8) Non-specific chain splitting during sequence analysis is considerably reduced by amidation of side-chain carboxyl groups[2]. 9) If C-terminal carboxyl groups are modified after digestion, the DMED forms a convenient end-of-sequence signal and its polarity can be exploited to help retain peptides in film method sequencing. Limitations imposed by amidation include the loss of fragmentation methods specific for acidic residues, such as partial acid hydrolysis and digestion with S. aureus protease, the need for a carrier of opposite charge from Polybrene in liquid phase sequencing, and the demand that analytical systems for PTHs resolve two or three new derivatives. However, as the new sequencing requirements can be met, the net result of amidation with DMED would seem overwhelmingly positive.

AMIDATION WITH DMED

The classical solvent for complete amidation of proteins via carbo-diimide (CDI) activation of carboxyl groups is 7.5 M urea or 5 M guanidine HCl[3]. These are effective with most proteins, but not with very hydrophobic membrane proteins, the class that has caused me the most trouble. While working with cytochrome c oxidase 10 yrs ago I "discovered" HFA, a general and gentle protein solvent, and soon found that it was compatible with CDI reactions. HFA exists as the gem-diol with a pK of 6.6, which means that it buffers around the pH of 5 generally recommended for amidation. I took this to be a positive feature and for a long time buffered the reaction mixture up with pyridine. However, recently I found that the reaction proceeds more reliably at low pH, presumably because the CDI is attacked by the alkoxide form of HFA. But attack by HFA has value, too, as side-reactions due to prolonged exposure to excess CDI are minimized: I have seen no indication of reaction of either Asn or Tyr. Besides eliminating pyridine and dropping the pH, I have over the last 3 yrs also decreased the final concentration of HFA, increased that of the attacking amine, and stopped adding CDI as a solid. While experiments support each of these changes, they were not always fully controlled and rigorous, so the procedure offered below is not necessarily optimal.

The best modification was judged from solubility in dilute acid, chemical stability, and sharpness on RP-HPLC of myoglobin reacted with a dozen low molecular weight amines (to minimize the MW increase). Glycine amide and methyl ester, the traditional choices[2,3], were not considered as they would only confuse compositional analysis and would be most unlikely to improve solubility. Three classes of derivatives were very soluble and gave sharp, polar peaks. The hydroxylamine and methoxylamine products were more stable to acid and base than expected, but were still rejected on this basis. The O-methylurea product also seemed stable, but was suspected of being capable of further reaction. That left DMED. For global purposes it is preferable to its parent, ethylenediamine, as its HCl salt was much more soluble in the reaction mixture, there was no possibility of cross-linking, and residues modified by it would remain polar during sequencing. However, ethylenediamine and its oligomers can be useful in solid-phase attachment procedures (see Discussion).

Procedure for amidation via CDI in HFA

All reactions are conveniently performed in 6X50 mm culture tubes. The previously published procedure[4] specifies that the protein be dissolved in HFA using 4 µl/nmol. A more general advice is to use 200 µl/mg, but this should be taken as extremely rough: the procedure has been observed to give excellent results at a concentration as high as 20 µg/µl and there is no reason to suspect a lower limit. HFA sesquihydrate is preferred (my sources have dried up) but the trihydrate now offered by Aldrich should work well enough. Make sure the protein is truly dissolved, not just solvated and dispersed, before adding the amine. The practical upper limit for concentration of DMED·HCl is 4 M, as that is produced by adding about 1.5 vol of conc HCl to 1 vol DMED (mix cautiously at 0° and adjust to roughly pH 3; anywhere between 1.5 and 4 seems to work; store under N_2 at -20°). Add 1/2 vol of amine per vol HFA and mix well; sometimes some protein precipitates, but it should readily redissolve. Add enough ethyldimethylaminopropyl CDI to 4 M amine in a separate tube to make a saturated (or nearly) solution, then add 1/4 vol (relative to HFA) of this with rapid mixing. Although the CDI apparently disappears in a few minutes, the reaction of the amine seems fairly slow, whatever the contributions of the two likely pathways (Fig. 1), so always allow at least 1 hr at 22°; overnight does no harm. The reaction mixture is diluted with 3 vol aqueous acid and purified by RP-HPLC in 0.1% trifluoroacetic acid (TFA), water vs MeCN or MeCN:propanol 3:1.[1]

Amidation via the peptide methyl ester

Small hydrophilic peptides are hard to generate by the preceding method as they will overlap the reagents on HPLC. These are efficiently made by first converting the peptide to the methyl ester in 1N HCl/methanol (add 0.3 ml acetyl Cl dropwise to 3.9 ml methanol at 0°) at 50° for 10 min, vacuum dry exhaustively. Add 10 μl DMED:hexafluoroisopropanol 4:1, mix, stand 24 hrs 22°, then vacuum dry again. The fluoroalcohol increases the reaction rate several fold. Raising the temperature instead increases unfavorably the side reaction of amide bond rupture, which is appreciable with large peptides even at room temperature. An alternative is to substitute the faster reacting homologue N,N-dimethyl-1,3-propanediamine and stop the reaction at 9 hrs. The product need not be purified if intended only for sequencing: the first cycle will be normal except for residual DMED.

CHROMATOGRAPHY

Structural data has appeared that employed amidation with DMED of hydrophobic or acidic proteins[5,6,7] and data for several other proteins awaits publication. These and modified standards all exhibit similar properties: they are very soluble in dilute acid, fairly soluble at neutrality, elute quantitatively from wide-pore RP-HPLC columns, and are attacked readily with trypsin to yield "limit" digests. Fig. 2 shows the characteristic chromatographic behavior of ovalbumin. After alkylation with

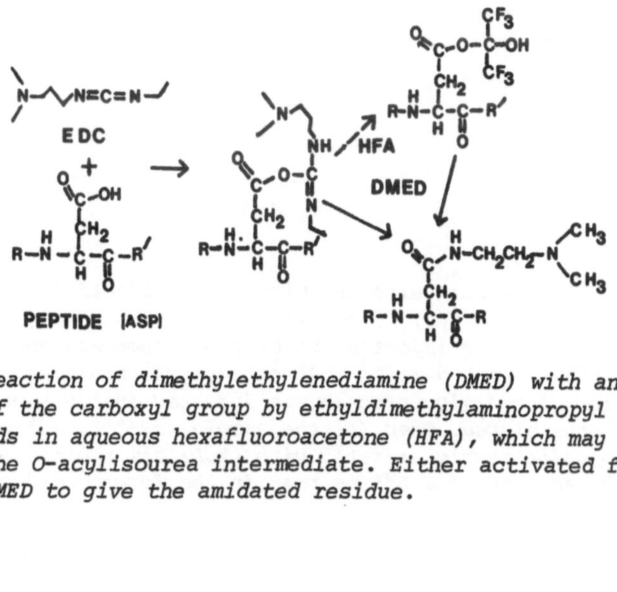

Figure 1: *Reaction of dimethylethylenediamine (DMED) with an Asp residue. Activation of the carboxyl group by ethyldimethylaminopropyl carbodiimide (EDC) proceeds in aqueous hexafluoroacetone (HFA), which may subsequently react with the O-acylisourea intermediate. Either activated form may then react with DMED to give the amidated residue.*

4-vinylpyridine in 6 M guanidine HCl[4] a portion was acidified, diluted and injected on the C3 reversed phase column. The gradient eluted virtually no protein. After reequilibration, 200 μl 88% formic acid was injected and the gradient rerun, which brought off most of the glycoprotein. This echo effect, though generally much less severe, has been observed with a majority of experimental proteins. On the other hand, pyridylethylated ovalbumin further modified with DMED, injected next onto this same column, eluted completely in the first run. (Note that non-DMED protein is still appearing in these next two runs.) The same results are obtained if the two samples are mixed and injected. To be fair, I should note that if alkylated ovalbumin is injected in dilute TFA rather than 1-2 M salt, most (70-80% in the test case) elutes the first time. This is more typical, although preparative runs with other proteins have given as much as 50% "echo", even when the loaded sample is "chased" with formic acid before the first run. Good chromatography was observed on other wide-pore columns, such as Vydac and Brownlee C18 and μBondapak C18, CN and phenyl.

RP-HPLC is the only separation method for amidated protein thoroughly investigated, but preliminary results indicate that acceptable size-exclusion chromatography can be obtained on wide-pore anion exchange columns. The theory here was that charge repulsion should render the support invisible to the modified protein. I have not been able to test suitable commercial columns, so cannot make recommendations. Electrophoresis in SDS is precluded because of precipitation, but reasonable results may be obtained with cationic detergents (Carlton Paul, personal communication).

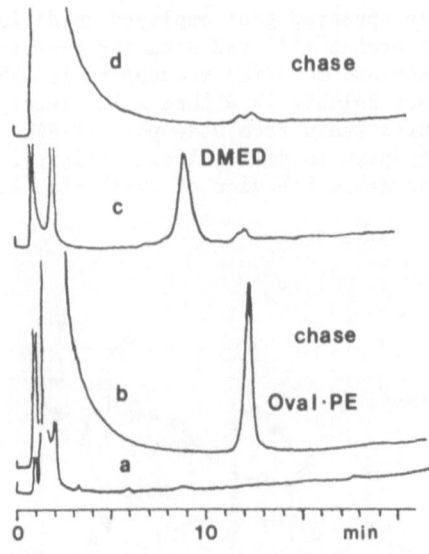

Figure 2. HPLC of ovalbumin with and without modification by DMED. The samples were run in order a,b,c,d on Ultrapore RPSC with a gradient of 0.1% aq TFA vs MeCN; after injection at 0% the column was developed at 2%/min from 20-60%, flow 0.5 ml/min. Absorbance was monitored at 230 nm. (a) The pyridylethylated glycoprotein was injected in 1.5 M guanidine HCl; virtually none eluted. Then (b) the sample was "chased" with formic acid. (c) A companion sample amidated with DMED; all protein elutes, as indicated by the absence of a DMED'd peak in the formic acid chase (d).

Enzymatic digestion

Even myoglobin, a small protein generally considered well-behaved, is resistant to attack by trypsin (TPCK from Worthington). As shown in Fig. 3, addition of urea is ineffective and while organic solvents help to open the substrate to attack, they also greatly reduce trypsin's activity. In contrast, digestion of myoglobin/DMED proceeds readily in the absence of denaturants and is simply slowed in their presence. Note the production of large incompletely cleaved fragments in the presence of propanol but not with urea. This effect, seen with both free and amidated proteins, seems to be general and could be exploited in sequencing efforts. Digestion of other unmodified proteins has been often variable, apparently because of varying conformation due to the history of treatment of particular samples. DMED'd proteins have given extremely repoducible digests, which is especially important when comparing peptide maps.

Results with endoproteinase Lys-C (Boeringer Mannheim) were similar, although the unmodified myoglobin digests more readily and the effect of

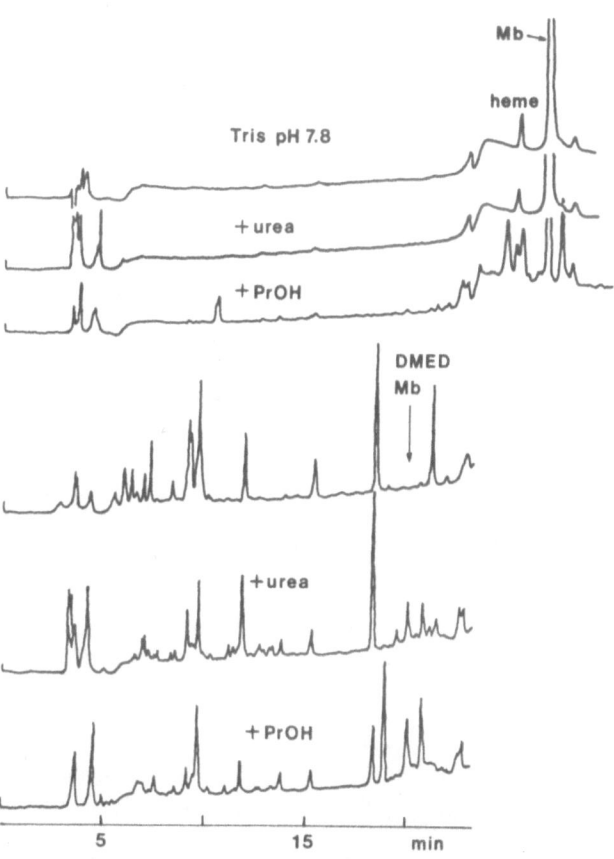

Figure 3. Effects of amidation and denaturants on tryptic digestion of myoglobin. Mb and DMED'd Mb were digested with trypsin 50:1 w:w for 77 min at 23°, the hydrolysis stopped by acidification. The middle trace in each set shows the effect of 2.9 M urea added to the 0.1 M buffer. The lower trace shows digestion in the presence of 20% 1-propanol. Chromatography conditions as in Fig. 2 except gradient from 10%, flow 1 ml/min.

propanol is not as striking (Fig. 4). Chymotrypsin and clostripain were found to fragment DMED'd material but, unsurprisingly, S. aureus V8 protease did not. The only confirmed case of splitting at a modified residue was with trypsin at high enzyme:substrate ratio where other anomalous splits were also observed.

Chemical Digestion

The standard solvent for CNBr fragmentation at Met is 70% formic acid[8]. One deleterious effect of this solvent is esterification of Ser and Thr, which must be reversed to obtain sharp single peaks on RP-HPLC[1]. Another effect is partial rupture of acid sensitive peptide bonds, particularly Asp-Pro. Obviously, this unwanted fragmentation is prevented by amidation. An example of the results with a large hydrophobic protein is shown in Fig. 5. The relative broadness of the amidated starting material (see also Fig. 2) and the trailing shoulders on most of the fragments suggests a multiplicity of forms. One's first suspicion would be that the reaction of DMED is incomplete, but I have no evidence from sequencing that this is so, nor have additional (and partial) modifications been detected — insulin chains, for instance, give one peak each with no minor forms. Also, many non-amidated fragments show splitting into several peaks. On the other hand, collected shoulders rechromatograph to their original positions, so the differences seem to be more than conformational. Adsorptive effects probably contribute to broadening, as uniformly modified peptides (e.g., insulin chains) are also noticably broader than their free counterparts and the DMED'd PTHs (see below) are much broader than uncharged PTHs even on

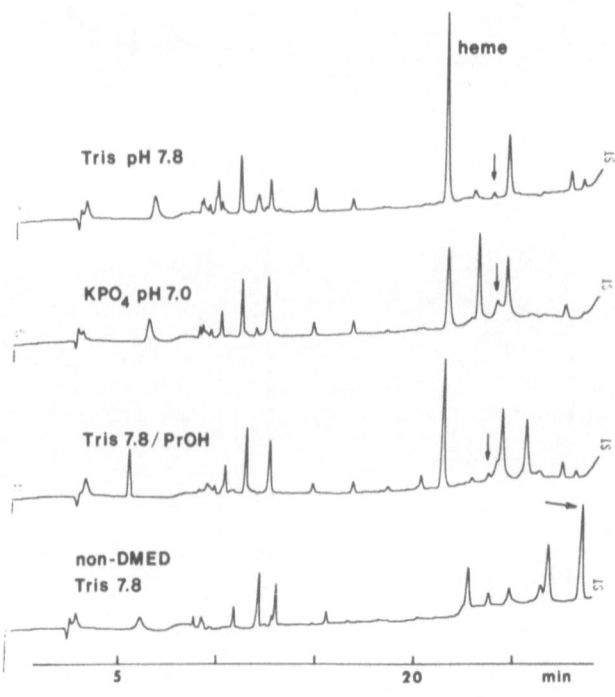

Figure 4. Digestion of myoglobin DMED by endoproteinase Lys-C. Substrate: enzyme 50:1, 5 hrs at 37°, 0.1 M buffer, concentration of 1-propanol in third trace 20%. The position of the DMED'd starting material is indicated by the arrow in the upper 3 traces; the non-amidated Mb used for the 4th trace was dehemed. HPLC as in Fig. 3.

columns showing minimal adsorption of Arg and His derivatives. This interpretation is supported by tests with casein/DMED which showed that peak
sharpening occurred when dimethyloctylamine was included in HPLC solvents.

Fragmentation at Trp with BNPS-skatole[9] proceeds readily with DMED'd
material, though, as the solvent here is acetic acid and cleavage at Asp is
not much of a problem, the advantage over non-amidated material is only that
the resulting fragments are more soluble. The other chemical cleavage of
interest is hydroxylamine cleavage of Asn-Gly bonds[10], which is theoretically extended by either amidation or CDI side-reaction to Asp-Gly.
Illuminating results are not yet available.

AMINO ACID ANALYSIS

Acid hydrolysis of DMED'd peptides isolated by HPLC yields one mole of
DMED per mole of carboxyl group, which includes the C-terminal group if the
amidated peptide is C-terminal or not a product of digestion. The HCl double

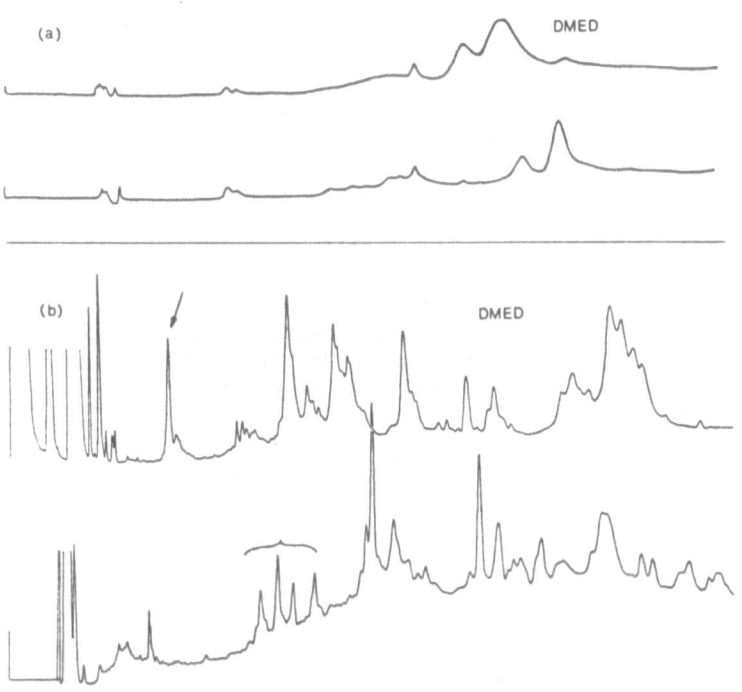

*Figure 5. HPLC comparison of intact and CNBr fragmented cytochrome P-450
with and without DMED modification. In (a) the pyridylethylated proteins
were run on 5μ Vydac C18; curves show fluorescence due to Trp. The double
peak pattern, as yet unexplained, survives both chemical modifications.
The main DMED'd peak was used for the CNBr digestion shown as the upper
trace in (b): Vydac, 0.1% aq TFA vs ethanol, gradient 30-80% at 0.33%/
min, flow 0.5 ml/min, absorbance at 229 nm. The lower digest was performed
and chromatographed a year earlier under different conditions, but the
main points of comparison are valid. For instance, the non-amidated protein
shows a cluster of 4 peaks where the DMED'd protein shows only one (arrow,
corresponding to the second of the cluster). The other 3 result from acid
rupture of Asp-Pro bonds, of which there are 5 in the sequence. Note the
superior recovery of late-eluting peaks from the DMED'd protein.*

salt is stable and non-volatile even when triethylamine (TEA) is used to drive off ammonia, as in the standard analytical system for PTC-amino acids[4] marketed by Waters as Pico-Tag. The separation on NovaPak columns resolves the PTC-DMED peak nicely after PTU (Fig. 6). Other columns or solvents that place PTC-Arg and His differently would be expected to also place DMED elsewhere. Quantification of DMED from standards and from unknowns subsequently sequenced has proven accurate (extinction at 254 nm is about that of most PTC-amino acids).

SEQUENCE ANALYSIS

The sequencing of DMED'd peptides and proteins presents difficulties for the usual methods, which are based on liquid partitioning between hydrated Polybrene, a polyquarternary amine[11], and organic solvents. The charge structure of the carrier might be expected to repel the similarly charged peptide, leading to extractive losses, and in practice it was difficult to sequence amidated peptides with Polybrene without severe washout. Adding an equal weight of the polysulfonic acid dye Ponceau S to the Polybrene gave good retention of DMED'd peptides and much "real" sequencing was accomplished with this mixed carrier[6,7]. However, balancing the polarity of the wash and extraction solvents for removal of byproduct and residue without also removing the peptide was difficult, and His was partially destroyed, for unexplained reasons. A new carrier was therefore considered that would bear sulfonic acid groups to associate with DMED; it was felt that inclusion of tertiary and quarternary amino groups would increase hydrophilicity and self-association. A carrier with these properties is (sulfopropyl)$_n$pentaethylenehexamine (SPPEXA), where n is about 8, synthesized from PEXA and propanesultone[4]. SPPEXA is so hydrophilic that it is precipitated from concentrated aqueous solution by methanol. This property enables the inclusion of water in sequencing wash and extraction solvents, which gives an enhanced partitioning effect and firm control of moisture content, particularly important in manual sequencing where the sequencing vessel is open to the air. Procedural details are given elsewhere[4]. An example of sequencing a short peptide with C-terminal DMED is shown in Fig. 7. No attempt has been made yet to incorporate SPPEXA and its associated solvents into a commercial sequenator.

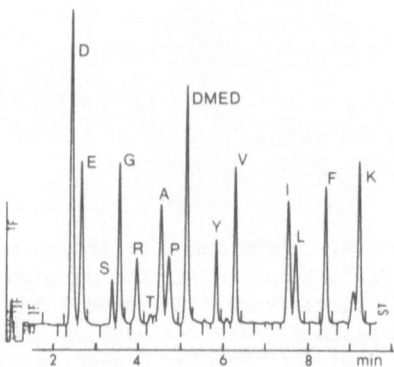

Figure 6. Composition of a DMED'd peptide by the PTC method. The 24 residue peptide with 4 modified carboxyl groups was purified by RP-HPLC from a BNPS-skatole digest of flavodoxin/DMED. The sample was hydolyzed 1 hr 150°, coupled with PITC, and analyzed at 40° on NovaPak C18 with a gradient of aq 140 mM K acetate pH 5.6 vs 84% MeCN, absorbance 254 nm, 0.1 OD. His and Met, absent from the peptide, elute after Gly and Val.

Peptides that lack the C-terminal DMED will not be held by SPPEXA as well when they near the end. Ideally, these should be covalently attached to amino resins or other solid surfaces according to the usual solid-phase techniques[12]. Small peptides (up to 20 residues) sequence well with the manual partitioning method[4], while peptides longer than 40-50 residues generally do not sequence far enough to present a problem. The ones in between are currently in limbo.

The PTHs from amidated residues do not chromatograph well on Ultrasphere C18, the column for our standard analytical system, eluting broadly near Tyr in "reverse" order (Glu-DMED first)[4]. Attempts to improve peak shape and resolution with higher ionic strength and the more aggresive amine TEA have been only partially successful. NovaPak columns show much less adsorptive effect and permit the elution of basic residues as reasonably sharp peaks near the front of the chromatogram. The solvent system given in Fig. 7 has been revised recently to 60 mM TEA phosphate pH 4.9 to bring them off sharper and earlier: Asp and Glu-DMED before PTU, Arg and His in that order between PTU and Asn. The position of the first three basics is determined by salt concentration alone, and His is positioned finally by pH adjustment. Incidently, the sequencing of completely amidated peptides does not mean that no free acids or methyl esters (depending on the conversion medium) will appear in the analyses: these are produced during conversion, particularly from Gln (the DMED derivatives are much more resistant than the natural amides). One deficiency of the NovaPak system is low resolution as compared with Ultrasphere. This is primarily due to column length and should be fixed by a tandem pair of columns if higher resolution is judged necessary. Another, related, problem is failure to separate the Edman byproduct diphenylthiourea (DPTU) from PTH-Trp. Adjustment of the gradient, solvent, and/or temperature could probably solve this problem, but the solution I most favor is elimination of Trp residues from the start by reduction[13]. Trp often causes a sudden drop in sequencing yield because of its tendency to oxidize and cyclize (Fig. 7 is an example of this, too).

DISCUSSION

The fundamental obstacles to the use of amidated proteins and peptides for structural determination have been overcome. The DMED procedure can be

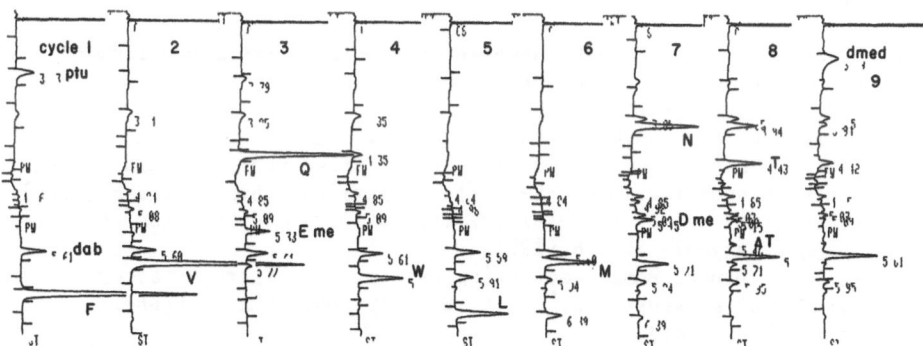

Figure 7. Sequencing of a small DMED'd peptide using SPPEXA as a carrier. About 800 pmol peptide was sequenced to the "end-of-sequence" DMED on cycle 9. PTH analyses employed NovaPak at 45°, 1 ml/min, gradient of 100 mM KPO₄ vs 84% MeCN, 9 min injection to injection. Elution order: D·DMED PTU E·DMED H N R S O T G A Y PEC M V P W K F I L. Details in ref. 4.

applied to materials of widely varying solubility characteristics to give products with desirable, uniform and predictable properties. Benefits are realized at every stage of the determination process: purification, fragmentation, amino acid analysis and sequencing. However, all of the procedures presented above need further investigation and refinement, and there are several areas largely or wholly unexplored. One already mentioned is hydroxylamine fragmentation at modified Asp-Gly linkages. Another is the possibility of cleavage at Asp or Glu modified to resemble the normal substrates of trypsin or endoproteinase Lys-C. The latter may be the best bet as it has already been shown to be much more active than trypsin toward the artificial residue aminoethyl-Cys[5], and trypsin was found to be inactive toward ethylenediamine and O-methylurea derivatives. Reversible modifications, say with N,N-dimethylaminoethanol, may make acidic residue cleavages possible without losing the benefits of solubility at earlier stages of digestion and fractionation.

Peptide modification may be effective in purifying peptide mixtures. An example using reaction of α-amino groups with PITC is given in Tarr et al.[5] and in one uncited case I was successful employing DMED this way. Other residues susceptible to specific, often reversible, modification are Lys, Arg, Met, Ser and Thr, Tyr, and Cys. This set provides for a powerful and rarely exploited strategy for difficult separations. The C-terminal peptide in the digest of a DMED'd protein is unique in having no free carboxyl groups. A special purification strategy employing anion exchange to capture the other peptides may yield this peptide in a general fashion.

One of the major benefits of protein amidation for sequencing is the facilitation of solid-phase attachment. A peptide produced by proteolysis will almost always possess one free carboxyl group strategically located at the C-terminus, by which it can be readily attached to solid-phase amino groups via CDI activation. No side chains are available for attachment, so no residues are invisible during sequencing. An alternative strategy is amidation with ethylenediamine, followed by an attachment procedure that is deliberately incomplete, e.g., cross-linking to solid-phase amines with a mixture of di- and mono-isothiocyanates. Yields will be low for reactive residues, but those recovered will necessarily have higher extinction coefficients and may therefore be just as visible as ordinary residues. Both strategies are under development. Prospects for liquid-phase sequencing are good, provided a suitable commercial carrier is or becomes available and the wash solvents in the gas phase or spinning cup sequenators are adapted.

REFERENCES

1. G. E. Tarr and J. W. Crabb, Reverse-phase high-performance liquid chromatography of hydrophobic proteins and fragments thereof, Anal. Biochem. 131:99 (1983).
2. W. F. Brandt, A. Henschen, and C. von Holt, Nature und extent of peptide bond cleavage by anhydrous heptafluorobutyric acid during Edman degradation, Hoppe-Seyler's Z. Physiol. Chem. 361:943 (1980).
3. K. L. Carraway and D. E. Koshland, Carbodiimide modification of proteins Methods Enzymol. 25:616 (1972).
4. G. E. Tarr, Manual Edman sequencing system, in: "Microcharacterization of Polypeptides," J. E. Shively, ed., Humana Press, Clifton NJ (1985).
5. G. E. Tarr, S. D. Black, V. S. Fujita, and M. J. Coon, Complete amino acid sequence of phenobarbital-induced cytochrome P-450 (isozyme 2) from rabbit liver, Proc. Natl. Acad. Sci. USA 80:6552 (1983).
6. V. S. Fujita, S. D. Black, G. E. Tarr, D. R. Koop, and M. J. Coon, On the amino acid sequence of cytochrome P-450 isozyme 4 from rabbit liver microsomes, Proc. Natl. Acad. Sci. USA 81:4260 (1984).

7. M. L. Ludwig, K. A. Pattridge, and G. Tarr, FMN:protein interactions in flavodoxin from _A. nidulans_, _in_: "Flavins and Flavoproteins," R. C. Bray, P. Engel, and S. G. Mayhew, eds., Gruyter, Berlin (1984).

8. E. Gross, The cyanogen bromide reaction, _Methods Enzymol._ 11:238 (1967).

9. A. Fontana, Cleavage at tryptophan by BNPS-skatole, _Methods Enzymol._ 25:419 (1972).

10. P. Bornstein and G. Balian, Cleavage at Asn-Gly bonds with hydroxyl-amine, _Methods Enzymol._ 47:132 (1977).

11. G. E. Tarr, J. F. Beecher, M. Bell, and D. J. McKean, Polyquarternary amines prevent peptide loss from sequenators, _Anal. Biochem._ 84:622 (1978).

12. R. A. Laursen, Coupling techniques in solid-phase sequencing, _Methods Enzymol._ 47:277 (1977).

13. W. S. D. Wong, D. T. Osuga, and R. E. Feeney, Pyridine borane as a reducing agent for proteins, _Anal. Biochem._ 139:58 (1984).

A METHOD FOR DIRECT AMINO ACID SEQUENCE ANALYSIS OF THE NH_2-TERMINAL REGIONS OF FIBRINOGEN

Chung Y. Liu and Francis J. Morgan

College of Physicians and Surgeons
Columbia University
New York, New York 10032

INTRODUCTION

Fibrinogen is one of the major plasma protein involved in blood coagulation and its primary structure has been completely determined[1-6] (Fig. 1). A number of abnormal fibrinogens have been purified and their structural defects have been found mainly in the NH_2-terminal regions[7,8]. However, fibrinogen is a macromolecule containing six polypeptide chains and the NH_2-terminal amino acids of the two Bβ-chains are pyroglutamates which are not susceptible to direct sequence analysis by automated Edman degradation. In this report we present a simple method for the direct sequence analysis of the NH_2-terminal regions of fibrinogen without prior separation of the polypeptide chains to investigate the structural abnormalities of fibrinogen.

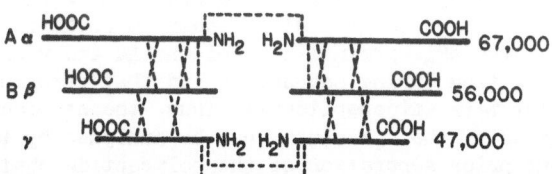

Fig. 1. Schematic representation of the major features of fibrinogen molecule. Fibrinogen molecule (Mr= 340,000) is a dimer with two identical subunits[1-6]. Each subunit contains three polypeptide chains (designed Aα, Bβ, and γ) crosslinked by disulfide bonds (----). The numbers are molecular weights of the respective polypeptide chains.

Materials and methods were as previously described[9],[10], with exceptions described in details in this report. Amino acid sequence analysis of fibrinogen was performed by automated Edman degradation (Beckman Sequencer 890C) using a 1.0 M Quadrol program[11]. 3-Phenyl-1-thiohydantoins (PTHs) were identified by HPLC by comparing to the standard materials[12]. NH_2-terminal pyroglutamate of the Bβ-chain is not susceptible to Edman degradation. In order to permit amino acid sequence analysis of the Bβ-chain alone, purified fibrinogen (24 mg) was first succinylated with succinic anhydride (100 mg) in 12 ml of borate buffer (0.1 M, pH 9.0) at $4°C$ for 60 min[13]. This procedure blocks the NH_2-termini of the Aα- and γ-chains without reaction with the NH_2-terminal pyroglutamate of the Bβ-chain. This sample was then digested with 45 μg of pyroglutamate aminopeptidase (E.C. 3.4.11.8) in 24 ml of sodium phosphate buffer (0.1 M, pH 8.0) containing EDTA (10 mM) and dithiothreital (5 mM) at $4°C$ for 9 hours under nitrogen to remove the NH_2-terminal pyroglutamate of the Bβ-chain. An additional aliquot of pyroglutamate aminopeptidase (45 μg) was added to the sample which was further digested for 16 hours at $25°C$ under nitrogen to complete the release of the NH_2-terminal pyroglutamate. This digest was dialyzed against acetic acid (50 mM, aldehyde-free) at $4°C$ for 18 hours[14] to remove the free pyroglutamic acid[15]. The resulting preparations were then subjected to 20 cycles of automated Edman degradation.

RESULTS AND DISCUSSION

Fibrinogen was purified from human blood plasma and its purity was confirmed by SDS-PAGE analysis of the three types of polypeptide chains and by measurement of the fibrinopeptides A and B contents in the sample[9],[10]. In order to analyze the Aα- and γ-chains, fibrinogen was directly sequenced for 20 steps. Two sequences (Fig. 2) were found which corresponded exactly to the normal sequences expected for the Aα- and γ-chains[1-6]. It appears from these data that the NH_2-terminal ends of the Bβ chains are still blocked by pyroglutamate as expected in normal fibrinogen (Fig. 1). These results suggest that the sequences of Aα- and γ-chains of the NH_2-terminal regions can be directly sequenced by automated Edman degradation without prior separation of these chains.

When fibrinogen was treated with succinic anhydride and then sequenced, Edman degradation showed no release of amino acids (data not shown), indicating that the NH_2-terminal ends of Aα- and γ-chains could be blocked from Edman degradation by succinylation and that the NH_2-terminal ends of Bβ-chains remained blocked by pyroglutamate. When succinylated fibrinogen was then treated with pyroglutamate aminopeptidase and sequenced for 20 steps, only one amino acid sequence was obtained, corresponding to the amino acid sequence of the normal Bβ-chain as shown in Fig. 3. These data indicate that the NH_2-terminal ends of the Aα and γ-chains remained blocked by succinylation during sequencing and that pyroglutamate was removed by pyroglutamate aminopeptidase. Thus, the sequences of the NH_2-terminal regions of the six chains can be analyzed by these simple procedures without prior separation of the polypeptide chains.

A high non-specific background was observed in the sequencing (Fig. 2A). This background increased with increase in the number of sequence cycles. However, this observation is not atypical in sequencing high molecular weight proteins such as fibrinogen. Clear identification of the respective amino acids could be performed.

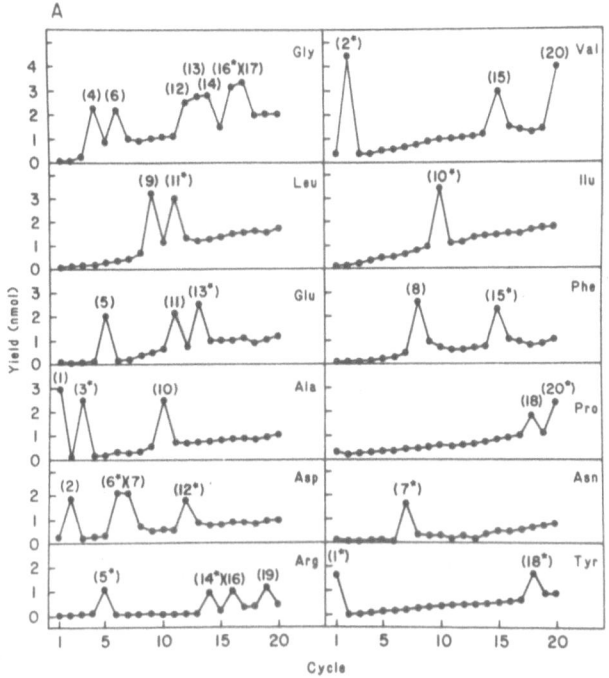

B

Aα–Chain	Ala Asp Ser Gly Glu Gly Asp Phe Leu Ala Glu Gly Gly Gly Val Arg Gly Pro Arg Val–
γ–Chain	Tyr Val Ala Thr Arg Asp Asn (Cys) (Cys) Ilu Leu Asp Glu Arg Phe Gly Ser Tyr (Cys) Pro–
Sequence Cycle	I 3 5 7 9 II 13 15 17 19

Fig. 2　Sequence analysis of Aα- and γ-chains of fibrinogen NY-1a.
NY-1a was directly sequenced for 20 steps. A, uncorrected
yields of PTH-derivatives detected by HPLC. Numbers in
parentheses without asterisks denote the position of a residue
in the first sequence (Aα-chain); those with asterisks indicate
the second sequence (γ-chain). PTH-Ser and PTH-Thr were
identified but not quantified, and the expected PTH-Cys at
cycles, 8, 9, and 19 were not identified under these
experimental conditions, but have been assumed from the known
sequence. B, NH$_2$-terminal sequences of Aα- and γ-chains.
Sequence Cycle refers to the numbering in A, and these numbers
correspond exactly to those residue positions of the Aα- and
γ-chains of normal human fibrinogen[9].

357

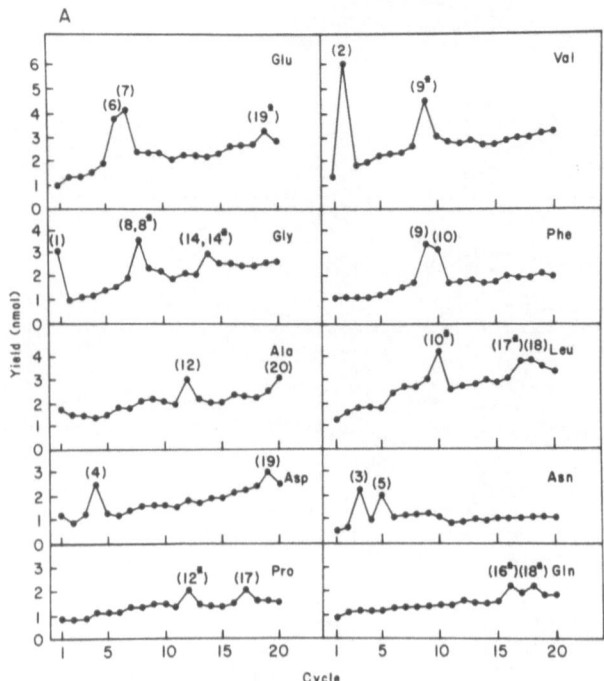

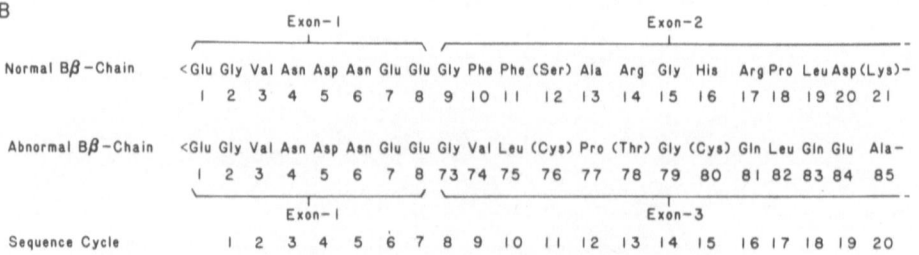

Fig. 3. Sequence analysis of Bβ-chain of fibrinogen NY-1a.
A, fibrinogen NY-1a was succinylated and then digested with
pyroglutamate aminopeptidase before direct sequencing.
Uncorrected yields of representative PTH-derivatives detected
by HPLC at each degradation cycle are given. Numbers in
parentheses without asterisks denote the positions of
residues in the first sequence (normal); those with asterisks
indicate the second sequence beginning at cycle 8 (abnormal).
PTH-His and PTH-Arg were identified but not quantified and
the expected PTH-Ser, PTH-Thr, PTH-Cys, and PTH-Lys at cycles
11, 13, 15, and 20, respectively were not clearly identified
under these conditions (see B).
B, NH$_2$-terminal amino acid sequence of the Bβ-chain of
fibrinogen NY-1a. The sequence of the abnormal chain is
given below the normal Bβ sequence for comparison; numbering
is that of the normal chain. Sequence Cycle refers to the
numbering in A and is out of phase as pyroglutamic acid
(<Glu) was removed enzymatically. The expected residues in
brackets at cycles 11, 13, 15, and 20 were not clearly
identified, but have been assumed from the known sequence[9].

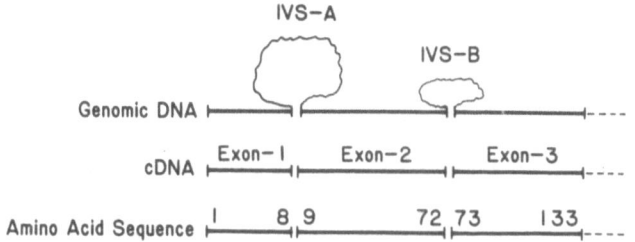

Fig. 4.   Identification of the abnormal Bβ-chain with a deletion of
          Bβ (9-72) corresponding exactly to exon 2 of the gene. A
          portion of the genomic DNA and cDNA structure corresponding
          to the NH$_2$-terminal region of the normal Bβ-chain[16] is shown.
          The numbers for the Bβ-chain are the amino acid positions
          corresponding to the respective exons. See refs. 9, 16, 17
          for further information.

These procedures were then tested by analyzing an abnormal fibrinogen
(Fibrinogen New York-1a) purified from a patient with a thrombotic
tendency[8],[9]. Two types of Bβ-chains were previously demonstrated in this
abnormal fibrinogen by SDS-PAGE analysis and Sephadex gel chromatography.
Sequence analysis by Edman degradation using the present procedure
demonstrated (Fig. 3) that this abnormal fibrinogen had two amino acid
sequences for the Bβ-chain: one corresponding to a normal Bβ-chain and one
corresponding to an abnormal Bβ chain with a deletion of the Bβ (9-72)
which corresponds exactly to exon-2 of the gene (Fig. 4). The NH$_2$-terminal
regions of Aα- and γ-chains were found normal (data not shown). Thus, the
results of the present studies demonstrate that amino acid sequences of the
NH$_2$-terminal regions of fibrinogen can be analyzed without prior separation
of the polypeptide chains, providing a simple and fast method for
investigating structural abnormalities in the NH$_2$-terminal regions of the
fibrinogen molecule.

ACKNOWLEDGEMENTS

This work was support by NHLBI grant HL-15486.

REFERENCES

1. F. Lottspeich, and A. Henschen, Amino Acid Sequence of Human Fibrin -
      Preliminary Notes on the Completion of the γ-Chain Sequence,
      Hoppe-Seyler's Z. Physiol. Chem. 358:935 (1977).
2. A. Henschen, and F. Lottspeich,   Amino Acid Sequence of Human fibrin
      - Preliminary Note on the Completion of the β-Chain Sequence,
      Hoppe-Seyler's Z. Physiol. Chem. 358:1643 (1977).
3. A. Henschen, F. Lottspeich, and B. Hessel, Amino Acid Sequence of Human
      Fibrin - Preliminary Note on the Completion of the Intermediate
      Part of the α-Chain Sequence, Hoppe-Seyler's Z. Physiol. Chem.
      360:1951 (1979).

4. R. F. Doolittle, K. W. K. Watt, B. A. Cottrell, D. D. Strong, and M. Riley, The Amino Acid Sequence of the α-Chain of Human Fibrinogen, Nature 280:464 (1979).

5. B. Hessel, M. Makino, S. Iwanaga, and B. Blomback, Primary Structure of Human Fibrinogen and Fibrin, Eur. J. Biochem. 98:521 (1979).

6. B. Blomback, B. Hessel, and D. Hogg, Disulfide Bridges in NH_2-Terminal Part of Human Fibrinogen, Thromb. Res. 8:639 (1976).

7. C. Southan, A. Henschen, and F. Lottspeich, The Search for Molecular Defects in Abnormal Fibrinogens, in: "Fibrinogen", A. Henschen, H. Graeff, and F. Lottspeich, eds., Walter de Gruyter, Berlin (1982).

8. C. Y. Liu, J.A. Koehn and H. L. Nossel, Abnormal Bβ-Chain in Fibrinogen New York (In Patients with a Thrombotic Tendency), Thromb. Haemost. 50:336 (1983).

9. C. Y. Liu, J.A. Koehn and F.J. Morgan, Characterization of Fibrinogen New York 1, J. Biol. Chem., 260:4390 (1985).

10. C. Y. Liu, D.A. Handley and S. Chien, Gold-Labelling of Thrombin and Ultrastructural Studies of Thrombin-gold Conjugate Binding by Fibrin, Anal. Biochem. 147:49 (1985).

11. P. Edman, and G. Begg, A Protein Sequenator, Eur. J. Biochem. 1:80 (1967).

12. C. L. Zimmerman, E. Appella, and J.J. Pisano, Rapid Analysis of Amino Acid Phenylthiohydantoins by High-Performance Liquid Chromatography. Anal. Biochem. 77:569 (1977).

13. I. M. Klotz, Succinylation, in: "Methods of Enzymology" 11:576 (1967).

14. D. N. Podell, and G.N. Abraham, A Technique for the Removal of Pyroglutamic Acid from the Amino Terminus of Proteins Using Calf Liver Pyroglutamate Amino Peptidase, Biochem. Biophys. Res. Commun. 81:176 (1978).

15. R. W. Armentrout, and R.F. Doolittle, Pyrrolidonecarboxylyl Peptidase: Stabilization and Purification, Arch. Biochem. Biophys. 132:80 (1969).

16. D. W. Chung, B. G. Que, M.W. Rixon, M. Mace, Jr., E.W. Davie, Characterization of Complementary Deoxyribonucleic Acid and Genomic Deoxyribonucleic acid for the β-Chain of Human Fibrinogen, Biochem. 22:3244 (1983).

17. C. Y. Liu, P. Wallen, and D.A. Handley, Fibrinogen New York 1: The Structural, Functional, and Genetic Defects and a Hypothesis of the Roles of Fibrin in Coagulation and Fibrinolysis, in: "Fibrinogen, Fibrin Formation and Fibrinolysis", D. Lane, A. Henschen, and K. Jasani, eds., Walter de Gruyter, Berlin, In press (1985).

VII. Analysis of PTH Amino Acids

PTH AMINO ACID ANALYSIS

Michael W. Hunkapiller

Research and Development
Applied Biosystems, Inc.
Foster City, CA

INTRODUCTION

The chemical process employed by automated protein/peptide sequencers is derived from the technique originated by Pehr Edman in the 1950s for the sequential degradation of peptide chains.[1,2]. The first step in this degradation is selective coupling of a peptide's amino-terminal amino acid with the Edman reagent, phenylisothiocyanate (PITC), a reaction catalyzed by an organic base delivered with the coupling reagent. The second step is cleavage of this derivatized amino acid from the remainder of the peptide, a reaction effected by treating the peptide with a strong organic acid. Each repeated coupling/cleavage cycle occurs at the newly-formed amino-terminal amino acid left by the previous cycle. Thus, repetitive cycles provide sequential separation of the amino acids which form the primary structure of the peptide.

The sequencing process is not completed by the Edman degradation alone. Once the amino acids are removed from the sample, they must be analyzed to determine their identity. Since the cleaved amino acid derivative, the anilinothiazolinone (ATZ), is not generally suitable for analysis, it is converted to a more stable phenylthiohydantoin (PTH) form before analysis is attempted. In modern sequencers[3,4], this conversion is accomplished automatically in a reaction vessel separate from that in which the Edman degradation occurs. The ATZ produced at each degradation cycle is extracted from the peptide with an organic solvent, transferred to the reaction vessel and treated with an aqueous solution of a strong organic acid to effect conversion to the PTH. The PTHs produced from each degradation cycle may be transferred to fraction collector vials until several are manually collected and prepared for analysis. Alternatively, the PTHs may be transferred directly and automatically from the sequencer conversion vessel to an on-line analysis system.[5,6]

Although a variety of analytical procedures have been used to identify the amino acids released during the Edman degradation, only high performance liquid chromatography (HPLC) is currently in widespread use. In fact, HPLC on reverse phase, silica-based packings has revolutionized peptide sequencing. It provides rapid, sensitive and quantitative analysis of PTH amino acids and is the only technique used for PTH analysis that can reliably resolve all of the PTH amino acids in a single chromatographic run. Moreover, because it provides quantitative data at the picomole level, HPLC is the only analytical method suitable for microsequencing by the latest generation of automated Edman sequencers.

The reliability, accuracy, and sensitivity of any Edman degradation scheme will ultimately be limited by the weakest of its components. In many cases, the PTH analysis protocol is the weakest link. Following are procedures recommended for optimal PTH analysis using HPLC.

ANALYTICAL PARAMETERS

Resolution

The first requirement of a PTH analysis protocol is to resolve the PTH amino acids derived from the sequencer and any contaminating compounds with similar spectral properties. Fulfilling this requirement can be difficult; there are likely to be 20 to 30 chemically similar compounds that must be separated.

Particular attention must be paid to providing good resolution of the PTH amino acids from any sequencer-derived contaminants. When small amounts of peptide are being sequenced, these contaminants may be present in much greater amounts than the PTH amino acids. Any overlap between a contaminant and a PTH amino acid can interfere with quantitation or identification of the PTH amino acid itself.

Sensitivity

The second requirement is to provide adequate detection sensitivity for the amount of sample being sequenced. The recoveries of certain PTH amino acids from the sequencer are typically low; the recoveries of all PTH amino acids decrease as the Edman degradation proceeds through the peptide. Thus, fulfilling this requirement generally means reaching a detection limit that is <10% of the starting amount of peptide.

Detection sensitivity is limited by a variety of factors:

1. Intrinsic noise level of the UV absorbance detector used to monitor the elution of compounds from the HPLC column: The best UV monitors have operating noise levels of <0.00002 absorbance units (AU), but others may exhibit noise levels substantially above this.

2. Elution volume of the PTH amino acids: Smaller elution volumes mean higher concentration and, hence, higher detection sensitivity. This is determined by:
    *   The particle size of the column packing (smaller particles generally give smaller volumes);
    *   The uniformity of the packing (columns with more

theoretical plates per unit length give smaller
volumes);
- Column size (shorter, narrower bore columns give
  smaller volumes); and,
- Extra-column effects (improper tubing connections
  and excessive tubing lengths between either the
  injector and the column or between the column and the
  detector give larger volumes).
- Elution mode - With isocratic elution, the retention
  time of a particular component also influences the
  elution volume (longer elution times give larger
  volumes), while with gradient elution the gradient
  steepness influences the elution volume (steeper
  gradients give smaller volumes).

3. Detection wavelength: The local UV absorbance maximum for
    most PTH amino acids is between 266 and 270 nm. Variable
    wavelength detectors that can be set in this range will
    give a 40 to 50% signal increase over that obtained with a
    fixed wavelength detector operating from a mercury
    emission line at 254 nm. This provides increased
    sensitivity as long as the noise level of the variable
    detector is equivalent to that of the fixed wavelength
    detector.

4. Chromatography artifacts: Random or periodic fluctuations
    in pump output or incomplete solvent mixing can produce
    baseline fluctuations that increase the effective detector
    noise. Contaminated solvents, column inlet frits, or
    column packings can give excessive baseline drift or
    produce artifact peaks that also increase the system noise
    level.

5. Sequencer artifacts: By-products of the Edman chemistry
    can coelute with some of the PTHs, thereby increasing the
    amount of PTHs required for reliable detection and
    quantitation.

6. Protein/peptide artifacts: Contaminants in the peptide
    sample being sequenced can produce chromatography
    background peaks that decrease the effective sensitivity.
    Contaminants that are also peptides are particularly
    troublesome. They will undergo Edman degradation
    themselves thus producing PTHs that interfere with
    identification of those produced by the primary sample.

Reliability

     Many peptides are being sequenced in such small amounts
that there is only enough PTH sample for one injection on the
HPLC. Thus, reliability of the analysis system is crucial.
With an automated injection system, malfunctions in the pump,
injector, or detector can be devastating because they may go
undetected while a number of cycles are analyzed.

     Retention time reproducibility is also important,
especially if the elution of the several PTHs is closely
spaced. Since there are numerous PTHs that share common
chemical structures, this is almost always the case. Moreover,
the occasional presence of chemically modified amino acids that
elute close to the common ones can often be detected if

retention times are quite reproducible. Generally, the system should have a relative standard deviation (RSD) for elution times of less than 0.3%.

Analysis of reproducible portions of each cycle's PTH product is important for accurate quantitation. Repeated treatment of the protein sample by the cleavage acid during the Edman degradation gradually fragments the sample into a series of smaller peptides. This process tends to produce a steadily increasing background level of PTHs from the fragments which is superimposed over the specific PTH released from the intact protein. Thus, the specific PTH, especially later in sequence runs, can only be identified by a quantitative distinction between the level of one PTH at a given cycle versus the background level of that PTH at preceeding and subsequent cycles.

Injection reproducibility is affected partly by the HPLC system, chiefly injector operation, and partly by the sample itself. In conventional PTH analysis, the PTHs must be quantitatively transferred from the sequencer and reconstituted in a reproducible volume of injection solution. Furthermore, they must remain stable between their formation and their injection onto the HPLC column. Practical limitations of these processes typically give a cycle-to-cycle variability (RSD) of 15 to 20%. With direct, on-line transfer of PTHs from the sequencer to the HPLC however, PTH instability can be minimized and an RSD of only 2% is possible (Table 1).

Columns

A variety of HPLC column packings have been used for PTH analysis. Although the octyldecylsilyl[7] and cyano-propylsilyl[8] silicas are currently the most widely used, octylsilyl,[9] phenylsilyl,[10] and mixed[11] supports have also been used successfully. However, selectivity of column packings of the same nominal type vary substantially from one manufacturer to another, and even from one batch to another from the same manufacturer. The reverse phase loading, residual silanol level, nature of any end-capping, partical diameter, and particle pore size all affect the selectivity and resolution of the packed column. The length and internal diameter of the column also affect resolution and detection sensitivity, with smaller diameter columns capable of providing higher sensitivity if used with suitable HPLC equipment.

Isocratic versus Gradient Elution

Both isocratic[7,12,13] and gradient[7-11] elution systems have been used successfully for PTH analysis. Isocratic systems are generally simpler, less expensive, and more easily transported from one set of HPLC equipment to another. They also place less stringent requirements on the purity of the mobile phase and the function of the solvent pump and mixing systems. However, they also have many drawbacks.

When used with complex samples containing many compounds that have widely differing retention behaviors, they typically exhibit the general elution problem[14]. Early-eluting peaks in the chromatogram are easily seen, but are not well separated.

Table 1. Reproducibility of On-Line PTH Analysis*

Amino Acid	Retention Time				Peak Height			
	Mean (min)	Low (min)	High (min)	RSD (%)	Mean (mV)	Low (mV)	High (mV)	RSD (%)
Aspartic Acid	6.08	6.05	6.12	0.30	8.22	7.40	8.80	4.22
Asparagine	6.55	6.52	6.60	0.34	7.97	7.65	8.20	1.66
Serine	7.41	7.38	7.45	0.24	4.76	4.58	4.92	1.88
Glutamine	7.84	7.80	7.90	0.26	7.19	6.91	7.39	1.67
Threonine	8.25	8.23	8.30	0.25	6.13	5.83	6.39	2.27
Glycine	8.60	8.58	8.65	0.25	5.86	5.56	6.08	1.93
Glutamic Acid	9.29	9.25	9.35	0.25	7.96	7.51	8.32	2.19
DMPTU	9.74	9.70	9.80	0.29	2.67	2.51	2.82	2.42
Alanine	11.87	11.83	11.93	0.23	6.37	6.03	6.58	2.14
Histidine	12.93	12.88	13.00	0.32	3.77	3.59	3.87	1.75
Tyrosine	14.90	14.85	14.95	0.13	7.36	6.93	7.67	2.25
Arginine	15.28	15.20	15.38	0.32	4.35	4.10	4.50	1.95
Proline	18.24	18.17	18.30	0.15	7.42	7.09	7.63	1.46
Methionine	18.85	18.80	18.90	0.12	7.66	7.19	7.97	2.35
Valine	19.25	19.20	19.30	0.11	8.34	7.89	8.58	1.79
DPTU	20.64	20.57	20.67	0.10	5.46	5.09	5.73	2.97
Tryptophan	21.96	21.88	22.00	0.10	9.09	8.65	9.36	1.88
Phenylalanine	22.84	22.75	22.88	0.10	8.11	7.73	8.33	1.79
Isoleucine	23.30	23.20	23.35	0.11	7.31	6.96	7.52	1.87
Lysine	23.65	23.55	23.68	0.11	9.99	9.58	10.40	2.38
Leucine	23.98	23.88	24.02	0.11	8.18	7.80	8.40	1.68
Mean				0.20				2.12

* Data from 38 consecutive analyses performed with on-line transfer from Applied Biosystems Model 470A Gas Phase Protein/Peptide Sequencer to Applied Biosystems Model 120A PTH Analyzer. Injection volumes were 50 microliters (40% of total transferred). Injections nominally contained 30 pmol of each PTH amino acid.

Later-eluting peaks are separated well, but they come off the column too slowly and are difficult to detect because of excessive band broadening. Pre-column dead volume, either in the injector, the tubing between the injector and the column, or the column inlet, tends to broaden peak elution volumes in isocratic elution, thereby decreasing both resolution and sensitivity. Use of larger sample volumes also tends to broaden peak elution volumes, particularly with smaller bore columns. The necessity of using smaller sample volumes to obtain good resolution may limit the effective sensitivity of samples that require larger volumes for proper dissolution and transfer onto the column. Finally, highly retained compounds injected with one sample may elute as "ghost" peaks after injection of a second or third sample and interfere with peak identification in those samples.

Gradient elution systems are more complex and expensive, but they provide more uniform sensitivity as well as flexibility in adjusting the parameters necessary for good resolution. They also permit injection of larger sample volumes without peak broadening of early-eluting components, and pre-column void volumes are relatively insignificant. As a result, they are well suited for use with smaller bore columns. Mobile phase purity is crucial to good results, since impurities can be concentrated on the column under starting solvent conditions and elute as artifact peaks or sharply rising baselines during gradient development.

Conventional versus On-line Analysis

With conventional HPLC analysis, the PTH samples are left in the sequencer either in solution or as a dried residue in fraction collector vials. Samples must be removed from the collector and manually prepared for analysis. Virtually any type of HPLC system can then be used, although a good HPLC autosampler is essential for reproducible results. One of the primary drawbacks of this technique is the significant delay between the Edman chemistry and obtaining the analytical results. This delay reduces the efficiency of the sequencer by as much as half. Moreover, the manual sample preparation often causes contamination, sample loss, and PTH degradation - problems that are particularly damaging to microsequencing results.

With on-line HPLC analysis, as each PTH sample is produced by the sequencer, a portion is automatically transferred into the HPLC system and analyzed. This provides rapid acquisition of results with minimal user intervention and none of the sample workup problems inherent in conventional PTH analysis. It requires an interface between the sequencer and the HPLC system that provides efficient and reproducible sample transfer and sampling as well as operational synchronization of the two units.

GRADIENT SEPARATION PROTOCOL USING PTH-C18 COLUMNS

The following protocol employs a gradient elution system that is suitable for PTH analysis using either a conventional or on-line HPLC unit. It has been optimized to provide >95% separation of the common PTH amino acids and the three primary artifacts from the gas phase sequencer. It uses the MPLC[R] concept of interchangeable cartridges and a reusable holder that allows easy replacement of the analytical cartridge with finger-tightened connections.

Each PTH-C18 cartridge is slurry packed with a porous, 5-micron, octyldecylsilyl-type sorbent specially selected for PTH analysis. The cartridges are 22 cm in length and are available with 4.6-mm ID (standard analytical bore) or 2.1-mm ID (narrow bore). The integral frits sealed in Tefzel[R] at each end of cartridge retain the sorbent and filter solids from the sample. The cartridges are shipped containing the mobile phase (40% aqueous acetonitrile) with which the columns were tested.

Column Installation and Maintenance

Use clean solvents and only those which are compatible with the column packing material and type 316 stainless steel. <u>Halogen acids and salts can corrode stainless steel</u>. The acceptable pH range is 2.2 to 8.0. All solvents should be free of particulate matter which might plug the 2-micron entrance frit in the cartridge. An in-line filter placed just before the injection system is also highly recommended.

The cleanliness of the injected sample greatly affects the column life. Samples containing components which are not eluted during the analysis cause a loss in efficiency and increase column back pressure. Column life can be extended by preparing samples free from retained components or by injecting smaller quantities. If performance loss due to retained sample components is suspected, reversing the column flow and pumping pure acetonitrile through the column may help restore performance.

High back pressure may be an indication of plugged frits. A plugged column can often be unplugged by reversing the direction of flow through the column (the column outlet should be temporary disconnected from the detector cell to prevent particulates from lodging there).

To install the column holder, connect the female compression fittings on each of the holder end nuts to the HPLC unit using the male fittings provided with the holder and 1/16-inch O.D. tubing. To install a cartridge, unscrew the holder end nuts from the cartridge jacket and carefully insert the cartridge. The preferred flow through the cartridge is from left to right as you read the label, but the design is symmetrical and the column may be backflushed. Reseal the holder to finger tightness using both end nuts. <u>Never use tools to tighten the end nuts to the jacket!</u> Once the cartridge is installed, leave it in the 55°C column heater or oven until temperature equilibrium is reached. Then, retighten the column holder end nuts <u>before flowing liquid through the column</u>. Loose end nuts can result in leakage of the column packing from the outlet end of the column. Leaked packing material can plug the column outlet line and/or the detector flow cell.

If the cartridge leaks at less than its rated pressure limit (7000 psi), do not try to force a seal with tools. A leak indicates that the sealing surface is either dirty, scratched, or deformed. Replace the cartridge with another and check again for leaks. If the leak persists, the high pressure seals inside the column holder end nuts must be replaced. Use the Snap Ring Seal Replacement Kit (Brownlee Part Number 140-260) for seal replacement.

To remove a cartridge, unscrew both holder end nuts. Then, using your thumb and forefinger together, pull the cartridge out. If you remove the cartridge from the holder and intend to reuse it later, store it with its ends wrapped with Parafilm[R] or similar material to protect the Tefzel frit assemblies from dirt and scratches.

Column Flushing

The column must be purged of the packing, testing, and shipping solvents before use for PTH analysis. This flush procedure will shorten the time required for optimizing the PTH separation and will significantly improve the peak shape of the charged PTH amino acids - Asp, Glu, His, and Arg. This purge procedure should always be performed when installing a new column.

Use only HPLC-quality water, tetrahydrofuran (THF), acetonitrile, sodium acetate, and acetic acid and sequencer grade trifluoroacetic acid (TFA). The HPLC grade acetonitrile should have a UV cutoff value of 188 nm or less. THF should be free of peroxides that destroy low levels of PTH amino acids and other UV-absorbing impurities that give high baselines. A 5% aqueous THF solution is more stable than neat THF.

All glassware used with the solutions must be scrupulously clean! Rinse it thoroughly with pure water or solvent, as appropriate, before using. Common laboratory detergents tend to leave deposits on the glassware that result in high gradient baselines or specific peaks in the chromatogram. Since this residue is difficult to remove by simple rinsing, you should not generally have HPLC glassware cleaned with detergent. Instead, have glassware dedicated only for HPLC use, rinse the glassware with water or solvent after use, and cover with aluminum foil for storage.

Start the purging process with 0.25% TFA (made by diluting Applied Biosystems Sequencer Reagent R3 or R4A with water) as solvent A and acetonitrile as solvent B. Set the HPLC to deliver 50% A/50% B at 1.5 mL/min (4.6-mm ID cartridge) or 0.4 mL/min (2.1-mm ID cartridge) for two hours. Discard the TFA solution, rinse the container with water, fill the container with water, and continue the purge for 20 minutes at 50% B. Discard the water, rinse the container with fresh water, fill the container with water, and continue the purge for another 20 min at 50% B. The system is now ready to install the separation solvents for PTH analysis.

Separation Solvents

A solvent kit for PTH separation should include two solvents (5% aqueous tetrahydrofuran and acetonitrile), two 3-$\underline{M}$ sodium acetate buffer concentrates (pH 3.8 and pH 4.6), and an oxidant scavenger (N,N-dimethyl-N'-phenylthiourea, DMPTU, available in 500-nmol vials, Applied Biosystems Part Number 400349). These reagents are used to form the mobile phases required for the gradient elution of PTHs from the analytical column. The solvents should be stored in a cool, dark place. The buffers should be stored at 4°C; and the DMPTU should be stored at -20°C.

A standard PTH mixture must include the PTHs and common sequencer-derived contaminants. The Applied Biosystems PTH standard (Part Number 400316) includes 19 PTH amino acids (no cysteine derivative), N,N'-diphenylthiourea (DPTU), dithiothreitol (DTT), and DMPTU. Vials containing dried PTH residues or stock solutions in acetonitrile should be stored at -20°C, and protected from contamination with water.

Separation Optimization

    Because of the wide variety of HPLC equipment, it is not
possible to provide a specific separation protocol for every
system. Therefore, start with the following procedure and
optimize it according to the accompanying stepwise procedures
described below until you have obtained the correct separation.
This optimization may include changes in the %B gradient,
buffer pH, buffer concentration, and column temperature.
Typical PTH separations obtained at Applied Biosystems are
shown in Figure 1.

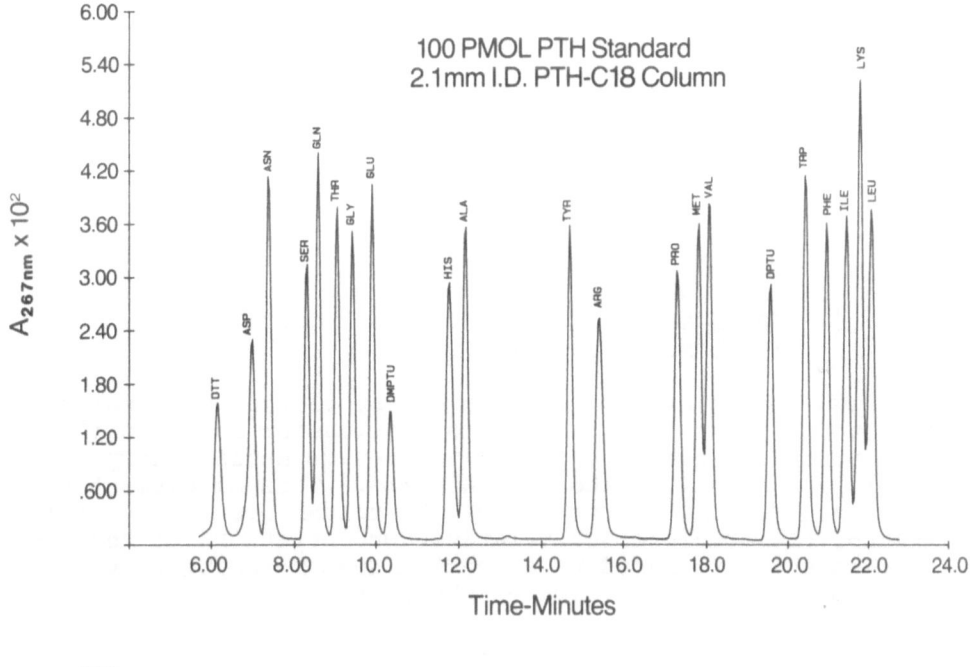

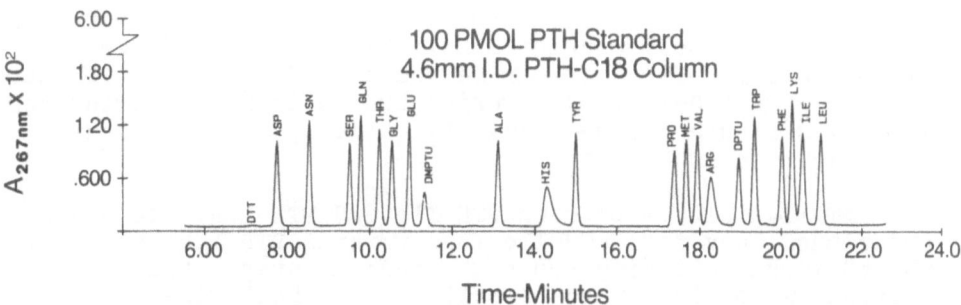

Fig. 1. Separation of PTH amino acids on PTH-C18 columns.

<u>Nominal starting solvent compositions are</u>:

Solvent A (per liter):   5% aqueous THF
                                   30 mL pH 3.8 buffer
                                   7 mL pH 4.6 buffer

Solvent A (per liter):   acetonitrile
                                   500 nanomoles DMPTU

<u>Nominal HPLC parameters are</u>:

Flow Rate: 1.0 mL/min (4.6-mm ID column)
                     0.2 mL/min (2.1-mm ID column)

Temperature: 55°C.

Detector Wavelength: 270 nm (with variable wavelength
                     detector)
                     254 nm (with fixed wavelength
                     detector)

Nominal gradient for a 4.6-mm ID column is as follows (all segments are programmed as linear gradient changes):

8% B at time 0 min
19% B at time 2 min   (8 to 19% B from 0 to 2 min)
51% B at time 16 min
51% B at time 19 min
8% B at time 19.1 min

Nominal gradient for a 2.1-mm ID column is:

10% B at time 0 min
14% B at time 2 min
40% B at time 20 min
60% B at time 25 min
10% B at time 25.1 min

Two stages of optimization of the elution protocol are required. First, the gradient of increasing acetonitrile concentration must be adjusted to provide satisfactory resolution of all of the neutral PTHs. This adjustment may also include some adjustment of the column oven temperature. Second, the absolute and relative concentrations of the two buffers that are added to solvent A must be adjusted to position the elution times of the charged PTHs.

Only two of the neutral PTH amino acids, PTH-Gln and PTH-Lys, are likely to require repositioning because of overlap with other, closely-eluting PTHs. PTH-Gln elutes between PTH-Ser and PTH-Thr. It can be positioned away from PTH-Ser (and towards PTH-Thr) by lowering the %B increase during the initial 2 minute of the gradient. Raising the %B increase moves it towards PTH-Ser.

PTH-Lys elutes between PTH-Ile and PTH-Leu. It can be positioned away from PTH-Ile (and towards PTH-Leu) by lowering the %B increase during the next gradient step. Raising the %B increase moves it towards PTH-Ile. With both PTH-Gln and PTH-Lys, small changes (a few %B) should be sufficient to provide the required separation, although HPLC units with very

372

large or very small mixer dead volumes may require larger changes.

The effect of acetonitrile concentration on the relative elution positions of all the PTHs is illustrated in Figure 2. The isocratic elution profiles shown in this figure can be used to estimate the acetonitrile gradient changes that might be required to position any of the neutral PTHs relative to the others.

Once all of the neutral PTHs are separated from each other, the positions of PTH-Asp and PTH-Glu should be adjusted. Their positions are determined primarily by the pH of the Solvent A buffer. Lowering the pH causes both PTHs to elute later; raising the pH causes both PTHs to elute earlier. PTH-Asp should be positioned just before PTH-Asn and well after oxidized DTT which will be present in both sequencer samples and the PTH standard. If there is insufficient spacing between DTT and PTH-Asn for PTH-Asp, then either decrease the column oven temperature by a few degrees or reduce the %B at time 0.0 of the gradient to provide the required spacing. These latter parameters may also affect the positioning of PTH-Gln and PTH-Lys.

Finally, position PTH-His and PTH-Arg by adjusting the Solvent A buffer concentration. Increasing the buffer concentration causes both PTHs to elute earlier; decreasing it causes both to elute later. They should be positioned somewhere between DMPTU and PTH-Pro so that they do not coelute with PTH-Ala, PTH-Tyr, or the DTT adduct of PTH-Ser (which elutes midway between PTH-Ala and PTH-Tyr). Ideal positions are just after PTH-Ala for PTH-His and just after PTH-Tyr for PTH-Arg.

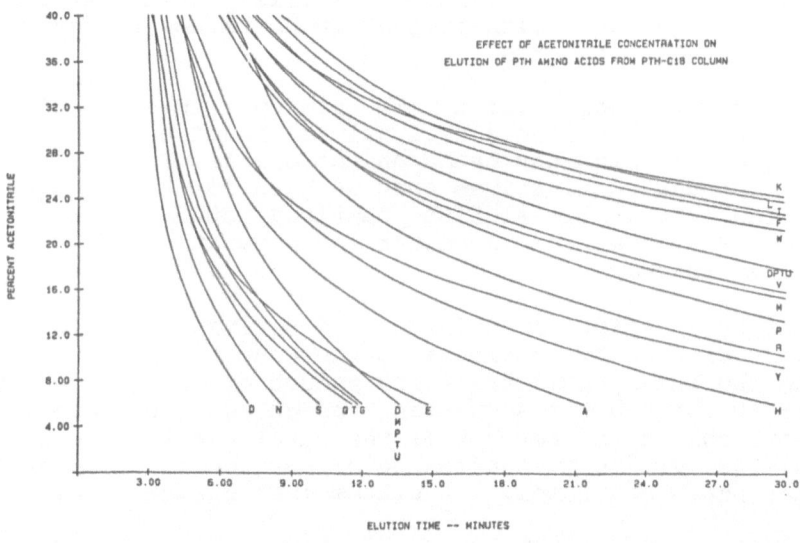

Fig. 2.   Isocratic elution profiles of PTH amino acids from PTH-C18 column.

by adding DMPTU to the Solvent B reservoir at 500 nmol DMPTU/L of Solvent B. It serves as a scavenger for elements on the column packing surface or in the mobile phases that might otherwise destroy the PTHs. This addition to Solvent B will cause a very slight increase in the baseline absorbance level at the end of the gradient, about 0.001 AU if the PTH elution is being monitored at 270 nm (0.002 AU at 254 nm).

AMINO ACID ASSIGNMENT

Figure 4 shows typical chromatographic data from the first portion of a sequence analysis of a small protein. The data was obtained using on-line PTH analysis in conjunction with the gas phase protein sequencer. The assignment of amino acid sequence at each cycle in data such as this must be made in relation to the background level of amino acids present in all cycles, and to the level of any carryover from the preceding cycles due to incomplete degradations at those cycles. The simplest qualitative method for making this assignment is to overlay chromatograms from succeeding cycles on a light box and compare the increase and decrease in the heights of specific peaks. The carryover of signal in cycles immediately after that in which a signal increase above background occurs can be used to confirm the assignment of the amino acid residue at that cycle.

Quantitation can be made either by peak height or peak area measurements. Manual peak height analysis is simple and requires no on-line computer system. It is probably more accurate than peak area analysis when measuring amounts of PTH amino acids near the limits of the HPLC system's sensitivity. It is, however, quite time consuming. Peak area analysis generally requires an automated integrator that will increase the cost of the HPLC system. However, some of the more sophisticated data systems greatly simplify and speed calculations, particularly those such as baseline subtraction, correction of injection volume variability using internal standards, and conversion of peak areas to molar quantities using external standards.

Many amino acids yield more than one PTH derivative during sequencing. All of these derivatives should be used to confirm the sequence assignment based on the primary PTH derivative. Those secondary derivatives that we have identified are described below. The elution positions noted are for gradient elution using the PTH-C18 column and separation protocol described above.

Aspartic Acid. An unidentified derivative forms upon exposure of aspartyl residues to the coupling reagent and base during the Edman chemistry. The amount of this derivative increases through the sequencing run at the expense of PTH-Asp. However, only a few percent of the total aspartic acid appears as this derivative even after 40 or more degradation cycles in the gas phase sequencer. It elutes just before DPTU.

Asparagine. About 10% of PTH-Asn is degraded by deamination to yield PTH-Asp in the conversion flask under typical conversion conditions. Additional PTH-Asp can result

Sample Considerations

The recommended solvent for sample injections is 10 to 20% aqueous acetonitrile. If higher solvent concentrations are used, especially with large injection volumes on the 2.1-mm ID column, early-eluting PTHs may show peak broadening and poor resolution. Some PTHs, notably PTH-Ser and PTH-Thr, are unstable in aqueous acetonitrile. Standards for manual injection should be made fresh daily by diluting with water a stock solution made up in neat acetonitrile containing 0.001% DTT. Standards stored in the sequencer for on-line analysis should be dissolved in neat acetonitrile containing 0.001% DTT.

Dissolution of the PTHs in the solvents is not instantaneous. As long as 20 to 30 minutes may be required. Failure to wait may result in apparent low recoveries of the PTHs during the HPLC analysis. Mixing the vial contents by vortexing periodically during this time can aid in dissolution.

When the total amount of all the PTHs in a mixture injected for analysis is less than a few hundred picomoles, the recovery of several of the PTHs from the column may be low (Figure 3). This loss of sample on the column can be minimized

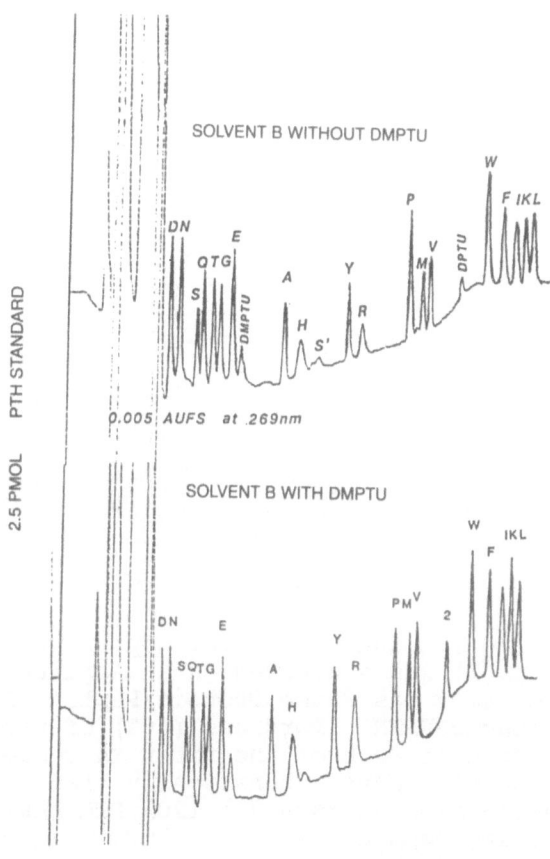

Fig. 3.   Effect of scavenger on recovery of pmol levels of PTH amino acids.

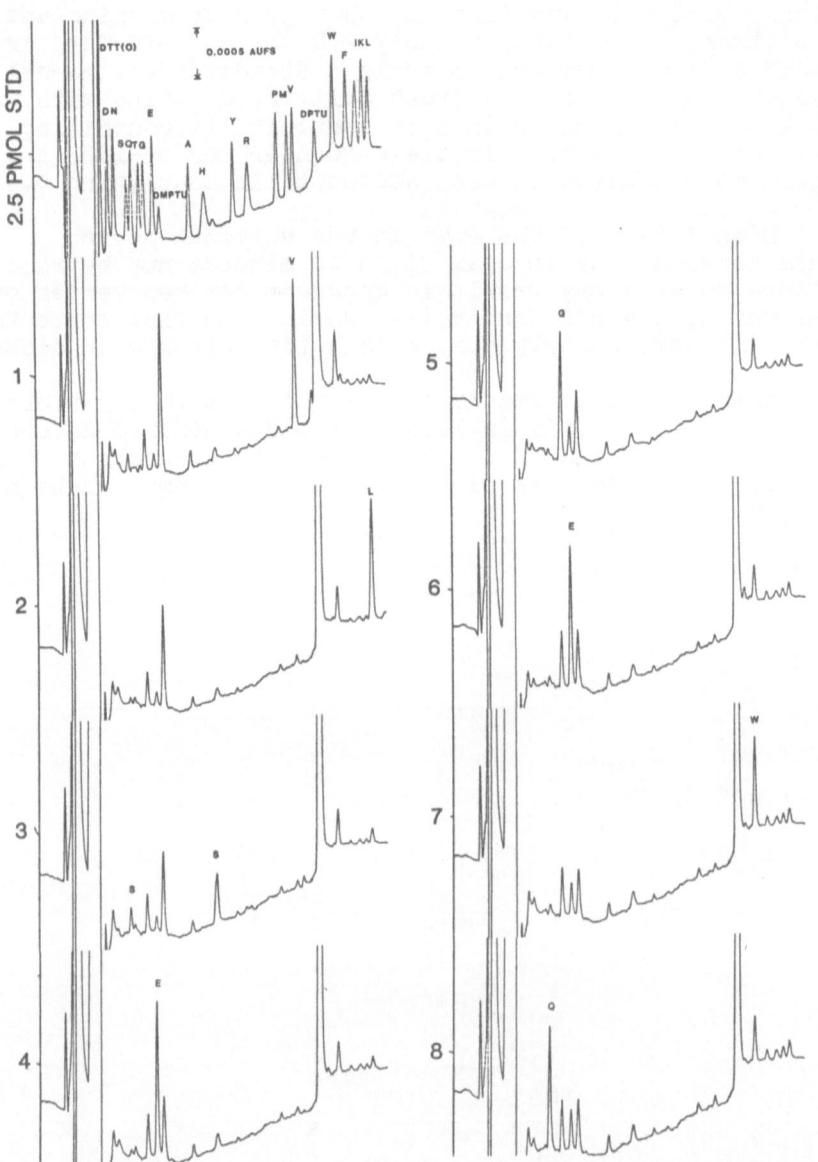

Fig. 4. On-line PTH analysis data. Sperm whale apomyoglobin (20 pmol) was sequenced on an Applied Biosystems Model 470A Gas Phase Protein/Peptide Sequencer using Program 03RPTH. Portions (40%) of the PTH solution produced at each of the first 24 cycles of the degradation were automatically transferred into an Applied Biosystems Model 120A PTH Analyzer and chromatographed.

from deamination of asparaginyl residues during purification or handling of the protein prior to sequencing.

Asparaginyl residues N-linked to complex carbohydrate moieties, produce ATZ derivatives that are insoluble in Applied Biosystems Sequencer Solvent S3 (1-chlorobutane). However, removal of all but the directly linked N-acetylgalactosamine (AGA), by treatment of the protein or peptide prior to sequencing with endoglycosidase H, results in formation of an S3-soluble ATZ. PTH-AGAAsn elutes between oxidized DTT and PTH-Asp[15].

Serine. The seryl hydroxyl group is esterified by trifluoroacetic acid during the cleavage reaction of the Edman chemistry. During subsequent conversion of the ATZ, loss of the trifluroacetyl group gives PTH-dehydroalanine. This derivative has frequently been used for identification of serine in Edman sequencing, although it is very reactive and unstable. It elutes near PTH-Tyr and can be monitored by its absorbance at 313 nm.

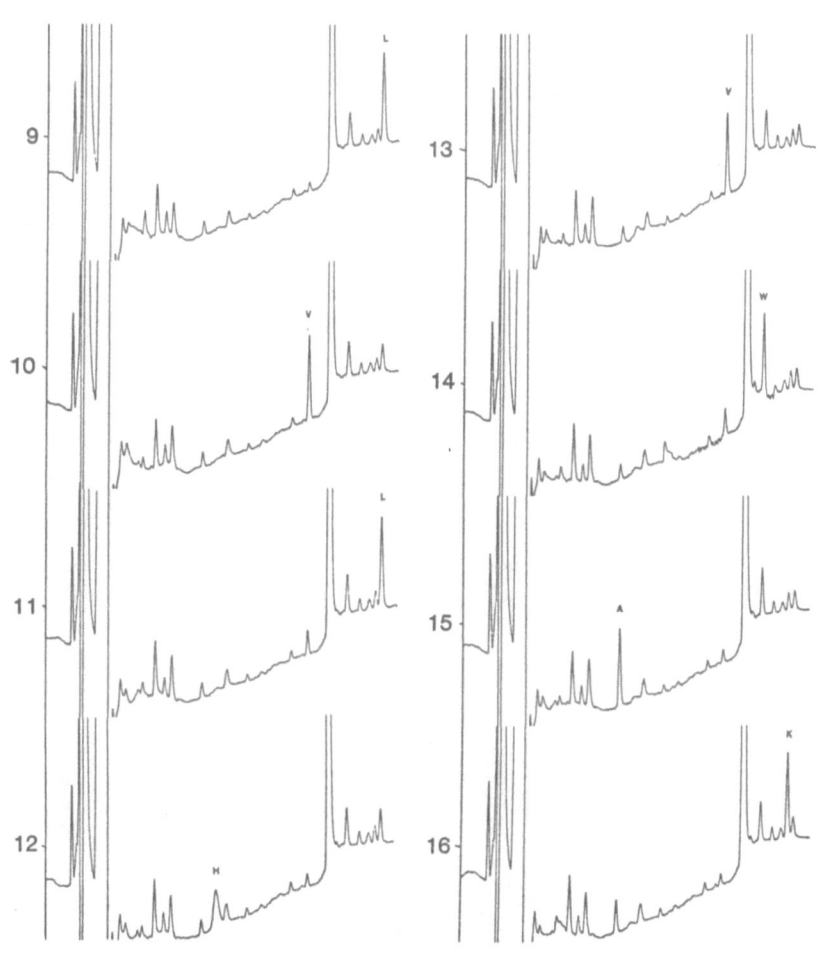

Fig. 4. (continued)

The standard gas phase sequencer programs are designed to trap and stabilize the dehydro product with DTT delivered to the conversion flask just before transfer of the ATZ from the cartridge. The DTT-trapped derivative elutes midway between PTH-Ala and PTH-Tyr and can be monitored by its absorbance at 254 to 270 nm. Recoveries of this derivative are typically 40-60%. Some authentic PTH-Ser, usually 10-20%, is also recovered.

Glutamine. About 10% of PTH-Gln is degraded by deamination to yield PTH-Glu in the conversion flask under typical conditions. Additional PTH-Glu can result from deamination of glutaminyl residues during purification or handling of the protein prior to sequencing.

Threonine. The threonyl hydroxyl group is esterified by trifluoroacetic acid during the cleavage reaction of the Edman chemistry. During subsequent conversion of the ATZ, loss of the trifluroacetyl group gives

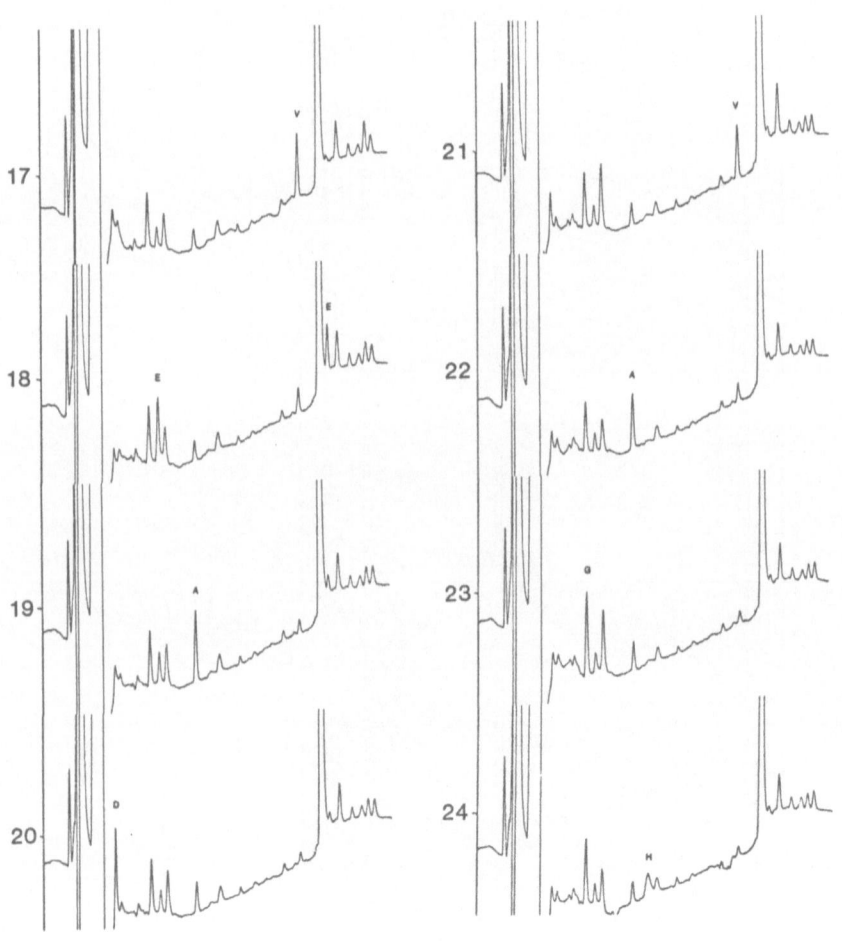

Fig. 4.   (continued)

PTH-dehydro-alpha-aminoisobutyric acid. This derivative has frequently been used for identification of threonine in Edman sequencing, although it is only moderately stable. It elutes near PTH-Pro and can be monitored by its absorbance at 313 nm.

The standard gas phase sequencer programs are designed to trap and stabilize some of the dehydro product with DTT delivered to the conversion flask just before transfer of the ATZ from the cartridge. The DTT-trapped derivative elutes as two to four peaks midway between PTH-Tyr and PTH-Pro and can be monitored by their absorbance at 254 to 270 nm. Recoveries of these derivatives are typically 5% each. Some authentic PTH-Thr, typically 20-30%, is also recovered.

Glycine. ATZ-Gly converts to PTH-Gly somewhat slowly, the reaction being only 80-85% completed during the standard conversion conditions in the gas phase sequencer. The remaining 15-20% is observed as phenylthiocarbamylglycine (PTC-Gly), which elutes near the end of the solvent front.

Glutamic Acid. An unidentified derivative forms upon exposure of glutamyl residues to the coupling reagent and base during the Edman chemistry. The amount of this derivative increases through the sequencing run at the expense of PTH-Glu. After 40 or more degradation cycles in the gas phase sequencer, a substantial portion of the glutamic acid is represented by this compound. It elutes midway between DPTU and PTH-Trp.

Proline. Hydroxyproline (HYDPro), if present in the protein or peptide, produces PTH-HYDPro that elutes as two peaks, one just before and one just after PTH-Ala.

Tryptophan. An unidentified derivative of tryptophan that elutes midway between DPTU and PTH-Trp can be observed in many samples of tryptophan-containing proteins and peptides. Presumably, it is the result of modification of the tryptophanyl indole ring during sample purification or handling prior to sequencing. It may account for 0 to 100% of the tryptophan signal, but it is generally only stable enough to be seen with on-line PTH analysis.

Lysine. Hydroxylysine (HYDLys), if present in the protein or peptide, produces PTH-HYDLys that elutes just after PTH-Val. Methyllysine (METLys), if present, produces PTH-METLys that elutes just after PTH-Leu, with its exact position being sensitive to the chromatography buffer concentration. Succinyllysine (SUCLys), if present, produces PTH-SUCLys that elutes midway between DMPTU and PTH-Ala, with its exact position being sensitive to the chromatography buffer pH.

Cysteine. Authentic PTH-Cys is not usually recovered in sufficient yield to be seen. PTH-dehydroalanine, generated by loss of H_2S from the side chain, can be observed directly by monitoring at 313 nm or indirectly as its DTT derivative by monitoring at 254 to 270 nm, although the recovery of this compound is generally less with cysteine than with serine (see discussion of serine above).

Cysteine is easily identified after modification of its side chain to give a form more stable to the Edman chemistry, and a variety of modification methods have been used.

Alkylation with 4-vinylpyridine to give
S-beta-(4-pyridylethyl)cysteine (PECCys) is the ideal method.[15]
PTH-PECCys, which is positively charged at pH 4, can be
positioned to elute midway between PTH-Val and DPTU by
adjusting the chromatography buffer concentration.

Alkylation of the protein with iodoacetic acid gives
S-carboxymethylcysteine (CMCys). PTH-CMCys elutes near PTH-Ser
and PTH-Gln, with the exact position being sensitive to the
chromatography buffer pH.

Alkylation of the protein with iodoacetamide gives
S-carboxamidomethylcysteine (CAMCys). PTH-CAMCys elutes just
before DMPTU. About 50% of PTH-CAMCys is degraded by
deamination to yield PTH-CMCys in the conversion flask under
typical conversion conditions. Additional PTH-CMCys can result
from deamination of CM-cysteinyl residues during purification
or handling of the protein prior to sequencing.

Oxidation of the protein with performic acid gives cysteic
acid (CysA). PTH-CysA elutes near the end of the solvent
front.

REFERENCES

1. P. Edman, Method for determination of the amino acid
   sequence in peptides, Acta Chem. Scand. 4:283 (1950).
2. P. Edman and G. Begg, A protein sequenator, Eur. J.
   Biochem. 1:80 (1967).
3. B. Wittmann-Liebold, H. Graffunder, and H. Kohls, Amino
   acid sequence studies on ten ribosomal proteins of
   Escherichia coli with an improved sequenator equipped with
   an automatic conversion device, Hoppe-Seyler's Z. Physiol.
   Chem. 354:1415 (1973).
4. R. M. Hewick, M. W. Hunkapiller, L. E. Hood, and W. J.
   Dreyer, A gas-liquid-solid phase peptide and protein
   sequenator, J. Biol. Chem. 256:7990 (1981).
5. H. Rodriguez, W. J. Kohr, and R. M. Harkins, Design and
   operation of a completely automated Beckman
   microsequencer, Anal. Biochem. 140:538 (1984).
6. W. Machleidt and H. Hofner, Fully automated solid phase
   sequencing with on-line identification of PTHs by high
   pressure liquid chromatography, in: "Methods in Peptide
   and Protein Sequence Analysis," C. Birr, ed.,
   Elsevier/North-Holland Biomedical Press, Amsterdam,
   (1980).
7. C. L. Zimmerman, E. Appella, and J. J. Pisano, Rapid
   analysis of amino acid phenylthiohydantoins by
   high-performance liquid chromatography, Anal. Biochem.
   77:569 (1977).
8. N. D. Johnson, M. W. Hunkapiller, and L. E. Hood, Analysis
   of phenylthiohydantoin amino acids by high-performance
   liquid chromatography on DuPont Zorbax cyanopropylsilane
   columns, Anal. Biochem. 100:335 (1979).
9. B. Wittmann-Liebold, An evaluation of the current status
   of protein sequencing, in: "Methods in Protein Sequence
   Analysis," M. Elzinga, ed., Humana Press, Clifton, New
   Jersey (1982).

10. L. E. Henderson, T. D. Copeland, and S. Oroszlan, Separation of amino acid phenylthiohydantoins by high-performance liquid chromatography on phenylalkyl support, Anal. Biochem. 102:1 (1980).

11. R. L. Cunico, R. Simpson, L. Correia, and C. T. Wehr, High-sensitivity phenylthiohydantoin amino acid analysis using conventional and microbore chromatography, J. Chromatog. 336:105 (1984).

12. F. Lottspeich, Identification of the phenylthiohydantoin derivatives of amino acids by high pressure liquid chromatography, using a ternary, isocratic solvent system, Hoppe-Seyler's Z. Physiol. Chem. 361:1829 (1980).

13. T. E. Tarr, Rapid separation of amino acid phenylthiohydantoins by isocratic high-performance liquid chromatography, Anal. Biochem. 111:27 (1981).

14. J. L. Glach, Gradient elution in liquid chromatography, LC Magazine 2:746 (1984).

15. R. J. Paxton, and J. E. Shively, Structural analysis of carcinoembryonic antigen (CEA) and a related tumor associated antigen (TEX), in: "Proceedings of Symposium of American Protein Chemists," J. L'Italien, ed., Plenum, New York, N. Y. (in press).

16. C. S. Fullmer, Identification of cysteine-containing peptides in protein digests by high-performance liquid chromatography, Anal. Biochem. 142:336 (1984).

AN ON-LINE ISOCRATIC HPLC SYSTEM FOR THE ANALYSIS OF PTH-AMINO ACIDS ON A GAS-PHASE SEQUENCER

J. E. Shively [*], D. Hawke [*], R. M. Kutny [&], B. Krieger [&], and J. L. Glajch [&]

[*]Division of Immunology, Beckman Research Institute of the City of Hope, Duarte, CA 91010

[&]Central Research and Development Department, E. I. Du Pont De Nemours & Co., Experimental Station, Wilmington, DE 19898

INTRODUCTION

Since the description of a gas phase sequencer in 1981 by Hewick et al.[1], the protein sequencing field has increasingly used this methodology for routine sequence analysis of peptides and proteins in the low nanomole to low picomole range. The commercially available instrument from Applied Biosystems has greatly simplified automated Edman chemistry, including the conversion of anilinothiazolinone (ATZ) to phenythiohydantoin (PTH) derivatives of amino acids. The analysis of the PTH derivatives is routinely performed by high performance liquid chromatography (HPLC) using various reverse phase columns. Investigators have utilized either gradient[2-4] or isocratic[5-6] modes for the separation of the 20 common PTH derivatives. Routinely, the sample collected from the sequencer is dried in a Speed-Vac, reconstituted in a small volume (10-50 uL of acetonitrile containing an internal standard), and injected onto an HPLC. Investigators may inject 10-100% of the sample, depending on the required sensitivity of the analysis. If, as is sometimes the case, the sample must be reanalyzed, it is recommended that no more than 40-60% of the sample be analyzed in a single HPLC run. Most laboratories allow a sequencer run to proceed overnight, workup the 10-15 cycles accumulated at one time, and load these samples onto a HPLC equipped with an autosampler. Since it is often impossible to determine if a sequencer run is successful by examining a single or even two cycles, it may be late morning before it is realized that the evening's run was successful or not. If successful, it may require many more hours of sample workup and HPLC time before it can be determined if the run is "over." Thus we ask, have we reached the end of a peptide, or can we interpet any more cycles on a protein? Because of this delay between sample production by the sequencer and sample analysis by the HPLC, the efficiency of sequence analysis is reduced by at least a factor of two. It is clear that an on-line HPLC which analyzes the cycles as fast as they are produced would greatly increase the efficiency of microsequence analysis. Although such a system is now available from Applied Biosystems, when this work was started in late 1984 no commercial system was available. This report may be of interest for those who wish to consider the parameters involved in the development of an on-line system, or who wish to construct their own.

Major considerations for the development of this on-line system was the need to have a totally reliable separation of the PTH amino acids, and to minimally disturb the instrument in terms of its normal operation. Since the relative positions of PTH amino acids may change or drift in a gradient HPLC system, and a gradient HPLC is inherently more complex to maintain in good operation condition over extended periods, we chose an isocratic separation described by Glajch et al.[7]. This report describes the on-line system and its performance with a protein and peptide standard. We also include here experiments determining the need for precycling polybrene and the first use of an all Teflon cartridge in an Applied Biosystems instrument.

The On-Line System

The HPLC comprised a four solvent Du Pont HPLC, a Du Pont Zorbax PTH column (dimensions: 0.41 x 25 cm), and a Waters Assoc. column heater and Model 440 UV detector. The injector was a pneumatically actuated Rheodyne Model 7126. The line from the sequencer valve block to the fraction collector was replaced with 0.007 in. Teflon tubing, and was rerouted to the injector. The connection to the injector was made by sleeving the stainless steel needle with sixteenth inch Teflon tubing and connecting the two lengths of sixteenth inch tubing with slightly oversized heat shrink Teflon tubing, followed by heating. The vent port of the injector was connected to the fraction collector via 0.007 in. Teflon tubing. The pneumatic actuators of the injector were connected to three-way solenoid valves supplying 40-60 psi of compressed air. The 24 VDC solenoids were controlled by an interface containing 5 and 24 VDC power supplies and a simple electronic driver circuit. The interface was connected to an Apple II plus computer which had the appropriate software to time the "load" and "inject" functions of the injector. The computer communicated to the interface through a VIA board available from John Bell Engineering. The signal taken from the sequencer was "prep S1", a 24 VDC signal piggybacked from a solenoid on the back of the instrument. It should be noted that other 24 VDC signals could have been utilized, but we chose this one since it wasn't being used on this instrument (the need to wash with S1-heptane- is not well established for gas phase sequencing).

The requirements for a reproducible delivery of sample from the conversion flask to the injector loop are 1) that a sufficient volume is present to load the loop and leave a margin on either side of the loop, and 2) that no bubbles are entrained in the sample. The size of the conversion flask (about 0.6 mL) and the volume of the conical portion (about 100 µL) further dictate the sample volume. A survey of different sample vs injection volumes rapidly convinced us that an ideal working range would be 90-120 uL of sample and 50 µL injected. These combinations would lead to 42-55% of the sample injected. In order to maximize the extraction of the PTH derivative from the conversion flask, since only a single extraction could be performed, we choose 120 uL as the routine sample volume. The need to inject 50 µL onto an isocratic HPLC system lead to other considerations. If the sample were dissolved in 100% acetonitrile, the maximum injected volume would be 10uL. Beyond that the peaks would be broadened due to the substantial amount of acetonitrile present. We circumvented this problem by including water in the acetonitrile; thus changing S4 to 50% aqueous acetonitrile. An internal standard (PTH aminoisobutyric acid) was added to S4 so that a 50uL injection resulted in 50 pmoles of standard. The flask pressure was adjusted so that a 7-12 sec delivery from the conversion flask to the injecteor loop resulted in a complete filling of the loop with equal amounts of excess liquid on either side. The injector was left in the load position during filling, turned to inject to allow the sample to enter the column, and then returned to the load position. The excess sample together

Table I. The on-line sequencer program.

Step	Cartridge	Flask	Time	Purpose
32	Del S2	Collect	60	Clear injector line
34	Pause	Del S4	14	120uL to flask
40	Prep S1	Pause	2	Start Apple computer
41	Ar Dry	Ar flush	30	Remove bubbles from flask
44	Ar Dry	Collect	8	Fill injector loop (load)
45	Ar Dry	Pause	40	Inject, return to load
46	Ar Dry	Collect	40	Empty loop to collector
47	Ar Dry	Pause	20	Allow droplets to form
48	Ar Dry	Collect	40	Remove droplets
49	Ar dry	Del R5	20	Rinse flask
51	Ar dry	Empty	40	Flask to waste
53	Ar dry	FC adv	2	Collector advance

Only portions of the sequencer program relevant to the on-line system are shown. The step numbers are arbitrary. Time is in seconds.

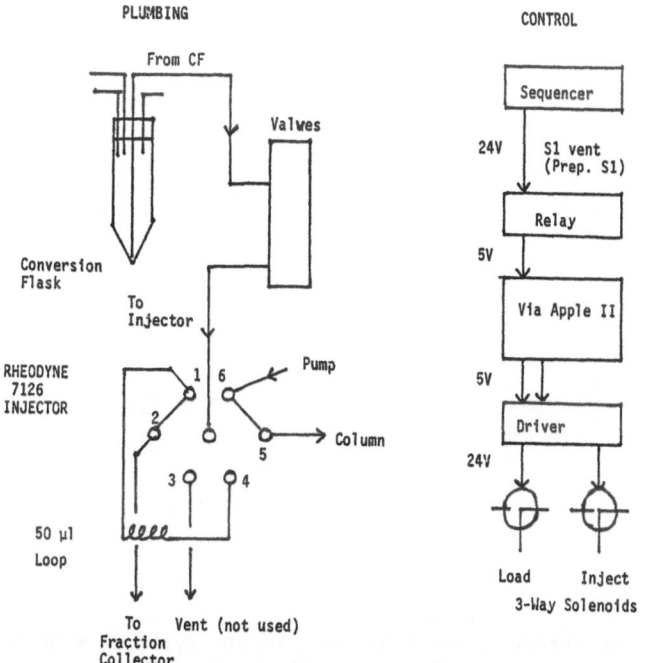

Figure 1. Schematic for on-line system. The schematic shows the plumbing for the connections between the conversions flask and injector, and from the injector to the HPLC and fraction collector. The control schematic shows the DC voltages between the various elements. The two three-way solenoids control the load and inject functions of the injector.

with 50 uL of HPLC buffer is delivered to the fraction collector. A schematic of the on-line system is shown in Figure 1. A summary of the program steps which accomplish the injection and delivery to the fraction collector are shown in Table I. The pause step at 47 reflect the need to wait before clearing the microbore line of all remaining liquid. If this step is omitted, the line may develop droplets which effectively can block the line. The delivery of R5 (acetonitrile) at step 49 is to rinse the conversion flask. Omission of this step results in a slow build-up of internal standard in the conversion flask.

<u>Performance of the On-Line System</u>

Before a sequencer run is initiated, a 50 pmole standard mixture of the 20 PTH amino acid derivatives is injected. The on-line needle is removed and 50 uL of the standard mixture is loaded into the loop. A short "inject" program is initiated which automatically turns the injector first to inject and then to load. At this point the on-line needle is reinserted and the program proceeds to clear the loop with R5 (acetonitrile). At this point the protein or peptide sample can be loaded and the sequencer run started. There is no need to wait for the standard run to finish since no injection will be made for at least the next 40 min of the sequencer cycle. A typical example of a 50 pmole standard mixture injected on-line is shown in Figure 2. In this work the results are recorded on a standard 10 mV recorder (10 in. paper) and the data is collected, processed, and stored on a Hewlett Packard Laboratory Automation System. The data is replotted on an arbitrary scale and normalized to any appropriate value full scale. The data can be accessed by modem to a home personal computer,

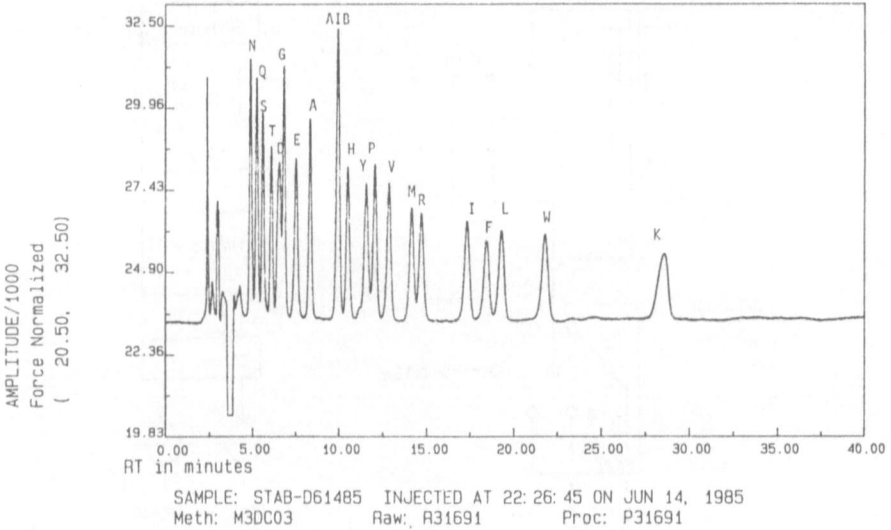

SAMPLE: STAB-D61485   INJECTED AT 22:26:45 ON JUN 14, 1985
Meth: M3DC03   Raw: R31691   Proc: P31691

Figure 2.   A PTH standard run from the on-line system. A 50 pmole standard mixture of the PTH amino acids (excluding Cys) was injected in 50 uL of 50% aqueous acetonitrile through the on-line system. The column is a DuPont Zorbax PTH column (0.46 x 25cm) run at 1.00 mL/min in the isocratic solvent system described by Glajch et al.[7]. The internal standard is PTH aminoisobutyric acid (AIB). The absorbance is arbitrary units at 254 nm using a Waters Assoc. Model 440 detector and recalculated using a Hewlett Packard Lab Automation System.

thus making it possible to evaluate the progress of a run at late hours or on weekends. Because of this convenience, it is possible to tend the sequencer and load consecutive samples earlier than previously possible. The increase in efficiency even without the modem is at least a factor of two.

A tridecapeptide with the following sequence was synthesized in order to rapidly test the performance of the on-line system: Val-Ala-Ser-Glu-Thr-Leu-His-Val-Ala-Leu-Ala-Val-Arg. The peptide is short enough to be sequenced in one day, allows repetitive yield calculations for Val, Ala, and Leu, and contains several of the amino acids which form unstable PTH derivatives (Ser and Thr). In addition, His and Arg are often poorly extracted as their PTH derivatives. Thirteen cycles from a sequencer run of 150 pmoles of the synthetic peptide are shown in Figure 3. The initial yield was about 100% based on amino acid analysis of an equivalent aliquot. The repetative yield from cycles 1 to 8 or 2 to 9 was about 82%. Similar results were obtained on a second instrument which was not configured as an on-line system. The large peak eluting just after 8 min on each chromatogram is the 50pmole internal standard. The size of the DPTU peak (diphenythiourea) is 4 pmoles injected by cycle 3 (the total is 10 pmoles, since only 40% of the sample is injected). The higher amount of DPTU on cycle 1 compared to the succeeding cycles is not unusual. Ser plus the the DTE adduct of dehydro-Ser (Ser*) is observed at cycle 3. Similarly, Thr plus Thr* is observed at cycle 5. The yield of His at cycle 7 is only 4 pmoles corrected for injection volume. This low yield is typical for the ABI instrument. By comparison the yield for Val at cycle 8 is 38 pmoles. The yield of Val at cycle 12 is 32 pmoles at least 10pmoles of which must be subtracted as carryover from the previous cycle. No Arg (indicated by arrow) is detected at cycle 13.

Once we established that the on-line system gave reproducible results that were equivalent to those we obtained by manually working up samples, we were interested in further increasing the efficiency of analysis by decreasing the polybrene precycle time. Since the Applied Biosystems polybrene sample contains large amounts of a peptide, it is necessary to precycle for up to five hours before the background is suitable for sequencing. We avoided this problem by using polybrene purchased from Aldrich. We next evaluated the levels of DPTU, the major background peak, for 3 precycles followed by loading 60 pmoles of sperm whale myoglobin and continuing the analysis for up to 13 cycles. These results, together with other controls such as sequence analysis with no polybrene, are shown in Table II. The lowest background levels for DPTU (precycle 3 or myoglobin cycle 3) were obtained when the disk contained no polybrene. Although it is tempting to avoid the use of polybrene, it should be noted that polybrene is absolutely required to immobolize peptides on the glass fiber disk (ie the peptide will wash out by cycles 2-3). Second, the absence of polybrene had a detrimental effect on the yields of PTH Ser at cycle 3, and PTH His at cycle 12. Although it is generally recommended to use 2 mg of polybrene on a glass fiber disk, we were not aware of studies to determine the optimum amount, nor the effect of amount of polybrene versus background or repetitive yields. The data in Table II suggest that there is very little difference in these parameters over the range of 1-3 mg of polybrene. The amount of polybrene per disk was obtained by directly weighing the dried disk before and after addition of polybrene. The dried disk was again weighed after the completion of the sequencer runs. Within an experimental error of 5%, the disks weighed the same, suggesting that no polybrene is lost in up to 15 cycles of Edman chemistry. Other research by us had suggested the possibility that the interaction of the glass fiber disk with polybrene may contribute to the background observed. In order to evaluate this possibility, we pretreated the disks with strong base or acid before adding polybrene, and in separate experiments, trimethylsilylated the disks after acid or base treatment. The results of these experiments,

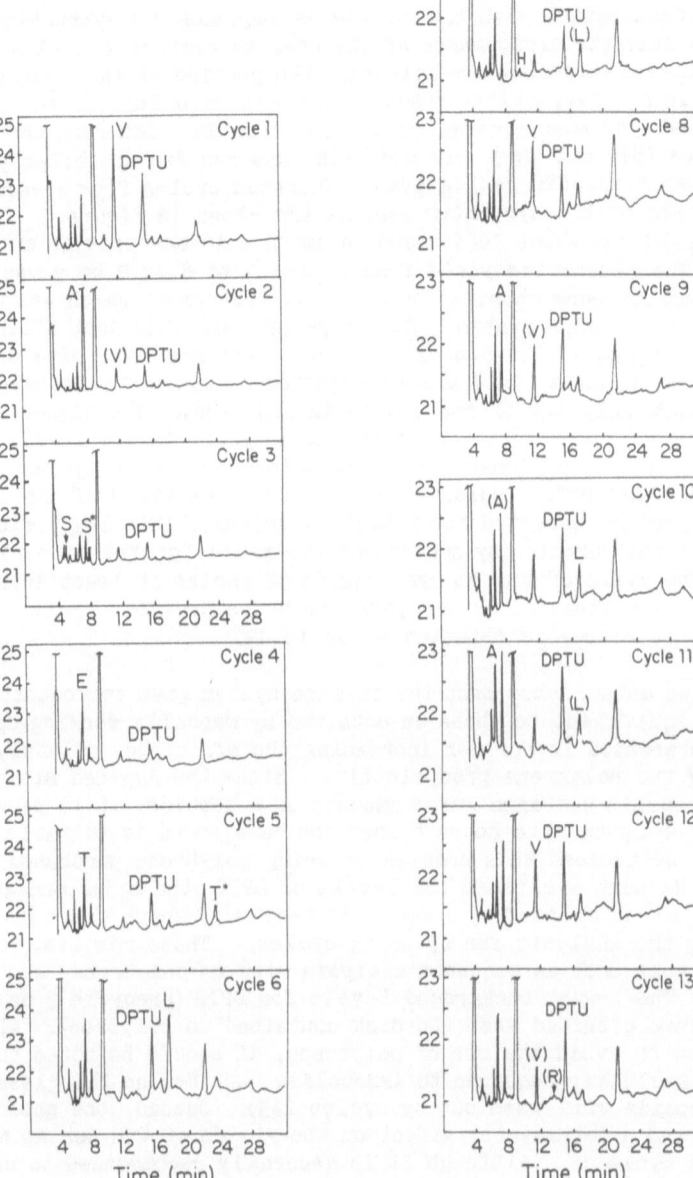

Figure 3. On-line sequence analysis of a synthetic tridecapeptide. The chromatographic conditions are as described in Figure 2. The peptide sequence is Val-Ala-Ser-Glu-Thr-Leu-His-Val-Ala-Leu-Ala-Val-Arg. The amount sequenced is 150 pmoles with an initial yield of 150 pmoles. The offscale peak eluting after 8 min. is the internal standard. The background peak, DPTU, is also indicated. The residue at cycle 13, Arg, is indicated by an arrow, but was not detected.

also shown in Table II, suggest that these treatments may increase initial
and repetitive yields, while at the same time increasing the background
levels of DPTU. The best results were obtained for disks pretreated with
ammonia and TMS. By cycle 3 of the myoglobin run the DPTU level was 16
pmole, an acceptable level considering the improvement in both initial and
repetitive yields.

These findings were consistent with the idea that the state of the
disk was an important parameter in optimizing sequence analysis. Since
these results were obtained for a protein, and it was already established
that polybrene is even more critical for a peptide, we were interested in
trying to find a disk derivative which could replace polybrene. We
reasoned that various silylating agents together with further chemical
modifications may create a polybrene-like mileu for peptide sequencing.
Accordingly, we derivatized disks with quaternary ammonium salts and aminopropylsulfo salts, and compared the results for the synthetic peptide with
polybrene. The results in Table III demonstrate that none of the derivatives were able to retain the peptide. Addition of polybrene back to the

Table II. On-line sequencer runs for myoglobin.

Disk Treatment	Polybrene	Init. Yield	Repet. Yield	DPTU at cycle P3	M3
None	None	43	83	0	3
None	1mg	28	89	4	32
None	2mg	41	91	149	15
None	2mg	32	90	66	25
None	3mg	25	89	71	17
NH_3	2mg	32	93	862	142
NH_3/TMS	2mg	50	92	453	16
HNO_3/TMS	2mg	45	92	834	132

In each case 60 pmoles of sperm whale apomyoglobin was sequenced. The disk
treatment refers to pretreated with dilute ammonia or nitric acid followed
by trimethylsilylation (TMS). The background peak (DPTU) was measured
after three precycles (P3), at which point the sample was applied, and at
the third myoglobin cycle (M3). The percent initial yield was calculated
at cycle one, and the percent repetitive yield from cycle 1 to 9.

Table III. On-line sequencer runs for a synthetic peptide.

Disk Treatment	Polybrene	Init. Yield	Repet. Yield	DPTU at cycle P3	S3
None	2mg	150	82	30	10
Quat. ammonium	none	15	0	21	21
Aminopropyl	none	92	0	25	18
APS/DMS	none	30	0	10	10
APS/propylsulfo	1mg	104	80	42	44

The synthetic peptide Val-Ala-Ser-Glu-Thr-Leu-His-Val-Ala-Leu-Ala-Val-Arg
was sequenced on disks treated under a variety of conditions including
silylation with dimethyloctyldecyl quaternary ammonium (quat. ammonium)
reagent, aminopropyl silylating reagent (APS), or APS followed by
dimethylsulfate (DMS) or 1,3-propanesultone (propylsulfo). The initial
yield are in pmoles, the repetetive yields in percent, and the DPTU at
precycle 3 (P3) or sample cycle 3 (S3) in pmoles. The amount sequenced was
150 pmoles.

aminopropylsulfo derivatized disk restored its peptide retention ability.
It is interesting to note that if five times the amount of peptide is
loaded on these derivatized disks, the peptide was often retained for up to
ten cycles before washout occurred. This point may be of some significance
for the analysis of nmole quantities of peptide, but it is clear that for
microanalysis of peptides, polybrene is a required carrier.

Teflon Cartridge

It was previously noted that the low yields of PTH His and Arg is a
significant problem in microsequence analysis. We and others have noted
that these derivatives tend to give broad peaks on elution from HPLC reverse phase columns. Columns which are highly "endcapped" or buffers which
contain ion-pair reagents partially overcome this problem. These observations suggest that these basic, positively charged derivatives interact
with the silanol groups on the silica supports. We reasoned that the glass
components of the microsequencer were responsible for binding PTH His and
Arg, thus resulting in low extractive yields. The glass components which
contact the ATZ or PTH derivatives include the cartridge, the glass fiber
disk, and the conversion flask. We have previously demonstrated that
better yields of PTH Ser and Thr are obtained with all Teflon cartridge and
conversion flask construction, together with a TMS derivatized glass fiber
disk[8]. In the Applied Biosystems instrument it was convenient to construct
a Teflon cartridge, but difficult to change the conversion flask. Thus our
experiments here were limited to substituting a Teflon cartridge for the
original glass one. A comparison of the two is shown in Figure 4. In
order to center the two halves a corresponding ridge and indentention was
incorporated in the design. The inlet and outlet lines were plumbed with
"pull through" fittings[8]. This feature eliminates the potential leak or
seal degradation inherent in the Teflon flare fit to glass seal required in
the original design. The Teflon surfaces are self sealing and do not
require a Zitex seal[8], but in this work, we maintained the Zitex seal.
The size of the glass fiber disk was reduced from 1.2 cm to 1.1 cm so that

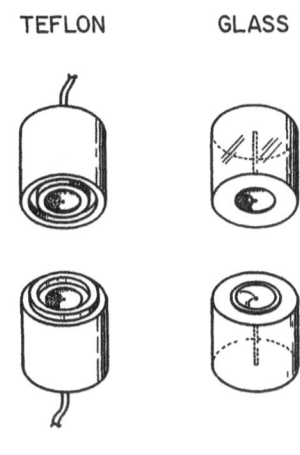

TEFLON   GLASS

2 cm

Figure 4. A Teflon cartridge for the gas phase sequencer. A Teflon
cartridge was constructed to fit in the cartridge holder of an
Applied Biosystems sequencer. A glass cartridge is shown for
comparison. The glass fiber disk size is 1.2 cm for the glass
cartridge and 1.1 cm for the Teflon cartridge. The Teflon
cartridge has a ridge and corresponding indentation to allign
the two halves. Teflon tubing is attached via a "pull through"
fitting on the Teflon cartridge and via a "flare fit " on the
glass cartridge.

the disk could lay flat (the glass cartridge induces a slight bow in the
oversized 1.2 cm disk). With this adjustment the flow properties through
either chamber was similar. A comparison of cycles from the glass versus
the Teflon cartridge for the sequence analysis of 150 pmoles of the synthetic peptide is shown in Figure 5 and in Table IV. Figure 5A demonstrates that the yield of PTH Ala at cycle 2 was similar for both cartridges. The higher DPTU background for the Teflon cartridge will be explained later. Figures 5B and 5C illustrate the increased yields of PTH
His and Arg from the Teflon cartridge. This data is summarized in Table
IV. In this experiment, little improvement in the yield of PTH Ser or Thr
was observed. We surmise that further improvements would be achieved with
the use of an all Teflon conversion flask.

After these experiments were completed, it was noted that the amount
of PITC (R1) delivered to the cartridge was excessive for the Teflon
cartridge. This may indicate that the Teflon cartridge offers less flow
resistance than the glass cartridge. Once the delivery was adjusted, the
background levels of DPTU decreased substantially. It was then noticed
that the excessive PITC delivery had led to the build-up a residue in the
conversion flask. The flask was replaced, resulting in a substantial improvement in the baseline and "breakthrough" area of the chromatographs.
This result is shown in Figure 6. At this point we conclude that it is
quite easy to replace the glass cartridge with a Teflon one, and that
sequencing results are improved provided the appropriate changes in disk
size and PITC delivery are made.

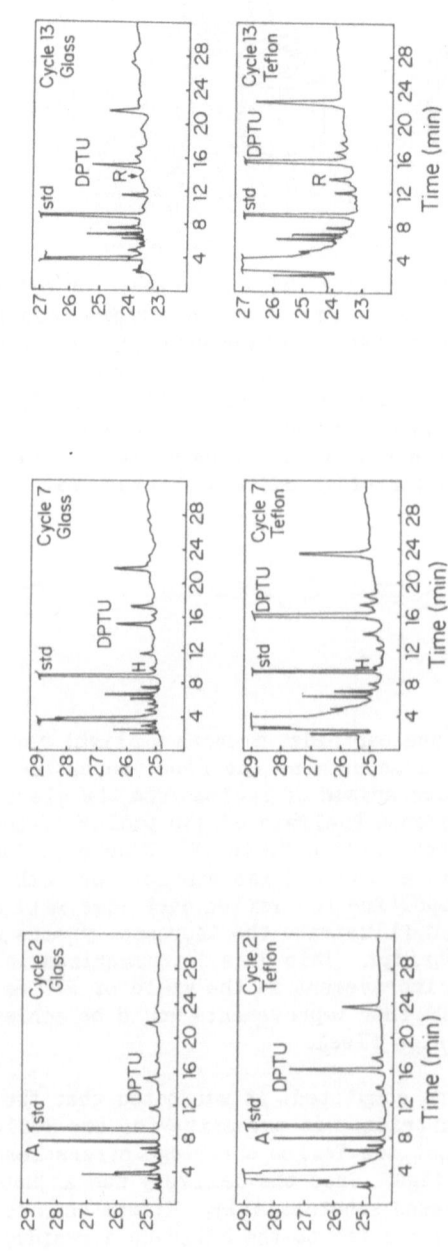

Figure 5A. Comparison of sequence results on a Teflon versus a glass cartridge. The synthetic peptide shown in Figure 3 was sequenced in the Teflon cartridge shown in Figure 4. A comparison of cycle 2 for the glass versus the Teflon cartridge is shown. In both cases 150 pmoles was sequenced. The yield at cycle 2 was 150 pmoles for the glass and 155 pmoles for the Teflon. **Figure 5B.** Comparison of sequence results on a Teflon versus a glass cartridge. Cycle 7, His, for the synthetic peptide is shown. The yield for glass is 4 pmoles and for Teflon 8 pmoles. See Figure 5A for details. **Figure 5C.** Comparison of sequence results on a Teflon versus a glass cartridge. Cycle 13, Arg, for the synthetic peptide is shown. The yield for glass is 0 pmoles and for Teflon is 4 pmoles. See Figure 5A for details.

Table IV. On-line sequencer results for a synthetic peptide on a Teflon versus a glass cartridge.

Cycle	Residue	Yield (pmoles) Glass	Teflon	Repetitive Yield Glass	Teflon
1	Val	150	136	81	86
2	Ala	150	155		
3	Ser	29	35		
5	Thr	36	32		
7	His	4	8		
13	Arg	0	4		

The synthetic peptide Val-Ala-Ser-Glu-Thr-Leu-His-Val-Ala-Leu-Ala-Val-Arg was sequenced on a Glass or Teflon cartridge. In each case 150 pmoles was sequenced. Yields for selected cycle are shown in pmoles. The repetitive yields were calculated from cycles 1 to 8. Similar results were calculated from cycles 1 to 12 and cycles 2 to 9.

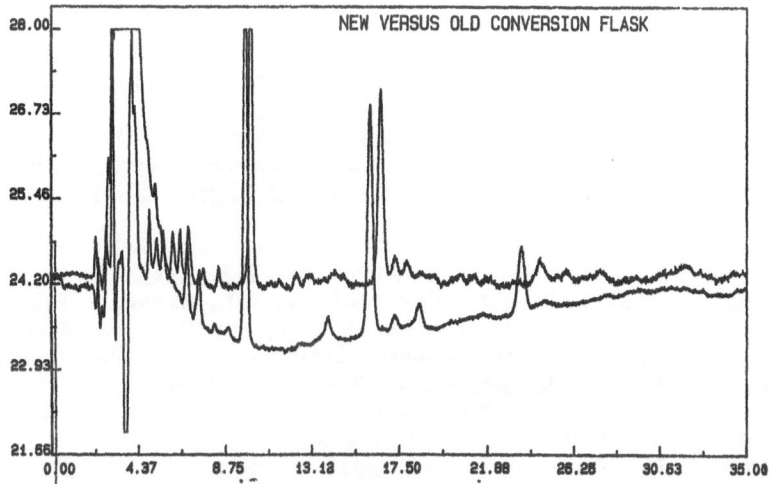

Figure 6. Effect of impurities in the conversion flask on HPLC behaviour. The lower tracing (similar to those in Figure 5) shows the chromatographic problems observed for a dirty conversion flask. The upper tracing shows the improved performance osberved with a new conversion flask. The build-up of impurities in the conversion flask was due to over-delivery of PITC (R1) to the. Teflon cartridge. This problem was corrected as shown in the upper trace, but only after the delivery of PITC was reduced and the conversion flask was replaced. See text for details.

The utility of the on-line system is best illustrated by the number of continuous injections made from Feb 1 to July 31, 1985, over 2300. This trouble-free performance and ease of setting up an online system prompts us to highly recommend it for the sequencing lab desiring greater efficiency.

ACKNOWLEDGEMENT

One of us (JES) is grateful for the opportunity to spend a 6 month sabbatical at DuPont Experimental Station, during which this work was performed.

•

REFERENCES

1. Hewick, R.M., Hunkapiller, M.W., Hood, L.E., and Dryer, W.J., A gas-liquid solid phase peptide and protein sequenator, J. Biol. Chem. 256:7990-7997 (1981).
2. Johnson, N.D., Hunkapiller, M.W., and Hood, L.E., Analysis of phenylthiohydantoin amino acids by high-performance liquid chromatography on Du Pont Zorbax cyanopropylsilane columns, Anal. Biochem. 100:335-338 (1979).
3. Zimmerman, C.L., Appella, C., and Pisano, J.J., Rapid analysis of amino acid phenylthiohydantoins by high-performance liquid chromatography, Anal. Biochem. 77:569-573 (1977).
4. Hawke, D., Yuan, P.-M., and Shively, J.E., Microsequence analysis of peptides and proteins. II. Separation of amino acid phenylthiohydantoin derivatives by high-performance liquid chromatography on octadecylsilane supports, Anat. Biochem. 120:302-311 (1982).
5. Lottspeich, F., Identification of the phenylthiohydantoin derivatives of amino acids by high pressure liquid chromatography, using a ternary, ioscratic system, Hoppe-Seyler Z. Physiol. Chem., 361:1829-1834 (1980).
6. Ashman, K., and Wittman-Liebold, B., A new isocratic HPLC separation for PTH amino acids based on 2-propanol, FEBS Lett., in press.
7. Glajch, J.L., Gluckman, J.C., Charikofsky, J.G., Minor, J. M., and Kirkland, J. J., Simultaneous selectivity optimization of mobile and stationary phase in reverse-phase liquid chrmatography of amino acid PTH derivatives, J. Chromatogr. 318:23-31 (1985).
8. Hawke, D.H., Harris, D.C., and Shively, J.E., Microsequence analysis of peptides and proteins. V. Design and performance of a novel gas-liquid-solid phase instrument, Anal. Biochem. 147:315-330, (1985).

ROUTINE ANALYSIS OF LOW-PICOMOLE-LEVEL PHENYLTHIOHYDANTOINS BY HPLC USING

A DIISOPROPYLETHYLAMINE-ACETATE/THF BUFFER AND ACETONITRILE GRADIENT

Carl J. March and Thomas P. Hopp

Immunex Corporation
51 University Street
Seattle, WA 98101

The commercial availability of the gas-phase protein sequencer (1,2) has greatly enhanced the ability of investigators to obtain information from subnanomolar amounts of polypeptide. To take full advantage of this new technology, however, it has been necessary to develop sensitive and efficient separation methods for phenylthiohydantoin (PTH) amino acids. Although reversed-phase high performance liquid chromatography (HPLC) has been the method of choice (3,4,5), few methods developed to date have provided reliable qualitative and quantitative information when analyzing just a few picomoles (pmol) of PTH amino acid. The most common problems in PTH analysis have been baseline fluctuations due to gradient elution, the lack of resolution of all PTH amino acids from each other and from frequently observed by-products of sequencing such as N,N'-diphenylthiourea (DPTU) and the elution of basic residues such as PTH-arginine in broad peaks which are difficult to quantify.

It has been recognized that the choice of buffer is critical in its effect on these problems. Indeed, the basic component of the aqueous phase has been widely varied, the most common of those investigated being sodium (5), ammonia and triethylamine (6). Because of improvements seen with triethylamine were thought to be due to its ability to act as a counter-ion to the free silicic acid groups on the solid phase and simultaneously to act as a hydrophobic species in binding to reversed-phase columns, we reasoned that N,N'-diisopropylethylamine (DIEA) might be an optimal choice. This compound, commonly used in peptide synthesis, has the advantage of being at the same time: cationic, hydrophobic (it is one of the largest water soluble alkyl-amines) and sterically hindered to prevent chemical reactivity. Thus, we have developed a binary gradient elution protocol based upon the use of a N,N'-diisopropylethylamine-acetate (DIEA-Ac) buffer containing 5% (v/v) tetrahydrofuran as the aqueous phase and acetonitrile as the organic phase, which eliminates the problems cited above. Here, we present standard chromatograms and sequence data to demonstrate the efficacy of this separation system in the routine analysis of low-picomole level phenylthiohydantoins.

HPLC Buffer Preparation

Buffer A. N,N'-diisopropylethylamine (DIEA) was purchased from Aldrich, dried over calcium hydride and then redistilled. The fraction boiling over from 125-127°C was collected and stored over 4A molecular sieves under argon at 4°C. Glacial acetic acid (Baker HPLC grade) was diluted with HPLC grade H_2O to 50 mM and adjusted to the desired pH with DIEA and then filtered through a 0.2 micron Nylon 66 filter. This buffer (DIEA-Ac) was adjusted to 5% (v/v) tetrahydrofuran (THF) (Burdick and Jackson, without preservative, UV grade) immediately before use.

Buffer B. The organic phase was acetonitrile (Burdick and Jackson, UV grade).

Preparation of PTH Standards

PTH amino acids (Sigma) were diluted in acetonitrile and standardized by molar extinction coefficient after the method of Edman and Henschen (7). The methyl esters of acid side chain PTH amino acids were prepared by resuspending 1 μmol of PTH amino acid in 1 ml methanolic HCl at 50°C for 15 min. N-dimethyl,N'-phenylthiourea (DMPTU) and N,N'-diphenylthiourea (DPTU) were purchased from Applied Biosystems. Stock solutions of all PTH amino-acids were diluted to the desired concentration for a 10 μl injection using equal volumes of DIEA-Ac and acetonitrile.

HPLC Columns

For all the results presented here, a 4.6 x 250 mm Shandon Hypersil CPS (Cyano) - 5 micron particle size HPLC column was used. However, identical results have been obtained using an IBM Cyano column of the same dimensions. The column was kept in a heated compartment at 40°C.

HPLC Hardware

All analyses were performed on a Hewlett Packard Model 1090A HPLC, equipped with a binary gradient system, heated column compartment, autosampler with autoinjector, HP Model 1040A diode array detector and an onboard DPU integrator. A 0.5 micron mechanical filter (SSI part number 05-0148) was placed immediately in front of the column inlet.

Protein Sequencing

Sequence analysis was performed on an Applied Biosystems Model 470A gas phase sequencer. All sequencing runs were made using the manufacturer-supplied programs designed for no vacuum drying and methanolic-HCl conversion. Sequencing reagents were purchased from Applied Biosystems. Sequencing solvents were purchased from Burdick and Jackson (Protein sequencing grade). Dithiothreitol (Cal Biochem-Behring, Ultrol grade) was added to S3 (n-butyl chloride) at a concentration of 2 mg/200 ml and to S4 (50/50 acetonitrile and methanol) at a concentration of 10 mg/200 ml. Filters were conditioned and samples applied according to manufacturer's specifications. Sequencer fractions were dried in vacuo in a Savant Speed-Vac and resuspended in 12 μl injection buffer (50/50 DIEA-Ac and acetonitrile).

Standard Chromatogram Analysis

Figure 1 shows a typical chromatogram of a standard containing five picomoles of each PTH amino acid. All peaks are well resolved from each other and the baseline is relatively flat, even though the scale is only 0.0015 absorbance units at full scale (AUFS). Chromatograms yielding useful information have been obtained as low as 0.0005 AUFS. Thus, the DIEA-Ac/THF buffer and acetonitrile gradient provide a means of reliably detecting 0.5 pmol of PTH amino acid. Some parameters useful in optimizing a particular HPLC system using the Shandon or IBM columns are presented in detail below.

Buffer pH. The use of DIEA to titrate the 50 mM acetate (before addition of THF) accomplishes a two-fold purpose. First of all, DIEA serves to drastically move PTH-arginine forward in the chromatogram with respect to other published methods using sodium acetate (5) and secondly, DIEA establishes the pH of the aqueous buffer. Of course, these two functions of DIEA are accomplished by simply titrating 50 mM acetate to a given pH. In our hands, the best pH for a new column is around 4.50. However, depending on batch to batch variability in functional group (CN) load on the silica, some differences may be observed between columns. As a column ages, the pH of the aqueous buffer needs to be adjusted upward. As a rule of thumb, the pH is adjusted about every 300 column runs. Easy adjustment can be maintained by having two buffer solutions, one adjusted to a pH slightly lower than the current operating pH and the other one higher, so that simple mixing of the two buffers in different ratios adjusts the pH. It has been our experience that if the pH is maintained to cause PTH-arginine to elute half-way between PTH-AspOMe and PTH-GluOMe (see Figure 1), that all other peaks will be in alignment. As the pH of the buffer is raised (i.e., the more DIEA added), PTH-histidine and PTH-arginine will elute closer to the front of the chromatogram. If the pH becomes too high, PTH-His and PTH-Gly begin to overlap as well as PTH-AspOMe and PTH-Arg. A low pH causes PTH-Arg to overlap with PTH-GluOMe, and PTH-His to overlap PTC-Gly (see below) which normally elutes between PTH-His and PTH-Ala.

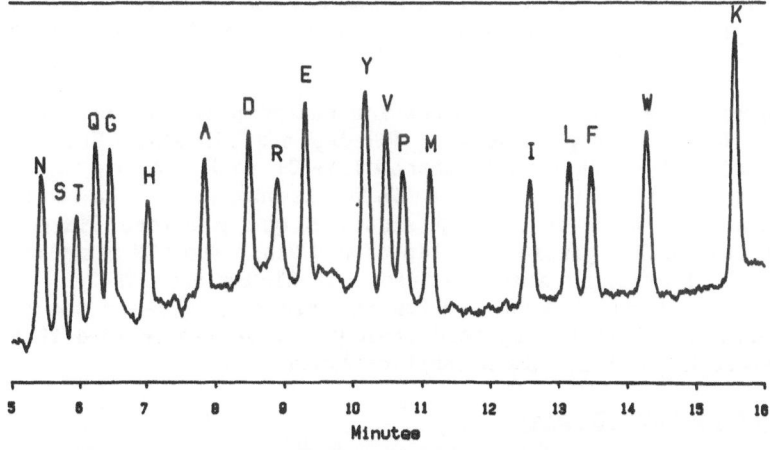

Fig. 1. *Chromatogram of PTH-amino acid standard. Each peak represents five picomoles of PTH-amino acid. Elution conditions are as described in the text. Full scale absorbance is 0.0015 AUFS at 265 nm with a 30 nm bandpass using Hewlett Packard Model 1040A Diode Array Detector.*

Table 1. Gradient Elution Conditions for DIEA-Ac/THF-Acetonitrile PTH Analysis on a Cyano Column.[a]

Time (minutes)	Percent Acetonitrile
0[b]	6
3	12
6	22
9	26
12	32
15	40
15.5	40
17	65
19[c]	6

[a]Buffers are as described in Methods. Flow rate is 1.2 ml/min and column temperature is 40°C. Acetonitrile concentrations are changed linearly between time points.
[b]Column is equilibrated at 94% A, 6% B.
[c]Column is re-equilibrated for 6 minutes prior to next injection.

        Gradient conditions. Table 1 provides the gradient elution conditions which have been optimized for the Hewlett Packard Model 1090A HPLC. These gradient conditions should only be taken as a guideline when using a different HPLC, especially one with a dynamic mixing chamber. We have achieved the best results using a flow rate of 1.2 mL/min and heating the column at 40°C. Using this gradient system, the cycle to cycle time (time between injections) is 25 minutes, with all PTH amino acids being eluted between 5 and 16 minutes after injection. Proper adjustment of the gradient is the most critical for the following pairs of PTH amino acids. Higher initial concentrations of acetonitrile result in a loss of resolution between PTH-Thr and PTH-Gln, as PTH-Gln moves forward. Increasing the percentage of acetonitrile too quickly at the 3, 6 or 9 minute time points results in PTH-Pro eluting at an earlier position, causing loss of resolution between it and PTH-Val. Similarly, too great of an increase at the 9 or 12 minute time points causes a loss of resolution between PTH-Leu and PTH-Phe.

        Free acid residues. The separation method presented here has been optimized for the methyl esters of PTH-Asp and PTH-Glu. These can be easily obtained by either using methanolic-HCl as R4 in the sequenator, or by conversion after the cycle if 25% TFA is used as R4. In the current system, the free acid of PTH-Asp elutes slightly before PTH-Asn, but PTH-Glu elutes with a partial overlap of PTH-Gln. The PTH-methylester of S-carboxymethyl cysteine elutes between PTH-GluOMe and PTH-Tyr, however the free acid elutes between PTH-Asn and PTH-Ser. Due to the overlap problem with the PTH-Glu free acid residue, it is recommended that only methyl esters be used in this separation system.

Analysis of Sequence Samples

        A method for analyzing sequencer residues is only useful if it clearly resolves all PTH amino acids from peaks arising from the expected byproducts of Edman chemistry. As discussed below, Figure 2 illustrates the separation of a five picomole PTH standard containing the most commonly observed sequencing byproduct, DPTU. Other frequently observed compounds are depicted by arrows at their corresponding elution times.

Expected peaks. DPTU is usually the sequencing byproduct present in the highest amount in each cycle. The addition of THF to a level of 5% (v/v) in the DIEA-Ac aqueous buffer achieves a baseline resolution between PTH-Met and DPTU. This is illustrated in Figure 2, where a 10 fold molar excess of DPTU (50 pmol) is included in the sample. The baseline resolution is maintained even at a 75-100 molar excess of DPTU over PTH-Met (data not shown). The DMA adduct with PITC (DMPTU) elutes between PTH-Gln and PTH-Gly (Figure 2). If present in very large amounts (> 25 pmol), DMPTU can cause either of the adjacent PTH amino acids to appear as a shoulder if they are present in low amounts (< 5 pmol). However, in our laboratory we rarely observe DMPTU above 10 pmol in the initial cycle and 5 pmol in subsequent cycles, and we have never had PTH-Gln or Gly obscured by DMPTU during a sequencing run. Due to the addition of DTT in S3 and S4, separation of DTT from residues in the front of the chromatogram is necessary. As shown in Figure 2, DTT (both the reduced and oxidized form) elutes well ahead of PTH-Asn and presents no problem in the chromatogram. N,N'-diphenylurea (DPU) is formed upon oxidation of DPTU resulting in the loss of the thiol group. Although typically present in very low amounts (< 5 pmol), DPU can, at times, exist at levels greater than DPTU. The elution position of DPU, immediately before PTH-Ile, is depicted by an arrow in Figure 2. When present in high amounts (probably due to sequencer malfunction), DPU can form a doublet peak with PTH-Ile. However, DPU never occludes and is easily distinguished from PTH-Ile by its retention time.

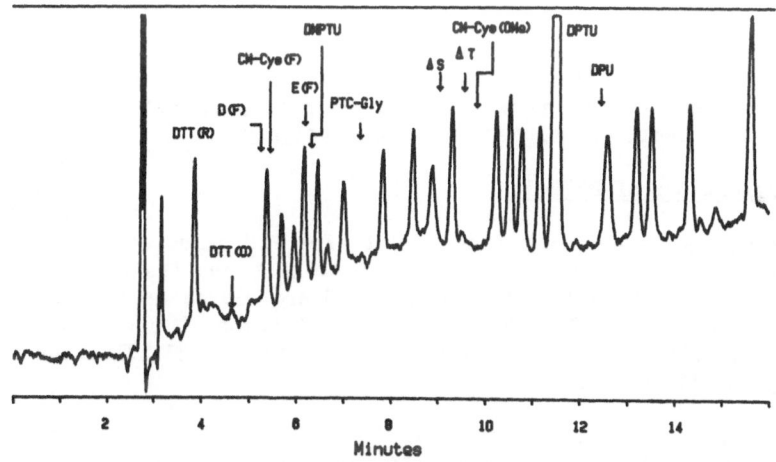

Fig. 2. Chromatogram of PTH-amino acid standard plus DTT and DPTU. The peaks eluting between 5 and 16 min are PTH-amino acids (five pmol each) as labeled in Fig. 1, with the exception of the large peak at 11.7 min which is 50 pmol DPTU. The first peaks at 2.7 and 3.3 min are due to solvent breakthrough. The peak at 3.9 min (DTT(R)) is the reduced form of DTT. Other components observed in practice, but not in the chromatogram are represented by arrows at their elution positions. They are: DTT(O): oxidized form of DTT. D(F): free acid of PTH-Asp. CM-Cys(F): free acid of PTH-S-carboxymethyl cysteine. E(F): free acid of PTH-Glu. DMPTU: N-dimethyl,N'-phenylthiourea. PTC-Gly: phenylthiocarbamyl-glycine. △S: dehydro serine. △T: dehydro threonine. CM-Cys(OMe): PTH-S-carboxymethyl cysteine methyl ester. DPU: N,N'-diphenylurea.

Other peaks. In the course of sequencing, many other peaks are observed. Some of these are normal derivatives of certain amino acids and some are due to less than perfect sequenator operation. The discussion of each peak is beyond the scope of this work. However, the positions of some of these peaks observed in our laboratory are indicated by arrows in Figure 2. In addition to the resolution of these components offered by the chromatogram at an absorbance wavelength of 265 nm, many of these peaks are more easily detected at other wavelengths if multiple channel monitoring or multiple detectors are used.

Sequence of Angiotensin II. Due to difficulties in sequencing the octapeptide angiotensin II, we chose to use this peptide in demonstrating the efficacy of the DIEA-Ac buffer system in the routine analysis of low levels of phenylthiohydantoin. Figure 3 shows all eight cycles from a sequencing run on 25 pmol of angiotensin II. Each panel is simply a plot of the raw data obtained from the indicated cycle.

Panel 2 shows the elution of PTH-Arg. The relatively early retention time provided a sharp peak with an easily quantifiable shape for integration. Like PTH-Arg, PTH-His (Panel 6, Figure 3) elutes relatively early in the DIEA-Ac system. The pH stability of the buffer provides reliable positioning for both these residues, and consequently, coupled with good peak shape due to early elution times, provide excellent integration data for quantitation of sequence runs.

Due to cyclization of the peptide at position 6 (Histidine), relatively poor yields have typically been observed for cycles 7 (Proline) and 8 (Phenylalanine). However, as shown in Panels 7 and 8 of Figure 3, both of these residues were easily detected, due to the flat baseline of the chromatograms. Insets, showing the replotting of the raw data at 0.002 AUFS for both cycles 7 and 8 highlight the appearance of proline and phenylalanine, respectively. The quantified yield of PTH-Pro was 3.8 pmol and that for PTH-Phe, 2.2 pmol. The ability to reliably detect as little as 1 pmol of PTH amino acid has been very useful, especially in analyzing very small quantities of unknown polypeptides.

Other Applications of DIEA-Ac

Although we have only presented data for PTH amino acid analysis on a cyano support, we have also obtained excellent results using a DuPont Zorbax ODS column. On the ODS column, PTH-Arg elutes as the first peak in the chromatogram, followed by PTH-His. However, because it is more difficult to obtain complete baseline resolution for all PTH amino acids, we have generally preferred the cyano support.

The separation method presented here also works nicely for amino acid analysis. After hydrolysis, the free acids are converted to PTH amino acids followed by methylation. The sample is then run using the same system. A standard hydrolysate can be used to compensate for differences in reaction rates and yields, thereby making integration possible. Thus, the same HPLC and buffer system can be used for both amino acid and sequencing analysis.

The transparency in the UV range of the DIEA-Ac buffer (without THF) makes it an ideal substitute for pyridyl-acetate systems in the fractionation of proteins and peptides on reversed-phase HPLC. We have observed similar retention times for both buffer systems, but prefer the DIEA-Ac because chromatograms can be obtained using absorbance at 210 nm (data not shown), and the fact that the DIEA-Ac buffer is stable as compared to pyridyl-acetate, which, upon degradation, can lead to the blocking of amino termini. Furthermore, the DIEA-Ac buffer is volatile, making it an ideal final HPLC step in preparation for sequencing because DIEA-Ac does not produce any artifacts in subsequent analysis, as has been observed for pyridine-based buffers.

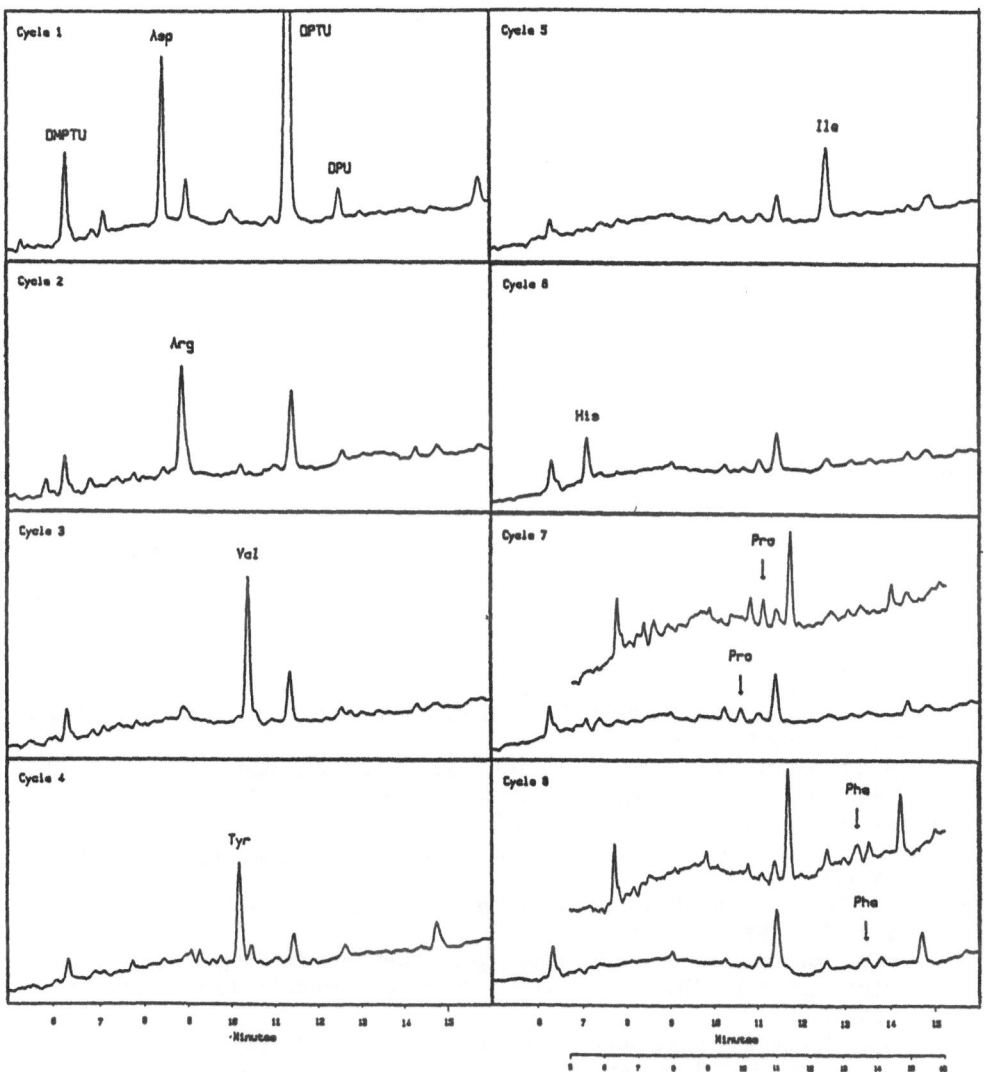

Fig. 3. N-terminal amino acid sequence analysis of 25 pmol human angiotensin II. Samples were prepared as described in the text. The HPLC chromatograms for each cycle are represented by the raw data plotted at 0.005 AUFS at 265 nm (30 nm bandpass). For cycles 7 and 8, insets of the same raw data replotted at 0.002 AUFS are shown. The inset plots are scaled to the time scale shown underneath that for the larger chromatograms. Initial yield for PTH-Asp was 16.3 pmol (65%).

SUMMARY

We have developed a PTH amino acid HPLC separation method using a novel buffer system on a cyano support. This elution protocol provides reliable qualitative and quantitative protein sequencing analysis at the 1-10 picomole level. In our hands, DIEA has outperformed all other basic compounds we have tried in the aqueous phase for PTH analysis, and thus, we consider DIEA close to the optimal base available for use in reversed-phase HPLC. Although the DIEA must be at least dried over CaH_2 and refluxed if not redistilled, many protein chemistry labs already routinely perform this function for use in peptide synthesis. Once redistilled, a one liter bottle will last several months with daily HPLC operation (150-200 liters of DIEA-Ac/liter DIEA). The DIEA-Ac buffer is stable over long periods of time if kept in amber bottles and inhibits microbial growth. The volatile nature of the buffer prevents salt residue build-up (as is observed with sodium-based systems) which extends column life to several thousand runs with little loss of resolution, provided the proper pH is maintained. The excellent signal to noise ratio afforded by the DIEA-Ac buffer system has been instrumental in our obtaining protein sequence at the low picomole level, leading to the cloning and expression of several rare proteins, including the human IL-2 receptor (8,9) and more recently, human IL-1β (10). The separation system presented here should be useful in developing femtomole sensitivity PTH analysis on small bore columns. Such studies are presently in progress.

ACKNOWLEDGEMENTS

We wish to thank Dr. Kathryn Prickett for useful discussions, Steve Hartman and Janet Merriam for excellent technical assistance, and Judy Byce for preparation of the manuscript.

REFERENCES

1. R.M. Hewick, M.W. Hunkapiller, L.E. Hood, and W.J. Dreyer, J. Biol. Chem. 256:7990 (1981).
2. M.W. Hunkapiller, R.M. Hewick, W.J. Dreyer, and L.E. Hood, in: "Methods in Enzymology," C.H.W. Hirs and S.N. Timasheff, eds., 91:399, Academic Press, New York (1983).
3. C.L. Zimmerman, E. Appella, and J.J. Pisano, Anal. Biochem. 77:569 (1977).
4. A.S. Bhown, J.E. Mole, A. Weissinger, and J.C. Bennett, J. Chromatogr. 148:532 (1978).
5. M.W. Hunkapiller and L.E. Hood, in: "Methods in Enzymology," C.H.W. Hirs and S.N. Timasheff, eds., 91:486, Academic Press, New York (1983).
6. S.D. Black and M.J. Coon, Anal. Biochem. 121:281 (1982).
7. Edman, P. and Henschen, A., in: "Protein Sequence Determination," S.B. Needleman, ed., Springer-Verlag, New York/Berlin (1975).
8. D.L. Urdal, C.J. March, S. Gillis, A. Larsen, and S.K. Dower, Proc. Natl. Acad. Sci. USA 81:6481 (1984).
9. D. Cosman, D.P. Cerretti, A. Larsen, L. Park, C. March, S. Dower, S. Gillis, and D. Urdal, Nature 312:768 (1984).
10. C.J. March, B. Mosley, A. Larsen, D.P. Cerretti, G. Braedt, V. Price, S. Gillis, C.S. Henney, S.R. Kronheim, K. Grabstein, P.J. Conlon, T.P. Hopp, and D. Cosman, Nature 315:641 (1985).

IDENTIFICATION OF SIDE-CHAIN PROTECTED L-PHENYLTHIOHYDANTOINS ON CYANO HPLC COLUMNS: AN APPLICATION TO GAS-PHASE MICROSEQUENCING OF PEPTIDES SYNTHESIZED ON SOLID-PHASE SUPPORTS

Daniel J. Mc Cormick, Benjamin J. Madden and Robert J. Ryan

Department of Cell Biology
Mayo Clinic and Mayo Research Foundation
Rochester, MN, 55905

INTRODUCTION

Recent advances in microsequencing procedures using gas-phase methods have provided for the sequence determination of relatively small amounts (50-1000 pmol) of proteins and large polypeptides (1,2), and the efficient sequence analysis of very small amounts (10-100 pmol) of short peptides containing 10 to 50 amino acids (3,4). Developments in improved High Performance Liquid Chromatography (HPLC) instrumentation have increased the sensitivity of detection (less than 5 pmol) of the phenylthiohydantoin (PTH) amino acids derived from Edman degradation (5,6).

In addition to the sequencing of small polypeptides and proteins immobilized on Polybrene discs, sequence information has also been obtained on relatively short (less than 40 residues) side-chain protected synthetic peptides prior to their cleavage from polystyrene supports (7,8). Gas-phase sequencing methodology is, therefore, an applicable tool in solid-phase synthesis for the assessment of peptide homogeneity, and for the quantitative determination of amino acid coupling during the synthesis of longer peptides (8,9). Reports describing the HPLC of side-chain protected PTH residues, derived from the sequencing of synthetic peptides, on Zorbax-ODS columns (Dupont) have been described by Simmons and Schlesinger (10,11).

In the present study, we report the use of gas-phase microsequencing on protected synthetic peptides linked to solid-phase supports, and the identification of the resultant side-chain protected PTH-L-amino acids by reversed-phase HPLC on Ultrasphere Cyano (CN) columns. We further report on the stability of the chemical linkages between the peptide (C-terminus) and various solid-phase supports for quantitative sequence analysis.

MATERIALS AND METHODS

The primary structures of the synthetic peptides used for gas-phase sequencing are given in Figure 1. Peptides I (res. 159-169), II (res. 185-199) and III (res. 131-147) represent sequences of the human acetylcholine receptor α-chain as reported by Noda et al. (12). The peptides IV (res. 26-41), V (res. 36-51) and VI (res. 81-92) were taken from the primary sequence of human chorionic gonadotropin α-chain

(13).   Each peptide sequence was synthesized on benzhydrylamine (BHA),
Merrifield hydroxymethyl (MHM), or 4-(oxymethyl)-phenylacetamidomethyl (PAM)
resin using tBoc-L-amino acid derivatives (see Figure 1).   Details of the
solid-phase synthetic procedures used are described in detail elsewhere
(14-16).

Coupling of the first tBoc-amino acid (C-terminus) to BHA or MHM
resins was accomplished by using 7.0 mmol excess of tBoc-amino acid and
N,N'-dicyclohexylcarbodiimide   in   50%   dichloromethane/dimethylformamide
(v/v) for 18 h.   The residues tBoc-aspargine or tBoc-glutamine were
coupled as their N-hydroxybenzotriazole active esters.   After coupling,
each resin was washed extensively, and 1.0 mg aliquots were hydrolyzed
(6 N HCL/propanoic acid, 1:1) for 3 h at 130°C.   Each hydrolysate was
subjected to amino acid analysis and the substitution was 0.31 mmol of
asparagine/g of BHA resin, and 0.43 mmol of glycine or leucine/g of MHM
resin.

```
            159                                              169
  I.  LYS-GLY-GLY-ASN-PRO-GLU-SER-ASP-GLN-PRO-ASP-LEU-SER-ASN-BHA
      ClZ                 OBz Bzl OBz         OBz       Bzl


      185            190              195           199
 II.  LYS-HIS-SER-VAL-THR-TYR-SER-CYS-SER-PRO-ASP-THR-PRO-TYR-LEU-MHM
      ClZ Dnp Bzl     Bzl ClB Bzl MeB Bzl     OBz Bzl     ClB


      131        135            139          143            147
III.  ILE-VAL-THR-HIS-PHE-PRO-PHE-ASP-GLU-GLN-ASN-CYS-SER-NLE-LYS-LEU-GLY-MHM
              Bzl Dnp         OBz OBz         MeB Bzl    ClZ


      26         30           34          38         41
 IV.  LEU-GLN-CYS-MET-GLY-CYS-CYS-PHE-SER-ARG-ALA-TYR-PRO-THR-PRO-LEU-PAM
              MeB         MeB MeB     Bzl Tos     BrZ     Bzl


      36         40           44          48         51
  V.  ALA-TYR-PRO-THR-PRO-LEU-ARG-SER-LYS-LYS-THR-MET-LEU-VAL-GLN-LYS-PAM
      BrZ     Bzl          Tos Bzl ClZ ClZ Bzl               ClZ


      81         85           89          92
 VI.  ALA-CYS-HIS-CYS-SER-THR-CYS-TYR-TYR-HIS-LYS-SER-PAM
      MeB Tos MeB Bzl Bzl MeB BrZ BrZ Tos ClZ Bzl
```

Figure 1.   Primary sequences of side-chain protected synthetic peptides
           coupled to solid-phase resins for gas-phase sequencing. Peptide I
           (159-169) was synthesized on BHA resin, peptides II (185-199) and
           III (131-147) were synthesized on MHM resins, and peptides IV
           (26-41), V (36-51) and VI (81-92) were synthesized on PAM resins.
           In peptide I, an additional sequence of Lys-Gly-Gly was added to
           the N-terminal end. For peptides II and III, the amino acids (see
           underline) serine (Bzl) and norleucine (Nle) were substituted for
           the residues Cys-193 and Met-144, respectively. Abbreviations of
           the side-chain protecting groups shown are: ClZ, ε-2-chloro-
           benzyloxycarbonyl for lysine; OBz, β or γ-benzyl ester for
           aspartic and glutamic acid, respectively; Bzl, O-benzyl for serine
           and threonine; Dnp, N^{IM}-2,4-dinitrophenyl for histidine; ClB,
           0-2,6-dichlorobenzyl for tyrosine; MeB, S-4 methylbenzyl for
           cysteine; Tos, p-toluenesulfonyl (tosyl) for arginine and
           histidine; BrZ, 0-2-bromobenzyloxycarbonyl for tyrosine; tBoc,
           tert-butyloxycarbonyl.

The coupling of the amino acids tBoc-leucine, tBoc-lysine (ClZ) and tBoc-serine (Bzl) to PAM resins was performed according to the procedures described by Mitchell et al. (17). Substitution of the PAM resins was determined to be 0.72 mmol of leucine or serine and 0.57 mmol of lysine/g resin.

Approximately 1.0 mmol of the tBoc-substituted BHA or MHM resins were used for the manual synthesis of peptides I, II and III. Completion of coupling after each residue was monitored by the ninhydrin method of Kaiser (18). An Applied Biosytems 430A solid-phase peptide synthesizer was used for the automated synthesis of peptides IV, V and VI on 0.5 mmol of the PAM resins. After synthesis, 50-mg aliquots of each protected peptide-resin were treated with anhydrous trifluoroacetic acid (TFA) for 15 min to remove the N-terminal tBoc group, and then saved for microsequencing analysis. The remaining amounts of each peptide were cleaved from the resin support by treatment with 8 ml anhydrous HF (distilled in CoF$_3$), containing an excess of 10 mmol anisole, at 0°C for 60 min. The crude peptides were extracted in TFA, precipitated in cold diethyl ether, redissolved in 10% acetic acid/water (v/v) and freeze-dried. Peptides were purified by chromatography methods described elsewhere (15,19). The primary structure and composition of each nonprotected peptide was also evaluated by microsequence analysis.

Approximately 50-100 μg of each peptide-resin (0.2-2.0 nmol peptide) was sequenced on an automated 470A gas-phase sequencer (Applied Biosystems, CA) using the 02nbgn and 02nvac programs supplied by the manufacturer. Protected peptide resins were suspended in 50% CH$_2$Cl$_2$/CH$_3$OH (v/v) and applied on a glass filter disc which was then sandwiched between two Teflon filters. The side-chain proteced PTH amino acids obtained from each cycle were analyzed by HPLC methods described below. Nonprotected synthetic peptides were analyzed by gas-phase sequencing in order to determine the degree of side-chain deprotection of each residue following HF treatment of the peptide resins. Approximately 1.0 nmol of each nonprotected peptide was applied to Polybrene-treated discs and sequenced using the programs above. The PTH amino acids were also analyzed by HPLC. Conversion of the 2-anilino-5-thiazolinone amino acids to their corresponding PTH derivatives was accomplished by the use of methanolic HCl (1 N) in the conversion chamber. After conversion, each PTH sample was dried in a Speed Vac concentrator (Savant Inst.), dissolved in 33% acetonitrile:water (v/v), and injected into the HPLC system described below.

The HPLC instrumentation (Beckman Instruments, Inc., CA) employed for the separation and identification of both standard (nonprotected) and side-chain protected PTH L-amino acids is described below in Figure 2. An example of a separation of PTH standards on an Ultrasphere Cyano column is illustrated. The conditions, program and buffers used for the set-up of the gradient is also given below (see Figure 2).

RESULTS AND DISCUSSION

The conditions and gradient elution of all PTH standards by HPLC on an Ultrasphere Cyano (CN) column are depicted in Figure 2 above. The average elution times of each standard PTH amino acid from several gradient runs is further summarized in Table 1 below. In our hands, the separation of PTH-His from PTH-Asp (O-methyl ester derivative) was determined to be critically dependent upon the pH of buffer A. Variation of the pH by as much as ± 0.02 pH units caused PTH-His to coelute with PTH-Asp. Aside from the occasional coelution of histidine with aspartic

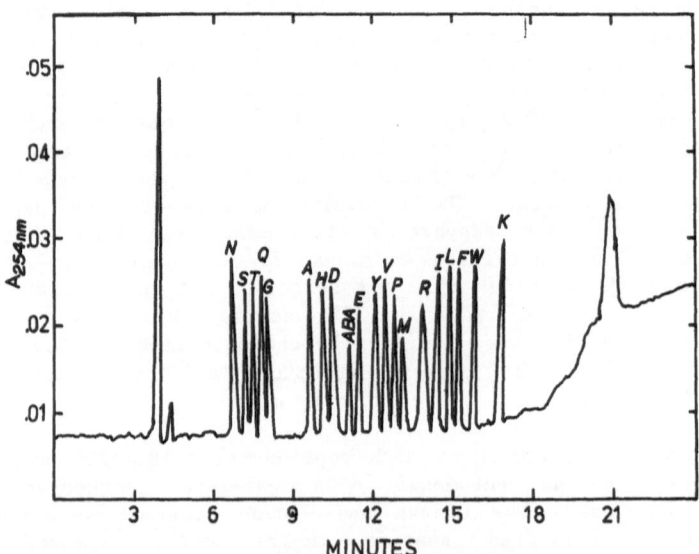

Figure 2. Reversed-phase HPLC of PTH L-amino acid standards (50 pmol in 20 µl). The instrumentation employed the following Beckman equipment: two Model 114M HPLC pumps, a Model 340 gradient mixing chamber, a Model 504 autosampler, a Model 160 fixed-wavelength UV detector, and a Model 450 data System integrator/controller. Chromatography was performed on a Beckman Ultrasphere-CN column (0.46 x 25 cm, 5 µ particle size) connected to a Waters Associates Guard-PAK precolumn. Gradient elution of standard and side-chain protected PTH amino acids was carried out at 37°C at a flow rate of 1.0 ml/min using Buffer A (21 mM sodium acetate, pH 4.96, containing 5% tetrahydrofuran) and Buffer B (100% acetonitrile). Injections by autosampler (20 µl) were programmed at 33 min intervals using the program below.

Time (min)	Function	Value	Duration (min)
0	Flow	1.0 ml/min	
0	%B	9%	
0.05	%B	51%	19.0
24.10	%B	100%	2.0
27.00	%B	9%	4.0
30.00	Flag	2	0.1
33.00	Flag	1	0.1

acid, the complete resolution of all other PTH amino acids was accomplished by using a linear gradient from 9% to 51% acetonitrile (buffer B) within 19 min.

A microsequence analysis of protected synthetic peptides I (159-169), II (185-199) and III (131-147), that were synthesized on BHA or MHM resins, was done. The elution times (E_t) of the resultant side-chain protected PTH amino acids by HPLC on a CN-column are listed in Table 2. The nonprotected PTH amino acids Ala, Asn, Gln, Gly, Ile, Leu, Met, Pro, Phe and Val, produced from the Edman degradation of each protected synthetic peptide, are not shown in this table. The elution of these nonprotected PTH amino acids, however, corresponded exactly with the elution times of the PTH amino acid standards given in Table 1.

The side-chain protected amino acids Cys(MeB), His(Dnp), Lys(ClZ), and Tyr(ClB) (key to abbreviations in Fig. 1) were exceptionally stable during the repetitive Edman degradation sequencing of each synthetic peptide. The O-benzyl ester protecting group of PTH-Asp(OBz) was less stable, yielding approximately 50% nonprotected PTH-Asp and 50% protected PTH-Asp(OBz) (average E_t's = 11.17 and 18.70 min, respectively) when this residue was encountered in the synthetic sequence. PTH-GLU(OBz) was completely unstable and routinely yielded a single nonprotected PTH-Glu O-methyl ester peak (average E_t = 12.38 min). The protected amino acids

Table 1. Separation of Standard Phenylthiohydantoin (PTH) Amino Acids by High Performance Liquid Chromatography form an Ultrasphere-Cyano (CN) Column[a]

PTH-Amino Acid	Average Elution Time (min)[b]
Asn	7.02 + 0.03
Ser	7.38 + 0.03
Thr	7.63 + 0.04
Gln	8.21 + 0.04
Gly	8.40 + 0.03
Ala	10.18 + 0.05
His	11.07 + 0.06
Asp	11.16 + 0.05
ABA[c]	11.85 + 0.04
Glu	12.34 + 0.06
Tyr	13.15 + 0.05
Val	13.56 + 0.05
Pro	13.94 + 0.04
Met	14.28 + 0.06
Arg	14.96 + 0.04
Ile	15.70 + 0.06
Leu	16.28 + 0.06
Phe	16.60 + 0.06
Trp	17.25 + 0.05
Lys	18.41 + 0.07

[a]Elution conditions of the PTH standards (50 pmol in 20 µl) from the CN-column (4.6 x 250 mm, Beckman Inst.) is described in Figure 2.

[b]Elution time of each standard PTH amino acid listed is given as the average of five runs (n=5), + standard deviation.

[c]Internal standard was 30 pmol of PTH-2-aminobutanoic acid (ABA).

Table 2. Identification of Side-Chain Protected PTH-Amino Acids from Gas-Phase Sequencing of Peptides Synthesized on BHA and MHM-Resins[a]

Cycle No.	Residue No.	PTH-Amino Acid Identified	Elution Times (min)[b]	Relative Percent Area of Peaks[c]
		Peptide I: 159-169(BHA)		
1	Cap	Lysine (ClZ)	20.78	100%
6	161	Glutamic (OBz)	12.43	100%
7	162	Serine (Bzl)	9.48, 12.78, 18.09 19.55, 20.20	11%, 25%, 36%, 12%, 16%
8	163	Aspartic (OBz)	11.22, 18.76	52%, 48%
11	166	Aspartic (OBz)	11.17, 18.70	55%, 45%
13	168	Serine (Bzl)	(see Cycle No. 7)	
		Peptide II: 185-199(MHM)		
1	185	Lysine (ClZ)	20.82	100%
2	186	Histidine (Dnp)	17.45	100%
3	187	Serine (Bzl)	9.41, 12.66, 17.92 19.41, 20.03	10%, 32%, 39%, 10%, 9%
5	189	Threonine (Bzl)	12.87, 18.40, 18.71, 20.64	26%, 19% 22%, 33%
6	190	Tyrosine (ClB)	22.67	100%
7	191	Serine (Bzl)	9.42, 12.67, 17.95, 19.44, 20.08	11%, 30%, 38%, 10%, 11%
8	192	Cysteine (MeB)	20.77	92%
9	193	Serine (Bzl)	(see Cycle No. 7)	
11	195	Aspartic (OBz)	11.14, 18.67	52%, 48%
12	196	Threonine (Bzl)	(see Cycle No. 5)	
14	198	Tyrosine (ClB)	22.73	100%
		Peptide III: 131-147(MHM)		
3	133	Threonine (Bzl)	12.90, 18.56, 18.78, 20.71	17%, 20% 28%, 35%
4	134	Histidine (Dnp)	17.50	100%
8	138	Aspartic (OBz)	11.13, 18.68	46%, 54%
9	139	Glutamic (OBz)	12.35	100%
12	142	Cysteine (MeB)	20.72, 21.65	89%, 19%
13	143	Serine (Bzl)	(see Cycle No. 3 for peptide II)	
14	144	Norleucine	16.40	100%
15	145	Lysine (ClZ)	20.87	100%

[a]Sequencing of synthetic peptides is described in Materials and Methods; elution of PTH-amino acids is described in Figure 2.

[b]Elution times of the ABA internal standard (30 pmol) was 11.83, 11.73 and 11.75 min for peptides I, II and III, respectively.

[c]Major and minor peak areas of each amino acid are expressed as a relative percentage of the total peak area (= 100%).

PTH-Cys(MeB), PTH-His(Dnp), PTH-Lys(ClZ) and PTH-Tyr(ClB) gave one elution peak by HPLC analysis, and had average E_t's of 20.75, 17.48, 20.82 and 2.69 min, respectively. At times, PTH-Cys(MeB) produced a second minor peak with an E_t of 21.65 min by HPLC analysis (peptide III, Table 2).

Sequencing of the amino acids Ser(Bzl) and Thr(Bzl) in the synthetic peptides produced several peaks by HPLC analysis (see Table 2). In the case of PTH-Ser(Bzl), HPLC analysis yielded four minor derivatives and one major derivative (average E_t = 17.98 min) presumed to be PTH-Ser(Bzl). The nature of the other minor peaks is unknown, although they may represent various sequencing artifacts of both protected and nonprotected PTH-Ser. Peptides containing protected Thr(Bzl) residues produced three minor peaks and one major peak (average E_t = 20.67 min) by HPLC on CN-columns. Of interest, was the elution time of PTH-Norleucine (an uncommon amino acid often used as a substitute for Met in peptide synthesis), that was determined to be about 16.40 min. This elution corresponds to a position after PTH-Leu and before PTH-Phe (see Table 1).

The sequencing of the protected peptides synthesized on PAM resins (i.e., peptides IV, V and VI), and the resultant separation of the side-chain protected PTH amino acids is given in Table 3. The elution patterns of the protected PTH amino acids for these peptides were essentially the same as that described for the PTH derivatives listed in Table 2. HPLC analysis of the tosyl (Tos) protected derivatives of PTH-His(Tos) and PTH-Arg(Tos), however, produced only one major peak with average E_t's of 11.37 and 16.57 min, respectively. In the sequencing conditions used for these studies, we found that protected tosyl-His was very unstable to Edman degradation. Generally, greater than 95% of PTH-His(Tos) gave a peak elution time (E_t = 11.37) which corresponded exactly to the elution of nonprotected PTH-His. The instability of the tosyl-His derivative to Edman degradation appears to be contrary to the findings reported by Schlesinger (11) for the sequencing of synthetic peptides containing this protected amino acid.

Microsequencing of peptides containing side-chain protected Tyr(BrZ) yielded PTH derivatives with two distinct elution times (see Table 3). The major elution time of PTH-Tyr(BrZ) was determined to be about 23.26 min. The characteristic elution times of all 12 of the side-chain protected PTH amino acids, analyzed by HPLC in this study, is summarized in Table 4. The following table presents the average expected elution time for each protected PTH derivative on CN-columns when using the gradient conditions specified above (see Figure 2).

The sequencing of each nonprotected synthetic peptide (obtained after HF cleavage of the protected peptide from its resin) was also undertaken. HPLC analysis of the PTH amino acids from each peptide confirmed the removal of the side-chain protecting groups used for synthesis (see Fig. 1). Some peptides, however, contained some residually protected residues [e.g., Cys(MeB), peptides IV and VI] which were identified in the sequence analysis. Therefore, the identification of side-chain protected PTH amino acids by HPLC on CN-columns has proven to be a useful analytical tool in determining the efficiency of a particular HF treatment (i.e., cleavage and side-chain deprotection) on synthetic peptide-resins.

Finally, a preliminary study on the stability of the chemical linkages between the protected synthetic peptide and the resin supports, for sequencing analysis was done. The results of our findings are summarized in Table 5 below. In general, amide linkages were highly susceptible to cleavage by one-step Edman degradation. Ester linkages, on the other hand, were more stable to the Edman degradation reactions in gas-phase sequencing. This stability was increased when the esterified

Table 3. Identification of Side-Chain Protected PTH-Amino Acids from
Gas-Phase Sequencing of Peptides Synthesized on PAM-Resin
Supports[a]

Cycle No.	Residue No.	PTH-Amino Acid Identified	Elution Times (min)[b]	Relative Percent Area of Peak
		Peptide IV: 26-41(PAM)		
3	28	Cysteine (MeB)	21.14, 21.88	79%, 21%
6	31	Cysteine (MeB)	21.16, 21.90	72%, 28%
7	32	Cysteine (MeB)	21.20, 21.92	68%, 32%
9	34	Serine (Bzl)	9.75, 13.04, 18.35, 19.77, 20.33	18%, 23%, 27% 21%, 11%
10	35	Arginine (Tos)	16.61	100%
12	37	Tyrosine (BrZ)	13.38, 23.15	35%, 65%
14	39	Threonine (Bzl)	12.97, 18.50, 18.75, 20.70	25%, 17%, 27%, 31%
		Peptide V: 36-51(PAM)		
2	37	Tyrosine (BrZ)	13.31, 23.05	27%, 73%
4	39	Threonine (Bzl)	12.88, 18.45, 18.67, 20.61	22%, 15%, 20%, 43%
7	42	Arginine (Tos)	16.52	100%
8	43	Serine (Bzl)	9.68, 12.98, 18.23, 19.65, 20.30	14%, 22%, 44%, 7%, 13%
9	44	Lysine (ClZ)	20.88	100%
10	45	Lysine (ClZ)	21.03	100%
11	46	Threonine (Bzl)	12.91, 18.49, 18.71, 20.64	24%, 18%, 20%, 38%
16	51	Lysine (ClZ)	20.88	100%
		Peptide VI: 81-92(PAM)		
2	82	Cysteine (MeB)	21.41, 22.20	72%, 28%
3	83	Histidine (Tos)	11.38	100%
4	84	Cysteine (MeB)	21.45, 22.22	70%, 30%
5	85	Serine (Bzl)	9.82, 13.21, 18.55, 20.01, 20.61	15%, 18%, 35%, 19%, 13%
6	86	Threonine (Bzl)	13.10, 18.69, 18.95, 20.88	20%, 21%, 24%, 36%
7	87	Cysteine (MeB)	21.50, 22.25	68%, 32%
8	88	Tyrosine (BrZ)	13.57, 23.38	26%, 74%
9	89	Tyrosine (BrZ)	13.60, 23.47	29%, 71%
10	90	Histidine (Tos)	11.35	100%
11	91	Lysine (ClZ)	21.26	100%
12	92	Serine (Bzl)	(see Cycle No. 5)	

[a]Elution of PTH-amino acids is described in Figure 2.

[b]Average elution times of the ABA internal standard (30 pmol) was 12.11,
12.05 and 12.26 min for peptides IV, V and VI, respectively.

C-terminal residue contained a branched side-chain (e.g., Leu, Lys and Ser ester bonds were more stable than Gly ester bonds). The increased stability of PAM-resin peptides to Edman degradation appeared to be due to the high electron withdrawing potential of the phenylacetamidomethyl bridge (17).

The present study clearly demonstrates the usefulness of gas-phase microsequencing in the analysis of protected synthetic peptides assembled by solid-phase methods. The identification of nonprotected PTH amino acids from side-chain protected PTH amino acids, using identical gradient conditions on CN-columns, provides for an accurate evaluation of: (1) the sequence of the synthetic peptide prior to cleavage and deprotection with HF; (2) the extent of side-chain deprotection after HF treatment and purification of the peptide.

Table 4. Summary of the Major and Minor Peak Elution Times for Side-Chain Protected PTH L-Amino Acids on Ultrasphere CN Columns[a]

Amino Acid	Side-Chain Protecting Group	Average Elution Time of Peaks (min)	Average Percent Area of Peaks
Arg	Toluenesulfonyl (Tos)	16.57 + 0.06 (n=2)	100%
Asp	β-Benzyl ester (OBz)	11.17 + 0.04 (n=4)[b]	51% + 3.7
		18.40 + 0.04 (n=4)	49% + 3.7
Cys	4-methyl benzyl (MeB)	21.17 + 0.30 (n=8)[b]	75% + 8.1
		21.96 + 0.24 (n=8)	25% + 8.1
Glu	γ-Benzyl ester (OBz)	12.38 + 0.06 (n=2)	100%
His	Dinitrophenyl (Dnp)	17.48 + 0.04 (n=2)	100%
His	Toluenesulfonyl (Tos)	11.37 + 0.02 (n=2)	100%
Lys	2-Chloro Z (ClZ)	20.93 + 0.16 (n=7)	100%
Nle	None	16.42 + 0.05 (n=4)	100%
Ser	O-Benzyl (Bzl)	9.57 + 0.18 (n=10)	13% ± 2.7
		12.87 + 0.22 (n=10)	26% ± 5.3
		18.16 + 0.25 (n=10)[b]	37% ± 4.3
		19.62 + 0.23 (n=10)	13% ± 4.8
		20.25 + 0.22 (n=10)	11% ± 2.5
Thr	O-Benzyl (Bzl)	12.93 + 0.08 (n=7)	23% ± 3.4
		18.50 + 0.10 (n=7)	18% ± 2.0
		18.75 + 0.09 (n=7)	23% ± 3.2
		20.69 + 0.09 (n=7)[b]	36% ± 4.0
Tyr	Dichlorobenzyl (CLB)	22.70 + 0.04 (n=2)	100%
Tyr	2-Bromo Z (BrZ)	13.47 + 0.14 (n=4)	29% ± 4.0
		23.26 + 0.20 (n=4)[b]	71% ± 4.9

[a]Elution conditions of each PTH-amino acid is described in Figure 2.

[b]Indicates the elution position of the major peak of each derivative; n = number of values used to determine the average elution time, ± standard deviation.

Table 5. Efficiency of the Removal of C-Terminal Residues from Solid-Phase Resin Supports During Gas-Phase Sequencing of Synthetic Peptides

Peptide	Resin Support	Residue & Linkage	Initial Yield (pmol)[a]	Final Yield (pmol)[b]	Percent C-Term Removed
159–169	BHA	Asn,amide	751	657	87.5%
			905	772	85.3%
185–199	MHM	Leu,ester	722	16	2.2%
			1027	155	15.1%
131–147	MHM	Gly,ester	1672	545	32.6%
			207	65	31.4%
26–41	PAM	Leu,ester	846	207	24.4%
36–51	PAM	Lys,ester	869	95	10.9%
			1770	173	9.8%
81–92	PAM	Ser,ester	931	100	10.4%

[a]Total yield of the first N-terminal residue sequenced assuming a 95% initial yield.

[b]Total yield of the last residue (C-terminus) linked to the resin assuming a sequencing repetitive yield of 93%.

ACKNOWLEDGEMENT

This work was supported in part by a grant from the Mellon Foundation and by funds of the Mayo Clinic/Mayo Research Foundation. Appreciation to Barbara Baldus and Sharon Jones for the preparation of the typescript is expressed.

REFERENCES

1. R. M. Hewick, M. W., Hunkapiller, L. E. Hood, and W. J. Dreyer, A gas-liquid solid phase peptide and protein sequenator, J. Biol. Chem. 256:7990 (1981).
2. M. W. Hunkapiller, R. M. Hewick, W. J. Dreyer, and L. E. Hood, High Sensitivity sequencing with a gas-phase sequenator, in: "Methods in Enzymology," Vol. 91, C. H. W. Hirs and S. N. Timasheff, eds., Academic Press, New York (1983).
3. M. W. Hunkapiller and L. E. Hood, Protein sequence analysis: Automated microsequencing, Science 219:650 (1983).
4. F. S. Esch, Polypeptide microsequence analysis with the commercially available gas-phase sequencer, Anal. Biochem. 136:39 (1984).
5. A. S. Bhown, and J. C. Bennett, The use of HPLC in protein sequencing, in: "Handbook of HPLC for the Separation of Amino Acids, Peptides and Proteins," Vol. II, W. S. Hancock, ed., CRC Press, Boca Raton (1984)
6. M. W. Hunkapiller and L. E. Hood, Analysis of phenyltiohydantoins by ultrasensitive gradient HPLC, in: "Methods in Enzymology," Vol. 91, C. H. W. Hirs and S. N. Timasheff, eds., Academic Press, New York (1983).

7. J. E. Strickler, M. W. Hunkapiller, and K. J. Wilson, Utility of the gas-phase sequencer for both liquid- and solid-phase degradation of proteins and peptides at low picomole levels, Anal. Biochem. 140:553 (1984).

8. G. R. Matsuedu, E. Haber, and M. N. Margolies, Quantitative solid-phase Edman degradation for evaluation of extended solid-phase peptide synthesis, Biochemistry 20:2571 (1981).

9. S. B. Kent, M. Riemen, M. LeDoux, and R. B. Merrifield, A study of the Edman degradation in the assessment of the purity of synthetic peptides, in: "Methods in Protein Sequence Analysis," M. Elzinga, ed., Humana Press, Clifton (1982).

10. J. Simmons and D. H. Schlesinger, HPLC of side-chain protected amino acid phenylthiohydantoins, Anal. Biochem. 104:254 (1980).

11. D. H. Schlesinger, HPLC of side-chain protected phenylthiohydantoins: Application to solid-phase peptide synthesis, in: "Methods in Enzymology," Vol. 91, C. H. W. Hirs and S. N. Timsheff, eds., Academic Press, New York (1983).

12. M. Noda, Y. Furutani, H. Takahashi, M. Toyosata, T. Tanabe, S. Shiisu, S. Kikyotani, T. Kayano, T. Hirose, S. Inayama, and S. Numa, Cloning and sequence of calf cDNA and human genomic DNA encoding alpha-subunit precursor of muscle acetylcholine receptor, Nature 305:818 (1983).

13. J. G. Pierce and T. F. Parsons, Glycoprotein hormones: Structure and function, Ann. Rev. Biochem. 50:465 (1981).

14. J. Koketsu and M. Z. Atassi, Immunochemistry of sperm-whale myoglobin-XVI: Accurate delineation of the single region in sequence 1-55 by immunochemical studies of synthetic peptides, Immunochemistry 11:1 (1974).

15. D. J. Mc Cormick and M. Z. Atassi, Localization and synthesis of the acetylcholine binding site in the alpha-chain of Torpedo californica acetylcholine receptor, Biochem. J. 224:995 (1984).

16. D. J. Mc Cormick, V. A. Lennon, and M. Z. Atassi, Synthesis of an antigenic site of native acetylcholine receptor peptide 159-156 of Torpedo acetylcholine receptor alpha-chain, Biochem. J. 226:193 (1985).

17. A. R. Mitchell, S. B. Kent, M. Engelhard, and R. B. Merrifield, A new synthetic route to t-Boc aminoacyl-4-(oxymethyl)phenylacetamido-methyl-resin, and improved support for solid-phase peptide synthesis, J. Org. Chem. 43:2845 (1978).

18. E. Kaiser, R. L. Colescott, C. D. Bassinger, and P. T. Cook, Color test for the detection of free terminal amino groups in the solid-phase synthesis of peptides, Anal. Biochem. 34:595 (1970).

19. D. J. Mc Cormick and M. Z. Atassi, Antigenic structure of human hemoglobin: Delineation of the antigenic site (Site 2) within region 41-65 of the alpha-chain by immunochemistry of synthetic peptides, J. Protein Chem. in press (1985).

SEPARATION OF AMINOACID PHENYLTHIOHYDANTOIN DERIVATIVES BY HIGH PRESSURE LIQUID CHROMATOGRAPHY

Joseph L. Meuth and J. Lawrence Fox

Department of Molecular Biology
Abbott Laboratories
Abbott Park, Illinois 60064

The Edman degradation is the preferred method for determining the primary structure of a protein (1). During this process, the amino terminal residue of the protein is sequentially removed and routinely converted to its PTH derivative. The PTH amino acid derivative can then be identified by thin-layer chromatography (2), gas-liquid chromatography (3), HPLC on reversed phase supports (4-17), or reconversion to the free amino acid and subsequent analysis on an amino acid analyzer (18). HPLC is the most commonly used method of identification.

In recent years, numerous reversed phase HPLC separations for PTH amino acids have been reported (4-17). Both isocratic and gradient elution systems have been utilized. Isocratic systems are usually very simple and fast, provide quiet baselines, and are quite reproducible since only one buffer and one buffer pump are needed for the analysis. However, the lack of adequate resolution is a serious problem in the isocratic system. Gradient elution systems can be complicated, have longer elution times, often possess sizable baseline shifts, and, if not controlled properly, the reproducibility can suffer. On the other hand, gradient elution provides more versatility and higher resolution.

Since this laboratory has the capability of performing microsequencing analysis on polypeptides at the picomole level, a rapid, high sensitive analysis of PTH amino acids was desired. HPLC offers the best performance at this time. A new protocol was developed for the separation of PTH amino acids using reversed phase HPLC since satisfactory results were not readily obtained using some of the methods reported in the literature. This method is simple to use and can routinely identify PTH amino acids in the low picomole range. This report outlines the methodology now being used in this laboratory.

MATERIALS AND METHODS

Analyses were performed using an Altex 344 binary liquid chromatograph equipped with two Model 112 pumps, a Model 421 controller, a Model 340 system organizer, and a Model 160 fixed wavelength detector set at 254 nm. Samples were injected using a Model 210 manual injector with a 10 μl injection loop or a Model 504 autosampler with a 20 μl injection loop. The column was an Altex Ultrasphere ODS:PTH column (5 μ, 4.6 x 250 mm) preceded by an Altex PTH Amino Acid (ODS) precolumn (4.6 x 45 mm). The column and precolumn were jacketed and held to a constant temperature of with a Lauda MGW Model S-1 recirculating water bath. The chromatographic response was recorded and integrated using a Hewlett-Packard Model 3390A integrator.

Ammonium hydroxide (reagent grade) was obtained from Fisher. Acetonitrile (MeCN) was obtained from Burdick and Jackson. High quality water was obtained by passing distilled water through a Sybron/Barnstead Nanopure II system equipped with an Organopure unit. PTH amino acid standards were obtained from Pierce and were redissolved in a 2:1 mixture of water and MeCN.

The elution of the PTH amino acids was achieved using a combination of isocratic and linear gradient separation steps. Both buffers A and B were prepared from water, MeCN, and a 1 M stock solution of acetic acid adjusted to pH 4.5 with concentrated ammonium hydroxide. In both buffers the final ammonium-acetate concentration was 17 mM. Finally, buffer A contained 10% MeCN and buffer B contained 90% MeCN. Each buffer was degassed using sonication and vacuum at the beginning of each day. Buffer preparations more than 48 hrs. old were not used.

Initial chromatographic conditions were 70% buffer A and 30% buffer B. A flowrate of 1.2 ml/min and a temperature of 50°C were used throughout each analysis. Beginning one minute after injection, the concentration of buffer B was increased linearly to 50% over a three minute period. After being held at this level for eight minutes, the concentration of buffer B was reduced to 30% over a one-minute period and, following a re-equilibration time of eight minutes, another sample was injected. Total time per analysis was approximately 21 minutes. Any slight baseline rise observed when the concentration of the second buffer was increased was eliminated by the addition of a few microliters of acetone per liter to buffer A to increase its ultraviolet absorbance. This introduction of acetone had no obvious effect on the separation or retention of the PTH amino acids.

RESULTS AND DISCUSSION

The isocratic procedure of Tarr (7) and its step-gradient variation as reported by Black and Coon (4) were used as the starting point for the development of the protocol presented here. Tarr's method (7) provided rapid analysis but lacked complete resolution between several of the PTH's, particularly in the early part of the chromatography. Black and Coon (4) reported an improvement in the resolution, but this methodology proved to be unusable in this laboratory because of excessive baseline shifts and undesirable baseline noise at high sensitivity.

In the present protocol the optimized separation of the PTH amino acids was accomplished by the use of a rapid linear gradient and the adjustment of the ionic strength and MeCN concentration of the buffers used to effect chromatography. Because of the extensive studies conducted by others (4,6,7) it was not necessary to use a temperature other than 50°C, except in one special case as reported later. Results, as shown later, were also obtained which indicated that a pH of 4.5 provided the best separation of the PTH's. Only an Altex Ultraphere ODS:PTH column was used because previous experience in this laboratory indicated that this column provided the best separation of the PTH amino acids. Finally, special attention was given to the observation that the various PTH amino acids elute in two different groups. The first group contained the early eluting, hydrophilic PTH derivatives and the second group contained the slower eluting, hydrophobic PTH derivatives. Adjustment of chromatographic parameters sometimes had a positive effect on one group and a negative effect on the other, therefore, this always had to be a consideration.

Initially, the use of a linear gradient accomplished two things. First, it eliminated the baseline problems seen with a step gradient and, second, it greatly improved the resolution between most, but not all, of the PTH's. The linear gradient produced a very quiet baseline with only a slight rise occurring as the concentration of the second buffer increased. A completely flat baseline was obtained by the addition of a few microliters of acetone per liter of the first buffer to increase its ultraviolet absorbance. After several experiments, it was observed that the timing for the beginning and the length of the gradient affected the separation of various PTH amino acids, particularly that of PTH-Ala and PTH-Tyr. The most acceptable separation of the PTH's was obtained by introducing, one minute after injection, a three minute linear gradient with the second buffer from initial to final conditions. It should also be noted that the gradient may be a function of the size of the gradient mixing chamber. The instrument used in this study has a 1.2 milliliter mixing chamber volume - a larger mixer (e.g. 3.0 ml) may work well with a shorter gradient time.

The pH and the ionic strength of the buffers were also considered. Illustrated in Figure 1 is the effect of pH. Only those PTH's whose elution was affected are shown. From this experiment it was obvious that a pH of 4.5 provided the desired separation. The ionic strength of the buffers is particularly important for the proper elution of PTH-His and PTH-Arg (4,6,7,14). At a 50 mM acetate concentration it was not possible to separate PTH-Glu from PTH-His and PTH-Thr from PTH-Arg. Reducing the acetate concentration, and thus lowering the ionic strength, resulted in PTH-His and PTH-Arg being retained longer on the column. It was eventually determined that the most ideal elution positions were to have PTH-His eluting after PTH-Gly and to have PTH-Arg eluting ahead of PTH-Met. Exactly how low the acetate concentration has to be in order to achieve this elution pattern must be determined for the particular column being used. In the present study an acetate concentration of 17 mM was needed, however, in some early work with another column an acetate concentration of 11.25 mM was best. Periodically, it became necessary to increase the acetate concentration because as the column aged PTH-His and PTH-Arg tended to absorb more tightly onto it. This phenomenon has been reported earlier by Fohlman et al. (14), and they too overcame the effect by increasing the concentration of the buffer salts. Just recently, though, it was determined that the occasional cleaning of the column with 70% methanol or with 0.1% trifluoroacetic acid in 50% MeCN (19) would slow down the aging process and extend the useful life of the column.

The desired positioning of PTH-His and PTH-Arg was accomplished without having triethylamine-acetate in the system. It has been observed (4) that triethylamine-acetate could greatly influence the elution of PTH-His and PTH-Arg, but experiments here indicated that

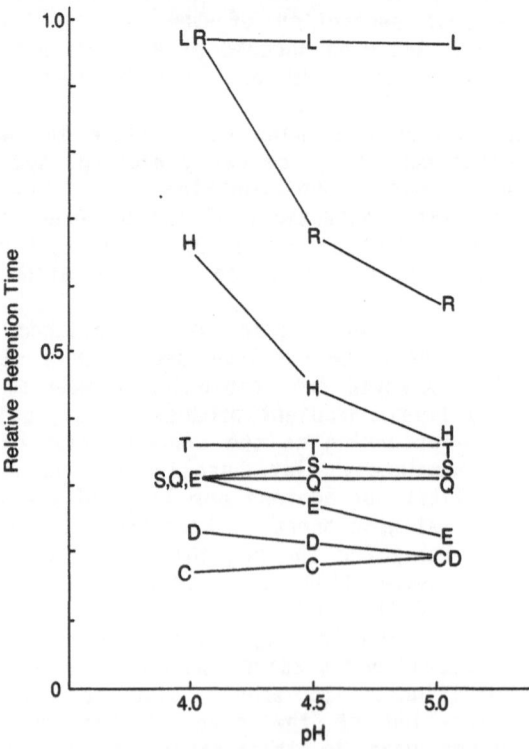

Figure 1.   Effect of pH on the separation of PTH amino acids.   Only those PTH's whose elution was affected are shown.

its presence was not absolutely necessary. By removing triethylamine-acetate from the system, the buffers became simpler to prepare, and the need for the redistillation of an organic solvent which degrades over time was eliminated.

Final optimization for the PTH separation was achieved by making adjustments to the MeCN concentration at either initial or final conditions or both. Resolution of pairs of certain PTH amino acids indicated if the proper MeCN concentration was being used. For example, the lack of resolution between PTH-Lys and PTH-Phe usually indicated that either the initial or final MeCN concentration was too low. If the initial MeCN concentration was too high, resolution was lost between PTH-Gln and PTH-Ser and also between PTH-Ala and PTH-Tyr. The resolution between PTH-Pro and PTH-Trp may be affected when the final MeCN concentration was too high. In the protocol presented here the best results were obtained with an initial MeCN concentration of approximately 34% and a final MeCN concentration of 50%. Again, other columns may require slightly different conditions. It should also be noted that the respective MeCN concentrations of the two buffers being used were 10% and 90%. This was established to facilitate the identification of the side-chain protected amino acids found in the synthetic peptides made in this laboratory. The elution of the PTH derivatives of some of these amino acids is accomplished more efficiently at an MeCN concentration higher than 50% (20). Other concentrations could be used if so desired.

Figure 2 illustrates the elution profile routinely obtained in this laboratory for the common PTH amino acids. Arrows indicate the elution positions of other frequently encountered PTH derivatives and certain contamination by-products of the Edman chemistry. The profile shows good peak sharpness and essentially baseline separation between all of the PTH amino acids. The last peak, PTH-Nle (internal standard), elutes from the column in about eleven minutes. Including re-equilibration, the total analysis time for each cycle requires approximately twenty-one minutes. The elution profile and retention times are very reproducible as was seen from three experiments consisting of fifteen consecutive injections conducted on separate days using different buffer preparations. For all three experiments the largest standard deviation in elution time was +0.047 minutes. Unlike other methods (5, 11, 14), the gradient and flow programming being used are very simple and produce a quiet baseline, thus allowing for high sensitivity. Detection of one to two picomoles of PTH amino acid is no problem. The system can be completely automated through the use of an autosampler. So far, no appreciable loss in the column selectivity has been observed when using the conditions outlined above.

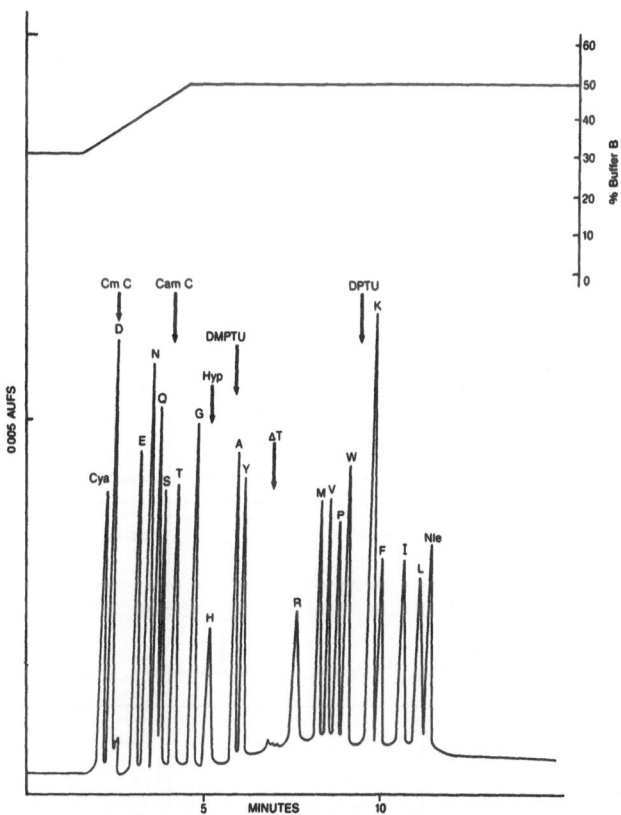

Figure 2. Separation of the common PTH amino acids (20 pmol each) via reversed phase HPLC on an Altex Ultrasphere ODS: PTH column (4.6 x 250 mm) preceeded by a precolumn (4.6 x 45 mm). Absorption was monitored at 254 nm. Vertical arrows indicate the eultion positions for a) carboxymethyl-cystiene, Cm-C; b) carboxyamidomethyl-cystiene, Cam-C; c) hydroxyproline, Hyp; d) dimethylphenylthiourea, DMPTU; e) dehydrothreonine, ΔT; and f) diphenylthiourea, DPTU. The protocol used is given in Materials and Methods. PTH-D and PTH Cm-C may be separated by isocratic rechromatography at 20% buffer B.

When performing actual sequence analysis some contamination peaks from the Edman chemistry are observed on the HPLC chromatogram. In particular, at 50°C DMPTU and PTH-Ala elute as a doublet and DPTU elutes near PTH-Trp. The effect of these contaminants depends upon how much of each is present. However, in accordance to the manufacturer's recommendations, a new, smaller conversion flask and a heptane wash after the phenylisothiocyanate coupling have been introduced into the Applied Biosystems vapor-phase sequenator. The two modifications have so decreased the amounts of DMPTU and DPTU observed with each cycle that these contaminants no longer significantly interfere with the analysis. It should be noted that lowering the analysis temperature to 45°C will result in an increased resolution of DMPTU and PTH-Ala which may prove to be advantageous. The trade-off here is that PTH-Trp will co-elute with DPTU at this lower temperature and, thus, it becomes important to know if the peptide being sequenced contains tryptophan or not.

A contamination peak resulting from the dithiothreitol (DTT) in the MeCN used as a solvent in the sequencer is also seen during the chromatography. This peak elutes under PTH-Glu, and, at times, can be quite large. The DTT is added to the MeCN to improve the yields of PTH-Ser, PTH-Thr, and PTH-His. Recent experience has indicated that sequence analysis proceeds very well without DTT and, therefore, its use has been discontinued, thereby resulting in the disappearance of the contamination peak on HPLC. Figure 3 illustrates the results of the sequence analysis of 500 pmol of sperm whale myoglobin without DTT in the MeCN. The appearance of PTH-Ser and its secondary peaks in cycle 3 are quite obvious and taken together probably indicate a 30% recovery yield. It should also be observed that there is not significant background peak at the elution position of PTH-Glu.

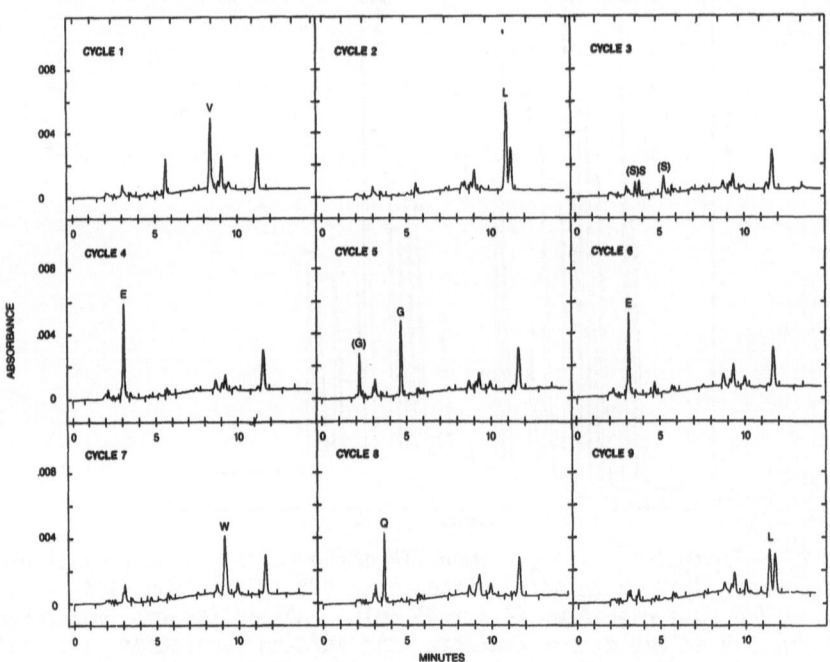

Figure 3. Sequence analysis of 500 pmol of sperm whale myoglobin. Absorbance was 0.010 AUFS at 254 nm. For each cycle 25% of the total sample was injected. Secondary peaks for Ser (cycle 3) and Gly (cycle 5) are indicated by ( ).

The chromatographic separation of the PTH-amino acids outlined in this report should be attractive to any laboratory actively involved in microsequencing because it is simple and reproducible, provides high resolution and peak sharpness with a quiet baseline, and is versatile. Detection of PTH amino acid below one picomole is possible, and the analysis time per cycle is relatively short. In addition, the entire system can be automated, thus providing more efficient use of personnel time. It is important to remember, though, that the differences between HPLC systems and columns may require the fine-tuning of a method to accomplish an optimal chromatographic separation.

REFERENCES

1. Edman, P., and Henschen, A. (1975) Mol. Biol., Biochem., and Biophys. 8, 232.

2. Inagami, T., and Murakami, K. (1972) Anal. Biochem. 47, 501.

3. Pisano, J.J., Brenzert, T.J., and Brewer, H.B. (1972) Anal. Biochem. 45, 43.

4. Black, S.D., and Coon, M.J. (1982) Anal. Biochem. 121, 281-285.

5. Hawke, D., Yaran, P.M., Shively, J.E. (1982) Anal. Biochem. 120, 302-311.

6. Kim, S.M. (1982) J. Chromat. 247 103-110.

7. Tarr, G.E. (1981) Anal. Biochem 111, 27-32.

8. Rose, S.M., and Schwartz, B.D. (1980) Anal. Biochem. 107, 206-213.

9. Somack, R. (1980) Anal. Biochem. 104, 464-468.

10. Greibrokk, T., Jensen, E., and Ostvold, G. (1980) J. Liq. Chromat. 3, 1277-1298.

11. Harris, J.U., Robinson, D., and Johnson, A.J. (1980) Anal. Biochem. 105, 239-245.

12. Chang, J.Y., Lehmann, A., and Wittman-Liebold, B. (1980) Anal. Biochem. 102, 380-383.

13. Henderson, L.E., Copeland, T.D., and Oroszlan S. (1980) Anal. Biochem. 102, 1-7.

14. Fohlman, J., Rask, L., and Peterson, P.A. (1980) Anal. Biochem. 106, 22-26.

15. Johnson, N.D., Hunkapiller, M.W., and Hood, L.E. (1979) Anal. Biochem. 100, 335-338.

16. Bhown, A., Mole, J.E., Weissinger, A., and Bennett, J.C. (1978) J. Chromat. 148, 532-535.

17. Frank, G., and Strubert, W. (1973) Chromatographia 6, 522-524.

18. Zimmerman, C.L., Appella, E., and Pisano, J.J. (1977) Anal. Biochem. 77, 569-573.

19. Hawke, D., personal communication.

20. Fox, J. L., Kim, Y., and Sarin, V. (1985) J. Cell. Biochemistry Supplement 9B, 131.

VIII. Computer Analysis of Protein Sequence Data

HYDROPHOBIC MOMENTS AS TOOLS FOR ANALYSIS OF PROTEIN SEQUENCES AND STRUCTURES

David Eisenberg, William Wilcox, and Steven Eshita

Molecular Biology Institute, Dept. of Chem. and Biochem.

University of California, Los Angeles, California 90024

INTRODUCTION

A challenge confronting protein chemists today is the interpretation of the structural and functional implications of DNA-derived amino acid sequences of proteins, which are so readily available as fruits of recombinant DNA methods. Knowing only an amino acid sequence, we would like to be able to predict the three-dimensional structure of the protein, also its function, its ligands and regulators, and even its sites for post-translational chemical modification. Of course we must presently settle for much less, as we seek new links between sequence and structure.

This paper describes several computational tools which relate amino acid sequence to structure. They all utilize hydrophobicities and hydrophobic moments, quantitative concepts that rest on measurements of free energies of transfer. The concepts are crude compared to those of the quantum chemist, and the parameters that define them are only roughly known compared to those from atomic physics. However, the concepts have an important feature for treating biochemical structures: they deal with the free energy of folding. The free energy is the physical property that determines the structure at constant temperature and pressure. For this reason, these concepts may present a useful alternative to conventional energetic methods in attempting to understand protein folding.

1. HYDROPHOBICITIES

Biophysical chemists at least since Cohn and Edsall[1] have recognized that the free energy change for transfer of a chemical group from an apolar solvent to water is to some extent independent of the larger molecule to which the group is attached. For amino acid residues, these free energies of transfer have been termed "hydrophobicities." An early determination of values for some of the amino acid residues was made by Nozaki and Tanford[2], who measured the free energy of transfer of amino acids from ethanol and other apolar solvents to water. These hydrophobicity values are given with others in Table I.

Each hydrophobicity represents the free energy of transfer of one mole of residue at a defined concentration from an apolar solvent to water. Apolar residues, such as Trp or Leu have positive values. Polar residues such as Glu and Lys have negative values. The magnitude of the hydrophobicity is a measure of how much the side chain seeks or avoids association with water. Because the spontaneous folding of a chain of amino acids into a compact globular protein must depend to some extent on association with water or avoidance of it by its residues, it might be expected that the sequence of amino acid hydrophobicities would be a determinant of protein structure. In fact, the extent of correlation of residue hydrophobicities in two protein sequences seems to reflect the similarity of their three-dimensional structures[3].

There have now been numerous determinations of residue hydrophobicities by a variety of methods[4-9], and much discussion about the proper interpretation[10-14] of them. Rose et al[13] have recently shown that the magnitude of the residue hydrophobicity as measured by Nozaki and Tanford[2] reflects the the tendency of a residue to be buried beneath the aqueous-accessible surface of the protein. Even with this reassurance that the free energies of transfer are related to structure, it is difficult to know exactly which scale of hydrophobicities to work with. Table I contains our normalized Concensus scale[20] which was synthesized from five earlier scales[2,4-7]. It also contains a scale determined by Fauchere and Pliska[9] from the transfer of amino acid analogues from octanol to water. Table I also contains a scale based on additive atomic contributions, the parameters for which were determined from the data of Fauchere and Pliska. The derivation and use of this ATOM scale will be described elsewhere[15].

2. HYDROPHOBIC MOMENTS

Hydrophobic moments describe asymmetry of hydrophobicity, or amphiphilicity[10,16]. The hydrophobic dipole moment is the exact analog of the electric dipole moment, except that it measures the asymmetric distribution of the hydrophobicity in a structure, rather than the asymmetry of electric charge. An alpha helix having apolar side chains protruding from one cylindrical surface and polar ones from the opposite cylindrical surface has a large hydrophobic moment. In this case, the hydrophobic moment is simply a quantitative expression of the familiar idea of helix amphiphilicity[17-19]. Hydrophobic moments can also be used to describe amphiphilicity in beta structures, irregular structures, entire molecules, or even small groups of atoms[10,15,27].

The hydrophobic moment of a protein, or of a segment of it, can be calculated from the atomic coordinates in cases where they are known. This is called the <u>structural hydrophobic moment</u>, and is given by

$$\vec{\mu} = \sum_{\text{residues},n} H_n \vec{s}_n \tag{1}$$

in which H_n is the hydrophobicity of the nth residue of the segment, and $\vec{s}_n$ is a unit vector pointing from its alpha carbon atom to the center of the residue's side chain. Thus the moment is a vector sum of the directions of the side chains, with each being weighted by its hydrophobicity. A very hydrophobic residue contributes strongly to the moment, whereas a very hydrophilic residue (having a negative value for H) contributes in the opposite direction. The largest possible moment for a given segment is achieved when hydrophobic residues protrude from one side, and hydrophilic residues protrude from the other, because then both sets contribute to the sum with the same sign.

Even if the atomic coordinates of a protein are not known, it is still possible to estimate a hydrophobic moment, called the <u>sequence hydrophobic moment</u>. Normally it is assumed that the segment has some particular regular secondary structure, and that the the side chains protrude perpendicular to the axis of the segment. The magnitude of the sequence hydrophobic moment is then given by

$$\mu = \{[\sum_{residues,n} H_n \sin(\delta n)]^2 + [\sum_{residues,n} H_n \cos(\delta n)]^2\}^{\frac{1}{2}} \qquad (2)$$

Again the sum is over all residues of the segment, and H_n is the hydrophobicity of the nth residue. Delta is the angle, measured in radians, at which successive side chains emerge from the central axis of the structure. For an alpha helix, delta is 100°; for a strand of beta sheet, delta is about 160°.

Table I. Amino Acid Hydrophobicities (Free Energies of Transfer). Observed and calculated free energies of transfer, are given in units of kcal mol^{-1}. ΔG_{FP} and ΔG_{TAN} are experimental free energies of transfer relative to glycine. ΔG_{ATOM} is a calculated value.

Residue	Normalized Consensus[+]	ΔG_{FP} [*]	ΔG_{ATOM} [#]	ΔG_{TAN} [%]
Gly	0.48	(0)	(0)	(0)
Ala	0.62	0.42	0.67	0.50
Val	1.08	1.66	1.5	1.50
Leu	1.06	2.32	1.9	1.80
Ile	1.38	2.46	1.9	
Pro	0.12	0.98	1.2	
Cys	0.29	1.34	0.38	
Met	0.64	1.68	2.4	1.30
Thr	-0.05	0.35	0.52	0.40
Ser	-0.18	-0.05	0.01	-0.30
Phe	1.19	2.44	2.3	2.50
Trp	0.81	3.07	2.6	3.40
Tyr	0.26	1.31	1.6	2.30
Gln	-0.85	-0.30	-0.22	
Asn	-0.78	-0.82	-0.60	
Glu	-0.74	-0.87	-0.76	
Asp	-0.90	-1.05	-1.2	
His	-0.40	0.18	0.64	0.50
Lys	-1.50	-1.35	-0.57	
Arg	-2.53	-1.37	-2.1	

+ An average of five hydrophobicity scales.[10,20]
* Free energy of transfer for amino acid analogues from 1-octanol to water determined by Fauchere & Pliska[9]
# A calculated, atom-based free energy of transfer[15], based on data from Reference 9.
% Free energy of transfer for amino acid side chains determined by Nozaki & Tanford[2]

3. THE HYDROPHOBIC MOMENT PLOT

The hydrophobic properties of an alpha helix can be classified on a hydrophobic moment plot[20,21], an example of which is given in Fig. 1. The vertical coordinate registers the magnitude of the hydrophobic moment per residue. The curved boundary shows the largest moment per residue that any alpha helix can have for a given average hydrophobicity. The curve intersects the horizontal axis at a value of 1.38, the hydrophobicity per residue of a hypothetical poly-isoleucine alpha helix. Other points on the curve represent hypothetical helical copolymers of isoleucine and arginine.

Alpha helices of different types tend to cluster in different regions of the hydrophobic moment plot. Helices from globular proteins tend to plot in the region of Fig. 1 labelled GLOBULAR, whereas probable transmembrane helices tend to plot in the two nearly triangular regions at the lower right. The alpha helix of the lytic protein melittin, and related molecules plot in the region labelled SURFACE, as discussed below.

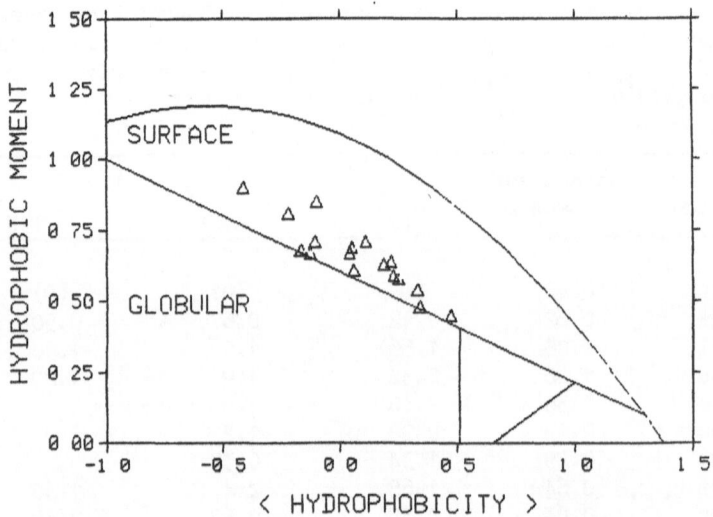

Figure 1. Hydrophobic moment plot for surface-seeking helices. Each triangular point characterizes one of the lytic peptides listed in Table II. The horizontal coordinate gives the hydrophobicity per residue of an 11-residue segment of the peptide. This is the segment having the largest hydrophobic moment. The hydrophobicities are on the normalized scale of Table I. The vertical coordinate gives the hydrophobic moment of the segment as calculated from Equation 2 of the text, using the same hydrophobicity scale. It is assumed that the peptide is an alpha helix, with side chains protruding normal to the helix axis. The boundary curve represents the largest hydrophobic moment that a helix of each average hydrophobicity can have. The equation of the boundary curve is: Moment = $-0.315\langle H\rangle^2 - 0.358\langle H\rangle + 1.087$.

To compare the hydrophobic properties of different alpha helices, it is convenient to choose a standard number of residues. In earlier work we have found that 11 residues is a useful standard, and for purpose of comparison, we select the 11 residue segment within a helix that has the largest hydrophobic moment. In computing the hydrophobic properties for a given protein, we move an 11 residue "window" through the amino acid sequence, and for each position, compute the mean hydrophobicity and hydrophobic moment.

For any given peptide, one usually displays only the point corresponding to the 11 residue segment having the largest moment.

Table II. Lytic peptides whose hydrophobic properties are represented by the triangular points of Figure 1. Each point refers to an 11-residue helix starting at the indicated residue.

Peptide	Starting Residue	Plot Coordinates		Reference
		$\langle H \rangle$	Moment	
Cecropin A	3 (Lys)	-0.22	0.80	36
Cecropin B	3 (Lys)	-0.41	0.89	36
Delta-hemolysin				
S. aureus	16 (Ile)	-0.11	0.70	37
S. aureus (canine strain)	16 (Ile)	-0.17	0.67	37
Mellitin				
A. mellifera	12 (Gly)	0.25	0.57	24
A. florea	12 (Gly)	0.34	0.47	38
Artificial cytotoxin	12 (Leu)	0.19	0.62	37
Artificial cytotoxin	2 (Leu)	-0.10	0.84	40
Bombolitin I	2 (Lys)	0.06	0.60	26
Bombolitin II	2 (Lys)	0.05	0.68	26
Bombolitin III	2 (Lys)	0.11	0.70	26
Bombolitin IV	2 (Asn)	0.04	0.66	26
Bombolitin V	6 (Ile)	0.47	0.44	26
Mastoparan	2 (Leu)	0.23	0.58	26
Crabrolin	2 (Leu)	0.33	0.54	26

4. SURFACE SEEKING HELICES

In earlier work we found that 8 hemolytic peptide toxins all plot in the SURFACE region of the hydrophobic moment plot, assuming they are helical[20]. Two of these are melittins from the venoms of two species of bee. Melittin is known to be an alpha helix from crystal structure studies[22,23], and is known to accumulate at a air-water surface[24]. The origin of melittin's surface-seeking activity is almost certainly that it forms a highly amphiphilic alpha helix, with an apolar side and a polar side. The helix aligns itself parallel to the air-liquid surface, with its apolar side toward air and its polar side in water. In other words, a protein segment with large hydrophobic moment tends to seek the surface between an aqueous and an apolar phase. This same principle is probably involved in melittin's lytic action, with the apolar phase now being lipid rather than air[25].

Recently a new class of 17-residue lytic peptides has been discovered in the venom of the bumblebee, the bombolitins[26]. Five such peptides lyse erythrocytes and liposomes, release histamine from rat peritoneal mast cells, and stimulate phospholipase A_2. All five peptides have large hydrophobic moments if assumed to be alpha helical, and plot in the SURFACE region of the hydrophobic moment plot, in positions similar to those of the other lytic peptides. Similarly mastoparan (from wasps, hornets, and yellow jackets), and crabrolin (from European hornets) plot in the same region.

These seven additional peptides bring to 15 the total of apparently amphiphilic peptides whose hydrophobic properties are similar. This strengthens the hypothesis that lytic, highly surface-seeking peptides can be recognized from their amino acid sequences by their positions on the hydrophobic moment plot[16].

Transmembrane protein pores may also contain highly amphiphilic helices. Finer-Moore and Stroud[35] and Eisenberg et al[20] have proposed algorithms for detecting these from amino acid sequences, both utilizing hydrophobic moments.

5. TRANSMEMBRANE HELICES

Can membrane-penetrating protein segments be recognized from the amino acid sequence alone? This question has been addressed by several research groups[8,20,28,29,30]. It is clear that membrane-penetrating segments are more apolar on the average than segments of globular proteins, when the average is taken over 20 to 30 residues, the expected length of a transmembrane helix. But from an amino acid sequence it is not a simple matter to distinguish transmembrane segments from non-membrane segments. Quite a few segments in globular proteins are more apolar on the average than one of the putative transmembrane helices in bacteriorhodopsin[31].

A possible solution to this dilemma has been proposed[20]. It may be that association of protein helices with membranes is cooperative. The first transmembrane association may require a very apolar helix, or two quite apolar helices. Additional transmembrane helices may associate with the others if only moderately apolar. This simple cooperative model was found to fit most, but not all, data on membrane associated proteins[20].

6. STRUCTURAL HYDROPHOBIC MOMENTS

The structural hydrophobic moment, defined by Equation 1 above, can help in visualizing the hydrophobic component of the free energy of protein folding. For each segment (alpha helix, strand of beta structure, or irregular stretch of sequence) the magnitude and direction of the hydrophobic moment is computed. Then the relative positions and directions of the moments are examined by computer graphics. Two general features are found in globular proteins[10]: (1) Hydrophobic moments of segments at the aqueous surface point inwards (towards greater hydrophobicity); and (2) Hydrophobic moments of neighboring segments tend to oppose each other. The result is that in smaller proteins formed from two layers of structure[32], the hydrophobic moments of virtually all segments point inwards toward the center. In a larger protein, triosephosphate isomerase, having four layers of polypeptide chains, the moments of the inner beta barrel tend to point outwards, opposing the moments of the surrounding alpha helices which themselves point inwards. In other words, the very center of the enzyme is less apolar on the average that the region midway between the center and the aqueous surface. In structures with segments of low amphiphilicity, hydrophobic moments may not reveal all hydrophobic forces.

7. OPPOSITION OF STRUCTURAL HYDROPOHOBIC MOMENTS IN MELITTIN

A fairly complete picture of the spatial organization of the hydrophobic free energy can be given for melittin, the peptide toxin of bee venom[24], whose three-dimensional structure is known[22,23]. The melittin monomer consists of 26 amino acid residues, folded as an alpha helix bent by 60° near the center. In aqueous salt solution, and in the crystalline state, melittin exists as a tetramer.

The hydrophobic interaction is the principal binding force among the four monomers within the melittin tetramer. There are no hydrogen bonds among the four monomers of the tetramer, and electrostatic charge interactions between melittin side chains are all repulsive. This is so because there are no negative charges in melittin: the C-terminus is amidated, and there are no acidic residues. Thus the tendency of the 40 large apolar side chains to avoid water must account in large measure for the stability of the tetramer.

A visual representation of the hydrophobic energy of melittin is provided by the structural hydrophobic moments in Fig. 2. Hydrophobic moments are shown by straight lines drawn from the center of each melittin monomer towards the direction of increasing hydrophobicity. Because melittin is more hydrophobic on the inside than the outside, the moments point inwards. Furthermore, their directions oppose each other. Another view of the melittin tetramer with its opposing moments is shown in Fig. 3. This opposition of moments from neighboring segments reflects the directional aspect of the hydrophobic energy of folding.

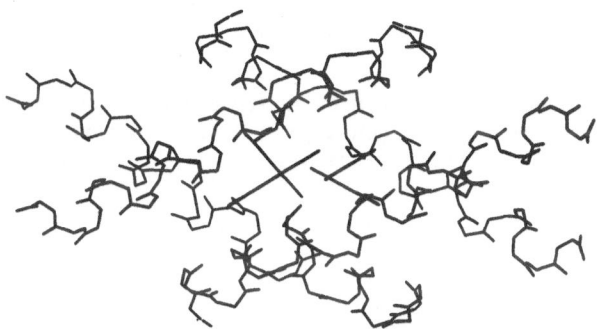

Figure 2. The melittin tetramer, with monomers 1 and 4 having their

A quantitative measure of the extent of opposition of moments of different protein segments in the folded structure is given by C

$$C = \frac{1 - \sum\limits_{\text{segments } i} \tilde{\mu}_i}{\sum\limits_{\text{segments } i} |\tilde{\mu}_i|} \tag{3}$$

Using values of the hydrophobicity from the atomic scale of Table I, we find that the hydrophobic moments of melittin monomers 1 and 3 are 13.2 kcal mol^{-1} Å, and those of monomers 2 and 4 are 12.0 kcal mole^{-1} Å. This difference is due to the slightly different conformation these monomers adapt. The vector sum of the moments for the four chains is 2.4 kcal mole^{-1} Å, showing that four moments are in nearly perfect opposition, as suggested by Figs. 2 and 3. The fractional cancellation determined from Equation 3 is 0.95.

carboxyl termini at the top. The hydrophobic moments are shown as straight lines drawn from the geometric center of each monomer in the direction of increasing hydrophobicity; that is, toward the center of the molecule. In this figure and the following ones, the lengths of the displayed vectors are all on the same relative scale.

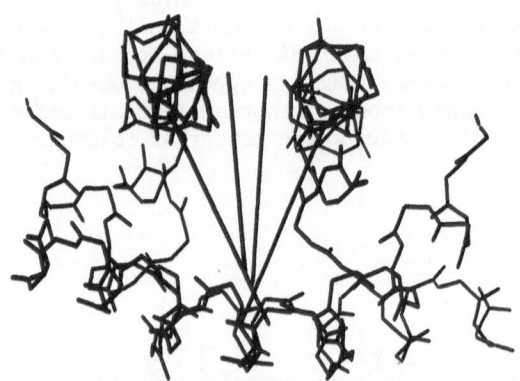

Figure 3. The melittin tetramer from another direction, roughly down the helix axes of monomers 3 and 4 (on top). Notice that the moments of 3 and 4 oppose each other only slightly, but are much more opposed in direction to the moments from monomers 1 and 2 (the nearly vertical lines directed upwards).

It is interesting to consider the melittin monomers in pairs (Table III). Monomers 1 and 2 run nearly antiparallel, and are bonded by only a few hydrophobic residues. Most of their apolar side chains extend towards monomers 3 and 4. This is shown in Fig. 3 by the directions of their hydrophobic moments. The moments form an angle of 62 with each other, and therefore only partly oppose each other: the fractional cancellation is 0.14 (Table III).

Table III. Opposition of Hydrophobic Moments in Melittin

Monomer(s)	Moment Length[#] (kcal Å)	Angle[*] (deg)	C (cancellation)[%]
1	13.2		
2	12.0		
3	13.2		
4	12.0		
1+2+3+4	2.4		0.95
1+2	21.6	62.	0.14
1+3	11.8	127.	0.55
1+4	7.6	145.	0.70

* Angle between two moments
\# Vector sum
% From Equation 3

In contrast, monomer pairs 1-3 and 1-4 have moments that more effectively oppose each other. They form angles nearer to 180°, and thus have much more substantial fractional cancellation. In other words, the hydrophobic forces are most directed between chains 1 and 3 and between chains 1 and 4. These cancellations are illustrated in Figs. 4 and 5.

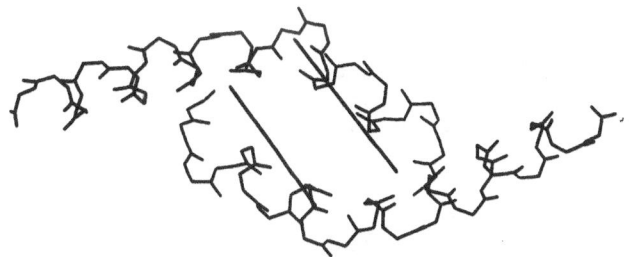

Figure 4. Stereo image of melittin monomers 1 and 3. The hydrophobic moments are more nearly opposed that those of monomers 1 and 2.

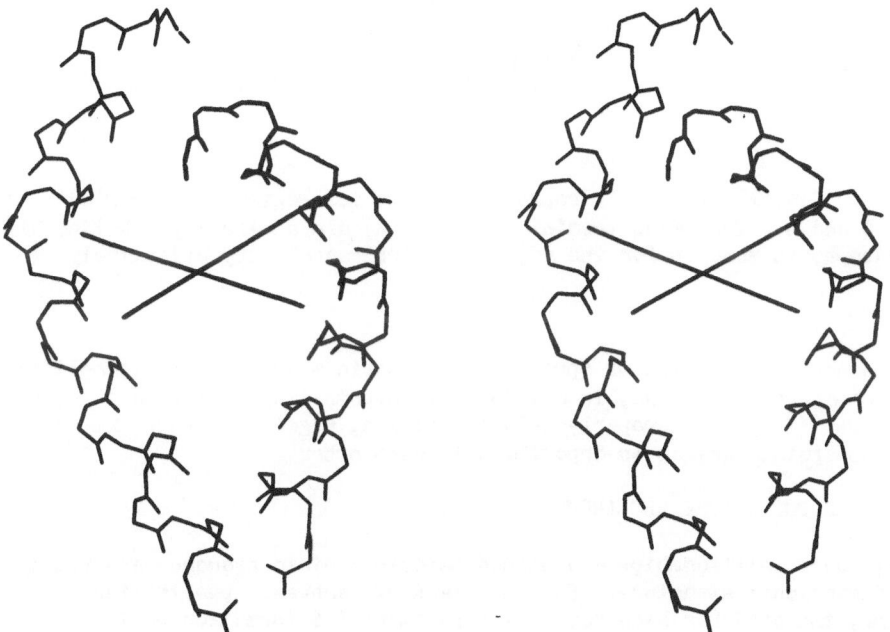

Figure 5. Melittin monomers 1 and 4, showing strong opposition of the hydrophobic moments.

A difference between hydrophobic moments and electrostatic moments is that hydrophobic moments do not interact over a distance. Their directions have significance only for adjacent segments of protein. There are regions of spatial separation between chains 1 and 3 of melittin (Fig. 4) as well as between chains 1 and 4 (Fig. 5). This raises the question of whether it is meaningful to discuss the opposition of the moments of these pairs of chains.

A more detailed analysis can be made by considering each melittin chain to be two segments: segment A running from residue 1 to residue 13, and segment B running from residue 15 to residue 26. Then hydrophobic moments can be calculated for each of these segments, and their magnitudes and directions analyzed. We have done this, and find that the moments of adjacent segments cancel even more effectively than do moments from the whole chains. For example, the angle between segments 1A and 3A (Fig. 6) is 121° and their fractional cancellation C is 0.51. The angle between segments 1B and 3B is 164°, and their fractional cancellation C is 0.86. Thus their average fractional cancellation is 0.68, considerably higher than the apparent cancellation of the entire chains 1 and 3 of 0.55 (Table III).

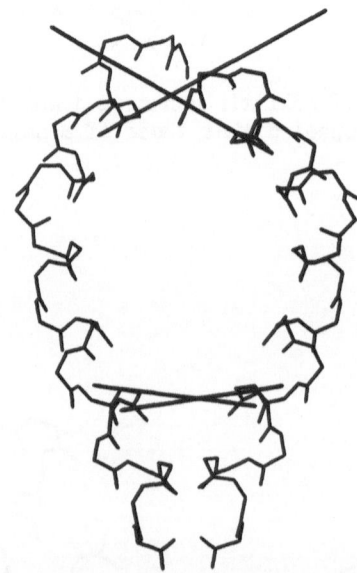

Figure 6. Cancellation of hydrophobic moments for segments A and B on chains 1 and 3. The amino termini (segments A) are at the top. Notice that the hydrophobic moments for the adjacent pairs cancel very effectively.

The coherent pattern of opposing hydrophobic moments found in melittin and in other real proteins, is not found in hypothetical, misfolded protein structures[32,33]. In improperly folded proteins, segment moments are small, and do not point inwards in opposition to each other.

8. AVENUES FOR FUTURE RESEARCH

Stronger relationships are needed between protein sequence and protein three-dimensional structure. For the class of membrane- penetrating proteins, the need has been for more experimental information on the structures. Pending work on the three-dimensional structures of membrane proteins will provide extremely important experimental systems against which algorithms can be calibrated.

In the wider area of protein folding studies, the need is for methods that take account of solvent effects[33], since water plays such an important role in protein folding. An atom based method for computing hydrophobic energy may help[15].

ACKNOWLEDGEMENTS

The authors gratefully acknowledge support from NIH research (USPHS GM 31299) and training (NIH GM 07185) grants; and NSF grants DMB 85-05867 and CHE-8509657. We thank G. Kriel for discussions.

REFERENCES

1. Cohn, E.J. & Edsall, J.T. in "Proteins, Amino Acids, and Peptides as Ions and Dipolar Ions", Chapter 9 (ACS Monograph Series, Reinhold Publishing Co., New York, 1943).
2. Nozaki, Y. & Tanford, C. J. Biol. Chem. 246, 2211-2217 (1971).
3. Sweet, R.M. & Eisenberg, D. J. Mol. Biol. 171, 470-488 (1983).
4. Chothia, C. J. Mol. Biol. 105, 1-14 (1976).
5. Janin, J. Nature (London) 277, 491-492 (1979).
6. Von Heijne, G. & Blomberg, C. Eur. J. Biochem. 97, 175 (1979).
7. Wolfenden, R., Anderson, L., Cullis, P.M. & Southgate, C.C.B. Biochemistry 20, 849-855 (1981).
7a. Hopp, T.P. & Woods, K.R. Proc. Natl. Acad. Sci. USA 78, 3824-3828 (1981).
8. Kyte, S. & Doolittle, R.F. J. Mol. Biol. 157, 105-132 (1982).
9. Fauchere, J.-L. & Pliska, V. Eur. J. Med. Chem.-Chem. Ther. 18, 369-375 (1983).
10. Eisenberg, D., Weiss, R.M., Terwilliger, T.C. & Wilcox, W. Faraday Symp. Chem. Soc. 17, 109-120 (1982).
11. Frommel, C. J. Theor. Biol. 111, 247-260 (1984).
12. Guy, H.R. Biophys. J. 47, 61-70 (1985).
13. Rose, G.D., Geselowitz, A.R., Lesser, G.J., Lee, R.H. & Zehfus, M.H. Science 229, 834-838 (1985).
14. Edsall J.T. & McKenzie, W.A. Adv. Biophys. 16, 53-183 (1983).
15. Eisenberg, D. & McLachlan, A.D. Nature (London), in press (1985).
16. Eisenberg, D., Weiss, R.M. & Terwilliger, T.C. Nature (London) 299, 371-374 (1982).
17. Perutz, M.F., Kendrew, J.C. & Watson, H.C. J. Mol. Biol. 13, 669-678 (1965).
18. Schiffer, M. & Edmundson, A.B. Biophys. J. 7, 121-135 (1967).
19. Kaiser, E.T. & Kezdy, F.J. Proc. Natl. Acad. Sci. USA 80, 1137-1143 (1983).
20. Eisenberg, D., Schwarz, E., Komaromy, M. & Wall, R. J. Mol. Biol. 179, 125-142 (1984).
21. Eisenberg, D. Ann. Rev. Biochem. 53, 595-623 (1984).
22. Terwilliger, T.C. & Eisenberg, D. J. Biol. Chem. 257, 6010-6015, (1982).
23. Terwilliger, T.C. & Eisenberg, D. J. Biol. Chem. 257, 6016-6022, (1982).
24. Haberman, E. Science, 177, 314-322 (1972).
25. Terwilliger, T.C., Weissman, L. & Eisenberg, D. Biophys. J. 37, 353-361 (1982).
26. Argiolas, A. & Pisano, J.J. J. Biol. Chem. 260, 1437-1444 (1985).
27. Eisenberg, D., Weiss, R.M. & Terwilliger, T.C. Proc. Acad. Sci. USA 81, 140-144 (1984).
28. Argos, P., Rao, J.K.M. & Hargrave, P.A. Eur. J. Biochem. 128, 565-575 (1982).
29. Kuhn, L.A. & Leigh, J.S.,Jr. Biochim. Biophys. Acta 828, 351-361 (1985).

30. Engelman, D.M., Goldman, A. & Steitz, T.A. <u>Meth. Enzymol.</u> 88, 81-88 (1981).

31. Engelman, D.M., Henderson, R., McLachlan, A.D. & Wallace, B.A. <u>Proc. Natl. Acad. Sci. USA</u> 77, 2023-2027 (1980).

32. Richardson, J. <u>Adv. Pro. Chem.</u> 34, 167-339 (1981).

33. Eisenberg, D., Wilcox, W. & McLachlan, A.D. <u>J. Cellular Biochem.</u>, in press (1985).

34. Novotny, J., Bruccoleri, R. & Karplus, M. <u>J. Mol. Biol.</u> 177, 787-818 (1984).

35. Finer-Moore, J. & Stroud, R.M. <u>Proc. Acad. Sci. USA</u> 81, 155-159 (1984).

36. Steiner, H., Hultmark, D., Engstrom, A., Bennich, H. & Bowman, H.G. <u>Nature</u> (London) 292, 246-248 (1981).

37. Fitton, J.E., Dell, A. & Show, W.V. <u>FEBS Letters</u> 115, 209-212 (1980).

38. Kriel, G. <u>FEBS Letters</u> 33, 241-244 (1973).

39. DeGrado, W.F., Kezdy, F.J. & Kaiser, E.T. <u>J. Amer. Chem. Soc.</u> 103, 679-681 (1981).

40. DeGrado, W.F. Personal communication (1983).

IDENTIFICATION OF PROTEIN SURFACES AND INTERACTION SITES

BY HYDROPHILICITY ANALYSIS

Thomas P. Hopp

Immunex Corporation
51 University Street
Seattle, WA 98101

The packing of hydrophobic amino acids into the interior of proteins has been recognized as a major factor in the creation of a three dimensional structure from a linear amino acid sequence. A related realization is that regions of protein sequences that contain clusters of charged and polar amino acids are usually highly exposed at the protein surface, and are therefore likely to be involved in protein interactions with other molecules, for example, binding to antibodies. Long hydrophobic segments have been recognized as probable membrane spanning sequences. In 1981, we published a method for averaging the hydrophilicity values of the amino acids in protein sequences (Hopp and Woods, ref. 1). The hydrophilicity profiles made by this procedure are useful in locating antigenic sites on native protein antigens, because major antigenicity is always found in the highest peaks (1,2). In 1982, Kyte and Doolittle (3) published a procedure that is virtually identical to ours, but made the additional observation that regions of wide hydrophobic maxima (valleys on our plots) are usually associated with membrane spanning segments of peptide chain. In this paper we compare our method to other similar procedures and describe several improvements to our method.

Comparison of hydrophilicity/hydrophobicity methods. Figure 1 shows the great similarity of the profiles generated using the different available scales of amino acid values. Almost without exception, the profiles show the same set of peaks and valleys, only differing in the magnitude of displacement from zero. Table 1 shows the success rate of each method in locating antigenic sites, determined as previously described (1). This is, to some extent, a measure of the correctness of a method in locating highly exposed segments. The methods all show some ability to locate antigenic sites, although our method, which was developed specifically for this purpose, is best. Furthermore, all procedures show a correlation of hydrophobic valleys with segments of secondary structure, as was pointed out earlier by Rose and Roy (4). A number of other such scales have been published, but were omitted from this comparison because many were very poor predictors of antigenic sites, and others did not have a full set of 20 amino acid values. The acrophilicity scale was developed by us (5) from direct observations on protein 3 dimensional structures. Interestingly, it locates secondary structure elements as well or better than any of the other methods, but is less successful in locating antigenic sites. This underscores our previous conclusions (1), that antigenic sites are a subset of surface locations, and that some highly exposed sites will not be immunogenic.

The major differences between scales occur at their top (hydrophilic) ends. Two groups of amino acids figure prominently: the <u>charged</u> amino acids, and the <u>small</u> amino acids. Our hydrophilicity method gives the four highly charged amino acids the same maximum value of +3.0, while the other scales spread these amino acids over a wide range. We found that the success rate for locating antigenic sites was improved by making all four values equal (1). This discrepency probably contributes to the lower prediction success rates of the other scales, because all of the charged residues are likely to contribute equally to protein-protein interactions by forming charge pairs. Furthermore, as seen in Table 1, the other scales are biased in favor of positively charged sites.

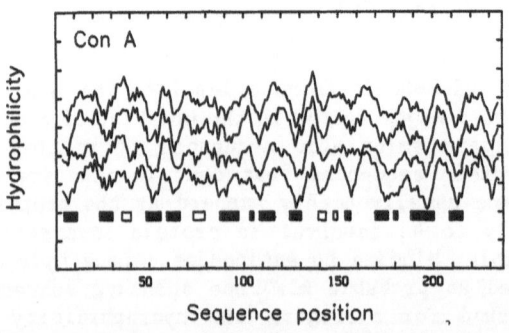

Fig. 1. Hydrophilicity analysis of concanavalin A by four sets of values. The scale has been compressed to facilitate comparisons, and each profile is offset by two hydrophilicity units from adjacent profiles. Top profile: values of Rose et al. (4); second profile: values of Kyte and Doolittle (3); third: Hopp & Woods (1); fourth: acrophilicity values (5). Strands of the two β pleated sheets are indicated at bottom. Solid bars, internal strands; open bars, edge strands.

<u>Relationship to secondary structure</u>. As mentioned earlier, all of the hydrophilicity/hydrophobicity scales show large hydrophobic (downward) deflections in regions of secondary structure. Our experience with a large number of proteins has led to the following observations: 1) the internal strands of β-pleated sheets are always correlated with deep hydrophobic valleys, while edge strands can have much higher values (cf. Figure 1). 2) similarly, large helices usually correlate with deep valleys because their central regions are packed against the core of the folded protein. In contrast the ends of large helices, and the whole extent of small helices are much less hydrophobic (cf. Figure 2). This is appropriate, because these segments, as well as the edges of β-pleated sheets, are necessarily highly exposed on protein surfaces. 3) even where valleys exist in the absence of secondary structure, the segment is usually buried. For example, the last valley of Con A is not involved

Table 1. Comparison of hydrophilicity methods.

A. Antigenic site selectivity. Percent correct among verifiable predictions.[a]

Peaks	Methods[b]							
	HYDRO4	H&W	Welling	Rose	K&D	Acro	C&F	Random
Highest	100	100	89	75	71	63	44	56
Top 3	88	75	76	68	65	59	43	57

B. Charge selectivity. Expressed as number of charges per six amino acids in peak hexapeptides.[c]

Peaks	Methods														
	H&W			K&D			Rose			Acro			HYDRO4		
	+	−	+/−	+	−	+/−	+	−	+/−	+	−	+/−	+	−	+/−
Highest	1.7	2.1	0.8	1.8	1.2	1.5	1.7	1.2	1.4	0.8	0.8	1.0	1.9	1.9	1.0
Top 3	1.4	1.8	0.8	1.5	1.3	1.2	1.5	1.2	1.3	0.7	0.8	0.9	1.6	1.7	0.9

[a]Determined as in reference 1.
[b]Methods are: HYDRO4, hydrophilicity with N,C and amino acid adjustments as described in this paper; H&W, original hydrophilicity (1); Welling, Welling et al. (8); Rose, Rose et al. (6); K&D, Kyte and Doolittle (3); Acro, acrophilicity (5); C&F, Chou and Fasman (9); Random, numbers from 3 to −3.4 assigned to the amino acids randomly.
[c]Averaged results for 70 proteins.

in secondary structure, but in fact represents a random coil segment that is indeed buried inside the molecule. In cytochrome c (Figure 3) valleys corresponding to residues 27-35 and 80-83 have no secondary structure, but actually represent critical regions of hydrophobic amino acids lining the heme binding pocket.

The observations above lead to the following generalization concerning hydrophilicity plots: the parts of the profile above the zero line represent the edge strands of β-pleated sheets, the ends of helices, and highly exposed loop regions connecting them, while the parts below zero represent the internal strands of β-pleated sheets, the central residues of large helices and occasional buried coils. It should also be emphasized that an averaging group length of 6 amino acids is optimal for extracting such information from protein sequences. This was established in our original hydrophilicity paper (1) and has been reconfirmed in our recent work in developing the acrophilicity scale (T. Hopp and J. Merriam, in preparation).

Improvements to the method. We have been investigating ways to improve the usefulness of hydrophilicity analysis, both in predicting antigenic sites and in locating structural elements of proteins. Several useful adjustments have been found: 1) because N and C termini are

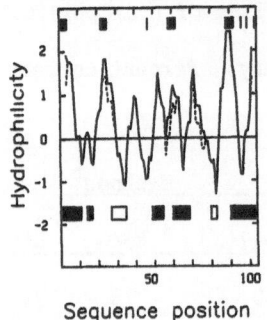

Fig. 2. Hydrophilicity profile for myoglobin. The solid profile was made by the HYDRO4 procedure; the dotted profile, by the original method (1). Bars below the profile represent the 8 helices of myoglobin. Slashes and bars above represent antigenic residues and segments.

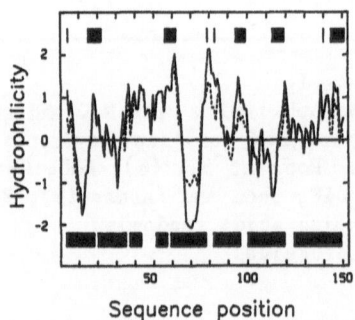

Fig. 3. Hydrophilicity profile for cytochrome c. Solid bars below profile represent helices. The open bars are two segments in contact with the heme moiety. Antigenic sites are indicated above the profile.

usually highly exposed, raising the value of the first and last hexapeptide averages increases prediction success rates. 2) the amino acids Gly, Ser, Thr, His, Tyr and Trp are somewhat ambiguous in their hydrophilic/hydrophobic behavior (6). We found that lowering the values of selected Gly, Ser and Thr residues (when four neighboring residues are hydrophobic) increases the ability to see transmembrane segments as broad valleys (Figure 4) while leaving the rest of the profile unchanged.

440

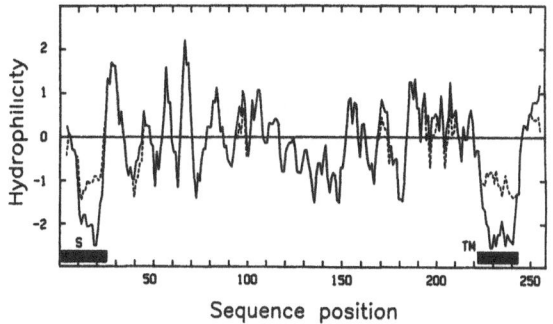

Fig. 4. *Hydrophilicity profile for the class II histocompatibility antigen chain, IAa^k. The solid profile was made by the HYDRO4 procedure; the dotted profile, by the original method. The Gly, Ser and Thr downward adjustments cause dramatic lowering of the profile in the regions of the signal peptide (S) and the transmembrane anchor segment (TM). The highest peaks probably represent sites of major antigenicity and other protein-protein interactions. The intervening valleys are probably the β-strands of this immunoglobulin-like molecule.*

Increasing the values of His, Tyr and Trp residues (when four neighboring residues are hydrophilic) on the other hand, increases success in predicting antigenic sites (Figures 2 and 3, Table 1).

Hydrophilicity in the future. The redundancy of the profiles made with four different scales shown in Figure 1 implies that the development of new hydrophilicity/hydrophobicity scales is not likely to produce much new information. Rather, new uses for the existing scales may lead to improvements like those seen for our N and C terminal and amino acid adjustments described above. At the same time, lessons already learned should not be forgotten. The use of hexapeptide averages has been advocated based upon experimental findings (1) but has not been generally applied. This is unfortunate because the use of other average sizes can obscure secondary structure information and, at the least, has served to obscure the redundancy of the methods developed to date.

In a recent survey of the immunological literature (5), we have uncovered an interesting fact. All of the major disease organisms for which detailed information is available, have proven to have major antigenicity associated with the most hydrophilic sites on their surface antigens. Thus, hydrophilicity analysis correctly predicted the location of neutralizing and/or strain specific antigenic sites on proteins from influenza, polio, foot and mouth disease, hepatitis B, herpes and common cold viruses, as well as on streptococcal M protein and gonococcal pilin. Most of these studies were carried out without using our method as a guide, so that the outcome strongly indicates that using hydrophilicity analysis in the future will lead to an accelerated success rate with other organisms. On the other hand, there recently have been a few reports of failed attempts at raising antiprotein responses using hydrophilic peptides. Experience in our own laboratories has demonstrated that it is indeed easy to fail to get a useful immunization using peptides, but also that minor changes in the structure of the immunogen can drastically alter the outcome in generating antiprotein antibodies. Therefore, it is clear that hydrophilicity analysis will only live up to its full potential to immunologists after ways of assuring the proper immunogenicity of the synthetic peptides have been found. Experimentation directed toward this goal is underway in our laboratories.

Some of the most exciting developments may arise simply through an increased understanding of the information already present in hydrophilicity profiles. We have recently observed that a number of important protein interactions other than antibody-antigen interactions can occur at the most hydrophilic sites (5,7). These include the compliment binding site on IgG, and receptor binding sites on apolipoprotein E, fibronectin and interleukin 2, to name a few (reviewed in reference 5). Sites of phosphorylation, proteolysis and other post-translational modifications also occur with great frequency at the most hydrophilic sites. In all probability, the acceptance of a standard hydrophilicity scale and averaging group length would go a long way toward fostering new discoveries, especially because it would enable investigators to communicate results and to understand each other more readily.

The hydrophilicity analysis methods described in this paper (HYDRO4 and ACRO3) are available from the author on request.

TABLE 2.

Comparison of hydrophilicity and acrophilicity values.

Hydrophilicity			Acrophilicity	
Asp	3.0		Gly	3.0
Glu	3.0		Pro	2.6
Lys	3.0		Asn	2.3
Arg	3.0		Asp	2.1
Ser	0.3		Ser	1.8
Asn	0.2		Lys	1.4
Gln	0.2		Glu	0.5
Gly	0.0		Arg	0.3
Pro	0.0		Thr	-0.1
Thr	-0.4		Gln	-0.2
His	-0.5		His	-0.4
Ala	-0.5		Ala	-0.5
Cys	-1.0		Val	-1.7
Met	-1.3		Met	-1.8
Val	-1.5		Tyr	-2.0
Leu	-1.8		Leu	-2.5
Ile	-1.8		Ile	-2.5
Tyr	-2.3		Cys	-2.6
Phe	-2.5		Phe	-2.7
Trp	-3.4		Trp	-3.0

REFERENCES

1. T.P. Hopp and K.R. Woods, Prediction of protein antigenic determinants from amino acid sequences, Proc. Natl. Acad. Sci. USA 78:3824 (1981).
2. T.P. Hopp and K.R. Woods, A computer program for predicting protein antigenic determinants, Molec. Immunol. 20:483 (1983).
3. J. Kyte and R.F. Doolittle, A simple method for displaying the hydropathic character of a protein, J. Mol. Biol. 157:105 (1982).
4. G.D. Rose and S. Roy, Hydrophobic basis of packing in globular proteins, Proc. Natl. Acad. Sci. USA 77:4643 (1980).
5. T.P. Hopp, Protein Antigen Conformation, in: "International Conference on Synthetic Antigens," P. Neri, ed., Sclavo, Siena (1984).
6. G.D. Rose, A.R. Geselowitz, G.J. Lesser, R.H. Lee, and M.H. Zehfus, Hydrophobicity of amino acid residues in globular proteins, Science 229:834 (1985).
7. T.P. Hopp, Computer Prediction of Protein Surface Features and Antigenic Determinants, in: "Molecular Basis of Cancer, Part B," R. Rein, ed., Alan R. Liss, Inc., New York (1985).
8. G.W. Welling, W.J. Weijer, R.v.d. Zee, and S. Welling-Wester, Prediction of sequential antigenic regions in proteins, FEBS Letters 188:215 (1985).
9. P.Y. Chou and G.D. Fasman, Prediction of the secondary structure of proteins from their amino acid sequence, Adv. Enzymol. 47:45 (1978).

THE PROTEIN IDENTIFICATION RESOURCE (PIR): AN ON-LINE COMPUTER SYSTEM
FOR THE CHARACTERIZATION OF PROTEINS BASED ON COMPARISONS WITH
PREVIOUSLY CHARACTERIZED PROTEIN SEQUENCES

David G. George, Winona C. Barker, and Lois T. Hunt

National Biomedical Research Foundation
Georgetown University Medical Center
Washington, D.C. 20007

This conference is dedicated to examining new methods for the
isolation and characterization of proteins. One extremely effective
method for the characterization of a new protein involves the comparison
of its amino acid sequence with the sequences of previously determined
proteins. Although this method is not new (but dates back to the early
days of protein sequencing methodology), the wealth of information
available is only recently being fully appreciated. The rapid increase
in the accumulation of sequence data, owing to recombinant DNA technology,
has greatly heightened interest in this area and has made large database
searching a much more fruitful enterprise. The primary structures of
well over 3,000 proteins containing almost three quarters of a million
residues are now known, more than double what was known just 5 years ago.

Sequence comparison methods allow the identity of proteins to be
established, e.g., proteins subsequently identified as ubiquitin were at
various times separately isolated from different tissues and called
thymosin polypeptide beta$_1$, protein S in trout testis, nonhistone
chromosomal protein HMG-20, and ATP-dependent proteolysis factor I.[1] In
addition, these methods can often provide an indication of a protein's
structure and/or function and suggest possible activities. For example,
the recently discovered similarity between the viral oncogene v-erbB and
the kinase-related domain of the epidermal growth factor receptor
suggested that the v-erbB protein might exhibit tyrosine kinase
activity.[2,3] The epidermal growth factor (EGF) receptor consists of three
domains: an extracellular domain, a transmembrane domain, and an
intracellular domain. Binding of EGF to the extracellular domain of the
receptor induces a tyrosine kinase activity in the intracellular domain,
which triggers the events that eventually lead to DNA synthesis and
mitosis. The catalytic region shows homology with a number of other
tyrosine-specific protein kinases.[4-6] As can be seen in Figure 1, the
v-erbB protein is homologous to the transmembrane and intracellular
domains of the EGF receptor but lacks the major portion of the EGF binding
domain. Along its entire length the v-erbB protein is less than 20%
different from the corresponding region of the EGF receptor; it differs by
only about 3% in the kinase-related region. This structural similarity
strongly indicates that the v-erbB protein functions analogously to the
receptor protein but lacks the regulatory domain responsible for
inhibiting kinase activity. Subsequently, kinase activity has been

445

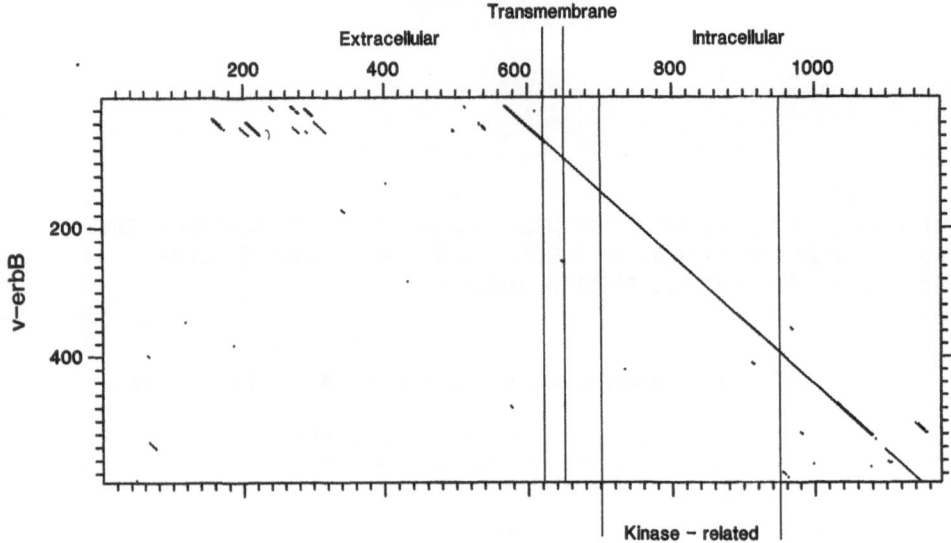

Fig. 1. A DOTMATRIX comparison of the human epidermal growth factor
receptor protein with the v-erbB protein from avian
erythroblastosis virus. A segment length of 25 and a minimum
score of 30 were used.

found,[7,8] confirming this supposition. These findings are important not
only from a biochemical point of view but they also provide a direct link
between normal cellular proteins and oncogenic proteins. This correlation
has made a fundamental contribution to the understanding of the general
mechanisms involved in cellular transformations induced by oncogenic
proteins.

    In order to make sequence comparison methods available to the
general scientific community, the National Biomedical Research Foundation,
in conjunction with the Division of Research Resources of the NIH, has
established the Protein Identification Resource (PIR), a Resource designed
specifically to aid the research community in the identification and
interpretation of protein sequence information. This Resource is the
direct outgrowth of an on-line protein and nucleic acid database system
that has been operational since 1980.[9-11] It consists of several protein
and nucleic acid databases and a set of interactive computer programs
integrated into a sophisticated computer system. The system provides fast
database searching methods, sequence comparison methods, and a number of
retrieval and sequence manipulation routines. The PIR on-line system is
accessible by direct telephone connection or through the TYMNET
communications network. All of the methods described below are available
through the PIR system.

THE NBRF PROTEIN SEQUENCE DATABASE

    The NBRF Protein Sequence Database had its beginnings in the early
days of protein sequence determination. At that time, under the
directorship of the late Margaret O. Dayhoff, our research group began
collecting sequences for use in protein evolutionary studies. The
collection was subsequently published in a series of volumes as the

Atlas of Protein Sequence and Structure.[12-15] The origin of the database as a data set for protein evolutionary studies has had a profound influence on its later development. In 1975, Margaret Dayhoff introduced the concept of a superfamily of proteins.[16-18] A protein superfamily is a group of proteins and/or protein domains that are believed on the basis on their sequence similarity to be evolutionarily related. Thus, in principle, each of the proteins in a superfamily can be interrelated by an evolutionary tree.

In the present NBRF Protein Sequence Database, all proteins are classified into superfamilies and each superfamily is further divided into families and subfamilies. Within a superfamily, proteins less than 50% different are assigned to the same family and those less than 20% different are assigned to the same subfamily. As these criteria are based on pairwise sequence comparisons, some latitude is given in their application, i.e., it is often the case that two sequences may differ by less than 20% and one of them differs from a third by less than 20%, whereas the other differs from the third by more than 20%. This classification scheme makes the database particularly amenable to database searching methods in that, once a similarity has been detected, the relationship between the selected sequence and other sequences in the database is already known.

In general, sequence comparison methods alone cannot prove that proteins are homologous (evolved from the same ancestor) but can only indicate that the similarity between proteins is unusual. Functional relationships must be demonstrated experimentally and homology must be inferred in conjunction with other biological reasoning. Sequence comparison methods are most effectively employed to identify unusual sequence similarities, the identification of which can spawn experimental studies aimed at discovering their biological significance. Although we rely heavily on sequence comparison methods in defining superfamilies there is no single criterion on which the classification is based.

As more sequence data have become available, we have become increasingly aware that similarity between protein sequences has profound biochemical implications irrespective of their evolutionary relationships. Although the evolutionary relationship between the v-erbB protein and the EGF receptor protein is of great interest, the importance to cancer research of the previously discussed findings is completely independent of any evolutionary arguments. The remainder of this paper will discuss protein similarity in terms of possible structural and functional relationships rather than of possible homology.

SEARCH COMPARISON METHODS

The method of choice for searching the protein sequence database depends upon the amount of sequence information the user has and the type of relationship the user wishes to detect (see references 19-22 for recent reviews of sequence searching and comparison methods). Successful use of any method depends upon a proper evaluation of the results. Figure 2 shows a plot of the observed percentage difference versus the estimated evolutionary distance between proteins, obtained in a simulation study. The curve can be divided into three general regions: a region of low percentage difference, which exhibits a sharp downward slope; an intermediate region (enclosed in the box), where the slope gradually decreases; and a region of high percentage difference, where the curve levels off. This curve reflects the results generally obtained by database searching methods (R. F. Doolittle, personal communication); the leveling off corresponds to the loss of information due to increasing evolutionary

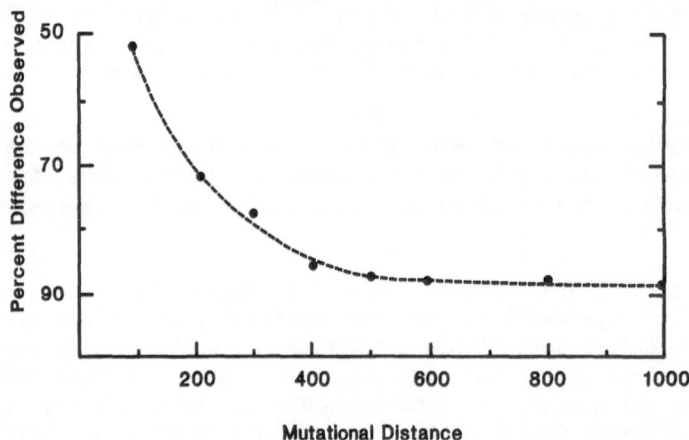

Fig. 2. Dependence of the observed percentage difference on the total
number of mutations per 100 residues. Data were compiled from
mutation simulation studies described in reference 28, from which
this figure was adapted with permission.

distance between the sequences. In the region of low (below about 75%)
percentage difference, virtually any method, even searches for strings of
identically matching residues, will detect the relationship; further
studies will generally indicate structural and functional relationships
between the proteins. As the curve levels off, more sensitive methods are
needed to detect the relationship and observed similarities in this region
often cannot be demonstrated to be biologically significant. In the region
above about 85% difference, only the most sophisticated of methods can
detect sequence similarities and often they cannot be detected at all. For
obvious reasons, most research in sequence comparison methods has been
directed toward the region enclosed in the box in Figure 2.

     Sequence comparison methods can be classified as global or local.
Global methods compare sequences along their entire lengths, whereas local
methods compare only local regions irrespective of the remainder of the

sequences. Needleman and Wunsch[23] described a global method that computes
a maximum scoring alignment between two sequences, allowing gaps to be
inserted to maximize matching. The alignment score is computed by summing
amino acid pair scores for each pair of the aligned residues and
subtracting a penalty for each region of gaps inserted in the alignment.
The amino acid pair scores are determined from an amino acid similarity

matrix. In our implementation,[18] a matrix bias can also be added to each
element of the amino acid scoring matrix; this is equivalent to adding a
penalty for each gap in the alignment. An alternative global alignment

algorithm was developed by Sellers,[24] but with certain restrictions it can
be shown to be mathematically equivalent to the Needleman—Wunsch
method.[25,26]

     Most other commonly used sequence comparison methods involve local
comparisons. The predominance in the number of local methods with respect
to global methods has its origin in the nature of the data. It has become
increasingly evident in recent years that many proteins have evolved by
gene fusion/splitting and transpositional events. Consequently, many
proteins are composed of several functionally and structurally distinct
domains having separate evolutionary origins. Moreover, although proteins
may remain highly conserved over large regions of their sequences and thus
are easily identified as being related, other regions of the same

molecules may no longer exhibit significant sequence similarity. For example, cytochrome b_6 from spinach chloroplast contains only about half as many residues as cytochrome b from baker's yeast mitochondrion and shows strong sequence similarity to a large portion of the amino end of the baker's yeast protein. There is another gene, 3' to the cytochrome b_6 gene in spinach chloroplast, that codes for a 17K protein of unknown function. This protein contains a region that shows strong sequence similarity to a portion of the carboxyl end of baker's yeast mitochondrial cytochrome b.[27] The cytochrome b gene in spinach chloroplast has apparently been split and has given rise to two separate genes. Global methods are not well suited to handle situations such as these. Furthermore, they require too much computational time to be routinely used in large database searches. Global methods are still extremely useful but produce reasonable results only when applied over regions of proteins that exhibit sequence similarity. These regions can best be located using local methods.

The simplest local searching method is to select a small subsequence from the test sequence and to search for identically matching subsegments in the database. A string of six consecutive amino acids is usually sufficient to identify closely related proteins. Only about 10% of all possible pentapeptides and less than 1% of all possible hexapeptides are present in the current database, making the occurrence of an identically matching hexapeptide a rare event. Because proteins less than 75% different often contain a number of regions of consecutive identities, closely related proteins can usually be identified using this approach. The SCAN command of our PSQ program[11] uses a precompiled peptide table and a simple look-up algorithm to locate all occurrences in the database of amino acid segments of 3 to 30 residues; the response time is instantaneous from the user's point of view. This should always be the first searching strategy tried; it requires only partial sequence information and can be used to check the progress of a sequencing project. We know of several instances in which it has been discovered that, rather than sequencing a protein of interest, the investigators were in fact sequencing another commonly occurring protein or a portion of the cloning vector. If detected early, a considerable amount of time and money could be saved.

A more sophisticated approach is to search the database, allowing for mismatches, using a similarity matrix to score amino acid pairs. Proteins can be between 75% and 85% different and can still be recognized to be related on the basis of the types of replacements observed at each position. A scoring system that gives a higher weight to amino acids with similar properties allows a more sensitive database search to be conducted. We have developed an empirical scoring matrix, the Mutation Data Matrix (MDM),[28] which is based on observed amino acid replacement frequencies. It has been found to be very effective for detecting distantly related proteins.[28,29]

Our SEARCH program[18] compares a subsegment of a specified length from a test sequence with all other subsegments in the database and computes a score equal to the sum of the amino acid pair scores of the aligned residues; the comparisons with the highest corresponding scores are selected. The amino acid pair scores are defined by the scoring matrix. This method has been in general use for the past 10 years and has been found to be reasonably effective when used to select candidates for further testing.

In any biologically realistic sequence alignment, provision must be made to allow for gaps in the sequences corresponding to insertions and/or deletions of genetic material. In most cases, however, these alignments exhibit substantial regions of consecutively aligned residues. Thus, the SEARCH program has been found effective even though insertion/deletion events are not explicitly taken into account. Even when gaps occur within the region of the alignment corresponding to the test segment, the similar residues contribute enough to give an overall high score. We have found segment lengths of 25 to be effective when MDM is used as the scoring matrix.

The similarity between two sequences can be more extensively analyzed by comparing all segments from one sequence with all segments from the other. The results of these comparisons can be conveniently evaluated by plotting the segment comparison scores in matrix form with one axis corresponding to each sequence (see Figure 1, for example). In most implementations, the points in the matrix where the corresponding segment comparison scores exceed a specified value are marked by dots. Regions of similarity appear as diagonals in the plot. This method is also extremely effective in locating sequence duplications when the sequences are plotted against themselves. A number of variations of this method have been published;[30-33] they differ in the method used to assign the scores. However, in most cases they produce equivalent results. In our version,[34] scores are computed as they are in the SEARCH program; the segment length and minimum score are adjustable. Using MDM as the scoring matrix, we have found values of 20 to 30 to be most useful for the segment length and minimum score.

Although the dotmatrix method is simple to apply and very useful, it obviously cannot be applied in large database searches. Wilbur and Lipman[35,36] have described a diagonal method for large database searching based on a dotmatrix type of approach. More recently, Lipman and Pearson[37] have described a modification, program FASTP, specifically developed for protein sequence comparisons. Essentially, the algorithm scores identities or k-tuples of identities (k contiguous identities) between sequences and selects the highest-scoring diagonals. These diagonals are rescored using MDM as the scoring matrix, and a top-scoring diagonal is chosen for each pair of sequences compared. From among these rescored values the overall top-scoring diagonals are selected. The alignments corresponding to each of these diagonals are optimized using a modified Needleman-Wunsch algorithm. This method is particularly well suited for large database searches and is generally the method of choice if an initial SCAN of the database is unsuccessful.

With the exception of SCAN, the methods so far discussed generally require that a significant portion of the test sequence be well established. There are several other methods that can be applied when less information is available. Often, in the initial stages of the isolation and characterization of a protein, sequenator analysis yields two or more amino acids at each step because of a mixed or partially purified sample. For example, in the early stages of analysis of baker's yeast alcohol dehydrogenase, sequenator analysis of an aliqout from a partially separated tryptic digest may yield the following information at each position:

1	2	3	4	5	6	7	8	9	10	11	12
Ser	Ile	Pro	Glu	Thr	Gln	Lys	Gly	Val	Ile	Phe	Tyr
Trp	Val	Val	Ser	Ala	Ala	His	Cys	Tyr	Lys	Ser	Gly
	Gly	Gly	Pro	Val	Val	Cys	Ser	Gly		Leu	Gln

A slight modification to the SEARCH program allows information such as this to be used in a database search. When the database was searched, only three exactly matching segments were found. These correspond to the first 12 residues of baker's yeast alcohol dehydrogenase and residues 40-52 and 183-194 of bovine trypsinogen. Of the segments that matched at 10 or more positions, all are related to either alcohol dehydrogenase or trypsinogen. These results would indicate that the major contaminants in the preparation are products of the self-digestion of trypsin.

In most cases, identification of only 7 or 8 not necessarily contiguous amino acids will uniquely identify a protein if it is in the database. Thus, microsequencing of peptides containing certain radiolabeled amino acids can yield sufficient information to identify a protein.[38] The SEARCH program will recognize 'X' as an ambiguous amino acid symbol that matches all other residues and this symbol can be used to indicate a lack of information at particular positions.

It is also possible to make limited identifications based on partial compositional data. It is surprisingly easy to recognize a protein in the database knowing only the molecular weight and the percentages for a few amino acid residues. For example, suppose we could reliably estimate the molecular weight within 20% and the amount of each amino acid within 2 residues or 20% (whichever is larger). Using the known values for bovine catalase and successively searching for sequences with matching properties, the following results were obtained:

	Range	Number of Sequences
Mol. wt.	46,068-69,102	302
% Cys	0.39-1.19	72
% Trp	0.79-1.59	20
% His	3.32-4.98	2
% Asn	4.74-7.12	1

After searching using the molecular weight and the first three amino acid percentages, only bovine catalase and E. coli diaminopimelate decarboxylase remained. With four amino acid percentages, only bovine catalase was found. In general, even with proteins of average composition and molecular weight, knowledge of the molecular weight and the percentage composition for only four or five amino acids is sufficient to reduce the list of possible candidates to a small group of related proteins.

The PIR on-line system also provides a full battery of retrieval operations that allow sequences to be located using combinations of protein name, biological source and its taxonomic classification, superfamily organization, author and journal citation, and other ancillary information (including keywords and data relating to the architecture of the protein). Thus, the PIR contains many of the necessary facilities for a complete analysis of protein sequence data.

ACKNOWLEDGMENTS

The Protein Identification Resource is supported by grant RR01821 from the Division of Research Resources of the National Institutes of Health.

REFERENCES

1. L. T. Hunt and M. O. Dayhoff, Evolution of chromosomal proteins, in: "Macromolecular Sequences in Systematic and Evolutionary Biology," M. Goodman, ed., Plenum, New York (1982).

2. J. Downward, Y. Yarden, E. Mayes, G. Scrace, N. Totty, P. Stockwell, A. Ullrich, J. Schlessinger, and M. D. Waterfield, Close similarity of epidermal growth factor receptor and v-erb-B oncogene protein sequences, Nature 307:521 (1984).

3. A. Ullrich, L. Coussens, J. S. Hayflick, T. J. Dull, A. Gray, A. W. Tam, J. Lee, Y. Yarden, T. A. Libermann, J. Schlessinger, J. Downward, E. L. V. Mayes, N. Whittle, M. D. Waterfield, and P. H. Seeburg, Human epidermal growth factor receptor cDNA sequence and aberrant expression of the amplified gene in A431 epidermoid carcinoma cells, Nature 309:418 (1984).

4. W. C. Barker and M. O. Dayhoff, Viral src gene products are related to the catalytic chain of mammalian cAMP-dependent protein kinase, Proc. Nat. Acad. Sci. USA 79:2836 (1982).

5. M. L. Privalsky, R. Ralston, and J. M. Bishop, The membrane glycoprotein encoded by the retroviral oncogene v-erb-B is structurally related to tyrosine-specific protein kinases, Proc. Nat. Acad. Sci. USA 81:704 (1984).

6. G. Naharro, K. C. Robbins, and E. P. Reddy, Gene product of v-fgr onc: hybrid protein containing a portion of actin and a tyrosine-specific protein kinase, Science 223:63 (1984).

7. T. Gilmore, J. E. DeClue, and G. S. Martin, Protein phosphorylation at tyrosine is induced by the v-erbB gene product in vivo and in vitro, Cell 40:609 (1985).

8. R. M. Kris, I. Lax, W. Gullick, M. D. Waterfield, A. Ullrich, M. Fridkin, and J. Schlessinger, Antibodies against a synthetic peptide as a probe for the kinase activity of the avian EGF receptor and v-erbB protein, Cell 40:619 (1985).

9. M. O. Dayhoff, R. M. Schwartz, H. R. Chen, W. C. Barker, L. T. Hunt, and B. C. Orcutt, Nucleic acid sequence database, DNA 1:51 (1981).

10. B. C. Orcutt, D. G. George, J. A. Fredrickson, and M. O. Dayhoff, Nucleic acid sequence database computer system, Nucl. Acids Res. 10:157 (1982).

11. B. C. Orcutt, D. G. George, and M. O. Dayhoff, Protein and nucleic acid sequence database systems, Annu. Rev. Biophys. Bioeng. 12:419 (1983).

12. M. O. Dayhoff, R. V. Eck, M. A. Chang, and M. R. Sochard, "Atlas of Protein Sequence and Structure 1965," National Biomedical Research Foundation, Silver Spring, MD (1965).

13. R. V. Eck and M. O. Dayhoff, "Atlas of Protein Sequence and Structure 1966," National Biomedical Research Foundation, Silver Spring, MD (1966).

14. M. O. Dayhoff, ed., "Atlas of Protein Sequence and Structure," Vol. 5, National Biomedical Research Foundation, Washington (1972).

15. M. O. Dayhoff, ed., "Atlas of Protein Sequence and Structure," Vol. 5, Suppl. 3, National Biomedical Research Foundation, Washington (1979).

16. M. O. Dayhoff, P. J. McLaughlin, W. C. Barker, and L. T. Hunt, Evolution of sequences within protein superfamilies, Naturwissenschaften 62:154 (1975).

17. M. O. Dayhoff, The origin and evolution of protein superfamilies, Fed. Proc. 35:2132 (1976).

18. M. O. Dayhoff, W. C. Barker, and L. T. Hunt, Establishing homologies in protein sequences, Methods Enzymol. 91:524 (1983).

19. B. C. Orcutt and W. C. Barker, Searching the protein sequence database, Bull. Math. Biol. 46:545 (1984).

20. M. S. Waterman, General methods of sequence comparison, Bull. Math. Biol. 46:473 (1984).

21. D. Sankoff and J. B. Kruskal, eds., "Time Warps, String Edits, and Macromolecules: The Theory and Practice of Sequence Comparison," Addison-Wesley, Reading, MA (1983).

22. J. B. Kruskal, An overview of sequence comparison: time warps, string edits, and macromolecules, SIAM Rev. 25:201 (1983).

23. S. B. Needleman and C. D. Wunsch, A general method applicable to the search for similarities in the amino acid sequence of two proteins, J. Mol. Biol. 48:443 (1970).

24. P. H. Sellers, On the theory and computation of evolutionary distances, SIAM J. Appl. Math. 26:787 (1974).

25. T. F. Smith, M. S. Waterman, and W. M. Fitch, Comparative biosequence metrics, J. Mol. Evol. 18:38 (1981).

26. M. S. Waterman, Sequence alignments in the neighborhood of the optimum with general application to dynamic programming, Proc. Nat. Acad. Sci. USA 80:3123 (1983).

27. W. R. Widger, W. A. Cramer, R. G. Herrmann, and A. Trebst, Sequence homology and structural similarity between cytochrome b of mitochondrial complex III and the chloroplast b_6-f complex: Position of the cytochrome b hemes in the membrane, Proc. Nat. Acad. Sci. USA 81:674 (1984).

28. R. M. Schwartz and M. O. Dayhoff, Matrices for detecting distant relationships, in: "Atlas of Protein Sequence and Structure," Vol. 5, Suppl. 3, M. O. Dayhoff, ed., National Biomedical Research Foundation, Washington (1979).

29. D. F. Feng, M. S. Johnson, and R. F. Doolittle, Aligning amino acid sequences: comparison of commonly used methods, J. Mol. Evol. 21:112 (1985).

30. A. J. Gibbs and G. A. McIntyre, The diagram, a method for comparing sequences, its use with amino acid and nucleotide sequences, Eur. J. Biochem. 16:1 (1970).

31. A. D. McLachlan, Tests for comparing related amino-acid sequences. Cytochrome c and cytochrome c_{551}, J. Mol. Biol. 61:409 (1971).

32. J. V. Maizel, Jr. and R. P. Lenk, Enhanced graphic matrix analysis of nucleic acid and protein sequences, Proc. Nat. Acad. Sci. USA 78:7665 (1981).

33. R. Staden, An interactive graphics program for comparing and aligning nucleic acid and amino acid sequences, Nucl. Acids Res. 10:2951 (1982).

34. D. G. George, L.-S. L. Yeh, and W. C. Barker, Unexpected relationships between bacteriophage lambda hypothetical proteins and bacteriophage T4 tail-fiber proteins, Biochem. Biophys. Res. Commun. 115:1061 (1983).

35. W. J. Wilbur and D. J. Lipman, Rapid similarity searches of nucleic acid and protein data banks, Proc. Nat. Acad. Sci. USA 80:726 (1983).

36. W. J. Wilbur and D. J. Lipman, The context dependent comparison of biological sequences, SIAM J. Appl. Math. 44:557 (1984).

37. D. J. Lipman and W. R. Pearson, Rapid and sensitive protein similarity searches, Science 227:1435 (1985).

38. M. O. Dayhoff and B. C. Orcutt, Methods for identifying proteins by using partial sequences, Proc. Nat. Acad. Sci. USA 76:2170

COMPUTER ANALYSIS OF PROTEIN SEQUENCING DATA

Norman Froelich, Lynn C. Williams[*], John T. Casagrande and Minnie McMillan

Norris Cancer Center, USC School of Medicine

2025 Zonal Avenue, Los Angeles, CA 90033

The concept of a Microchemical facility, in which sophisticated instruments for the analysis and synthesis of genes and proteins of biological significance are centralized in one laboratory and are available to principal investigators on a service basis, serves as a powerful, effective method of interfacing the highly efficient and complex chemistries utilized by these machines with fundamental problems in biology (1). The cornerstone of such a facility is the gas-phase sequenator service which is used by investigators who are cognizant of the fundamental importance and clinical relevance of their proteins, but who are often not familiar with the techniques of protein purification or protein sequencing. In order to assist these investigators in interpretation of their results we have developed a computer program for the analysis of protein-sequencing data. This program plots peak area for each PTH amino acid for every cycle in the sequencing run. From such plots, one can readily establish the sequences of pure proteins by inspection; in addition, one can analyse minor species and monitor the technical performance of the sequenator and autosampler, for example. This system is likely to be but a primitive prototype of one in which a protein, after being loaded on the sequenator, will be automatically analyzed and the sequence(s) directly printed out by computer.

METHODS

The Microchemical Core Laboratory in the Norris Cancer Center is equipped with a Model 470A Applied Biosystems Model gas-phase sequenator (2). PTH derivatives are analysed on a Perkin-Elmer HPLC system comprised of two Series 4 pumps, a ISS-100 autosampler, a LC95 variable wavelength detector and a Chromatographics 2 Data Station. PTH-derivatives from the sequenator are separated on an IBM or Zorbax CN column at 34°C using a 20 mM sodium acetate, pH 5.08 5% THF buffer/acetonitrile solvent system. The chromatograph resulting from each sequenator cycle is stored by the Data station on a $5\frac{1}{4}$ inch floppy disc. It can be graphically presented as a plot of absorbance at 254 nm versus time (see Figure 1A); alternately a table of peak areas/retention times can be printed (see Figure 1B). It is this tabulation which is used in our computer analyses.

We have interfaced the Data station with a Digital Equipment Corporation MINC 11-23 computer and printer and written a program to enable us to plot peak area of a particular PTH-amino acid for every cycle in the sequencing run.

The DEC MINC 11-23 computer has 64 K of memory, two RX02 8 inch floppy discs and a DEC VT 125 graphics terminal with a DEC LA50 printer. The MINC interface to the Perkin-Elmer Data Station was achieved using an RS-232 cable.

The program consists of three segments:

(1)   the menu program which controls program flow to the various modules

(2)   the Perkin-Elmer Data station to MINC data transfer program

(3)   the analysis program which graphs the peak area for a specific retention time over all the cycles of a particular protein sequencing run.

The menu program allows the user to choose between transferring data and/or graphing results. The MINC transfer program is designed to intercept the tabular data, to extract the retention time and peak area, and to store each parameter separately in two files on the floppy drive for future processing by the MINC amino acid analysis program. This graphing program displays a chart in which peak area for a specific retention time is plotted for all the cycles for a particular sequenator run.

The following parameters must be specified by the user:

(1)   retention time for a specific amino acid

(2)   "correction factor" or retention time window. This feature allows for any slight drift of retention times from cycle to cycle caused by column compression, change in buffer pH, etc. If the "correction factor" results in overlap of peaks then the program compensates by making the error factor 1% less than the next amino acid retention time.

(3)   the maximum peak area chosen by inspection of the data for the first few cycles.

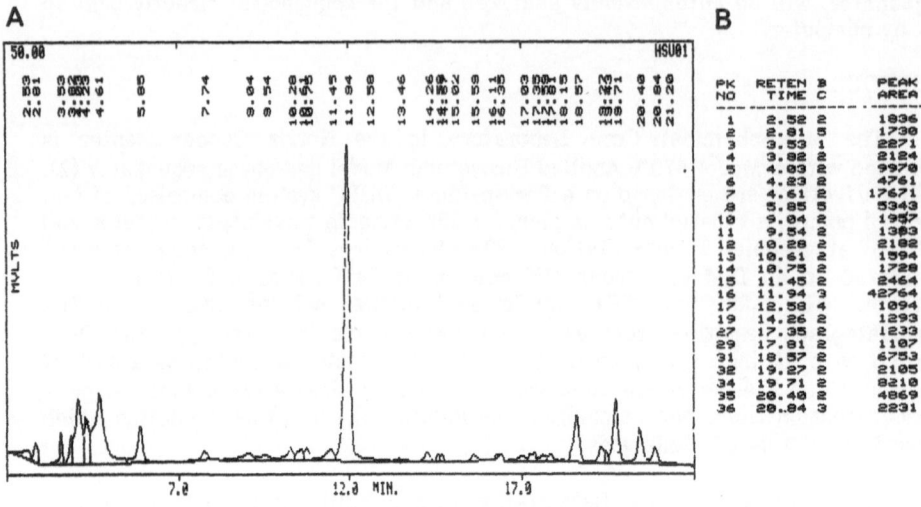

Figure 1. HPLC separation of PTH–derivatives from one sequenator cycle plotted graphically (A) and tabulated (B).

456

Sample	Reten. Time	Area
1	18.10	4422
2	18.13	1892
3	18.11	3118
4	18.10	123907
5	18.11	9045
6	18.09	4657
7	18.10	3743
8	18.08	5143
9	18.10	4654
10	18.10	3590
11	18.10	4827
12	18.09	5569
13	18.09	6989
14	18.09	11332
15	18.09	8168
16	18.10	6162
17	18.06	6899
18	18.06	5100
19	18.08	7019
20	18.09	6154
21	18.15	11021
22	18.15	11021
23	18.09	7571
24	18.10	30715
25	18.11	10557
26	18.13	10938

Figure 2. Computer tabulation of cycle number (sample), retention time and peak area for PTH-leucine. Graphic display is given in Figure 3.

For each amino acid the program steps through the data files searching for retention times that fall into the retention time window specified for that amino acid. A table is then printed for each amino acid showing cycle number, retention time for that cycle, and the peak area value (See Figure 2). After the table is printed, the graph of peak area versus residue number is also given.

At the end of the report the program prints a tentative amino acid sequence. This sequence is generated by picking the largest area in each cycle and identifying the correct amino acid by retention time.

RESULTS

This method of computer analysis in which peak area of a particular amino acid is plotted for every cycle in a sequencing run has three major uses:

(1) **identification of amino acids by inspection** instead the overlaying of pen recordings on chart paper. This is illustrated in Figure 3 which shows a partial computer analysis of the N-terminus of the light chain of high molecular weight kininogen. The following assignments can be made with ease:

Asparagine at cycle number 3

Histidine at cycle numbers 2,6,8,10,16,18

Leucine at cycle numbers 4,24,34

Arginine at cycle numbers 12,20

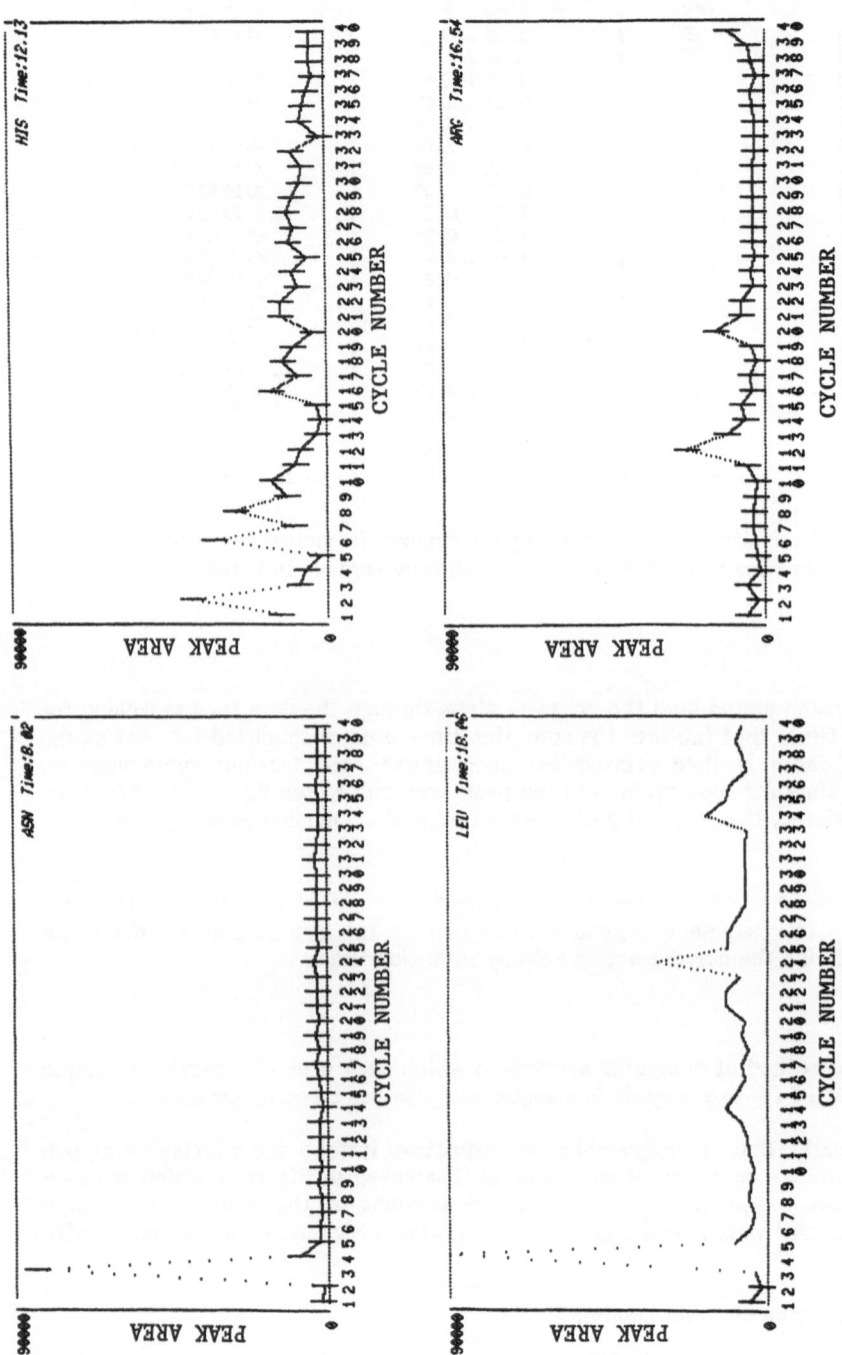

Figure 3. Partial computer analysis of the N-terminal sequence of the light chain of high molecular weight kininogen showing peak area plotted for each sequenator cycle for the PTH-amino acids asparagine, histidine, leucine and arginine. Retention time is indicated (top right).

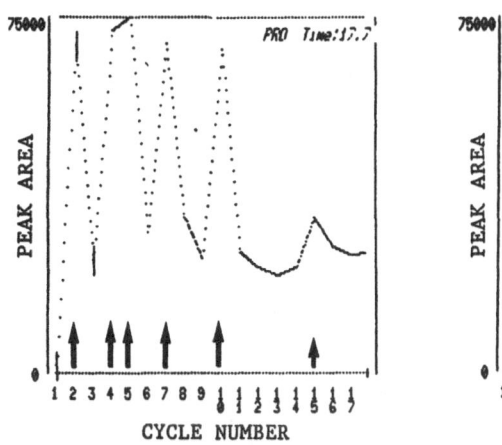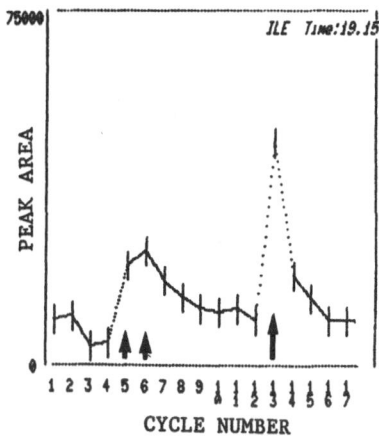

Figure 4. Partial computer analysis for the N-terminal sequences obtained from a sample of dental enamel protein. Peak areas of PTH-amino acids proline and isoleucine are plotted for each cycle. Assignments for the major species are indicated by large arrows, for the minor species by small arrows.

(2) **identification of minor sequences from contamination species.** This is illustrated in figure 4 which shows the partial computer analysis of a dental enamel protein. The following assignments can be made by inspection:

Major Species: Pro at cycle numbers 2,4,5,7,10
                     Ile at cycle number 13

Minor Species: Pro at cycle number 15
                     Ile at cycle numbers 5,6

(3) **monitoring the technical performance** of the sequenator and accuracy of the autosampler. Figure 5 shows very clearly how "lag" of PTH amino acids increases during a sequenator run. We have used our computer program to detect both malfunctioning of the autosampler and also poor performance of the sequenator.

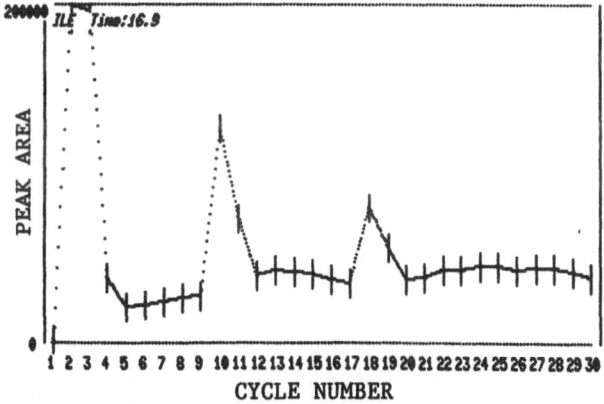

Figure 5. Partial computer analysis of the N-terminus of the PII protein of Borrelia Hermisii showing peak areas for PTH-isoleucine plotted for each sequenator cycle.

We have also used our computer program to predict the amino acid sequence of a protein from the stored data and because of the relative simplicity of our analysis were astonished when 27 out of 35 assignments were correct. Several histidine residues were "missed" because histidine is difficult to detect and gives low peak area values.

DISCUSSION

The introduction of the gas-phase sequenator with its increased sensitivity has resulted in the ability to analyse proteins of major biological importance which previously would have been impossible (2). The introduction of on-line HPLC analyses of PTH derivatives is a major step in the further automation of protein sequencing. Another phase of this automation is computer analysis of data and a program such as that presented here will greatly enhance the reliability and speed of analysis of protein sequencing data and assignments of amino acids.

We are indebted to Drs. W. Hsueh, S. Schiffman, M. Simon and M. Snead for allowing us to use their data prior to publication. We are most grateful to Ms. Margaret Cameron and Ms. Sarah Olivo for their outstanding efforts word processing this manuscript. This work was supported by grant 5P30 CA 14089-10 from the National Institutes of Health.

REFERENCES

(1)   M. Hunkapiller, S. Kent, M. Caruthers, W. Dreyer, J. Firca, C. Giffin, S. Horvath, T. Hunkapiller, P. Tempst, and L. Hood, Nature 310:105 (1984).

(2)   R. Hewick, M. Hunkapiller, L. Hood and W. Dreyer, J. Biol. Chem. 256:7990 (1981).

IX. Miscellaneous Methods of Polypeptide Characterization

SIMULTANEOUS MULTIPLE PEPTIDE SYNTHESIS: THE

RAPID PREPARATION OF LARGE NUMBERS OF PEPTIDES

Richard A. Houghten, Sarah R. Hoffmann,
Mairead K. Bray, Nicole Frizzell, J.M. Ostresh,
Suzanne M. Pratt and John Sitarik

Scripps Clinic and Research Foundation
10666 North Torrey Pines Road
La Jolla, California 92037

The method of simultaneous multiple peptide synthesis (SMPS-as illustrated in Figure 1), developed recently in this laboratory (1, 2, 3) has made the synthesis of large numbers (>100) of specific peptides possible in amounts useful for virtually all studies (>10mg) and in a greatly shortened time frame (weeks). These methods address the very rapid escalation in the need for synthetic peptides for the development of synthetic peptide vaccines (4-5), the detailed study of antigen-antibody interactions (1, 6-11, 28), the preparation of optimal analogs of biologically active peptides (12-14), the optimization of peptide antigens of clinical diagnostic utility (15-17), the mapping of protein products of brain specific genes (18,19), and the study of peptide and protein conformational parameters (20-22). The limiting factor in the majority of these studies has been the availability and cost of the desired peptides.

The detailed methods used to carry out SMPS and a specific example of the utility of the methods in the study of antibody-antigen inter-action is presented herein.

SYNTHETIC METHODS

Preparation Of Multiple Peptide Resins

General Methods. Standard Boc amino acid resin (50-100 mg, 0.2-0.8 meq/gm) was contained in polypropylene mesh packets (74µ), having approximate dimensions of 15x20mm. After placing a number at the top of the unsealed bag with a black marking pen, the packet was closed and the number was permanently sealed into the polypropylene, giving an easily readable label for each bag. The general procedures apply, after adjustment for different swelling properties, to standard benzyl linked polystyrene resins (23), spacer linked styrene resins (24), polyamide resins (25), or macroreticular resins (6). These resin packets can be used for simultaneous multiple peptide syntheses (SMPS), that is, syntheses in which many different peptides are produced concurrently; or multiple analog peptide synthesis (MAPS), syntheses in which many analogs of a particular peptide are produced concurrently. The standard deprotecting, neutralization, coupling, and wash protocols used in this laboratory are variations of Merrifield's original solid phase procedures (23) and are described in detail elsewhere (26, 27).

Synthesis of Multiple Peptide Resins. Using completely manual methods, up to two hundred and fifty individual packets containing the desired starting resin were readily carried through their common Boc removal, washing, and neutralization steps. It must be emphasized that adequate agitation of the resin packets must exist to ensure adequate solvent transfer at these various steps. Following methylene chloride washes, to remove excess base (See Figure 1, step II 6), the packets containing the neutralized peptide resins were removed from the reaction vessel and added to solutions of the desired activated amino acid. Computer generated check lists ensure that the correct packet is put in the appropriate activated amino acid (30). In situ coupling or preformed symmetrical anhydrides of the next protected amino acid were used with equal success. The individual coupling steps were carried out for 60 minutes while shaking vigorously on an Eberbach reciprocating shaker at room temperature. After completion of the coupling steps, the resin packets were returned to the reaction vessel and the synthesis continued through additional cycles of common wash, deprotection, neutralization, and coupling steps until the syntheses were completed. Specific variations of analogous peptides such as residue replacement or omission analogs, or chain lengthened or shortened analogs were easily accomplished by removing the individual, coded packets at the point of variation during the synthesis, carrying out the desired variation separately and, if appropriate, returning the packet to the common reaction vessel for completion. Following the synthesis of a series of peptides, the resin filled packets were washed thoroughly, dried and weighed to give an initial indication of yield. The protected peptide resins, still contained within their packets, were then cleaved using conventional hydrogen fluoride/anisole as described below (23).

Cleavage of Multiple Peptide Resins

The apparatus for the cleavage of multiple peptide resins is described in detail elsewhere (3) and is commercially available (30). Briefly, this apparatus consists of three polypropylene plates with a polypropylene chamber separating the top and middle plates. The bottom plate serves as a safety support for the 24 individual reaction vessels and is attached through a threaded post on the middle plate. The top plate has two threaded ports. The first port serves as a vent, while the second port leads to a polypropylene radial distribution manifold having 24 ports. Polypropylene tubes are connected to the radial manifold in the central chamber leading through the middle plate and terminating 8cm

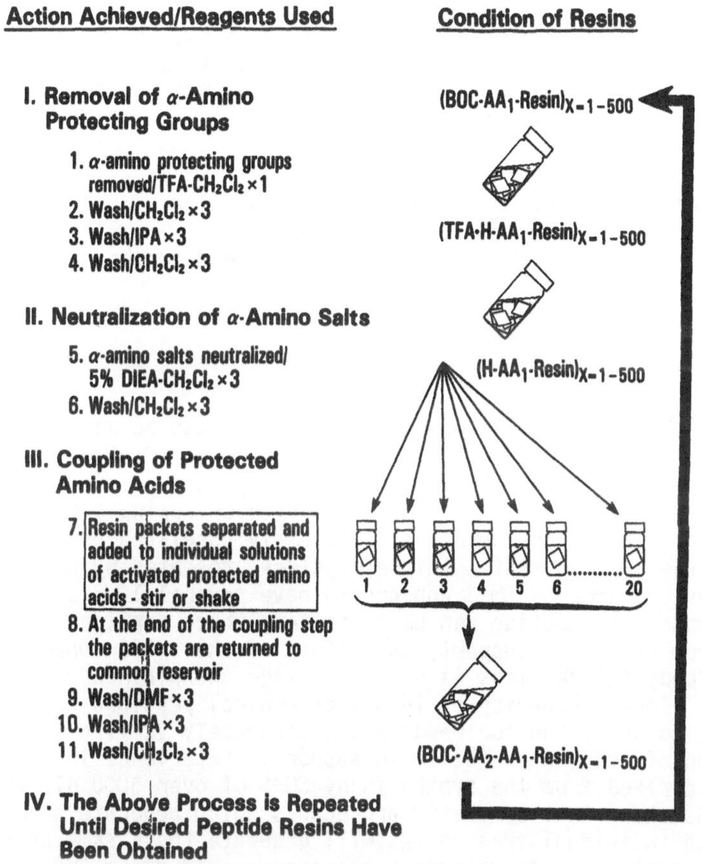

Action Achieved/Reagents Used

Condition of Resins

I. **Removal of α-Amino Protecting Groups**

$(BOC-AA_1-Resin)_{X=1-500}$

    1. α-amino protecting groups removed/TFA-CH$_2$Cl$_2$ ×1
    2. Wash/CH$_2$Cl$_2$ ×3
    3. Wash/IPA ×3
    4. Wash/CH$_2$Cl$_2$ ×3

$(TFA-H-AA_1-Resin)_{X=1-500}$

II. **Neutralization of α-Amino Salts**

    5. α-amino salts neutralized/ 5% DIEA-CH$_2$Cl$_2$ ×3
    6. Wash/CH$_2$Cl$_2$ ×3

$(H-AA_1-Resin)_{X=1-500}$

III. **Coupling of Protected Amino Acids**

    7. Resin packets separated and added to individual solutions of activated protected amino acids - stir or shake
    8. At the end of the coupling step the packets are returned to common reservoir
    9. Wash/DMF ×3
    10. Wash/IPA ×3
    11. Wash/CH$_2$Cl$_2$ ×3

1 2 3 4 5 6   20

$(BOC-AA_2-AA_1-Resin)_{X=1-500}$

IV. **The Above Process is Repeated Until Desired Peptide Resins Have Been Obtained**

Figure 1. Illustration of the principle of simultaneous multiple peptide synthesis (SMPS). Packets refer to individual packets or bags made from polypropylene mesh that contain the resins necessary for the peptides being prepared.

from the bottom of each reaction vessel. The middle plate is bored to allow the attachment of 0-ring pressure fitting vessels.

Into each of the 24 reaction vessels was added either a free resin (100-500mg) or, more conveniently, resin contained in a polypropylene synthesis packet (1,2,3). To the resin was added an amount of anisole equaling 10% of the expected volume of hydrogen fluoride along with a small teflon coated magnetic stir bar. After cooling in a dry ice/acetone solution, the vessels were purged with N2 for 15 minutes and HF (5-10ml) was condensed simultaneously and equally into all 24 reaction chambers. When the desired amount of HF was condensed, the dry ice/acetone bath was replaced with an ice water bath and the reaction solutions stirred for 1.0 hr at 4°C. Substantially improved peptide purity was obtained when the cleavage was carried out at -15°C. After 1.0 hr the HF was removed rapidly (<15 min) by a precooled N2 flow. Nitrogen purging was continued for a further 15 minutes and the vessels were then removed from the apparatus and the parties extracted from the residual anisole and resin by conventional methods.

Peptide Antigen-Monoclonal Antibody Interaction

Immunoprecipitation and ELISA procedures used are described in detail elsewhere (1, 28).

DISCUSSION

We have shown that large numbers of peptides can be made using the technique of simultaneous multiple peptide synthesis (SMPS-ref 1,2,3). It is now practical to make large numbers of peptides (>100) for any study requiring a specific series of analogs, overlapping series of nested peptides, or large numbers of completely unrelated peptides. The methods for the synthesis of multiple peptides presented by Geysen et. al. (9,10) as compared to SMPS appears to have several limitations: only microgram amounts of peptide can be generated, these peptides are not removable from the solid support, the effect of the solid support on antigen-antibody interactions is unclear, large unexplained variations in the antibody binding behavior of identical control peptides were found, and there is no provision for readily and accurately varying the concentration of peptide on the solid support. Experience in this laboratory, derived from the synthesis by SMPS of over 5000 different peptides, has shown that synthetic procedures which generate good peptides with individual free resins will generate these same peptides in identical or better purity when they are synthesized using SMPS. This is true for all synthetic procedures and resins normally used in standard solid phase peptide synthesis. Even when preparing single peptides it was found convenient to have the resin confined to a mesh packet since no resin is lost to reaction vessel walls or storage bottles, resin is more completely coupled and washed because it cannot stick to solvent inaccessible areas of some reaction vessels, removal of DNP groups from histidine can be carried out simultaneously for all resins containing DNP-histidine, storage of the peptide resin is simplified, and cleavage of the peptide resin is made easier as the spent resin remains in the packet following removal of the newly generated free peptide. One aspect of SMPS which must not be minimized is the necessity for sufficient agitation of the resin filled packets to ensure complete solvent transfer throughout all of the resin in the packets. This is best accomplished by mechanical stirring, or shaking. SMPS can be used successfully with a small number of resin packets on the Applied Protein Technologies PSS-80, Beckman 990B, Biosearch SAM II and Peptide International's automated peptide synthesizers, and the Applied Biosystems synthesizer.

The importance of complete characterization and purification prior to the use of very large numbers of peptides must be considered in light of the specific problem being investigated. If for example, 400 analogs of a biologically active peptide are desired, which require complete characterization as well as 95% or greater purity, then the method of SMPS coupled with simultaneous multiple HF procedures offers a greatly reduced preparation time for the desired crude peptides. If complete characterization and purification is required, this would have to be accomplished whether the crude peptides were synthesized by standard solid phase procedures over a period of one to two years, or over one to two months by SMPS. The coupling completions at any single point, or at every point in a given synthesis can readily be monitored by non-destructive methods such as Gisen's picric acid procedure (29) on the entire contents of the bag and additional coupling steps carried out if necessary. We have written computer programs (30) in order to keep an account of the various coupling steps when large numbers (i.e. 500) of individual peptide resins are being generated simultaneously. Without these computer programs it is difficult to eliminate human error when very large numbers (i.e., 500) of peptides are being generated simultaneously.

If the complete characterization and purification of the peptides generated is not essential, and if the purity of the specific peptides is adequate for the study being carried out, as is true for many immunological and chromotographic studies. (1-11,5-22,28), then the technique of SMPS very greatly simplifies the preparation of large numbers of peptides. These conditions apply to many initial studies involving the screening of large numbers of peptides for biological activity, peptide antibody-antigen interactions, or the synthesis of multiple overlapping segments of entire protein sequences and their examination as possible synthetic vaccine candidates.

The use of the simultaneous multiple HF apparatus used herein (3, 30) offers a very large convenience factor compared to conventional HF procedures since now fourty eight individual peptide resins can be cleaved in a single day versus 2 to 4 peptide resins per day with other apparatuses. It is also felt that this apparatus is substantially safer to use than a conventional HF apparatus since it greatly reduces the number of times that HF must be handled. This apparatus also provides more uniform cleavage conditions for a given set of 24 resins.

The utility of SMPS has been illustrated by the preparation of 260 individual analogs of a specific peptide antigen derived from the influenza hemagluttinin protein and the detailed study of their interactions with a single monoclonal antibody (1,28). Figure 2 shows

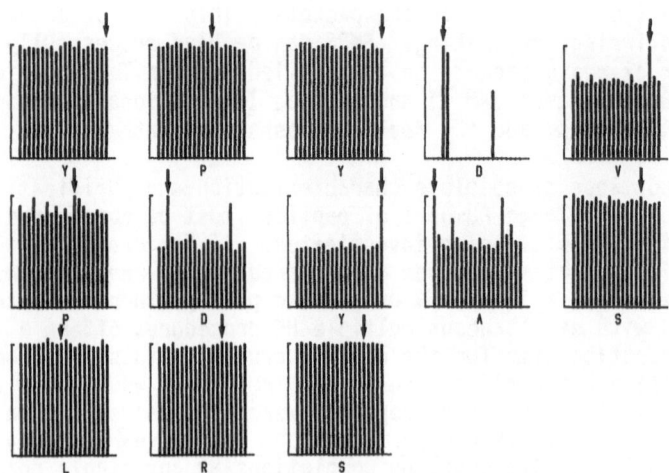

Figure 2. Each vertical line on the graph represents the concentration of a given analog which inhibited 50% of the binding of ^{125}I labelled control peptide (YPYDVPDYASLRS). The tick mark on each graph represents a concentration of 10^{-1} mole and the axis intersect represents a concentration of 10^{-5} mole. Each individual position of each graph is comprised of the replacement analogs along the X-axis in the following order: A, C, D, E, F, G, H, I, K, L, M, N, P, Q, R, S, T, V, W, Y.

ELISA: GLYCINE INSERTION ANALOGS

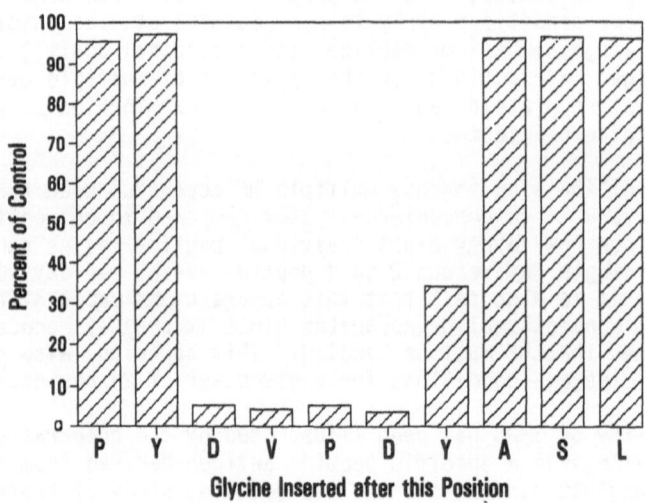

Figure 3. The effect on the interaction of monoclonal antibody (20C01-ref 1) with its known antigenic peptide sequence Y-P-Y-D-V-P-D-Y-A-S-L-R-S upon insertion of a glycine residue between successive pairs of residues (i.e. between residue P and Y; Y and D; D and V, etc.).

468

the IC-50's for the inhibition of ^{125}I radiolabeled control peptide by the various analogs using an immunoprecipitation assay. As can be seen, the relative positional variation was found to be quite large and was best represented logarithmically. The IC-50 displacement ability for the average of the substitutions at positions 101(D), 102(V), 103(P), 104(D), 105(Y) and 106 (A) caused reductions in displacing ability of >10,000; 140; 5; 360; 840 and 720 respectively over controls. Figure 3 shows the effect of insertion of a single glycine residue into the sequence of the antigenic region of this same peptide. This clearly demonstrates that the insertion, which puts the sequence out of phase by a single residue, greatly reduces the ability of this antibody to recognize this peptide antigen. The possible importance of relative positional variation in antigen "drift" and antigenic "shift" has been discussed (28).

In conclusion we feel that the technique of SMPS and the use of a simultaneous multiple HF cleavage apparatus greatly facilitates the preparation of large numbers of peptides. The peptides generated have purities which are as high as the chemistry permits for an individual free resin. These procedures enable large numbers of peptides (>100) to be generated rapidly (4-6 weeks) and in amounts (>10 mg) which will satisify the needs of most investigators.

REFERENCES

1. R.A. Houghten, Proc. Natl. Acad. Sci., USA 82, 5131-5135 (1985).
2. R.A. Houghten, S.T. DeGraw, M.K. Bray, S.R. Hoffman, and N.D. Frizzell, Biotechniques, 4, 522-529 (1986).
3. R.A. Houghten, M.K. Bray, S.T. DeGraw, C.J. Kirby, Inter. J. Pept. Prot. Res., 27, 673-678 (1986).
4. R.A. Lerner, Nature (London) 299, 592-596 (1982).
5. J.L. Bittle, R.A. Houghten, H. Alexander, T.M. Shinnick, J.G. Sutcliffe, R.A. Lerner, D.J. Rowlands, and F. Brown, Nature (London) 298, 30-33 (1982).
6. I.J. East, P.E. Todd, and S.J. Leach, J. Immunol. 17, 519-525 (1980).
7. I.A. Wilson, H.L. Niman, R.A. Houghten, A.R. Cherenson, M.L. Connolly, and R.A. Lerner, Cell, 37, 767-778 (1984).
8. S. McMillan, M.V. Seiden, R.A. Houghten, B. Clevinger, J.M. Davie, and R.A. Lerner, Cell, 35, 859-863 (1983).
9. H.M. Geysen, R.H. Meloen, and S.J. Barteling, Proc. Natl. Acad. Sci. USA 81, 3998-4002 (1984).
10. H.M. Geysen, S.J. Barteling, and R.H. Meloen, Proc. Natl. Acad. Sci. USA 82, 178-182 (1985).
11. H.L. Niman, R.A. Houghten, L.E. Walker, R.A. Reisfeld, I.A. Wilson, J.M. Hogle, and R.A. Lerner, Proc. Natl. Acad. Sci. USA 80, 4949-4953 (1983).
12. D. Yamashiro, and C.H. Li, The Peptides, Vol. 6, ed. H. Meienhofer, Academic Press, NY; pp. 191-217 (1984).
13. V. Hruby, in Conformationally Directed Drug Design. eds. J.A. Vida and M. Gordon, Am. Chem. Soc. pp. 9-27 (1984).
14. D.F. Veber, F.W. Holly, W.J. Paleveda, R.F. Nutt, S.J. Bergstrand, M. Torchiana, M.W. Glitzer, R. Saperstein, and R. Hirschman, Proc. Natl. Acad. Sci. USA, 75, 2636-2640 (1978).
15. P.P. Chen, R.A. Houghten, S. Fong, G.H. Rhodes, T.A. Gilbertson, J.H. Vaughan, R.A. Lerner, and D.A. Carson, Proc. Natl. Acad. Sci. USA 81, 1784-1788 (1984).

16. F.A. Klipstein, R.F. Engert, R.A. Houghten, and B. Rowe, J. Clin. Microbiology 19, 798-803 (1984).
17. P.P. Chen, R.A. Houghten, and S. Fong, J. Exp. Med., in press (1985).
18. F.E. Bloom, E.L. Battenberg, R.J. Milner, and J.G. Sutcliffe, J. Neurosci., in press (1985).
19. J.G. Sutcliffe, R.J. Milner, T.N. Shinnick, and F.E. Bloom, Cell, 33, 671-682 (1983).
20. R.A. Houghten and S.T. DeGraw, J. Chromat. 386, 223-228 (1987).
21. R.A. Houghten and J.M. Ostresh, Biochromatography, 2, 80-84 (1987).
22. R.M. Freidinger, D.F. Veber, D.S. Perlow, J.R. Brooks, and R. . Saperstein, Science, 210, 656-658 (1980).
23. R.B. Merrifield, J. Amer. Chem. Soc. 85, 2149-2154 (1963).
24. A.R. Mitchell, S.B. Kent, M. Engelhard, and R.B. Merrifield, J. Org. Chem. 43, 2846-2852 (1978).
25. E. Atherton, D.L. Clive, and R.C. Sheppard, J. Amer. Chem Soc. 97, 6584-8585 (1975).
26. R.A. Houghten, W.C. Chang, and C.H. Li, Int. J. Pept. Prot. Res. 16, 311-320 (1980).
27. R.A. Houghten, J.M. Ostresh, and F.A. Klipstein, Eur. J. Biochem. 145, 157-162 (1984).
28. R.A. Houghten, S.R. Hoffman, H.L. Niman, in Modern Approaches to Vaccines. Cold Spring Harbor Symposium, (R.M. Chanock and R.A. Lerner eds.). Cold Spring Harbor, New York, pp. 21-25.
29. B.F. Gisen, Anal. Biochem. 58, 248-252 (1972).
30. Available from Multiple Peptide Systems, La Jolla, California.

MAPPING FUNCTIONAL DOMAINS OF HUMAN PLATELET THROMBOSPONDIN WITH ELECTRO-BLOTTING AND HIGH SENSITIVITY SEQUENCING

Gregory A. Grant, Vishva M. Dixit, Nancy J. Galvin,
Karen M. O'Rourke, Samuel A. Santoro and William A.
Frazier

Washington University School of Medicine
St. Louis, MO 63110

INTRODUCTION

Thrombospondin (TSP) is secreted from platelet alpha-granules at sites of platelet activation and aggregation and is also produced by a variety of cell types in vitro and in vivo. Our laboratory has recently selected a monoclonal antibody (Mab) against TSP which blocks platelet aggregation and has provided direct evidence that TSP is required in this important and highly regulated process (1). The protein is a trimer of subunits of 180 KDa linked by disulfide bonds (2). The only practical source for its purification is the secretion products of fresh platelets stimulated with thrombin from which it can be purified in quantities of aobut 2 to 5 mg per preparation (3). Thus any characterization of the structure of this large protein requires high sensitivity techniques.

DOMAIN STRUCTURE OF TSP

Like the more thoroughly studied adhesive glycoprotein fibronectin (FN), TSP binds a variety of macromolecules. These include heparin (3), fibrinogen (4), collagen (5), sulfated galactosyl lipids or sulfatides (6), plasminogen (7), histidine-rich glycoprotein (7) and fibronectin (8). We initially sought to determine whether TSP was composed of domains which might contain these binding sites within compact regions of protein structure separated by more flexible regions of connecting peptide. Using the same approach employed by others for fibronectin, we digested TSP with a variety of proteases to determine which gave rise to relatively stable fragments which might then be subjected to purification by affinity methods. Figure 1A shows a 0.5% w/w digest of TSP with chymotrypsin in the presence of Ca^{++}. A 25 KDa peptide is rapidly liberated with the generation of a 140 KDa species which then decays to 120 KDa with the cleavage of an 18 KDa peptide which remains disulfide linked to the 120 KDa fragment. It is this 18 KDa peptide which contains the epitope for the monoclonal antibody that blocks platelet aggregation (1). Using affinity chromatography, we have determined that the 25 KDa peptide contains the site of heparin binding (3), both the 120 and 140 KDa fragments bind to fibrinogen (4) and trimers composed of 140 KDa chains bind preferentially to red cells (9) and fixed, activated platelets (10).

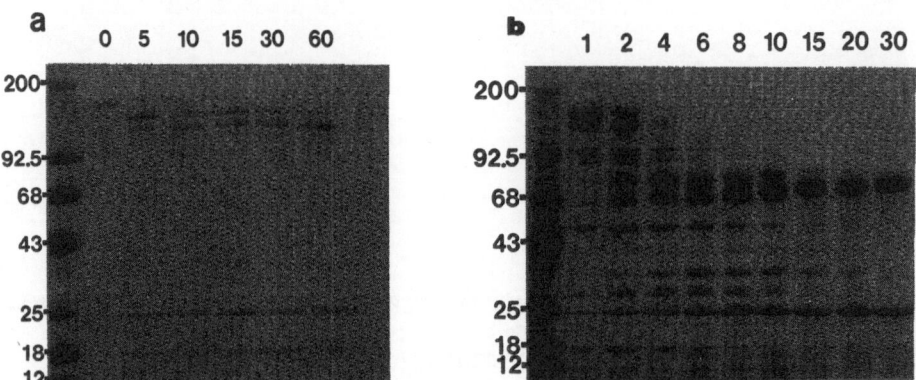

Fig. 1. Digestion of TSP with chymotrypsin in 1 mM Ca^{++} (a) and 5 mM EDTA (b). Numbers at the top of each lane indicate the time of digestion in min. after which an aliquot was removed, boiled in SDS sample buffer and run on an SDS slab gel. Standards are labeled with M_r in kiloDaltons. Staining was with Coomassie blue.

When Ca^{++} is removed from TSP it undergoes dramatic conformational changes which have been best illustrated by changes in the pattern of its proteolysis (11). As shown in Figure 1B, only two stable fragments are produced by chymotryptic digestion in the absence of Ca^{++}, a 25 KDa monomeric species and a trimeric 70 KDa fragment. The 25 KDa peptide is the same heparin binding domain obtained in the presence of Ca^{++} while the 70 KDa peptide appears to be derived from the 140 and 120 KDa species (12) and has been shown by us and others (5) to contain one or more binding sites for collagen, preferentially binding type V.

SEQUENCING OF TSP AND ITS FRAGMENTS

The first sequencing experiments were aimed at determining if the intact peptide chains of the TSP trimer were unblocked at their N termini and if all three were identical as had been supposed (3). Reduced and carboxymethylated peptide chains were subjected to automated Edman degradation in an Applied Biosystems gas phase sequencer. In several independent runs initial yields of up to 85% were obtained for the sequence:

Asn-Arg-Ile-Pro-Glu-Ser-Gly-Gly-Asp-Asn-Ser-Val-Phe-

Coligan and Slayter (13) obtained the same sequence and reported the additional residues:

-Asp-Ile-Phe-Glu-Leu-Thr-Gly-Ala-Ala-Trp-Lys-Gly-

It was clear from the high initial yields obtained in our analyses that all chains had the same sequence and it was unlikely that any were blocked (3). Within a short time three groups including our own reported that the 25 KDa heparin domain isolated by affinity chromatography on heparin-agarose (3), isoelectric focussing (13) or gel filtration (14) had a N-terminal sequence identical to that of the intact peptide chain. This result locates the heparin domain at the very amino terminus of the TSP peptide chain. Since the different groups all used different proteases of varying specificity to cleave off this domain, it also points out the extreme stability of this region of the TSP molecule, and further suggests that the extreme amino terminal region of the peptide chain is protected from proteolysis by sequestration within the domain.

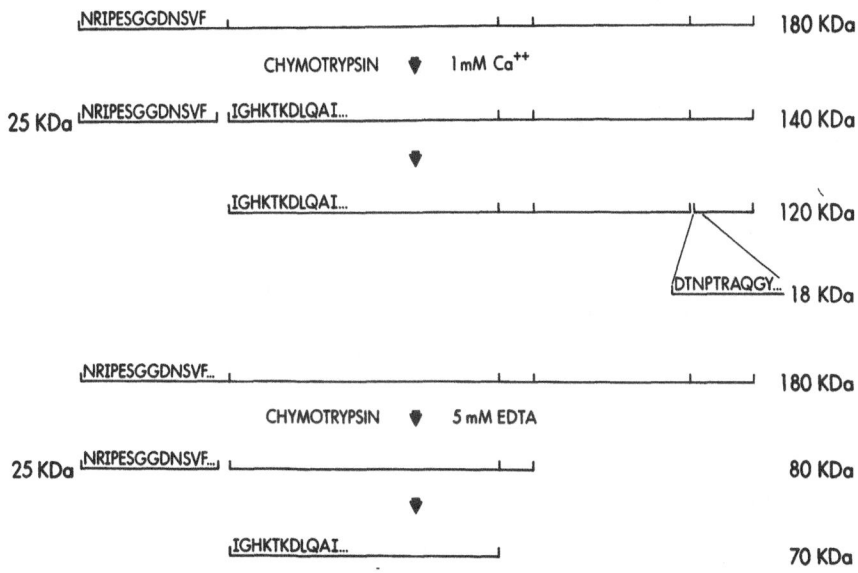

Fig. 2. Schematic representation of the chymotryptic digestion of TSP with the amino terminal amino acid sequence of each fragment indicated in the single letter amino acid code.

We reasoned that if the 18 KDa peptide were cleaved from one end of the 140 KDa fragment to yield the 120 KDa peptide, that one of these two fragments might have the same N-terminal sequence as the 140 KDa peptide. To test this hypothesis, we were faced with the problem of sequencing large fragments which could only be obtained as bands on SDS gels. At the time we began, the only available method of effectively recovering the protein band from the gel was the electroelution method of Hunkapiller et al. (15). However, soon thereafter, the electroblotting method of Aebersold et al. (16) became available and we employed this technique as well. For both methods, samples of 200 pmol or less were analyzed in the Applied Biosystems gas phase sequencer. Thus preparative digests of TSP were prepared and separated on SDS PAGE. For electroelution the gels were stained with Coomassie blue, the bands excised and placed in the elution apparatus where electroelution and electrodialysis were carried out over a period of several days. The sample was then recovered, concentrated and spotted on the sample disk of the gas phase sequencer. For electroblotting the gels were sandwiched against acid cleaned glass fiber paper and the peptides were transferred to the paper at acid pH as described by Aebersold et al. (16). In this technique the glass fiber paper is then stained for protein and the bands are cut out. The protein-impregnated paper is loaded directly into the reaction cartridge of the gas phase sequencer.

Figure 2 summarizes the results of these experiments (12). Both the 140 and the 120 KDa species have the N-terminal sequence:

Ile-Gly-His-Lys-Thr-Lys-Asp-Leu-Gln-Ala-Ile-

indicating that the 18 KDa piece must be removed from the C-terminal end of the 140 KDa peptide to generate the 120 KDa fragment. The 18 KDa fragment was also sequenced and found to have a distinct N-terminal sequence as shown in the figure. We then electroblotted and sequenced the 25 KDa and 70 KDa peptides obtained by chymotryptic digestion in the absence of Ca++. This confirmed that the heparin binding domain obtained in EDTA was in fact the same fragment as that produced in Ca++. Sur-

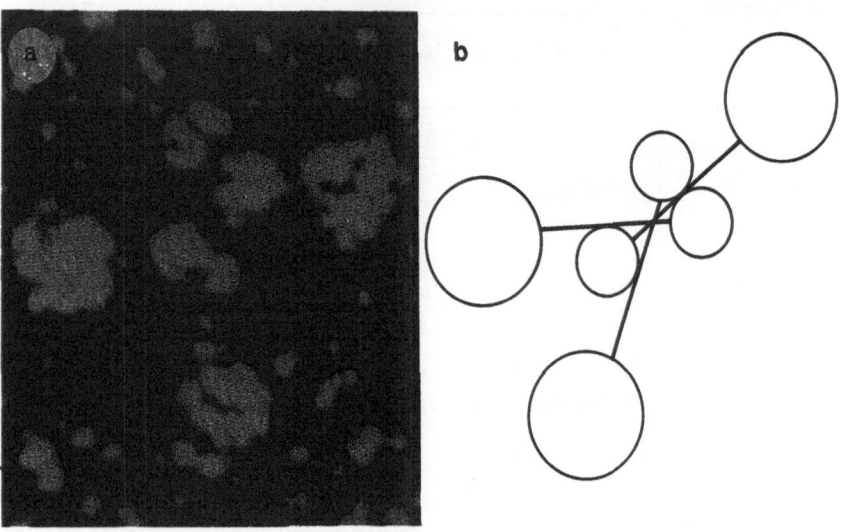

Fig. 3. Image of the TSP trimer obtained with electron microscopy of a
rotary shadowed, carbon coated replica (a). The final magnification is
about 1,000,000 X. (b) A model of the trimer which indicates the structure
of each asymetric monomer. The orientation of each monomer is about the
same as that in part (a).

prisingly, the 70 KDa fragment had exactly the same N-terminal sequence
as the 140/120 KDa fragments. Thus all of the peptide material digested
from the large 140 KDa fragment of TSP is removed at the carboxyl terminal
end while the amino terminal end adjacent to the heparin binding domain
is stable to digestion even in the absence of Ca^{++} (12).

ELECTRON MICROSCOPY OF TSP

    Using transmission electron microscopy of rotary shadowed, carbon
coated replicas of TSP we (12) and Lawler et al. (11) have obtained high
resolution images of the TSP trimer. A model derived from these data
shown in Figure 3 reveals the structure of the TSP subunit to be an
asymmetric dumbbell. That is a small globular region is connected by a
thin strand to a larger globular region. With the ability to visualize
this degree of detail in the TSP structure we went on to use monoclonal
antibodies specific for different domains or fragments of TSP to identify
these regions in the three-dimensional images of TSP obtained with electron microscopy. Mab A2.5 immunoprecipitates the 25 KDa amino terminal
heparin binding domain of TSP (17) while Mab C6.7 which blocks platelet
aggregation (1) immunoprecipitates the 18 KDa C-terminal fragment. We
prepared complexes of TSP and these Mabs at 1:1 molar ratios and examined
the products in the electron microscope (12). Mab A2.5 was seen to bind
to the small globular domains often crosslinking two of them causing a
bunching together of the large domains at the other end of the molecule.
In contrast, Mab C6.7 bound as far away from the small domains as possible,
on the face of the large globular domain opposite the point at which the
connecting strand enters it. Thus the N-terminal heparin binding domain
resides in the small globular domain and the C-terminal 18 KDa fragment
is on the outer face of the large globular domain the maximum possible
distance from the N-terminus. Lawler et al. (11) have also shown that
removal of the heparin domains by limited proteolysis results in a trimer
lacking the small globular domains. We have mapped two other Mabs which
react with the 120 KDa and the 70 KDa fragments and these bind to a different region of the large domain and the connecting strand region respectively.

474

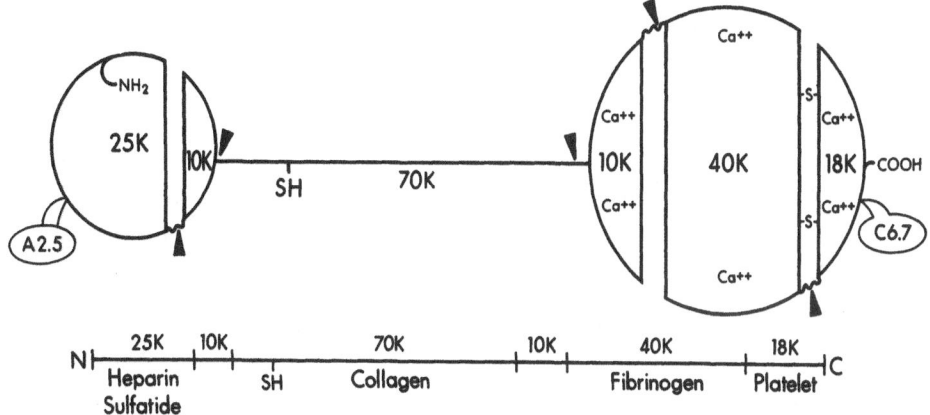

Fig. 4. Schematic representation of the TSP subunit drawn above the linear map of the proteolytically generated fragments discussed in the text. Solid arrowheads mark the points of proteolytic cleavage. A2.5 and C6.7 refer to the two monoclonal antibodies specific for the heparin binding domain and the 18 KDa domain respectively.

In conjunction with the amino acid sequencing data, these antibody data indicate that, to a large extent, the subunit of TSP is a colinear representation of the peptide chain (12). This is an interesting result since the fibronectin subunit is a strictly colinear map of the peptide chain, and the same may be true of von Willebrand factor as well. Our current picture of the TSP subunit is shown schematically in Figure 4. The N-terminus is probably buried within the 25 KDa heparin binding domain, a region that contains little or no cysteine, yet is remarkably stable to proteolysis by a variety of proteases. The peptide chain leaves this domain and courses through a ca. 10 KDa region which contains cleavage sites for a variety of proteases resulting in heparin domains of variable sizes with different proteases. The chain then enters the ca. 70 KDa extended connecting region which contains the interchain disulfides and the collagen binding domain. At the end of this region the chain enters the large C-terminal globular domain where its folding requires Ca^{++}. This region is competely degraded by proteases in the absence of Ca^{++} and we have identified Mabs which bind here only in the absence of Ca^{++}. At the end of the 120 KDa fragment, a cleavage site for chymotrypsin occurs which marks the N-terminus of the 18 KDa domain. This C-terminal domain contains the epitope for the Mab that blocks platelet aggregation. It is ideally situated on the outer face of the large globular domain for interaction with other platelet proteins which take part in the aggregation reaction. Perhaps one of these is fibrinogen (18). Other candidates include the platelet surface glycoproteins IIb/IIIa (19) and fibronectin (20), since antibodies against any of these proteins will block the aggregation of washed platelets.

REREFERENCES

1. V. M. Dixit, D. M. Haverstick, K. M. O'Rourke, S. W. Hennessy, G. A. Grant, S. A. Santoro, and W. A. Frazier, A monoclonal antibody against human thrombospondin inhibits platelet aggregation, Proc. Natl. Acad. Sci. USA, 82:3472 (1985).

2. J. W. Lawler, H.S. Slayter, and J. E. Coligan, Isolation and charac-
    terization of a high molecular weight glycoprotein from human
    blood platelets, J. Biol. Chem. 253:8609 (1978).
3. V. M. Dixit, G. A. Grant, S. A. Santoro, and W. A. Frazier, Isolation
    and characterization of a heparin-binding domain from the amino
    terminus of platelet thrombospondin, J. Biol. Chem. 259:10100
    (1984).
4. V. M. Dixit, G. A. Grant, W. A. Frazier, and S. A. Santoro, Isolation
    of the fibrinogen-binding region of platelet thrombospondin,
    Biochem. Biophys. Res. Commun. 119:1075 (1984).
5. S. M. Mumby, G. J. Raugi, and P. Bornstein, Interactions of thrombo-
    spondin with extracellular matrix proteins: selective binding
    to type V collagen, J. Cell Biol. 98:646 (1984).
6. D. D. Roberts, D. M. Haverstick, V. M. Dixit, W. A. Frazier, S. A.
    Santoro, and V. Ginsburg, The platelet glycoprotein thrombospondin
    binds specifically to sulfated glycolipids, J. Biol. Chem. 260:
    9405 (1985).
7. R. L. Silverstein, L. L. K. Leung, P. C. Harpel, and R. L. Nachman,
    Platelet thrombospondin forms a trimolecular complex with
    plasminogen and histidine-rich glycoprotein, J. Clin. Invest.
    75:2065 (1985).
8. J. Lahav, J. Lawler, and M. A. Gimbrone, Thrombospondin interactions
    with fibronectin and fibrinogen, Eur. J. Biochem. 145:151 (1984).
9. D. M. Haverstick, V. M. Dixit, G. A. Grant, W. A. Frazier, and S. A.
    Santoro, Localization of the hemagglutinating activity of
    platelet thrombospondin to a 140,000 Dalton thermolytic fragment,
    Biochemistry 23:5597 (1984).
10. D. M. Haverstick, V. M. Dixit, G. A. Grant, W. A. Frazier, and S. A.
    Santoro, Characterization of the platelet agglutinating activity
    of thrombospondin, Biochemistry 24:3128 (1985).
11. J. L. Lawler, H. Derick, J. E. Connolly, J. H. Chen, and F. C. Chao,
    The structure of human platelet thrombospondin, J. Biol. Chem.
    260:3762 (1985).
12. N. J. Galvin, V. M. Dixit, K. M. O'Rourke, S. A. Santoro, G. A. Grant,
    and W. A. Frazier, Mapping of epitopes for monoclonal antibodies
    against human platelet thrombospondin with electron microscopy
    and high sensitivity amino acid sequencing, J. Cell Biol. in
    press.
13. J. E. Coligan and H. S. Slayter, Structure of thrombospondin, J.
    Biol. Chem. 259:3944 (1985).
14. G. J. Raugi, S. M. Mumby, C. A. Ready, and P. Bornstein, Location
    and partial characterization of the heparin-binding fragment
    of platelet thrombospondin, Thromb. Res. 36:165 (1984).
15. M. W. Hunkapillar, E. Lujan, F. Ostrander, and L. E. Hood, Isolation
    of microgram quantities of proteins from polyacrylamide gels
    for amino acid sequence analysis, Methods in Enzymol. 91:227
    (1983).
16. R. H. Aebersold, D. B. Teplow, L. E. Hood, and S. B. H. Kent,
    Electroblotting onto activated glass: high efficiency prepara-
    tion of proteins from analytical SDS-polyacrylamide gels for
    direct sequence analysis, J. Biol. Chem. in press.
17. V. M. Dixit, D. M. Haverstick, K. M. O'Rourke, S. W. Hennessy, G. A.
    Grant, S. A. Santoro, and W. A. Frazier, Effects of anti-thrombo-
    spondin monoclonal antibodies on the agglutination of erythro-
    cytes and fixed, activated platelets by purified thrombospondin,
    Biochemistry 24:4270 (1985).
18. T. K. Gartner, J. M. Gerrard, J. G. White, and D. C. Williams Fibrin-
    ogen is the receptor for the endogenous lectin of human
    platelets, Nature (London) 189:688 (1981).

19. R. P. McEver, E. M. Bennett, and M. N. Martin, Identification of two structurally and functionally distinct sites on human platelet membrane glycoprotein IIb-IIIa using monoclonal antibodies, J. Biol. Chem. 258:5269 (1983).

20. V. M. Dixit, D. M. Haverstick, K. M. O'Rourke, S. W. Hennessy, T. J. Broekelmann, J. A. McDonald, G. A. Grant, S. A. Santoro, and W. A. Frazier, Inhibition of platelet aggregation by a monoclonal antibody against human fibronectin, Proc. Natl. Acad. Sci. USA 82:3844 (1985).

IMMUNOPROTECTION - A NOVEL APPROACH FOR MAPPING

EPITOPES ON AN ANTIGEN

Betsy J. Bricker, Robert R. Wagner and Jay W. Fox*

Department of Microbiology, University of Virginia Medical
School, Charlottesville, VA 22908

INTRODUCTION

Since the introduction of hybridoma technology, monoclonal antibodies have proven to be very useful tools for the study of specific
sites on proteins (1,2,3 and 4). The site specificity of monoclonal
antibodies offers new ways to link areas of protein function with structure. For some proteins, this mapping has been a simple matter of
digesting the antigen and analyzing which fragment(s) the antibodies
interact with (5,6,7 and 8). However, a large population of monoclonal
antibodies (MAbs) are highly dependent on the protein conformation of the
antigen (9,10). These MAbs often lose their reactivities with their
antigenic determinants when the target protein is digested or during
protein preparation for cleavage. This has been a serious barrier for
studying many complex proteins.

It has been shown that the surface glycoprotein (called the
G-Protein) of Vesicular Stomatitis virus (VSV) is intimately involved in
host cell infection (11). To better understand this important molecule,
a panel of 25 MAbs was developed against the purified G-Protein (details
to be published elsewhere). However, attempts to use these MAbs to map
the functional regions of the molecule were confounded by the high
dependence of most of the MAbs on the tertiary structure of the
G-Protein.

The following report defines a unique mapping technique which can
accomodate the structural integrity required of the target protein.
Using this method it was possible to map a conformationally dependent MAb
to a limited region of the published G-Protein sequence (12). Placement
of this epitope could not have done using the more conventional mapping
methods.

MATERIALS AND METHODS

Origin of the Monoclonal Antibodies

A panel of monoclonal antibodies was made as the result of a fusion
between <u>Balb c</u> mouse spleen cells immunized against purified G-Protein,

and SP 2/0 myeloma cells as described by Oi and Herzenberg (13). A detailed characterization of these monoclonal antibodies will be published elsewhere.

Isolation of G-Protein

G-Protein was obtained from VSV harvested from BHK cells as published by Petri and Wagner (14). G-Protein was extracted from the virus by incubation in 0.01 M Tris, pH 8.0, containing 0.03 M β-D-octyl glucoside for 1 hour at room temperature. The remaining viral particles were removed by centrifugation at 150,000 g for 1.5 hours into a small 50% glycerol cushion.

Covalent Cross-linking of MAbs to Protein A Sepharose

Cross-linking was performed by the method developed by Schneider et al (15) (Fig. 1). Borate buffer (0.1 M, pH 8.1) was used to wash 1.5 ml (packed volume) of Protein A Sepharose (Pharmacia) three times. Approximately 13 mg MAb #19 (purified by affinity chromatography) in a volume of 13 ml (0.1 M Borate buffer, pH 8.1) was added to the washed beads and incubated overnight with agitation at either room temperature or 4°C. The pelleted beads were washed twice in 0.2 M triethanolamine (pH 8.2) and then incubated in 20 volumes (30 ml) of 20 mM Dimethyl Pimelidate (DMP, Pierce Chems), prepared in 0.2 M triethanolamine (pH 8.2) immediately before use. The pH was readjusted to 8.2 after bound to the beads during the 1 hour incubation with agitation at room temperature. The beads were then pelleted and washed in borate buffer twice. After a final wash in 0.14 M phosphate buffer (pH 8.9) containing 0.05% sodium azide, the beads were resuspended in 5.0 ml of this same buffer for storage at 4°C. For screening purposes, proportionately smaller preparations were used.

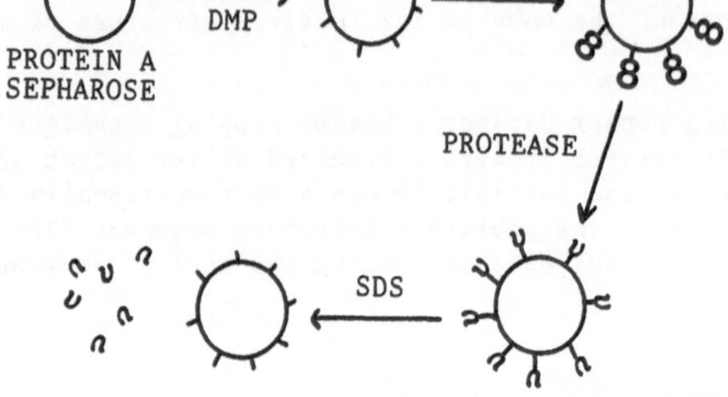

Figure 1: *Outline of the steps comprising the immunodigestion procedure. Clockwise: Protein A sepharose is used as the substrate for covalent binding of MAb; antigen is bound, followed by proteolytic digestion. The bound antigenic determinants are finally released in the presence of SDS.*

Immunodigestion

MAb covalently cross-linked to Protein A sepharose (1.5 ml packed volume) was washed twice in PBS (pH 7.6) before resuspension in 10 ml PBS containing 5 mg G-Protein antigen. The antigen was allowed to bind overnight with agitation at 23°C. The antigen-bound beads were then centrifuged and washed twice in PBS. At this time the appropriate protease was added (at a concentration, volume and pH determined empirically for each antigen). For 2.5 mg G-Protein actually bound to MAb, 250 µg Staph. aureus protease (Worthington Enzymes) in a volume of 2.75 ml PBS (pH 7.6) was used (giving a protein to enzyme ratio of 10 to 1 wt/wt).

Digestion continued at 23°C for 18 hours with mild shaking. The reaction was stopped by 3 washes in PBS followed by a final rinse in 10 mM sodium phosphate, pH 7.6. As much fluid as possible was removed from the beads, followed by the addition of 0.5 ml of either SDS-PAGE sample buffer (16), or 2% SDS, both without 2-mercaptoethanol or other reducing agents, and incubation for 3-5 minutes in a boiling water bath. The G-protein fragments eluted into the supernatant were then reduced with 2-mercaptoethanol and used immediately for further purification procedures.

For analyses on SDS-PAGE, smaller preparations of sample were appropriate. These were obtained by following the preceding procedure with a few modifications. First, 5.5 µl (packed volume) of MAb covalently linked to Protein A sepharose were reacted with 50 µg of the antigen in 100 µl PBS, pH 7.6. Digestion was initiated with the addition of 30 µg Staph. aureus protease in 100 µl PBS (pH 7.6). After digestion and washing, 20 µl of SDS-PAGE sample buffer (without reducing agents) was used to elute the fragments.

Isolation of Digestion Fragments

Immunodigests for both fragment recovery and for screening purposes were electrophoresed on SDS polyacrylamide gels (16) using a 10-17% acrylamide gradient for high resolution of fragments. Samples used for screening were separated on a 12 cm running gel then stained by the silver stain of Oakley et al (17).

Alternatively, a sample containing fragments to be isolated and purified was layered across the entire width of a 28 cm long running gel. Sodium thioglycolate (0.1 mM) was included in the upper tank buffer to protect the side groups of tryptophan, methionine and histidine. After separation, the gel was lightly stained with Coomassie Brilliant Blue and a narrow band containing the fragment of interest was excised. The fragment-containing strips were washed for 2 hours against several changes of distilled water at 4°C to return the pH to neutral, then stored at -20°C overnight or longer.

For electorelution, each frozen acrylamide block was cut into 1 mm cubes with a razor blade and transferred to an electroelution cell (model ECU 40, CBS Scientific Co., San Diego, CA.) fitted with a Spectra/Por membrane (2000 MW cut off). The conditions for electroelution were taken from Hunkapillar et al (18). The sample was covered with 0.4 M ammonium carbonate containing 2% SDS and 0.1% Dithiothreitol, then overlayed with elution buffer (0.1% SDS in 0.05 M ammonium carbonate). After soaking for 3-5 hours, electroelution was carried out in elution buffer overnight; then the sample was electrodialyzed in a 1:5 dilution of the elution buffer for 24 hours. The sample was collected in a minimal volume and lyophilized. SDS and other contaminants were removed by a chloroform-methanol protein extraction step performed exactly as published by Wessel and Flugge (19). The remaining precipitate was lyophilized and stored at -20°C for amino acid sequencing.

Amino Acid Sequence Analysis

The peptide samples were sequenced with an Applied Biosystem 470A Sequencer. The glass filters were previously treated with TFA and the polybrene (3 mg) contained 2 mg NaCl. Resultant PTH amino acids were analyzed by HPLC using an IBM cyano column.

RESULTS

Digestion of the G-Protein of VSV followed by Western Blot analysis of the resulting fragments was found to be inappropriate for epitope mapping of anti-G-Protein MAbs due to a strong dependence on the conformation of the native antigen. A technique was needed which could retain the tertiary structure of the target protein long enough to allow MAb binding. With this in mind, a technique was developed in which the MAb was first bound to the antigenic protein, allowing the protein-protein interaction to stabilize the epitope structure.

MAb 19 was covalently cross-linked to Protein A sepharose beads. This was necessary to eliminate MAb-derived fragments from contaminating the G-protein digestion products. Analysis of covalently cross-linked MAb by SDS-PAGE using silver stain detection methods indicated that all detectable MAb was covalently bound and could not be removed by boiling in 2% SDS for 5 minutes (data not shown).

The cross-linked MAb was next used to immunoprecipitate a partially purified extract of G-Protein. Again, using SDS-PAGE analysis, the covalent cross-linking procedure had no apparent effect on the antibody's ability to specifically bind the G-Protein.

With the target protein bound to the immobilized MAb, the appropriate protease was added. Theoretically, the epitope on the G-protein was sterically protected by the G-protein MAb. The rest of the G-protein, not immediately adjacent to the binding site, was accessible to the protease and digested. At the end of the incubation period, the protease and the unbound MAb and G-protein fragments were completely washed away from the beads. An advantage to this approach is shown in Figure 2. If an epitope is defined by 2 discontinuous regions of the protein juxtapositioned into a tertiary structure, both regions would remain bound to the MAb and be available for analysis. Fragments resulting from two discontinuous regions must be distinguished by N-termini sequencing from fragments formed by incomplete digestion (multiple cleavage sites) of a single linear stretch along the G-protein sequence (dotted line). The digestion cannot be carried to completion since the MAb becomes increasingly vulnerable to proteolytic cleavage as the surrounding antigen is removed. Eventually the MAb is cleaved from the sepharose bead.

Figure 3 shows the digestion pattern for the epitope defined by MAb 19. The G-protein was immunodigested with Staph. aureus protease at a concentration of 250 μg per 2.5 mg bound G-Protein in 2.5 ml PBS (pH 7.5) for 18 hours. The appropriate concentration and digestion conditions for a given protease must be determined empirically for each antigen.

Figure 2: *Diagrammatic representation of immunodigestion. Left side: globular protein is attached to the binding site of a covalently cross-linked MAb. Right side: only those regions of the protein directly interacting with the binding region of the MAb remain after digestion. Note that two discontinuous regions of the protein form the antigenic determinant and both remain bound after digestion.*

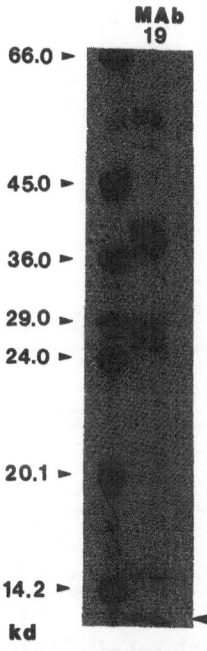

Figure 3: *Immunodigested G-protein. After elution of the bound digest fragments from the MAb, the sample was reduced and analyzed by SDS-PAGE. The arrow at the right marks the smallest major fragment at 12 kd.*

Since the smallest fragment (12 kd) would best limit the location of the epitope, further analysis centered on this fragment. Although HPLC initially seemed the best method for purifying the digestion fragments, solubility problems made this approach inappropriate. Purification of the 12 kd fragment (and several larger fragments) was achieved by separation on SDS-PAGE followed by electroelution of the excised band. This method resulted in a sample pure enough for amino acid sequencing.

The 12 kd fragment was sequenced and the first 10 N-terminal residues identified. The sequence matched <u>identically</u> a central region of the published G-Protein sequence shown in Figure 4. The terminus of the fragment was estimated from the molecular weight of the fragment and the specificity of the protease. Sequencing of select fragments with larger molecular weights are supportive of this placement. Although we have not yet studied all the fragments produced by immunodigestion, there is no indication at this time of a second discontinuous region of the protein involved in the epitope structure.

It would have been very useful to demonstrate that MAb 19 would rebind the 12 kd fragment on a Western Blot. Unfortunately, as might be expected from the strong dependence on conformation, MAb 19 no longer bound the 12 kd fragment or any of the digestion fragments after electrophoresis. Surprisingly, a MAb specific for another epitope on the G-Protein did bind the 12 kd fragment on a Western blot. This MAb was known to be less dependent on the tertiary structure of the G-protein than most of the other MAbs in our panel (data to be published elsewhere). This result is highly suggestive that these two epitopes are located in the same region of the molecule.

DISCUSSION

The technique presented in this paper has been useful in the mapping of an epitope unapproachable by conventional methods. It should be a useful means for the initial characterization of not only MAb-antigen interactions, but also the mapping of other protein-protein binding sites dependent on the tertiary conformation of the reactants.

```
215
...-G-L-P-E-T-G-I-R-S-N-Y-F-P-Y-I-S-T-E-G-I-C-K-M-P-F-C-R-K-
                         225

            250
Q-G-Y-K-L-K-N-D-L-W-F-Q-I-M-D-P-D-L-D-K-T-V-R-D-L-P-H-I-K-D-

    275                                                  300
C-D-L-S-S-S-I-I-T-P-G-E-H-A-T-D-I-S-L-I-S-D-V-E-R-I-L-D-Y-A-

                                                   325
L-C-Q-N-T-W-S-K-I-E-S-G-F-P-T-T-P-V-D-L-S-Y-L-G-...
        ▲         ▲              ▲
```

Figure 4: Partial copy of the published G-Protein sequence (12). The wavy underline marks the 10 N-terminal amino acids sequences after immunodigestion of the 12 kd fragment. The straight underline marks the remaining portion of the sequence believed to contain the epitope to MAb 19. Arrows show the possible termination (cleavage) sites for the fragment.

There are several conditions which must be met for this technique to be useful. First, the sequence of the target protein must be known, if mapping is the goal. Second, to obtain sufficient quantities of a fragment for sequencing, adequate amounts of the reactants are needed. For antigens difficult to isolate in large, stable quantities, this will be a serious problem. Third, the procedure must be modified for use with mouse IgG1 type MAbs since these do not bind Protein A under normal conditions (20). Finally, several bands for each digest may need to be sequenced to identify epitopes composed of discontinuous regions.

The results from an immunodigest can only be considered suggestive since conformationally dependent MAbs usually do not rebind the localized fragment for conformation. However, it does provide a starting point for other mapping methods. Synthetic peptides are currently being produced to further pinpoint the epitope for MAb 19 by way of anti-peptide antibodies (21). Working within the boundaries of a 12 kd fragment will be much more efficient than trying to use a shotgun approach across the entire 65,000 dalton sequence. Additionally, the epitope-containing fragment can be further trimmed by using several successive protease treatments. Also, carboxypeptidase sequencing could be useful in concisely defining the carboxyl termini of fragments. Finally, the digestion fragments themselves could be useful immunogens to produce antibodies which can either inhibit protein function directly, or compete with those MAbs which do.

REFERENCES

1. D. E. Yelton and M. D. Scharff, Monoclonal antibodies: a powerful new tool in biology and medicine, Ann Rev Biochem., 50:657-680 (1981).
2. H. Harris, Monoclonal antibodies to enzymes, in "Monoclonal Antibodies and Functional Cell Lines. Progress and Applications". R. H. Kennett, K. B. Bechtal, and T. J. McKearn, eds., pp. 33-65, Plenum Press, New York (1984).
3. M. F. Greaves, "Monoclonal Antibodies to Receptors: Probes for Receptor Structure and Function. Receptors and Recognition Series B Volume 17". Chapman and Hall, London (1984).
4. J. W. Yewdell and W. Gerhard, Antigenic characterization of viruses by monoclonal antibodies, Ann Rev Microbiol., 35:185-206 (1981).
5. H. Towbin and J. Gordin, Immunoblotting and dot immunobinding - current status and outlook, J Immunol Methods, 72:313-340 (1984).
6. B. T. Atherton, D. M. Taylor and R. O. Hynes, Structural analysis of fibronectin with monoclonal antibodies, J of Supramolec Struct and Cell Biochem., 17:153-161 (1981).
7. M. D. Pierschbacher, E. G. Hayman and E. Ruoslahti, Location of the cell-attachment site in fibronectin with monoclonal antibodies and proteolytic fragments of the molecule, Cell 26:259-268 (1981).
8. W. J. Gullick, S. Tzartos, and J. Lindstrom, Monoclonal antibodies as probes of acetylcholine receptor structure. I. Peptide mapping, Biochem., 20:2173-2180 (1981).
9. P. Parham, M. J. Androlewicz, F. M. Brodsky, N. J. Holmes and J. P. Ways, Monoclonal antibodies: purification, fragmentation and application to structural and functional studies of class I MHC antigens, J Immunol Methods, 53:133-173 (1982).
10. J. A. Berzofsky, G. K. Buckenmeyer, G. Hicks, F. R. N. Gurd, R. J. Feldmann and J. Minna, Topographic antigenic determinants recognized by monoclonal antibodies to sperm whale myoglobin, JBC, 257(6):3189-3198 (1982).
11. J. M. Kelly, S. U. Emerson, and R. R. Wagner, The glycoprotein of

vesicular stomatitis virus is the antigen that gives rise to and reacts with neutralizing antibody, J Virol, 10:1231-1235 (1972).

12. C. J. Gallione and J. K. Rose, Nucleotide sequence of a cDNA clone encoding the entire glycoprotein from the New Jersey serotype of vesicular stomatitis virus, J Virol., 46:162-169 (1983).

13. V. T. Oi and L. A. Herzenberg, Immunoglobulin-producing hybrid cell lines, in: "Selected Methods in Cellular Immunology", B. B. Mishell and S. M. Shiigi, eds., pp. 356-359, W. H. Freeman and Co., San Francisco (1980).

14. W. A. Petri and R. R. Wagner, Reconstitution into liposomes of the glycoprotein of vesicular stomatitis virus by detergent dialysis, JBC, 254:4313-4316 (1979).

15. C. Schneider, R. A. Newman, D. R. Sutherland, U. Asser and M. F. Greaves, A one-step purification of membrane proteins using a high efficiency immunomatrix, JBC, 257:10766-10769 (1982).

16. U. K. Laemmli, Cleavage of structural proteins during the assembly of the head of bacteriophage T4, Nature, 227:680-685 (1970).

17. B. R. Oakley, D. R. Kirsch and N. R. Morris, A simplified ultra-sensitive silver stain for detecting proteins in polyacrylamide gels, Anal Biochem., 105:361-363 (1980).

18. M. W. Hunkapillar, E. Lujan, F. Ostrander, and L. E. Hood, Isolation of microgram quantities of protein from polyacrylamide gels for amino acid sequence analysis, Methods of enzymology, 91:227-236 (1983).

19. D. Wessel and U. I. Flugge, A method for the quantitative recovery of protein in dilute solution in the presence of detergents and lipids, Anal Biochem., 138:141-143 (1981).

20. Problems with Protein A binding my mouse IgG1 type MAbs can be circumvented by binding the MAbs to Protein A sepharose in Biorad's MAPS buffer followed by 2 rinses in 0.2 M triethanolamine, pH 8.2 and covalently cross-linking as previously described. Biorad MAPS buffer contains components which inhibit cross-linking by DMP and so it must be removed completely. Even with these modifications, some IgG1 MAbs cannot be bound to Protein A sepharose in adequate amounts.

21. I. A. Wilson, H. L. Niman, R. A. Houghton, A. R. Cherenson, M. L. Connolly and R. A. Lerner, The structure of an antigenic determinant in a protein, Cell, 37:767-778 (1984).

MICRO METHOD FOR THE DETECTION OF HEPARIN-BINDING PROTEINS AND

PEPTIDES

Nobuyoshi Hirose, Lilian Socorro, Richard L. Jackson,
and Alan D. Cardin

Merrell Dow Research Institute, Cincinnati, OH 45215
and Division of Lipoprotein Research, Department of
Pharmacology and Cell Biophysics, University of
Cincinnati College of Medicine, Cincinnati, Ohio
45267-0575

INTRODUCTION

Acid mucopolysacchrides, such as heparin, are known to bind to
many proteins and mediate various biological processes (1). For
example, heparin precipitates plasma low density lipoproteins (LDL)
and very low density lipoproteins in the presence of Ca^{2+}. Heparin
binds to antithrombin-III (AT-III) and potentiates the inhibition of
thrombin (Th). Recently, it has been reported that heparin binds to
several growth factors (2,3). For these reasons, there is interest
in the development of methods for the detection and purification of
heparin-binding proteins. In the present report we describe a sensitive and quantitative dot-blot method for heparin-binding.

MATERIALS AND METHODS

Crude heparin from bovine lung was a generous gift of Hepar
Industries (Franklin, Ohio). Bovine serum albumin (BSA) was from
Sigma. Thrombin was kindly given by Dr. J. W. Fenton (Division of
Laboratories and Research, New York Department of Health, Albany, New
York). Rabbit thymus histones (Hn) was from Dr. R. C. Krueger
(Department of Biological Chemistry and Molecular Genetics,
University of Cincinnati College of Medicine, Cincinnati, Ohio). The
3-(p-hydroxyphenol)-propionic acid N-hydroxysuccinimide ester (Tagit)
was from Calbiochem; carrier free $Na[^{125}I]$(17 Ci/mg) was from New
England Nuclear; X-OMAT AR film (XAR-2) was from Kodak; AffiGel-10,
nitrocellulose paper, and Bio-Dot™ Microfiltration apparatus were
from Bio-Rad. Bovine milk lipoprotein lipase (LpL) was purified as
previously described (4). Human AT-III and fibronectin (Fn) were
purified from lipoprotein deficient plasma by heparin- and
gelatin-Sepharose chromatography (5). The human apolipoproteins A-I,
A-II, C-III and E were purified from the plasma of normal subjects
or patients with type V hyperlipoproteinemia (6). LDL were purified
from the plasma of subjects with type II hyperlipoproteinemia by
sequential ultracentrifugation in KBr (7). Purification and
radioiodination of high reactive heparin (HRH) was done as previously
described (8). In brief, crude heparin (100 mg uronic acid) was

fractionated on 50 ml of LDL-AffiGel-10 equilibrated with 10 mM Tris-HCl, pH 8.0 (standard buffer), 10 mM CaCl$_2$. HRH was eluted from the column with 0.5 M NaCl. HRH (2 mg uronic acid) in 1 ml of 0.1 M borate buffer, pH 9.0, was reacted twice with 1 mg of Tagit at 0°C for 20 min, then radioiodinated in 0.15 M glycine, pH 9.0, 0.4 M NaCl with 1 mCi of Na[125I] and ICl. Radioiodinated HRH was purified on 10 ml of LDL-AffiGel-10. The specific radioactivity of HRH was 1-2 dpm/ng uronic acid in the experiments. A Bio-Dot™ Microfiltration apparatus was assembled according to the manufacturer's manual (9). In a typical experiment, 0.5 nmole of protein in 250 μl of 50 mM ammonium bicarbonate was applied to nitrocellulose paper (0.45 μm). After the protein was applied and filtered, each spot was washed 3 times with 200 μl of 10 mM Tris-HCl, pH 8.0, 0.15 M NaCl. The nitro-cellulose paper was then cut into strips and placed into plastic conical tubes (Corning™, 15 ml centrifuge tube with screw cap). The strips were washed once with 10 ml of incubation buffer without heparin for 5 min. Then the strips were incubated as described in Results with radioiodinated heparin with gentle shaking at room tem-perature (Bellco rocker platform). After incubation, each strip was transferred to a new conical tube and washed 4 times, 1 min each, with 10 ml of incubation buffer. The strips were blotted on filter paper (Whatman #1) and air dried overnight. Air dried strips were placed in the x-ray cassette and fixed with scotch tape. Several X-ray films and an intensifying screen (Cronex Dupont) were overlayed on the nitrocellulose and stored at -70°C. The spots were visualized by radioautography. Each spot of the radioautograph was scanned by a Helena Labs soft laser densitometer using the transmittance mode. The strips were then stained with 0.1% Amido Black and the protein-staining spots were scanned by the densitometer using the reflectance mode.

RESULTS

Effect of Incubation Time on [125I]-HRH Binding

To determine the effect of incubation time on heparin-binding, 0.5 nmole of LpL, LDL, AT-III and BSA was spotted on nitrocellulose paper. The paper was washed, cut into strips and incubated with 5 ml of 10 mM Tris-HCl, pH 8.0, 10 mM CaCl$_2$ containing [125I]-HRH (2,000 dpm/ml, 1 μg uronic acid/ml) for 1, 3, 4, 6, and 9 h. After washing, the nitrocellulose paper was overlayed with X-ray film and the film was developed after 48 h. Each spot of the radioautograph was scanned and the % relative intensity was determined. Figure 1 shows the time course of binding; the inset shows the radioautograph. The binding of [125I]-HRH to LpL, LDL and AT-III was time-dependent; BSA showed no visible spot even after 9 h. For LDL and AT-III, [125I]HRH binding increased linearly and was maximal after 4 h incubation. For LpL, maximal [125I]-HRH binding required 9 h.

Effect of Cations on [125I]-HRH Binding

To determine the specificity of HRH-binding and the possible role of cations, seven heparin-binding proteins (LDL, apoE, LpL, Fn, AT-III, Th and Hn) and 4 non-heparin-binding proteins (BSA, apoC-III, apoA-I and apoA-II) were spotted on nitrocellulose. The paper was then cut into strips and incubated for 3 h with [125I]-HRH (20,000 dpm/ml, 10 μg/ml) in 10 mM Tris-HCl, pH 8.0 in either the absence or the presence of 10 mM Ca^{2+}, Mg^{2+} or La^{3+} (Figure 2). In the absence of Ca^{2+}, LpL, Fn, Th and Hn bound [125I]-HRH; no binding was evident

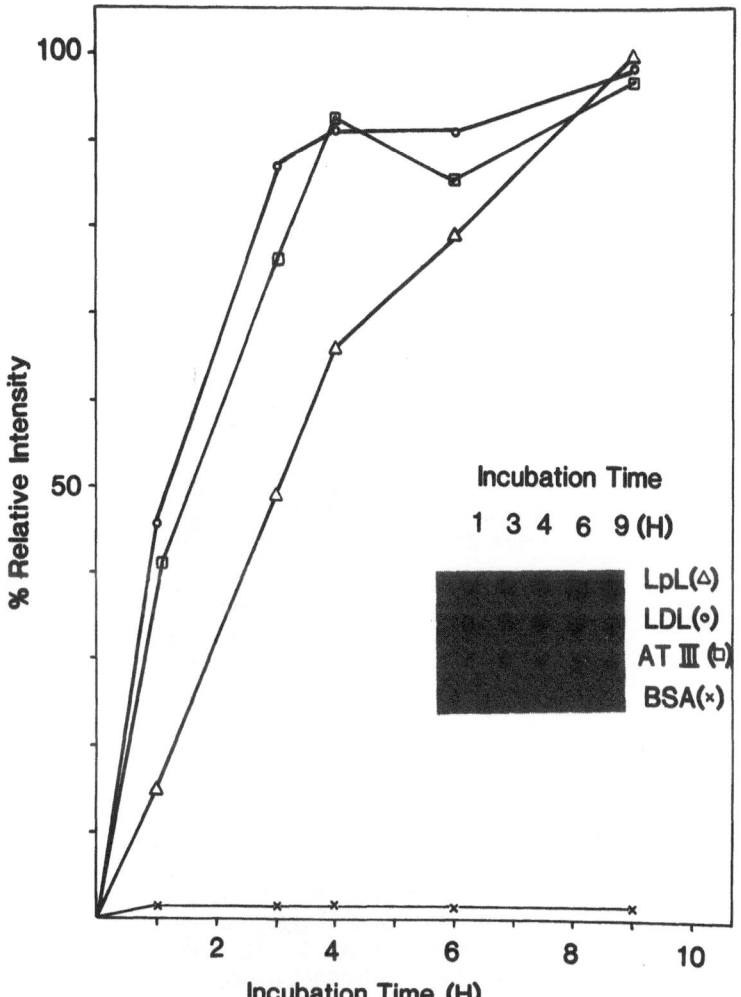

Figure 1: Effect of time on the binding of [^{125}I]-HRH to LpL, LDL, AT-III and BSA as detetermined by the dot-blot method. Proteins (0.5 nmole) were applied to nitrocellulose paper and incubated with 2,000 dpm/ml of [^{125}I]-HRH for the times indicated. The nitrocellulose paper was subjected to radioautography (48 h). The intensity of each spot was determined at each time point, and normalized to that at 9 h (% relative intensity). The inset shows the radioautograph.

in the absence of Ca^{2+} with LDL, apoE and AT-III. However in the presence of Ca^{2+}, LDL, apoE and AT-III bound $[^{125}I]$-HRH. In addition, the presence of Ca^{2+} enhanced the binding of $[^{125}I]$-HRH to LpL, Fn, Th and Hn. BSA, apoC-III, apoA-I and apoA-II did not bind to $[^{125}I]$-HRH in either the presence or absence of Ca^{2+}. Equivalent

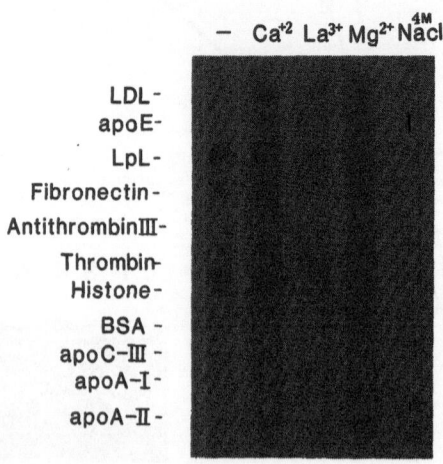

Figure 2: The effect of cations on the binding of $[^{125}I]$-HRH to LDL, apoE, LpL, Fn, AT-III, Th and Hn. Proteins (0.5 nmole) were spotted on nitrocellulose and incubated with $[^{125}I]$-HRH (20,000 dpm/ml) either in the absence (standard buffer alone) or in the presence of 10 mM Ca^{2+}, Mg^{2+}, or La^{3+}. Radioautography was for 48 h. Non-heparin-binding proteins (BSA, apoA-I, A-II and C-III) were included as controls. The effect of 4 M NaCl wash on the binding of $[^{125}I]$-HRH is also shown.

results were obtained with Mg^{2+}. La^{3+}, as well as 4 M NaCl wash abolished the binding of $[^{125}I]$-HRH. Scanning of Amido Black stained strips showed that the washing and incubation conditions did not decrease the amount of protein on the paper. Therefore the lack of binding is not explained by the removal of the proteins from the paper.

Quantitation of Heparin-Binding

The correlation between radioactivity and spot intensity of the radioautograph was examined with the same heparin binding proteins as described in Figure 2 . The amount of protein spotted was 0.5 nmole. The strips were incubated with $[^{125}I]$-HRH in order to determine the relation between intensity and the amount of $[^{125}I]$-HRH bound to the paper. Each spot was scanned by a densitometer using the intensity of Hn in the presence of Ca^{2+} as 100%. Then the strips were cut into spots and radioactivity was determined. Figure 3 shows a linear cor-

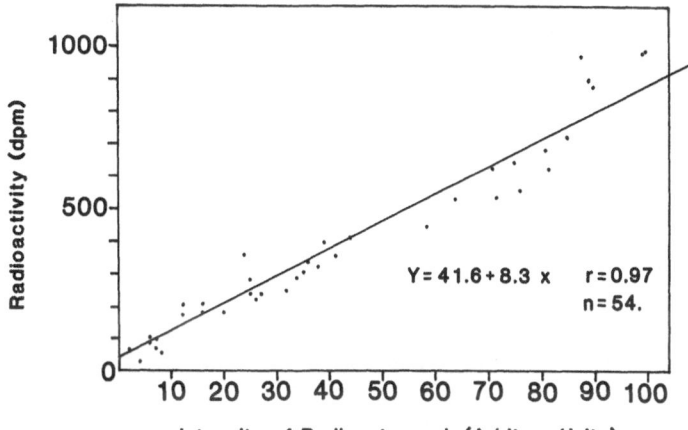

Figure 3: Correlation between radioactivity and intensity of radioautograph. Spots of the radioautograph of LDL (n=7), apoE (n=7), LpL (n=7), Fn (n=7), AT-III (n=7), Th (n=7), and Hn (n=12) were scanned by a laser densitometer using the intensity of Hn in the presence of 10 mM Ca^{2+} as 100%. After radioautography, the spots were cut from the nitrocellulose and the radioactivity was determined.

relation of intensity and radioactivity with a correlation coefficient (r) of 0.97 (P < 0.001). With this standard curve and knowledge of the specific radioactivity of $[^{125}I]$-HRH, the amount of HRH mass bound to LDL, apoE, LpL, Fn, AT-III, Th and Hn was determined as 300, 220, 670, 240, 60, 400 and 970 ng uronic acid, respectively.

DISCUSSION

The method most commonly used to assess heparin-binding is chromatography of proteins and peptides on a heparin-affinity column. However, this method has several disadvantages which include the following: non-specific binding; limitations in the buffer systems for chromatography; modification of important regions in the heparin with its immobilization to the affinity column; inability to determine the stoichiometry of binding; and large amounts of protein are required. The dot-blot method described in this report overcomes several of these limitations. The method requires small amounts of protein or peptide with pmole quantities being sufficient. When applying the sample to the nitrocellulose, guanidine-HCl or urea can be used. Finally, the amount of $[^{125}I]$-HRH bound can be determined. To use this method effectively, several factors need to be considered. Free $[^{125}I]$ should be completely removed from the radioiodinated heparin. If free $[^{125}I]$ is present, the background intensity of the radioautograph becomes high. If the specific activity of radioiodinated heparin is high (> 100 dpm/ng), background intensity also increases. In this case, the specific activity of the radioiodinated heparin should be diluted to < 5 dpm/ng by the addition of unlabeled heparin. If the molecular weight of proteins or peptides is less than 20 Kdalton, 0.1 µm pore size nitrocellulose should be used in order to trap the proteins on the paper. Alternatively, nylon sheets may be used to enhance protein binding but the background intensity of nylon is higher than that of nitrocellulose. Although proteins are detected by Amido Black staining, their quantitation could not be accurately determined; proteins differ in their chromogenicities and are difficult to stain and destain uniformly.

ACKNOWLEDGEMENTS

This research was supported by the United States Public Health Grants HL-30999 and HL-31387 and by the the American Heart Association (A.D.C.). We gratefully acknowledge the assistance of Mrs. Liz Wendelmoot and Mrs. Robin Wright in preparing the manuscript for publication.

REFERENCES

1. U. Lindahl and M. Hook, Glycoaminoglycans and their binding to biological macromolecules. Ann. Rev. Biochem., 47:385-417, 1978.
2. Y. Shing, J. Folkman, R. Sullivan, C. Butterfield, J. Murray, and M. Klagsbrun, Heparin affinity: Purification of a tumor-derived capillary endothelial cell growth factor. Science, 223: 1296-1299, 1984.
3. T. Maciag, T. Mehlman, and R. Friesel, Heparin binds endothelial cell growth factor, the principal endothelial cell mitogen. Science, 225: 932-935, 1984.
4. L. Socorro and R. L. Jackson, Monoclonal antibodies to bovine milk lipoprotein lipase. Evidence for proteolytic degradation of the native enzyme. J. Biol. Chem., 260: 6324-6328, 1985.
5. A. D. Cardin, R. L. Barnhart, K. R. Witt, and R. L. Jackson, Reactivity of heparin with the human plasma heparin-binding proteins thrombin, antithrombin III, and apolipoproteins E and B-100. Thromb. Res., 34: 541-550, 1984.
6. A. D. Cardin, G. Holdsworth, and R. L. Jackson, Isolation and characterization of plasma lipoproteins and apolipoproteins. In: "Methods in Pharmacology", pp. 141-166, Plenum Publishing Co., New York (1984).
7. A. D. Cardin, K. R. Witt, C. L. Barnhart, and R. L. Jackson, Sulfhydryl chemistry and solubility properties of human plasma apolipoprotein B. Biochemistry, 21: 4503-4511, 1982.
8. A. D. Cardin, K. R. Witt, and R. L. Jackson, Visualization of heparin-binding proteins by ligand blotting with ^{125}I-Heparin. Anal. Biochem. 137: 368-373, 1984.
9. Bio-Dot™ Microfiltration Apparatus Instruction Manual.

RAPID AND SENSITIVE DETERMINATION OF PROTEIN DISULFIDE BONDS

Hsieng S. Lu, Michael L. Klein, Richard R. Everett and Por-Hsiung Lai

Amgen, Thousand Oaks, CA 91320

INTRODUCTION

Therapeutically important polypeptides such as lymphokines and growth factors often contain disulfide bonds. The intact disulfide structures in these proteins are formed post-translationally and are often essential for biological activity. Such proteins produced by genetically modified E. coli cells are usually recovered using various processes including the denaturation-renaturation step. This step also allows formation of correct disulfide bonds. When foreign protein is secreted by genetically modified yeast, the effect of passage through its secretory pathway on the fidelity and efficiency of protein folding, particularly the disulfide formation in the recombinant protein, has to be determined. Different disulfide structures may cause changes in gross protein conformation, biological activity and stability and alter antigenicity. It is important to be able to determine and quantify correct disulfide bonds as well as incorrect ones in such molecules. A number of chemical methods are available for use in such characterization. So far, diagonal electrophoresis (1,2) has been found to be most useful. However, characterization and quantitation of protein disulfide structures using this method usually involves multi-step procedures, requires a considerable amount of materials and is time-consuming. Here we describe a rapid and sensitive procedure for the direct determination of protein disulfide bonds in several rDNA-derived proteins. The procedure developed is based upon the current advancements in protein chemistry techniques and instrumentation. It includes selective enzymatic fragmentation of intact proteins, high-recovery isolation of peptide fragments by HPLC, compositional determination of peptides by an improved analysis of phenylthiocarbamyl (PTC-) amino acids and quantitative microsequencing of disulfide-paired peptides. This procedure allows complete determination of protein disulfide bonds at the subnanomole level within several days. A step-by-step procedure is summarized in Table 1.

MATERIALS AND METHODS

Protein Purification, Peptide Fragmentation and HPLC Separation

Recombinant consensus interferon (IFN) secreted by Saccharomyces cerevisiae is isolated according to Zsebo et al. (3). Recombinant interleukin-2 [rMetHuIL-2(ala125)] analog is produced in E. coli and purified according to purification schemes which will be published elsewhere.

Tryptic peptide fragments are generated by incubating the protein sample in 0.2M NH_4HCO_3, pH 8.0 containing 1.0 mM $CaCl_2$ with TPCK-treated trypsin (substrate-to-enzyme = 50:1, w/w) at 37°C for 18 hours. The tryptic digest is injected directly and separated by HPLC using a C-4 column (Vydac or Synchropak, 300 A wide pore) and a gradient of acetonitrile in 0.1% trifluoroacetic acid.

Amino Acid Analysis Using PITC-Derivatization and HPLC

Hydrolysis of the peptide sample either in Wisp vials (Millipore) for regular HPLC or in micro vials (Hewlett Packard) for microbore HPLC is performed in Pico-Tag (Millipore) vacuum vials (4), which contain 1.5 to 2.0 ml of 5.7 N HCl (constant boiling) with 0.2% phenol. Performic acid oxidation is carried out in either Wisp or micro vials using half of the acid hydrolysate.

Standard amino acids and peptide hydrolysates are coupled with phenylisothiocyanate (PITC) using a modified coupling buffer (ethanol: trimethylamine: water; 7:1:1, v/v) (5). Trimethylamine is specially purified to be free of any contaminating primary or secondary amines (6). Coupling of peptide hydrolysates with PITC is performed directly in the same hydrolysis vial to prevent sample loss caused by transferring.

Quantitation and separation of PTC-amino acid derivatives is achieved at picomole level with a regular HPLC using an improved procedure modified from Heinriksen and Meredith (7). The separation of PTC-amino acids is carried out with a SP8700 HPLC system (Spectra Physics) in a Rainin 3 micron C-18 column (0.46 x 10 cm) using sodium acetate (NaOAc)-methanol-acetonitrile gradient elution. Mobile phase A is prepared as 50 mM NaOAc, pH 6.0 (adjusted with phosphoric acid), and mobile phase B is 100 mM NaOAc-H_3PO_4 (pH 6.0)/MeCN/MeOH (40:50:10). The column is equilibrated at 25°C in 94% A/6% B at a flow rate of 1 ml/min, and is developed using multiple gradient conditions (5%-18% B in 7.5 min, 18%-20% B in 2 min, 20%-40% B in 4 min and 40%-60% in 5 min). PTC-amino acids could also be analyzed by reverse-phase microbore HPLC (HP1090) using Altex narrow bore C-18 columns (0.2 x 25 cm) with an adaptation of the similar solvent elution system used in regular HPLC. Mobile phases C and D are identical to A and B

Table 1: Strategy of Disulfide Bond Determination

1. Choice of a specific cleavage of intact protein for isolation of individual disulfide pairs.
2. Preparative isolation of peptides via RP-HPLC under optimized conditions.
3. Identification of disulfide-containing peptides by compositional analysis using regular or microbore HPLC of PTC-amino acids derived from peptide hydrolysate.
4. Estimation of cystine in the form of PTC-cysteic acid by compositional analysis of performic acid-oxidized disulfide peptides.
5. Direct sequencing of polybrene-immobilized disulfide-containing peptides using a gas-phase seqeuncer.
6. HPLC of PTH-amino acids and identification of PTH-cystine released from expected sequencer cycle.

respectively, except that the pH of NaOAc solution in C and D is adjusted to pH 6.5 with H₃PO₄. The column is equilibrated in 95% C/5% D and operated at higher temperature (40°C) at a lower flow rate (0.25 ml/min), using a gradient of 5%-50% D in 15 min, followed by an isocratic elution with 50% D for 5 min. Data acquisition and processing are performed with a Nelson Analytical 4400 system and XTRACHROM software.

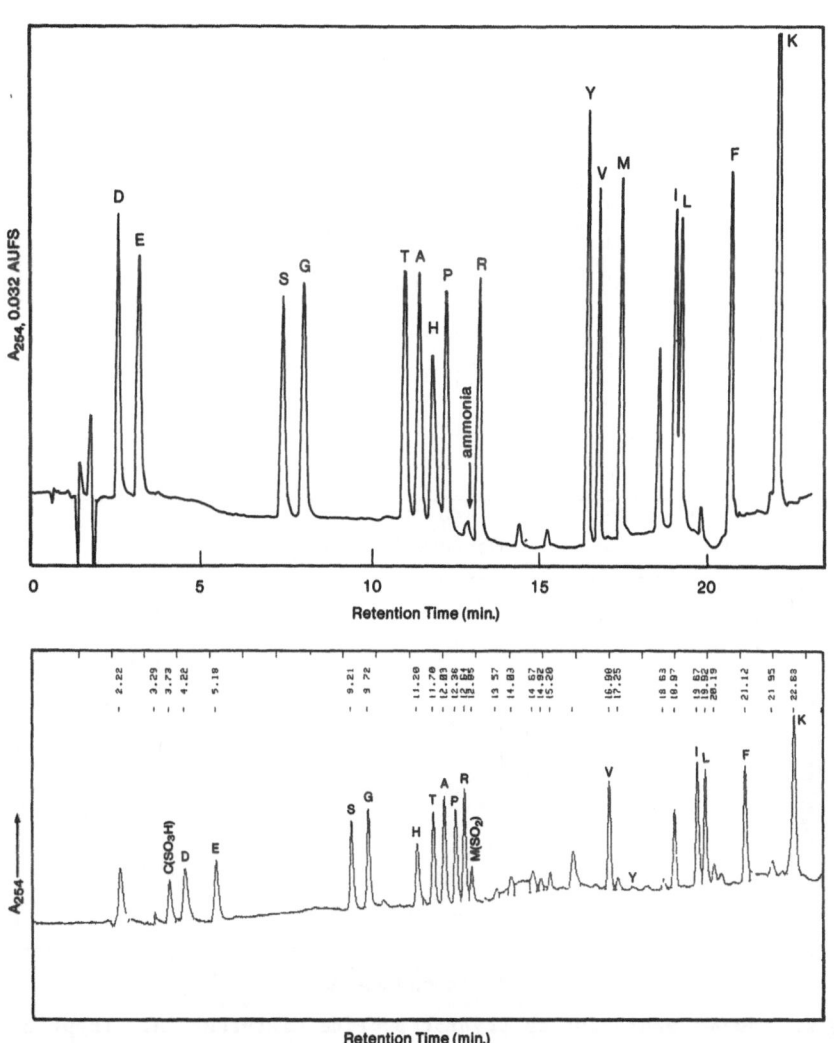

Figure 1. A. HPLC separation of PTC-amino acids (200 pmol each) using a Rainin 3 micron C-18 column. B. HPLC separation of PTC-oxidized amino acid standards (10 pmol each) in an Altex narrow bore C-18 column.

Sequence Analysis

Automated sequence analysis (6,8) is performed with a gas-phase sequencer using either the standard program or a new program designated MHNVAC supplied by Applied Biosystems. The peptide, dissolved in 50% formic acid, is loaded onto a glass fiber disc containing precycled polybrene. The PTH-amino acid obtained from each sequencer cycle is identified by RP-HPLC (9). To prevent reduction of PTH-cystine during transfer and drying, methanol and acetonitrile containing no dithiothreitol in S-4 is used. PTH-cystine is promptly analyzed after conversion.

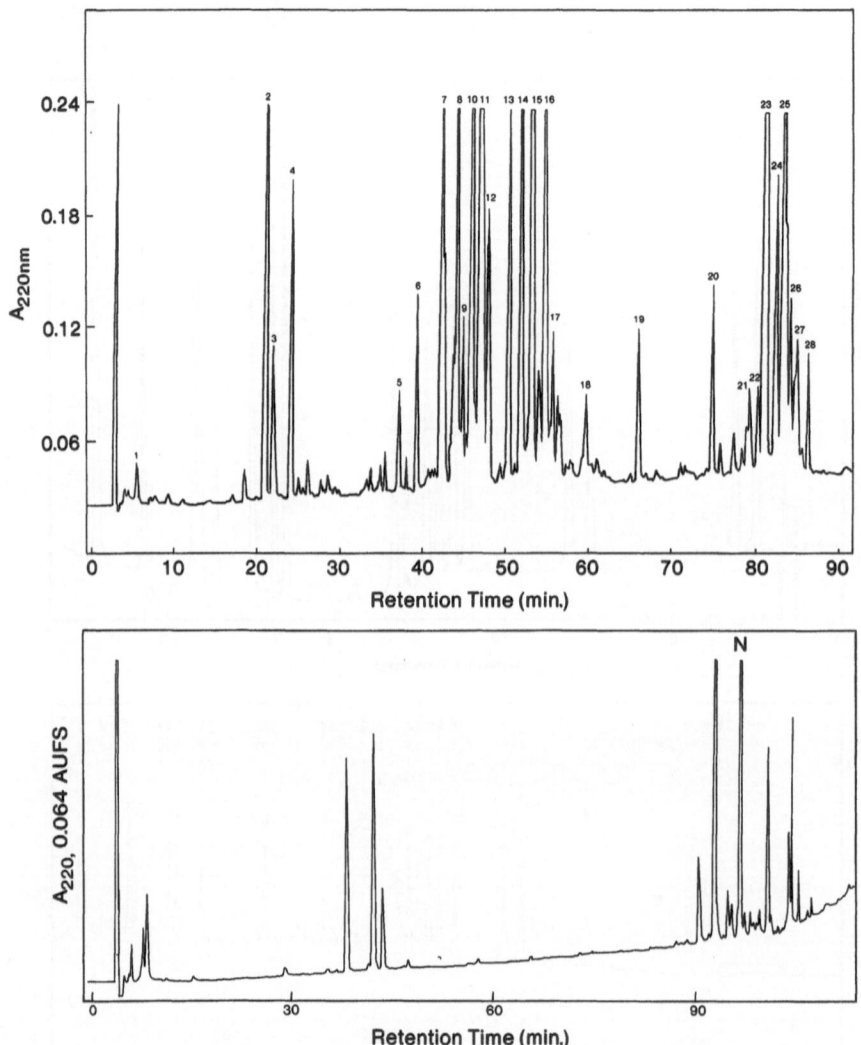

Figure 2. HPLC separation of tryptic peptide mixtures. A. Tryptic digest of consensus IFN (10 nmol). The separation starts with a 5 min isocratic elution at 97% of mobile phase A (0.1% TFA) and 3% of B (90% MeCN containing 0.1% TFA), then develops with a gradient of mobile phase B from 3% to 55% in 90 min. B. Tryptic digest of rMetHuIL-2(ala125) (2 nmol). Elution conditions are identical to A except the 90-min gradient is developed from 3% to 45% B.

RESULTS AND DISCUSSION

Figure 1A illustrates the separation profile of PTC-derivatives of all common amino acids obtained with a Rainin 3 micron C-18 column using NaOAc-MeOH-MeCN gradient elution. Baseline resolution is achieved for all of the PTC-amino acid derivatives except the Ile-Leu pair which is 80% resolved. This method allows the detection of PTC-amino acids at low picomole

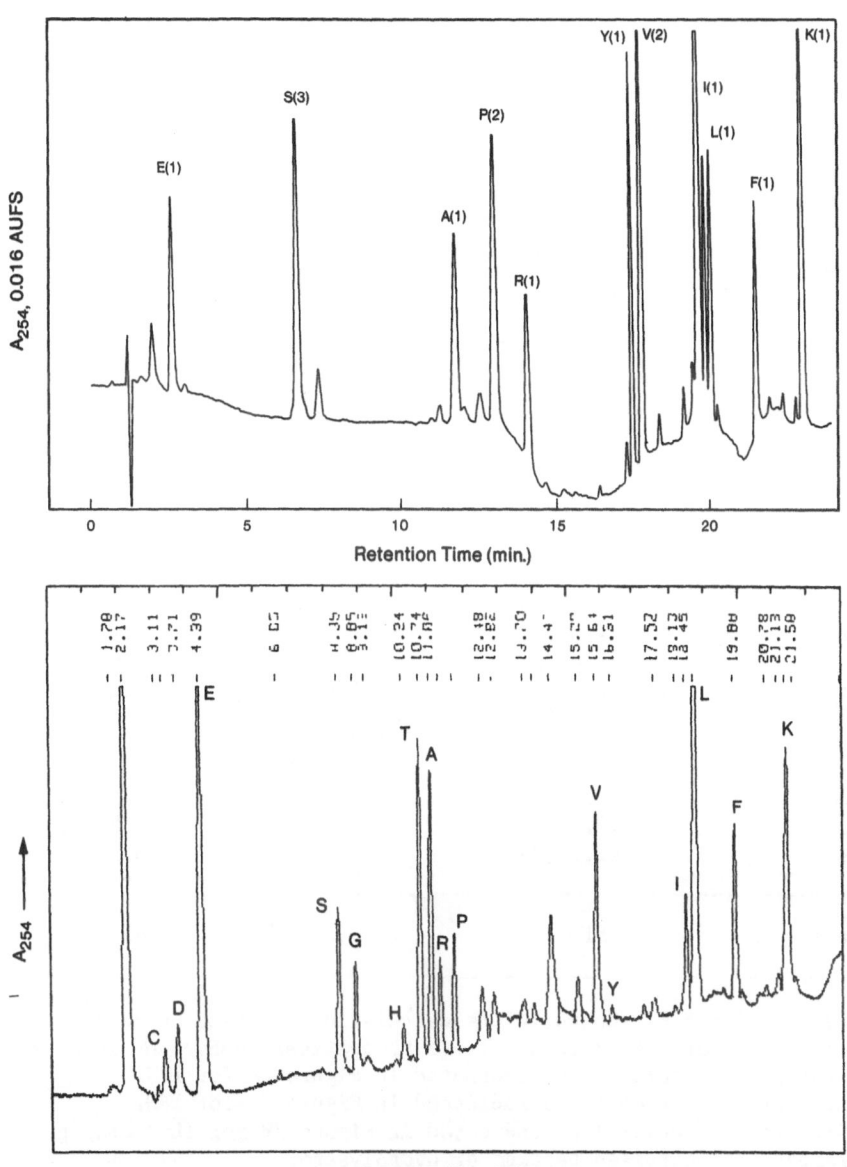

Figure 3. Compositional analyses of the disulfide peptides collected in fraction 15 of consensus IFN (see Figure 2A) using regular HPLC (A, 50 pmol), and fraction N of rMetHuIL-2(ala125) (see Figure 2B) using microbore HPLC (B, 2-3 pmol).

levels. As shown in Figure 1B, PTC-amino acids can also be separated by RP-HPLC using a narrow bore C-18 column. The high efficiency of the column has allowed the complete resolution of all amino acid derivatives including PTC-cysteic acid and methionine sulfone using a simple gradient elution. The use of a lower flow rate (0.25 ml/min) in microbore LC system gives a quieter baseline at high senstivity (<0.005 AUFS). The increased resolution and improved signal-to-noise ratio allows highly sensitive analysis at subpicomole levels.

RP-HPLC has been broadly used for protein and peptide separations (10). In these practices, RP-C18 columns are widely employed. However, isolation of larger disulfide peptides using HPLC requires a method which allows separation of complex mixtures with high recovery, acceptable

Table 2. Amino Acid Composition of Disulfide Peptides Derived from the Tryptic Digestion of Consensus IFN and rMetHuIL-2(ala125)[1].

PTC Amino Acid	Disulfide Pair		
	A[2]	B[3]	C[4]
	pmol	pmol	pmol
Cys(SO$_3$H)[5]	37.8 (2)	27.0 (2)	4.1 (2)
Asp	126.0 (6)	3.3 (0)	5.6 (3)
Glu	184.8 (9)	20.2 (1)	25.2 (11)
Ser	48.3 (2)	50.6 (3)	3.6 (2)
Gly	48.3 (2)	3.4 (0)	2.4 (1)
Thr	50.4 (3)	1.7 (0)	10.1 (4)
Ala	48.0 (2)	16.7 (1)	8.4 (3)
His	10.5 (1)	0 (0)	1.7 (1)
Pro	37.8 (2)	35.4 (2)	2.9 (1)
Arg	33.6 (2)	13.4 (1)	2.9 (1)
Tyr	48.5 (2)	13.4 (1)	2.6 (1)
Val	111.3 (5)	34.0 (2)	5.5 (2)
Met	16.8 (1)	1.7 (1)	0.7 (1)
Ile	47.9 (2)	15.2 (1)	3.1 (1)
Leu	132.3 (7)	15.2 (1)	18.7 (7)
Phe	29.4 (1)	13.4 (1)	5.8 (2)
Lys	42.0 (2)	13.4 (1)	5.5 (2)
Total	51	18[6]	45

[1] See Figure 2A and B. Compositions calculated are based on pmol recovered. Integers in parentheses are theoretical numbers of amino acids.
[2] Obtained from fraction 23 as indicated in Figure 2A for IFN.
[3] Obtained from fraction 15 as indicated in Figure 2A for IFN.
[4] Obtained from fraction N as indicated in Figure 2B for IL-2 analog.
[5] Determined from oxidized peptide or hydrolysate.
[6] An extra trp is not analyzed.

resolution and reproducibility. We found that the RP-C4 (wide pore) columns are superior to C18 columns as judged by these criteria. Figure 2A demonstrates the reverse phase separation of a tryptic peptide mixture derived from purified intact consensus IFN produced in S. cerevisiae. Complete resolution of major tryptic fragments (28 peaks) is achieved using a 90-min gradient elution. The tryptic peptide mixture of rDNA derived rMetHuIL-2(ala125) is also separated by a C-4 column using a similar elution condition (Figure 2B). Yields estimated by amino acid analysis for each identified peptide are higher than 70%.

These peptides are further characterized by high sensitivity compositional analysis and microsequencing. Figure 3A shows the PTC-amino acid analysis of the peptides in fraction 15 obtained from the tryptic map of consensus IFN (Figure 2A). Fraction 15 contains a lysyl peptide and an arginyl peptide cross-linked by the disulfide bond, Cys(29)-Cys(139), in consensus IFN. Once the disulfide-containing peptide is identified, cystine can be detected as cysteic acid in PTC-form by compositional analysis of the performic acid-oxidized hydrolysate. The results for the detection of cystine in the disulfide peptides of fraction

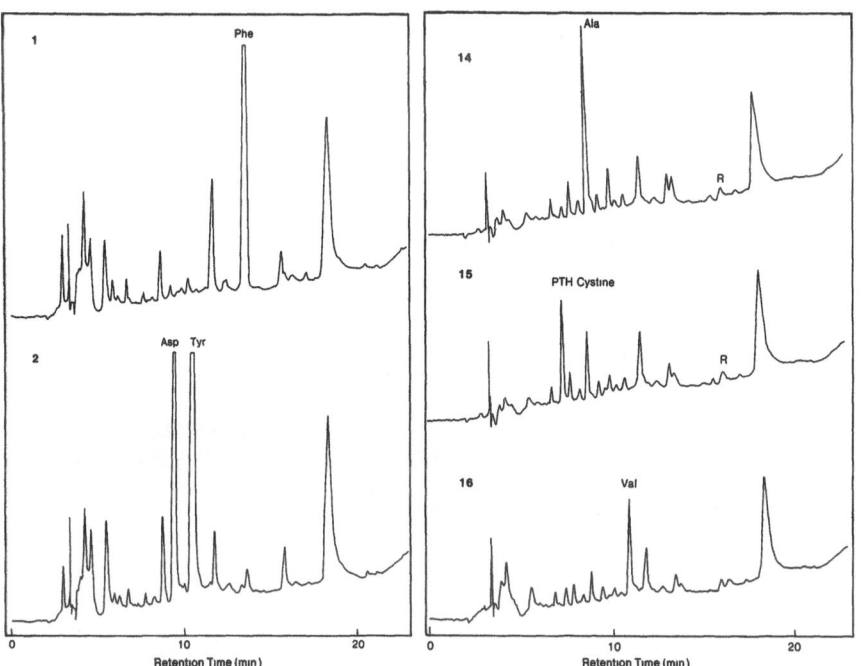

Figure 4. Sequence analysis of disulfide peptide, fraction 23, of consensus IFN. One-third of the sample containing released PTH-amino acids at each degradation cycle is analyzed by RP-HPLC using a cyano column (9). Amount of sample loaded: 500 pmol. Initial yield: 60%.

15 are shown in Table 2. An example using narrow bore column PTC-amino acid analysis of an oxidized hydrolysate of the disulfide-containing tryptic peptides from rMetHuIL-2(ala125) is shown in Figure 3B. Cysteic acid in PTC-form is recovered in a molar ratio of 1.8 per peptide molecule. Since only a small amount of the sample is needed for each analysis, it is possible to analyze both oxidized and unoxidized samples from the same batch of hydrolysate. Table 2 lists the amino acid composition of some disulfide-containing peptides prepared according to methods described for Figure 2A and 2B.

Disulfide-paired peptides which are detected by amino acid analysis are further identified by microsequencing. Initially, quantitative amounts of two PTH-amino acids are recovered from each Edman degradation cycle during sequencing of the disulfide paired peptides until the shorter peptide fragment is sequenced through. After this point, a single peptide sequence is observed. Figure 4 demonstrates the sequence analysis of the disulfide-paired peptides found in fraction 23 of the chromatogram shown in Figure 2A. The peptides are linked by the disulfide bond, Cys(1)-Cys(99), in consensus IFN. Repetitive yield for both sequences is about 90%. The first cycle shows only PTH-phe since the other residue is disulfide-linked to another unsequenced cysteine on the other polypeptide chain. From cycle two, PTH-Asp and PTH-Tyr at approximately equal quantities are obtained. The shorter peptide chain (position 1-12 of consensus IFN) is sequenced through after 12 cycles of degradation.

Table 3. Amino Acid Sequence of Disulfide Peptides Isolated from the Tryptic Digestion of Consensus IFN and rMetHuIL-2(ala125).[1]

Fraction	Position	Sequence
A. **Consensus IFN:**		
13	14-33/135-145	ALILLAQMRRISPFSCLKDR
		KYSPCAWEVVR
14	24-33/136-145	ISPFSCLKDR
		YSPCAWEVVR
15	24-31/136-145	ISPFSCLK
		YSPCAWEVVR
23	1-12/85-122	CDLPQTHSLGNRR
		FYTELYQQLNDLEACVIQEVGVEETPLMNVDSILAVKK
20	1-13/85-122	(See footnote 2)
24	1-12/85-121	(See footnote 3)
B. **IL-2 Ala Analog:**		
N	55-76/98-120	HLQCLEEELKPLEEVLNLAQSK
		GSETTFMCEYADETATIVEFLNR

[1] For details see Figures 2A, B.
[2] Arg at position 13 is cleaved as compared to fraction 23.
[3] Lys at position 122 is cleaved as compared to fraction 23.

The distinct PTH-cystine is detected at cycle 15, which is assigned for PTH derivative of Cys(1)-Cys(99). This peptide is sequenced through cycle 35 and a single sequence is confirmed after cycle 14.

Only one single disulfide pair of peptides is obtained from RP-HPLC of the complete digestion of rMetHuIL-2(ala125) (Figure 2B and Table 3). However, a number of disulfide peptides are isolated from the tryptic digest of consensus IFN. This indicates that incomplete digestion had occurred due to the effect of penultimate amino acids. As shown in Table 3, fractions 13, 14 and 15 consist of the disulfide [Cys(29)-Cys(139)] paired peptides of different length, while fractions 20, 23 and 24 account for length varying components containing Cys(1)-Cys(99). The recovery yields of these peptides range from 70% to 85%. No mis-paired disulfide or unpaired (free SH) components are observed with proteins used in this study, suggesting the correct disulfide formation in these molecules.

In summary, a much more rapid and sensitive procedure employing the combination of several analytical techniques for protein characterization is developed as a powerful tool for measuring the fidelity and efficiency of protein production by rDNA technique, as well as for analyzing physiologically important proteins such as lymphokines and growth factors. The analysis time is greatly reduced (3-5 working days) and the characterization can be performed at subnanomole levels. Furthermore, the characterization at low picomole to subpicomole levels can be achieved if a microbore HPLC system is adapted.

REFERENCES

1. Brown, J.R. and Hartley, B.S., Biochem J., 101, 214 (1966).
2. Creighton, T.E. in Methods in Enzymology (F. Wold and K. Moldave, eds.) Vol. 107, pp. 305 (1984).
3. Zsebo, K.M., Lu, H.S., Fieschko, J.C., Goldstein, L., Davis, J., Suggs, S., Lai, P.H. and Bitter, G.A., J. Biol. Chem., manuscript submitted for publication.
4. Cohen, S.A., Tarvis, T.L. and Bidlingmeyer, B.A., American Lab., August, 48-59 (1984).
5. Lu, H.S. and Lai, P.H., manuscript submitted for publication.
6. Lai, P.H., Analytical Chim. Acta. 163, 243-248 (1984).
7. Heinrikson, R.L. and Meredith, S.C., Anal. Biochem. 136, 65-74 (1984).
8. Hewick, R.W., Hunkapiller, M.W., Hood, L.E. and Dreyer, W.J., J. Biol. Chem. 256, 7990-7997 (1981).
9. Hunkapiller, M.W. and Hood, L.E., Science 219, 650-659 (1983).
10. Heftmann, E. (Ed.) 2nd Int'l. Symposium on HPLC of Proteins, Peptides and Polynucleotides, J. Chromatogr., 266, 3-665 (1983).

SIMULATION OF PROTEIN HYDRODYNAMIC CHANGES OBSERVED BY UREA GRADIENT GEL

ELECTROPHORESIS

William Shalongo and Earle Stellwagen

Department of Biochemistry
University of Iowa
Iowa City, IA 52242

INTRODUCTION

The kinetics of protein conformational changes are normally monitored by absorbance or fluorescence measurements since spectral techniques are amenable to data collection on a rapid time scale. Unfortunately, these spectral techniques often reflect local changes in tertiary structure and cannot be assumed to reflect the initial hydrodynamic collapse of a randomly coiled denatured polypeptide into a
compact structure. Since such a collapse is likely an early event in
protein folding, it would be useful to have a procedure available to
observe the kinetics of the hydrodynamic change accompanying protein
folding and unfolding. Creighton (1979, 1980) has demonstrated that
zone electrophoresis in polyacrylamide slab gels containing a urea
gradient can provide information regarding the kinetics of the hydrodynamic changes accompanying unfolding or refolding in urea, the presence of multiple forms of the protein, and the transient accumulation
of intermediate forms. Unfortunately, the protein profiles observed
upon staining of urea gradient gels after electrophoresis can only be
qualitatively interpreted. While Creighton has employed the equation of
Mitchell (1976) to simulate such profiles, we find that this equation is
awkward to use and can give unreliable results. In this report we
illustrate the application of the equation of Endo et al. (1983) to such
simulations. We find this equation to be easy to use and the results
obtained to be sensible and readily displayed in an easily grasped
format.

UREA GRADIENT ELECTROPHORESIS

As described by Creighton (1979), a slab gel is poured using solutions of acrylamide and of acrylamide containing 8 M urea to create a
gel containing a linear gradient of urea. The gel is oriented in a
electrophoresis apparatus so that electrophoresis of the protein is
normal to the urea gradient. Either native or denatured protein is
applied to the top of a gel and the gel is stained following completion
of electrophoresis. In the case of a rapid two state transition, the
mobility of the compact native protein is indicated by an unfeatured
staining band observed at low urea concentrations, the mobility of the
randomly coiled denatured protein by an unfeatured staining band in high

urea concentrations and the transition zone in which both forms of the protein are populated is indicated by a sigmoidal staining band connecting the two unfeatured band. Creighton (1980) has sketched the profiles expected for two state transitions in which the halftimes for exchange are comparable and are slow relative to electrophoresis time, for transitions having transiently stable intermediate forms and for transitions involving a multiplicity of native or denatured forms.

SIMULATIONS

The mathematical formulation of Endo et al. (1983) was adapted to the sieving features of a polyacrylamide gel as opposed to a porous exclusion column. Input information includes the length of the gel in the direction of electrophoresis, the concentration of urea at which the simulation is being done, the distances migrated by the compact and unfolded forms of the protein, the electrophoresis time, the halftime for the hydrodynamic change, the urea concentration in which the two forms are equally populated at equilibrium, the urea concentrations in which 10% and 90% of the transition has occurred, the form and the volume of the protein applied to the gel and the number of theoretical plates in the gel. The distributions of protein at discrete urea concentrations are individually calculated and a multiplicity of such calculations are diplayed in an offset mode to emulate the gel as illustrated in Fig. 1.

RESULTS

Protein distribution patterns characteristic for hydrodynamic changes associated with protein folding involving two states, the denatured and native protein, whose halftimes for exchange are fast, comparable and slow relative to electrophoresis time are shown in Fig. 1. In this discussion we denote comparable as an exchange halfitme equal to electrophoresis time, fast as an exchange halftime ten times shorter than the electrophoresis time, and slow as an exchange halftime ten times longer than the electrophoresis time. As expected, fast exchange between the two forms generates a smooth sigmoidal shaped staining band spanning the equilibrium transition concentration range in urea, a comparable exchange rate generates a reaction boundary beginning in the equilibrium transition concentration range and extending to low urea concentration, while slow exchange gives little evidence of reformation of native protein at any urea concentration. However, the halftimes for protein conformation changes are strongly dependent on urea concentration. In the simulations shown in Fig. 2, we have assumed that the reference exchange halftime (fast, comparable or slow) occurs at the midpoint of the equilibrum transition, 5 M urea, and that the exchange halftime decrease by an order of magnitude per 2 M urea. As can been seen, the sigmoidal nature of the fast exchange example sharpens, the comparable exchange in 5 M urea becomes fast in 3 M urea and below while the slow exchange in 5 M urea becomes comparable in 3 M urea and fast in 1 M urea and below. The result is that all urea dependent exchange rates can facilitate complete transformation to native in low urea concentration as was previously only correct for fast exchange. However, unlike fast exchange, the transition zones displayed in Fig. 2 occur at lower concentrations than the equilibrum transition and the transition zones are not described by a relatively sharp sigmoidal curve but rather by a smeared zone of low staining intensity.

The advantages of these simulations are at least three-fold. Firstly, if a staining profile is recognized to be characteristic for either slow, comparable or fast, simulations can place boundary condi-

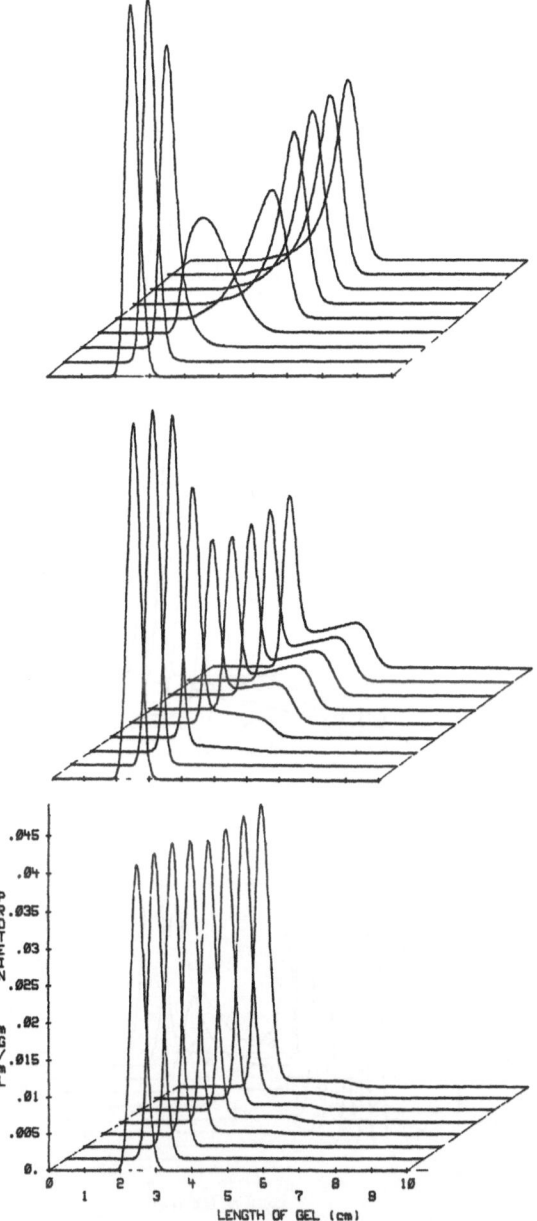

Fig. 1.   Fig. 1.   Simulation of fast, comparable, and slow exchange
between denatured and native protein during urea gradient
electrophoresis. It was assumed that exchange had an equilibrium transition centered at 5 M urea and that 90% of the
transition occurred over a range of 2 M urea. The simulations
were constructed assuming that denatured protein was added to
the left edge of each gel, that denatured protein would migrate
2 cm and that native would migrate 5 cm. The gels are oriented
so that each profile represents the protein distribution at an
integral value of urea molarity with 8 M urea nearest the
viewer and 0 M most distant. Vertical distances represent the
concentration of protein in each profile. The top, middle and
bottom panels represent fast, comparable and slow exchange
each of which is urea independent.

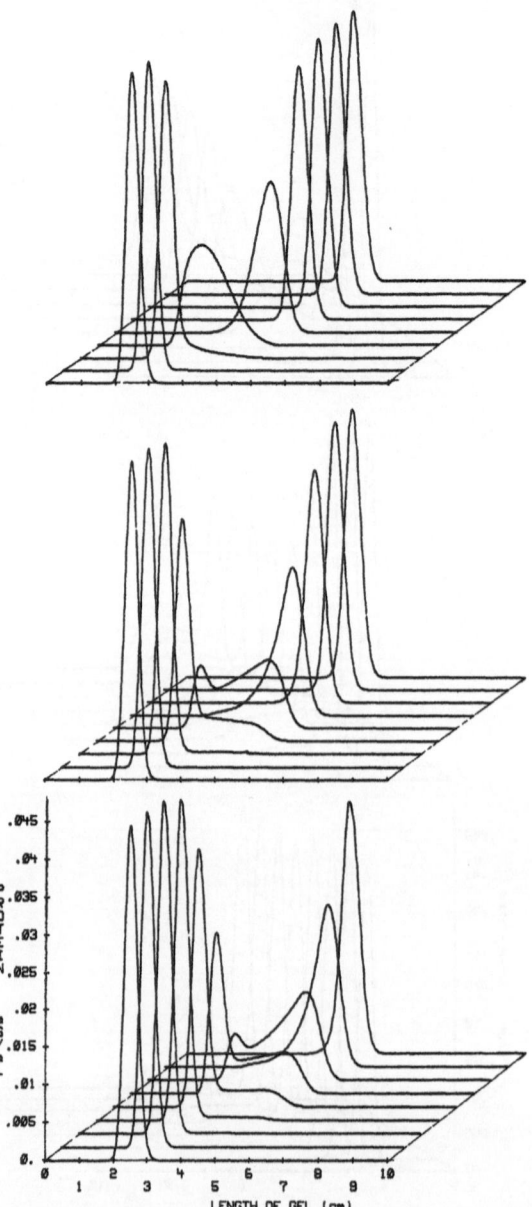

Fig. 2.   Simulation of exchange between denatured and native protein
which is urea dependent during urea gradient electrophoresis.
All other considerations are the same as described for Fig. 1.

tions on the magnitudes of exchange rates which would generate a characteristic pattern. If the electrophoresis time can be varied, the boundary conditions can become more narrowed. Secondly, if exchange rates having been measured for a conformational change in urea using a different probe such as fluorescence, the hydrodynamic and fluorescence changes can be compared. Thirdly, if the observed staining pattern represents the occurrence of two or more discrete kinetic events, then the simulated distributions may be combined in different proportions to emulate the observed staining pattern. For example in the case of thioredoxin, the fluorescence changes accompanying refolding can be resolved into three kinetic phases. Simulation of the gel staining pattern resulting from the exchange rate for each of the three kinetic events clearly indicates that the slowest kinetic phase does not involve a hydrodynamic change and that the observed hydrodynamic change is principally generated by the fastest kinetic phase. Accordingly, the fast kinetic phases represents the major compaction of the denatured protein and the slow phase must represent the annealing of the compacted protein into the native conformation.

ACKNOWLEDGMENT

This research was supported by a Public Health Service research grant GM22109 and program project grant HL14388.

REFERENCES

Creighton, T.E., 1979, Electrophoretic analysis of the unfolding of proteins by urea, J. Mol. Biol. 129:235.

Creighton, T.E., 1980, Kinetic study of protein unfolding and refolding using urea gradient electrophoresis, J. Mol. Biol. 137:61.

Endo, S., Mairo, Y. and Wada, A., 1983, Denaturant gradient chromatography for the study of protein denaturation: principle and procedure, Anal. Biochem. 131:108.

Mitchell, R.M., 1976, The spatial distribution of moving macromolecules undergoing isomerization, Biopolymers, 15:1717.

B. ANALYSIS OF POLYPEPTIDE STRUCTURE AND FUNCTION

X. Site-Directed Mutagenesis

SITE-SPECIFIC MUTATIONS IN DIHYDROFOLATE REDUCTASE AT THE DIHYDROFOLATE

BINDING SITE

Ruth J. Mayer, Jin-Tann Chen, Kazunari Taira, and
Stephen J. Benkovic

The Pennsylvania State University
Department of Chemistry
152 Davey Lab.
University Park, PA 16802

INTRODUCTION

The high resolution X-ray crystal structure of an enzyme provides an opportuntiy for precise understanding of structure-functions relationships in that enzyme. Dihydrofolate reductase, which catalyzes the reduction of dihydrofolate (H_2folate) to tetrahydrofolate (H_4folate) with NADPH as cofactor, is an excellent choice for such studies. The methotrexate-binary complex structure of the *Escherichia coli* enzyme identifies several close interactions of methotrexate (1,2), and by analogy, dihydrofolate, with conserved amino acid side chains. The specific roles of these amino acids in binding and catalysis can be studied using the technique of site-specific mutagenesis by replacing the particular residue postulated to be essential with one lacking the specific interaction.

The effects of such mutations can be considered in the context of current mechanistic proposals based on the X-ray structure and the following kinetic scheme proposed by Stone and Morrison (3):

$$
\begin{array}{c}
\ce{E <=>[k_9(DHF)][k_8] E\cdot DHF} \\
\Big\updownarrow K_1 \\
\ce{E\cdot H <=>[k_1(DHF)][k_2] EH\cdot DHF <=>[k_3][k_4] E\cdot DHFH ->[k_5] E\cdot THF ->[k_7] E}
\end{array}
\tag{1}
$$

The protonation of enzyme, manifest as a pK_a of 8.1 in the V/K_{DHF} pH profile, was proposed to be specifically the protonation of the active site Asp-27, based on the crystal structure. The identity of this amino acid has been confirmed by replacing the Asp with Asn or Ser (4). Neither the Asn-27 or Ser-27 enzymes exhibit a pK_a in the V/K_{DHF} pH profile, and both have low k_{cat} values relative to the wild type (wt) enzyme at pH 7. The Asp-27 is also thought to be responsible for the protonation of dihydrofolate, either by direct proton transfer to N-5 or by stabilization of the less predominant enol tautomer, allowing protonation of N-5 from the 4-OH group. The properties of the Asn and Ser-27 mutant enzymes do not distinguish between these possible mechanisms. Proton transfer or an associated conformational change is probably largely rate-limiting, as experiments with NADPD show a minimal isotope effect, indicating that hydride transfer is not entirely rate-limiting. Note that proton transfer does not occur from the medium to the ternary complexes.

The effects on substrate binding and catalytic efficiency of mutations at three other conserved residues in the dihydrofolate binding site are reported here. The mutations that have been made are Thr-113 →Val, Leu-54 →Gly; Phe-31 → Val, Tyr. Thr-113 interacts with the substrate via hydrogen bonding to the 2-NH$_2$ group and also hydrogen bonds to Asp-27. Van der Waal's contacts are made by Leu-54 with the p-aminobenzoyl moiety and by Phe-31 with both the p-amino benzoyl moiety and the pteridine. The roles of these three amino acids were analyzed with regard to several questions.

First, how important are these interactions in specific binding energy? Second, is the hydrophobic binding energy required to facilitate either protonation or hydride transfer? Specifically, do either Phe-31 or Leu-54 destabilize the N5-C6 imine to facilitate reduction of this bond? Third, can we affect the precise mechanism of proton transfer by changing residues within this region?

MATERIALS AND METHODS

Plasmid TY1 (5,6) (a construction of pBR322 lacking an EcoRI site with the fol gene cloned into the BamHI site) was grown in *E. coli* strain HB101 and purified by standard procedures. Mismatches were created by the method of Dalbodie-McFarland *et al.* (7) using the unique EcoRI site in the fol gene to generate nicked plasmid DNA. Oligonucleotides were synthesized on an Applied Biosystems synthesizer and were purified either by reverse-phase HPLC or polyacrylamide gel electrophoresis. Mutants were identified after transformation either by colony hybridization (8) (Val-113), by recloning of a new restriction fragment (Tyr, Val-31) or isolation and religation of linear plasmid generated from a unique restriction site at the mutation site (Gly-54). The mutation was confirmed by nucleotide sequencing of the fol gene (9,10).

Mutant dihydrofolate reductases were purified by methotrexate affinity chromatography (11). In some cases (Val-31, Gly-54) the lower affinity of the mutant enzyme for methotrexate permitted elution of the mutant enzyme from the resin under conditions (no folate in elution buffer) which left any wild type enzyme still bound to the resin. In no case was wild type contamination from chromosomal DHFR found to be a problem in data analysis (12).

The conformation of the Val-113 and Gly-54 mutants was studied by stopped-flow fluorescence quenching at 340 nm according to the method of Cagley, *et al.* (13) using equipment constructed in the laboratory of K.A. Johnson (14). Kinetic parameters as a function of pH were determined under conditions similar to those used by Stone and Morrison (3) in the same buffer system (50 mM Mes), (125 mM Tris), 25 mM ethanolamine, 0.1 M sodium chloride-MTEN buffer.

RESULTS AND DISCUSSION

Two methods were used to characterize the Val-113 and Gly-54 enzymes to ascertain that no gross conformational changes had occurred as a result of the mutation. The equilibrium distribution between two principal conformational states as established by Cayley *et al.* (13) was measured as well as the rate of NADPH binding to ensure that the binding site remote from the mutation site was not affected.

Data for the binding of NADPH to wt is biphasic and has been described by the following Scheme:

$$E_1 + NADPH \overset{k_{on}}{\underset{k_{off}}{\rightleftharpoons}} E_1 \cdot NADPH \qquad (2)$$

$$\downarrow k_2$$

$$E_2$$

The equilibrium concentration of the conformers E_1 and E_2 sets the relative amplitude of the two phases, where k_2 is the rate of conversion of $E_2 \longrightarrow E_1$ as measured by the decrease in fluorescence at 340 nm of the slow phase. The rate constant for NADPH binding (k_{on}) is measured from the NADPH dependence of the rate of the fluorescence change for the fast phase. The values for k_2, k_{on} and the relative amplitudes are listed in Table 1 for the wt, Val-113 and Gly-54 mutants. The only deviation from wt parameters is in the relative amplitudes of the phases in the Val-113 mutants. The overall agreement of the data with wt parameters suggests that the conformational effects of both mutations are local.

Table II summarizes the V/K, V and pK_a values observed in pH profiles determined for each mutant. All of these mutants have V values at least as large at that of wt, while the V/K values are always smaller, but of varying magnitude depending on the mutation. Also, each mutation results in a pK_a shift to lower pH for the Asp-27 ionization in both V/K and V profiles.

Table 1: Effect of mutations Val-113 and Gly-54 on enzyme conformation.

	k_{on}[a] $(M^{-1} s^{-1})$	k_2[a] (s^{-1})	rel amp[a]
wild type	1.6×10^7	0.030	1.2:1
Val-113	2.0×10^7	0.035	1:2.5
Gly-54	1.9×10^7	0.029	1.2:1

a) *The k_{on}, k_2 and relative amplitude values were determined according to Scheme 2. Conditions were: 0.46 - 4 μM DHFR, 2 - 50 μM NADPH, MTEN buffer (pH 7.0) 25°C, 340 nm.*

The effect of the Val-113 mutation is simply a 25-fold increase in K_{DHF}, with no decrease in V. This increase in K_{DHF} is probably a reflection of a higher K_D value for H_2folate, by analogy with a parallel increase in the dissociation constant for methotrexate. The slight decrease in the observed pK_a value is most reasonably a result of removing the hydrogen bonding interaction between Thr-113 and Asp-27, destabilizing the acid form of Asp-27. The loss of a single hydrogen bond (1-2 kcal/mol) also can account for the effect on K_M for H_2folate.

The mutation Leu-54 ⟶ Gly, which removes the hydrophobic interaction, also results in an increase in K_{DHF} with no change in V. The increase in K_{DHF} is now a factor of 140, but at pH's below the new apparent pK_a, the velocity is completely unaffected. The pH dependence of K_i for the competitive inhibitor, 2,4-diamino-6,7-dimethylpteridine, indicates that the apparent pK_a is actually an intrinsic pK_a for the E·NADPH complex. The large shift in pK_a is thought to be due to an increased accessivility of solvent water to the active site. The pK_a of 5.8 in the Gly-54 mutant agrees well with that observed in free enzyme, determined from the pH-dependence of the binding of methotrexate or 2,4-diamino-6,7-dimethyl pteridine (15).

Similar effects are observed in the Val-31 mutant. The V/K_{DHF} is 25-fold smaller than wild type, solely as a result of the 25-fold increase in K_{DHF}. The apparent pK_a value likewise is shifted 1.2 pH units to lower pH (Fig. 1), probably due to the introduction of water in the active site to fill the volume originally occupied by the Phe-31.

The Tyr-31 mutant has only a 4-fold smaller value for V/K_{DHF} in the pH independent region, which results from the combination of a 6-fold increase in K_{DHF}, and a 50% increase in V. The small increase in K_{DHF} at low pH values is consistent with the minimal changes in the amino acid side chain. At higher pH values, however, the K_{DHF} value increases to 300 μM (pH 9.5), possibly due to the ionization of Tyr, drastically decreasing the hydrophobicity of the side chain. The 50% increase in V may be the result of the introduction of another hydrogen bonding interaction which may facilitate reduction.

Table 2: Summary of pH-independent values of V/K_{DHF} and V and apparent pK_a values for Thr-113, Gly-54, Val-31 and Tyr-31 mutants.[a]

	V (s^{-1})	V/K (M^{-1} s^{-1})	pK_1
wt	16	1.5×10^7	8.1
Thr-113	14	5.6×10^5	7.7
Gly-54	14	1.0×10^5	5.8
Val-31	14	6.5×10^5	6.9
Tyr-31	24	4.2×10^6	6.8

a) *Conditions: 1 - 700 nm DHFR, 3 - 200 μM H_2folate, 60 μM NADPH, MTEN buffer, 25°C, 340 nm. the pK_1 values were determined according to equation 5 in (3).*

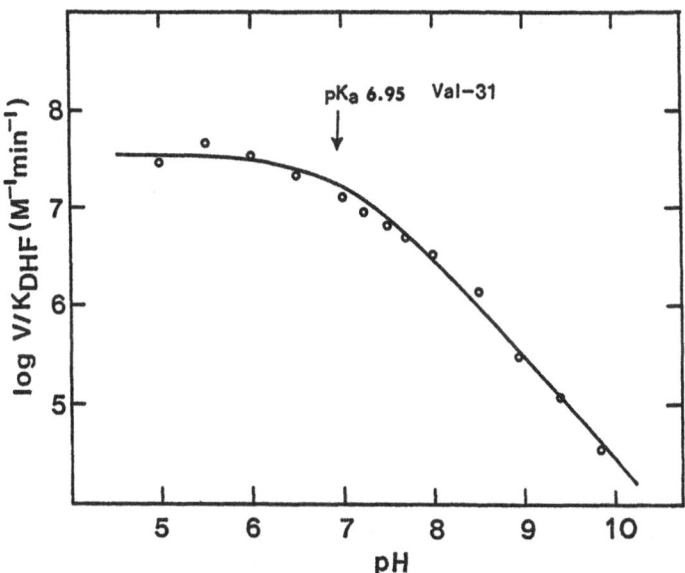

Figure 1: The pH dependence of V/K for the Val-31, where the solid line is the theoretical curve for the pK_a value indicated.

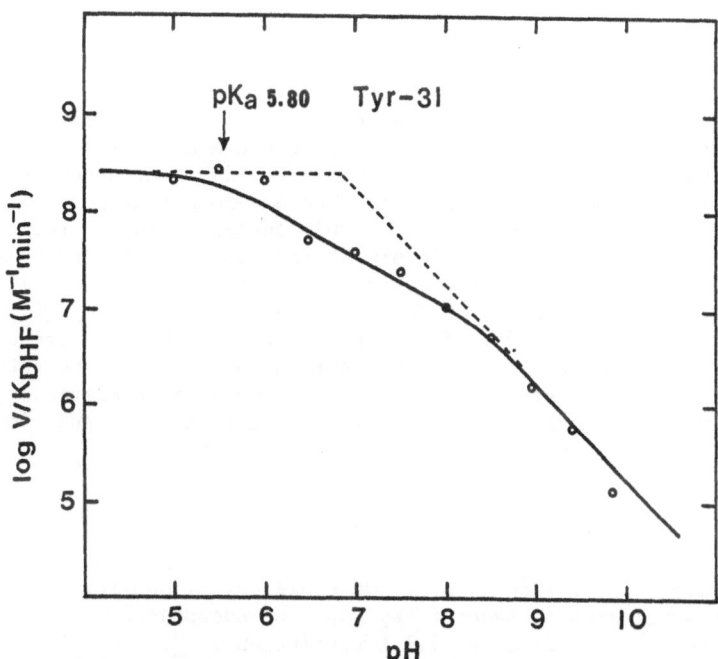

Figure 2: The pH dependence of V/K for the Tyr-31, where the solid line is the theoretical curve for the intrinsic pK_a of 5.8 and the conditions of $k_1 > k_9$, $k_2 = 0.1 k_3$, and $k_{11}H > k_8$. The intercept of the dashed line corresponds to the apparent pK_a of 6.8.

The apparent pK_a value of 6.8 (Fig. 2) is based on a theoretical fit assuming the kinetic sequence proposed by Stone and Morrison in which there is no proton equilibrium between E·DHF and E·DHFH, so that the apparent pK_a is the intrinsic pK_a. A closer examination of the data shows, however, that this sequence is indequate. A better fit to the "hollow" phenomenon is obtained if the following kinetic scheme is used:

$$
\begin{array}{ccccccccc}
E & \underset{k_8}{\overset{k_9(DHF)}{\rightleftharpoons}} & E\cdot DHF \\[1em]
K_1 \Big\updownarrow H^+ & k_{10}\Big\updownarrow & \Big\updownarrow k_{11}(H) \\[1em]
EH & \underset{k_2}{\overset{k_1(DHF)}{\rightleftharpoons}} & EH\cdot DHF & \underset{k_4}{\overset{k_3}{\rightleftharpoons}} & E\cdot DHFH & \overset{k_5}{\longrightarrow} & E\cdot THF & \overset{k_7}{\longrightarrow} & E
\end{array}
\tag{3}
$$

Appropriate choice of rate constants ($k_2 < k_3$, $k_8 < k_{11}H$ and $k_1 > k_9$) (16) results in the theoretical curve shown in Fig. 2. In this sequence, the observed pK_a is displaced about 1 pH unit above the intrinsic pK_a. Note that this mechanism requires that solvent water have access to the active site, allowing protonation of the ternary E·DHF·NADPH complex. The mechanism can be confirmed by measurement of pH dependence of the deuterium isotope effect which is in progress.

In conclusion, the effect of the Thr-113 ⟶ Val mutation seems to be a simple case of weakening substrate binding by removing hydrogen bonds required to keep the K_M for dihydrofolate low, but not required in any way for catalysis. The effects of the remaining mutations appear to be more complex; the major effects are on substrate binding and solvent accessibility to the active site. The Val-31 mutant exhibits both weaker substrate binding and a lower pK_a value for Asp-27, attributable to an increase in the number of water molecules within the active site. Replacement by Tyr-31 maintains tight substrate binding but introduces some degree of solvent equilibration of the two ternary complexes (Eq. 3) as evidence by the apparent "hollow" in the V/K pH profile. The Gly-54 mutant possibly introduces a second "channel" that permits rapid equilibration of the ternary complexes and lowers the intrinsic pK_a of the Asp-27 by two pH units. In all four cases the mutant enzyme achieves the same V as wt, despite decreased dihydrofolate binding and/or a marked drop in the pK_a of the catalytic group. The lack of dependence of V on pK_a can be interpreted in several ways. The proton transfer may not be reflected in the rate-limiting step. Alternatively, Asp-27 may not be functioning as a simple proton donor to N-5 but might be acting bifunctionally to stabilize the enol form of the substrate by hydrogen bonding at N-3 and at the 4-OH.

REFERENCES

1. J. T. Bolin, D. J. Filman, D. A. Matthews, R. C. Hamlin, J. Kraut, Crystal structures of *Escherichia coli* and *Lactobacillus casei* dihydrofolate reductase refined at 1.7 Å resolution I, J. Biol. Chem. 257: 13650 (1982).

2. D. J. Filman, J. T. Bolin, D. A. Matthews, J. Kraut, Crystal structures of *Escherichia coli* and *Lactobacillus casei* dihydrofolate reductase refined at 1.7 Å resolution II, J. Biol. Chem. 257:13663 (1982).

3. S. R. Stone and J. F. Morrison, Catalytic mechanism of the dihydro-folate reductase reaction as determined by pH studies, Biochemistry 23:2753 (1984).

4. E. E. Howell, J. E. Villafranca, M. S. Warren, S. J. Oatley, J. Kraut, The functional role of aspartic acid 27 in dihydrofolate reductase revealed by mutagenesis, submitted for publication.

5. D. R. Smith and J. M. Calvo, Nucleotide sequence of the E. coli gene coding for dihydrofolate reductase, Nuc. Acids. Res. 8:2255 (1980).

6. T. Yaegashi, C-. P. D. Tu, unpublished results.

7. G. Dalbadie-McFarland, L. W. Cohen, A. D. Riggs, C. Morin, K. Itakura, J. H. Richards, Oligonucleotide-directed mutagenesis as a general and powerful method for studies of protein function, Proc. Natl. Acad. Sci. USA 79:6409 (1982).

8. J. P. Gergen, R. H. Stein, P. C. Wensink, Filter replicas and perman-ent collections of recombinant DNA plasmids, Nuc. Acids. Res. 7:2115 (1979).

9. A. M. Maxam and W. Gilbert, Sequencing end-labeled DNA with base-specific chemical cleavages, Methods Enzym. 65:499 (1980).

10. D. de Laurmier and D. Bryant, Dideoxy sequencing in a double-stranded template, unpublished results.

11. B. T. Kaufman, Methotrexate-agarose in the purification of dihydro-folate reductase, Methods Enzym. 34:272 (1974).

12. G. Spears, J. G. T. Sneyd, E. G. Loten, A method for deriving kinetic constants for two enzymes acting on the same substrate, Biochem. J. 125:1149 (1971).

13. P. J. Cayley, S. M. J. Dunn, R. W. King, Kinetics of substrate, co-enzyme and inhibitor binding to Escherichia coli dihydrofolate reduc-tase, Biochemistry 20:874 (1981).

14. M. E. Porter, K. A. Johnson, Characterization of the ATP-sensitive binding of Tetrahymena 30S Dynein to bovine-brain microtubules, J. Biol. Chem. 258:6575 (1983).

15. S. R. Stone and J. F. Morrison, The pH dependence of the binding of dihydrofolate and substrate analogues to dihydrofolate reductase from Escherichia coli, Biochem. Biophys. Acta 745:247 (1983).

16. W. W. Cleland, Determining the chemical mechanisms of enzyme-catalyzed reactions by kinetic studies, in "Advances in Enzymology", A. Meister, ed., Interscience, New York (1975).

SITE SATURATION MUTAGENESIS OF ACTIVE SITE RESIDUES OF β-LACTAMASE

Steve C. Schultz, Steven S. Carroll and John H. Richards

Division of Chemistry and Chemical Engineering
California Institute of Technology
Pasadena, California 91125

INTRODUCTION

Techniques such as site-specific *in vitro* mutagenesis that enable one to make specific changes in amino acid sequence by altering the DNA sequence of a corresponding structural gene provide new approaches for studying the relationship between structure and function in proteins (1, 2). Detailed knowledge of the three-dimensional structure and the mechanism of action of a particular protein greatly assist in choosing which mutant proteins to produce and study. Even when such information is available, particular amino acid substitutions often affect the behavior of a protein in unanticipated ways. Therefore, to study thoroughly the role of an important residue, substitution with all 19 possible amino acids (site saturation) becomes necessary.

The site saturation approach to protein structure/function studies requires a procedure to generate efficiently the appropriate mutants and is particularly applicable when a rapid assay for protein activity is available. An effective method for phenotypic screening that avoids purification of each individual protein greatly simplifies these studies and allows this approach to be extended to the simultaneous saturation of two or more sites in the protein to assess the effect of combinations of amino acids on protein function, or perhaps to create a protein with a novel activity. The increase in the number of mutant proteins essentially by a factor of 20^n implies an upper limit on the number of sites, n, that can be simultaneously saturated.

We have used site saturation to study the pBR322 coded β-lactamase (EC 3.5.2.6) (3). This is an RTEM-1 enzyme originally taken from R-factor R1 (4-6) that catalyzes the hydrolysis of the β-lactam ring of penam and cephem antibiotics (7). β-Lactamase confers resistance to these antibiotics on cells that produce the enzyme and thereby provides a convenient screening procedure to assess the activities of mutants.

β-Lactamase has been studied extensively due to its clinical importance. Several β-lactamases have been isolated and characterized; these have been classified into three groups: A, B and C. The class A β-lactamases include *S. aureus* PC1, *B. licheniformis* 749/c, *B. cerus* 569/H 1 as well as TEM-1 and TEM-2 enzymes (8). The catalytic mechanism of this class is the most clearly understood. The class A β-lactamases generally hydrolyze penam antibiotics more rapidly than cephem antibiotics. Class B consists of a *B. cereus* enzyme that requires zinc for catalytic activity (9) Class C enzymes are the chromosomal β-lactamases from *E. coli* K12 and *Ps. aeruginosa*. These enzymes have a nucleophilic serine like class A, but hydrolyze cephalosporins more rapidly than penicillins (10, 11).

The class A lactamases may have evolved from D-Ala-D-Ala carboxypeptidases that are the target enzymes for penicillins (8, 12); β-lactamases have very low levels of peptidase activity for D-Ala-D-Ala substrates (13). Both β-lactamases and D-Ala-D-Ala carboxypeptidases contain a serine residue at their active sites (8, 14). This serine hydroxyl group reacts with the carbonyl carbon of the β-lactam ring of the antibiotic forming an acyl-enzyme intermediate in both types of enzymes. However, the acyl-enzyme intermediate of β-lactamase deacylates rapidly whereas the acyl-enzyme intermediate of the carboxypeptidase is stable (15, 16).

The class A β-lactamases contain a conserved triad [Ser-Thr-Xaa-Lys, residues 70-73 in a consensus numbering of these enzymes (8)] that includes the active site serine (residue 70) (15, 17). The substitution Ser 70 → Cys produces an active mutant (18) whereas Ser 70 → Thr produces an inactive mutant (19) suggesting that a primary nucleophile is required in this position. The Lys residue at position 73 is conserved in class A lactamases, class C lactamases and D-Ala-D-Ala carboxypeptidases (8, 20). Recently, a solution has been reported for the structure of a penicillin sensitive D-alanyl carboxypeptidase-transpeptidase from *Streptomyces* R61 with a bound cephalosporin C molecule (21). In this structure, the residues between Ser and Lys are near the binding site of the antibiotic. The nucleophilic Ser lies near the carbonyl carbon of the β-lactam ring and the Lys residue is near the larger of the two rings and could interact with a C3 substituent or with the C4 carboxylate of cephalosporin C. At the resolution presently available, the side chain of residue 71 is difficult to orient.

Using site saturation, we have generated mutants with all of the 19 amino acid substitutions at Thr 71 and Lys 73. For residue 71, 14 of the substitutions resulted in active β-lactamase (3) suggesting that this residue is not involved directly in binding or catalysis. However, all of these mutants are degraded by cellular proteases more rapidly than the wild-type enzyme indicating that residue 71 is important for structural stability of the protein. In contrast, none of the mutants at residue 73 were active suggesting that this residue is essential for binding or catalysis. For site saturation at residue 73 the mutant proteins were present in quantities identical to wild-type enzyme indicating that mutations at this residue do not significantly affect stability.

SITE SATURATION TECHNIQUE

Restriction sites that flank the codons for residue 71 and 73 were introduced into pBR322 by oligonucleotide directed mutagenesis: an Ava I site at position 3972 and a Sca I site at 3937. The DNA fragment between 3972 and 3937 (corresponding to the Pro 62 to Val 74 segment of β-lactamase) was removed and replaced with a mixture of synthetic, double-stranded oligonucleotides that included 32 codons for either residue 71 or 73 (see Fig. 1); these mixtures coded for all 20 amino acids and an amber codon. The resulting mixture of plasmids was used to transform *E. coli* and the colonies were screened for their sensitivity to penam and cephem antibiotics. Plasmids derived from individual colonies were sequenced to identify the mutants (3).

SATURATION AT RESIDUE 71

Table 1 lists the maximum concentration of antibiotic at which cells containing the various mutant β-lactamases were able to grow. Mutants with His, Cys and Ser at residue 71 produce resistance to high levels of all of the penam antibiotics tested. Mutants containing very large or positively charged amino acid side chains in this position are inactive (Trp, Tyr, Lys and Arg); the

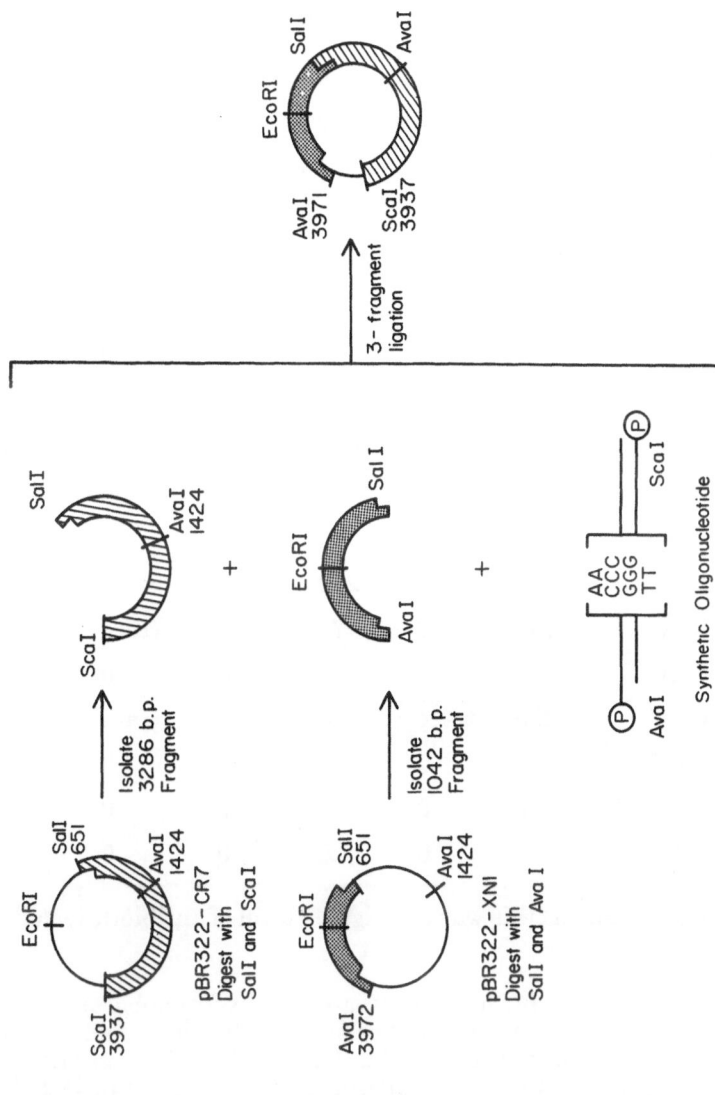

Fig. 1. Design of three-fragment ligation for insertion of the mixture of oligonucleotides.

Table 1: Maximum level of resistance of mutations at residue 71

Amino Acid	Codons	Ampicillin		Benzylpenicillin		6-Aminopenicillanic Acid	
		30°	37°	30°	37°	30°	37°
Gly	GGC, GGG	500+	trace	500+	50	35	20
Ala	GCG,GCC	500+	75	500+	100	trace	trace
Val	GTG	500+	150	500+	175	75	trace
Leu	TTG, CTC, CTG	500+	100	500+	125	80	trace
Ile	ATC	500+	500+	500+	500+	90	40
Met	ATG	500+	125	500+	150	30	0
Pro	CCG, CCC	500+	200	500+	350	50	40
Phe	TTC	60	0	0	0	0	0
Trp	TGG	0	0	0	0	0	0
Tyr	TAC	0	0	0	0	0	0
His	CAC	500+	500+	500+	500+	250+	250+
Cys	TGC	500+	500+	500+	500+	250+	100
Ser	TCC, AGC, TCG	500+	500+	500+	500+	250+	250+
Thr	ACG, ACC	500+	500+	500+	500+	250+	250+
Asn	AAC	270	trace	210	trace	55	trace
Gln	CAG	130	20	70	trace	trace	0
Asp	GAC	0	0	0	0	0	0
Glu	GAG	200	trace	70	trace	trace	trace
Lys	AAG	0	0	0	0	0	0
Arg	CGC, AGG	0	0	0	0	0	0
Stop	TAG	0	0	0	0	0	0

a) Units are mg/l; + means this was the highest level of antibiotic tested.

Thr 71 → Asp mutant is also inactive. Mutants with hydrophobic amino acids at residue 71 provide high levels of resistance for benzylpenicillin and ampicillin, but low levels for 6-aminopenicillanic acid. The Thr 71 → Glu and Thr 71 → Gln mutants confer higher resistance to ampicillin than to benzylpenicillin. None of the mutants produced resistance to cephalothin or cephalexin at 37°C, although the Thr 71 → Ser mutant showed trace resistance at reduced temperature.

Fig. 2 shows a western blot of all 19 mutants at 37°C and 30°C. The mutant enzymes appear much less stable to cellular proteases than the wild-type enzyme. A mutant (P2) analogous to Thr → Ile has been reported for β-lactamase PC1 from *Staphyloccocus aureus* (8). Unlike the Thr 71 → Ile mutant of the pBR322 coded β-lactamase, the Thr 71 → Ile mutant of the *S. aureus* β-lactamase is inactive; incomplete folding has been observed for this mutant

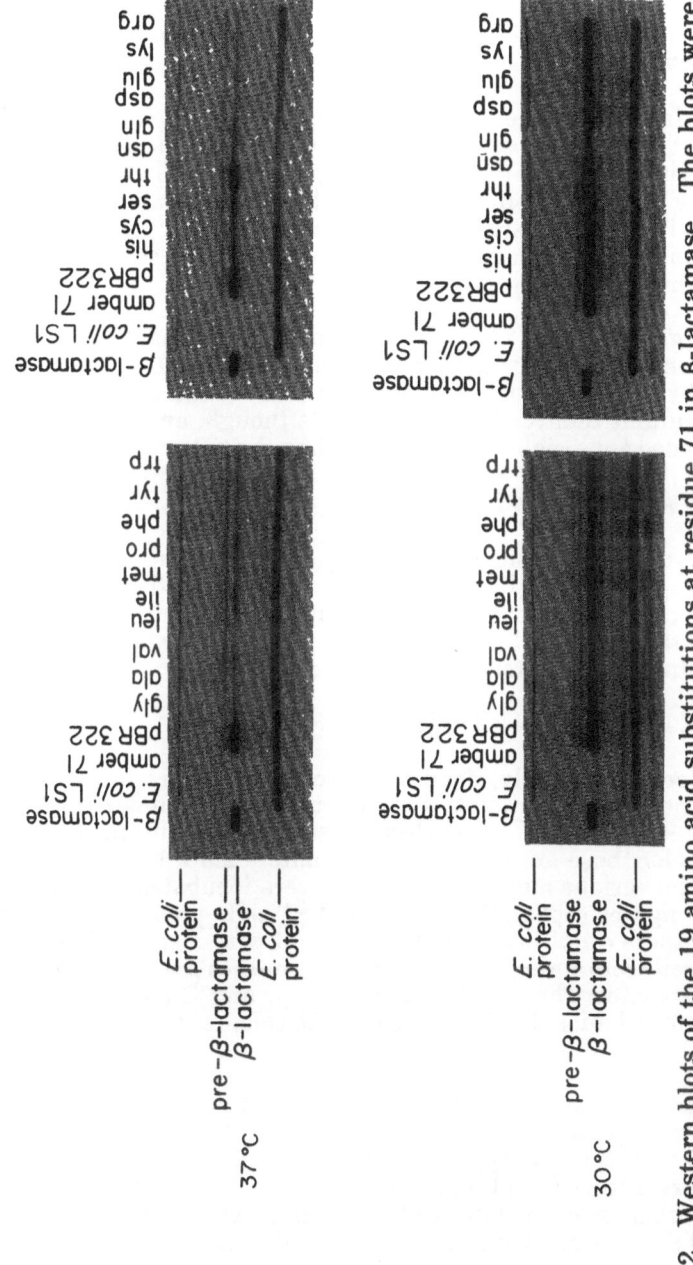

Fig. 2. Western blots of the 19 amino acid substitutions at residue 71 in β-lactamase. The blots were treated with rabbit anti-β-lactamase antibody and visualized using the horse radish peroxidase assay.

protein (22). The difference in behavior between these two mutants possibly reflects the presence of a disulfide bond between Cys 77 and Cys 123 in the pBR322 β-lactamase. The *S. aureus* β-lactamase lacks any disulfide bonds and may, therefore, be more susceptible to mutations that affect the folding of these proteins into the active conformation.

The results of these experiments indicate that residue 71 is very important for structural stability of the protein, but is not essential for either binding or catalysis. An apparent change in specificity suggests that this residue may affect the region of the protein that accommodates the R-group on the β-lactam ring of the antibiotic.

Another interesting result of these experiments is the increasing antibiotic resistance for the series Gly<Ala<Leu<Val (see Table 1) which emphasizes the importance of the methyl group of Thr 71. The reduced stability observed for the Thr 71 → Ser mutant (23) further demonstrates this point since only the methyl group is absent in this mutant. The Thr 71 → Ile mutant is somewhat anomalous, but may indicate the presence of a strong hydrophobic interaction in this region of the protein.

Surprisingly, the Thr 71 → His mutant shows considerable catalytic activity. Also noteworthy is the stability of the Thr 71 → Pro mutant as demonstrated by the western blots. The rigid nature of this residue may help to stabilize this mutant relative to other mutants though, even in this case, the protein is degraded appreciably more readily than the wild-type enzyme.

SATURATION AT RESIDUE 73

All 19 mutants at residue 73 in β-lactamase are unable to confer resistance to ampicillin. Western blot analysis shows that the mutant proteins are present in cells in amounts similar to that of wild-type enzyme. These results suggest that residue 73 is important for binding or catalysis but does not affect the structure of the protein significantly.

SECOND-SITE REVERSION EXPERIMENTS

As mentioned previously, none of the mutations at residue 71 provided resistance to the cephem antibiotics cephalothin and cephalexin, except the mutant Thr 71 → Ser at 30°C. To select for second site mutations that might restore activity for these antibiotics, approximately 2.5 x 10^7 cells containing plasmid with the various mutant lactamases were incubated on L-agar plates containing 50 mg/l of cephalothin. *E. coli* LS1 cells produced 2 individual colonies under these conditions as did the mutant with a stop codon at residue 71. However, cells containing the mutant β-lactamases produced more: Thr 71 → Ile gave 151 colonies; Thr 71 → Trp, Thr 71 → Tyr and Thr 71 → His each gave approximately 50 colonies; Thr 71 → Asp gave 26 colonies and all others gave 1-10 colonies.

These results suggested that many possible second-site mutations are capable of restoring cephalothin resistance. To select for mutants with possibly higher activity than the wild-type enzyme, approximately 10^10 cells of each of the mutants were incubated in L-broth containing 1000 mg/l of cephalothin (500 mg/l is the maximum level at which cells containing wild-type enzyme normally grow). None of the mutants grew under these conditions.

To allow for more than a single mutation to restore cephalothin resistance, the mutants were grown sequentially at increasing levels of cephalothin: 50, 100, 250, 500 and 1000 mg/l. Plasmid was prepared from the final 1000 mg/l culture and used in transformation of LS1 cells. These cells were incubated in L-broth with 500 mg/l of cephalothin. Mutants with Thr 71 → Lys, Ala, Pro and Ser all yielded revertants that were resistant to this concentration of cephalothin. However, these cells now grew in the presence of 1500 mg/l of this antibiotic. For each of these mutants, a single base change in the codon for residue 71 can and apparently does, lead to a codon for threonine so that all revertants with this high level of resistance to cephalothin have regained the wild-type amino acid sequence in the region of the active site.

These results suggest that mutants with amino acids other than threonine at residue 71 can recover low levels of activity toward cephalothin though none approach the activity of the wild-type enzyme under the conditions of this experiment. The increased level of resistance for the revertants that have threonine at residue 71 may result from overproduction of the enzyme, perhaps due to mutations that increase the plasmid copy number or the activity of the promoter, or may be caused by mutations in the structural gene that give rise to enzymes that are inherently more active against cephalosporins, examples of which have been previously observed (24).

REFERENCES

1. G. Dalbadie-McFarland, and J. H. Richards, *In vitro* mutagenesis: Powerful new techniques for studying structure-function relationships in proteins, Ann. Rep. Med. Chem. 18:237-245 (1983).
2. G. K. Ackers, and F. R. Smith, Effects of site-specific amino acid modification on protein interactions and biological function, Ann. Rev. Biochem. 54:597-629 (1985).
3. S. C. Schultz, and J. H. Richards, Site saturation studies of β-lactamase. Production and characterization of mutant β-lactamases with all possible amino acid substitutions at residue 71, Proc. Natl. Acad. Sci. USA 83:1588-1592 (1986).
4. M. Matthew, and R. W. Hedges, Analytical isoelectric focusing of R factor-determined β-lactamases: Correlation with plasmid compatibility, J. Bacteriol. 125:713-718 (1976).
5. F. Bolivar, R. L. Rodriguez, P. J. Greene, M. C. Betlach, H. L. Heyneker, and H. W. Boyer, Construction and characterization of new cloning vehicles. II. A multipurpose cloning system, Gene 2:95-113 (1977).
6. M. So, R. Gill, and S. Falkow, The generation of a Col E1-Ap^r cloning vehicle which allows detection of inserted DNA, Mol. Gen. Genet. 142:239-249 (1975).
7. N. Datta, and P. Kontomichalou, Penicillinase synthesis controlled by infectious R factors in enterobacteriaceae, Nature 208:239-241 (1965).
8. R. P. Ambler, The structure of β-lactamases, Phil. Trans. R. Soc. Lond. B. 289:321-331 (1980).
9. S. Kuwabara, and E. P. Abraham, Some properties of two extracellular β-lactamases from *Bacillus cereus* 569/H, Biochem. J. 103:27c-30c (1967).
10. R. B. Sykes, and M. Matthew, The β-lactam antibiotics, J. Antimicrob. Chemother. 2:115-157 (1976).
11. B. Jaurin, and T. Grundström, Amp C cephalosporinase of *Escherichia coli* K-12 has a different evolutionary origin from that of β-lactamases of the penicillinase type, Proc. Natl. Acad. Sci. USA 78:4897-4901 (1981).
12. D. J. Tipper, and J. L. Strominger, Mechanism of action of penicillins: A proposal based on their structural similarity to acyl-D-alanyl-D-alanine, Proc. Natl. Acad. Sci. USA 54:1133-1141 (1965).
13. R. F. Pratt, and C. P. Govardham, β-Lactamase-catalyzed hydrolysis of acyclic depsipeptides and acyl transfer to specific amino acid receptors, Proc. Natl. Acad. Sci. USA 81:1302-1306 (1984).
14. R. A. Nicholas, R. Ishino, W. Park, M. Matsuhashi, and J. Strominger, Purification and sequencing of the active site tryptic peptide from penicillin-binding protein 5 from the dacA mutant strain of *Escherichia coli* (TMRL 1222), J. Biol. Chem. 260:6394-6397 (1985).
15. J. Fisher, J. G. Belasco, S. Khosla, and J. R. Knowles, β-Lactamase proceeds via an acyl-enzyme intermediate. Interaction of the *Escherichia coli* RTEM enzyme with cefoxitin, Biochemistry 19:2895-2901 (1980).

16. D. J. Waxman, and J. L. Strominger, Penicillin-binding proteins and the mechanism of action of β-lactam antibiotics, <u>Ann. Rev. Biochem</u>. 52:825-869 (1983).

17. V. Knott-Hunziker, S. G. Waley, B. S. Orlek, and P. G. Sammes, Penicillinase active sites: Labelling of serine-44 in β-lactamase I by 6 β-bromopenicillanic acid, <u>FEBS Lett</u>. 99:59-61 (1979).

18. I. S. Sigal, W. F. DeGrado, B. J. Thomas, and S. R. Petteway, Jr., Purification and properties of thio β-lactamase, <u>J. Biol. Chem</u>. 259:5327-5332 (1984).

19. G. Dalbadie-McFarland, L. W. Cohen, A. D. Riggs, C. Morin, K. Itakura, and J. H. Richards, Oligonucleotide-directed mutagenesis as a general and powerful method for studies of protein function, <u>Proc. Natl. Acad. Sci. USA</u> 79:6409-6413 (1982).

20. V. Knott-Hunziker, S. Petursson, G. S. Jayatilake, S. G. Waley, B. Jaurin, and T. Grundstrom, Active sites of β-lactamases, <u>Biochem. J</u>. 201:621-627 (1982).

21. J. A. Kelly, J. R. Knox, P. C. Meows, J. H. Gilbert, J. B. Bartolone, and H. Zhao, 2.8-Å Structure of penicillin-sensitive D-alanyl carboxypeptidase-transpeptidase from *Streptomyces* R61 and complexes with β-lactams, <u>J. Biol. Chem</u>. 260:6449-6458 (1985).

22. S. Craig, M. Hollecker, T. E. Creighton, and R. H. Pain, Single amino acid mutations block a late step in the folding of β-lactamase from *Staphylococcus aureus*, <u>J. Mol. Biol</u>. 185, in press.

23. G. Dalbadie-McFarland, J. Neitzel, and J. H. Richards, Active site mutants of β-lactamase/Use of inactive double mutant to study requirements for catalysis, <u>Biochemistry</u> 25:332-338 (1986).

24. A. Hall, and J. R. Knowles, Directed selective pressure on a β-lactamase to analyze molecular changes involved in development of enzyme function, <u>Nature</u> 264:803-804 (1976).

ENZYME THERMOSTABILITY AND ITS ENHANCEMENT

BY GENETIC ENGINEERING

Tim J. Ahern and Alexander M. Klibanov

Department of Applied Biological Sciences
Massachusetts Institute of Technology
Cambridge MA 02139

ENZYME THERMOSTABILITY

Enzymes are observed to lose their catalytic activity as temperature is increased. Such thermal inactivation can be either reversible or irreversible, depending on whether return to ambient temperature restores enzymatic activity (1). Reversible conformational transitions that result in the denaturation (partial unfolding) of the enzyme can account for the loss of enzyme activity that is regained upon cooling; this phenomenon has been the subject of extensive investigation and is now well characterized (2-7). We have recently elucidated the major processes accounting for the irreversible thermoinactivation of enzymes. In studies of hen egg-white lysozyme (8), bovine pancreatic ribonuclease (9), and yeast triosephosphate isomerase (10), we have shown that the processes of irreversible enzyme inactivation at 90 - 100°C are deamidation of asparagine residues, hydrolysis of peptide bonds at aspartic acid residues, destruction of S-S bonds, and formation of incorrect (scrambled) structures. The relative contribution of each process depends on the nature of the protein and the pH.

This knowledge of why and how proteins lose their biological activity at high temperatures is crucial for understanding thermophilic behavior and should demarcate the upper limit of protein thermostability and hence of life. Such findings also make possible the formulation of a rational strategy for the enhancement of enzyme thermostability. From the biotechnological standpoint, thermal inactivation is by far the most frequently encountered mode of enzyme inactivation in practice (1). There are a number of compelling reasons for conducting industrial enzymatic processes at elevated temperatures, e.g. enhanced reaction rates, reduced likelihood of microbial contamination, increased solubility of substates, and reduced viscosity. Therefore, nearly all existing commercial processes employing enzymes are carried out at elevated temperatures: production of fructose by glucose isomerase at 60-65°C, hydrolysis of starch by alpha-amylase at 85-110° C, resolution of D,L-amino acids by aminoacylase at 50°C, etc. (1). It would be highly advantageous to perform most such processes at yet higher temperatures; however, the enzymes used are not sufficiently thermostable.

Conversely, other enzymes are "too stable" for the desired application. For example, the persistance of the proteolytic activity of microbial rennet hinders its adoption for use in the food industry. Therefore the means to both _reduce_ as well as _enhance_ enzyme thermostability are prominent subjects of research.

As a result of our findings, several approaches to enzyme thermostabilization are now apparent. The first is disarmingly straightforward: exclude all excess water from the system. Since water is a reactant in peptide hydrolysis and deamidation, the mediator of alkaline destruction of cystine, and the solvent facilitating the mobility leading to incorrect structure formation, it follows that substitution of another medium for water could stabilize an enzyme with respect to these processes. As expected, the rates of irreversible thermoinactivation of enzymes are reduced by several orders of magnitude when incubated as dry powders (11,12) or in organic solvents such as cyclohexane, hexadecane, and 1-heptanol (11). This rationale also explains why the stability of lipase activity at 100° C in organic media decreases with increasing water content (13). Since a number of enzymes including not only lipase but tyrosinase, peroxidase, cholesterol oxidase, and xanthine oxidase also exhibit activity in nearly anhydrous media, the elimination of water may prove to be a general, practical strategy of thermostabilization.

THERMOSTABILIZATION OF ENZYMES BY MEANS OF SITE-DIRECTED MUTAGENESIS

The second strategy involves genetic engineering aimed at replacing the "weak links" in proteins, i.e., the amino acid residues implicated in irreversible thermoinactivation (Asp, Asn, cystine, and to a lesser extent Gln) or any neighboring groups found to accelerate the inactivating process. These approaches can be expected to result in decreased rates of irreversible thermoinactivation and thus more thermostable enzymes. For example, deamidation of Asn occurs at a significant rate in enzymes at elevated temperatures in aqueous media throughout the range of pH relevant to enzymatic catalysis (8). Therefore, replacement of Asn with residues that will not affect adversely the native conformation or catalytic activity should stabilize against thermoinactivation those enzymes that lose their activity due to deamidation.

The technique to be employed, site-directed mutagenesis (Fig. 1), is a proven, powerful tool for the selective substitution of amino acids within sequences of proteins (14, 16-18). By synthesizing oligonucleotide primers coding for amino acid replacements at specific codons and annealing such primers to the gene of interest, one can construct mutant genes that, when expressed by a suitable plasmid vector in the appropriate host, will result in mutant proteins that differ from the wild type protein by as few as one amino acid substitution.

For the purposes of this study, we have chosen triosephosphate isomerase (TIM) of the yeast _Saccharomyces cerevisiae_ as a model system. As a result of work in the laboratory of Professor Gregory Petsko at MIT, the yeast enzyme has been cloned and expressed in _E. coli_ (19), its complete three-dimensional structure has been determined by X-ray diffraction at 0.19 nm resolution (20; T. Alber, M. Rose & G.A. Petsko, unpublished work), and a method of site-directed mutagenesis has been successfully devised and implemented for the study of the enzyme's catalytic properties (21). TIM is a relatively small (53,000 MW) enzyme consisting of two identical single-chain subunits composed solely of the twenty common amino acid residues. It is ideal for the purposes of our study because deamidation alone appears to satisfactorily account for the irreversible thermoinactivation of TIM by covalent mechanisms at near-neutral pH.

Our method of incubation and assay for irreversible thermoinactivation is the following: yeast TIM is incubated at either pH 6 or 8 in concentrated denaturant (e.g., 6 M GuHCl) at 100° C; at intervals aliquots are removed,

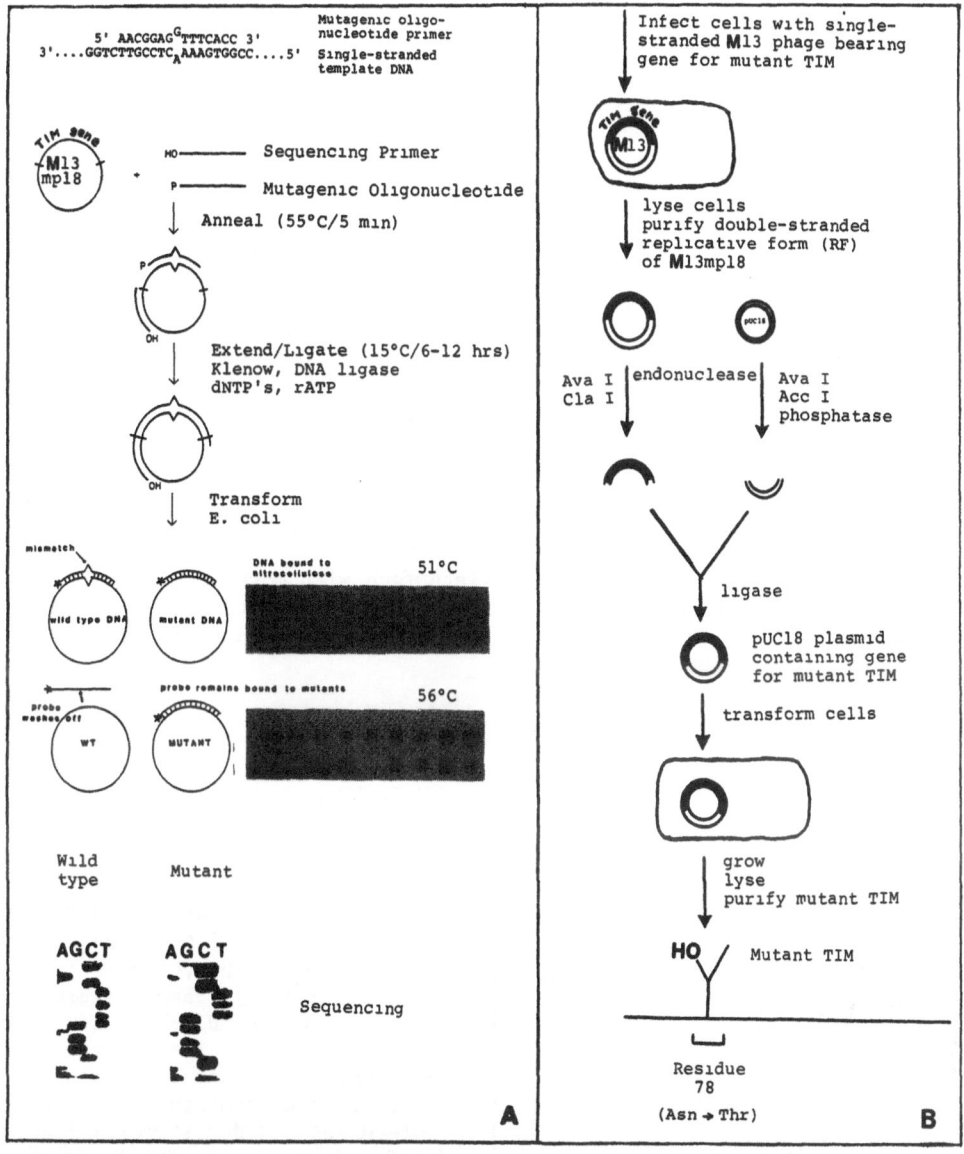

Figure 1. Site-directed mutagenesis (A) and expression (B) to produce mutant triosephosphate isomerase (TIM). In this example, Asn-78 is replaced by Thr according to the double-primer method of Zoller and Smith (14). Both a non-mutagenic sequencing primer and a pentadecamer oligonucleotide priming the synthesis of a DNA strand encoding the amino acid Thr in place of Asn at position 78 within yeast TIM were synthesized and annealed to a single-stranded template of DNA from M13mp18 phage containing the gene encoding TIM of the yeast <u>Saccharomyces cerevisiae</u> (20). The primer was then elongated, ligated, identified by dot-blot hybridization, and the entire TIM gene was sequenced by the chain-termination method of Sanger (15) to confirm that only one change had occurred. The mutated gene for yTIM was then excised, inserted into an expression vector (pUC 18) and used to transform competent <u>E. coli</u> cells that were deficient in the gene for bacterial TIM. The mutant yTIM enzyme was obtained from cultures of the transformed cells and purified by ion-exchange and immunoadsorption chromatography to yield mutant yTIM of >95% purity having a specific activity equivalent to that of native yTIM.

cooled, the denaturant is diluted, and the enzymatic activity is assayed. It was found (10) that deamidation (resulting in the evolution of free ammonia and a decrease in the enzyme's isoelectric point) can account for the irreversible thermoinactivation of TIM at pH 6 (Fig. 2) and 8. Monodeamidated and dideamidated TIM isolated by means of preparative isoelectofocusing exhibit approximately one-half and one-third of native TIM's specific activity, respectively.

These findings helped us formulate the following strategy:

(1) The dissociation of TIM to monomers appears to be the rate-limiting step in the loss of catalytic activity under destabilizing conditions such as dilute denaturant solutions (22). Stabilization against such dissociations may also have a stabilizing effect at elevated temperatures or promote the reversible regeneration of activity upon cooling.

(2) Deamidation of Asn residues located at the subunit interface appear to destabilize the enzyme by promoting dissociation (23). Such a covalent change, whose rate is accelerated at elevated temperature, is observed to be the rate-limiting covalent cause of irreversible inactivation of TIM (10). There are three Asn residues, numbers 14, 65, and 78, per monomer at the subunit interface of yeast TIM (Fig. 3).

(3) Replacement of Asn at the subunit interface with more stable hydrophobic residues by means of genetic engineering should prevent the deleterious effects of deamidation and therefore stabilize enzyme yeast TIM when exposed to elevated temperatures.

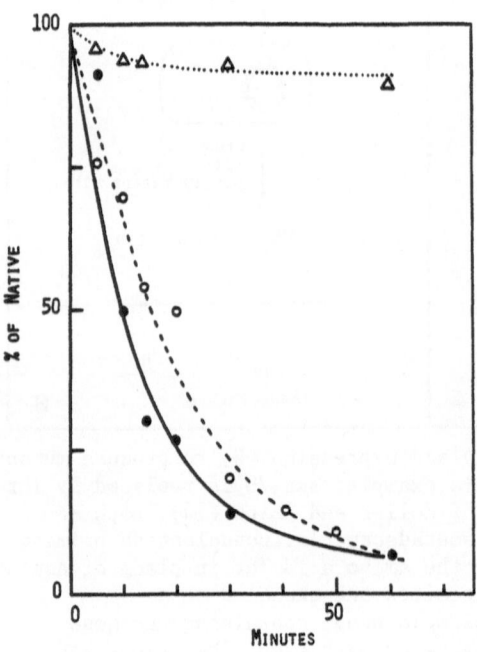

Figure 2. The time course of irreversible inactivation of yeast TIM vs. the rates of deamidation and peptide chain hydrolysis. o, The residual activity of yeast TIM (1.5mg/mL) after heating at 100° in K-phosphate (0.1 M, pH 6) containing guanidinium chloride (6 M) and EDTA (8.4 mM). Samples were cooled (23°C) and diluted (1:400) prior to assay. ●, The fraction of non-deamidated yeast TIM during heating, as determined by isoelectrofocusing and densitometry (see (8) for details). Δ, The fraction of yeast TIM that did not undergo peptide bond hydrolysis, as determined by SDS polyacrylamide gel electrophoresis (see (8)).

We reason that the substitutions for Asn in TIM that have the greatest probability of conserving the enzyme's conformation and catalytic activity should (1) approximate the size and shape of Asn, and (2) conserve the hydrophobicity at the subunit interface. It is also useful to ask: are certain amino acid substitutions for Asn avoided throughout the evolution of the family of TIM? The relative probability of random substitutions of the amino acids for Asn by spontaneous point mutations can be approximated by assigning equal probability to every nucleotide replacement (Table 1). The observed frequency of amino acid substitutions at Asn codons of yeast TIM within the family of TIM shown in the same table indicates that any substitution – with the possible exception of Tyr – is allowed as a random event when the twelve Asn residue sites are examined together. However, when one examines only those substitutions that have occurred at the residue sites 14, 65 and 78 (the Asn's found at the interface of yeast TIM), then one finds that only Ile, Thr, Val, Ser, and His are substituents. The most probable substitution, Lys, which would result in the introduction of a charged hydrophilic group at the subunit interface, is not observed (Table 2). Its exclusion as a replacement strengthens the assumption that addition of charge at the interface is destabilizing and therefore lends support to our hypothesis.

Ile, Thr, and Val satisfy the substitution criteria described above. For these reasons, we are in the process of replacing Asn residues 14, 65, and 78 by means of site-directed mutagenesis by both Ile and Thr in an attempt to increase the enzyme's resistance to irreversible thermoinactivation. As a control, mutant enzymes bearing Asp in place of Asn at the same loci are being constructed in order to determine the effect deamidation at single sites has on the enzyme's stability. The mutations resulting in the greatest degree of stabilization of TIM against irreversible thermoinactivation can be subsequently combined by means of serial, cumulative mutagenesis. Preliminary results involving substitution of Asn-78 have demonstrated that pure mutant TIM is more stable against irreversible thermoinactivation when Thr or Ile are the substituents, and that replacement of Asn by Asp has a destabilizing effect (10) and thus support our strategy of amino acid replacement by genetic engineering to increase the thermostabilization of enzymes.

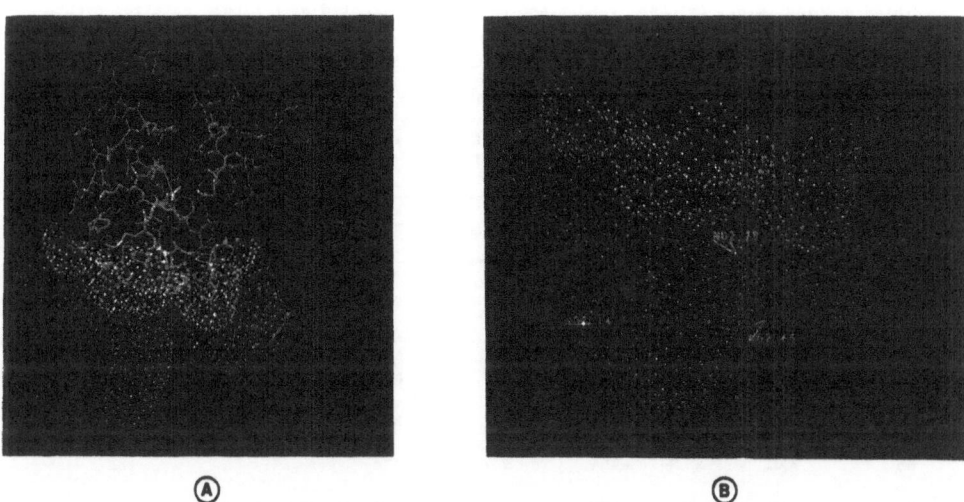

Ⓐ　　　　　　　　　　　　　　Ⓑ

Figure 3. Interfacial contacts between the surface of the subunit of yeast TIM and the asparagines of its partner subunit. A, The interfacial surface (dots) 0.3 nm distant from the atomic nuclei (interconnected by white lines) of a subunit of TIM. Only the surface within 0.4 nm of the atomic nucloi of the partner subunit (not shown) is depicted. B, The same interfacial surface represented in (A) in relation to the three asparagines (14, 65, and 78) of the partner subunit. The delta-nitrogen and oxygen plus the gamma-carbon of the residues are delineated in white. Each residue is in close contact with the other subunit. Residue 14 is partly obscured by the large loop in the foreground. The graphical representations were generated by Evans and Sutherland PS 300 graphics system by means of the Frodo system, version 6.0.

Table 1. Calculated probability and observed frequency of mutations at Asn codons within the family of triosephosphate isomerases (TIM).

Residue due to mutation	Single point mutations[a]	Double point mutations[b]	Calcuated probability[c]	Observed frequency[d]	Number of locations[e]
LYS	2		.158	.24	7
THR	1	3	.120	.17	5
ASP	1	1	.092	.10	3
ILE	1	2	.105	.034	1
HIS	1	1	.092	.10	3
SER	1	2	.105	.034	1
TYR	1	1	.092	0	0
ARG		3	.040	.034	1
GLN		2	.026	.14	4
GLU		2	.026	.07	2
GLY		1	.013	.07	2
VAL		1	.013	.03	1
MET		1	.013	0	0
PHE		1	.013	0	0
CYS		1	.013	0	0
LEU		1	.013	0	0
PRO		1	.013	0	0
ALA		1	.013	0	0

[a] The number of mutations requiring only one oligonucleotide substitution to convert the predominant Asn codon (AAC) to codons specifying the residue indicated.

[b] The number of mutations requiring two oligonucleotide substitutions to convert the predominant Asn codon (AAC) to codons specifying the residue indicated.

[c] The probability of a given amino acid substitution, assuming an equal probability for all oligonucleotide substitutions.

[d] The observed frequency of amino acid substitutions within the family of TIMs at all sites coding for Asn in yeast TIM. The same substitution within TIMs of relatively closely related species (e.g., chicken, rabbit and human) were treated as a single mutation event. Sequences used for comparison are those of human (24), rabbit (25), chicken (26), coelacanth (27 Eschericia coli (28), and Bacillus stearothermophilus (29).

[e] The number of single mutation events observed at Asn codons.

Table 2. Observed substitutions at specific Asn residues within the family of triosephosphate isomerases (TIM).

Residue[a]	Asn codon	X –ASN– Y		Conserved	Replacements	Interface
10	AAC	GLY–	–PHE	6		no
14	AAC	LEU–	–GLY	5	HIS	yes
28	AAC	LEU–	–THR	2	GLU, LYS, ARG, GLN	no
35	AAT	GLU–	–VAL	0	ASP (4), THR, GLY	no
65	AAC	GLN–	–ALA	5	THR	yes
78	AAC	GLU–	–SER	0	ILE (4), VAL, SER	yes
148	AAC	LEU–	–ALA	0	LYS (4), GLU, ASP	no
159	AAC	THR–	–VAL	0	LYS (5), GLY	no
213	AAC	ALA–	–GLY	1	THR (4), LYS	no
216	AAC	SER–	–ALA	3	THR (3)	no
245	AAC	ILE–	–SER	3	ASP, LYS, GLN	no
248	AAC	ARG–	–OH	0	GLN (4), HIS (2)	no

[a] Asparagines exist at these residues in yeast TIM

The methods outlined above are designed to eliminate the major general mechanism resulting in irreversible thermoinactivation at pH 6 and 100° C, namely deamidation. Once eliminated, however, it is likely that yet another, although much slower process will then demarcate a new, higher upper limit of thermostability common to all or most enzymes. Future work shall concentrate on their elimination as well. Probable secondary mechanisms of irreversible thermoinactivation at pH 6 include (1) deamidation of Gln, (2) hydrolysis of peptide bonds adjacent to Asp, and (3) destruction of cystine residues (8). Another mechanism, i.e., racemization of amino acid residues, has been shown to occur in peptides at 100°C within the pH range of biotechnological interest (30,31). Although the reported rates of racemization are much slower than the rates of irreversible thermoinactivation observed by us, once the upper limit of thermostability is raised by site-directed mutagenesis, the role of racemization will be greater; therefore, it is being presently evaluated in collaboration with Professor Jeffrey L. Bada of the University of California at San Diego.

Depending on whether or not these processes are sequence-dependent, the subsequent strategy would involve the replacement by site-directed mutagenesis of either the new "weak link" itself (e.g., replacing Gln with Leu and replacing Asp with Glu) or the neighboring residues that accelerate the destruction of the weak links, should such be found to be the case.

Non-covalent, conformational processes may be found to be rate-limiting as well. At present, these are eliminated by incubation in the presence of strong denaturant, but this method is not applicable to practical catalysis at high temperatures. Another method to eliminate such conformational mechanisms may be to shift the enzyme's isoelectric point away from the pH of incubation. Such shifts, brought about by means of post-translational modifications introducing new charged groups on the surface of the enzyme, has been shown to prevent incorrect folding of ribonuclease (9).

SUMMARY

A thorough understanding of both <u>reversible</u> and <u>irreversible</u> thermoinactivation of enzymes should provide insights into the nature of thermophilic behavior and help clarify the upper temperature limit of life. From the biotechnological standpoint, it is recognized that the use of enzymes as industrial catalysts is severely restricted by their instability at elevated temperatures. The strategies outlined above are a set of general guidelines describing how to make enzymes more stable against irreversible thermoinactivation by means of site-directed mutagenesis. These strategies are entirely different yet complementary to those (32,33) derived from comparative investigations of proteins from mesophilic vs. thermophilic organisms in which thermal denaturation has been the focal point. This difference is due to the fact that while the approach for preventing <u>reversible</u> thermal unfolding is to introduce additional stabilizing interactions (hydrogen bonds, hydrophobic contacts, salt bridges) in the enzyme molecule, the strategy for stabilizing against <u>irreversible</u> thermoinactivation (occurring even prior to and frequently independently of thermal unfolding) is to eliminate chemical reactions that lead to damage of the enzyme molecule. The principles elaborated above should have general applicability and are of both basic and practical interest.

REFERENCES

1. A.M. Klibanov, Stabilization of enzymes against thermal inactivation, <u>Advan. Appl. Microbiol.</u> 27:1 (1983).
2. W. Kauzmann, Some factors in the interpretation of protein denaturation, <u>Advan. Protein Chem.</u> 14:1 (1959).

3. C. Tanford, Protein denaturation, <u>Advan</u>. <u>Protein</u> <u>Chem</u>. 23:121 (1968) and 24:1 (1970).

4. S. Lapanje, "Physicochemical Aspects of Protein Denaturation," Wiley, New York (1978).

5. P.L. Privalov, Stability of proteins, <u>Advan</u>. <u>Protein</u> <u>Chem</u>. 33:167 (1979).

6. W. Pfeil, The problem of stability of globular proteins, <u>Molec</u>. <u>Cell</u>. <u>Biochem</u>. 40:3 (1981).

7. T. E. Creighton, "Proteins," Chapter 7, Freeman, New York (1983).

8. T. J. Ahern and A. M. Klibanov, The mechanism of irreversible enzyme inactivation at 100°C, <u>Science</u>, 228:1280 (1985).

9. S. E. Zale and A. M. Klibanov, Mechanisms of irreversible thermoinactivation of enzymes, <u>Ann</u>. <u>N</u>. <u>Y</u>. <u>Acad</u>. <u>Sci</u>. 434: 20-26 (1984).

10. T. J. Ahern, I. J. Casal, R. C. Davenport, G. A. Petsko, and A. M. Klibanov, (in preparation).

11. T. J. Ahern and A. M. Klibanov, Why do enzymes irreversibly inactivate at high temperatures?, <u>J</u>. <u>Cell</u>. <u>Biochem</u>. (in press).

12. P. F. Mullaney, Dry thermal inactivation of trypsin and ribonuclease, <u>Nature</u> 210:953 (1966).

13. A. Zaks and A. M. Klibanov, Enzymatic catalysis in organic media at 100°C, <u>Science</u> 224:1249 (1984).

14. M. J. Zoller and M. Smith, Oligonucleotide-directed mutagenesis: a simple method using two oligonucleotide primers and a single-stranded DNA template, <u>DNA</u> 3:479 (1984).

15. F. Sanger, S. Nicklen, and A. R. Coulson, DNA sequencing with chain-terminating inhibitors, <u>Proc</u>. <u>Nat</u>. <u>Acad</u>. <u>Sci</u>. USA 74:5463 (1977).

16. D. Botstein and D. Shortle, Strategies and applications of in vitro mutagenesis, <u>Science</u> 229:1193 (1985).

17. C. A. Hutchinson III et al., Mutagenesis at a specific position in a DNA sequence, <u>J</u> . <u>Biol</u>. <u>Chem</u>. 253:6551 (1978).

18. A. Razin, T. Hirose, K. Itakawa, and A. D. Riggs, Effective correction of a mutation by use of chemically synthesized DNA, <u>Proc</u>. <u>Nat</u>. <u>Acad</u>. <u>Sci</u>.<u>USA</u> 75:4268 (1978).

19. T. Alber and G. Kawasaki, Nucleotide sequence of the triose phosphate isomerase gene of Saccharomyces cerevisiae, <u>J</u>. <u>Mol</u>. <u>Appl</u>. <u>Genetics</u> 1:419 (1982).

20. G.A. Petsko et al.,Probing the catalytic mechanism of yeast triose phosphate isomerase by site-specific mutagenesis, <u>Biochemical</u> <u>Society</u> <u>Transactions</u> 12:229 (1984).

21. T. Alber et al., On the three-dimensional structure and catalytic mechanism of triose phosphate isomerase, <u>Phil</u>. <u>Trans</u>. <u>R</u>. <u>Soc</u>. <u>London</u> B293:159 (1981).

22. S. G. Waley, Refolding of triose phosphate isomerase,<u>Biochem</u>. <u>J</u>. 135:165 (1973).

23. P. M. Yuan, J. M. Talent, and R. W. Gracy Molecular basis for the accumulation of acidic isozymes of triosephosphate isomerase on aging, <u>Mechanisms</u> <u>of</u> <u>Ageing</u> <u>and</u> <u>Development</u> 17:151 (1981).

24. H. S. Lu, P. M. Yuan, and R. W. Gracy, Primary structure of human triosephosphate isomerase, <u>J</u>. <u>Biol</u>. <u>Chem</u>. 259:11958 (1984).

25. P. H. Corran and S. G. Waley, The amino acid sequence of rabbit muscle triose phosphate isomerase, <u>Biochem</u>. <u>J</u>. 145:335 (1975).

26. M. J. Furth, J. D. Milman, J. D. Priddle and R. Offord, Studies on the subunit structure and amino acid sequence of triose phosphate isomerase from chicken breast muscle, <u>Biochem</u>. <u>J</u>. 139:11 (1974).

27. E. Kolb, J. I. Harris, and J. Bridgen, Triose phosphate isomerase from the coelacanth, <u>Biochem</u>. <u>J</u>. 137:185 (1974).

28. E. Pichersky, L. Gottlieb, and J. F. Hess, Nucleotide sequence of the triose phosphate isomerase gene of Eschericia coli, <u>Mol</u>. <u>Gen</u>. <u>Genet</u>. 195:314 (1984).

29. S. Artavanis-Tsakonas and J. I. Harris, Primary structure of triosephosphate isomerase from Bacillus stearothermophilus, Eur. J. Biochem. 108:599 (1980).

30. J. L. Bada, Racemization of amino acids, in: "Chemistry and Biochemistr of Amino Acids," Chapman and Hall, London, pp. 399-413 (1985).

31. S. M. Steinberg, P. M. Masters, and J. L. Bada, The racemization of fre and peptide-bound serine and aspartic acid at 100°C as a function of pH implications for in vivo racemization, Biorg. Chem. 12:349 (1984).

32. H. Zuber, ed., "Enzymes and Proteins from Thermophilic Organisms," Birkhauser Verlag, Basel (1976).

33. M. F. Perutz, Electrostatic effects in proteins, Science 201:1187 (1978).

EXPRESSION OF HUMAN C5a IN E. COLI

Chen-Chen Kan, Yoshihiro Fukuoka, Tony E. Hugli and Georg H. Fey

Department of Immunology, Research Institute of Scripps Clinic 10666 North Torrey Pines Road, La Jolla, California 92037

INTRODUCTION

Host defense mechanisms against tissue damage, bacterial and viral infections, fungi, parasites and other foreign intruders rely on several systems. The best known are specific cellular and humoral immune responses, but other non-specific mechanisms are of comparable importance. After a first encounter with a novel infectious agent, mammalian organisms need one to two weeks to raise sufficient neutralizing antibody titers. In this time non-specific, broad-range systems such as the complement proteins play a critical role for the survival of the organism. The complex set of reactions characterized as the acute inflammatory response is another mechanism to control damage inflicted on the organism from the outside environment. Not surprisingly, some of these defense systems are interrelated. The focus of this study is the classical C5a anaphylatoxin peptide, which regulates both a number of important inflammatory processes and, indirectly, aspects of the immune response reaction.

In a localized infection or an inflammatory site, C5a is generated from complement component C5 by specific proteolytic cleavage of a single peptide bond. This cleavage is catalyzed by C5 convertase, a multicomponent enzyme of the classical and alternative pathways of complement activation (1,2). C5b, the larger fragment of C5, participates in late events of the complement cascade. The smaller fragment C5a, a 74 amino acid polypeptide, participates in inflammatory and immune-regulatory reactions (3,4,5). C5a exerts its functions by interacting with specific C5a receptors present on a number of cell-types, including mast cells, neutrophils, monocytes/macrophages and possibly others (6,7,8). Binding of C5a to mast cell receptors leads to the release of histamines, other preformed vasoactive amines and stimulates intracellular biosynthesis of arachidonate products. These substances locally regulate the blood vessel diameter and permeability. Binding of C5a to the neutrophil C5a receptor leads to changes in the surface adhesiveness of neutrophils and promotes emigration of these cells from the lumen of blood vessels, across the vessel wall, into the surrounding tissues. This directed movement of neutrophils against a gradient of the ligand is referred to as neutrophil chemotaxis. Binding of C5a to the neutrophil receptor also triggers the respiratory burst, a sequence of reactions leading to the release of oxygen free radicals including superoxide anion which mediate the potent bactericidal effect of neutrophils. Finally, binding of C5a to macrophage/monocyte C5a receptors is believed to trigger the release of

Table I: Biological Activities of C5a

Pulmonary Injury
Enhancement of vascular permeability
Contraction of smooth nuscle
Granulocyte chemotaxis
Neutrophil aggregation
Induction of cellular release of stored:
    enzymes
    vasoactive amines
    oxygen metabolites
    arachidonate metabolites
Indirect regulation of lymphocyte immune responses.

monokines, which in turn interact with lymphocytes and participate in the
regulation of immune responses (9,10). C5a is known to elicit responses from
yet other tissues, including smooth muscles, cardiac muscle and lung tissues,
which are probably also mediated through receptors. However, specific
saturable C5a receptors have until now only been demonstrated for the cell
types mentioned above. The most important biological responses mediated by
C5a are summarized in Table I and in references 3 to 5.

Thus, C5a is an important mediator in inflammation and host defense. It
can be expected that by modifying the peptide or the receptor, or by designing ligand-analogues, it is possible to create substances of pharmacologic
value in regulating inflammatory and immunologic responses. In view of these
long range objectives, it is essential to understand in greater detail the
three dimensional structure of the ligand and receptor, as well as the
stereochemistry of the binding interaction. Until now these studies have
been hampered by availability of C5a, which is a trace component in human
plasma. Therefore, we have attempted to generate C5a by genetic engineering
from cloned cDNA. By this approach it should be possible both to make
available sufficient quantities of C5a for structural analysis (crystallization, X-ray diffraction, nuclear magnetic resonance), and to create C5a
mutants. Specific mutagenesis will be useful in identifying amino acid
residues that are essential for receptor-ligand interactions, and thus for
C5a biological function. Here we report the production of recombinant human
C5a from cloned cDNA in E. coli.

MATERIALS AND METHODS

Cloning and Sequence Analysis of Human C5a cDNA

A synthetic oligonucleotide probe (a mixed 17mer with 32 sequence
permutations) was constructed from residues 67-72 of the published human C5a
sequence (11), which contains a methionine residue in this region. A human
liver cDNA library was screened with this probe. One cDNA clone representing
the coding region for C5a and several hundred base pairs (bp) on both sides
was isolated. The DNA sequence analysis was performed by subcloning random
fragments of 200-300 bp in length into the phage vector M13 and by the
dideoxy sequencing method (12-15).

Preparation and Purification of Synthetic Oligonucleotides

Oligonucleotides were synthesized on an Applied Biosystems DNA synthesizer, using phosphoramidites purchased from the same supplier. Synthetic
oliognucleotides were labeled with $[\gamma-^{32}P]$ ATP and polynucleotide kinase (PL
Laboratories), and analyzed by electrophoresis in polyacrylamide gels and
autoradiography. For screening cDNA libraries it was sufficient to use the

oligonucleotides without further purification steps. For use as mutagenesis-primers, oligonucleotides were enriched by preparative elution from poly-acrylamide gels.

Mutagenesis by the Double Primer Method

We have used two methods for oligonucleotide directed mutagenesis: a) the gapped plasmid approach (16), allowing mutagenesis to be carried out on the plasmid itself, which is used as the expression vector. This approach does not require intermediate transfer into an M13 vector; and b) the double primer method after transfer of sequences to be mutated into M13 phage vectors (17-19). The gapped plasmid approach was used successfully to introduce a 15 bp insertion, coding for an enterokinase cleavage site (see: Results). However, for deletions and small alterations the double primer method in M13 was used exclusively. One of the two primers was the universal M13 17mer primer LMB2 (13), the other was the specific mutagenesis primer.

Enrichment of Mutants by Kunkel's Method

In the early stages of this work mutants were identified in a conventional manner by plaque-hybridization using the radiolabeled mutagenesis primer as a hybridization probe (17-19). Later we found it advantageous to enrich mutants by Kunkel's procedure (20). In this procedure the M13 mutagenesis template DNA is propagated in E. coli BW313 (ung⁻), which produces progeny phage DNA containing uracil instead of thymine bases. Using this template mutagenesis is then carried out with the double primer method (see above), including dATP, dCTP, dGTP and TTP in the copying reaction. This creates a double stranded heteroduplex with the template strand containing uracil (U-DNA) and the copied strand thymine (T-DNA). The heteroduplex is used for transformation of E. coli CSH 50 (ung⁺) by the standard calcium chloride technique. This host strain degrades U-DNA, and enriches for T-DNA progeny. Random white plaques were picked after transformation of this host and subjected to sequence analysis by standard dideoxy sequencing. In two experiments we obtained close to 40% of the random progeny as mutants, in agreement with the frequencies obtained in the original report (20).


```
 T L Q K K I E E I A A K Y K H S V V K K C C Y D G
ACGCTGCAAAAGAAGATAGAAGAAATAGCTGCTAAATATAAACATTCAGTAGTGAAGAAATGTTGTTACGATGGA

                                          S
 A C V N N D E T C E Q R A A R I I L G P R C I K A
GCTTGCGTTAATAATGATGAAACATGTGAGCAGCGAGCTGCACGGATTATTTTAGGGCCAAGATGCATCAAAGCT

     Y
 F T E C C V V A S Q L R A N I S H K D M Q L G R
TTCACTGAATGTTGTGTCGTCGCAAGCCAGCTCCGTGCTAATATCTCTCATAAAGACATGCAATTGGGAAGG
```

Figure 1. cDNA Sequence and Derived Amino Acid Sequence of the Anaphylatoxin Peptide C5a. Amino acid residue 42 was reported to be a serine coded by AGT in one other published sequence (30). Residue 57 was reported as a tyrosine by sequence analysis of C5a isolated from pooled human sera (11), but found to be a valine in both published sequences derived from cDNA (12,30). Cloning and sequencing of cDNA was performed as described in Materials and Methods and reference 12. The underlined part of the sequence was used for the construction of synthetic oligonucleotides.

The cDNA sequence and derived amino acid sequence of C5a is shown in Figure 1. With the exception of one residue (valine instead of tyrosine at position 57) the sequence confirmed the published amino acid sequence of human C5a (11,12). This finding probably reflects the existence of a sequence polymorphism of C5a in the human population. The published protein sequence was determined from pooled plasma samples of many different donors, while the cDNA library was prepared with liver mRNA from one single donor. If the C5a type of this donor was a rare variant, then this variation would not have been detected in the pooled protein sequence. In a first series of experiments (first row in Figure 2), the cDNA sequences coding for C5a plus flanking sequences from the C-terminal portion of the C5β-chain and the N-terminal part of the C5α'-chain (labeled C5β' and C5α' in Figure 2) were

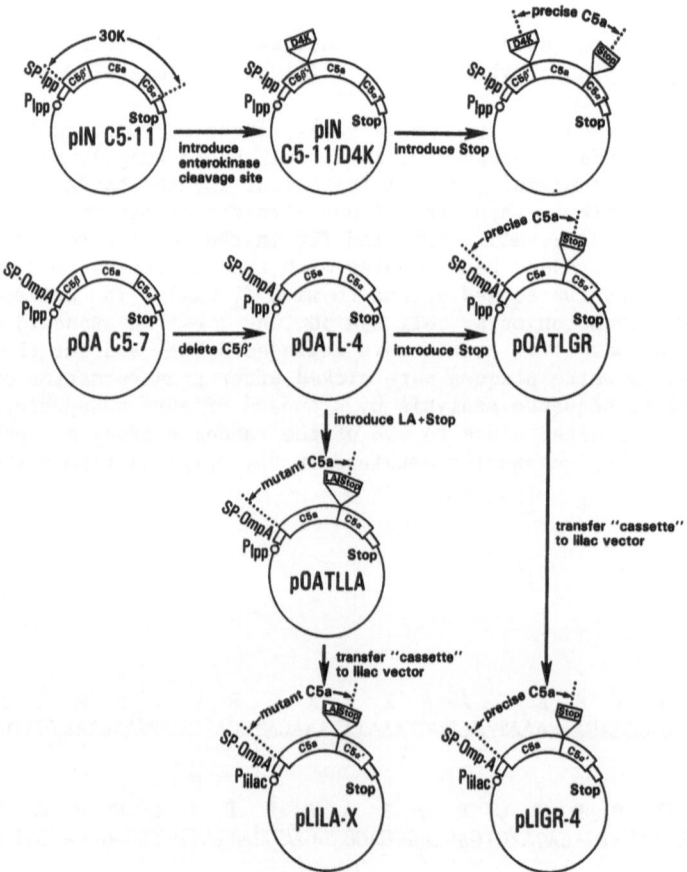

Figure 2. Engineering of Precise Human C5a in E. coli. Three series of constructs were prepared as described in the text. The top row shows constructs made in the first-generation vectors pINIIIC3, using the lipoprotein promoter and signal peptide (21). The second row shows the second series of constructs made in the second generation vectors. These vectors contained the lipoprotein promoter in combination with the ompA signal-peptide coding sequences and translational control sequences (24). The bottom row shows the third series of constructs, prepared in the latest generation of vectors, containing the stronger Lilac-promoter, a hybrid promoter between the lipoprotein and the lac-promoter (M. Inouye, personal communication).

inserted into the expression plasmid pINIII C3 (21). This plasmid vector carries the strong promoter P_{lpp} of the E. coli lipoprotein gene, followed by the translation-initiation sequences and the coding sequences for the signal peptide of the lipoprotein gene (SP-lpp) and a polylinker sequence as a cloning site. Downstream from this insertion site the vector contains coding sequences for the remainder of the lipoprotein. The promoter is preceded by a lac-operator sequence, inducible with isopropyl-thio-galactoside (IPTG) (21). A 618 bp SauIIIA cDNA fragment was inserted in both orientations into the BamHI site of the polylinker, creating constructs pINC5-10 and pINC5-11 (Figure 2, top row, left construct). From the sequence of pINC5-11 a 30 kdal recombinant peptide was expected, containing C5a in its central portion. Figure 3 demonstrates that the expected recombinant polypeptide was produced and that it was reactive with anti-human C5a serum. The recombinant polypeptide was only produced from pINC5-11 carrying the cDNA insert in the forward orientation (Figure 3, tracks 4 and 5), but not from pINC5-10 carrying the same insert in the reverse orientation (Figure 3, tracks 2 and 3). Although not well resolved, the 30 kdal polypeptide was visible on the coomassie brilliant blue stained SDS-polyacrylamide gel loaded with total protein extracts from E. coli cells harboring pINC5-11. We estimated this species to be in the 0.1 to 0.5 per cent range of total E. coli protein. From a quantative evaluation of the Western blot analysis, we estimated the content

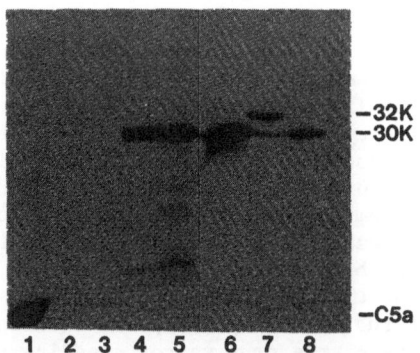

Figure 3. *Immunologic Reactivity of Recombinant C5a Polypeptides with anti-C5a Sera. E. coli DH 1 cultures containing constructs pINC5-10 (tracks 2,3) pINC5-11 (tracks 4,5) and pOAC5-7 (three independent isolates, tracks 6,7,8) were grown to mid-logarithmic phase. Then some cultures were induced with IPTG at a final concentration of 5 mM (tracks 3,5,6 to 8) and others (tracks 2,4) were left without the inducer. After an additional growth of 4 hours in presence of the inducer, the cells were pelleted, freeze-thawed and sonicated. Cells from 1 ml of culture were then resuspended in SDS-sample buffer, containing β-mercaptoethanol, and boiled for 2 minutes. After removal of insoluble debris by centrifugation, the protein extracts were electrophoresed in an 11% SDS-polyacrylamide gel. After electrophoresis, the proteins were electrophoretically transferred to nitrocellulose filters (Western blots). The blots were developed by incubation with rabbit anti human C5a serum and goat anti rabbit immunoglobulin (IgG-fraction), conjugated with horse radish peroxidase. Color development was achieved with BioRad HRP color development reagent, containing 4 chloro-1-naphthol. Track 1 contains 2 μg of authentic human C5a, purified from plasma.*

of this 30 kdal polypeptide to be 2-3 milligrams per liter of E. coli culture in late logarithmic phase, after induction with 5 mM IPTG for 4 hours.
In order to create C5a with an authentic N-terminus, we planned to insert
coding sequences for an enterokinase cleavage site, Asp-Asp-Asp-Asp-Lys (D4K,
22,23) in front of the C5a coding sequence, and to process the purified 30
kdal protein with enterokinase in vitro. To this effect, a 15 bp insertion
was successfully placed into the plasmid pINC5-11, creating pINC5-11/D4K
(Figure 2, top row, middle construct). This mutagenesis was performed by
using the gapped plasmid approach (16). The plan was to further introduce a
stop codon behind the coding sequences for C5a, as shown in the right part of
the top row of Figure 2. However, this construct has not yet been engi-
neered, because more promising second generation expression plasmids had
become available from Dr. Inouye's laboratory in the meantime. These new
vectors (Figure 2, second row) are derived from the pIN series, and still
carry the strong P_{1pp} promoter. As a new feature they contain the translation initiation sequences and the coding sequences for the signal peptide
(SP-ompA) of the E. coli outer membrane protein ompA (24). These vectors
were chosen because they promised two important advantages: a) It was known
that the signal peptidases responsible for the removal of the ompA signal
peptide in E. coli could recognize and cleave at hybrid cleavage sites. The
C-terminus of the ompA signal peptide was apparently sufficient to determine
specificity of these proteases. The N-terminus of other sequences following
downstream from the signal peptide appeared not to influence specificity
(24). The expectation therefore was that these signal peptidases would
recognize the designed cleavage site between the ompA signal peptide and the
N-terminus of mature C5a, thus leaving authentic C5a without an additional
methionyl residue at its N-terminus. Other engineering approaches would have
to solve this problem, because C5a is an internal peptide fragment of C5, and
thus is not normally preceded by a translation initiating methionyl residue.
b) It was expected that the vectorial transport, mediated by the ompA signal
peptide, should lead to accumulation of recombinant C5a in the periplasmic
space of E. coli. This was a desirable property, because authentic C5a
contains three internal disulfide bridges. The cytosol of E. coli is a
reducing environment, not likely to allow correct formation of these bridges.
Thus, C5a produced in the cytosol of E. coli would probably need additional
treatment to introduce these bonds correctly. The periplasmic space is less
reducing and it was therefore anticipated that the disulfide bridges might be
formed properly in this compartment without the need for additional preparative steps. Finally, it was also expected that the periplasmic space may
contain less proteinases then the cytosol, and that it may be easier to
recover substantial quantities of intact C5a from this compartment. Thus,
the cassette coding for the 30 kdal polypeptide was transferred from pINC5-11
into the polycloning site of the ompA vector, leading to the construct pOA
C5-7 (Figure 2, second row, left construct). From this construct two forms
of a recombinant, C5a-containing polypeptide, were obtained: a 30 kdal and a
32 kdal form (Figure 3, tracks 6-8). These forms were observed in E. coli
cells carrying several independent isolates of the same pOA C5-7 construct.
It is therefore thought that the predominance of one form over the other is
not a function of the construct, but of the state of the E. coli cells in
these different cultures at the time of harvest. The interpretation is that
the 32 kdal form probably represents the recombinant polypeptide still containing the ompA signal peptide at its N-terminus, and the 30 kdal form
represents the protein after successful removal of the signal peptide. This
interpretation needs verification by direct N-terminal sequence analysis on
the purified peptide. We have not attempted to perform this analysis on the
30 kdal recombinant polypeptide, because we wish to perform it at a later
stage, when authentic C5a is obtained in E. coli. However, the result was
suggestive that the signal peptide could indeed be removed at or very near
the desired position. The sequences coding for the C5β' portion and the
polylinker sequences were removed by deletion mutagenesis with synthetic
oligonucleotides complementary in one part to the ompA signal peptide coding

sequences and in the other part to the coding sequences for the N-terminus of
C5a. Mutagenesis was performed with the double primer method and mutants
were enriched with Kunkel's procedure (see: Material and Methods). The
resultant constructs were sequenced (pOATL-4, second row in Figure 2, middle
construct) to verify the correct construction. The critical part of the
sequence, showing the newly generated precise junction between the ompA
signal peptide and the N-terminus of C5a is shown in Figure 4. The final
step in this second series of constructions was to introduce a translation
termination stop codon following the triplet for the C-terminal Arg74 of C5a.
This was achieved by oligonucleotide directed insertion mutagenesis with the
double primer method and enrichment of mutants by Kunkel's procedure, leading
to construct pOATLGR (Figure 2, second row, right construct). From this construct an authentic sized, 9 kdal recombinant C5a polypeptide, specifically
immunoreactive on Western blots with anti C5a-sera, was obtained (data not
shown). However, the content of this peptide in E. coli cultures was very
low, estimated to be less than 5 µg per liter of E. coli culture. In the
meantime, a third generation of expression plasmids had become available from
Dr. Inouye's laboratory. This new series carried a stronger promoter, the
LILAC-promoter, a hybrid between the lipoprotein and the E. coli lac-operon
promoter, which was reported to allow overproduction of foreign polypeptides
in E. coli at five to ten times higher concentrations than the vectors
discussed so far (Inouye, personal communication). Therefore, the cassette
of DNA-sequences leading to the expression of authentic C5a from pOATLGR was

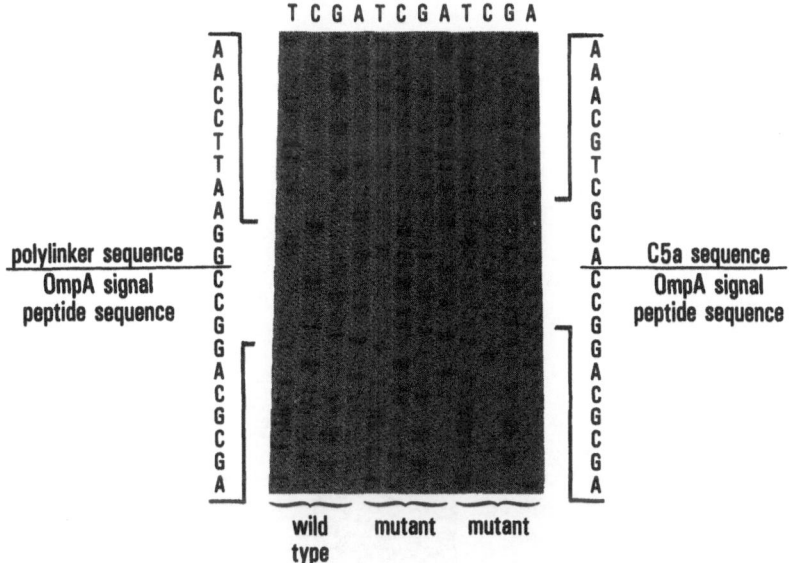

Figure 4. Verification of Mutagenesis by DNA-Sequence Analysis. The left
four tracks, (labeled wild-type at the bottom) show the sequence of construct
pOAC5-7 (Figure 2, second row, left construct) at the transition between the
sequence coding for the C-terminus of the ompA signal peptide and the polylinker sequence, preceding the C5β' coding sequence. The next two groups of
four tracks, labeled mutant at the bottom, show the corresponding region of
construct pOATL-4, (Figure 2, second row, middle construct), after successful
deletion of the polylinker sequence and the C5β'-sequences. The last
residues coding for the ompA signal peptide are AGGCC. Then follow the
triplets ACG CTG CAA coding for the first three amino acids TLQ of C5a (see
Figure 1). Deletion mutagenesis was performed with the synthetic 36mer GCT
ACC GTA GCG CAG GCC ACG CTG CAA AAG AAG ATA by the double primer method
followed by Kunkel's enrichment procedure. DNA sequence analysis with the
chain termination method was as described in Materials and Methods.

inserted into a LILAC-vector, leading to the construct pLIGR-4, shown on the bottom right of Figure 2. As expected, from this construct, expression of a 9 kdal (authentic size) C5a recombinant polypeptide was obtained in E. coli at several-fold higher concentrations than from pOATLGR (Figure 5). A series of experiments has been initiated, but not yet completed, to optimize the yield of recombinant human C5a from pLIGR-4 carrying E. coli cells. As judged by quantitative Western blot analysis with authentic human C5a as a standard, the yield of the 9 kdal polypeptide was 16 μg per liter of culture, harvested in the late logarithmic growth phase after an induction with 2.5 to 5 mM IPTG for 4 hours.

The extraction of this polypeptide from E. coli, the purification procedure, its amino acid sequence and its biological properties are currently under study. This peptide has until now been identified as recombinant C5a on the basis of a) its specific reactivity with anti-C5a serum; b) its electrophoretic mobility in SDS-polyacrylamide gels, which is indistinguishable from authentic C5a; and c) its expression in an orientation-dependent, IPTG inducible manner from a plasmid construct which was shown by DNA sequencing to contain the precise coding sequences for C5a.

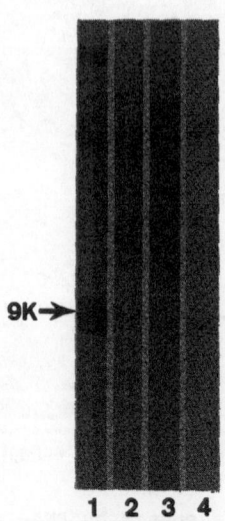

9K→

1 2 3 4

Figure 5. Immunologic Reactivity of Authentic Sized Recombinant Human C5a. E. coli CAG 456 cultures containing the Lilac-vector as a control (track 2) or the construct pLIGR-4 (tracks 3,4) (see Figure 2, bottom row, right construct) were grown to mid-logarithmic phase. Then one culture (track 4) was induced with 5 mM IPTG, the other (track 3) was left without the inducer. After an additional growth for 4 hours in the presence of the inducer, the cultures were harvested and processed for Western blot analysis as described in Figure 3. Track 1 contains 2 μg of authentic human C5a, purified from human blood plasma. Note the presence of two C5a-reactive molecular species in the high-molecular weight region of tracks 3 and 4. These species were repeatedly observed and are believed to originate from attachment of C5a to other cellular proteins by disulfide interchange from the unpaired cysteinyl residue (residue 27 in Figure 1).

DISCUSSION

In order to produce substantial amounts of small peptides (less than 100 amino acids), several choices are available ranging from chemical peptide synthesis over expression in E. coli, and expression in yeast to expression in animal cells. At the time when this project was initiated, chemical synthesis and purification of peptides over 50 residues in length was not sufficiently advanced to allow elimination of synthesis products differing from the desired product in only a few residues. This situation has improved over the last two years, and chemical synthesis is now a prime choice. However, in the size range of 70 to 100 residues, genetic engineering presently still is a very viable option, in particular, if it is intended to engineer a large number of mutants. Only a few laboratories are presently able to synthesize and purify fault-free 74 amino acid polypeptides. It was known that a typical obstacle encountered upon expression studies in E. coli is the propensity of E. coli proteinases to degrade foreign polypeptides. The severity of this effect is greater with shorter polypeptides. Therefore, the initial constructs were designed to produce recombinant polypeptides of over 100 amino acid residues in length (the 30 kdal polypeptide). These constructs were designed to optimize the choice of the vector, promoters, control signals, growth conditions and E. coli host strains. Once a good expression system was found, it was intended to trim the polypeptide to its final shorter size. This plan was executed with partial success. We have tested several expression plasmids first, including pUC7 (25), pAS1 (26) and the pIN-vector series. Expression was obtained only from the pIN-vectors, although pAS1 is an excellent general expression vector, successfully employed in many other laboratories. This exemplifies the current state of knowledge in the field. For any given sequence to be expressed, it is at present not possible to predict on theoretical grounds, which is the most suitable expression system. Several systems need to be tried, and it is common to optimize the system which works. Similarly, in order to reduce the problems caused by proteolytic degradation, several E. coli strains were tested, including DH1 (27), JM101 (28) and E. coli CAG 456 (29). Optimal results were obtained with CAG 456, and this strain is currently used for our pilot studies at the 50 liter scale. Still, the yields are not satisfactory. As shown in Table II, the yields decreased from 2-3 mg per liter for the 30 kdal polypeptide to 16 µg per liter for the 9 kdal polypeptide with intermediate yields for intermediate size products. In order to facilitate the purification from E. coli and the recovery of sufficient amounts for structural and functional studies, it would be an advantage, to improve the yields by one to two orders of magnitude. Current experiments are underway to achieve this goal. However, if within a reasonable time these goals cannot be reached, alternatives will be considered, including the expression in eucaryotic cells. If the yields can be substantially improved, then the constructs discussed here should provide a useful starting point for mutagenesis experiments and the elucidation of the structure/function relationship of this important effector molecule.

Table II. Content of C5a-Recombinant Polypeptides in E. coli [a]

Construct	Size of recombinant polypeptide	Content per liter of E. coli culture
pINC5-11	30 kdal	2-3 mg
pLIGR-4	9 kdal	16 µg

Contents evaluated by comparison with authentic C5a standards from Western blot analysis.

ACKNOWLEDGEMENTS

We thank Drs. T. Belt, M.C. Carroll, R.R. Porter and D. Stafford for
kindly making human liver cDNA libraries available, and Drs. M. Inouye, A.
Shatzman and M. Rosenberg for generously providing us with expression
plasmids. The expert technical assistance of M. Gehring, J. Wagner and M.
Kawahara is gratefully acknowledged. We thank K. Dunn for the preparation of
this manuscript. C.-C. K. was supported by a training grant from the
National Institute of Health, awarded to Dr. H. Müller-Eberhard (grant number
HL 07915). This work was supported by NIH research grants number AI19651 and
number AI17354 awarded to G.H.F. and T.E.H. This is publication number
4126-IMM from the Department of Immunology, Research Institute of Scripps
Clinic.

REFERENCES

1. T. E. Hugli and H. J. Müller-Eberhard, Anaphylatoxins: C3a and C5a,
   Adv. Immunol. 26:1 (1978).
2. U. R. Nilsson, R. J. Mandle, Jr. and J. A. McConnell-Mapes, Human C3 and
   C5: Subunit structure and Modification by Trypsin and C42-C423, J.
   Immunol. 114:815 (1975).
3. T. E. Hugli, The Structural Basis for Anaphylatoxin and Chemotactic
   Functions of C3a, C4a, and C5a, CRC Reviews in Immunol. 1:32 (1981).
4. T. E. Hugli, Bioactive Factors of the Blood Complement System, in:
   "Proteins in Biology and Medicine," R. A. Bradshaw, R. L. Hill and J.
   Tang, eds., Academic Press, New York (1982).
5. T. E. Hugli, Structure and Function of the Anaphylatoxins, Springer
   Semin. Immunopathol. 7:193 (1984).
6. D. E. Chenoweth and T. E. Hugli, Demonstration of a Specific C5a
   Receptor on Intact Human Polymorphnuclear Leukocytes, Proc. Natl.
   Acad. Sci. USA 75:3943 (1978).
7. D. E. Chenoweth, M. G. Goodman and W. O. Weigle, Demonstration of a
   Specific Receptor for Human C5a Anaphylatoxin on Murine Macrophages,
   J. Exp. Med. 156:68 (1982).
8. T. E. Rollins and M. S. Springer, Identification of the C5a Receptor of
   Human Polymorphnuclear Leukocytes, Fed. Proc. 44:1767 (1985).
9. W. O. Weigle, E. L. Morgan, M. G. Goodman, D. E. Chenoweth and T. E.
   Hugli, Modulation of the Immune Response by Anaphylatoxin in the
   Microenvironment of the Interacting Cell, Fed. Proc. 41:3099 (1982).
10. W. O. Weigle, M. G. Goodman, E. L. Morgan and T. E. Hugli, Regulation of
    the Immune Response by Components of the Complement Cascade and their
    Activated Fragments, Springer Semin. Immunopathol. 6:173 (1983).
11. H. N. Fernandez and T. E. Hugli, Primary Structural Analysis of the
    Polypeptide Portion of Human C5a Anaphylatoxin, J. Biol. Chem.
    253:6955 (1978).
12. G. H. Fey, M. H. L. de Bruijn, C.-C. Kan, A. Chain, M. R. Gehring, E.
    Solomon and H. Leffert, C3, C5 and Alpha$_2$-macroglobulin: cDNA
    Sequences and Regulation of mRNA Concentrations in Acute Inflamma-
    tion, in: "Advances in Gene Technology: Molecular Biology of the
    Immune System," J. W. Streilein, ed., Cambridge University Press,
    Cambridge (1985).
13. A. T. Bankier and B. G. Barrell, Shotgun DNA Sequencing, in: "Techniques
    in Nucleic Acid Biochemistry," R. A. Flavell, ed., Elsevier/North
    Holland Scientific Publishers, Ltd., Ireland (1983).
14. F. Sanger, S. Nicklen and A. R. Coulson, DNA Sequencing with Chain-
    Terminating Inhibitors, Proc. Natl. Acad. Sci. USA 74:5463 (1977).
15. F. Sanger, A. R. Coulson, B. G. Barrell, A. J. H. Smith and B. A. Rose,
    Cloning in Single-stranded Bacteriophage as an Aid to Rapid DNA
    Sequencing, J. Mol. Biol. 143:161 (1980).

16. Y. Morinaga, T. Franceschini, S. Inouye and M. Inouye, Improvement of Oligonucleotide-Directed Site-Specific Mutagenesis Using Double-Stranded Plasmid DNA, Biotechnology, July (1984).

17. M. J. Zoller and M. Smith, Oligonucleotide-Directed Mutagenesis of DNA Fragments Cloned into M13 Vectors, in: "Methods in Enzymology," S. P. Colowick and N. O. Kaplan, eds., 100:468 (1983).

18. K. Norris, F. Norris, L. Christiansen and N. Fiil, Efficient Site Directed Mutagenesis by Simultaneous Use of Two Primers, Nucl. Acids Res. 11:5103 (1983).

19. M. J. Zoller and M. Smith, Oligonucleotide-Directed Mutagenesis: A Simple Method Using Two Oligonucleotide Primer and a Single-Stranded DNA Template, DNA 3:479 (1984).

20. T. A. Kunkel, Rapid and Efficient Site-Specific Mutagenesis Without Phenotypic Selection, Proc. Natl. Acad. Sci. USA, 82:488 (1985).

21. Y. Masui, J. Coleman and M. Inouye, Multipurpose Expression Cloning Vehicles in Escherichia Coli, in: "Experimental Manupulation of Gene Expression," M. Inouye, ed., Academic Press, New York (1983).

22. E. W. Davie and H. Neurath, Identification of a Peptide Released During Autocatalytic Activation of Trypsinogen, J. Biol. Chem. 212:515 (1955).

23. J. W. Brodrick, C. Largman, M. W. Hsiang, J. H. Johnson and M .C. Goekas, Structural Basis for the Specific Activation of Human Cationic Trypsinogen by Human Enteropeptidase, J. Biol. Chem. 253:2737 (1978).

24. J. Ghrayeb, H. Kimura, M. Takahara, H. Hsiung, Y. Masuri and M. Inouye, Secretion Cloning Vectors in Escherichia Coli, The EMBO Journal, 3:2437 (1984).

25. J. Messing and J. Vieira, A New Pair of M13 Vectors for Selecting either DNA Strand of Double-Digest Restriction Fragments, Gene 19:269 (1982).

26. M. Rosenberg, Y.-S. Ho and A. Shatzman, The Use of pKC30 and its Derivatives for Controlled Expression of Genes, in: "Methods in Enzymology," S. P. Colowick and N. O. Kaplan, eds., Academic Press, New York, 100:123 (1983).

27. T. Maniatis, E. F. Fritsch and J. Sambrook, in: "Molecular Cloning, A Laboratory Manual," Cold Spring Harbor Laboratory, Cold Spring Harbor (1982).

28. J. Messing, New M13 Vectors for Cloning, in: "Methods in Enzymology," S. P. Colowick and N. O. Kaplan, eds., Academic Press, New York 101:20 (1983).

29. T. A. Baker, A. D. Grossmann and C. A. Gross, A Gene Regulating the Heat Shock Response in Escherichia Coli also Affects Proteolysis, Proc. Natl. Acad. Sci. USA 81:6779 (1984).

30. A. B. Lundwall, R. A. Wetsel, T. Kristensen, A. S. Whitehead, D. E. Woods, R. C. Ogden, H. R. Colten and B. F. Tack, Isolation and Sequence Analysis of a cDNA Clone Encoding the Fifth Complement Component, J. Biol. Chem. 260:2108 (1985).

XI. Active Site Studies

USE OF TRINITROBENZENE SULFONATE TO DETERMINE THE pK_a VALUES OF TWO ACTIVE-SITE LYSINES OF RIBULOSEBISPHOSPHATE CARBOXYLASE/OXYGENASE[1]

Fred C. Hartman, Sylvia Milanez, and Eva H. Lee[2]

Biology Division, Oak Ridge National Laboratory and the University of Tennessee—Oak Ridge Graduate School of Biomedical Sciences, Oak Ridge, Tennessee 37831

INTRODUCTION

Ribulose-P_2 carboxylase[3] (E.C. 4.1.1.39), which catalyzes the carboxylation of ribulose-P_2 to yield two molar equivalents of D-3-phosphoglycerate, provides the only significant route by which atmospheric CO_2 is converted to carbohydrate and hence is absolutely essential to all higher life forms (see ref. 1 for a thorough review). The carboxylation reaction is suppressed by O_2, which is both a competitive inhibitor (2) and a substrate of ribulose-P_2 carboxylase (3). In the presence of O_2, the enzyme catalyzes the conversion of ribulose-P_2 to one molar equivalent each of phosphoglycolate and 3-phosphoglycerate. Although multiple substrate specificities among enzymes are not unusual, the bifunctionality of ribulose-P_2 carboxylase is perhaps unprecedented in that the two reactions catalyzed are the initial steps in competing metabolic pathways — photosynthetic assimilation of CO_2 and photorespiration, an energy wasteful process without a known function which results in the release of previously fixed CO_2. Suppression of the carboxylase reaction by O_2 clearly accounts for the long-recognized inhibition of photosynthesis by O_2, i.e. the Warburg effect (4-8). Minimization of the Warburg effect by cultivating plants under controlled atmospheres of reduced O_2 or elevated CO_2 levels results in substantially enhanced growth rates and yields (9). Obviously, if the carboxylase/oxygenase ratio of ribulose-P_2 carboxylase could be enhanced either by genetic engineering or by chemical treatment, agronomic benefits would ensue.

[1] The Research from the authors' laboratory was sponsored jointly by the Office of Health and Environmental Research, U.S. Department of Energy under contract DE-AC05-84OR21400 with the Martin Marietta Energy Systems, Inc. and by Grant PCM-8207516 from the National Science Foundation.

[2] Postdoctoral investigator supported by Grant PCM-8207516 from the National Science Foundation.

[3] Abbreviations used are: ribulose-P_2 carboxylase, D-ribulose 1, 5-bisphosphate carboxylase/oxygenase; TNBS, trinitrobenzene sulfonate; TNP, trinitrophenyl; HPLC, high-performance liquid chromatography.

A consensus concerning the feasibility of improving the carboxylase/ oxygenase ratio is lacking. Encouraging observations include differential effects by divalent cations (however, in all cases the carboxylase activity relative to the oxygenase is suppressed) (10,11) and the increased efficiency as a carboxylase that has occurred during evolution (12). The formidability of "engineering" a superior carboxylase is emphasized by the realization that both activities are catalyzed by the same active site and that both carboxylation and oxygenation pathways proceed from a common reaction intermediate, the C2-C3 enediol derived from ribulose-P_2 (1) (see Discussion). Evidence is lacking for an essential role of any active-site residue exclusively to one of the two activities.

Like many investigators, we believe that the central importance of ribulose-P_2 carboxylase to photosynthate yields justifies a thorough exploration of its structure and mechanism and that without such fundamental information the feasibility of modulating the enzyme's activities by external means cannot be rigorously evaluated.

As one facet of elucidating the enzyme's mechanism, we have focused on active-site characterization through the use of affinity labels. The structures of five different reagents, which exhibit the usual characteristics of affinity labels, and the primary targets of their reactions with ribulose-P_2 carboxylase from both spinach and R. rubrum are shown in Fig. 1. Given the structural complexity of ribulose-P_2 carboxylase and the lack of absolute specificity of the affinity labels used, we have relied on comparative sequence analyses to reveal whether residues

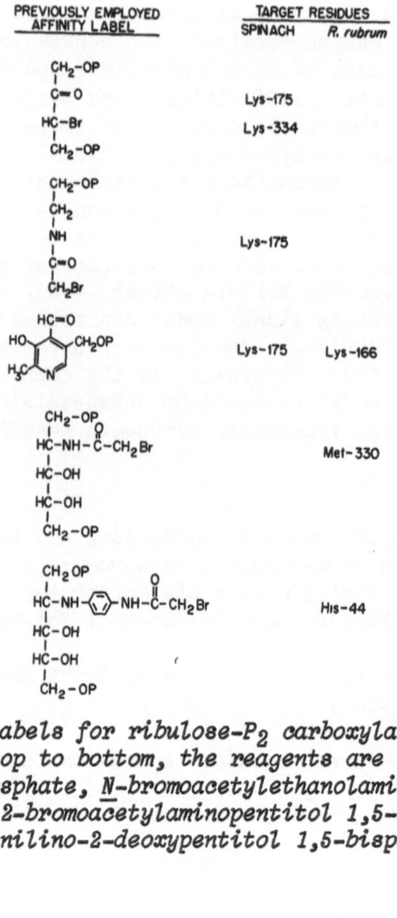

Figure 1: Affinity labels for ribulose-P_2 carboxylase and the targets of modification. From top to bottom, the reagents are 3-bromo-1,4-dihydroxy-2-butanone 1,4-bisphosphate, N-bromoacetylethanolamine phosphate, pyridoxal phosphate, 2-bromoacetylaminopentitol 1,5-bisphosphate, and 2-(4-bromoacetamido)anilino-2-deoxypentitol 1,5-bisphosphate.

implicated at the active-site by affinity labeling are indeed species invariant and thus likely to be essential to function (13).

Because of their evolutionary diversity and structural dissimilarities, the carboxylases from spinach and R. rubrum provide a stringent test of structural conservation. Whereas the quaternary structure of ribulose-P_2 carboxylase from spinach is typical of that observed among all higher plant and most bacterial carboxylases in that it is a hexadecamer with eight large (53,000-Da) and eight small (14,000-Da) subunits (14-16), the functionally analogous enzyme from the purple, non-sulfur photosynthetic bacterium R. rubrum is a homodimer of 53,000-Da subunits (13,17,18). Furthermore, despite >80% sequence homology among most ribulose-P_2 carboxylases (1), the homology between the R. rubrum enzyme and the large subunit of the spinach carboxylase is only 31% (13).

Studies (see ref. 13 for a review) with the affinity labels shown in Fig. 1 have provided evidence for at least three regions of primary structure comprising the catalytic site. In each case, the residues, for which sound chemical evidence has suggested functional importance, are located within highly conserved regions among diverse species (Fig. 2). Lys-175 in the spinach enzyme is selectively labeled by three different reagents — 3-bromo-1,4-dihydroxy-2-butanone 1,4-bisphosphate, N-bromoacetylethanolamine phosphate, and pyridoxal 5'-phosphate. The corresponding residue in the R. rubrum enzyme, Lys-166, is modified by pyridoxal phosphate with a high degree of specificity. An identical seven-residue sequence encompasses the reactive lysyl residue in the two species. A second active-site region contains Lys-334 in spinach

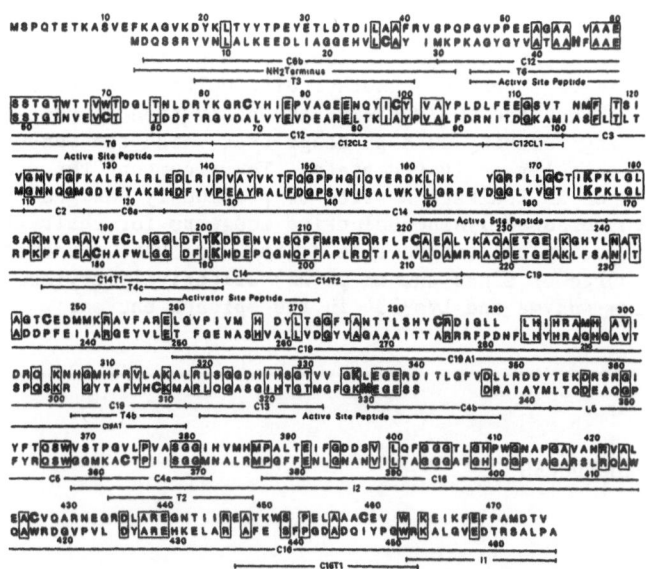

Figure 2: Amino acid sequences of ribulose-P_2 carboxylase from spinach (15) (upper) and R. rubrum (13) (lower). Alignments were made by visual inspection. Residues identical in both enzymes are enclosed in boxes. Gaps, attributed to deletions or insertions, appear as blank spaces. Peptide notations are explained in ref. 13. Cysteinyl residues and residues identified at the active site or activator site by selective chemical labeling are illustrated with larger type. Abbreviations: Asp, D; Asn, N; Glu, E; Gln, Q; Thr, T; Ser, S; Pro, P; Gly, G; Ala, A; Cys, C; Val, V; Met, M; Ile, I; Leu, L; Tyr, Y; Phe, F; His, H; Lys, K; Trp, W; Arg, R. (Reprinted from ref. 13 with permission of publisher.)

carboxylase and Met-330 in the R. rubrum enzyme. The former is labeled by 3-bromo-1,4-dihydroxy-2-butanone 1,4-bisphosphate and the latter by 2-bromoacetylaminopentitol 1,5-bisphosphate. Although Met-330 is replaced by Leu-335 in the spinach enzyme and hence cannot function in catalysis, Lys-334 and eight other residues in its immediate vicinity are conserved. Thus, two different lysyl residues (Lys-175 and Lys-334 in the spinach enzyme, which correspond to Lys-166 and Lys-329 in the R. rubrum enzyme) appear essential. A region encompassing His-44 in R. rubrum carboxylase is implicated by the selective labeling of this residue with 2-(4-bromoacetamido)anilino-2-dexoypentitol 1,5-bisphosphate. The residue that is subject to selective alkylation by the affinity label apparently does not function in catalysis, because it is deleted in the enzyme from spinach. However, 11 of the 14 residues from positions 40 through 54 are conserved, strongly suggestive of functional participation of at least one of the invariant residues.

Another region of high homology is seen at the activator site (labeled "activator-site" peptide in Fig. 2). All ribulose-P_2 carboxylases require CO_2 and Mg^{2+} for activation; the process entails condensation of CO_2 with a lysyl ε-amino group (Lys-201 in the spinach enzyme) to form a carbamate that is stabilized by Mg^{2+} (19,20). Chemically, the process is the same for the R. rubrum enzyme (21) with the lysyl residue at position 191 as the CO_2 acceptor.

Because assignment of a residue to the active site does not define its function (if any), we have sought to determine the pK_a values of the two essential lysines (Lys-166 and Lys-329 in the R. rubrum enzyme and Lys-175 and Lys-334 in the spinach enzyme) and thereby place constraints on their possible roles. In principle, the pH-dependence of inactivation by the affinity labels would provide the pK_a values of the targeted residues, but this approach is complicated by the pH-dependence of the reagents' binding affinities due to ionization of the phosphate groups over the pH-range of interest. Another difficulty, irrespective of reagent used, is that the CO_2/Mg^{2+}-induced activation is pH-dependent (22) necessitating focusing on a pH-range over which the enzyme can be maintained in its fully activated state. To circumvent the problem imposed by changes in ionization state of previously designed affinity labels, we have screened lysine-selective reagents for their ability to preferentially modify the active-site lysines and have discovered that TNBS (with a charge of minus one at pH >2) selectively arylates Lys-166 in the R. rubrum enzyme and Lys-334 in the spinach enzyme.

EXPERIMENTAL PROCEDURES

Materials

Ribulose-P_2 carboxylases from spinach and R. rubrum were purified as described earlier (23,18) and were stored at -70°C as concentrated stocks (>40 mg/ml) in activation buffer (50 mM Bicine/66 mM NaHCO$_3$/10 mM MgCl$_2$/1 mM EDTA, pH 8.0) that contained 20% (v/v) glycerol. As needed, aliquots of the stocks were dialyzed at room temperature against either the activation buffer or deactivation buffer (50 mM Bicine/1 mM EDTA, pH 8.0) to which was added Chelex-100 resin (10 g/liter of buffer) to ensure complete removal of trace metal ions. Protein concentrations were based on absorbancies at 280 nm using $\varepsilon^{1\%} = 12.0$ for the R. rubrum enzyme (18) and $\varepsilon^{1\%} = 16.4$ for the spinach enzyme (23). Their respective specific activities were 5.5 units/mg and 1.6 units/mg.

Methods

Assays for Ribulose-P_2 Carboxylase. Slight modification (18) of the spectrophotometric method of Racker (24) was used to monitor carboxylase activity during treatments with TNBS. The direct $^{14}CO_2$-fixation assay (25) was used in the determination of the degree of activation (i.e. carbamate stability) as a function of pH. All assays were performed at 25°C.

Kinetics of Reactions of TNBS with Ribulose-P_2 Carboxylase and with N-α-Acetyllysine. Reactions with enzymes were carried out in septum-sealed glass reaction vials (capacity = 4.2 ml) maintained at 25 ± 0.01°C. Final protein concentrations in the reaction mixtures (4.0 ml) were 0.3 mg/ml (5.7 μM subunit in the case of the R. rubrum enzyme and 4.55 μM large·small subunit pair in the case of the spinach enzyme); final TNBS concentrations were 0.2 mM and 0.1 mM for the R. rubrum and spinach enzymes, respectively. Stocks of TNBS were standardized by reaction with excess N-α-acetyllysine and observing product formation spectrophotometrically at 367 nm ($\varepsilon = 1.1 \ 10^4$) (26).

An aliquot of the concentrated enzyme stock (>40 mg/ml in either activation or deactivation buffer, see "Materials") was delivered by microsyringe into the sealed reaction vial which contained the buffer of desired pH. The reaction was initiated by the addition, with a microsyringe, of 40 μl or 80 μl of 10 mM TNBS. Identical reaction mixtures that lacked TNBS served as controls. The TNBS-inactivations of the activated forms of both enzymes were monitored by periodic withdrawal of 5-25 μl aliquots of the reaction mixtures and direct transfer into assay cuvettes. In the case of deactivated spinach enzyme, aliquots of the reaction mixtures were combined with an equal volume of activation buffer that contained 0.01 M 2-mercaptoethanol and assayed one hour later. This approach provided instantaneous quenching of the TNBS and subsequent activation of the native enzyme that remained at the time of sampling. A ± 5% reproducibility was assumed for each given assay, and standard least-squares fitting was used in all calculations.

Reactions of TNBS with both of the activated enzymes were carried out in the following buffer systems: pH 6.5-7.5, 50 mM Pipes/66 mM NaHCO$_3$/10 mM MgCl$_2$; pH 7.5-9.5, 50 mM Bicine/66 mM NaHCO$_3$/10 mM MgCl$_2$; pH 9.5-10.5, 66 mM NaHCO$_3$/10 mM MgCl$_2$. Reactions of TNBS with the deactivated spinach enzyme were carried out in the following buffer systems: pH 6.5-7.5, 50 mM Pipes/0.1 mM EDTA; pH 7.5-9.5, 50 mM Bicine/ 0.1 mM EDTA; pH 9.5-10.5, 50 mM Ches/0.1 mM EDTA. Prior to addition of EDTA to these buffers, they were treated with Chelex-100 resin (10 g/ liter) and filtered as a precaution against trace metal contamination. The ionic strengths of all buffers used to evaluate the pH-dependencies of the enzyme inactivations by TNBS were adjusted with 3 M NaCl to μ = 0.13. The pH 6.5-7.5 buffers that contained bicarbonate were prepared just prior to use because the pH was unstable over long periods due to evolution of CO$_2$. However, with freshly prepared buffers and sealed reaction vials with little air space, the pH values were unaltered during the short time-courses of the reactions.

Reactions of TNBS (1 mM) with N-α-acetyllysine (0.1 mM) at 25°C were continuously monitored at 367 nm. Stock solutions of N-α-acetyllysine were standardized by amino acid analysis following hydrolysis. Buffer compositions and ionic strengths were the same as those in which the enzyme inactivations were carried out. Neither the rates of arylation of N-α-acetyllysine by TNBS nor the extinction coefficient at 367 nm of the reaction product was altered by the presence of Mg^{2+} in the buffers.

Preparative Modification of Ribulose-P$_2$ Carboxylase by TNBS. The enzymes from spinach and R. rubrum, at 5 mg/ml (76 μM and 89 μM, respectively) in either activation or deactivation buffer, were inactivated at room temperature with TNBS (300 μM) in order to obtain samples for identification of site(s) of modification and for examining the interactions of the derivatized enzymes with carboxyarabinitol-P$_2$. For chemical characterization, reactions were terminated at 70–80% inactivation by the addition of 2-mercaptoethanol (0.01 M); the derivatized proteins were carboxymethylated with iodoacetic acid and digested with trypsin. Reactions were allowed to proceed until >95% inactivation had occurred in those cases in which binding of carboxyarabinitol-P$_2$ was to be studied; quenching was again achieved with 2-mercaptoethanol (0.01 M).

TNP-labeled peptides were isolated from the tryptic digests by successive chromatography on DEAE cellulose, Sephadex G-25, and Lichrosorb. During purification, these peptides were detected by monitoring column effluents at 367 nm. A digest (20 mg) was first chromatographed on a column (1 × 25 cm) of Whatman DE-52 equilibrated with 0.01 M NH$_4$HCO$_3$ (pH 8.0) and eluted with a 400-ml linear gradient comprised of 200 ml of equilibration buffer and 200 ml of 0.2 M NH$_4$HCO$_3$ (pH 8.0). Fractions containing a TNP-peptide were pooled and concentrated by lyophilization; the concentrate was then gel filtered on a 1.7 × 220 cm column of Sephadex G-25 in 0.05 M NH$_4$HCO$_3$ (pH 8.0). Following these preparative steps, an aliquot of the partially purified TNP-peptide (~5 nmols) was subjected to HPLC on a 250 × 4.6 mm column of Lichrosorb RP8 (5 μm) from laboratory Data Control. The solvent system consisted of 0.1% (v/v) aqueous trifluoroacetic acid (equilibration) and 0.1% (v/v) trifluoroacetic acid in acetonitrile (limit) as described previously (27). Fractions were collected for subsequent identification of peptides by amino acid analyses.

Amino Acid Analysis. Total acid hydrolysis of peptides and N-α-acetyllysine was achieved in evacuated (<50 μm Hg) sealed tubes with 6 N HCl/0.01 M 2-mercaptoethanol at 110°C for 21 h. Hydrolysates were dried on a Speed Vac concentrator (Savant Instruments Inc.) and subjected to chromatography on a Beckman 121 M amino acid analyzer using Beckman's "3-hour, single-column system." The analyzer was hard wired to a PDP data acquisition system. Computer programs used in data collection and analysis were provided by S. S. Stevens and J. T. Holderman of the Oak Ridge National Laboratory.

RESULTS

Reaction of TNBS with Ribulose-P$_2$ Carboxylase (Fig. 3)

The fully-activated forms of both the R. rubrum and spinach enzymes are rapidly inactivated by TNBS in pseudo first-order fashion. Saturating levels of carboxyribitol-P$_2$ (0.1 mM), a competitive inhibitor with K_i = 2 μM (18,28), afford only partial protection. The rates of inactivation are directly proportional to the TNBS concentration up to 2 mM (Fig. 3, inset); hence, there is no indication that the enzymes have affinity for TNBS.

Characterization of TNBS-Inactivated Ribulose-P$_2$ Carboxylase

A. Identification of sites of arylation. The enzymes from R. rubrum and spinach were treated with TNBS under both activation and deactivation conditions (50 mM Bicine, pH 8.0, with or without 10 mM MgCl$_2$/66 mM NaHCO$_3$); the inactivated samples were then carboxymethylated and digested with trypsin (see "Methods"). Unfractionated tryptic digests of the four samples were inspected by reverse-phase HPLC monitored at either 215 nm for detection of all peptides or 367 nm for detection of TNP-peptides. Fig. 4A shows that only one TNP-peptide of significance is present in digests of the inactivated spinach carboxylase, irrespective of whether the enzyme was modified under activation or deactivation conditions. Although the digest of the R. rubrum enzyme that was treated with TNBS under activation conditions also contains only one major TNP-peptide, the corresponding digest of TNBS-treated deactivated enzyme contains four major labeled peptides (Fig. 4B). The TNP-peptides from digests of both enzymes were purified by DEAE chromatography, gel filtration, and HPLC in succession (see "Methods"). Amino acid analyses of these peptides permit the following assignments of their locations within the respective polypeptide chains (see Fig. 2).

(a) peptide from spinach enzyme (Fig. 4A2 and 4A3) — fragment 320-339 with TNP-Lys at position 334.

(b) peptide from R. rubrum enzyme (modified under activation conditions, Fig. 4B2) — fragment 149-168 with TNP-Lys at position 166.

(c) peptides from R. rubrum enzyme (modified under deactivation conditions, Fig. 4B3);

peptide #1 — fragment 289-301 with TNP-Lys at position 300.
peptide #2 — fragment 1-6 with N-α-TNP-Met at position 1.
peptide #3 — fragment 314-337 with TNP-Lys at position 329.
peptide #4 — fragment 149-168 with TNP-Lys at position 166.

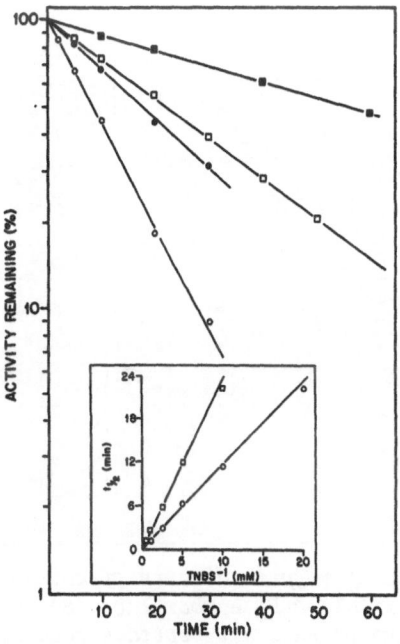

Figure 3: Inactivation of R. rubrum ribulose-P_2 carboxylase (0.3 mg/ml) by TNBS (0.2 mM) in the absence (□) and presence (■) of carboxyribitol-P_2 (0.1 mM) and inactivation of spinach ribulose-P_2 carboxylase (0.3 mg/ml) by TNBS (0.2 mM) in the absence (○) and presence (●) of carboxyribitol-P_2 (0.1 mM). All reactions were carried out at 25°C and at pH 8.0 in 50 mM Bicine/66 mM $NaHCO_3$/10 mM $MgCl_2$.

Separate samples of the two activated enzymes were also inactivated by TNBS at pH 9.5 and at pH 6.5. Subsequent to carboxymethylation and trypsin digestion, aliquots were inspected by HPLC. The specificities of arylation were the same as those observed at pH 8.0 (data not shown).

B. <u>Interaction of trinitrophenylated ribulose-P_2 carboxylase with carboxyarabinitol-P_2.</u> The transition-state analogue 2-carboxyarabinitol-P_2 interacts strongly with the carboxylase in the activated state to form a stable quaternary complex (Enz·CO_2·Mg^{2+}·carboxyarabinitol-P_2), readily isolable by gel filtration (29). Thus, a facile method is provided for determining whether a chemically derivatized, catalytically inactive carboxylase is capable of binding the transition-state analogue, and by inference the substrate. The <u>R. rubrum</u> enzyme, inactivated as a consequence of arylation of Lys-166, is incapable of tight complexation with carboxyarabinitol-P_2 (Fig. 5A). The spinach enzyme, inactivated as a consequence of arylation of Lys-334, does bind carboxyarabinitol-P_2 with a stoichiometry (0.60 moles/mole large·small subunit pair) somewhat less than observed for the native enzyme (0.95 moles/mole large·small subunit pair) (Fig. 5B). In the absence of CO_2/Mg^{2+}, the extent of carboxyarabinitol-P_2 binding to TNP-carboxylase is reduced by 75% (data not shown).

<u>pH-Dependencies of the Reactions of TNBS with Ribulose-P_2 Carboxylase and with N-α-Acetyllysine.</u> In the presence of 66 mM NaHCO$_3$ and 10 mM MgCl$_2$ from pH 6.5-9.5, both the spinach and <u>R. rubrum</u> enzymes retained >90% of their maximally attainable activity, observed at pH 8.0, during a 2-hr period. Hence, over this pH range, the pH-dependency of a chemical modification can be examined without complications arising from an altered activation state of the carboxylase.

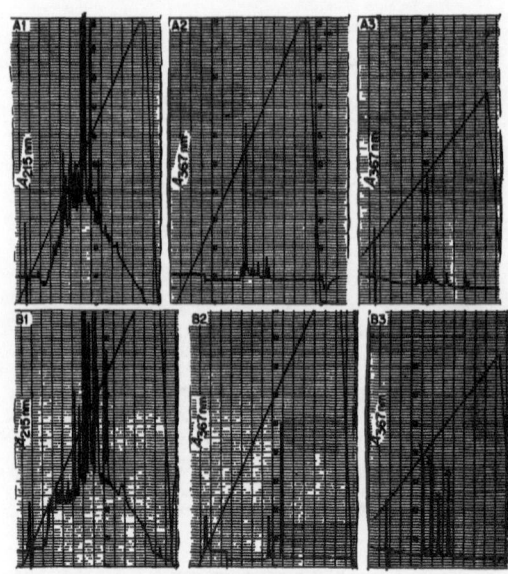

Figure 4: HPLC profiles of tryptic digests of spinach ribulose-P_2 carboxylase (A) and <u>R. rubrum</u> ribulose-P_2 carboxylase (B) that had been treated with TNBS. Panels A1 and B1 show all peptides detected at 215 nm, whereas the other four panels show only TNP-peptides detected at 367 nm. Panels A2 and B2 represent samples of enzymes (duplicates of those shown in panels A1 and B1, respectively) inactivated by TNBS under activation conditions; panels A3 and B3 represent samples of enzymes inactivated by TNBS under deactivation conditions. Note that the gradient is not the same in every panel, and thus elution positions cannot be compared directly.

560

The reaction of TNBS with amines is second-order, and for most compounds examined the pH-dependencies show that only unprotonated amino groups are reactive (30,31). The observed second-order rate constant $k_{2(obsd)}$ will thus reflect the degree of ionization so that the ionization constant (K) can be determined from the equation

$$k_{2(obsd)} = k_o \, K/(K + [H^+]) \qquad (1)$$

where k_o is the intrinsic reactivity of the unprotonated group (32). In the present study, the instability of the activated enzymes (i.e. the carbamates) at pH >9.5 precluded direct determination of k_o. However, a linear form of equation 1 (32)

$$1/k_{2(obsd)} = [H^+]/K \, k_o + 1/k_o \qquad (2)$$

permits determination of k_o by extrapolation of the data upon plotting $1/k_{2(obsd)}$ vs. $[H^+]$, whereby the ordinate intercept equals $1/k_o$ and the slope equals $1/K \, k_o$. Such plots are shown in the insets of Fig. 6 for the reactions of TNBS with N-α-acetyllysine (panel A), activated R. rubrum carboxylase (panel B), activated spinach carboxylase (panel C), and deactivated spinach carboxylase (panel D). Also, shown in Fig. 6 are plots of log $k_{2(obsd)}$ vs. pH for the entire pH-range examined with the solid lines representing theoretical curves established by k_o and pK_a values determined from the linear plots in the insets.

The calculated pK_a values and intrinsic reactivities (k_o) for acetyllysine, Lys-166 of the activated R. rubrum enzyme, and Lys-334 of the spinach enzyme (under both activation and deactivation conditions) are provided in Table 1. Particularly noteworthy are the increased

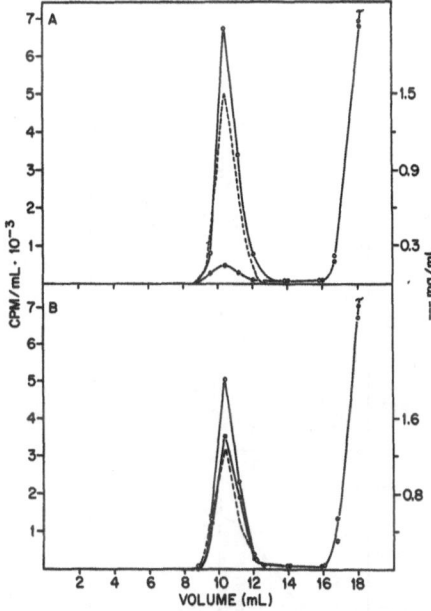

Figure 5: Gel filtration of native (O) and TNBS-inactivated (●) R. rubrum ribulose-P2 carboxylase (A) and spinach ribulose-P2 carboxylase (B) and after mixing with [14C]carboxyarabinitol-P2 (sp. act. = 325 cpm/ nmol). Enzyme solutions (1 ml containing 5 mg of protein) in 50 mM Bicine/ 66 mM NaHCO3/10 mM MgCl2 (pH 8.0) were incubated at room temperature for 30 min with [14C]carboxyarabinitol-P2 (0.5 mM) prior to chromatography on a 1 × 24-cm column of Sephadex G-50 equilibrated with the same buffer. Fractions were assayed for radioactivity and protein concentration.

Table 1: Observed second-order rate constants for the reactions of TNBS with N-α-acetyllysine and with ribulose-P$_2$ carboxylase

Sample	pK$_a$	k$_0$(M^{-1} min^{-1})	k$_2$ (M^{-1} min^{-1}) pH 7.0	pH 8.0
N-α-acetyllysine	10.8	1,250	0.2	2.0
Lys-166 of activated R. rubrum	7.9	670 (0.54)[a]	73 (365)	364 (182)
Lys-334 of activated spinach enzyme	9.0	4,500 (3.6)	46 (230)	409 (205)
Lys-334 of deactivated spinach enzyme	9.8	26,000 (20.8)	41 (205)	406 (203)

[a] Values in parentheses are relative to those of acetyllysine.

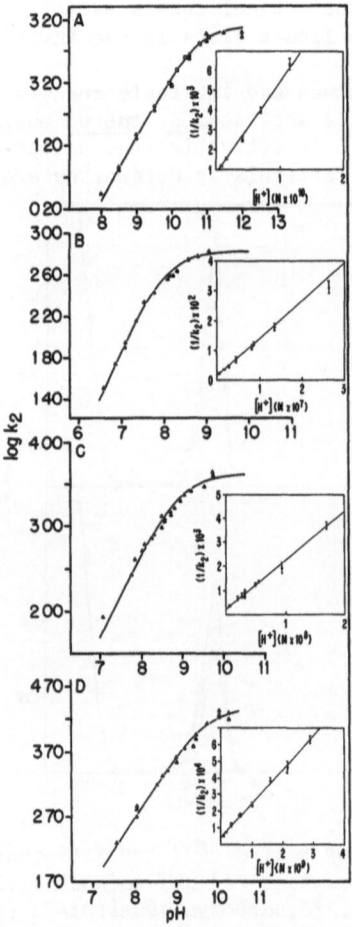

Figure 6. pH-Dependencies of the reactions of TNBS with N-α-acetyllysine (A), activated R. rubrum ribulose-P$_2$ carboxylase (B), activated spinach ribulose-P$_2$ carboxylase (C), and deactivated spinach ribulose-P$_2$ carboxylase (D).

acidities of the two protein ε-amino groups relative to that of acetyllysine and their striking nucleophilicities ($\underline{k}_o$) despite these increased acidities. The differences in these parameters for Lys-166 and Lys-334 are counterbalanced so that their observed reactivities ($\underline{k}_2$) at pH 7.0 and at pH 8.0 are quite similar (Table 1).

DISCUSSION

To determine the $p\underline{K}_a$ value of any amino acid residue within an enzyme from the pH-dependency of its inactivation by a chemical reagent, a precise correlation between site of modification and loss of catalytic activity must be established. This criterion is most easily met if the modification is specific, i.e. restricted to a single amino acid side-chain. TNBS fulfills this requirement beautifully in its reaction with ribulose-P_2 carboxylase (Fig. 4). Although unanticipated, the finding that TNBS selectively arylates Lys-166 in the $\underline{R}$. $\underline{rubrum}$ enzyme in contrast to Lys-334 in the spinach enzyme provides an avenue for ascertaining the $p\underline{K}_a$ values of both essential residues with a single reagent.

Another rather obvious, but frequently overlooked, criterion for assigning an inflection point to a particular residue is the demonstration of unaltered reagent specificity throughout the pH-range examined. In the cases of the arylations reported herein, the tryptic patterns presented in Fig. 4 were not significantly changed when the enzymes were derivatized at either pH 6.5 or pH 9.5, the extremes of pH used for modifications under activation conditions.

A third criterion that must be satisfied with the carboxylase, due to its pH-dependent activation by CO_2/Mg^{2+}, is verification that changes in the activation state do not occur over the pH-range examined. The activation of ribulose-P_2 carboxylase by CO_2/Mg^{2+} entails conformational changes, which can be detected by physical methods (29,33), and are reflected in altered ligand binding properties (34,35) and altered accessibility of some amino acid residues to chemical reagents (13,36). By diluting fully-activated stocks of the enzyme (in pH 8.0 buffer containing saturating levels of CO_2 and Mg^{2+}) into buffers at various pH (also containing the concentrations of CO_2 and Mg^{2+} as before), the deactivation process could be monitored with the $^{14}CO_2$-fixation assay (see "Methods"). This assay readily distinguishes the activated from the deactivated carboxylase, because the time required to measure the enzyme activity (30 $\underline{s}$) is short compared to the rate of activation, which is greatly slowed by ribulose-P_2 (25,34). The deactivated enzyme exhibits <10% of the activity of the fully activated enzyme in this assay (25). These experiments demonstrate that from pH 6.5 to pH 9.5, the fully-activated state is maintained for at least 2 hr.

From the foregoing, the assignment of the inflections, observed in the pH-dependencies of the TNBS-induced inactivations, to Lys-166 in the $\underline{R}$. $\underline{rubrum}$ carboxylase and to Lys-334 in the spinach carboxylase are unequivocal.

Because the site-specificity of TNBS for the spinach enzyme is insensitive to its activation state, an opportunity is provided to ascertain whether the ionization constant for active-site Lys-334 of the spinach enzyme is altered as a consequence of the conformational changes associated with the activation/deactivation process. The present study clearly demonstrates that in the activated enzyme, as compared to the deactivated counterpart, Lys-334 is considerably more acidic ($p\underline{K}_a$ of 9.0 vs. 9.8) and less reactive ($\underline{k}_o$ of 4500 vs. 26,000 M^{-1} min^{-1}). During activation, movement of the Lys-334 side-chain into a more hydrophobic

563

environment having less accessibility to solvent could explain the increased acidity.

Previous correlation between basicity of amino acids (and small peptides) and reactivity toward TNBS would predict that amino groups with pK_a values of 7.9 (Lys-166 of the R. rubrum enzyme), 9.0 (Lys-334 of the activated spinach enzyme), and 9.8 (Lys-334 of the deactivated spinach enzyme) will have k_o values of 120, 280, and 600 $M^{-1} min^{-1}$, respectively (37). By contrast, the respective k_o values determined experimentally are 670, 4500, and 26,000 $M^{-1} min^{-1}$. These two active-site lysyl residues are thus considerably more nucleophilic than anticipated from their pK_a values, and their selective arylation by TNBS is favored by both enhanced acidity and enhanced nucleophilicity. This unusual situation might reflect the microenvironment at the catalytic site. An alternative explanation could be provided by selective reversible binding and orientation of TNBS so as to increase the rates of arylation. However, any such affinity must be quite weak based on the linear relationship between rate of inactivation and reagent concentration up to 2 mM (Fig. 3).

Although the unusual acidity and nucleophilicity can account for the specificity of TNBS for Lys-166 and Lys-334, the similar reactivities at pH 8.0 observed for the two residues (see Table 1) raise the question as to why only one of them is modified in each species of carboxylase investigated. The most obvious explanation for the failure of both to react is that the two lysines are juxtaposed within the catalytic site so that derivatization of both is precluded on steric grounds. Preliminary experiments in our laboratory indicate that Lys-166 and Lys-329 in the R. rubrum enzyme can be bridged with 4,4'-diisothiocyano-2,2'-disulfonate stilbene, a covalent cross-linking agent that spans ~12 Å. A positive charge on one of two proximal active-site lysines could account, at least in part, for the other's very low pK_a (7.9) as observed presently. A weak noncovalent interaction between TNBS and the enzyme, albeit undetected by the experiments carried out, in which the reagent is rather precisely oriented could explain the restriction of arylation to only one of the two potentially reactive lysines. The species dependence of the lysyl residue targeted would then be a consequence of rather subtle differences in active-site geometries. Interestingly, both Lys-329 and Lys-166 in the R. rubrum enzyme are arylated by TNBS under deactivation conditions (Fig. 4B3), an observation which could be interpreted to indicate that the two residues are farther apart in the deactivated enzyme.

Given a knowledge of the pK_a values for the two essential lysines of ribulose-P_2 carboxylase, can one exclude from consideration some of their possible roles? In addressing the question of a role in substrate binding vs. a role in catalysis, several observations argue against the former. The incomplete protection of both enzymes against TNBS-inactivation afforded by saturating levels of carboxyribitol-P_2 is inconsistent with the lysines, which are targets for arylation, participating in salt linkages with phosphate groups of ribulose-P_2. Secondly, the inactivated spinach carboxylase, in which Lys-334 has been arylated, is still able to form the quaternary complex with CO_2, Mg^{2+}, and carboxyarabinitol-P_2. The catalytically inactive, derivatized protein can thus undergo carbamylation and bind the transition-state analog resulting in a conformational change that renders the ligands slowly exchangeable. Thirdly, with respect to Lys-166 in the R. rubrum carboxylase, its extreme acidity (pK_a = 7.9) appears incompatible with effective utilization as a phosphate binding site. However, this is not true with Lys-334 in the spinach enzyme whose pK_a is very close to that of Lys-41 of RNAase (pK_a = 8.8) (38), a residue involved in phosphate binding (39). The inability of the inactivated R. rubrum enzyme to form an isolable quaternary complex may be due to interference by the TNP-moiety of the conformational change necessary for tight binding or to prevention of substrate binding per se for steric reasons. Lastly, the enhanced nucleophilicities of both lysines are more consistent with a catalytic involvement than with a function in binding.

For a consideration of the catalytic roles in which lysyl residues could participate, it is instructive to consider the carboxylation reaction pathway (1,40) (Fig. 7). The reaction is a complex one entailing enolization, carboxylation, hydration, carbon-carbon scission, and protonation. A protonated lysine might (a) polarize the carbonyl group of ribulose-P_2 (however, Mg^{2+} is a more likely candidate for this role, see ref. 41), (b) polarize CO_2 to facilitate its addition to the enediolate intermediate, (c) stabilize either of the two carbanion intermediates, or (d) protonate the C2 carbanion of 3-phosphoglycerate to complete the reaction. With a pK_a of 9.0 (established in the present study), Lys-334 could mediate any one of the preceding steps and correspond to the essential acid with a pK_a of ~8.4 revealed by the pH-dependency of V_{max}/K_m (42).

Steps reasonably subject to mediation by an unprotonated lysine are abstraction of the C3 proton of ribulose-P_2 to initiate enolization or hydration of the 2-carboxy-3-keto intermediate. Based on the primary deuterium isotope effect with [3-^{2}H]ribulose-P_2 (42), the proton abstraction step that initiates enolization appears to be at least partially rate limiting and would thus be expected to contribute to the pH profile of enzyme activity. Data illustrating the pH-dependence of V_{max}/K_m suggest that this catalytically essential base in the free enzyme has a pK_a of 6.8-8.1 (42,43). This broad range represents two independent studies and in part may reflect species variation as well as differences in conditions. Given the additional complication of disparities between kinetic and intrinsic pK_a values, the possibility of Lys-166 (pK_a = 7.9) as the essential base cannot be excluded. Although the precise roles of the two active-site lysines of ribulose-P_2 carboxylase cannot yet be pinpointed, their strong acidities and enhanced nucleophilicities demonstrated in this study provide further evidence of catalytic functionality.

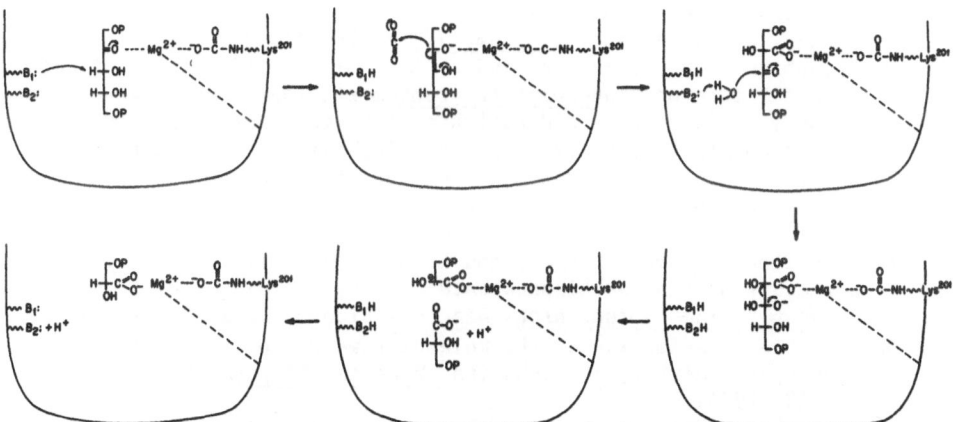

Figure 7. Reaction pathway for carboxylation of ribulose-P_2 as catalyzed by ribulose-P_2 carboxylase. The oxygenation pathway (not shown) presumably entails addition of oxygen to the enediolate intermediate to form a hydroperoxide which undergoes cleavage to phosphoglycolate and 3-phosphoglycerate.

We assume that the species invariance and the adjacent sequence homologies reflect properties and functions of the two active-site lysines that are similar in ribulose-P$_2$ carboxylases from all photosynthetic organisms. Although the specificity of TNBS allows the determination of the pK_a of only Lys-166 in the R. rubrum enzyme and only Lys-334 in the spinach enzyme, it is reasonable to predict that the relative acidities and reactivities of the two active-site lysines are comparable among species despite the likelihood of quantitative differences.

REFERENCES

1. H. M. Miziorko and G. H. Lorimer, Ribulose-1,5-bisphosphate carboxylase/oxygenase, Ann. Rev. Biochem. 52:507 (1983).
2. G. Bowes and W. L. Ogren, Oxygen inhibition and other properties of soybean ribulose 1,5-diphosphate carboxylase, J. Biol. Chem. 247: 2171 (1972).
3. G. H. Lorimer, T. J. Andrews, and N. E. Tolbert, Ribulose diphosphate oxygenase. II. Further proof of reaction products and mechanism of action, Biochemistry 12:18 (1973).
4. J. T. Bahr and R. G. Jensen, On the activity of ribulosediphosphate carboxylase with CO$_2$ and O$_2$ from leaf extracts of Zea mays, Biochem. Biophys. Res. Commun. 57:1180 (1974).
5. T. J. Andrews, M. R. Badger, and G. H. Lorimer, Factors affecting interconversion between kinetic forms of ribulose diphosphate carboxylase-oxygenase from spinach, Arch. Biochem. Biophys. 171:93 (1975).
6. W. A. Laing, W. L. Ogren, and R. H. Hageman, Bicarbonate stabilization of ribulose 1,5-diphosphate carboxylase, Biochemistry 14:2269 (1975).
7. R. Chollet and W. L. Ogren, Regulation of phostorespiration in C$_3$ and C$_4$ species, Bot. Rev. 41:137 (1975).
8. R. Chollet, The biochemistry of photorespiration, Trends Biochem. Sci. 2:155 (1977).
9. R. W. F. Hardy, U. D. Havelka, and B. Quebedeaux, The opportunity for and significance of alteration of ribulose 1,5-bisphosphate carboxylase activities in crop production, in "Photosynthetic Carbon Assimilation," H. W. Siegelman and G. Hind, eds., Plenum Press, New York (1978).
10. J. T. Christeller and W. A. Laing, Effects of manganese ions and magnesium ions on the activity of soya-bean ribulose bisphosphate carboxylase/oxygenase, Biochem. J. 183:747 (1979).
11. P. D. Robinson, M. N. Martin, and F. R. Tabita, Differential effects of metal ions on Rhodospirillum rubrum ribulosebisphosphate carboxylase/oxygenase and stoichiometric incorporation of HCO$_3^-$ into a cobalt(III)-enzyme complex, Biochemistry 18:4453 (1979).
12. D. B. Jordan and W. L. Ogren, Species variation in the specificity of ribulose bisphosphate carboxylase/oxygenase, Nature 291:513 (1981).
13. F. C. Hartman, C. D. Stringer, and E. H. Lee, Complete primary structure of ribulosebisphosphate carboxylase/oxygenase from Rhodospirillum rubrum, Arch. Biochem. Biophys. 232:280 (1984).
14. A. C. Rutner, Estimation of the molecular weight of ribulose diphosphate carboxylase subunits, Biochem. Biophys. Res. Commun. 39:923 (1970).
15. G. Zurawski, B. Perrot, W. Bottomley, and P. R. Whitfeld, The structure of the gene for the large subunit of ribulose 1,5-bisphosphate carboxylase from spinach chloroplast DNA, Nucleic Acids Res. 9:3251 (1981).

16. P. G. Martin, Amino acid sequence of the small subunit of ribulose-1,5-bisphosphate carboxylase from spinach, _Aust. J. Plant Physiol._ 6:401 (1979).

17. F. R. Tabita and B. A. McFadden, D-Ribulose 1,5-diphosphate carboxylase from _Rhodospirillum rubrum_, _J. Biol. Chem._ 249:3459 (1974).

18. J. V. Schloss, E. F. Phares, M. V. Long, I. L. Norton, ·C. D. Stringer, and F. C. Hartman, Ribulosebisphosphate carboxylase/oxygenase from _Rhodospirillum rubrum_, _Methods Enzymol._ 90:522 (1982).

19. G. H. Lorimer and H. M. Miziorko, Carbamate formation of the ε-amino group of a lysyl residue as the basis for the activation of ribulosebisphosphate carboxylase by CO_2 and Mg^{2+}, _Biochemistry_ 19:5321 (1980).

20. G. H. Lorimer, Ribulosebisphosphate carboxylase — amino acid sequence of a peptide bearing the activator carbon dioxide, _Biochemistry_ 20:1236 (1981).

21. M. I. Donnelly, C. D. Stringer, and F. C. Hartman, Characterization of the activator site of _Rhodospirillum rubrum_ ribulosebisphosphate carboxylase/oxygenase, _Biochemistry_ 22:4346 (1983).

22. G. H. Lorimer, M. R. Badger, and T. J. Andrews, The activation of ribulose-1,5-bisphosphate carboxylase by carbon dioxide and magnesium ions. Equilibria, kinetics, a suggested mechanism, and physiological implications, _Biochemistry_ 15:529 (1976).

23. M. Wishnick and M. D. Lane, Ribulose diphosphate carboxylase from spinach leaves, _Methods Enzymol._ 23:570 (1971).

24. E. Racker, D-Ribulose-1,5-diphosphate, in: "Methods of Enzymatic Analysis," H. U. Bergmeyer, ed., Academic Press, New York (1963).

25. G. H. Lorimer, M. R. Badger, and T. J. Andrews, D-Ribulose-1,5-bisphosphate carboxylase-oxygenase: Improved methods for the activation and assay of catalytic activities, _Anal. Biochem._ 78:66 (1977).

26. B. V. Plapp, S. Moore, and W. H. Stein, Activity of bovine pancreatic deoxyribonuclease A with modified amino groups, _J. Biol. Chem._ 246:939 (1971).

27. W. C. Mahoney and M. A. Hermodson, Separation of large denatured peptides by reverse phase high performance liquid chromatography, _J. Biol. Chem._ 255:11199 (1980).

28. J. Pierce, N. E. Tolbert, and R. Barker, Interaction of ribulosebisphosphate carboxylase/oxygenase with transition-state analogues, _Biochemistry_ 19:934 (1980).

29. H. M. Miziorko and R. C. Sealy, Characterization of the ribulosebisphosphate carboxylase-carbon dioxide-divalent cation-carboxypentitol bisphosphate complex, _Biochemistry_ 19:1167 (1980).

30. R. B. Freedman and G. K. Radda, The reaction of 2,4,6-trinitrobenzenesulphonic acid with amino acids. Peptides and Proteins, _Biochem. J._ 108:383 (1968).

31. A. R. Goldfarb, A kinetic study of the reactions of amino acids and peptides with trinitrobenzenesulfonic acid, _Biochemistry_ 5:2570 (1966).

32. W. R. Welches and T. O. Baldwin, Active center studies on bacterial luciferase: Modification of the enzyme with 2,4-dinitrofluorobenzene, _Biochemistry_ 20:512 (1981).

33. M. I. Siegel and M. D. Lane, Interaction of ribulose diphosphate carboxylase with 2-carboxyribitol diphosphate, an analogue of the proposed carboxylated intermediate in the CO_2 fixation reaction, _Biochem. Biophys. Res. Commun._ 48:508 (1972).

34. D. B. Jordan and R. Chollet, Inhibition of ribulose bisphosphate carboxylase by substrate ribulose 1,5-bisphosphate, _J. Biol. Chem._ 258: 13752 (1983).

35. D. B. Jordan, R. Chollet, and W. L. Ogren, Binding of phosphorylated effectors by active and inactive forms of ribulose-1,5-bisphosphate carboxylase, _Biochemistry_ 22:3410 (1983).

36. J. V. Schloss, C. D. Stringer, and F. C. Hartman, Identification of essential lysyl and cysteinyl residues in spinach ribulosebisphosphate carboxylase/oxygenase modified by the affinity label N-bromoacetylethanolamine phosphate, _J. Biol. Chem._ 253:5707 (1978).

37. R. Fields, The measurement of amino groups in proteins and peptides, _Biochem. J._ 124:581 (1971).

38. A. L. Murdock, K. L. Grist, and C. H. W. Hirs, On the dinitrophenylation of bovine pancreatic ribonuclease A. Kinetics of the reaction in water and 8 M urea, _Arch. Biochem. Biophys._ 114:375 (1966).

39. N. Borkakoti, The active site of ribonuclease A from the crystallographic studies of ribonuclease-A-inhibitor complexes, _Eur. J. Biochem._ 132:89 (1983).

40. A. Jaworowski and I. A. Rose, Partition kinetics of ribulose-1,5-bisphosphate carboxylase from _Rhodospirillum rubrum_, _J. Biol. Chem._ 260: 944 (1985).

41. H. M. Miziorko and R. C. Sealy, Electron spin resonance studies of ribulosebisphosphate carboxylase: Identification of activator cation ligands, _Biochemistry_ 23:479 (1984).

42. J. V. Schloss, Primary deuterium isotope effects with [3-^{2}H]ribulose 1,5-bisphosphate (RuBP) in the carboxylase reaction of RuBP carboxylase/oxygenase: Effect of pH and CO_2, _Fed. Proc._ 42: 1923 (1983).

43. J. T. Christeller, Effects of divalent cations on the activity of ribulosebisphosphate carboxylase: Interactions with pH and with D_2O as solvent, _Arch. Biochem. Biophys._ 217:485 (1982).

ADVANCES IN AFFINITY LABELING OF

PURINE NUCLEOTIDE SITES IN DEHYDROGENASES

Roberta F. Colman

Chemistry Department
University of Delaware
Newark, Delaware 19716

An important goal of many biochemists is the identification of amino acids within the active and allosteric sites of enzymes and the elucidation of the role of those amino acids. One approach is to modify chemically critical amino acids and to rely on the specificity of the enzyme for its substrate or normal regulatory compound to limit the extent of chemical modification to the region of the active or allosteric sites. This is the strategy termed affinity labeling, which can potentially lead to specific but irreversible attack within purified enzymes, or even of particular enzymes when present in a mixture of proteins. In the case of dehydrogenases and kinases, custom designed purine nucleotide analogues which have reactive functional groups located at key positions of the natural structure can be effective in mapping amino acid residues in different subsites of the nucleotide binding regions (1).

Examples of such compounds are the fluorosulfonylbenzoyl derivatives of nucleosides shown in Fig. 1. Fig. 1a is 5'-p-fluorosulfonylbenzoyl adenosine (5'-FSBA), which is prepared by reaction of p-fluorosulfonylbenzoyl chloride with adenosine (2). This compound might reasonably be considered as an analogue of ADP, ATP or DPNH. In addition to the adenine and ribose moieties, it has a carbonyl group adjacent to the 5'-position which is structurally similar to the first phosphoryl group of the naturally occurring purine nucleotides. If the molecule is arranged in an extended conformation, the sulfonyl fluoride moiety may be located in a position analogous to the terminal phosphate of ATP or to the ribose proximal to the nicotinamide ring of DPNH. This sulfonyl fluoride moiety can act as an electrophilic agent in covalent reactions with several classes of amino acids, including tyrosine, lysine, histidine, serine and cysteine.

Fig. 1b is 5'-p-fluorosulfonylbenzoyl guanosine (5'-FSBG) in which guanine replaces the adenine moiety in the first derivative. Synthesized by reaction of guanosine with p-fluorosulfonyl chloride (3,4), this purine nucleotide alkylating agent would be expected to be specifically directed toward GTP sites in proteins.

The structure shown in Fig. 1c is the fluorescent compound 5'-p-fluorosulfonylbenzoyl-1,N^6-ethenoadenosine (5'-FSBεA) (5,6). This nucleotide analogue, with a fluorescence emission maximum at 412 nm, may provide a means of introducing a covalently bound fluorescent probe into

nucleotide sites in proteins. We have also synthesized a related
fluorescent compound, 5'-p-fluorosulfonylbenzoyl-2-aza-1,N^6-ethenoadenosine
(5'-FSBaϵA), in which a nitrogen replaces the carbon atom at the 2-position
(7). This compound, illustrated in Fig. 1d, is structurally very similar
to 5'-FSBϵA but has different spectral properties, with a fluorescence
emission peak at 490 nm. In this paper the fluorosulfonylbenzoyl
nucleosides will initially be used to exemplify the types of structure-
function studies that can be conducted with purine nucleotide affinity
labels, including the use of the fluorescent probes to estimate distances
between labeled sites.

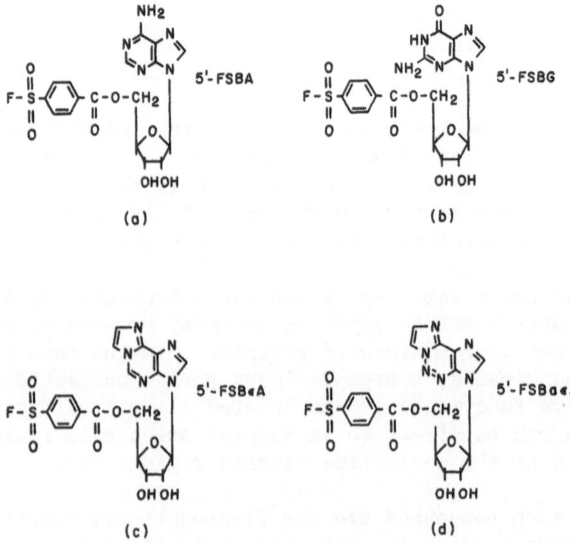

Fluorosulfonylbenzoyl Nucleosides

Figure 1: 5'-p-Fluorosulfonylbenzoyl Derivatives of Adenosine,
Guanosine, 1, N^6-Ethenoadenosine and 2-Aza-1,N^6-ethenoadenosine.

A different class of nucleotide analogues is represented by the three
compounds of Fig. 2 which we have recently synthesized: the
bromo-2,3-dioxobutyl nucleotides (8-10). These compounds are closely
related to the adenine nucleotide structure, are water soluble and are
negatively charged at neutral pH. The bromoketo group is potentially
reactive with several nucleophiles found in proteins, including cysteine,
histidine, glutamic and aspartic acid. In addition, the dioxo group lends
the possibility of reaction with arginine residues. In the case of
6-(4-bromo-2,3-dioxobutyl)thioadenosine 5'-diphosphate (6-BDB-TADP) shown
in Fig. 2a (8), the compound might be expected to react with amino acid
residues in the purine region of the adenine nucleotide binding sites of
proteins because of the location of the functional groups adjacent to the

6-position. This compound should thus be complementary to the
FSB-nucleosides shown in Fig. 1. The 2-(4-bromo-2,3-dioxobutyl)-
thioadenosine 5'-diphosphate (2-BDB-TADP) pictured in Fig. 2c (9),
retains the 6-amino group of the purine ring, thus more closely resembling the
natural nucleotides; this factor may be important in specifically directing
reaction within the nucleotide binding sites of some enzymes. Furthermore,
the position of the reactive functional group adjacent to the 2-position of
the purine ring contrasts with that of Fig. 2a. The middle compound of
Fig. 2, 2-(4-bromo-2,3-dioxobutyl)-thio-1,N^6-ethenoadenosine
2',5'-diphosphate (2-BDB-TɛADP) features a 2'-phosphate moiety and thus
should be directed to TPN binding sites in enzymes (10). In addition, it
is fluorescent, with an emission peak at 428 nm and an excitation maximum
at 302 nm. This property allows easy measurement of the extent of reagent
incorporation into protein, as well as providing a convenient means of
introducing a fluorescent probe into a putative TPN binding site of a
protein. The range of information that can be obtained by the use of
purine nucleotide affinity labels will be illustrated by studies conducted
in my laboratory on bovine liver glutamate dehydrogenase and pig heart
TPN-dependent isocitrate dehydrogenase.

BOVINE LIVER GLUTAMATE DEHYDROGENASE

Glutamate dehydrogenase catalyzes the oxidative deamination of the
amino acid glutamate to form α-keto glutarate and ammonia using either DPN
or TPN as its coenzyme. The activity of the bovine liver enzyme is
regulated by several purine nucleotides, the most effective of which are
ADP, which activates, and guanosine triphosphate, which inhibits. The
enzyme is also inhibited by high concentrations of DPNH which bind to a
second non-catalytic site. Glutamate dehydrogenase is composed of six
identical subunits with several nucleotide sites per subunit, including a
site for ADP, two for GTP and two for DPNH (one catalytic and one
regulatory). We thought that affinity labeling might provide a useful
approach to distinguish among these several binding sites.

Incubation of glutamate dehydrogenase with 1.4 mM 5'-p-fluorosulfonyl-
benzoyl-1,N^6-ethenoadenosine (Fig. 1c) for 200 min at pH 8 and 30° results
in retention of full catalytic activity as measured in the absence of any

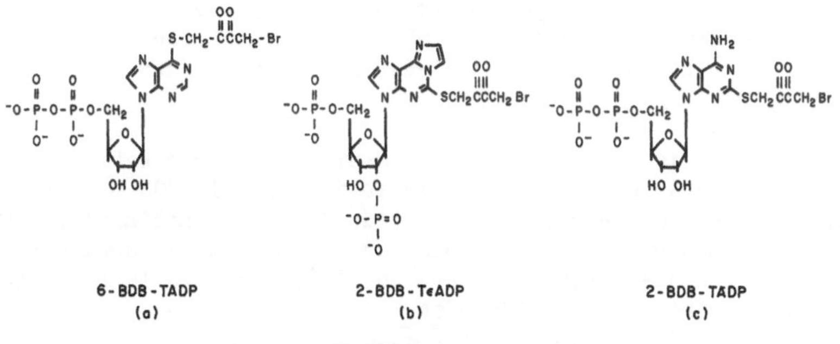

Bromo-2,3-Dioxobutyl Nucleotides

Figure 2: Bromo-2,3-Dioxobutyl Nucleotides. (a) 6-(4-Bromo-2,3-
dioxobutyl)thioadenosine 5'-diphosphate. (b) 2-(4-Bromo-2,3-
dioxobutyl)-thio-1,N^6-ethenoadenosine 2',5'-diphosphate. (c)
2-(4-Bromo-2,3-dioxobutyl)thioadenosine 5'-diphosphate.

regulatory compounds; that is, the reagent does not inactivate the enzyme. It thus appears that 5'-FSBεA does not react at the active site (6). Native enzyme is inhibited about 90% by 1 μM GTP, and during the incubation with 5'-FSBεA, a time-dependent increase is observed in the activity of the enzyme as assayed in the presence of 1 μM GTP. The rate of reaction of 5'-FSBεA with the enzyme can be monitored by this observed time-dependent desensitization to GTP inhibition. The reaction follows pseudo first order kinetics and a rate constant can be calculated from the semilogarithmic plot of the data.

The pseudo first order rate constant for reaction of glutamate dehydrogenase with FSBεA exhibits a non-linear dependence on reagent concentration, suggesting that the reagent binds reversibly to the enzyme (with K_I = 1.4 mM) prior to irreversible covalent modification (6). This behavior is characteristic of an affinity label.

The effects of added substrates and regulatory compounds on the rate of reaction of glutamate dehydrogenase with 1.4 mM 5'-FSBεA have been evaluated (6). The rate constant is unaffected by inclusion in the reaction mixture of the substrate α-ketoglutarate or of the coenzyme DPNH when these are present at concentrations high relative to their known dissociation constants. This indicates that modification by 5'-FSBεA does not occur at the active site or the DPNH inhibitory site. The activator ADP also does not affect the rate constant appreciably.

The k_{obs} is decreased with increasing concentrations of GTP alone, but is most markedly altered by combinations of GTP and DPNH. Indeed, complete protection is provided by saturating concentrations of these ligands, such as 25 μM GTP in the presence of 100 μM DPNH. Since GTP is known to bind more tightly to glutamate dehydrogenase in the presence of reduced coenzyme, these results are consistent with reaction of 5'-FSBεA occurring at a GTP site.

The amount of reagent incorporated into the enzyme was determined from its fluorescence. The extent of covalent incorporation increases with time of incubation with 5'-FSBεA and is directly proportional to the percent decrease in GTP inhibition (6,11). The data extrapolate to 1.28 moles of SBεA incorporated per subunit at 100% change in sensitivity to GTP inhibition. The catalytic and regulatory properties of modified enzyme containing about 1 mole reagent/mole subunit were compared with those of native enzyme to ascertain the functional site of modification.

As indicated above, glutamate dehydrogenase is inhibited by high concentrations of DPNH by binding to a site distinct from the catalytic site. The 5'-FSBεA-modified and native enzymes have essentially the same affinity for coenzyme, and both are inhibited by high concentrations of DPNH. These results indicate that 5'-FSBεA does not react at the DPNH regulatory site.

Native enzyme is inhibited 96% by saturating concentrations of GTP, whereas the 5'-FSBεA-modified enzyme is inhibited only 70% by saturating concentrations of the inhibitory nucleotide (6). The affinity for GTP is also about 15 times weaker in the modified enzyme. It is apparent that the sensitivity of glutamate dehydrogenase to GTP inhibition is decreased but not eliminated by reaction of 5'-FSBεA.

To determine whether 5'-FSBεA reacts at a GTP site of glutamate dehydrogenase, we measured the reversible binding of radioactive GTP to native and modified enzyme using an ultrafiltration technique. Native enzyme was found to bind 2 moles of GTP in the presence of DPNH, while only 1 mole of GTP was bound by modified enzyme (6). These results indicate that upon modification of glutamate dehydrogenase by 5'-FSBεA, one of the two GTP sites is eliminated. 5'-FSBεA has presumably reacted at this GTP site.

In order to identify the site on the protein which reacts with 5'-FSBεA, we have prepared stable derivatives of tyrosine and lysine by reaction with the fluorosulfonylbenzoyl nucleosides (12). Upon acid hydrolysis, the ester linkages between the benzoyl and the nucleoside moieties are hydrolyzed and the acid-stable products are carboxybenzenesulfonyl tyrosine (or CBS-Tyr) and carboxybenzenesulfonyl lysine (or CBS-Lys). As shown in Fig. 3, these derivatives can be separated and quantified on an ion exchange amino acid analyzer. CBS-Lys is eluted just ahead of tyrosine, and CBS-Tyr is eluted after phenylalanine; these derivatives are thus eluted at positions distinct from each other and from standard amino acids. For glutamate dehydrogenase incubated with 5'-FSBεA, the amount of CBS-Tyr is directly proportional to the change in GTP inhibition, suggesting that the tyrosine modified is in a GTP binding site (11).

To further characterize the site of 5'-FSBεA reaction, the modified glutamate dehydrogenase was digested successively with trypsin and then chymotrypsin, and the resultant proteolytic digest was fractionated by gel filtration followed by high performance liquid chromatography. A single type of peptide containing modified tyrosine was isolated, with amino

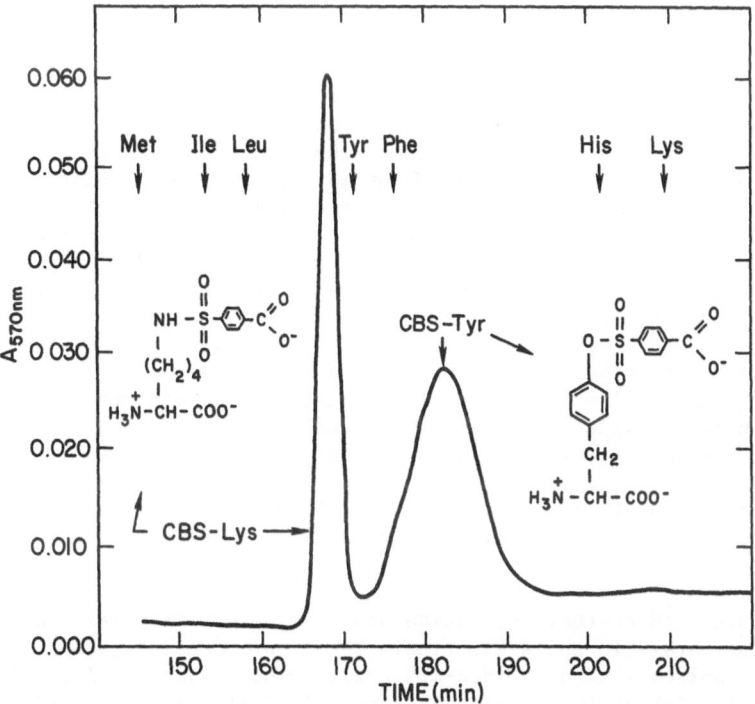

Figure 3: Structures of Nᵉ-(4-Carboxybenzenesulfonyl)-Lysine (CBS-Lys) and O-(4-Carboxybenzenesulfonyl)-Tyrosine (CBS-Tyr) and their elution positions from an ion exchange amino acid analyzer. The elution positions of Met (145 min), Ile (153 min), Leu (158 min), CBS-Lys (168 min), Tyr (171 min), Phe (176 min), CBS-Tyr (182 min), His (201 min) and Lys (209 min) are indicated.

573

terminal CBS-tyrosine as determined by dansylation. This peptide has been identified as amino acids 262-265:

$$Tyr^{262}-Leu-His-Arg^{265}$$

on the basis of the known amino acid sequence of glutamate dehydrogenase (14). Therefore the tyrosine residue modified by 5'-FSBεA is Tyr-262.

Having established the functional site and the specific amino acid at which 5'-FSBεA labels glutamate dehydrogenase, we set out to use the fluorescent probe to measure the distance between the GTP site and another defined site on the enzyme using the technique of fluorescence energy transfer. The distance between two chromophores on a protein can be determined by fluorescence energy transfer provided that the emission spectrum of the donor chromophore overlaps the absorption spectrum of the acceptor chromophore and that the two species are sufficiently close to each other (15-17).

For glutamate dehydrogenase, the fluorescent -SBεA at the GTP site was used as the energy donor; and we chose 2'(3')-O-(2,4,6-trinitrophenyl)-adenosine 5'-diphosphate (TNP-ADP) as the energy acceptor. We have found that TNP-ADP serves as an effective probe of the natural ADP activating site of glutamate dehydrogenase. TNP-ADP activates the enzyme two-fold, and competes for the binding of radioactive ADP (11).

The spectral overlap is excellent between the -SBεA-modified protein, with its fluorescence emission maximum at about 418 nm, and the TNP-ADP absorption spectrum, with its maximum absorption at about 405 nm. An R_0 value of 22 Å was calculated for the distance at which the efficiency of energy transfer would be 50% (11).

The fluorescence at 405 nm of FSBεA-modified glutamate dehydrogenase is progressively quenched by increasing concentrations of TNP-ADP bound reversibly to the ADP activating site. The efficiency of energy transfer is 77%, which allows the calculation of 18 Å as an average distance between the -SBεA and TNP-ADP sites (11). This result indicates that although the GTP and ADP regulatory sites are not identical, they are relatively close to one another. This may provide an explanation for earlier reports which indicate that GTP and ADP compete kinetically in the regulation of glutamate dehydrogenase.

We have used other fluorescent probes to extend our estimates of the distance relationships between additional sites of glutamate dehydrogenase. Iodoacetamidosalicylic acid was shown by Holbrook to react with lysine-126 at the α-ketoglutarate binding site (18,19). The ISA-modified enzyme is fluorescent with an emission peak at about 400 nm (7). This fluorescence overlaps the absorption spectrum of TNP-ADP bound reversibly at the ADP activating site, allowing a distance of 33 Å to be calculated between the two sites.

We have used another nucleotide analogue to estimate the distance between the catalytic and a GTP site. The fluorescent compound 5'-p-fluorosulfonylbenzoyl-2-aza-ethenoadenosine (shown in Fig. 1d) is structurally similar to FSB-ethenoadenosine and also reacts covalently at a GTP regulatory site of glutamate dehydrogenase; however, it has different spectral properties (7). The absorption spectrum of the -SBaεA-modified enzyme overlaps the corrected fluorescence emission spectrum of the ISA-modified enzyme, allowing the estimation of 23 Å between the GTP and catalytic sites in the doubly labeled enzyme (7).

Table I summarizes the distance relationships among these several sites. The distance of 33 Å between the ADP and catalytic sites and 23 Å between the GTP and catalytic sites suggests that the regulatory sites are closer to each other as compared with their distances from the catalytic site. These distance estimates are consistent with the allosteric model of regulation of glutamate dehydrogenase whereby the regulatory nucleotides are bound at distinct sites and indirectly influence the catalytic reaction.

We have used two different purine nucleotide affinity labels to covalently modify the NADH inhibitory site of glutamate dehydrogenase. The first of these is the new nucleotide analogue that we have recently synthesized: 6-(4-bromo-2,3-dioxobutyl)thioadenosine 5'-diphosphate (6-BDB-TADP), shown in Fig. 2a. Glutamate dehydrogenase reacts covalently with 6-BDB-TADP, and upon reaction, the modified enzyme completely loses its normal ability to be inhibited by high concentrations of DPNH (20). A plot of initial velocity versus DPNH concentration for the modified enzyme, in contrast to the native enzyme, follows normal Michaelis-Menten kinetics.

The time dependence of incorporation of 6-BDB-TADP into the enzyme reveals that only about one mole is incorporated per peptide chain. A plot of the percentage of maximum loss in DPNH inhibition (measured at 600 μM DPNH) as a function of the incorporation of reagent indicates that the DPNH inhibition is completely lost when one mole is incorporated per enzyme subunit (20).

The reaction rate of 6-BDB-TADP with glutamate dehydrogenase is not affected by the substrate α-ketoglutarate, or by GTP alone. However, it is decreased markedly by concentrations of NADH sufficient to bind to the inhibitory site (20). GTP is known to tighten the binding of NADH, and with 5 mM NADH plus 50 or 100 μM GTP little or no effect of 6-BDB-TADP on NADH inhibition is observed. We think there is good evidence that 6-(4-bromo-2,3-thioadenosine 5'-diphosphate acts as a specific affinity label of the NADH inhibitory site of glutamate dehydrogenase.

We have recently isolated a 19-member peptide from a tryptic and chymotryptic digest of enzyme labeled at the NADH site by 6-BDB-TADP. The modified residue in this case is one of the 6 cysteines in glutamate dehydrogenase, Cysteine-319 (21). Because of the location of the reactive bromodioxobutyl moiety adjacent to the purine portion of the nucleotide, it is likely that the cysteine is near the adenine binding site of NADH.

Table I. Summary of Energy Transfer Measurements Between Energy Donor-Acceptor Pairs on Glutamate Dehydrogenase

Donor	Location	Acceptor	Location	Distance (Å)	Ref.
-SBεA	GTP Site	TNP-ADP	ADP Site	18	(11)
ISA	Catalytic Site	TNP-ADP	ADP Site	33	(7)
ISA	Catalytic Site	-SBaεA	GTP Site	23	(7)

The second nucleotide analogue we have used to specifically modify the NADH inhibitory site of glutamate dehydrogenase is 5'-p-fluorosulfonyl-benzoyl adenosine (5'-FSBA), shown in Fig. 1a, in which the reactive functional group is in a region structurally equivalent to the nicotinamide ribose of the coenzyme. In that case, equal amounts of CBS-Lysine and CBS-Tyrosine are found throughout the course of the reaction, although no more than 1 mole of total reagent per peptide chain is incorporated (12). Two nucleosidyl peptides have been isolated from modified enzyme digested with thermolysin by chromatography on a dihydroxyboryl-substituted agarose column followed by HPLC (22).

From a comparison of the amino acid composition and amino terminals of these peptides with the known amino acid sequence of glutamate dehydrogenase (14), the amino acids labeled by 5'-FSBA have been identified as tyrosine-190 and lysine-420 (22). The reaction of 5'-FSBA with glutamate dehydrogenase is limited and specific, and it appears that the modified residues Tyr-190 and Lys-420 must both be located in or near the NADH inhibitory site. The modification of equal amounts of these two residues suggests that both are close to the reactive sulfonylfluoride of the enzyme-bound reagent and have an equal probability of reacting. Although these residues are separated by almost half the protein in the linear sequence, they appear to be quite close to each other in the folded structure.

Thus, affinity labeling has strongly implicated three amino acid residues from different parts of the linear protein molecule:

$$\text{Tyr}^{190} \text{ (5'-FSBA)}$$

$$\text{Cys}^{319} \text{ (6-BDB-TADP)}$$

$$\text{Lys}^{420} \text{ (5'-FSBA)}$$

as participants in the binding site of one regulatory nucleotide, DPNH. This type of data may provide important constraints for models of the folding of glutamate dehydrogenase.

PIG HEART TPN-DEPENDENT ISOCITRATE DEHYDROGENASE

Another example of the effectiveness of affinity labeling of purine nucleotide sites in dehydrogenases is provided by some recent work we have conducted on the TPN-dependent isocitrate dehydrogenase (10). This enzyme catalyzes the oxidative decarboxylation of isocitrate to form α-ketoglutarate, which is one of the critical steps in the energy producing Citric Acid Cycle. The pig heart TPN-specific isocitrate dehydrogenase has a single type of polypeptide chain of molecular weight 58,000 and under most conditions exists as a dimer.

The binding of coenzymes and coenzyme fragments by the enzyme has been measured by a variety of techniques including ^{31}P (23) and ^{1}H-NMR (24), UV difference and protein fluorescence spectroscopy (25). The enzyme binds tightly TPN and TPNH, as well as the coenzyme fragments 2'-phosphoadenosine 5'-diphosphoribose and adenosine 2',5'-diphosphate. DPN, which lacks the 2'-phosphate, does not function as a coenzyme for this enzyme-catalyzed reaction, and when measured as an inhibitor, DPN binds several orders of magnitude more weakly than do nucleotides with a 2'-phosphate group on the adenosine ribose. Thus in designing an affinity label which will react at the coenzyme binding site of isocitrate dehydrogenase, it is critical to include a 2'-phosphate moiety.

We have synthesized 2-(4-bromo-2,3-dioxobutyl)-thio-1,N^6-ethenoadenosine 2',5'-diphosphate (2-BDB-TεADP), shown in Fig. 2b, which contains the required 2'-phosphate (10). Starting with TPN, 2'-phosphoadenosine 5'-diphosphoribose (PADPR) was generated enzymatically. The PADPR was converted to the fluorescent εPADPR by reaction with chloroacetaldehyde. Treatment with NaOH, followed by reaction with carbon disulfide, yielded 2-thio-1,N^6-ethenoadenosine 2',5'-diphosphate (TεADP). Condensation of TεADP with 1,4-dibromobutanedione gave the final product, 2-BDB-TεADP.

Incubation of pig heart TPN-specific isocitrate dehydrogenase with 100 μM BDB-TεADP at pH 7.0 and 25° results in a time dependent inactivation of the enzyme. Biphasic inactivation kinetics is observed which can be described in terms of a fast initial phase of inactivation resulting in partially active enzyme of 8-10% residual activity, followed by a slower phase leading to total inactivation (10).

Complete protection against the fast phase is provided by TPNH, but not by DPN or DPNH, suggesting that the site attacked during the fast phase is a natural coenzyme binding site of the TPN-specific isocitrate dehydrogenase. In contrast, the addition of isocitrate and manganous ion causes total protection against the slow phase of inactivation, and this phase may result from reaction at a substrate binding site.

The incorporation of BDB-TεADP into the enzyme was determined by comparison of the fluorescence of modified enzyme (denatured in 5M guanidine hydrochloride) to that of BDB-TεADP standards. Only 0.99 moles reagent per mole enzyme subunit is incorporated with only 2.5% activity remaining (10). It is apparent that there is a limited extent of incorporation of BDB-TεADP into isocitrate dehydrogenase, which is an essential criterion of an affinity label.

CONCLUDING REMARKS

This paper has summarized a representative sampling of the types of studies being conducted in my laboratory using purine nucleotide affinity labels. We are hopeful that these various analogues will not only be useful for our own experiments, but that they will also be valuable to other laboratories in exploring nucleotide sites in a variety of proteins. The 5'-p-fluorosulfonylbenzoyl adenosine which was the first of this class of compounds that we described, has already been found to yield specific labeling of NAD sites in several dehydrogenases and reductases, as indicated in Table II.

As summarized in Table III, 5'-FSBA has also labeled ATP or ADP binding sites in a large number of kinases and synthetases, as well as providing an effective handle for examining an ADP receptor protein of platelet membranes. Furthermore, it has modified specific nucleotide sites in such diverse proteins as the ATPases, actin, myosin, luciferase and oxo-prolinase. We anticipate that the guanosine, the ethenoadenosine and aza-ethenoadenosine analogues, as well as the new 6-bromodioxobutyl-thioadenosine 5'-diphosphate, 2-bromodioxobutyl-thioadenosine 5'-diphosphate and the fluorescent 2-BDB-thio-ethenoadenosine 2',5'-diphosphate will similarly have widespread applications to the elucidation of purine nucleotide sites in enzymes.

Table II. Proteins Covalently Modified by 5'-FSBA (Partial List)

Dehydrogenases/Reductases

Glutamate Dehydrogenase	(12,22)[a]
Malate Dehydrogenase	(26)
3α,20β-Hydroxysteroid Dehydrogenase	(27)
17β-Estradiol Dehydrogenase	(28)
Xanthine Dehydrogenase	(29)
Aldehyde Reductase	(30)
NADH-Cytochrome b_5 Reductase	(31)

[a]Literature references

Table III. Proteins Covalently Modified by 5'-FSBA (Partial List)

Kinases and Other ADP/ATP Binding Proteins

Pyruvate Kinase (32-34)[a]	ADP Receptor Protein of Platelet Membranes (50-52)
Phosphofructokinase (35-37)	Mitochondrial F_1-ATPase (53-55)
cAMP-Dependent Protein Kinase (38-40)	Chloroplast ATPase (56)
cGMP-Dependent Protein Kinase (41-42)	Na^+,K^+-Transport ATPase (57)
Casein Kinase II (43)	Actin (58)
Epidermal Growth Factor-Stimulated Protein Kinase (44,45)	Myosin (58)
Glutamine Synthetase (46,47)	Luciferase (59)
Carbamyl Phosphate Synthetase (48,49)	5-Oxo-Prolinase (60)

[a]Literature references

References

1. Colman, R. F. (1983) Ann. Rev. Biochem. 52, 67-91.
2. Colman, R. F., Pal, P. K. and Wyatt, J. L. (1977) Methods in Enzymology 46, 240-249.
3. Pal, P. K., Reischer, R. J., Wechter, W. J. and Colman, R. F. (1978) J. Biol. Chem. 253, 6644-6646.

4. Pal, P. K. and Colman, R. F. (1979) Biochemistry 18, 838-845.
5. Likos, J. J. and Colman, R. F. (1981) Biochemistry 20, 491-499.
6. Jacobson, M. A. and Colman, R. F. (1982) Biochemistry 21, 2177-2186.
7. Jacobson, M. A. and Colman, R. F. (1984) Biochemistry 23, 3789-3799.
8. Colman, R. F., Huang, Y-C., King, M. M. and Erb, M. (1984) Biochemistry 23, 3281-3286.
9. Kapetanovic, E., Bailey, J. M. and Colman, R. F. (1985) Biochemistry 24, in press.
10. Bailey, J. M. and Colman, R. F. (1985) Biochemistry 24, in press.
11. Jacobson, M. A. and Colman, R. F. (1983) Biochemistry 22, 4247-4257.
12. Saradambal, K. V., Bednar, R. A. and Colman, R. F. (1981) J. Biol. Chem. 256, 11866-11872.
13. Jacobson, M. A. and Colman, R. F. (1984) Biochemistry 24, 6377-6382.
14. Julliard, J. H. and Smith, E. L. (1979) J. Biol. Chem. 254, 3427-3438.
15. Forster, T. (1959) Discuss. Faraday Soc. 27, 7-17.
16. Fairclough, R. H. and Cantor, C. R. (1978) Methods Enzymol. 48, 347-379.
17. Stryer, L. (1978) Annu. Rev. Biochem. 47, 819-846.
18. Holbrook, J. J., Roberts, P. A. and Wallis, R. B. (1973) Biochem. J. 133, 165-171.
19. Wallis, R. B. and Holbrook, J. J. (1973) Biochem. J. 133, 173-182.
20. Batra, S. P. and Colman. R. F. (1984) Biochemistry 23, 4940-4946.
21. Batra, S. P. and Colman, R. F. (1985) Biochemistry 24, 3359.
22. Schmidt, J. A. and Colman, R. F. (1984). J. Biol. Chem. 259, 14515-14519.
23. Mas. M. T. and Colman. R. F. (1984) Biochemistry 23, 1675-1683.
24. Ehrlich, R. S. and Colman. R. F. (1985) Biochemistry 24, in press.
25. Mas. M. T. and Colman, R. F. (1985) Biochemistry 24, 1634-1646.
26. Roy, S. and Colman, R. F. (1979) Biochemistry 18, 4683-4690.
27. Sweet, F. and Samant, B. R. (1981) Biochemistry 20, 5170-5173.
28. Tobias, B. and Strickler, R. C. (1981) Biochemistry 20, 5546-5549.
29. Nishino, T., Nishino, T. and Tsushina, K. in Flavins and Flavoproteins, Walter de Gruyter and Co., Berlin, 1984, pp. 319-322.
30. Flynn, T. G. and Cronin, C. N. (1985) Feb. Proc. 44, 1619.
31. Chen, S., Haniu, M., Shively, J. E. and Iyanagi, T. (1985) Fed. Proc. 44, 1619.
32. Likos, J. J., Hess, B. and Colman, R. F. (1980) J. Biol. Chem. 19, 9388-9398.
33. Annamalai, A. E. and Colman, R. F. (1981) J. Biol. Chem. 256, 276-283.
34. DeCamp, D. L. and Colman, R. F. (1985) Biochemistry 24, 3360.
35. Mansour, T. E., Colman, R. F. (1978) Biochem. Biophys. Res. Commun. 81, 1370-1376.
36. Pettigrew, D. W. and Frieden, C. (1978) J. Biol. Chem. 253, 3623-3627.
37. Weng, L., Heinrickson, R. L. and Mansour, T. E. (1980) J. Biol. Chem. 255, 1492-1496.
38. Zoller, M. J. and Taylor, S. S. (1979) J. Biol. Chem. 254, 8363-8368.
39. Hixson, C. S. and Krebs, E. G. (1979) J. Biol. Chem. 254, 7509-7514.
40. Zoller, M. J., Nelson, N. C. and Taylor, S. S. (1981) J. Biol. Chem. 256, 10837-10842.
41. Hixson, C. S. and Krebs, E. G. (1981) J. Biol. Chem. 256, 1122-1127.
42. Hashimoto, E., Takio, K. and Krebs, E. G. (1982) J. Biol. Chem. 257, 727-733.
43. Hathaway, G. M., Zoller, M. J. and Traugh, J. A. (1981) J. Biol. Chem. 256, 11442-11446.
44. Buhrow, S. A., Cohen, S., Staros, J. V. (1982) J. Biol. Chem. 257, 4019-4022.
45. Buhrow, S. A., Cohen, S., Garbers, D. L. and Staros, J. V. (1983) J. Biol. Chem. 258, 7824-7827.
46. Foster, W. B., Griffith, M. J. and Kingdon, H. S. (1981) J. Biol. Chem. 256, 882-886.

47. Pinkofsky, H. B., Ginsburg, A., Reardon, I. and Heinrikson, R. L. (1984) J. Biol. Chem. 259, 9616-9622.
48. Boettcher, B. R. and Meister, A. (1980) J. Biol. Chem. 255, 7129-7133.
49. Powers, S. G., Muller, G. W. and Kafka, N. (1983) J. Biol. Chem. 258, 7545-7549.
50. Bennett, J. S., Colman, R. F. and Colman, R. W. (1978) J. Biol. Chem. 253, 7346-7354.
51. Figures, W. R., Niewiarowski, S., Morinelli, T. A., Colman, R. F., and Colman. R. W. (1981) J. Biol. Chem. 256, 7789-7795.
52. Mills, D.C.B., Figures, W. R., Scearce, L. M., Stewart, G. J., Colman, R. F. and Colman, R. W. (1985) J. Biol. Chem. 260, 8078-8083.
53. Esch. F. S. and Allison, W. S. (1978) J. Biol. Chem. 253, 6100-6106.
54. DiPietro, A., Godinot, C., Martin, J.-C. and Gautheron, D. C. (1979) Biochemistry 18,1738-1745.
55. DiPietro, A., Godinot, C. and Gautheron, D. C. (1981) Biochemistry 21, 6312-6318.
56. DeBenedetti, E., and Jagendorf, A. (1979) Biochem. Biophys. Res. Commun. 96, 440-446.
57. Cooper, J. B. and Winter, C. G. (1980) J. Supramol. Struct. 13, 165-174.
58. Bennett, J. S., Vilaire, G., Colman, R. F. and Colman, R. W. (1981) J. Biol. Chem. 256, 1185-1190.
59. Lee, Y., Esch, F. S. and DeLuca, M.A. (1981) Biochemistry 21, 1253-1256.
60. Williamson, J. M. and Meister, A. (1982) J. Biol. Chem. 257, 9161-9172.

ESCHERICHIA COLI ADPGLUCOSE SYNTHETASE SUBSTRATE-INHIBITOR BINDING SITES
STUDIED BY SITE(S) DIRECTED CHEMICAL MODIFICATION AND MUTANT ENZYME
CHARACTERIZATION

Young Moo Lee, Charles E. Larsen and Jack Preiss

Department of Biochemistry
Michigan State University
East Lansing, Michigan 48824

INTRODUCTION

ADPglucose synthetase catalyzes the synthesis of ADPglucose from
α-glucose 1-P and ATP (reaction 1)

(1)   ATP + α-glucose 1-P $\rightleftharpoons$ ADPglucose + PP$_i$

ADPglucose is the glucosyl donor for starch synthesis in plants and for
glycogen synthesis in bacteria and ADPglucose synthetase has been shown
to be the major regulatory enzyme for α 1,4 glucan sythesis in bacteria
as well as in plants.

Most ADPglucose synthetases are activated by glycolytic
intermediates and inhibited by either AMP, P$_i$, or ADP[1-4]. Of interest is
that the activator specificity of the ADPglucose synthetases can differ
and the difference in activator specificity is dependent on the source of
the enzyme. As shown in Table I the ADPglucose synthetases may be
grouped into 7 different classes based on their activator specificity.
The prevalent activators seen in the various classes are fructose 6-P,
fructose 1,6-P$_2$, 3-P-glycerate and pyruvate and there is some overlap
with respect to the activators. This overlapping specificity suggests
that the activator binding site(s) of the enzyme in the different groups
are related. It is believed that the specificity of the activator for
the organism's ADPglucose synthetase is coordinated with the prevalent
carbon assimilatory pathway utilized by the organism. This hypothesis
has been extensively discussed in a number of reviews[3,4]. Suffice to say
here that organisms that metabolize their carbon via glycolysis generally
have fructose 1,6-P$_2$ as an activator while those aerobic organisms that
assimilate carbon via the ribulose bisphosphate carboxylase pathway have
an ADPglucose synthetase with 3-P-glycerate as the activator.

It was therefore of interest to us to determine and compare the
structure and function of these various activator sites. Because we were
able to obtain derepressed mutants of the glycogen biosynthetic enzymes
in E. coli and therefore obtain high expression of the structural genes,
we were able to study the ADPglucose synthetase in considerable detail.
Preparation of pure ADPglucose synthetase was facilitated by using
affinity chromatography[5] and large amounts were obtained after we were
able to clone the glycogen biosynthetic structural genes, glg A (glycogen

synthase), glg B (branching enzyme) and glg C (ADPglucose synthetase)[6]. Moreover, via DNA sequencing of the glg C gene we were able to determine the deduced amino acid sequence of the ADPglucose synthetase[7]. This deduced amino acid sequencing was consistent with the previous peptide sequencing[8-10], the amino acid analysis as well as the molecular weight determination[8].

Via chemical modification studies with pyridoxal phosphate we were able to identify the activator binding site of the ADPglucose synthetase as amino acid residue 38 from the N-terminal end of the polypeptide[9]. Other studies also suggested that epsilon amino acid groups of a lysyl residue and arginine residues were involved in the binding of the activator[11,12].

Some insight was also obtained about the catalytic or substrate binding site as a lysyl residue number 194 from the N-terminal end was identified as the amino acid residue that was covalently modifed when

Table I. Classes of ADPglucose synthetases based on allosteric activator specificity.

Activator(s)	Bacteria
1. Pyruvate	Rhodospirillum sp.; R. fulvum, R. molischianum, R. photometricum, R. rubrum, R. tenue; Rhodocyclus purpureus
2. Pyruvate Fructose 6-P	Agrobacterium tumefaciens, Arthrobacter viscosus, Chlorobium limicola, Chromatium vinosum, Rhodopseudomonas sp.; R. acidophila, R. blastica, R. capsulata, R. palustris
3. Fructose 6-P Fructose 1,6-P_2	Aeromonas hydrophila, Micrococcus luteus, Mycobacterium smegmatis, Rhodopseudomonas viridis

4. Fructose 1,6-P_2
    Pyridoxal-P
    NADPH
	Citrobacter freundii, Edwardsiella tarda, Enterobacter aerogenes, Enterobacter cloacae, Escherichia aurescens, Escherichia coli, Klebsiella pneumoniae, Salmonella enteriditis, Salmonella typhimurium, Shigella dysenteriae

5. Pyruvate
    Fructose 6-P
    Fructose 1,6-P_2
	Rhodopseudomonas sp.; R.gelatinosa, R. globiformis, R. sphaeroides

6. 3-P-glycerate	Aphanocapsa 6308, Synechococcus 6301, Plants
7. None	Clostridium pasteurianum, Enterobacter hafniae, Serratia sp.; S. liquifaciens, S. marcescens

inactivation of enzyme catalysis occurred due to reductive phosphopyridoxylation[9]. This residue and enzyme activity could be protected by the presence of ADPglucose and Mg^{++} ions during the reductive phosphopyridoxylation.

In order to obtain more information on the catlytic site(s) as well as the inhibitor site we decided to use the photoaffinity probes, 8-azido ATP and 8-Azido AMP[13]. AMP is a potent ihibitor of the E. coli ADPglucose synthetase and thus 8-Azido AMP could be considered as an inhibitor analogue.

Further insight into the structure-function relationships of the various ligand binding sites could also obtained by sequencing of mutant structural genes of the ADPglucose synthetase and determining where the amino acid substitutions have occurred. We were fortunate in having mutants of the ADPglucose synthetase that were affected in their allosteric properties[1-4,14]. Thus efforts were made to clone one of the allosteric mutants glg C gene. This report describes the results we have obtained in these studies.

Studies involving 8-N$_3$ATP and 8-N$_3$ADPglucose

In the course of our studies with 8-N$_3$ATP it was found that it was a substrate for the ADPglucose synthetase. Synthesis of 8-N$_3$ADPglucose occurred at 0.3% the rate seen for ADPG synthesis with fructose 1,6-P$_2$ as the activator and at 0.7% the rate with 1,6 hexanediol-P$_2$ as the activator. The photoaffinity sugar nucleotide was purified and isolated via paper chromatography. Spectral analysis showed a typical spectra of an azido nucleotide (8-N$_3$ATP) with an absorption maximum at 281 nm which diminished upon photolysis[13]. Using a molar extinction coefficient of 1.33×10^4 which has been determined for azido adenosine nucleotides[15]. The product contained 1 mol of glucose per mol of azido adenosine. The product was also shown to transfer glucose to glycogen when incubated with pure E. coli glycogen synthase thus indicating it was a sugar nucleotide (unpublished results).

The 8-azido ADPglucose product could be pyrophosphorolyzed to 8-azido ATP utilizing ^{32}PP and the ADPglucose synthetase. The reaction rate was about 1.0% the rate observed for ADPglucose. Thus we were able to use two photoaffinity probes for the substrate site, 8-N$_3$ATP and 8-N$_3$ADPglucose. Since we could only obtain 8-N$_3$ATP labeled only with [^{32}P] and we were able to synthesize the 8-azido sugar nucleotide labeled with [^{14}C] glucose, most of our studies were done with 8-azido ADPglucose. This was mainly due to the longer half-life of the radioactivity and the highly hydrophilic nature of the ATP preventing satisfactory peptide sequencing.

Some kinetic studies are summarized in Table II. Whereas 8-N$_3$ADP-glucose has the same apparent affinity as the natural substrate, 8-N$_3$ATP seemed to have a lower apparent affinity than ATP. The greatest effect seen is on V_{max} where the catalytic rates are much lower with the azido substrates. Other kinetic experiments have shown that 8-N$_3$ADPG is a competitive inhibitor with ADPG.

Irradiation of ADPglucose synthetase in the presence of 8-N$_3$ATP, fructose 1,6-P$_2$ and Mg^{2+} resulted in covalent modification accompanying the loss of the enzyme activity only if the reaction mixture was exposed to UV light. The enzyme retained all its activity if the irradiation was done in the absence of 8-N$_3$ATP. The ADPG synthetase becomes progressively inactive with time as increasing amounts of 8-N$_3$ATP are incorporated into the enzyme. A linear incorporation of label was observed up to 70% inactivation. At this point, 0.35 mol of label was bound per mol of the enzyme subunit. This extrapolated to 0.5 mol per mol of enzyme required for complete inactivation for ATP in previous studies[16] where half site binding was observed.

Protection from covalent modification by $8\text{-}N_3ATP$ can be seen in Table III. The most effective ligands in the protection experiments were the substrates, ATP and ADPglucose. The presence of either substrate in the photolysis reaction mixture virtually completely prevented the incorporation of $8\text{-}N_3ATP$ and concomitantly protected the enzyme from inactivation. The inhibitors AMP and ADP also prevented the loss of enzymatic activity but to a lesser extent. Other nucleotide ligands, such as UTP and UDPglucose, very poor substrates for E. coli ADPglucose synthetase, provided some protection. The protection experiments were also performed at various concentrations (500 to 1500 μM) of ligands. The natural substrates (ATP and ADPglucose) protected the enzyme significantly from photoinactivation at the lower concentration and near the Km values, 50-500 μM. However, UTP and UDPglucose gave noticeable protection only at high concentrations (1500 μM). A smaller ligand, P_i did not protect the photoinactivation process to any significant extent. Only slight protection was observed with the activator fructose 1,6-P_2 and Mg^{2+}, the divalent cation cofactor. Their presence in the reaction mixture however, made ATP a better protecting ligand.

As observed with $8\text{-}N_3ATP$, irradiation of the enzyme in the presence of $8\text{-}N_3ADPglucose$ resulted in a loss of the enzyme activity. The nucleotide specificity of photoaffinity labeling by $8\text{-}N_3ADPglucose$ was demonstrated by protection experiments with various ligands (Fig. 1). The rate of photoinactivation of ADPglucose synthetase is considerably reduced in the presence of ADPglucose and certain ligands which may bind competitively at the substrate site. The best protection was noted for ADPG and ATP. AMP, UDPG or UTP gave good protection only at higher concentrations. Fig. 2 shows the dependence of the protection on the ADPG and ATP concentrations. The hyperbolic shape of the curves indicates that the nucleotides protect by binding to the enzyme and not

Table II. Kinetic Parameters of ADPglucose Synthetase with the substrates, ATP, ADPglucose, $8\text{-}N_3ATP$ and $8\text{-}N_3ADPG$.

Substrate	Activator	$S_{0.5}$[a] mM	ñ	V_{max} μmol-min^{-1}-mg^{-1}
ATP	Fructose 1,6-P_2	0.40	2.0	100
ADPG		0.07	2.2	90
$8\text{-}N_3ATP$		0.76	1.2	0.3
$8\text{-}N_3ADPG$		0.075	2.0	0.9
ATP	1,6 Hexanediol-P_2	0.15	1.1	100
ADPG		0.075	1.0	130
$8\text{-}N_3ATP$		0.27	1.2	0.7
$8\text{-}N_3ADPG$		0.075	--	0.9

The assays used for pyrophosphorolysis and for synthesis of ADPglucose[8] have been described. [a]$S_{0.5}$ indicates the concentration needed for 50% of maximal velocity. ñ is the Hill Constant.

Table III. Protection of ADPglucose synthetase (5 μM) by various ligands from photoinactivation by $8\text{-}N_3ATP$.

Compound	Conc. mM	% activity
None	--	10
Phosphate	5	20
ATP	1.5	94
ADPglucose	1.5	100
AMP	1.5	36
ADP	2.5	36
None: $Fru\text{-}P_2$ omitted	--	14
None: Mg^{2+} omitted	--	4
UTP	1.5	30
UDPglucose	1.5	46
ATP; Mg^{2+} $Fru\text{-}P_2$ omitted	1.5	51
ATP; Mg^{2+} omitted	1.5	29
ATP; $FrupP_2$ omitted	1.5	53

Ten nmol of enzyme subunit was incubated in the dark for 5 min with a 20 molar excess of $8\text{-}N_3ATP$ in 50 mM Tris-HCL, pH 7.2 containing 1 mM fructose $1,6\text{-}P_2$ and 5 mM $MgCl_2$. The reaction mixture, total volume, 0.2 to 1 ml, was placed in a "Coors" spot plate well and irradiated with a Mineral-light, model UVS-54 lamp (254 nm) at a distance of 30 cm at room temperature. Aliquots of the reaction mixture were taken out for the enzyme assay before irradiation and at various time intervals during the photolysis reaction, diluted in 50 mM Tris-HCl buffer (pH 8.5) containing 0.1mg/ml BSA. In protection experiments, a nucleotide ligand was included in the reaction mixture to give the indicated final concentration.

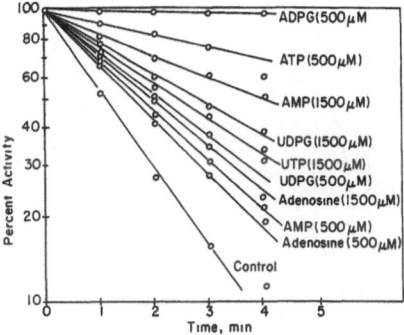

Figure 1. The effects of nucleotide and sugar nucleotide ligands on the kinetics of photoinactivation of $8\text{-}N_3ADPglucose$. ADPglucose synthetase (5μM) was inactivated at various time intervals with 10-fold molar excess of $8\text{-}N_3ADP[^{14}C]glucose$ as described in Table III. In the protection experiments, a ligand was included in the mixture at the indicated concentration. For enzyme assays, 10 μl portions was taken out before irradiation and at various time intervals during the photolysis reaction and diluted in 1.5 ml of 50 mM-Tris-HCl, pH 8.5, buffer containing 0.1 mg/ml BSA.

by "screening by UV absorption" (decreasing nitrene generation by the screening effect on activating photons). ADPglucose gives far better protection than ATP in the photoinactivation of the enzyme by 8-N_3ADPglucose. This observation is expected since ADPglucose has a higher apparent affinity for the enzyme ($S_{0.5}$, 75 μM) than ATP ($S_{0.5}$, 450 μM).

As observed with 8-N_3ATP photoinactivation there was a correlation between the extent of covalent labeling of the ADPglucose synthetase by [^{14}C]8-N_3ADPglucose and loss of enzyme activity. With loss of 80% activity there was an incorporation of 0.4 mol of 8-azido ADPglucose per mol of enzyme. ADPglucose at 1.5 mM gave complete protection of enzyme and virtually no incorporation of label was observed. With protection by ATP and AMP (1.5 mM) losses of enzyme activity were 30% and 50%, respectively. Only 0.1 to 0.14 mol of 8-azido analogue were incorporated in the ATP and AMP protection experiments.

Fig. 3 shows the HPLC fractionation of the tryptic peptides of the 8-azidoADP[^{14}C]glucose labeled ADPglucose synthetase on a C-4 Vydac column. The peptide peaks were monitored for radioactivity and as can be seen in the lower panel 7 radioactive peaks were observed. These radioactive peaks were purified further via HPLC on a RPC-18 column, large pore size from Varian or the RPC-4 column from Vydac.

It should be noted that the tryptic peptide map of the ADPglucose synthetase modified with 8-N_3[α-^{32}P]ATP was essentially similar to that observed in Fig. 3 and thus it is believed that the same sites on the enzyme were modified by 8-N_3ATP and 8-N_3ADPglucose.

Isolation of seven radioactive peptides suggested that some covalent modification by 8-N_3ADPG may have been non-specific in being at other sites other than the substrate binding area. However, with ADPglucose being present during photolysis no 8-N_3ADPG labeled peptides were found. With ATP and AMP as protectants labeling of all the peptides were decreased proportionately. Thus ADPG, AMP and ATP provided protection for all the labeled peptides and not for any specific one.

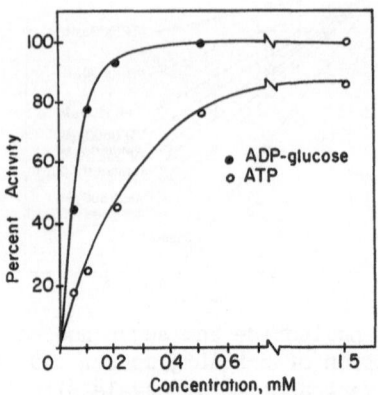

Figure 2. The effects of ADPglucose and ATP concentrations on protection from photoinactivation by 8-N_3ADPglucose. The experiment was carried out as described in the legend to Fig. 2. The photolysis reaction was terminated after 4 min by turning off the UV light and the enzyme was assayed for residual activity. Specific activity of fully active enzyme (100% activity) was 89 μmol/mg/min.

Peptide sequencing of the labeled peptides (Fig. 4) as well as amino acid analyses (Table IV) showed however that four of the seven peptides (peptides IA, IIIC, IVB and VB) were related with each other comprising the sequence from amino acid residues 107 to 128 of the enzyme. Sixty five percent of the radioactivity were found in the four peptides.

A second major binding region of $8-N_3ADP[^{14}C]$glucose (20%) is the sequence from residue 163 to 208 (tryptic peptides VIC and VIID). As indicated before, Lys^{194} is the site that was previously identified as being involved in the catalytic site binding of ADPglucose[9].

The third region where $8-N_3ADP$glucose is incorporated into the enzyme is amino acid residues 380-385 corresponding to peptide IIB. This peptide contained 15% of the total radioactivity in the seven peptides.

There are some interesting features of these 3 peptide regions. The peptide segment 107-128 which is the major binding region of the $8-N_3ADP$glucose is rich in basic amino acids, Lys at 108, Arg at 106, 114, 128 and 129. These may be essential for nucleotide substrate binding. Basic amino acids particularly arginine residues are considered to be essential for many enzymes whose substrates are phosphate esters[17].

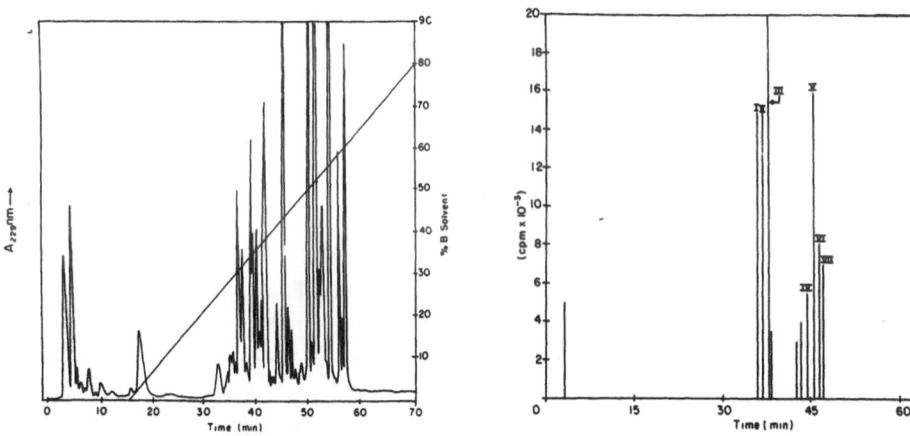

Figure 3. Peptide mapping of tryptic peptides derived from $8-N_3ADP[^{14}C]$-glucose incorporated ADPglucose synthetase.

Upper panel: HPLC chromatogram of tryptic peptides fractionated on a C-4 column. The $8-N_3ADP[^{14}C]$glucose incorporated enzyme was digested with a trypsin to substrate ratio (W/W) of 1:100 in 0.1 M ammonium bicarbonate buffer, pH 8.0 at $37°$ for 6 hrs. Following the digestion, the buffer was removed by repeated lyophilization from water.

Lower panel: Elution profile of $8-N_3ADP[^{14}C]$ incorporated peptides in chromatogram shown in the Upper panel.

About 30 nmol of tryptic peptides were applied to the Vydac 214 TP 54 reverse phase column (C-4, 0.4 X 25 cm) and the column was washed with 25 mM ammonium acetate, pH 6.0 (solvent A) for 15 minutes before a gradient elution, from solvent A to 80% of solvent B (60% acetonitrile, 40% 50 mM ammonium acetate, pH 6.0) was applied. The flow rate was 1 ml/min (a 1.5%/min increase in solvent B) at room temperature. Peptides were collected manually and concentrated in a speed vacuum concentrator (Savant).

```
              109                 114
IA^a          Gly Glu Asn Trp Tyr Arg


              107 ↓               114
IIIC          Met Lys Gly Glu Asn Trp Tyr Arg
              - - - - - - - - - - - →


                                  115                                               128
IVB^b                             Gly Thr Ala Asp Ala Val Thr Gln Asn Leu Asp Ile Ile Arg
                                  _____→


              109                         ↓                                         128
VB            Gly Glu Asn Trp Tyr Arg Gly Thr Ala Asp Ala Val Thr Gln Asn Leu Asp Ile Ile Arg
              _ _ _ _ _ _ _ _ _ _ _ _ →


              162                                                                                   187
VIID^a        Cys Thr Val Val Cys Met Pro Val Pro Ile Glu Glu Ala Ser Ala Phe Gly Val Met Ala Val Asp Glu Asn Asp Lys


              188
              Ile                                                      207
VIC^a,c       (Thr)Ile Glu Phe Val Gly Lys Pro Ala Asn Pro Pro Ser Met Pro Asn Asp Pro Ser Lys
              _ _ _ _ →


              380             385
IIB           Arg Cys Val Ile Asp Arg
              _ _ _ _ _ _ _ →
```

Figure 4. Sequences of 8-N$_3$ADP[14C]glucose incorporated peptides. An arrow (↓) indicates trypsin resistant sites created by covalent modification with the photoaffinity analogue. Underlined sequences are a portion that has actually been sequenced. a; sequences deduced from amino acid composition data (Table IV), b; characterized both by amino acid composition and -NH$_2$ terminal sequence analyses. Also, the sequence that was obtained from a major 8-N$_3$[α-32P]ATP incorporated peptide, c; two PTH amino acids (Ile and Thr) were detected at the -NH$_2$ terminus by HPLC analysis.

Table IV. Amino Acid Composition of 8-N$_3$ADP[14C]Cglucose Incorporated Tryptic Peptides

	Radioactive Peak			
	IA	IVB	VIC	VIID
CM-Cys				1.4 (2)
Asx	1.2 (1)	3.1 (3)	3.4 (3)	3.7 (3)
Thr		1.9 (2)	0.3 (1)	0.9 (1)
Ser			1.5 (2)	0.9 (1)
Glx	1.1 (1)	1.1 (1)	2.6 (1)	3.2 (3)
Pro			4.7 (5)	2.0 (2)
Gly	1.0 (1)	1.1 (1)	0.5 (1)	1.3 (1)
Ala		2.0 (2)	1.1 (1)	3.0 (3)
Val		1.0 (1)	1.3 (1)	3.8 (5)
Met			0.9 (1)	1.7 (2)
Ile		2.1 (2)	1.9 (1)	1.4 (1)
Lue		1.1 (1)		
Tyr	0.5 (1)			
Phe			1.0 (1)	1.1(1)
His				
Lys			2.0 (2)	1.0 (1)
Arg	1.0 (1)	1.0 (1)		
Trp	N.D. (1)	N.D.	N.D.	N.D.

All integral numbers in parentheses are theoretical values deduced from the known complete amino acid sequence of E. coli ADP-glucose synthetase[7]. N.D. = Not determined.

Another possible important aspect of this region is the predicted β-turn secondary structure when the structure is analyzed by the Chou and Fasman model[18]. Two strong β turns are predicted corresponding to the sharp peaks as seen in Fig. 5A for residues 109-112 (Gly-Glu-Asn-Trp) and the other is at 114-117 (Arg-Gly-Thr-Ala). A lesser possibility for a reverse turn is region 121-124 (Thr-Glu-Asn-Leu). It has been shown that reverse β-turns are abundant in globular proteins comprising about one-fourth of all residues[18,19] and that β-turns are located on the surface of proteins[20]. Reverse β-turns do not have a stable conformation and are regions of least resistance in the folding process of peptide chains. Thus, it would be of interest to see how this part of the peptide is associated with the allosteric binding between the substrates, ATP and ADPglucose and interacts with the allosteric inhibitor, AMP and activator, fructose 1,6-P$_2$ sites.

The second binding region for azidoADPG, residues 163-208, in contrast to the major peptide region, residues 107-128, has a predominance of negatively charged amino acids (Glu at 172, 173, 184 and 190 and Asp at 183, 186 and 204) and proline (residues at 168, 170, 194, 198, 199, 202 and 205). Similar to the residue region 107-128 the predicted secondary structure shows a strong potentiality of reverse β turns as seen in Figure 5B. The sharp peaks at Pro 198, Met 201 and Asp 204 could also be considered as a poly Pro conformation due to prolines at 198, 199, 202 and 205. Lysine 194, the putative binding site for ADPglucose is located between a predicted β-sheet (residues 188-192) and a reverse β-turn structure. Thus it is at the conformational boundary and the lys side chain could be at the surface of the enzyme well exposed to the solvent.

Residues 380-385, the third peptide region where 8-N$_3$ADPG was incorporated has a predominace of Arg residues at 379, 380 and 385. This could be essential for the electrostatic attraction of substrate to the enzyme. Arg is also present at residues 374 and 377. The predicted secondary structure of the residues 380-385 is a β-sheet conformation.

These 3 photoaffinity probe incorporated regions must be proximal to each other in the tertiary structure of the ADPglucose synthetase. Thus the substrate binding locus is possibly composed of regions of distal amino acids in the primary structure.

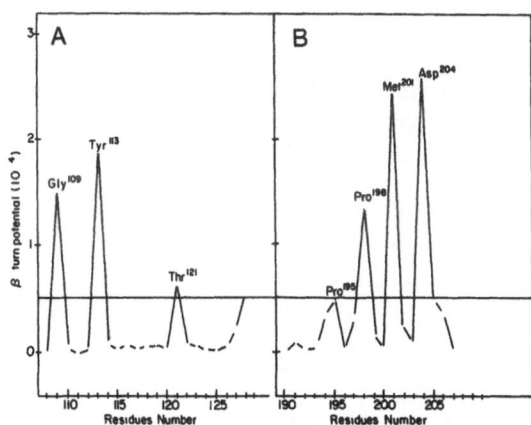

Figure 5. Prediction of reverse β-turn secondary structure by conformational analysis of two peptide segments where 8-N$_3$ADP[$_{14}$C]glucose binds covalently. A; peptide, sequence residues 108-128, B; peptide, sequence residues 189-207.

Studies involving the use of 8-N$_3$AMP

ADPglucose synthetase activity is inhibited by 8-N$_3$AMP. At lower concentrations of the activator, fructose 1,6-P$_2$, the enzyme is more sensitive to inhibition to 8-N$_3$AMP and as shown in table V this increased sensitivity to inhibition is similar to what was observed for the natural inhibitor AMP. However, 8-N$_3$AMP is a poorer inhibitor than AMP. At 50 μM Fructose 1,6-P$_2$, 12.6 times more 8-N$_3$AMP is required for 50% inhibition while at 1.5 mM fructose 1,6-P$_2$, 4.4 times more of the azido analogue is needed for 50% inhibition. Nevertheless, 8-N$_3$AMP appears to be similar to AMP as an inhibitor.

Photoinactivation of the ADPglucose synthetase by 8-N$_3$AMP concentrations 2-to 20-fold over the subunit concentration followed pseudo first order kinetics at the various azido analogue concentrations. The inactivation required UV light as incubations with the azido analogue in the dark did not cause inactivation of the enzyme. Table VI shows that ATP and ADPglucose but not 5'AMP could effectively protect the enzyme from photoinactivation. However if P$_i$ was added along with 5'AMP protection of the activity was 57% and therefore was as effective as ATP. Both P$_i$ and adenine could not significantly protect the enzyme from inactivation and thus the protections observed for the adenine nucleotides was not due to a shielding effect caused by UV absorption of the adenine ring.

TABLE V.   Inhibition Kinetics of ADPG Synthetase

Inhibitor	Fructose-P$_2$ Conc.	$^a I_{0.5}$ μM	Hill Slope ñ
AMP	50 μM	3.4	1.4
	1.5 mM	70	1.8
8-N$_3$AMP	50 μM	43	1.1
	1.5 mM	307	1.8

$^a I_{0.5}$ is the concentration required for 50% inhibition.

TABLE VI.   Effect of Ligands on 8-AzidoAMP Photoinactivation of ADPG Synthetase.

Ligand	Concentration (mM)	% Activity
none	–	2
ATP	2.1	50
ADPglucose	2.0	92
5' AMP	1.9	33
P$_i$	10	0
Adenine	2.0	20
5' AMP + P$_i$	1.9 + 10	57

Enzyme (25 μM subunit) was photoinactivated in the presence of 0.47 mM 8-N$_3$AMP and 50 μM fructose 1,6-P$_2$ for 15 minutes.

Since ATP and ADPglucose were the most efficient protectors and that P_i addition to AMP enhanced AMP protection it is possible that not only is 8-N_3AMP binding to the inhibitor site but that it is also binding to the substrate site or that there is some overlap between the two ligand sites.

Fig. 6 shows the relationship between inactivation by [^{3}H]8-N_3AMP and the extent of incorporation. Loss of activity follows pseudo first order kinetics up to at least 65% inactivation and extrapolates to 0.4 mol of 8-N_3AMP bound per mol of subunit. Complete inactivation requires about 0.8 to 1.0 mol of 8-N_3AMP bound per mol of subunit. There are at least two explanations for these results. Inactivation may occur initially at one site per dimer causing a conformational change resulting inactivation at all 4 subunits or non-specific labeling is occurring in addition to the specific site labeling per dimer.

Experiments were designed to identify the sites covalently modified. Enyzme, was irradiated in the presence of 0.5 mM [^{3}H]8-N_3AMP (1.5 x 10^6 cpm/μmol), Tris-Cl buffer, pH 7.2, 50 μM fructose 1,6-P_2, for 10 minutes. The enzyme was inactivated about 70%. The photolabelled enzyme was caboxymethylated with iodoacetic acid[8] and dialyzed against 6M guanidine-HCl or chromatographed on Sephadex G-50 in the presence of 6M guanidine-HCl in 50 mM Tris-Cl buffer, pH 7.2.

Fig. 7 shows that 5 major radioactive peaks were obtained after digestion by trypsin and reverse phase HPLC on a C18 column. The figure also shows the inhibition of incorporation by the addition of either 5 mM AMP, ATP or ADPG. Although the amount of [^{3}H]8-N_3AMP was significantly reduced, ATP and ADPG protected peaks ^{3}H-3 and ^{3}H-4 better than did AMP. ATP and ADPG protected enzyme retained 100% activity following photolysis, while AMP protected enzyme retained 76% activity as compared to 30% activity for the unprotected enzyme.

The finding that the tryptic digestion HPLC showed the same retention time for most [^{3}H]8-N_3AMP, [^{32}P]8-N_3ATP, and [^{14}C]8-N_3ADPG incorporated peptides on a C-18 column suggests strongly that the binding

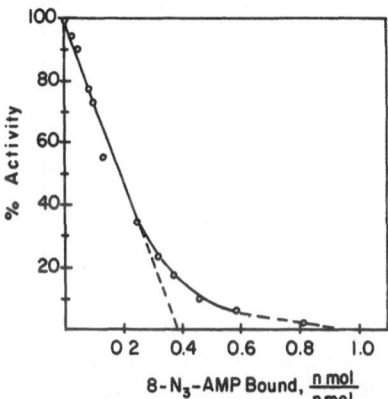

Figure 6. The effect of 8-N_3AMP incorporation on ADPglucose synthetase activity. 50 μM subunit of enzyme was photoinactivated with 0.5 mM [^{3}H]8-N_3AMP (1.5 x 10^7 cpm/μmol) in the presence of 5 mM MgCl$_2$ and 1 mM fructose 1,6-P_2 for 0 to 45 minutes. After aliquots were assayed for synthesis activity, the samples were exhaustively dialyzed against 6M guanidine HCl in 50 mM Tris-Cl buffer, pH 7.2, 1 mM EDTA and 50 mM KCl to remove non-covalently bound label. Aliquots were then counted for radioactivity.

site(s) of the substrates, ATP and ADPG and the allosteric inhibitor AMP overlap or may be identical.

We thus have undertaken studies to determine the sequence of the various 8-N$_3$AMP labeled peptides. Two approaches have been used to isolate the purified peptides. After covalent modification with 8-N$_3$AMP the enzyme was degraded by CNBr[21]. The generated peptides were subjected to Sephadex G-50 chromatography and then to phosphocellulose chromatography using a pH gradient with ammonium acetate buffer from pH 4.0 to 6.0. Four peaks of radioactivity was obtained and one of these peaks which accounted for 33% of the total label was treated with alkaline phosphatase in 4M urea with Tris-Cl buffer, pH 8.5 for 4 hours to remove the terminal phosphate. The alkaline phosphatase was separated from the peptide via gel filtration on a Sephadex G-75 column equilibrated with 20 mM ammonium acetate, pH 4.0, and 8M urea. The peptide fraction was rechromatographed on a phosphocellulose column. This peak was then purified via HPLC on a C-18 column and one radioactive peak was resolved from a number of UV absorbing peaks. Rechromatography of this peak gave just one A_{229nm} absorbing peak. The peptide was sequenced by Automated Edman degradation and each residue measured for radioactivity. The peptide sequence corresponded to a known sequence ranging from amino acid residue 109 to 152 in the ADPG synthetase subunit. Sequencing was continued to residue 127 and the sequence is shown below.

^{109}Lys-Gly-Glu-Asn-Trp-Tyr-Arg-Gly-Thr-Ala-Asp-Ala-Val-Thr-Gln-Asn-Leu-Asp-Ile^{127}

The tyrosine residue at 114 contained the [^{3}H]8-N$_3$AMP label.

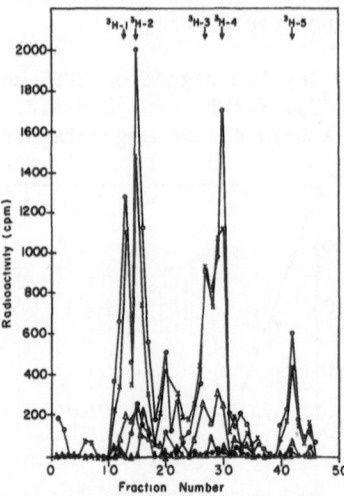

Figure 7. Elution profile of [^{3}H]8-N$_3$AMP incorporated peptides. Four nmol of tryptic digest peptides obtained from 70% photoinactivated enzyme were chromatographed on a protein C-18 (4.0 x 300 nm) Varian column. The column was washed for 5 min with solvent A (0.1% TFA in water) prior to a gradient elution of 0-60% solvent B (0.075% TFA in acetonitrile) in 60 min. The peptides were monitored at 214 nm. Each peak was collected manually to monitor radioactivity. o-o represents enzyme photoinactivated in the absence of protecting ligands. X-X represents photoinactivation in the absence of 5 mM MgCl$_2$. Δ-Δ, ●-● and ▲-▲ represent conditions where 5 mM AMP, 5 mM ATP and 5 mM ADPG were added as protectants. The tryptic digestion procedure is described in Fig. 3.

In order to purify all the labelled peptides a complete purification of the CNBr cleaved peptides was done using HPLC. However CNBr degradation gave us a soluble fraction containing 30% of the total radioactivity and an insoluble fraction containing 70% of the total radioactivity. The insoluble fraction was then digested with mouse sub-maxillary arginyl specific protease (MSAP).

The soluble fraction gave us two major radioactive peptides after HPLC chromatography and they are shown in Fig. 8. Also observed in the figure is the protection from labeling by 5 mM ADPG and 5 mM AMP. Both peaks are protected from labeling by ADPG while AMP gives better protection of peak ^{3}H-II than ^{3}H-I. In another experiment 5 mM ATP protected ^{3}H-I better than ^{3}H-II from labelling with 8-N$_3$AMP. In the experiment depicted in Fig. 8, The unprotected enzyme had only 15% of the original activity. ADPG-protected enzyme had 100% of the activity while AMP-protected enzyme retained 55% of the original activity. Since ^{3}H-II appeared to be an AMP-protected peak it was purified further by HPLC. Amino acid analysis of the peak showed that its composition corresponded to a cyanogen bromide fragment of amino acid residues 12 to 69. This corresponded to the region where the allosteric activator binding site for fructose 1,6-P$_2$ was situated (Lys 38).

HPLC analysis of the CNBr insouble fraction which was digested with MSAP yielded five peaks of radioactivity. The dominant radioactive peak when purified further it yielded two radioactive peaks. Thus far amino acid analysis has been done on the peaks and the analysis indicates that the peptides corresponds with a peptide of residues 116-129 and a peptide with residues 116-130. The difference between the two peptides are that one has two arginine residues at the C-terminal end. It is important to note that these peptides are part of the CNBr-5 peptide which had been sequenced before.

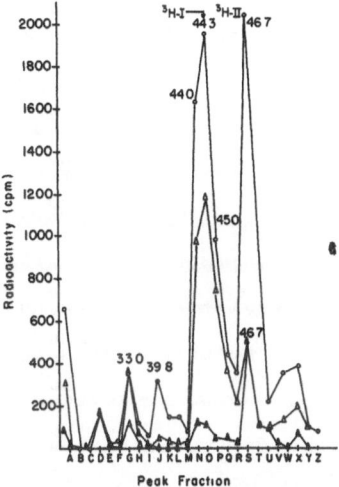

Figure 8. HPLC of the soluble fraction of CNBr peptides. Ten nmol of CNBr cleaved peptides obtained from 80% photoinactivated enzyme (the photoinactivation procedure is described in Figure 1 and Table III) were chromatographed on a protein C-4 (4.0 x 250 nm, Vydac) column. HPLC was done as indicated in Fig. 7 except that solvent B was (0.075% TFA in propanol) and the gradient elution was 0-70% of solvent B in 80 minutes. o-o, unprotected sample. ▲-▲, 5 mM ADPG protected sample, Δ-Δ, 5 mM AMP protected sample. The numbers in the figure indicate the elution time (min).

Fig. 9 shows the complete deduced amino acid sequence of the *E. coli* ADPG Synthetase. Asterisks at lys 39 and 195 correspond to the fructose 1,6-P_2 (activator) binding site and putative substrate binding site, respectively. Three [^{3}H]8-N_3AMP photolabelled peptides located in two regions of the enzyme have been observed. A major CNBr peptide that is labeled contains residues 109-152. A specifically modified Tyr-114 was found in the purified peptide and contained 33% of the total radioactivity. In that same region a MSAP peptide, residues 116-130 was found to contain 25% of the total label which indicated amino acid residues 114 to 130 contained about 60% of the total label of [^{3}H]8-N_3AMP incorporated into ADPG synthetase.

A second CNBr peptide (residues 12-69) labeled with [^{3}H]8-N_3AMP photoinactivated enzyme (containing 20% of the total radioactivity) is located near the N-terminus and contains also the allosteric activator site.

With the exception of the N-terminus site the 8-N_3ADPG major binding site, where 65% of the total label of the azido sugar nucleotide is incorporated, is very similar to the major 8-N_3AMP binding site; namely residues 109-128. Distinctive site(s) for 8-N_3ADPG are situated between residues 162-207 and residues 380-385.

The results seem to indicate that there is an overlapping between the substrate and inhibitor sites and also possible an interaction between the inhibitor and activator binding sites. Previous studies of the binding of activator, fructose 1,6-P_2 and inhibitor AMP have shown that although there are four sites for each ligand per tetrameric enzyme, the total number of activator plus inhibitor site filled with ligand when

10 20 30
MET VAL SER LEU GLU LYS ASN ASP HIS LEU MET LEU ALA ARG GLN LEU PRO LEU LYS SER VAL ALA LEU ILE LEU ALA GLY GLY ARG GLY
40 50 60
THR ARG LEU LYS ASP LEU THR ASN LYS*ARG ALA LYS PRO ALA VAL HIS PHE GLY GLY LYS PHE ARG ILE ILE ASP PHE ALA LEU SER ASN
70 80 90
CYS ILE ASN SER GLY ILE ARG ARG MET GLY VAL ILE THR GLN TYR GLN SER HIS THR LEU VAL GLN HIS ILE GLN ARG GLY TRP SER PHE
100 110 120
PHE ASN GLU GLU MET ASN GLU PHE VAL ASP LEU LEU PRO ALA GLN GLN ARG MET LYS GLY GLU ASN TRP TYR ARG GLY THR ALA ASP ALA
130 140 150
VAL THR GLN ASN LEU ASP ILE ILE ARG ARG TYR LYS ALA GLU TYR VAL VAL ILE LEU ALA GLY ASP HIS ILE TYR LYS GLN ASP TYR SER
160 170 180
ARG MET LEU ILE ASP HIS VAL GLU LYS GLY VAL ARG CYS THR VAL VAL CYS MET PRO VAL PRO ILE GLU GLU ALA SER ALA PHE GLY VAL
190 200 210
MET ALA VAL ASP GLU ASN ASP LYS THR ILE GLU PHE VAL GLY LYS*PRO ALA ASN PRO PRO SER MET PRO ASN ASP PRO SER LYS SER LEU
220 230 240
ALA SER MET GLY ILE TYR VAL PHE ASP ALA ASP TYR LEU TYR GLU LEU LEU GLU GLU ASP ASP ARG ASP GLU ASN SER SER HIS ASP PHE
250 260 270
GLY LYS ASP LEU ILE PRO LYS ILE THR GLU ALA GLY LEU ALA TYR ALA HIS PRO PHE PRO LEU SER CYS VAL GLN SER ASP PRO ASP ALA
280 290 300
GLU PRO TYR TRP ARG ASP VAL GLY THR LEU GLU ALA TYR TRP LYS ALA ASN LEU ASP LEU ALA SER VAL VAL PRO LYS LEU ASP MET TYR
310 320 330
ASP ARG ASN TRP PRO ILE ARG THR TYR ASN GLU SER LEU PRO PRO ALA LYS PHE VAL GLN ASP ARG SER GLY SER HIS GLY MET THR LEU
340 350 360
ASN SER LEU VAL SER GLY GLY CYS VAL ILE SER GLY SER VAL VAL VAL GLN SER VAL LEU PHE SER ARG VAL ARG VAL ASN SER PHE CYS
370 380 390
ASN ILE ASP SER ALA VAL LEU LEU PRO GLU VAL TRP VAL GLY ARG SER CYS ARG LEU ARG ARG CYS VAL ILE ASP ARG ALA CYS VAL ILE
400 410 420
PRO GLU GLY MET VAL ILE GLY GLU ASN ALA GLU GLU ASP ALA ARG ARG PHE TYR ARG SER GLU GLU GLY ILE VAL LEU VAL THR ARG GLU
430
MET LEU ARG LYS LEU GLY HIS LYS GLN GLU ARG

Figure 9. The deduced amino acid sequence of *E. coli* ADPG synthetase. The underlined portions correspond to sequences determined by Edman degradation. The asterisks at Lysines 39 and 195 correspond to the activator binding site and the postulated ADPG binding site, respectively.

fructose 1,6-P_2 and AMP are present is never greater than four[16]. The binding data plus the 8-N_3AMP covalent modification data would certainly strongly suggest interaction of the two allosteric sites via conformational change.

It would certainly be of interest to obtain X-ray crystallographic data on the three-dimensional structure of the ADPglucose synthetase to understand the possibilities and modes of interaction of the allosteric activator, inhibitor and substrate catalytic sites.

DNA recombinance studies with E. coli K12 618; an allosteric mutant

Another approach that we have taken to learn more about the structure-function relationships of the various allosteric sites has been to determine the amino acid substitutions that have occurred in some of the allosteric mutants of the E. coli ADPG synthetase. One such mutant is E. coli K12 strain 618[14]. As seen in Table VII the kinetic constants of the mutant enzyme have been altered. It has a higher apparent affinity for the activator, fructose 1,6-P_2 and has less sensitivity towards AMP inhibition than the wild type. The mutant E. coli containing this altered enzyme accumulates glycogen at a higher level and at a faster rate than the parent strain.

Since we had developed procedures to clone the glycogen biosynthetic genes of E. Coli K12[6], it was relatively easy for us to apply the same procedures to clone the glg genes of the E. coli K12 mutant 618, particularly glg C, the structural gene for the ADPG synthetase. The glg and glg B genes of mutant 618 were cloned onto plasmid pBR322 by insertion of the chromosomal DNA at the Pst I site by cotransformation with the Asd gene into an E. coli mutant, strain 6281, defective in Asd as well as in branching enzyme activity (glg B). Cells were then selected by growth on agar with enriched media containing 20 µg of tetracycline per ml. Strain 6281 requires diaminopimelic acid for growth even in enriched media because of the defective Asd gene. The Asd gene is very near the glg genes and thus the Asd gene was selected for by growth in media not containing diaminopimelic acid. Two transformants were isolated which displayed the phenotype Asd$^+$ Tet^R Amp^S. When the transformed cells were grown in enriched media containing 1% glucose and stained with I_2 they stained brownish-black indicating that they contained glycogen and branching enzyme activity.

The restriction endonuclease cleavage map of the plasmid containing the mutant glg C gene, pEBL 1, with 9 restriction nucleases, was similar to the restriction map of the cloning vehicle containing the "wild type" glg C gene, pOP12[6] except that a 1.6 Kb portion of the glg A gene (glycogen synthase) was missing as well as a 0.9 Kb fragment upstream from the Asd gene.

Fig. 10 shows the subcloning of the glg C gene and the DNA sequencing strategy of the mutant structural gene. A 1.9 Kb Hinc II fragment of pEBL 1 was cloned onto pUC8 and the subclone was designated as pLP228. Fig. 10 also shows the various DNA fragments sequenced by the Maxam and Gilbert technique. Since the functionally important sites of ADPG synthetase are located within 210 residues from the amino-terminus of the known primary structure, the DNA corresponding to that region of the glg C gene (about 600 Kb) was sequenced first to compare with the known wild type sequence.

Fig. 11 shows the nucleotide sequence obtained for the anti-sense strand where base changes were noted. At Bases 480, 482, 489 and 497, the mutagenesis with nitrosoguanidine caused a substitution of cytosine for thymine. In two of the four cases the change caused no change in amino acid. However in the other two cases residues 161 and 166 the amino acid valine was replaced by alanine in the mutant.

It is of interest that where the valine to alanine substitutions occur in the mutant enzyme is also near the site involved in substrate binding, residues 162-207. Thus the nature of amino acids in the substate binding site area may also play a role in the determination of the interactions of the activator and inhibitor sites with the catalytic sites. Further studies that will also involve oligonucleotide directed mutagenesis should provide more information in the nature and function of the amino acids involved in the regulation of ADPglucose synthesis and whether the alanine amino acid replacements are the cause for the alteration in the kinetic properties of the mutant enzyme.

Table VII. Kinetic Parameters of E. coli K12 and E. coli K12 mutant, 618 ADPG Synthetase

Bacterium	Ligand	$A_{0.5}$	$I_{0.5}$	Hill Slope
E. Coli K12	Fructose 1,6-P_2	62 μM	–	1.7
	AMP	–	66 μM	1.9
Mutant 618	Fructose 1,6-P_2	15 μM	–	1.0
	AMP	–	860 μM	1.9

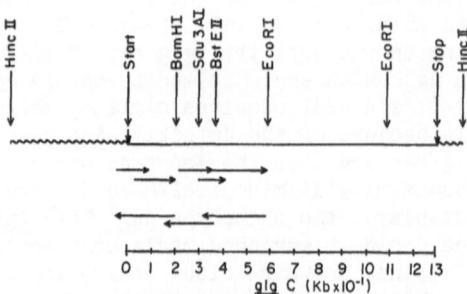

Figure 10. Sequencing strategy for the glg C region. The above arrows are oligonucleotide fragments of the anti sense strand while the arrows in the lower part of the figure are the oligonucleotide fragments of the sense strand. The arrows indicate the direction and extent of the sequence. The labeling of the oligonucleotide fragment is at the base of the arrow.

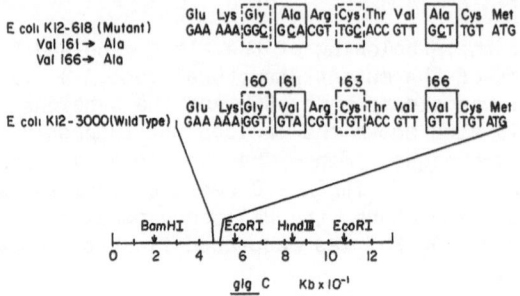

Figure 11. Mutation sites and the surrounding DNA sequences of ADPglucose synthetase structural gene from the wild type and mutant.

REFERENCES

1. J. Preiss, S.-G. Yung, and P. A. Baecker, Regulation of bacterial glycogen synthesis, Molec. Cell Biochem. 57:61-80 (1983).
2. J. Preiss, Bacterial glycogen synthesis and its regulation, Ann. Rev. Microbiol. 38:419-458 (1984).
3. J. Preiss, Regulation of adenosine diphosphate glucose pyrophosphorylase, Adv. Enyzmol. Related areas Mol Biol. 46:317-381 (1978).
4. J. Preiss, and D. A. Walsh, The Comparative biochemistry of glycogen and starch, In V. Ginsberg (ed). Biology of Carbohydrates, Vol. 1, John Wiley and Sons, Inc., New York (1981) pp 199-314.
5. T. H. Haugen, A. Ishaque, A. K. Chatterjee, and J. Preiss, Purification of Escherichia coli ADPglucose pyrophosphorylase by affinity chromatography, FEBS letters, 42:205-208 (1974).
6. T. W. Okita, R. L. Rodriguez, and J. Preiss, Biosynthesis of bacterial glycogen: cloning of the glycogen biosynthetic enzyme structural genes of Escherichia coli, J. Biol. Chem., 256:6944-6952 (1981).
7. P. A. Baecker, C. E. Furlong, and J. Preiss, Biosynthesis of Bacterial Glycogen: Primary structure of Escherichia coli ADPglucose synthetase as deduced from the nucleotide sequence of the glg C gene. J. Biol. Chem., 258:5084-5086 (1983).
8. T. H. Haugen, A. Ishaque, and J. Preiss, Biosynthesis of Bacterial Glycogen: Characterization of the subunit structure of Escherichia coli B glucose-1-phosphate adenylyltransferase (ec 2.7.7.27), J. Biol. Chem., 251:7880-7885 (1976).
9. T. F. Parsons, and J. Preiss, Biosynthesis of bacterial glycogen. Isolation and characterization of the pyridoxal-P allosteric activator site and the ADPglucose-protected pyridoxal-P binding site of Escherichia coli B ADPglucose synthase, J. Biol. Chem., 253:7638-7645 (1978).
10. W. K. Kappel, and J. Preiss, Biosynthesis of Bacterial glycogen: purification and characterization of ADPglucose pyrophosphorylase with modified regulatory properties from Escherichia coli B mutant CL 1136-504, Arch. Biochem. Biophys. 209:15-28 (1981).
11. C. A. Carlson, and J. Preiss, Modification of the allosteric activator site of Escherichia coli ADPglucose synthetase by trinitrobenzenesulfonate, Biochemistry, 20:7519-7528 (1981).
12. C. A. Carlson, and J. Preiss, Involvement of Arginine residues in the Allosteric Activation of Escherichia coli ADPglucose synthetase, Biochemistry, 21:1929-1934 (1982).
13. R. Potter, and B. E. Haley, Photoaffinity labeling of nucleotide binding sites with 8-azidopurine analogs: Techniques and applications, Methods Enzymol. 91:613-633 (1982).
14. N. Creuzat-Sigal, M. Latil-Damotte, J. Cattoneo, and J. Puig, Genetic analysis and biochemical characterization of mutants impairing glycogen metabolism in Escherichia coli K12. In Biochemistry of the Glycosidic linkage (R. Piras, and H. G. Pontis, eds.), pp. 647-680, Academic Press, New York (1972).
15. K. Muneyama, R. J. Bauer, D. A. Shuman, R. K. Robins, and L. N. Simon, Chemical synthesis and biological activity of 8-substituted adenosine 3',5'-cyclic monophosphate derivatives, Biochemistry, 10:2390-2395 (1971).
16. T. H. Haugen, and J. Preiss, Biosynthesis of Bacterial Glycogen: The nature of the binding of substrates and effectors to ADPglucose synthase, J. Biol. Chem., 254:127-136 (1979).
17. J. G. Riordan, K. D. McElvany, and C. L. Borders, Arginyl residues: anion recognition sites in enzymes. Science, 195:884-886 (1977).

18. P. Chou, and G. D. Fasman, Conformational parameters for amino acids in helical, β sheet and random coil regions calculated from proteins. <u>Biochemistry</u>, 13:211-222 (1976).

19. J. L. Crawford, W. N. Lipscomb, and C. G. Schellman, The reverse turn as a polypeptide conformation in globular proteins. <u>Proc. Natl. Acad. Sci. USA</u>, 70:538-542 (1973).

20. I. D. Kuntz, <u>J. Amer. Chem. Soc</u>., protein folding. 94:4009-4012 (1972).

21. E. Gross, The cyanide bromide reaction, <u>Methods Enzymol</u>. 11:238-255.

ACTIVE SITE AND OTHER SEQUENCE DATA FROM TORPEDO CALIFORNICA

ACETYLCHOLINESTERASE

Kathleen MacPhee-Quigley, Thomas S. Vedvick, Palmer Taylor, and
Susan S. Taylor

Department of Chemistry and Division of Pharmacology
University of California, San Diego
La Jolla, CA 92093

INTRODUCTION

Two distinct molecular forms of acetylcholinesterase are found in the
electric organ of Torpedo californica: a dimensionally asymmetric species
localized in the basal lamina of the synapse and a globular, hydrophobic
species associated with the plasma membrane (1-3). The asymmetric species
are comprised of two or usually three tetrametric sets of catalytic subunits
linked to both collagenous and non-collagenous structural subunits. These
elongated species sediment as discrete 13S and 17S species respectively and
as an 11S species after limited proteolytic treatment (4-6). The hydrophobic
globular species is a dimer of catalytic subunits with a sedimentation constant of 5.6S in the presence of a neutral detergent. The catalytic subunits
of these two forms exhibit small differences in electrophoretic migration,
amino acid composition, monoclonal antibody affinity, and peptide mapping
(3,4,7). Primary structural studies were undertaken in order to define
specific functional sites and, in addition, to elucidate whether the observed
structural differences between the two acetylcholinesterase species are the
result of multiple genes, alternate mRNA transcripts, or post-translational
modifications.

CHARACTERIZATION OF THE ACTIVE SITE PEPTIDE

The procedures for purifying both forms of acetylcholinesterase have
been described previously (3,6). A more detailed account of the isolation
and characterization of some of the active site peptides described here has
also been recently reported (8). In order to identify the reactive serine
residue, both the 5.6S and 11S forms of acetylcholinesterase were covalently
labeled with [^{3}H]diisopropyl fluorophosphate (DFP) and aged. The enzyme
(25-40 mg at a concentration of 1-2 mg/ml) was then allowed to react with a
slight molar excess of [^{3}H]diisopropyl fluorophosphate (specific activity,
4 Ci/mmol) in the presence of 0.02% NaN$_3$ for approximately 24-48 hours to
achieve complete inhibition and aging. ^{3}Following removal of the unreacted
DFP by dialysis, the sample was lyophilized and resuspended in 6M guanidine
HCl in 0.1 M Tris·HCl, at pH 8, to a concentration of 2-4 mg/ml. Following
the addition of a two-fold molar excess of dithiothreitol over the total
protein cysteine residues and bubbling with nitrogen, the preparation was
allowed to incubate at 50°C for 3 hours. The cysteines were then labeled

This work was supported in part by United State Public Health Service Grant
#GM18360 and DAMD Contract #17-83C-3202 Basic to P.T.

with [^{14}C]iodoacetic acid (specific activity, 10–50 mCi/mmol) by incubating for 1 hour at 25°C in the dark under nitrogen with a two-fold molar excess of IAA over total thiols. The reaction was stopped with a ten-fold excess of dithiothreitol and then dialyzed against 50 mM NH$_4$HCO$_3$, pH 8.3. Trypsin (1% wt/wt) was added to the carboxymethylated acetylcholinesterase and incubated overnight at 37°C. The resulting peptides were applied to a Sephadex G-50 (-40) super-fine column (1.5 x 200 cm) that had been equilibrated in 50 mM NH$_4$OH. The eluant was collected in 3 ml fractions at a flow rate of 30 ml/hr. Fractions were pooled, lyophilized, and then dissolved in 0.1% trifluoracetic acid. Pooled peptides were separated by reverse-phase HPLC on a Waters μBondapak C-18 or a Vydac C-18 column using an aqueous 0.1% trifluoracetic acid –acetonitrile gradient of 0–50% in 2 hours. Absorbance at 219 nm, [^{3}H] and [^{14}C] radioactivity were monitored. One major [^{3}H] peak was observed following elution from Sephadex G-50 (7), and Figure 1 illustrates the HPLC profile of this pooled fraction from the 11S species. This Sephadex fraction, which was pooled based on the DFP-labeling, also fortuitously contained both the NH$_2$- and COOH-terminal peptides as shown. The tritium was associated with a single well-resolved peptide and eluted at an identical position for both proteins. Amino acids analyses were carried out using an LKB 4400 amino acid analyzer. Peptides dissolved in 6N HCl were hydrolyzed <u>in vacuo</u> at 110°C for 20–24 hours. Table 1 shows the amino acid composition of the active site peptide as well as the NH$_2$- and COOH-terminal peptides. There were no significant differences in the amino acid compositions for the DFP-labeled peptide and the NH$_2$-terminal peptide isolated from both the 11S and 5.6S forms of acetylcholinesterase.

Dansyl-Edman degradations were performed manually. Solid phase sequencing was done on an automated sequenator modeled after the instrument described by Doolittle et al. (9), employing standard methods (10). Peptides (10–40 nmol) were coupled to phenyl diisothiocyanate glass beads (30 mg) in a

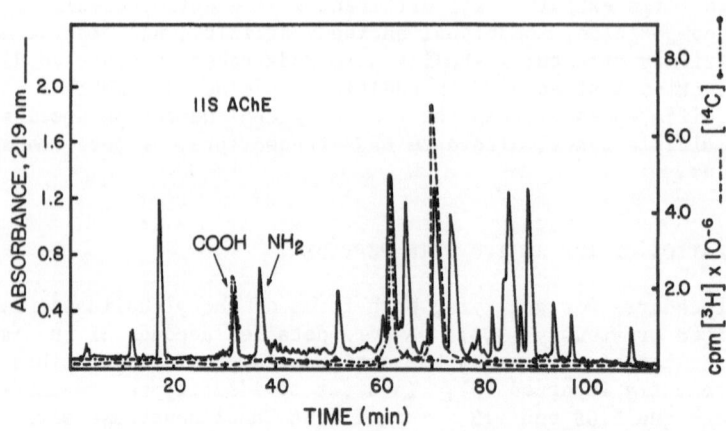

Figure 1. HPLC Purification of Sephadex fraction IV from the 11S species containing the [^{3}H]isopropylphosphonyl-labeled peptide.

The region of the Sephadex profile containing the DFP-label was pooled, lyophilized and separated on a reverse-phase C$_{18}$-μBondapack column using a gradient with aqueous 0.1% trifluoroacetic acid and acetonitrile between 0 and 50%. Fractions of 1 ml were collected over 2 hours. Absorbance at 219 nm, ———; [^{14}C]carboxymethyl-labeled peptides, -•-··-; [^{3}H]isopropylphosphoryl-labeled peptides, - - -. The NH$_2$-terminal peptide is identified with NH$_2$, the COOH-terminal is identified with COOH.

TABLE 1

Amino Acid Compositions of 11S AChE Tryptic Peptides

Amino Acid	DFP-labeled	NH$_2$-terminal	COOH-terminal
CM-Cys			1
Asx		3	
Thr	2		
Ser	4	1	1
Glx	1	1	2
Pro	1		
Gly	5	2	
Val	2	1	
Met	1		
Ile	2		
Leu	1	2	1
Phe	1		
His	1	1	1
Lys		1	
Arg	1		
NH$_2$-terminal	THR	ASX	HIS

Numbers are experimental values, rounded to the nearest whole number, from the averages of two or more determinations. Asx and Glx include both the acid and the amide. NH$_2$-terminal residues indicated were determined by dansylation.

buffer containing 0.1 M NaHCO$_3$, 10% 1-propanol, pH 9.3, and incubated overnight at 25°C (11). Thiazolinone amino acids were converted to their phenyl thiohydantoin derivatives by incubating at 80°C for 10 minutes with 0.2 ml of 1.5N HCl in methanol and drying under N$_2$. The phenylthiohydantoin amino acids were identified by HPLC according to the method of Bhown et al. (12) with

11S AChE

TRYPTIC

SPINNING CUP: THR-VAL-Thr-ILE-PHE-GLY-GLU- X -ALA-GLY-GLY-ALA-Ser-VAL-GLY-MET-HIS-ILE-LEU-Ser-PRO-GLY(Ser,Arg)

MANUAL: THR-VAL-THR-ILE-PHE-GLY(Gly, Ser, Ala, Gly, Gly, Ala, Ser, Val, Gly, Met, His, Ile, Leu, Ser, Pro, Gly, Ser, Arg)

CHYMOTRYPTIC

MANUAL: THR-VAL-THR-ILE -PHE-GLY(Glx, Ser, Ala, Gly, Gly, Ala, Ser, Val, Gly, Met, His) ILE-LEU-SER-PRO-GLY-SER-ARG

peptide A   peptide B

5.6S AChE

TRYPTIC

GAS PHASE: THR-VAL-THR-ILE-PHE-GLY-GLU- X -ALA-GLY-GLY-ALA-Ser-VAL-GLY-MET-HIS-ILE-LEU-Ser-PRO-GLY(Ser,Arg)

SOLID PHASE: THR-VAL-GLY-ILE-PHE-Gly-Glu-

Figure 2.   Sequence analyses of the 5.6S and 11S species of the active site tryptic peptides and 11S chymotryptic peptide (c.f. Reference 8).

The [³H]diisopropylphosphoryl-labeled serine is indicated by the large solid arrow. Residues that were identified unambiguously are indicated by a slash beneath the residue. Isolation and characterization of the chymotryptic peptides have been described elsewhere (8).

Buffer A at pH 3.9. Liquid phase sequencing was done using a Beckman Spinning Cup Sequencer (Model 890C). Gas phase sequencing was carried out using an Applied Biosystems Protein Sequencer (Model 470A).

The active site peptides were sequenced by a combination of manual dansyl-Edman degradation and automated sequencing using solid phase, spinning cup, and gas phase methodologies as summarized in Figure 2. The [^{3}H]labeled peptide from the 11S species was further characterized by a chymotryptic (1% wt/wt) digestion. The resulting peptides were separated by HPLC (profile not shown)(8). The reactive serine was localized at position 8 by the radioactivity [^{3}H] associated with that residue. The reactive serine peptide was isolated three times from the 5.6S species and twice from the 11S species.

IDENTIFICATION AND SEQUENCE OF NH$_2$-TERMINAL REGION

As shown in Figure 3, the amino terminal sequences of both enzyme forms of acetylcholinesterase were determined initially by spinning cup sequencing of the whole protein. These sequences were subsequently confirmed and extended by gas-phase sequencing of the whole proteins. Since the first four residues of the 11S enzyme form showed two amino acids at each step and, in addition, to elucidate and further confirm the NH$_2$-terminal sequences, the tryptic peptides following HPLC separation of all the pooled Sephadex fractions were screened by dansylation for peptides having an NH$_2$-terminal aspartic acid. One such peptide having the appropriate amino acid composition was fortuitously found in Fraction IV for the 11S enzyme and is identified in Figure 1. This peptide was sequenced by solid phase, gas phase, and manual dansyl-Edman methods, and is included in Figure 3. The amino terminal tryptic peptide of the 5.6S form is also indicated in Figure 3. This peptide from the 5.6S protein was found in the adjacent Sephadex fraction compared to the 11S enzyme. The composition of the 11S peptide is shown in Table 1. In addition, the tryptic peptide corresponding to residues 20-44 was subsequently sequenced and placed in this amino terminal region of the polypeptide chain.

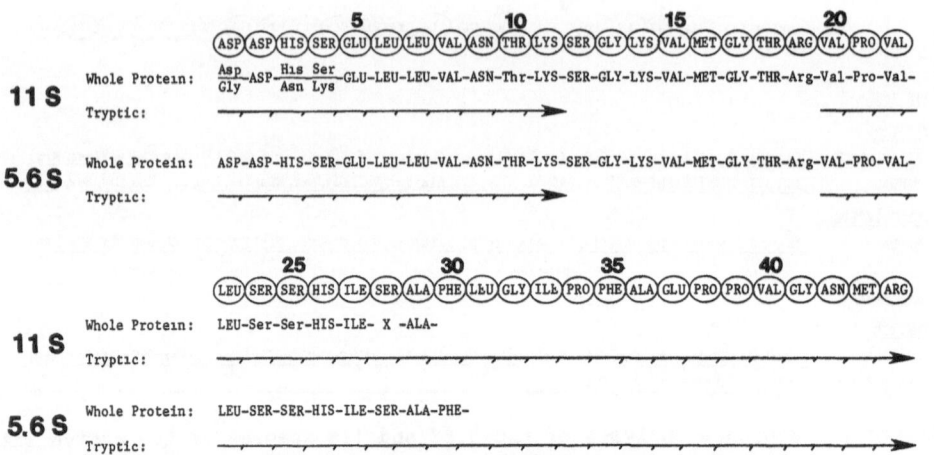

Figure 3. Sequence analysis of the NH$_2$-terminal sequences for the 5.6S and 11S species of acetylcholinesterase.

The consensus sequence indicated on the top is circled.

The tryptic peptides were also screened for a peptide that lacked both arginine and lysine. In addition, since there are preliminary indications that the COOH-terminal peptide contains a cysteine residue (13), all peptides in the 11S species containing the carboxymethyl [^{14}C] label were specifically screened by amino acid analysis. One candidate, labeled COOH in Figure 1, was identified with a composition devoid of either lysine or arginine amino acids (Table 1). This peptide was sequenced by gas phase and is presented in Figure 4. The comparable peptide has not yet been recovered from the 5.6S species. It is conspicuously missing from the same and neighboring pooled Sephadex fractions. One possible explanation for the apparent absence of this peptide is that it may be the target site for the post-translational modification that conveys the hydrophobic properties to the 5.6S enzyme. With this in mind, the hydrophobic peptides that bind very tightly to the reverse-phase HPLC column are being screened on the basis of amino acid composition and NH$_2$-terminal residue. This search has so far still failed to locate the COOH-terminal peptide from the 5.6S enzyme.

CONCLUSIONS

In addition to summarizing some of the sequences described above, Figure 4 also shows some sequence comparisons with human butyrylcholinesterase. The Torpedo enzyme exhibits a striking homology with human butyrylcholinesterase, particularly when the evolutionary distance which separates the two species and the different acyl group specificities of the two enzymes are considered

ACTIVE SITE PEPTIDE

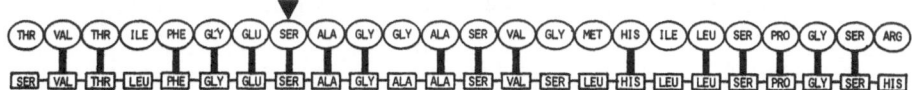

AMINO TERMINAL PEPTIDE

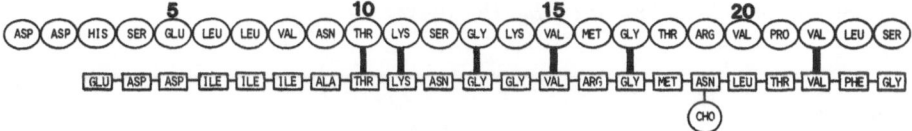

CARBOXY TERMINAL PEPTIDE

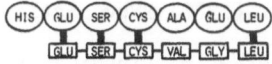

Figure 4. Sequence comparison of the active sites, NH$_2$-terminal, and COOH-terminal regions of Torpedo acetylcholinesterase and human serum butyrylcholinesterase.

The sequences for Torpedo acetylcholinesterase are circled, while those from butyrylcholinesterase are in boxes. Solid bars connect identical residues. The triangle above serine indicates the active site. The CHO on residue 19 of the NH$_2$-terminal sequence of human butyrylcholinesterase indicates a carbohydrate attachment site (15; and personal communication from O. Lockridge).

(15). Although the homologies are the most striking in the sequence surrounding the reactive serine residue, the homologies appear to extend
throughout the molecule, including even the COOH-terminal sequence.

Sequencing of the 11S and 5.6S forms of acetylcholinesterase has so far
not revealed any differences between the two enzymes, with the exception of
the COOH-terminal peptide which has not yet been found for the 5.6S enzyme.
This obvious descrepency may reflect a difference in the amino acid sequence.
On the other hand, it may simply reflect a difference in post-translational
modification. Monoclonal antibodies, which distinguish the two forms (7) and
which do not appear to be carbohydrate specific, are being used to localize
specific antigenic sites. These regions will be sequenced from both enzymes
in order to definitively establish whether or not the 5.6S and 11S forms of
acetylcholinesterase represent unique gene products.

Figure 5 localizes the NH$_2$-terminal, COOH-terminal, and active site
peptide with respect to the inferred primary sequence structure
recently obtained from recombinant DNA methodology (16). Also visualized are
the cysteine residues and carbohydrate attachment site candidates. The
carbohydrate-containing tryptic peptides have also been identified independently from the tryptic digest (data not shown). The molecule contains 575
amino acids with the active site serine occurring at residue 200.

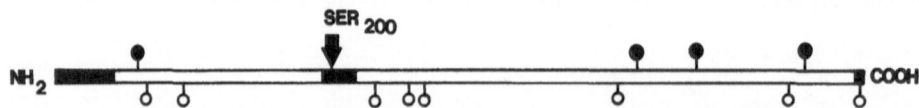

Figure 5. Line representation of how peptide placement aligns with the
inferred protein sequence based on the DNA sequence (16).

*Peptide sequences referred to in this paper are indicated by solid bars.
Possible carbohydrate sites are identified by solid circles, while cysteine
residues are localized with open circles. The active site serine is at
position 200 and is identified by the arrow.*

REFERENCES .

1. Vitatelli, O.M., and Bernhard, S.A. (1980) Biochemistry 19:4999-5007.
2. Bon, S., and Massoulie, J. (1980) Proc. Natl. Acad. Sci. U.S.A. 77:4464-
   4468.
3. Lee, S., Camp, S., and Taylor, P. (1982) J. Biol. Chem. 257:12302-12309.
4. Lwebuga-Mukasa, J.S., Lappi, S., and Taylor, P. (1982) Biochemistry
   16:1425-1434.
5. Reiger, F., Bon, S., Massoulie, J., Cartaud, J., Picard, B., and Bender,
   P. (1976) Eur. J. Biochem. 68:513-521.
6. Lee, S.L., Heinemann, S., and Taylor, P. (1982) J. Biol. Chem. 257:12283-
   12291.
7. Doctor, B.P., Camp, S., Gentry, M.K., Taylor, S.S., and Taylor, P. (1983)
   Proc. Natl. Acad. Sci. U.S.A. 80:5767-5771.

8. MacPhee-Quigley, K., Taylor, P., and Taylor, S. (1985) J. Biol. Chem.
   260, in press.

9. Doolittle, L.R., Mross, G.A., Fotheringill, L.A., and Doolittle, R.F. (1977) Anal. Biochem. 78:491-505.

10. Laursen, R.A., Bonner, A.G., and Horn, M.J. (1975) in Instrumentation in Amino Acid Sequence Analysis (Perhan, R.N., ed.) pp. 73-110, Academic Press, London.

11. Potter, R.L., and Taylor, S.S. (1979) J. Biol. Chem. 254:9000-9005.

12. Bhown, D.S., Mole, J.E., Weissenger, A., and Bennett, J.C. (1978) J. Chromatogr. 148:532-535.

13. Futerman. A.H., Fiorini, R.M., Roth, E., Low, M.G., and Silman, I. (1985) Biochem. J. 226:369-377.

14. Lee, S.L., Camp, S.J., and Taylor, P. (1982) J. Biol. Chem. 257:12283-12291.

15. Lockridge, O. (1984) in Cholinesterases, Fundamental and Applied Aspects (Brzin, M., Barnard, E.A., Sket, D., eds.) pp. 5-11, Walter De Gruyter and Co., Berlin.

16. Schumacher, M., Camp, S., Maulet, Y., Newton, M., MacPhee-Quigley, K., Taylor, S.S., Friedman, T., and Taylor, P. (1985), submitted.

ACETYLCHOLINE RECEPTOR α-BUNGAROTOXIN INTERACTIONS STUDIED BY CHEMICAL MODIFICATION, ENZYME DIGESTION AND HPLC

Allan L. Bieber, Jose Carlos Garcia-Borron*, and Marino Martinez-Carrion*

Ariz. St. Univ.	*Med. College of VA
Dept. of Chem.	Dept. of Biochem., VCU
Tempe, AZ	Richmond, VA

INTRODUCTION

The acetylcholine receptor (AchR), in high concentrations in the electric organ of certain fish, is a complex multisubunit structure (1,2,3). It is composed of four discrete polypeptides in a stoichiometry of 2α, 1β, 1γ, 1δ. Membrane permeability responses of the receptor are triggered by binding of acetylcholine to specific ligand binding sites on the α-subunits of the receptor complex. It has been well established that α-neurotoxins found in a number of snake venoms exert their effect by inhibiting binding of agonists, such as acetylcholine, to the receptor complex (4,5). Like the other α-neurotoxins from snake venoms, α-bungarotoxin (α-BTx) inactivates the acetylcholine receptor by specific interaction with the binding domains of the receptor complex. The α-neurotoxins, because of their tight, almost irreversible binding to the receptor, have been receiving considerable attention as exquisite probes for the study of structure and function of this important membrane complex. Published results demonstrate that methylated α-BTx retains high affinity for AchR and that the degree of methylation of α-BTx is lowered when the protein is bound to the receptor enriched membranes. The goal of our research was to use methylation studies of soluble and membrane bound α-BTx to locate the structural regions of the toxin that are masked by binding to AchR. Our approach was to use the established sequence of α-BTx (6) to predict protease cleavage patterns and HPLC to isolate the resulting cleavage peptides for analysis of radioactivity and composition. Some results derived from these studies are presented here.

MATERIAL AND METHODS

Preparation of alkaline extracted acetylcholine enriched membranes: Two hundred grams of frozen Torpedo marmorata electric organ tissue were homogenized in 300 ml of buffer A (10 mM Tris, 0.05% NaN_3, 5 mM EDTA, 400 mM NaCl, 5 mM iodoacetamide, 0.5 mM phenylmethylsulfonylfluoride, pH 7.4). Tissue disruption was accomplished at 4°C with a Polytron homogenizer in two 90 second steps with a 5 minute interval between steps. The homogenate was centrifuged at 3500 rpm in a Sorvall GSA rotor. The pellet was extracted with 100 ml of buffer in one step. The homogenate was centrifuged as before and the combined supernatants were passed through eight layers of cheese cloth. The filtrate was centrifuged at 30,000 rpm for 30 min at 4°C in a Beckman 35 rotor. The pellet was suspended in 100 ml of buffer B (10 mM NaH_2PO_4, 0.2% $NaNa_3$, 400 mM NaCl, 1 mM EDTA, pH 7.4). Aliquots

(25 ml) were layered over 15 ml of 50% sucrose in buffer B. The samples were loaded into a 35 rotor and subjected to 30,000 rpm for 2 hr at 4°C. The interfacial layer was carefully collected from each centrifuge tube, diluted six-fold with buffer C (10 mM NaH_2PO_4, 100 mM NaCl, 0.02% NaN_3, pH 7.4) and the suspension was centrifuged at 30,000 rpm for 30 min in a 35 rotor. The pellet was resuspended in 15 ml of buffer C, diluted six-fold with water and the pH was adjusted to 11 by dropwise addition of 1 N NaOH to the constantly stirred suspension. After 60 min at 4°C, the non-gradient centrifugation step was repeated. The pellet was resuspended in 12 ml buffer (10 mM Hepes, 100 mM $NaNO_3$) to give a final protein concentration approximately 4 mg/ml. The suspension was divided into aliquots of convenient volume for use and these were stored in liquid N_2 until needed. The samples had specific activities ranging from 20-35 μg α-BTx/mg protein by filter disc assay (7). Gel electrophoresis studies (8) showed four receptor bands were present along with a small amount of 90K band but the preparations were essentially free of the 43K band.

Reductive methylation of α-BTx was achieved under conditions used previously in this laboratory (9,10). The protocol for the heating-release studies of ^{14}C-Methyl-α-BTx is outlined in detail. Four hundred μl of alkaline extracted membrane suspension (3.6 mg/ml protein with binding capacity for 22 μg α-BTx per ml) was diluted ten-fold by addition of 3.6 ml of 54 mM Hepes, pH 7.4. Aliquots of 300 μl were dispensed and 10 μl of ^{14}C-methylated α-BTx (8.6 μM, 63,700 cpm/μg) was added. The mixture was incubated at room temperature for 30 min. Samples were transferred to a heating bath at 64°C. Aliquots (200 μl) were removed at the requisite time and were centrifuged at 26 psi in a Beckman Airfuge. Supernatant solution (100 μl) was counted in 5 ml of Biocount liquid scintillation solution. A blank sample without membranes was carried through the entire process and the radioactivity was used to represent 100% release. Comparable procedures were used in the acid release studies.

Protease digestion of methylated α-BTx: Methylated α-BTx was dissolved in 0.05 M Na_2HPO_4, pH 7.8 at concentrations ranging from about 0.5 mg/ml to 2.0 mg/ml. A volume of V8 protease (1 mg/ml in H_2O) was added to give a α-BTx/protease ratio of approximately 50/1. Incubation was carried out at room temperature for varying periods of time.

HPLC separation of V8 protease digests: Aliquots of the digestion mixture were injected directly into an Altex gradient HPLC system that was connected to a BioRad HiPore 318 reverse phase column (25 cm x 4.6 mm). The equilibration solvent, 0.1% TFA, served as the first solvent for the gradient and 0.1% TFA in acetonitrile was the second solvent. Detection of peptides was achieved by absorbance at 214 nm. Gradient conditions are delineated in the figure legends. Samples were collected manually when required for amino acid analysis and radioactivity measurements.

Amino acid analysis: The procedure used involved preparation and separation of phenylthiocarbamyl amino acids (PTC-amino acids) as described by Heinrikson and Meredith (11) HPLC separations of the PTC-amino acids were done at 45°C on an Altex reversed phase PTH-amino acid column (25 cm x 4.6 mm). Solvent system III of Heinrikson and Meredith was used. Detection was achieved by absorbance at 254 nm using the Altex detector.

RESULTS AND DISCUSSION

The major premise upon which this series of experiments rests is that one can separate membrane bound methylated α-BTx from soluble methylated α-BTx. Subsequent analyses of the isolated materials can then provide data relevant to the specific membrane binding.

Results from V8 protease cleavage: At pH 7.8, V8 protease is expected to cleave at each glutamic acid and aspartic acid linkage in a protein (12). In figure 1, the sequences of the peptides derived from V8 protease hydrolysis of fully methylated α-BTx are shown. Note that cleavage at asp 63 is not indicated. This point will be discussed later.

```
         5        10       15       20
    I V C H T T A T I P S S A V T C P P G E
    *                                     o

        25       30              35       40
    N L C Y R K M W C D        A F C S S R G K V V E
                  *   o                  *   o

        45       50       55                60       65       70
    L G C A A T C P S K K P Y E        E V T C C S T D K C N H P P K R Q P G
    L G C A A T C P S K K P Y E E      V T C C S T D K C N H P P K R Q P G
                  *  *   o  o             o  *                  *
```

* Expected methylation sites   °Expected cleavage sites

FIGURE 1: *Sequence of α-bungarotoxin peptides*

Figure 2 shows a typical separation of the products resulting from V8 digestion of methylated α-BTx. The better resolution obtained in the lower panel indicates that good resolution of five peaks is possible. Manual collection of samples permitted sufficient quantities of the peptides to be obtained for subsequent amino acid analysis. Both standard ninhydrin post column development and precolumn PTC-amino acid formation with subsequent HPLC separation were used. The peak fractions were easily recognized in terms of predicted cleavage sequences. The results are presented in TABLE 1.

TABLE 1

peak	Elution (min)	*sequence
1	21.4	56-74
2	27.0	31-41
3	27.5	42-56
4	35.6	1-20
5	36.9	21-30

* Consistent with amino acid analysis

As indicated earlier it is not likely that cleavage occurred at asp 63. The amino acid analysis data are consistent with sequence 57-74. An alternative explanation that peptides 57-63 and 64-74 are present but coelute is also possible. This seems less likely since a number of different gradients were used in attempts to isolate two fragments but resolution was not observed. Failure of V8 protease to cleave at asp sites adjacent to amino acids with large, bulky side chains has been noted in the literature. Dimethyllysine at position 64 could easily qualify as a large bulky group.

Figure 3 illustrates several related points that have a bearing on the membrane binding studies. First, the PTC-amino acid analysis system of Heinrikson and Meredith (11) can resolve monomethyllysine and dimethyllysine from the PTC-amino acids derived from the Pierce amino acid standard kit. Compare the small peaks (standard amino acids) with the three large peaks which are dimethyllysine, lysine and monomethyllysine at 24.68 min, 39.07 min and 39.54 min, respectively, (top panel). Second, the lower trace, derived from an acid hydrolyzed sample of methylated α-BTx shows that lysine and monomethyllysine are virtually absent whereas a large peak corresponding to dimethyllysine appears between arginine and proline. From this it appears that HPLC analysis of PTC-amino acids can be used to detect this specific chemical modification of α-BTx. Other applications of this type

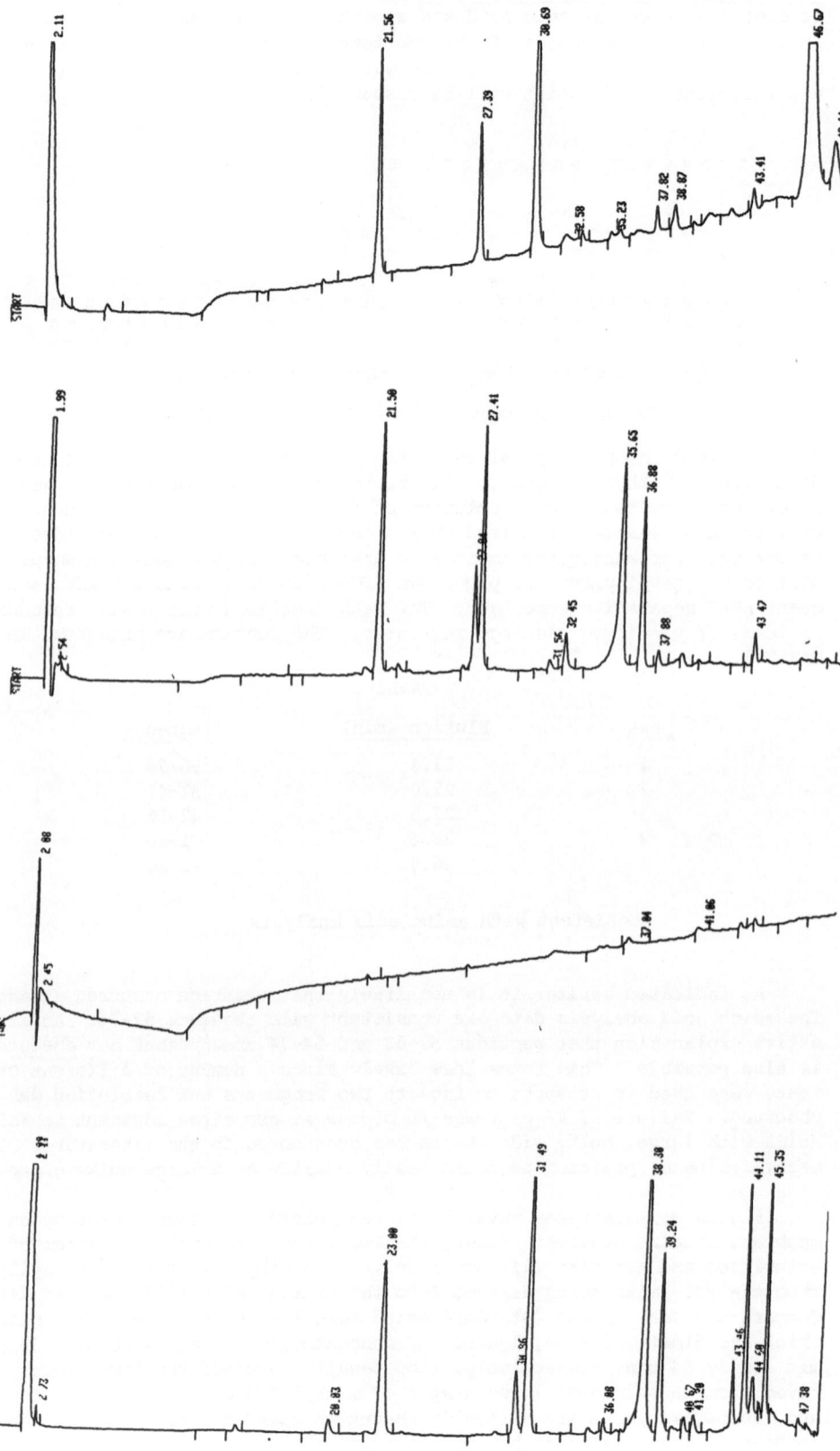

are likely to follow. Table 2 shows the composition derived from the PTC-amino acid analyses and compares the data to the known composition of α-BTx based on the sequence. The value for val is low but this may be the result of incomplete hydrolysis in 24 hrs of the val-val sequence in the α-BTx structure. Overall, the composition agrees quite well with the established structure of α-BTx.

Binding and release of $^{14}CH_3$-α-BTx from AchR membranes: In order to study membrane bound $^{14}CH_3$-α-BTx, it is necessary to have an effective means to release toxin from the receptor preparation. When pre-labelled $^{14}CH_3$-α-BTx was added to membrane suspensions, the solution incubated and subsequently subjected to the heating-release procedure described in the methods, toxin release was effected as shown Figure 4. When reductive methylation was done on receptor enriched membrane with α-BTx in excess of binding sites,

TABLE 2

PTC-AMINO ACID ANALYSIS OF CM-α-BTx

| Amino Acid | Mole amino acid / Mole methionine | | | |
| | Experimental | | Known | |
	Actual	Integer	Integer	Difference
D	3.8	4	4	
E	5.4	5	5	
CMC	10.3	10	10	
S	5.0	5	6	-1
G	3.8	4	4	
H	1.7	2	2	
T	4.9	5	7	-2
A	5.1	5	5	
R	3.7	4	3	+1
DMK*	7.4	7	0	+7
P	9.2	9	8	+1
Y	1.9	2	2	
V[†]	3.5	3	5	-2
M	1.0	1	1	
I	1.1	1	2	-1
L	2.1	2	2	
F	1.0	1	1	
K	.4	0	6	-6

* calculation based on proline standard
† extra peak at 31.22 min in chromatographic profile, may be V·V.

a much lower level of α-BTx was released form the membranes. The data in TABLE 3 illustrate the problems encountered in the release experiments under different sets of conditions. It is apparent that the reductive methylation reaction conditions alone produce an effect that markedly reduces the amount of toxin released. This large affect results from the combined influence of NaCNBH3 and formaldehyde since either reagent alone is much less inhibitory towards release. Preliminary studies indicate that mild acid treatment may be a more efficacious means for releasing the reductively methylated toxin from receptor enriched membranes.

FIGURE 2. *Separation of V8 protease digests*. A linear gradient from solvent A (0.1% TFA) to 30% solvent B (0.1% TFA) in acetonitrile over 44 min was used at a flow rate of 1.5 ml/min. A 4 min ramp to 100% B, 4 min ramp to 100% A followed by 8 min equilibration conditioned the column for the next injection. Panels from top to bottom: t = 0, 5 µl, 0.05 AUFS; t = 24 hr, 100 µl, 1.0 AUFS; V8 protease blank, 5 µl, 0.05 AUFS; altered gradient, 100 µl, 1.0 AUFS.

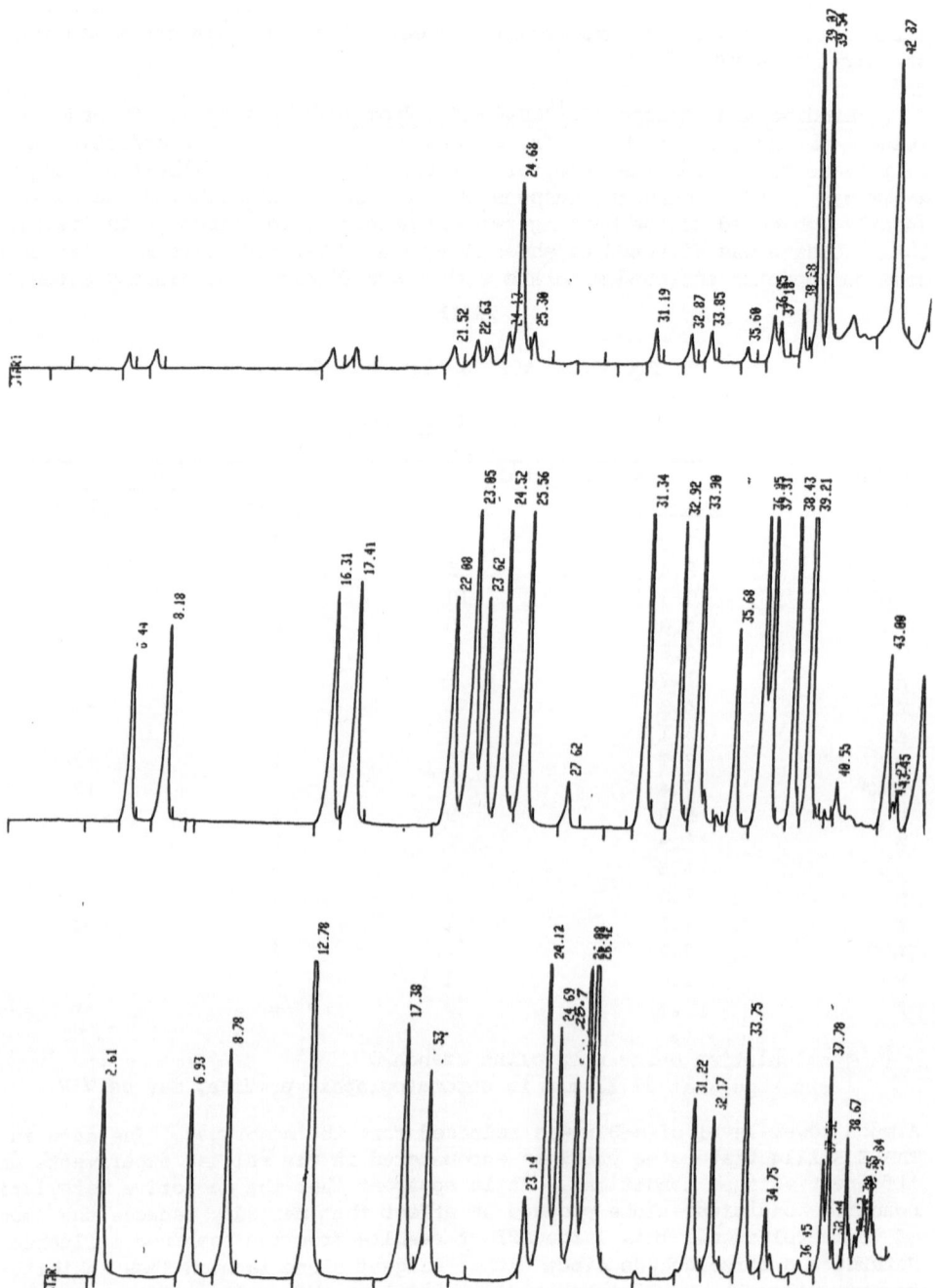

FIGURE 3. _PTC-amino acid analysis_. Gradient system III described by Heinrikson and Meredith (11) was used with flow rate at 1 ml/min, 45°C, and 0.05 AUFS.

Top: PTC-derivatives of dimethyllysine, lysine and monomethyllysine and a small amount of standard mixture of PTC-amino acids.

Center: PTC-amino acid standard mixture. 1.25 nmole of each amino acid.

Bottom: Carboxymethylated-α-BTx hydrolysate, PTC-amino acids.

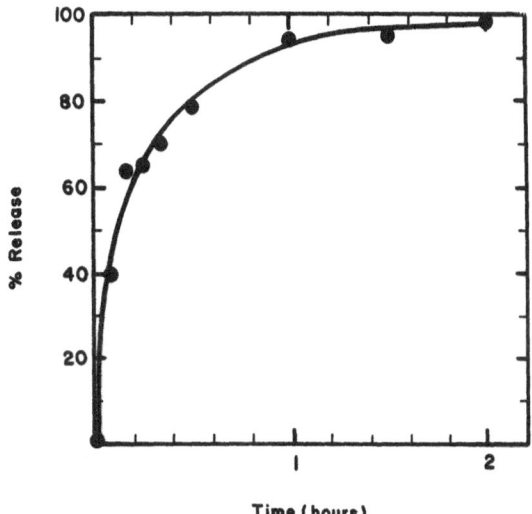

FIGURE 4. _Release of ^{14}C-α-BTx from AchR by heating._ See text for specific conditions.

TABLE 3

RELEASE OF α-BTx FROM TOXIN-RECEPTOR COMPLEX

Treatment	% Release
64°C, 30 min	41 ± 3, (3)
Reductive methylation followed by 64°C, 30 min	16 ± 2, (2)
Reductive methylation conditions without NaCNBH$_3$, followed by 64°C, 30 min	32 , (1)
Reductive methylation conditions without formaldehyde, followed by 64°C, 30 min.	30 ± 3, (2)
Incubation at pH 2.4, 120 min	57 ± 22, (2)
Reductive methylation followed by incubation at pH 2.4, 120 min	46 , (1)

* mean ± SD, (number of determinations)

Preliminary results pertaining to the radioactivity in peptides derived from recovered samples of soluble and membrane bound $^{14}CH_3$-α-BTx are presented in TABLE 4. It is apparent that the degree of methylation of

613

membrane bound α-BTx is less than that of the soluble fraction. This result agrees with earlier published results (10). From the peptide radioactivity data one can infer that the reactivities of free amino groups in the N-terminal and C-terminal regions of α-BTx are largely unaffected by binding to the receptor since they remain susceptible to methylation. The free amino groups in the internal regions of the α-BTx sequence are methylated to a considerably lower degree in the membrane bound α-BTx. These data are consistent with results of Tsetlin et al. (13) who used fluorescent labels and spin labels to study the interaction of AchR with specific amino groups of neurotoxin III from N.n. oxiana. Their results with a homologous neuro-toxin showed that spin label at lys 27 was most prominently altered by membrane binding and that the tail region, evidenced by studies of his 71, did not participate in the binding. Continuation of the studies using the general approach outlined in this paper will provide considerable data about the surface interaction of α-BTx and its receptor site on acetylcholine receptor enriched membranes.

TABLE 4

COMPARISON OF SOLUBLE AND MEMBRANE BOUND $^{14}CH_3$-α-BTx PEPTIDES

peak	sequence	area soluble peptide / area membrane peptide	cpm soluble peptide* / cpm membrane peptide
1	57-74	1.7	0.95
2	31-41	1.9	4.1
3	42-56	0.9	3.0
4	1-20	1.9	1.2
5	21-31	2.4	3.4

* Corrected to normalize areas.

REFERENCES

1. B.M. Conti-Tronconi and M.A. Raftery, Ann. Rev. Biochem. 51, 491 (1982).
2. F.J. Barrantes, Internat. Rev. Neurobiol. 24, 259 (1983).
3. J.P. Changeaux, A. Devillers-Thiery and P. Chemoulli, Science 225, 1335 (1984).
4. C.C. Chang and C.Y. Lee, Arch. Internat. Pharmacodyn. 144, 241 (1963).
5. E.X. Albequerque, A.T. Eldefrawi and M.E. Eldefrawi in Handbook of Experimental Pharmacology Vol. 52., 158-212 (1979) (C.Y. Lee, Editor) Springer-Verlag.
6. D. Mebs, K. Narita, S. Iwanaga, Y. Samejima & C.Y. Lee. Hoppe-Seyler Zeit Physiol. Chem. 353, 243 (1972).
7. J. Schmidt and M.A. Raftery, Anal. Biochem. 52, 349 (1973).
8. G. Soler, J.R. Mattingly, Jr. & M. Martinez-Carrion. Biochemistry 23, 4630 (1984).
9. P. Calvo-Fernandez and M. Martinez-Carrion, Arch. Biochem. Biophys. 208, 154 (1981).
10. G. Soler, M.C. Farach, H.A. Farach, Jr., J.R. Mattingly, Jr. and M. Martinez-Carrion, Arch. Biochem. Biophys. 225, 872 (1983).
11. R.L. Heinrickson and S.C. Meredith, Anal. Biochem. 136, 65 (1984).
12. G.R. Drapeau, Meth. Enzymol. 47, 188 (1977).
13. V.I. Surin, V.V. Kondakov, V.F. Bystrov, V.T. Ivanov and Yu. A. Ovchinnikov. Toxicon 20, 83 (1982).

IMIDOESTERS AND THE MECHANISM OF LIVER ALCOHOL DEHYDROGENASE

Bryce V. Plapp

Department of Biochemistry
The University of Iowa
Iowa City, IA 52242

INTRODUCTION

Protein chemists have just begun to exploit the full potential of imidoesters, which react specifically with amino groups (1,2), for studying protein structure and function. In this article, the uses of imidoesters to investigate the role of a lysine residue at the active site of liver alcohol dehydrogenase and to prepare derivatives for other mechanistic studies are illustrated. Furthermore, the kinetics of reactions of ethyl acetimidate and the optimal conditions for protein modification are discussed. The results presented here also support several theses of protein chemistry, which can apply to other types of chemical modification.

ARE AMINO GROUPS INVOLVED IN ACTIVITY?

The imidoester, methyl picolinimidate, was introduced by Benisek and Richards (3) for the purpose of attaching metal chelating sites to proteins for use in x-ray crystallography. When all ten of the amino groups of pancreatic DNase were modified with methyl picolinimidate, introducing the positively-charged picolinimidyl group, there was no change in activity, and we concluded that amino groups were not involved in catalysis (4). Such a conclusion is as valid as when one concludes that an essential residue is modified when a reagent inactivates. This is in accord with one thesis of protein chemistry, *If a group can be modified without affecting activity, the group is probably not involved in catalysis*. In contrast, when seven of the amino groups are carbamoylated, so as to remove positive charges, activity is retained, but neutralization of all ten amino groups inactivates. Likewise, four to five trinitrophenyl groups can be added without loss of activity, but seven of these neutral, hydrophobic groups inactivate, probably by changing the structure of the enzyme. These results demonstrate the utility of imidoesters for surveying the importance of amino groups for enzyme activity.

When methyl picolinimidate was tested on horse liver alcohol dehydrogenase, it was surprising to find a 20-fold increase in enzymatic activity (5). We could show that methyl picolinimidate reacted with a residue at the active site by differential modification. When all of the amino groups outside the active site were modified with ethyl acetimidate while the active site was protected with NAD$^+$ and pyrazole, then these reagents were removed, methyl picolinimidate activated the enzyme about 45-fold. Differential labeling was used to identify Lys-228 as the residue modified (6,7).

As shown in Table 1, alcohol dehydrogenase modified with different substituents on Lys-228 has different activities relative to the enzyme partially acetimidylated in the presence of NAD$^+$ and pyrazole (6-9). Maximum velocities, binding constants for coenzymes, and the Michaelis constants are generally increased by the different substituents, but the phosphopyridoxyl derivative has less than 20% of the activity of unmodified enzyme. This activity is due to the modified enzyme since the kinetic constants for the substrates are significantly increased as compared to those for the unmodified enzyme. Furthermore, the enzyme modified with pyridoxyl--without the phosphate--has greatly increased activity. Thus, *When an enzyme has residual activity, the kinetics should be studied in order to determine if the residual activity is due to some unmodified enzyme or to the modified enzyme itself.*

The change in kinetic characteristics can be explained in terms of the Ordered Bi Bi mechanism for the enzyme:

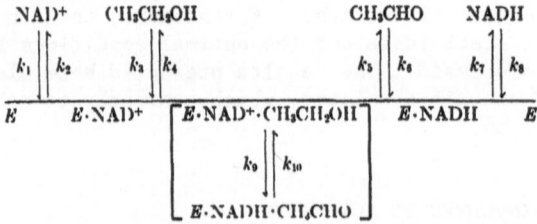

For native enzyme, release of NADH with a rate constant of k_7 is the rate-limiting step in turnover. In contrast, with picolinimidylated enzyme, the turnover number is increased by a factor of 10 and k_7 has become too fast to measure by stopped-flow techniques. With this modified enzyme, an isotope effect with deuterated ethanol indicates that the

Table 1: Activity of modified alcohol dehydrogenase

| Substituent on Lys-228 | Relative activity[a] 0.5 M EtOH, pH 9 | Kinetic constants[b] | | |
		V_1,s^{-1} K_{ia},μM	K_a,μM	K_b,mM
Native	1	3.5 26	3.9	0.35
Acetimidyl	18	9.0 30	25	7.4
Picolinimidyl	45	32 640	420	18
Methyl	29	29 220	36	7.9
Carbamoyl	3.9	1.2 1500	200	4.3
Pyridoxyl	15	21 3300	1600	61
Phosphopyridoxyl	0.2	0.2 170	390	6.6

[a] *Activity in a standard assay relative to enzyme acetimidylated on amino groups except the one at the active site.*

[b] *At pH 8, 25 °C. V_1 is the turnover number, K_{ia} and K_a are dissociation and Michaelis constants for NAD; K_b for ethanol.*

transfer of hydrogen from ethanol to NAD[+] in the ternary complex is at least partially rate-limiting (10). Such a modified enzyme was used to study the effect of pH and substrate structure on the hydrogen transfer step (11).

It appears that the amino group of Lys-228 is involved in the binding of coenzyme, and determination of the enzyme structure by x-ray crystallography (12) subsequently placed the amino group close to the 3'-hydroxyl group of the adenosine ribose and next to the carboxyl group of Asp-223 (Figure 1). We suggest that a substituent on the amino group interferes with the coenzyme binding. However, the complete explanation is probably not so simple, as discussed below.

CONFORMATIONAL CHANGES IN ALCOHOL DEHYDROGENASE

Alcohol dehydrogenase is a dimer of two identical subunits, each of which has a coenzyme binding domain and a catalytic domain, and between which the substrates bind. When the enzyme binds coenzyme and substrate, it changes conformation, so that the domains move closer, the active site cleft is closed up, and water in the substrate binding pocket is totally excluded (12).

Since modification of the amino groups of the enzyme could affect this conformational change, the enzyme was isonicotinimidylated and crystallized, and its structure determined by x-ray crystallography at a resolution of 3.2 Å (13). The electron density map showed that 23 of the 30 amino groups per subunit of the protein were modified; three of the amino groups were buried and did not appear to be modified; and four had no obvious electron density but were exposed to solvent and thus probably were modified.

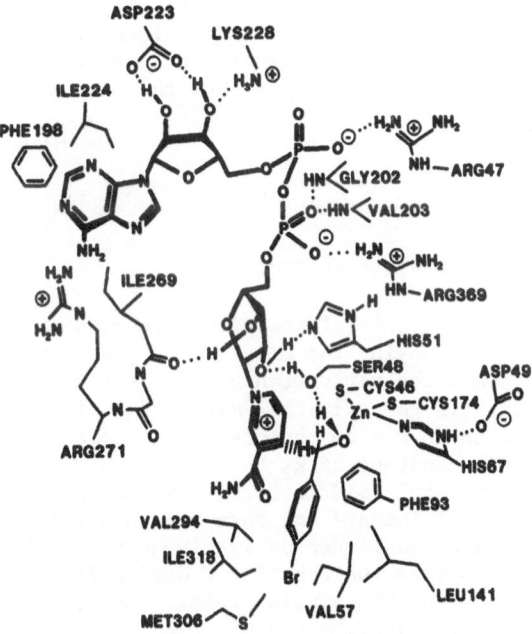

Figure 1: Active site of liver alcohol dehydrogenase (12,13)

When the structures of the isonicotinimidylated and native apoenzyme were superimposed as well as possible, the average of the differences in positions of alpha-carbon atoms was only 0.6 Å, which indicates that the two structures are almost identical. These results support another thesis of protein chemistry, *The extensive modification of amino acid side chains need not alter the overall conformation of a protein.*

Even though modification of the amino groups does not appear to affect the conformation of the apoenzyme, it may hinder the conformational change that occurs when coenzyme and substrate bind. Some evidence for this comes from the observation that NADH can diffuse into crystals of the modified enzyme without causing them to crack. With native apoenzyme, NADH cracks the crystals, as the enzyme changes conformation. Furthermore, pH dependency studies show that binding of NAD^+ to the modified enzyme or dissociation of the enzyme·NAD^+ complex is fastest when a group with a pK of 8 is unprotonated. The pH dependence is almost the opposite of that found with native enzyme, where the rates of reactions are fastest when a group with a pK value in this range is protonated.

The kinetic arguments required to explain these results are quite complex, but we propose that the enzyme·NAD^+ complex isomerizes, that is, changes conformation as seen by x-ray crystallography, as an obligatory step in the enzyme mechanism (14).

$$E \underset{k_2}{\overset{k_1}{\rightleftharpoons}} E \cdot NAD^+ \underset{k_4}{\overset{k_3}{\rightleftharpoons}} {}^*E \cdot NAD^+ \underset{k_6}{\overset{k_5}{\rightleftharpoons}} E \cdot NAD^+ \cdot CH_3CH_2OH \underset{k_8}{\overset{k_7}{\rightleftharpoons}} E \cdot NADH \cdot CH_3CHO$$

We think that for modified enzyme, k_3, the rate constant for isomerization, has become at least partially rate-limiting in the overall mechanism. If this is the case, then the rate constants calculated for the "on" and "off" velocities of NAD^+ binding are directly dependent upon the magnitude of k_3 and thus give the same pH dependency results. With native enzyme, however, the isomerization is faster, and the pH dependency results probably reflect k_1 and k_2. Thus, the chemical modification has uncovered a step that is normally hidden from the kinetic observations. This affords an opportunity to determine directly the rate of the conformational change and to identify the group with the pK of 8 that is involved in the isomerization. Protein fluorescence studies with the acetimidylated enzyme show that the group is not the lysine itself (15).

KINETICS OF REACTIONS OF ETHYL ACETIMIDATE

The reaction of ethyl acetimidate with alcohol dehydrogenase is a convenient system for studying the kinetics of reaction of imidoesters with amino groups since the activation of the enzyme can be used to follow the reaction. In general, the amino group of a protein reacts with an acetimidate to form a tetrahedral intermediate, which can then decompose with a rate constant of k_2 to form an amidine, or ammonia can be lost so that the N-alkyl imidate forms. Browne and Kent determined by ^{13}C-NMR that the N-alkyl imidate can form and then react with another amino group to form a cross-linked derivative with another amino group, or it can hydrolyze to form the acetylated derivative or to release the free amino group (16). In this simple kinetic model, a complex rate constant, k_3, is used to describe the kinetics of hydrolysis.

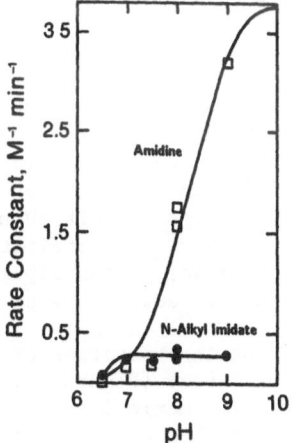

Reactions of acetimidate ester and protein

As shown previously (5,16), modification of liver alcohol dehydrogenase with ethyl acetimidate rapidly increases activity by 8 to 10-fold, but after this transient phase, there is some inactivation with the activity eventually leveling off at a factor of about 6-fold. Browne and Kent proposed that the transient activation was due to the N-alkyl imidate, which could slowly hydrolyze during the course of the reaction to regenerate free enzyme. Thus the enzyme could not be completely modified with one addition of reagent. On the assumption of the simple kinetic scheme, activation kinetics were used to determine rate constants for formation of the various products, as shown in the scheme above (17).

The kinetic constants for reaction at different pH values are shown in Figure 2. The rate constants for the formation of the N-alkyl imidate are much smaller than the rate constants for the formation of the acetamidine derivative, but the N-alkyl imidate seems to form even at pH 9 or above. (In the presence of sodium phosphate buffer at pH 8 or 9.5, the level of transient activation, and k_1, is somewhat higher than in triethanolamine buffers at about the same pH.) The formation of N-alkyl imidate at high pH values is also evidenced by electrophoresis of protein

Figure 2: pH Dependencies of amidine and N-alkyl imidate formation.

619

products. At pH 10 or so, in phosphate or phosphate and carbonate buffers, the activation of the enzyme occurs very rapidly, without a transient phase of activation, and the rate constants could not be determined accurately.

It appears that most of the amino groups of a protein can be modified with methyl or ethyl acetimidate in the pH range between 8 and 11 if four additions of reagent at hourly intervals are used. After an hour, the N-alkyl imidate intermediate should have hydrolyzed so that the next addition of reagent would modify some of the remaining amino groups. Nevertheless, even at pH 9, N-alkyl imidates may form and react to give some undesired by-products.

ROLE OF HISTIDINE RESIDUES IN ACTIVITY

Figure 1 shows the structure of the enzyme·NAD$^+$·bromobenzyl alcohol complex, which represents a real enzyme-substrate complex (18). The hydroxyl group of the alcohol is directly ligated to the zinc ion and forms a hydrogen bond with the hydroxyl group of Ser-48, which is hydrogen bonded to the 2'-hydroxyl group of the nicotinamide ribose, which is hydrogen bonded to the imidazole group of His-51. This hydrogen bond network may function as a proton relay system so that when the alcohol is oxidized to the aldehyde, the proton is released to the solvent. The zinc is buried in the structure whereas the imidazole group of His-51 is on the surface and could function as a base. Indeed, the maximal rate of hydrogen transfer depends upon a group with a pK of about 6.4, but this is thought to be due to the ionization of the zinc alcohol to form the zinc alkoxide (19).

It may also be noted that other studies have shown that modification of the cysteine residues 46 or 174 will inactivate the enzyme. Although these might be "essential" sulfhydryl groups, the three-dimensional structure suggests no obvious function for them. Instead, it appears that modification of either one of these residues will simply interfere with the proper binding of the substrate.

In order to obtain some evidence for the role of the histidine residue in the mechanism of the enzyme, we used diethyl pyrocarbonate, which is relatively specific for imidazole groups. However, the reagent can also modify amino groups, and this would cause some difficulties with alcohol dehydrogenase because modification of Lys-228 would probably activate. (See carbamoyl derivative in Table 1.) Thus, the amino groups were first blocked with the acetimidyl moiety (20).

Acetimidylated liver alcohol dehydrogenase is very rapidly inactivated by diethyl pyrocarbonate at pH 8. Neither NAD$^+$ nor NADH significantly protects against the inactivation, but formation of the ternary complex with NAD$^+$ or NADH and substrate analogs can almost completely protect. Thus, it seems that a histidine residue is "essential" for activity. Unfortunately, the ethoxyformyl group slowly hydrolyzes from the imidazole groups, and it is not possible to identify which of the 7 histidines per subunit are modified. During this reaction, 5 histidines are modified, presumably including His-51, since this residue is exposed on the surface of the protein.

The pH dependencies for the incorporation of ethoxyformyl groups and for inactivation are quite different (Figure 3). Groups with pK values of 7 must be unprotonated for the maximal rate of incorporation, whereas the inactivation occurs fastest when a group with a pK of about 9.6 is unprotonated. Furthermore, the residue that is involved in the

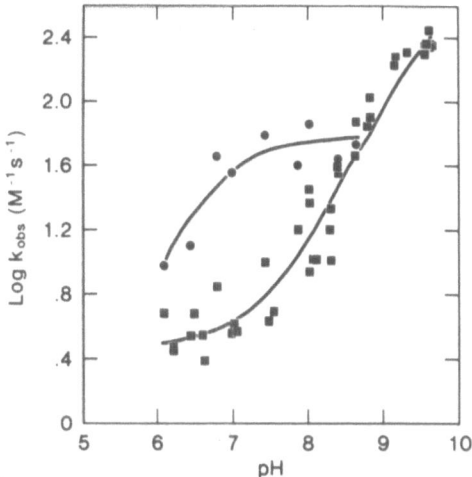

Figure 3: *pH Dependencies for reactions of diethyl pyrocarbonate with alcohol dehydrogenase: incorporation (●), inactivation (■); ref. 20.*

inactivation has a ten-fold higher rate of reaction at high pH than other histidine residues that have the pK values of 7.

A scheme for interpreting these results is as follows:

We assume that His-51 has a pK value of about 7, but when it is protonated, it reacts very slowly with diethyl pyrocarbonate. After the histidine deprotonates, it can participate in forming the hydrogen bond network. In this state, the imidazole still has a hydrogen that hinders reaction of diethyl pyrocarbonate. Thus at pH 7, the rate of inactivation might be considerably slower than one would expect if the imidazole were freely accessible. However, as the pH is raised, the water molecule attached to the zinc can ionize, with a pK value of about

621

9.6, to form the zinc-hydroxide complex. When the proton dissociates from the ionic species, the imidazole nitrogen is no longer protected and it can react with the diethyl pyrocarbonate. Furthermore, we suggest that the interaction of the imidazole through the hydrogen bond system to the zinc hydroxide enhances the reactivity of the histidine about ten-fold. This illustrates the thesis that, *Enzymes have groups that are hyperreactive due to the microenvironment*. These chemical modification studies suggest that an imidazole group is directly involved in the catalytic mechanism, but it would be important to determine how much base catalysis contributes to the overall rate enhancement furnished by the enzyme.

REFERENCES

1. Hunter, M. J., and Ludwig, M. L. (1972) *Methods Enzymol. 25*, 585-596.
2. Inman, J. K., Perham, R. N., DuBois, G. C., and Appela, E. (1983) *Methods Enzymol. 91*, 559-569.
3. Benisek, W. F., and Richards, F. M. (1968) *J. Biol. Chem. 243*, 4267-4271.
4. Plapp, B. V., Moore, S., and Stein, W. H. (1971) *J. Biol. Chem. 246*, 939-945.
5. Plapp, B. V. (1970) *J. Biol. Chem. 245*, 1727-1735.
6. Sogin, D. C., and Plapp, B. V. (1975) *J. Biol. Chem. 250*, 205-210.
7. Dworschack, R. T., Tarr, G., and Plapp, B. V. (1975) *Biochemistry 14*, 200-203.
8. Zoltobrocki, M., Kim, J. C., and Plapp, B. V. (1974) *Biochemistry 13*, 899-903.
9. Fries, R. F., Bohlken, D. P., Blakley, R. T., and Plapp, B. V. (1975) *Biochemistry 14*, 5233-5238.
10. Plapp, B. V., Brooks, R. L., and Shore, J. D. (1973) *J. Biol. Chem. 248*, 3470-3475.
11. Dworschack, R. T., and Plapp, B. V. (1977) *Biochemistry 16*, 2716-2725.
12. Eklund, H., Samama, J.-P., Wallén, L., Brändén, Ć.-I., Åkeson, Å, and Jones, T. A. (1981) *J. Mol. Biol. 146*, 561-587.
13. Plapp, B. V., Eklund, H., Jones, T. A., and Brändén, C.-I. (1983) *J. Biol. Chem. 258*, 5537-5547.
14. Plapp, B. V., Sogin, D. C., Dworschack, R. T., Bohlken, D. P., Woenckhaus, C., and Jeck, R., manuscript in preparation.
15. Parker, D. M., Hardman, M. J., Plapp, B. V., Holbrook, J. J., and Shore, J. D. (1978) *Biochem. J. 173*, 269-275.
16. Browne, D. T., and Kent, S. B. H. (1975) *Biochem. Biophys. Res. Commun. 67*, 126-132, 133-138.
17. Hennecke, M., and Plapp, B.V., manuscript in preparation.
18. Eklund, H., Plapp, B. V., Samama, J.-P., and Brändén, C.-I. (1982) *J. Biol. Chem. 257*, 14349-14358.
19. Kvassman, J., and Pettersson, G. (1978) *Eur. J. Biochem. 87*, 417-427.
20. Hennecke, M., and Plapp, B. V. (1983) *Biochemistry 22*, 3721-3728.

INVESTIGATION OF GRAMICIDIN CHANNEL FUNCTION BY SINGLE AMINO ACID

REPLACEMENT USING NON-GENETIC CODE AMINO ACIDS

Roger E. Koeppe II[*] and Olaf S. Andersen[+]

[*]Department of Chemistry
University of Arkansas
Fayetteville, Arkansas   72701

[+]Department of Physiology and Biophysics
Cornell University Medical College
New York, New York   10021

ABSTRACT

Oligonucleotide-directed mutagenesis is limited to amino acids which can be specified by the genetic code. By contrast, chemical synthesis of "mutant" polypeptides offers the possibility of introducing non-genetic code amino acids for particular investigations, e.g., comparisons of the functional effects of essentially isosteric side chains of different polarities (such a Val vs. trifluoro-Val). We have used a semisynthetic approach, consisting of degradation from the formyl-N-terminal followed by re-synthesis, to introduce variant amino acid side chains into the selective channel-forming pentadecapeptide, gramicidin A. Even though the side chains are not in direct contact with the permeating ions, the single-channel conductances for Na^+ and Cs^+ through gramicidin channels are markedly affected by changes in the physico-chemical characteristics of the side chains. The maximal single-channel conductance for Na^+ is decreased up to ten-fold when a polar side chain is present at position #1 in gramicidin. Furthermore, the selectivity for Cs^+ over Na^+ is increased when a polar side chain is at position #1. The transmembrane channels are dimers of gramicidin, and it is possible to observe hybrid channels formed between the natural and modified gramicidins. These hybrid channels suggest that the observed conductance changes are not due to gross changes in channel structure, but rather to subtle ion-side chain interactions which occur over distances of about 5-10 Å.

INTRODUCTION

The gramicidin family of peptides is a set of well characterized transmembrane cation channels (1,2). The parent Val_1- gramicidin A molecule has the sequence formyl-L-$\underline{Val}_1$-Gly-L-Ala-D-Leu-L-Ala-D-Val-L-Val-D-Val-L-Trp-D-Leu-L-$\underline{Trp}_{11}$-D-Leu-L-Trp-D-Leu-L-Trp-ethanolamine. Naturally occurring analogues include Ile_1-gramicidin A, Val_1- and Ile_1 gramicidin B which have Phe_{11}, Val_1- and Ile_1-gramicidin C which have Tyr_{11}, and the gramicidin K family which may have a fatty acid esterified to the ethanolamine (3). These analogues, together with chemically synthesized site-specific "mutants" of gramicidin A, provide valuable insight regarding the structure of the channel in lipid bilayer membranes (4, 5) and the mechanism of ion transport (6).

The dimeric single-stranded β helix (7) is now well established as the structure of the ion-conducting gramicidin channels (4,5). In this structure, each gramicidin monomer roughly spans a lipid monolayer, and the N-formyl ends of two monomers meet in the center of a bilayer to form a dimer stabilized by six hydrogen bonds (Fig. 1). This structure for the dimer is consistent with the known channel lifetimes of ~1 sec (e.g., ref. 6). The hydrophobic side chains project from the exterior surface and allow the channel to be incorporated into a lipid bilayer. The polar peptide groups form the wall of a pore with a diameter of ~4 Å, appropriate for the single-file transport of partially dehydrated monovalent alkali cations. Gramicidin channels do not transport divalent ions or negatively charged ions (1).

We report here structure-function studies of gramicidin channels having altered side chains at position #1. These side chains are at the junction of the two monomers near the center of the bilayer, and extend from ~5-10 Å away from the ion permeation pathway (Fig. 1). The side chains are not in physical contact with the passing ions. We have investigated channels that have either one or both position-1 side chains altered. Our results indicate that these side chains can significantly modulate the single-channel conductance, lifetime and ion selectivity, while not perturbing the channel conformation.

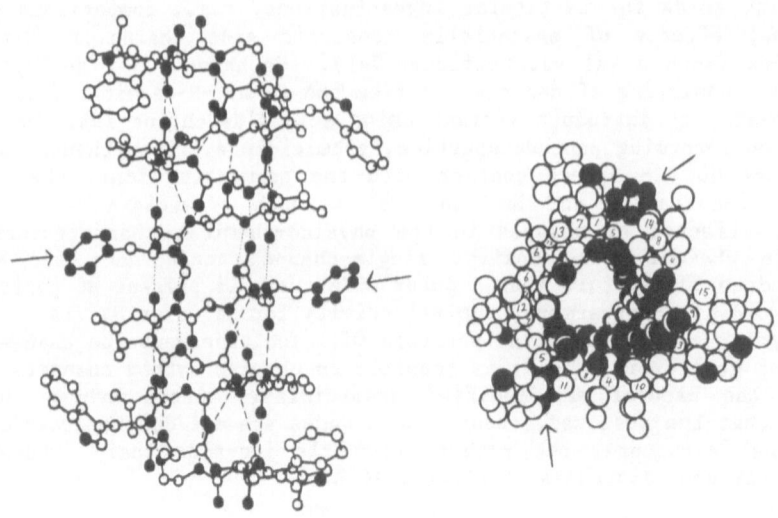

Fig. 1. Schematic drawing of Phe_1 gramicidin A in the channel conformation. The Phe_1 side chains (see arrows) are included to indicate the approximate positions of the side chain variations that are discussed in the text. The side view (left) shows the intramolecular (---) and six intermolecular (•••) hydrogen bonds. The end view (right) serves to roughly show the channel from the perspective of an approaching ion. Oxygen and nitrogen atoms and the phenyl rings are black.

METHODS

Gramicidin analogues were prepared by the semisynthetic procedure of Weiss and Koeppe (8), which is capable of dealing with the formyl blocking groups and with the hydrophobic nature of the gramicidin peptides. Briefly, the formyl group of gramicidin was removed in anhydrous 2 N methanolic HCl. The resulting desformyl gramicidin was purified by molecular sieve and ion exchange chromatography and treated with phenylisothiocyanate (PITC). Excess PITC was quenched by reaction with aniline, the PTH-valine cleaved in anhydrous HCl, and the products separated by column chromatography, as above.

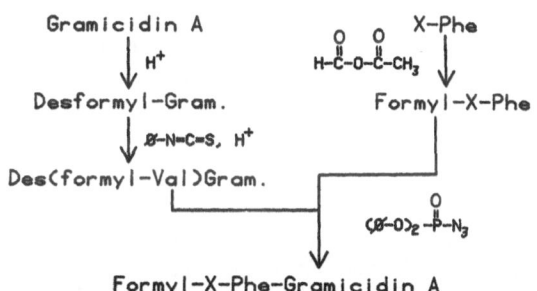

Synthetic Scheme

Fig. 2. Flow diagram illustrating the major reactions involved in replacing the natural Val$_1$ of gramicidin with a substituted phenylalanine (X-Phe) or other amino acid.

Amino acids were formylated using formic-acetic anhydride at 4°C, recrystallized from ethyl acetate, and coupled in dimethylformamide to des(formylvalyl)-gramicidin using a 2-fold molar excess of diphenyl-phosphorazidate at 0°C in the presence of N-methyl-morpholine. Highly purified gramicidins for single-channel studies were prepared directly from the coupling reaction mixtures by a "double passage" reversed phase HPLC technique (8). A summary of the semisynthetic reactions is shown in Fig. 2.

Single-channel measurements for gramicidin analogues at 25°C in planar bilayer membranes of diphytanoylphosphatidylcholine dissolved in n-decane were made using the bilayer-punch technique described by Andersen (9). For each analogue the conductance data in 1 M NaCl or 1 M CsCl were based on more than 300 current transitions measured at 25 and 50 mV applied potential.

RESULTS AND DISCUSSION

Table 1 lists the conductances of several pairs of gramicidin A analogues that are identical except for the first amino acid, and that have side chains of similar size and shape at position #1. In each case, a more polar side chain leads to a lower single-channel conductance. Data at other sodium ion concentrations (not shown) indicate that the effect is primarily a lowering of the maximal conductance, while the sodium ion activity for half maximal conductance is not significantly affected. Exceptions to this rule have been observed when Tyr or hexafluoro-Val is at position #1 of gramicidin A; channels formed by these analogues have higher apparent sodium affinities. We note that the position 1 side chains are relatively distant from the tight cation binding sites that are near the carbonyls of residues #11 and #13 (10).

Qualitatively similar results were observed with Cs^+ or Na^+ as the permeating ion (Table I). The relative conductance changes induced by polar side chains are, however, less for Cs^+ than for Na^+. Thus the cesium selectivity is enhanced for the compounds that have the more polar side chains at position #1. The smaller conductance changes observed with Cs^+ could be due to the higher intrinsic channel permeability to Cs^+ and/or to ion-side chain interactions that are less favorable for Na^+ than for Cs^+.

The average channel lifetime is shorter when a polar side chain is at position #1, e.g., 0.2 sec for trifluoro-Val vs. 0.6 sec for Val. This result is consistent with a β-helical structure that places both position-1 side chains near the center of the bilayer (Fig. 1) in a low dielectric medium. Channels with polar side chains at position #1 are expected to be less stable than channels with nonpolar side chains at position #1 because of the low dielectric lipid environment.

The question of whether the amino acid substitutions could substantially alter the channel conformation was addressed in experiments where we take advantage of the dimeric structure of the parent Val_1 gramicidin A channel. This provides a functional test for possible structural perturbations that may be induced by the variant side chains (11). Hybrid channels can form between the parent molecule and an analogue or between two analogues (Fig. 3). The hybrids appear to result from the random and independent association of two different gramicidin monomers, and generally have conductances and lifetimes that are between those of the respective symmetric channels (6,11). The intermediate and predictable

Table 1. Conductances of Gramicidin A Analogues

First Amino Acid	Conductance[a] (NaCl)	Conductance[a] (CsCl)
L-valine	12.4 pS	50 pS
4,4,4-trifluoro-L-valine	1.9	16
L-phenylalanine	10.2	45
p-fluoro-L-phenylalanine	5.9	32
L-norleucine	14.0	49
L-methionine	8.1	38

[a]1.0 M salt, diphytanoylphosphatidylcholine, 25-50 mV, 25°C.

properties of the hybrid channels suggest that the different monomers have similar overall structures and are at least compatible at their formyl-N-terminals. Once again, exceptions have been noted for Tyr_1- and hexa-fluoro-Val_1 gramicidin A. These monomers form hybrids with Val_1-gramicidin A that have shorter lifetimes and lower conductances than the symmetric channels.

We conclude that polar side chains at position #1 of gramicidin A can alter the channel conductance without either directly contacting the passing ions or changing the overall channel structure. The polar groups we have inserted lower the rate constant for one or more of the elementary steps of the transport process without affecting the ion affinity of the channel. The side chain effects presumably could be due to ion-dipole interactions or to electron inductive effects. Structural information about the channel has been important in ruling out direct ion-side chain contact. Our results with position-1 analogues in both symmetric and hybrid channels also lend support to the basic N-terminal-to-N-terminal structural motif. The ability to chemically introduce halogenated side chains through site-specific modification was crucial for these studies of the ion transport mechanism.

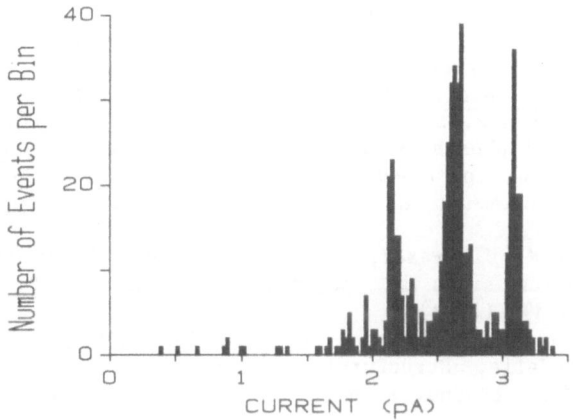

Fig. 3. Amplitude histograms of single-channel current steps obtained with mixtures of $norLeu_1$- and Met_1-gramicidin A. The three peaks in the histogram represent, from left to right: Met_1-gramacidin A channels; channels between $norLeu_1$- and Met_1-gramicidin A; and $norLeu_1$-gramicidin A channels. The two pure channel types were identified on the basis of the amplitudes of their single-channel currents. The central peak denotes the existence of hybrid channels. The average currents for the three peaks were, from left to right (with numbers of channels in parentheses): 2.156 ± 0.082 pA (119); 2.618 ± 0.076 pA (252); and 3.070 ± 0.051 pA (126). The three peaks account for 85% of the 586 channel transitions in the histogram. 1.0 M NaCl, 200 mV, 0.0503 pA bin width.

ACKNOWLEDGMENTS

We thank Judy Parli, Ruth Corder, Peter Landa and Julie Weiss for technical assistance. This work was supported in part by NIH grants NS-00648, GM-21342 and GM-34968.

REFERENCES

1. O. S. Andersen, Gramicidin channels, Ann. Rev. Physiol. 46:531 (1984).
2. J. F. Hinton and R. E. Koeppe II, The complexing properties of gramicidins, in: "Metal Ions in Biological Systems," H. Sigel, ed., Marcel Dekker, New York (1985).
3. R. E. Koeppe II, J. A. Paczkowski, and W. L. Whaley, Gramicidin K: a new linear channel-forming gramicidin from Bacillus brevis, Biochemistry 24:2822 (1985).
4. S. Weinstein, B. A. Wallace, E. R. Blout, J. S. Morrow, and W. R. Veatch, Conformation of gramicidin A channel in phospholipid vesicles: a ^{13}C and ^{19}F nuclear magnetic resonance study, Proc. Natl. Acad. Sci. USA 76:4230 (1979).
5. D. W. Urry, T. L. Trapane, and K. U. Prasad, Is the gramicidin A transmembrane channel single-stranded or double-stranded helix? A simple unequivocal determination, Science 221:1064 (1983).
6. J.-L.. Mazet, O. S. Andersen, and R. E. Koeppe II, Single-channel studies on linear gramicidins with altered amino acid sequences: a comparison of phenylalanine, tryptophane, and tyrosine substitutions at positions 1 and 11, Biophys. J. 45:263 (1984).
7. D. W. Urry, The gramicidin A transmembrane channel: a proposed π (L, D) helix, Proc. Natl. Acad. Sci., USA 68:672 (1971).
8. L. B. Weiss and R. E. Koeppe II, Semisynthesis of linear gramicidins using diphenyl phosphorazidate, Int. J. Peptide Prot. Res., in press.
9. O. S. Andersen, Ion movement through gramicidin A channels: single-channel measurements at very high potentials, Biophys. J. 41:119 (1983).
10. D. W. Urry, K. U. Prasad, and T. L. Trapane, Location of monovalent cation binding sites in the gramicidin channel, Proc. Natl. Acad. Sci. USA 79:390 (1982).
11. J. T. Durkin, O. S. Andersen, E. R. Blout, F. Heitz, R. E. Koeppe II, and Y. Trudelle, Structural information from functional measurements: single-channel studies on gramicidin analogues, Biophys. J., in press.

XII. Domain and Topographical Studies

IDENTIFICATION OF FUNCTIONAL DOMAINS OF THE INHIBITOR PROTEIN OF

cAMP-DEPENDENT PROTEIN KINASE

Heung-chin Cheng[+], Bruce E. Kemp[‡], Alan J. Smith[+], Richard B. Pearson[‡], Scott, M. Van Patten[+], Lufti Misconi[‡], and Donal A. Walsh[+]

[+]Department of Biological Chemistry, School of Medicine University of California, Davis, CA 95616 USA
[‡]Department of Medicine, University of Melbourne, Repatriation General Hospital, West Heidelberg, Victoria, 3081, Australia

INTRODUCTION

Phosphorylation-dephosphorylation of proteins is now recognized as a dynamic regulatory process in many cellular functions. cAMP-dependent protein kinase is one of the major enzymes involved in this process. The kinase mediates most, if not all, of the cellular functions of cAMP. Physiologically, the series of events leading to cAMP-dependent protein phosphorylation begins with the binding of a hormone or other agent to a cell surface receptor. In many cases, the binding leads to the activation of adenylate cyclase and a consequential increase in the intracellular level of cAMP. In eukaryotes, the major, if not sole, cellular receptor of cAMP is the regulatory subunit of cAMP-dependent protein kinase. The kinase is activated upon binding of cAMP to its regulatory subunits. The activated kinase then phosphorylates cellular proteins and such modifications lead to changes in cellular functions.

Structurally, cAMP-dependent protein kinase is a tetrameric multi-subunit enzyme. The holoenzyme contains a dimeric regulatory subunit (R_2) and two identical catalytic subunits (C) (Beavo et al., 1975). The holoenzyme is the inactive form of the protein kinase. Upon binding of cAMP to the regulatory subunits, the holoenzyme is dissociated into the dimeric regulatory subunits and monomeric free catalytic subunits according to the following equation (Corbin et al., 1978).

$$R_2C_2 + 4 \text{ cAMP} \rightleftharpoons R_2 \text{ cAMP}_4 + 2 \text{ C} \qquad \text{(equation 1)}$$

The free catalytic subunit is the active form of the enzyme and mediate most, if not all, of the physiological effects of cAMP by phosphorylating a number of cellular proteins.

Kinetically, the free catalytic subunits of cAMP-dependent protein kinase has been shown to follow an ordered BiBi mechanism.

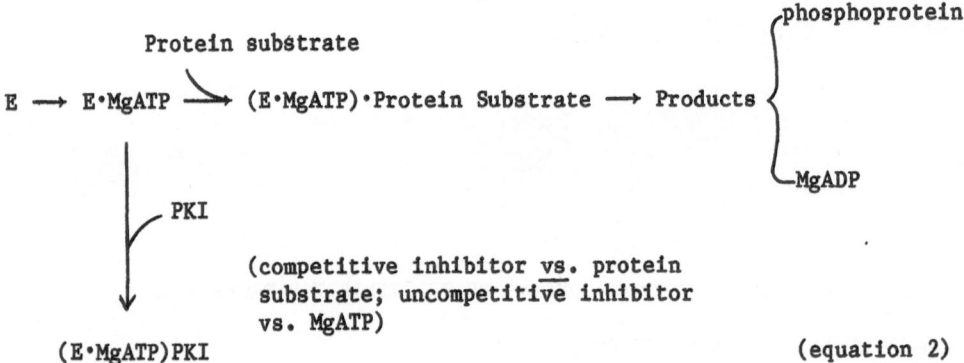

Protein substrate

E ⟶ E·MgATP ⟶ (E·MgATP)·Protein Substrate ⟶ Products ⟨ phosphoprotein / MgADP

↓ PKI

(competitive inhibitor vs. protein
substrate; uncompetitive inhibitor
vs. MgATP)

(E·MgATP)PKI   (equation 2)

The enyzme first binds to MgATP to form an enzyme·MgATP complex and this is
followed by the interaction of the enzyme·MgATP complex with the protein
substrate and the catalytic transfer of the γ-phosphate group from ATP to
the substrate protein. Kinetic analyses show that the K_m values of the
enzyme for both MgATP and protein substrates are in micromolar range (Whitehouse et al., 1983).

In addition to the regulatory subunits of the cAMP-dependent protein
kinase, another regulator protein also possibly acts as a physiological
modulator of the enzyme. This protein is the heat stable protein inhibitor
of the cAMP-dependent protein kinase. It exerts its action by binding to
the free catalytic subunit but not to the holoenzyme form of the protein
kinase (Ashby and Walsh, 1973). Kinetically, the inhibitor protein is competitive with respect to protein substrate and uncompetitive with respect to
MgATP (Whitehouse and Walsh, 1983). The inhibitor protein inhibits the
enzyme with a K_i value of 0.1 nM, a value of affinity four orders of magnitude greater than that of any substrate protein of the enzyme.

What are the structural features dictating the avid binding of the
inhibitor protein to the enzyme and making the inhibitor protein such a
potent inhibitor of the kinase? This question is what is being addressed by
experiments described in this report.

EXPERIMENTAL PROCEDURES

Materials

TLCK-treated α-chymotrypsin and histone (type II_A) were obtained from
Sigma Chemical Company. Staphylococcus aureus V8 protease was from Miles
Laboratories. Bovine heart cyclic AMP-dependent protein kinase was purified
to homogeneity as described previously (Whitehouse and Walsh, 1983). The
inhibitor protein of cAMP-dependent protein kinase was purified to homogeneity as described in Cheng et al. (1985a).

Procedures for Proteolytic Digestion of Protein and Peptide Inhibitors

Purified inhibitor protein, at a concentration of 1.4 mg/ml, was incubated in 0.1 M NH_4HCO_3, pH 7.8 with 0.16 mg/ml of Staphylococcus aureus V8
protease for five hours at 37°C. The reaction was terminated by a 60-fold
dilution of the reaction mixture into 10% trifluoroacetic acid. The
reaction was then injected to HPLC for separation of proteolytic fragments.

Chymotryptic digestion of the synthetic inhibitory peptide IP$_{20}$-amide was performed by incubating IP$_{20}$-amide at a concentration of 0.67 mg/ml and a peptide substrate to protease ratio of 700:1 (molar ratio). The reaction carried out in 10 mM Tris chloride (pH 8.0) at 23°C for two hours. The products were separated by HPLC.

HPLC Separation of Proteolytic Products

After digestion, samples were adjusted to 10% trifluoroacetic acid and applied to a reverse-phase µbondapak C$_{18}$ column. The column was washed with solvent A (0.1% trifluoroacetic acid in water) for 10 minutes at 1 ml/min and then eluted with a linear gradient of solvent A and solvent B (0.1% trifluoroacetic acid and 90% acetonitrile in water). The gradient started with 100% solvent A/0% solvent B and proceeded up to 40% solvent A/60% solvent B in 60 minutes at a flow rate of 1 ml/min. Fractions (1 ml) were collected.

Amino Acid Composition Analyses, Amino Acid Sequence Determination, Chemical Synthesis of Peptides, and Activity and Kinetic Analysis of Inhibitor Protein and Peptides

Procedures for amino acid composition and sequence analyses have been described in Cheng et al. (1985a). Model peptides were chemically synthesized as carboxy terminal amides by manual solid phase synthesis. Peptides were cleaved from the solid support and initially purified as described in Kemp (1979). Final purification of the peptides was performed by reverse phase HPLC as described in Cheng et al. (1985b). Procedures for activity and kinetic analyses of peptide and protein inhibitors were described in Cheng et al. (1985a).

RESULTS AND DISCUSSION

Isolation of Active Fragments from Staphylococcal Protease Digest of Native Inhibitor Protein

To find out which parts of the inhibitor protein contain the functional domains essential for the potent inhibitory activity of the protein, the inhibitor protein was cleaved into smaller fragments using a number of proteases and the fragments retaining significant inhibitory activity were isolated. Among the different proteases tried, Staphylococcus aureus V8 protease digestion of the inhibitor protein was found to generate two active fragments. After digestion by the protease for five hours, the products of proteolytic digestion were applied to a reverse phase HPLC column. Peptides were eluted with a solvent gradient of increasing acetonitrile as described in "Experimental Procedures". Fig. 1 shows the HPLC run of the digest. As indicated, digestion of the inhibitor protein by Staphylococcal protease for five hours resulted in its complete conversion to a number of proteolytic fragments. Among them, two exhibited inhibitory activity. The active fragment Peptide I was shown to be a peptide of twenty amino acids by amino acid composition analyses. The other active fragment, Peptide II, was found to contain thirty to forty amino acids. Peptide II was the precursor of Peptide I. Since Peptide I had twenty amino acids, it has been termed IP$_{20}$, meaning "inhibitory peptide with twenty amino acids."

Amino acid sequence determination of IP$_{20}$ confirmed that it is a twenty amino acid peptide with the sequence:

Thr·Thr·Tyr·Ala·Asp·Phe·Ile·Ala·Ser·Gly·Arg·Thr·Gly·Arg·Arg·Asn· Ala·Ile· His·Asp.

A peptide having the same amino acid sequence of IP_{20}, except with the addition of a C-terminal amide group, was chemically synthesized and tested for inhibitory activity. Kinetic analyses revealed that the native IP_{20} isolated from proteolytic digests of the inhibitor protein had a Ki value of 0.3 nM, the synthetic IP_{20}-amide a Ki of 2.3 nM (Table 1), and the inhibitor protein a Ki value of 0.1 nM (Cheng et al., 1985b). Thus, the native IP_{20} and synthetic IP_{20}-amide are potent inhibitors of cAMP-dependent protein kinase; their activities are nearly equivalent to that of the native inhibitor protein. The discrepancy between the inhibitory activity of the native and synthetic peptide is not yet understood. Possibly, it is due to the amide group at the C-terminal end of the synthetic IP_{20}-amide, although this seems unlikely (see below). A more likely explanation is that the native IP_{20} may contain a substituent group not detected by amino acid composition analyses and sequence determination and this possibility is under exploration. Data at this stage nevertheless indicates that this twenty amino acid sequence contains most, if not all, of the functional domains of the inhibitor protein essential for its potent inhibitory activity.

It has been well established that most, if not all, substrate proteins of cAMP-dependent protein kinase contain a basic subsite with a pair of basic amino acid residues (Lys-Arg or Arg-Arg) located on the N-terminal side of the target serine or threonine residue. Based on this, Kemptide with the sequence, -Leu-Arg-Arg-Ala-Ser-(PO_4)-Leu-Gly-, derived from the

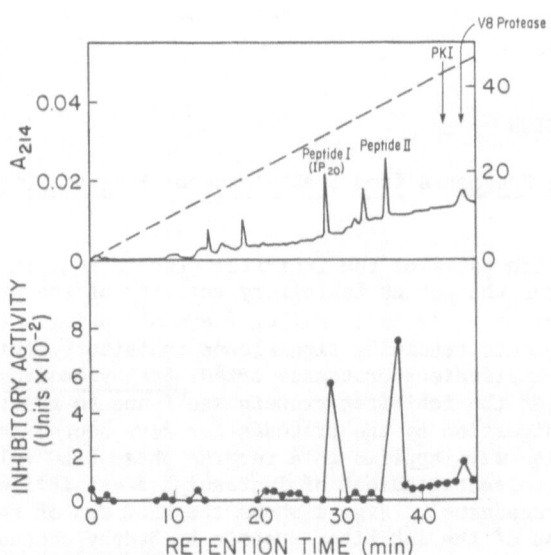

Figure 1. Hplc separation of the proteolytic products obtained by Staphylococcus aureus V8 protease digestion of the inhibitor protein.

The upper trace (——) is the A_{214} absorbance, the lower trace (—·—·—) is inhibitory activity, the gradient of solvent B (see "Experimental Procedures") is denoted by -------. Retention time is indicated from the start of the gradient. The arrows indicate the points of elution of native inhibitor protein and protease.

site of cAMP-dependent phosphorylation of porcine liver pyruvate kinase, has been used extensively as an artificial substrate of cAMP-dependent protein kinase. Extensive kinetic studies by Kemp et al. (1977) have demonstrated that the two arginine residues in Kemptide are indispensable for effective binding of the substrate peptide to the enzyme. Comparison of the phosphorylation site sequence (-Arg-Arg-X-Ser-(PO$_4$)-) with the C-terminal portion of IP$_{20}$ (-Gly-Arg-Arg-Asn-Ala-Ile-His-Asp) suggests that the pair of arginine residues in IP$_{20}$ is most likely a functional domain required for interaction with the protein kinase.

Using the sequence of Kemptide as the parent sequence, Feramisco et al. (1980) synthesized a peptide with alanine substituting the target serine residue. This peptide is termed Ala-peptide (Table 1). Activity analyses revealed that Ala-peptide was a poor inhibitor with a Ki value of 0.2 mM. D-Ser Kemptide, a Kemptide with the target serine substituted by a D-Serine residue, was found to be neither a substrate nor an inhibitor (Reed et al., 1985). Further analysis has, however, suggested that the hydroxy group of serine in Kemptide, besides acting as the phospho-acceptor, is important in enhancing effective binding of substrate protein to the protein kinase (Whitehouse et al., 1983; Reed et al., 1985). Moreover, the fact that Ala-peptide is a poor inhibitor means that the -Arg-Arg-X-Ala- domain in the IP$_{20}$ sequence is far from sufficient to make the inhibitor protein an inhibitor of such high potency. Therefore, other portions in the twenty amino acid sequence must be involved. In the absence of cAMP, the regulatory subunit is also a very potent inhibitor of the enzyme with an apparant dissociation constant for regulatory subunit-catalytic subunit interaction of between 0.1 to 0.6 nM (Hofmann, 1980). The type II regulatory subunits (R$_{II}$) is also phosphorylated by the catalytic subunit of the kinase. Comparison of the autophosphorylation site sequence of R$_{II}$ (Table 1) and IP$_{20}$ sequence suggests that the third arginine of IP$_{20}$ may also be involved in binding.

Chymotryptic Digestion of the Synthetic IP$_{20}$-Amide

First insights of the structural features of functional importance in IP$_{20}$-amide came when the peptide was digested with α-chymotrypsin (Fig. 2). After two hours of digestion, the majority of the parent peptide, IP$_{20}$-amide, was converted into peptide Ile^7-Asp^{20}-amide and a mixture of two peptides as indicated by arrows in Figure 2. Amino acid sequence determination revealed that the mixture contains peptides Ala^4-Asp^{20}-amide and Thr^1-Phe^6. When aliquots from the HPLC fractions collected were assayed for inhibitory activity, we found that Ile^7-Asp^{20}-amide, Ala^4-Asp^{20}-amide had activities very substantially lower than that of IP$_{20}$-amide.

Kinetic analysis of Ile^7-Asp^{20}-amide showed that its K$_i$ value (115 nM) was substantially higher than that of IP$_{20}$-amide (Table 1). No further analysis has yet been done to evaluate the inhibitory potency of Ala^4-Asp^{20}-amide due to its contamination by the Thr^1-Phe^6 peptide, but the data from HPLC activity profile (not shown) suggest that Ala^4-Asp^{20}-amide is at best no more inhibitory than Ile^7-Asp^{20}-amide.

Results from this experiment suggested that the Thr^1-Phe^6 portion of the twenty amino acid sequence of IP$_{20}$ contained one or more functional domains contributing to potent inhibitory activity of the inhibitor protein.

Table 1. *Protein kinase inhibitory peptide and analogous sequences. N.D. stands for not determined, NI stands for no inhibitory activity. *IP_{20} is a peptide derived from proteolytic digestion of the native inhibitor. All the other peptides shown in this table are synthetic peptides.*

Nomenclature	Sequence	K_i or K_m	Reference
Phosphorylation Site	$\overset{\displaystyle (PO_4)}{\underset{\displaystyle RRXS}{\mid}}$	—	Beavo et. al. 1975
Kemptide	$\overset{\displaystyle (PO_4)}{\underset{\displaystyle LRRASLG}{\mid}}$	4–6 µM	Kemp et. al. 1977
Autophosphorylation site of R_{II}	$\overset{\displaystyle (PO_4)}{\underset{\displaystyle PIPGRFDRRVSVC}{\mid}}$	—	Takio et. al. 1982
Ala–Peptide	LRRAALG	0.2 mM	Feramisco et. al. 1980
IP_{20}*	TTYADFIASGRTGRRNAIHD	0.3 nM	This Study
IP_{20}-amide	TTYADFIASGRTGRRNAIHD–NH_2	2.3 nM	"
Thr^1–Ile^{12}-amide	TTYADFIASGRTHRRNAI–NH_2	3.1 nM	"
Ala^4–Asp^{20}-amide	ADFIASGRTGRRNAIHD–NH_2	N.D.	"
Thr^1–phe^6	TTYADF	N.D.	"
Ile^7–Asp^{20}-amide	IASGRTGRRNAIHD–NH_2	115 nM	"
Tyr^3–ile^{18}-amide	YADFIASGRTGRRNAI–NH_2	27.5 nM	"
Phe^6–Asp^{20}-amide	FIASGRTGRRNAIHD–NH_2	73 nM	"
Gly^{10}–Asp^{20}-amide	GRTGRRNAIHD–NH_2	57 nM	"
Gly^{10}–Lys^{11}–Asp^{20}-amide	GLTGRRNAIHD–NH_2	0.37 µM	"
Gly^{10}–Lys^{14}–Asp^{20}-amide	GRTGLRNAIHD–NH_2	36 µM	"
Gly^{10}–Lys^{15}–Asp^{20}-amide	GRTGRLNAIHD–NH_2	4.2 µM	"
Ile^7–Ala^{26}	IASGRTGRRNAIHDILVSSA	0.8 µM	Scott et. al. 1985
Gly^{10}–Ala^{26}	GRTGRRNAIHDILVSSA	75 µM	"
Arg^{14}–Ala^{26}	RRNAIHDILVSSA	1.5 mM	"
Asn^{16}–Ala^{26}	NAIHDILVSSA	N.I.	"

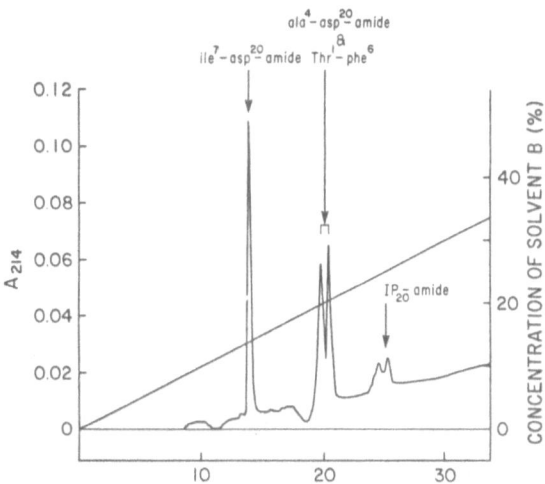

Fig. 2. *HPLC separation of the proteolytic products obtained by α-chymotryptic digestion of IP_{20}-amide. For explanations of symbols and lines, refer to legend of Figure 1.*

The Use of Synthetic Model Peptides to Identify the Structural Requirements for the Potent Inhibitory Activity of the Inhibitor Protein.

Based upon the sequence of IP_{20}, the synthesis of a variety of peptides has been initiated, aimed at probing the functional domains required for full interaction with the catalytic site of the protein kinase. In brief the results obtained to date are as follows:

C-terminal. Deletion of the two C-terminal residues (His, Asp) to yield the peptide, Thr^1-Ile^{18}-amide, did not appreciably modify inhibitory activity (Table 1).

N-terminal. Deletion of the first two threonine residues to yield the peptide Tyr^3-Ile^{18}-amide modified the K_i value 9-fold. The peptide Phe^6-Asp^{20}-amide, prepared synthetically, and Ile^7-Asp^{20}-amide (prepared by chymotrypsin digestion, see above) had K_i values of 73 nM and 115 nM, respectively. A shorter peptide $Gly^{10}-Asp^{20}$-amide, however, was slightly more inhibitory (K_i = 57 nM) (Table 1). Taken collectively, these data clearly indicate that within the first six N-terminal residues, and quite possibly within the first three, there is a critical site for interaction with the protein kinase. The marked reduction in activity upon removal of the first two threonines may suggest that one or both play an important role, however, such a deletion would result in the tyrosine residue having a free N-terminal, which might compromise any role in which it was involved. A similar reservation should also be applied in considering the reduced activity of the Phe^6-Asp^{20}-amide peptide.

Role of arginine residues. To study the functional importance of the three arginine residues, peptides with each of these substituted by lysine were synthesized using the $Gly^{10}-Asp^{20}$-amide peptide as the parent sequence. Substitution of either Arg^{14} or Arg^{15} by lysine (peptides $Gly^{10}-Lys^{14}-Asp^{20}$-amide and $Gly^{10}-Lys^{15}-Asp^{20}$-amide) drastically reduced inhibitory activity by in excess of of 80-fold, confirming the importance of the Arg-Arg basic subsite as a recognition domain. In addition, substitution of Arg^{11} with lysine also promoted a major decrease (~8-fold) in activity, demonstrating that rather than simply the pair of arginines, it was the full arginine cluster site that was critical as also appears to be the case with the regulatory subunit (Table 1).

In a concurrent but independent study, Scott et al. (1985) isolated an inhibitory peptide from mast cell proteinase II digest of the inhibitor protein. This peptide also contains 20-amino acids but is distinct from IP_{20} in not having the first six N-terminal residues but having an additional six C-terminal residues. Its sequence was determined to be IleAlaSerGlyArgThr-GlyArgArgAsnAlaIleHisAspIleLeuValSerSerAla. Based on this sequence Scott et al. (1985) have synthesized the series of peptides: Ile^7-Ala^{26}, $Gly^{10}-Ala^{26}$, $Arg^{14}-Ala^{26}$, and $Asn^{16}-Ala^{26}$ (Table 1). Kinetic analysis showed that $Asn^{16}-Ala^{26}$ had no activity but $Arg^{14}-Ala^{26}$ had weak inhibitory activity (K_i = 1.5 mM). These data likewise show the indispensability of the pair of arginine residues ($Arg^{14}Arg^{15}$) for the inhibitory action of the inhibitor protein. Peptides Ile^7-Ala^{26} and $Gly^{10}-Ala^{26}$ had significant inhibitory activity, with reported K_i values being 0.8 μM and 75 μM respectively, providing further evidence that Arg^{11} also contributes to the potent inhibitory activity of the inhibitor protein. In comparison of our results with those of Scott et al. (1985), the K_i value of $Gly^{10}-Ala^{26}$ appears to be markedly higher (0.8 μM) than that of $Gly^{10}-Asp^{20}$-amide (57 nM). Whether this might be due to the presence of the $Ile^{21}-Ala^{26}$ sequence at the C-terminal decreasing the inhibitory potency, is under study.

Functional Specificities of IP_{20}-Amide

Synthetic IP_{20}-amide has been shown to be a specific inhibitor of cAMP-dependent protein kinase. Its inhibitory action on other kinases has been tested with peptide concentration in the range of 1-5 μM and the appropriate protein substrates, at or below the K_m values for the corresponding kinases. IP_{20}-amide does not inhibit casein kinase I, casein kinase II, proteolytically-activated kinase I, Ca^{++}-phospholipid dependent protein kinase, cGMP-dependent protein kinase, Ca^{++}-calmodulin dependent protein kinase, or myosin light chain kinase (data not shown). Since arginine residues are also substrate specificity determinants for several of these kinases, this remarkably high degree of functional specificity of IP_{20}-amide suggests that the various key amino acid residues in the peptide so far elucidated, and also possibly others, are critical for the functional specificity to IP_{20}-amide and the inhibitor protein.

Summary

The existing data reveals that the inhibitor protein contains the following functional domains.

1) The inhibitor protein contains an arginine cluster with three functionally important arginines ($-Gly^{10}Arg^{11}Thr^{12}Gly^{13}Arg^{14}Arg^{15}Asn^{16}-Ala^{17}-$).

2) The N-terminal critical region (ThrThrTyrAlaAspPhe) in the IP_{20} sequence is suggested to contain one or more functional domains. Of particular interest are the first three residues ($Thr^1Thr^2Tyr^3$) having hydroxy groups as their side chains. As implicated by studies of ala-peptide (Whitehouse et al., 1983; Reed et al., 1985), the hydroxy group of serine in Kemptide and in cAMP-dependent phosphorylation sites of substrate proteins is important, not only as the phosphoacceptor, but also to enhance binding. A possible role for hydroxy group of any of the first three residues of IP_{20} (ThrThrTyr) might be postulated. If the inhibitor protein and other inhibitory peptides (e.g. IP_{20}) can assume a special conformation such that the hydroxy group of one of them can mimic the role of serine hydroxy group in terms of binding, but not as a phospho-acceptor, then this portion of the IP_{20} sequence could account in part for the high potency of inhibitory activity of the inhibitor protein.

3) Structural features conferring functional specificity to the inhibitor protein must be present in the twenty amino acid sequence of IP$_{20}$. With these structural features, IP$_{20}$-amide and most likely the inhibitor protein is able to specifically inhibit the cAMP-dependent protein kinase but not other protein kinases tested.

REFERENCES

Ashby, C. D., and Walsh, D.A., 1973, J. Biol. Chem. 248:1255.

Beavo, J. A., Bechtel, P. J., and Krebs, E. G., 1975, Adv. Cyclic Nucleotide Res. 5:241.

Cheng, H. C., Van Patten, S. M., Smith, A. J., and Walsh, D. A., 1985a, Biochem. J. 231:in press.

Cheng, H. C., Kemp, B. E., Pearson, R. B., Smith, A. J., Misconi, L., Van Patten, S. M., and Walsh, D. A., 1985b, J. Biol. Chem. submitted.

Corbin, J. D., Sugden, P. H., West, L., Flockhart, D. A., Lincoln, T. M., and McCarthy, D., 1978, J. Biol. Chem. 253:3997.

Feramisco, J. R., Glass, D. B., and Krebs, E. G., 1980, J. Biol. Chem. 255:4240.

Glass, D. B., and Krebs, E. G., 1979, J. Biol. Chem. 254:9728.

Glass, D. B., and Krebs, E. G., 1982, J. Biol. Chem. 257:1196.

Hofmann, F., 1980, J. Biol. Chem. 255:1559.

Hodges, R. S., and Merrifield, R. B., 1975, Anal. Biochem. 65:241.

Hathaway, G. M., Tuazon, P. T., and Traugh, J. A., 1983, in: "Methods in Enzymology, Vol. 99," J. C. Corbin and J. G. Hardman, eds., Academic Press, New York.

Henderson, P. J. F., 1972, Biochem. J. 127:321.

Kemp, B. E., Graves, D. J., Benjamini, E., and Krebs, E. G., 1977, J. Biol. Chem. 252:4888.

Kemp, B. E., and Pearson, R. B., 1985, J. Biol. Chem. 260:3355.

Kemp, B. E., 1979, J. Biol. Chem. 254:2638.

Pearson, R. B., Woodget, J. R., Cohen, P., and Kemp, B. E., 1985, J. Biol. Chem. submitted.

Reed, J., Kinzel, V., Kemp, B. E., and Walsh, D. A., 1985, Biochemistry 24:2961.

Scott, J. D., Fischer, E. H., Demaille, J. G., and Krebs, E. G., 1985, Proc. Natl. Acad. Sci. USA 82:4379.

Scott, T. R., Kemp, B. E., Su, H. -de., and Kuo, J. F., 1985, J. Biol. Chem. J. Biol. Chem. in press.

Takio, K., Smith, S. B., Krebs, E. G., Walsh, K., and Titani, K., 1982, Proc. Natl. Acad. Sci. USA 79:2544.

Tahara, S. M., and Traugh, J. A., 1981, J. Biol. Chem. 256:11558.

Whitehouse, S., Feramisco, J. R., Casnellie, J. E., Krebs, E. G., and Walsh, D. A., 1983, J. Biol. Chem. 258:3693.

Whitehouse, S., and Walsh, D. A., 1982, J. Biol. Chem. 257:6028.

Whitehouse, S., and Walsh, D. A., 1983, J. Biol. Chem. 258:3682.

ATRIAL NATRIURETIC FACTOR RECEPTOR IN ADRENAL PLASMA MEMBRANE:

IDENTIFICATION BY PHOTO-AFFINITY LABELING

Kunio S. Misono

Department of Biochemistry
Vanderbilt University School of Medicine
Nashville, TN 37232

INTRODUCTION

Atrial myocytes of mammalian heart contain a peptide hormone, termed atrial natriuretic factor (ANF), which has potent natriuretic (1) and vaso-relaxant activities (2,3). ANF is released into the circulation responding to blood volume expansion, where it is believed to play an important role in restoring and maintaining a constant volume/pressure of the cardiovascular system. ANF has been purified in multiple peptide of different sizes whose amino acid sequences have been determined by a number of investigators (4-9). These peptides are apparently derived from a single precursor protein whose sequence has been deduced from the nucleotide sequence of cloned cDNA (10,11). However, the mechanism of ANF processing yet remains largely unknown.

The natriuretic and vasorelaxant effects of ANF involve direct inter-actions of circulating ANF with the target tissues, the kidney, adrenal gland and vascular beds. This interaction apparently is mediated through a membrane bound receptor specific for ANF (12-15). In this report I present synthesis and purification of two photo-reactive radioiodinated ANF deriva-tives. Using these derivatives as a photo-affinity ligand and utilizing an in vitro receptor binding system with adrenal plasma membrane preparations, I obtained specific labeling of a membrane bound protein which most likely is a receptor protein or its subunit for ANF.

METHODS

Separation of Peptides and Peptide Derivatives

Separations of peptides and their derivatives were carried out by high pressure liquid chromatography (HPLC) on a Vydac C_{18} column (10μ, 0.46 x 25 cm, Separations group). Elution was carried out using a linear gradient of acetonitrile from 0% to 60% applied over a period of 90 min in 0.1% triflu-oroacetic acid at a flow rate of 1 ml/min. These elution conditions were used throughout as standard conditions.

---

*This study was supported by grant 84-1291 from the American Heart Association. I thank Mr. David Sullins for preparation of this manuscript.

‡Abbreviations: ANF, atrial natriuretic factor; AZB-, 4-azidobenzoyl.

Analytical Procedures

Amino acid analyses were carried out by reverse phase HPLC of phenyl-thiocarbamyl amino acid derivatives of peptide hydrolysate (16) using Waters Pico-Tag system. ^{125}I-Radioactivity was measured in a Micromedic gamma-counter with an efficiency of 80%.

Radioiodination of ANF-IV

ANF-IV (9 µg) with the sequence (4)

```
      1                  10                     20
      R-S-S-C-F-G-G-R-I-D-R-I-G-A-Q-S-G-L-G-C-N-S-F-R-Y
```

was radioiodinated in 50 µl of 0.5 M sodium phosphate buffer, pH 7.6, by reacting with 1 mCi of carrier-free [^{125}I]NaI and Chloramine-T (17) for 20 sec. The reaction was terminated by dilution with 5 ml of cold water. The mixture was then passed through a SepPak C_{18} cartridge (Waters) twice. After washing the cartridge with 10 ml of 0.1% trifluoroacetic acid, the bound material was eluted with 80% methanol. The eluant was concentrated to about 200 µl by evaporation under a nitrogen stream. The concentrated solution was diluted 2-fold with water and injected onto a Vydac C_{18} column. The column was eluted under the standard conditions described above. Fractions were collected every 30 sec. Mono-iodinated ANF-IV (^{125}I-ANF-IV) was eluted in a predominant radioactivity peak, coinciding with a discrete UV-absorption peak at 215 nm, at about 46 min after the beginning of gradient, while unlabeled peptide was eluted at 42 min. The radioactive peak fraction was diluted 4 times with distilled water and stored at 4°C. The purified ^{125}I-ANF-IV has a theoretical specific activity of 2.125 µCi/pmol.

Preparation of 4-Azidobenzoyl-^{125}I-ANF-IV

ANF-IV 1.35 mg (0.5 µmol), was reacted with 1.3 mg (5 µmol) of N-hydroxysuccinimidyl 4-azidobenzoate (18) obtained from Pierce in 200 µl of redistilled dimethylformamide containing 1 µl of triethylamine in the dark at room temperature for 5 h. After the reaction, the peptide was extracted in 3 ml of 1 M acetic acid. The acetic acid layer was washed twice with equal volumes of ethyl acetate and then lyophilized. 4-Azidobenzoyl (AZB-) ANF-IV was purified from this material by HPLC. AZB-ANF-IV was aliquotized, dried under vacuum and stored at -20°C.

Iodination of AZB-ANF-IV (19.5 µg) was carried out in a manner similar to that described above for ANF-IV. AZB-^{125}I-ANF-IV was purified by HPLC under the standard conditions, aliquotized, dried under vacuum and stored at -20°C in the dark.

Preparation of 4-Azidobenzoyl-^{125}I-ANF-P

ANF-P with the sequence from residues 4-25 was prepared by processing ANF-IV (2.4 mg) through 3 complete cycles of Edman degradation in a Beckman 890B Sequencer performed as described (19). The peptide was recovered from the reaction cup with 5 ml of 1 M acetic acid and lyophilized. ANF-P was purified by HPLC under the standard conditions. The structure of ANF-P (yield of 30%) was confirmed by amino acid analysis. ANF-P, 0.65 mg (0.27 µmol), was reacted with 3 mg (11.5 µmol) of N-hydroxysuccimidyl 4-azidobenzoate in 200 µl of dimethylformamide containing 1 µl of triethylamine in the dark overnight. The resultant AZB-ANF-P was extracted, purified by HPLC, and then radioiodinated in a manner similar to that described above for AZB-^{125}I-ANF-IV.

Preparation of Adrenal Cortical Membranes

Bovine adrenal glands were obtained from a local slaughterhouse. The adrenal cortex was sliced (0.5 mm thick) using a Stadie-Riggs microtome. Only the outermost slices were used for the preparation of plasma membrane which was carried out by the method of Glossmann et al. (20). The membranes were suspended in a buffer (assay buffer), consisting of 50 mM Tris-HCl buffer, pH 7.4, 0.15 M NaCl, 5 mM $MgCl_2$, 0.2% bovine serum albumin

(Miles Lab.) and 0.05% bacitracin (P.L. Biochemicals) at a membrane protein concentration of 1.7 mg/ml, divided into small aliquots and stored at -70°C.

Rat adrenal capsules (about 2 g wet weight) from 50 rats were separated from medulla by manual compression and homogenized in 40 ml of 20 mM NaHCO$_3$ in a Polytron at setting of 8 for 10 sec twice. The homogenate was centrifuged at 1,500 x g for 10 min at 4°C. The supernatant was recentrifuged. The final supernatant was then centrifuged at 104,000 x g for 30 min. The pellet was suspended in assay buffer (0.8 mg membrane protein/ml), divided into aliquots and stored at -70°C.

Competition Binding Analysis

Binding studies were carried out at varying concentrations of unlabeled ANF peptides or their analogues (10^{-12} to 10^{-5} M) in 0.5 ml of the assay buffer containing about 150,000 cpm of ^{125}I-ANF-IV (about 70 pM). Aliquots of bovine adrenal membrane (17 µg protein) were incubated in the mixtures for 30 min at 0°C. The binding reached plateau after 30 min. Bound ^{125}I-ANF-IV was determined by filtering 0.4 ml aliquots of the incubation mixture through Whatman GF/B filters. After washing two times with 3 ml of the same buffer without bovine serum albumin and bacitracin, ^{125}I-Radioactivity trapped on the filter was counted. Binding data were analyzed using the LIGAND computer program (21).

Photo-labeling Reaction of Adrenal Plasma Membranes with ^{125}I-labeled 4-Azidobenzoyl ANF Derivatives

Adrenal plasma membrane from bovine (340 µg membrane protein) or rat (150 µg membrane protein) was suspended in 200 µl of assay buffer and was incubated with AZB-^{125}I-ANF-IV (25 nM) in the absence or presence of various concentrations of unmodified ANF-IV. The incubation was carried out in a polypropylene microfuge tube at 0°C for 30 min in the dark. The mixture was then photolyzed for 20 min with a 250-W General Electric sun lamp at a distance of about 15 cm, while maintaining the temperature at 4°C using a jacketed glass vessel connected to a circulating cold water bath. After photolysis, the membrane was collected by centrifugation. The membrane pellet was washed by resuspending in 1 ml of assay buffer followed by centrifugation. The washing process was repeated twice more with the same buffer but without serum albumin and bacitracin. The pellet was suspended and boiled in 1% SDS containing 0.1 M dithiothreitol for 3 min and was subjected to SDS-gel electrophoresis in 7.5% gel by the method of Laemmli (22). For autoradiography, gels were dried using a Bio-Rad gel-dryer and exposed to Kodak XAR-5 X-ray film. The binding and photolysis with AZB-^{125}I-ANF-P (13 nM) were carried out under the same conditions.

RESULTS AND DISCUSSION

Preparation of radio-iodinated photo-affinity ligands

A photo-reactive 4-azidobenzoyl (AZB-) moiety was introduced to the NH$_2$-terminus of ANF-IV by reacting with N-hydroxysuccinimidyl 4-azidobenzoate in the dark. Since ANF does not contain Lys or His residue, the reaction is limited to the terminal α-amino group. The reaction products were separated by HPLC as shown in Figure 1A. AZB-ANF-IV was eluted in a peak having significant absorption at both 214 and 313 nm at 48 to 49 min after the beginning of the acetonitrile gradient. The structure of AZB-ANF-IV was confirmed by observing a characteristic absorbance peak at 340 nm is its UV-spectrum and by amino acid analysis which gave a composition identical to the parent molecule. By amino acid analysis, the yield of AZB-ANF-IV was 14%. The peak eluted at 42 min after the beginning of the gradient contained unreacted ANF-IV based on its elution time and amino acid composition. No other peak contained a significant amount of amino acids.

The ^{125}I-radiolabel was incorporated into AZB-ANF-IV by radioiodination at the COOH-terminal Tyr residue by the chloramine T method. The iodination

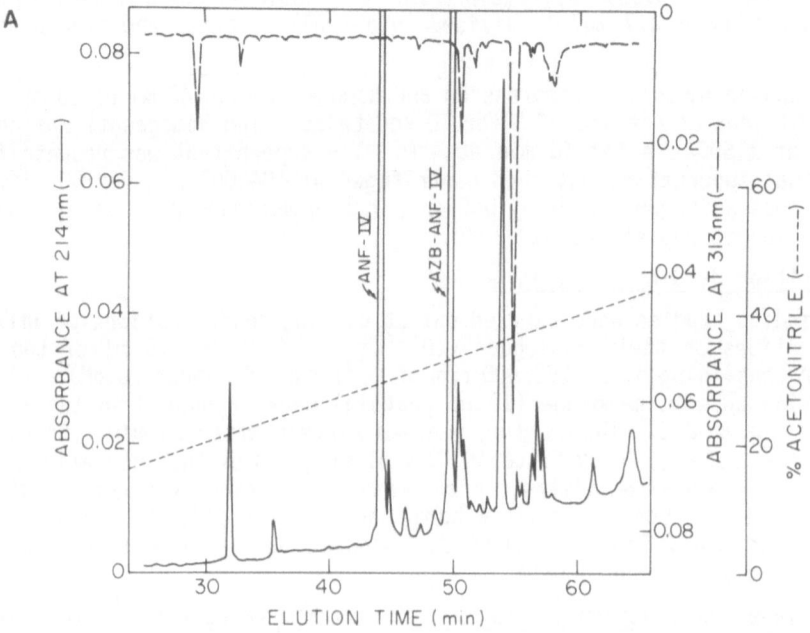

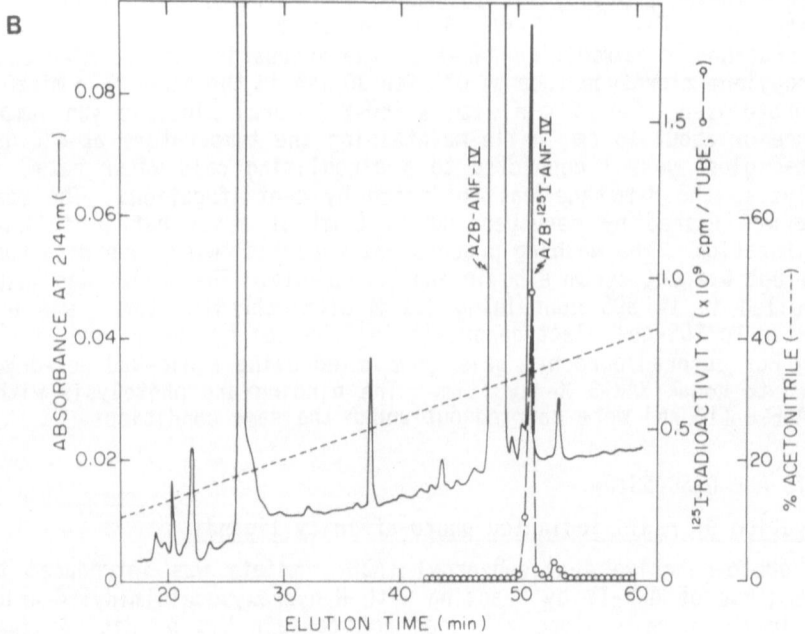

Figure 1. Preparation of mono-iodinated 4-azidobenzoyl ANF-IV. (A) ANF-IV was reacted with N-hydroxysuccinimidyl 4-azidobenzoate in the dark. The reaction products were extracted with 1 M acetic acid and was chromatographed on a Vydac C18 column (see text for details). Absorbance at both 214 nm and 313 nm were monitored by connecting two fixed wave length detectors in a series. (B) 4-aziobenzoyl ANF-IV (AZB-ANF-IV) from (B) was radioiodinated using chloramine T. The reaction products were extracted on a SepPak C18 cartridge, recovered from the cartridge and chromatographed on a Vydac C18 column (see text for details). Fractions of 0.5 ml (30 sec) were collected.

reaction was terminated by solid-phase extraction of peptides on a SepPak C18 cartridge. Addition of a reductant such as $Na_2S_2O_5$ was avoided in order to preserve the disulfide linkage in the ANF peptide which is known to be essential for the activity (4). The bound material was recovered and chromatographed by HPLC (Figure 1B). The major UV-absorbance peak eluting at about 48 min contained unreacted AZB-ANF-IV. A small but discrete UV-absorbance peak followed the major peak, eluting at 52 min which contained the majority of the radioactivity retained by the column. Based on the UV-absorbance and radioactivity contained in the peak fractions, the specific radioactivity of this material was estimated to be 1.5 to 2 µCi/pmole. This specific activity is consistent with mono-iodinated peptide which has a maximum theoretical specific activity of about 2.1 µCi/pmole. The formation of diiodinated AZB-ANF-IV apparently was not significant. It may have been prevented by the excess AZBANF-IV over [^{125}I]NaI (approximately 8-fold excess) added in the iodination mixture. The yield of purified AZB-^{125}I-ANF-IV ranged from 80 to 150 pmoles based on the radioactivity, representing recoveries of 17% to 30% of ^{125}I added in the reaction mixture.

AZB-^{125}I-ANFP was synthesized for the purpose of obtaining a potentially more geo-specific affinity labeling reagent. In this affinity reagent, the photo-reactive AZB-moiety was incorporated at Cys-4 position which is in the immediate proximity to the central and functionally essential domains of ANF localized within the sequence from residues 4 through 24. The NH_2-terminal Arg-Ser-Ser- sequence of ANF-IV was removed by 3 complete cycles of Edman degradation, yielding a 22-residue peptide designated ANF-P. ANF-P was recovered from the sequencer and purified by HPLC (not shown). The attachment of AZB-moiety to the terminal α-amino group and purification of resultant AZB-ANF-P were carried out in a manner similar to that used for ANF-IV. AZB-ANF-P was eluted from HPLC at 51 min and unreacted ANF-P at 44 min after the beginning of the gradient (not shown). The synthetic yield of AZB-derivative from ANF-P was low (2.5%) probably due to steric hindrance from the nearby peptide ring. Iodination of AZB-ANF-P on the other hand proceeded with yields comparable to those obtained in the iodination of AZB-ANF-IV. The elution time of AZB-^{124}I-ANF-P was 54 min.

<u>Receptor binding</u>

The receptor bindings of ANF and its analogues were studied using plasma membrane prepared from bovine adrenal cortex. The dissociation constants (K_d) of the peptides determined by competition binding assay are summarized in Table I. The biologically fully active peptide ANF-IV had K_d of 7.4×10^{-10} M, which is consistent with the IC_{50} reported for the ANF inhibition of aldosterone secretion from isolated rat adrenal glomerulosa cells stimulated by angiotensin I, ACTH or potassium (14). ANF-II had binding affinity nearly equal to ANV-IV, showing that the NH_2-terminal extension of the sequence in ANF-II has little contribution to the binding. The result again is consistent with the fact that the two peptides are essentially equipotent in causing smooth muscle relaxation and natriuresis (7). The substitutions of the residues in the ring portion of the sequence in [Tyr^5] ANF-IV and [Met^9] ANF-IV resulted in several fold reductions in the binding affinity. Atriopeptin II (6), lacking both NH_2-terminal Arg and COOH-terminal Tyr residues of ANV-IV, showed significantly reduced affinity. Further removal of COOH-terminal -Phe-Arg sequence to yield atropeptin I essentially abolished the binding to the membrane receptors. ANF-R peptide with residues 4-20, which represents only the ring portion of ANF, similarly showed little binding. The binding affinities of these peptides to the adrenal plasma membrane show close parallelism with their potencies in vasorelaxant and natriuretic activities, suggesting strongly that the present <u>in vitro</u> binding system is a viable representation of the membrane receptor system operative <u>in vivo</u>.

TABLE I.   Dissociation constants of atrial natriuretic peptides and analogues determined by competition binding against ^{125}I-ANF-IV to bovine adrenal cortex plasma membrane.

Peptide	Residues	Dissociation Constant (M)
ANF-IV	1-25	7.4×10^{-10}
ANF-II	-6-25	8.1×10^{-10}
[Tyr^5] ANF-IV	1-25	3.0×10^{-9}
[Met^9] ANF-IV	1-25	2.8×10^{-9}
Atriopeptin II	2-24	2.6×10^{-9}
Atriopeptin I	2-22	$>10^{-5}$
ANF-R	4-20	$>10^{-5}$

Competition bindings of AZB-ANF-IV and AZB-ANF-P against ^{125}I-ANF-V in the dark gave K_d of 1.3×10^{-9} M and 2.1×10^{-9} M, respectively, indicating that the introduction of the photoreactive AZB-moiety at the NH_2-termini of the molecules did not significantly affect the binding.

Photo-affinity labeling of ANF receptor

The results of photo-affinity labeling of adrenal membranes with AZB-^{125}I-ANF-IV and with AZB-^{125}I-ANF-P are shown in Figure 2. Incubation of AZB-^{125}I-ANF-IV with bovine adrenal plasma membrane followed by photo-activation resulted in radiolabeling of a number of proteins as analyzed by SDS gel electrophoresis and autoradiography (lane a). Of these, the major radioactive band with an apparent M_r of 124,000 was prevented from the

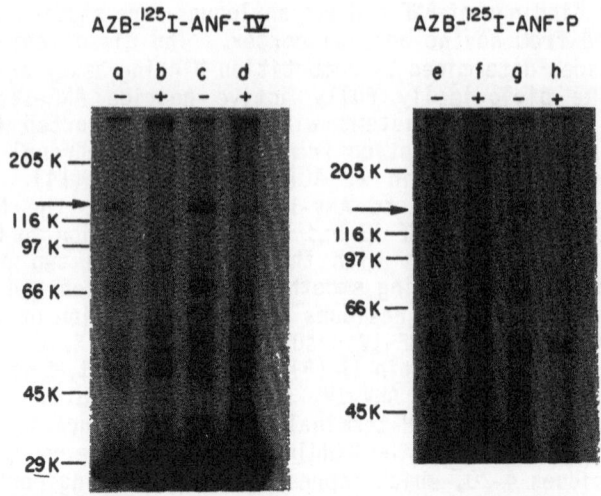

Figure 2. Autoradiograph showing the labeling of bovine (lanes a and b) and rat (lanes c and d) adrenal cortex plasma membranes by photolysis with 4-azidobenzoyl-^{125}I-ANF-IV in the absence (-) or presence (+) of 1 μM ANF-IV. The same experiments carried out with 4-azidobenzoyl-^{125}I-ANF-P with bovine (lanes e and f) and rat (lanes g and h) membranes. The arrows indicate the radioactive bands specifically displaced by ANF-IV.

labeling by inclusion of unmodified ANF-IV in the reaction mixture (lane b), demonstrating specificity of the labeling. Essentially identical results were obtained with rat adrenal membrane (lanes c and d), in which the specifically labeled protein had an apparent M_r of 126,000. The nature of other radio-labeled protein bands are not known. Bovine serum albumin and bacitracin as well as tris base present in the reaction mixture should have served as scavengers for non-specific labeling reactions. Some of the labeled bands not displacable by cold ANF may be a result of such scavenge reactions.

The specificity of the labeling of M_r 124,000 band was further confirmed by dose-dependent displacement by ANF-IV (Figure 3, lanes a - f). ANF-IV at concentrations above 10^{-8} M completely prevented the labeling by AZB-^{125}I-ANF-IV. The result is consistent with the K_d of 7.4 x 10^{-10} M for ANF-IV (Table I). On the contrary, a mixture of unrelated peptides angiotensin II, ACTH and [Arg^8] vasopressin (Figure 3, lane g), or atriopeptin I (lane h) at 1 µM concentrations showed no effect. These results together suggest that the labeling indeed involved a specific receptor-ligand interaction.

Photo-labeling reaction of AZB-^{125}I-ANF-P with bovine and rat adrenal membranes gave results essentially identical to those obtained with AZB-^{125}I-ANF-IV (Figure 2, lanes e - h) indicating that the reaction of AZB-^{125}I-ANF-P involved the same specific interaction with the membrane receptors. However, it should be noted that because of the difference in the positions of AZB-moieties in the two ANF derivatives, the point of affinity labeling in these labeled proteins are potentially different.

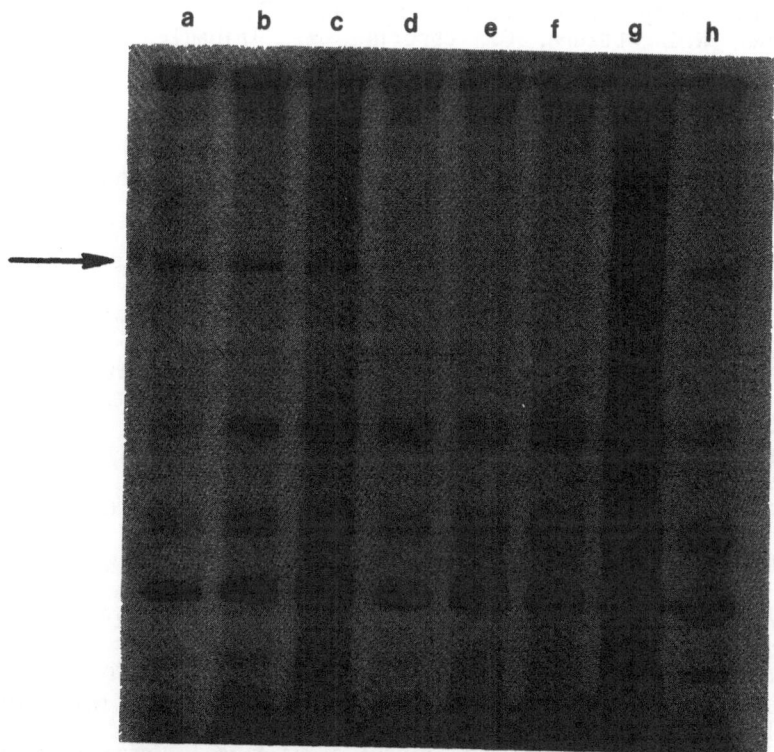

Figure 3. Autoradiograph showing the labeling of bovine adrenal cortex plasma membrane by photolysis with 4-azidobenzoyl-^{125}I-ANF-IV in the presence of increasing concentrations of ANF-IV (lane a, zero; b, 100 pM; c, 1 nM; d, 10 nM; 3, 100 nM; f, 1 µM), a combination of 1 µM each of angiotensin II, ACTH and [Arg^8] vasopressin (lane g), or 1 µM atriopeptin I (lane h). The arrow indicates radioactive band specifically labeled.

In summary, the results presented in the present report indicate that the protein with an apparent M_r of 124,000 in bovine adrenal membrane or that with M_r of 126,000 in rat adrenal membrane was specifically labeled by the two photo-reactive, radioiodinated ANF peptides synthesized here. This protein most likely is the ANF receptor protein or its subunits which are responsible for specific recognition and interaction of ANF with the target adrenal cells. The photoaffinity labeling reagents presented here may prove to be useful in molecular characterizations of ANF receptors.

REFERENCES

1. de Bold, A.J., Borenstein, H.B., Veress, A.T., and Sonnenberg, H. (1981) Life Science 28, 89-94.
2. Currie, M.G., Geller, D.M., Cole, B.R., Boylan, J.G., YuSheng, W., Holmberg, S.W., and Needleman, P. (1983) Science 221, 71-73.
3. Grammer, R.T., Fukumi, H., Inagami, T., and Misono, K.S. (1983) Biochem. Biophys. Res. Commun. 116, 696-703.
4. Misono, K.S., Fukumi, H., Grammer, R.T., and Inagami, T. (1984) Biochem. Biophys. Res. Commun. 119, 524-529.
5. Flynn, T.G., de Bold, M.L., and de Bold, A.J. (1983) Biochem. Biophys. Res. Commun. 117, 859-865.
6. Currie, M.G., Geller, D.M., Cole, B.R., Siegel, N.R., Fok, K.F., Adams, S.P., Eubanks, S.R., Galluppi, G.R., and Needleman, P. (1984) Science 223, 67-69.
7. Misono, K.S., Grammer, R.T., Fukumi, H., and Inagami, T. (1984) Biochem. Biophys. Res. Commun. 123, 444-451.
8. Seidah, N.G., Lazure, C., Chretien, M., Thibault, G., Garcia, R., Cantin, M., Genest, J., Nutt, R.F., Brady, S.F., Lyle, T.A., Paleveda, W.J., Colton, C.D., Ciccarone, T.M., and Veber, D.F. (1984) Proc. Nat. Acad. Sci. 81, 2640-2644.
9. Atlas, S.A., Kleinert, H.D., Camargo, M.J., Januszewicz, A., Sealey, J.E., Laragh, J.H., Schilling, J.W., Lewicki, J.A., Johnson, L.K., and Maack, T. (1984) Nature 309, 717-719.
10. Yamanaka, M., Greenberg, B., Johnson, L., Seilhamer, J., Brewer, M., Friedemann, T., Miller, J., Atlas, S., Laragh, K., Lewicki, J., and Fiddes, J. (1984) Nature 309, 719-722.
11. Maki, M., Takayanagi, R., Misono, K.S., Pandey, K.N., Tibbetts, C., and Inagami, T. (1984) Nature 309, 722-724.
12. Napier, M.A., Vandlen, R.L., Albers-Schönberg, G., Nutt, R.F., Brady, S., Lyle, T., Winquist, R., Faison, E.P., Heinel, L.A., and Blaine, E.H. (1984) Proc. Nat. Acad. Sci. 81, 5946-5950.
13. Hirata, Y., Tomita, M., Yoshimi, H., and Ikeda, M. (1984) Biochem. Biophys. Res. Commun. 125, 562-568.
14. Chartier, L., Schiffrin, E., Thibault, G., and Garcia, R. (1984) Endocrinology 115, 2026-2028.
15. De Lean, A., Gutkowska, J., McNicoll, N., Schiller, P.W., Cantin, M., and Genest, J. (1984) Life Sci. 35, 2311-2318.
16. Tarr, G.E., Black, S.D., Fujita, V.S., and Coom, M.J. (1983) Proc. Nat. Acad. Sci. 80, 6552-6556.
17. Hunter, W.M., and Greenwood, F.C. (1962) Nature 194, 670-677.
18. Galardy, R., Craig, L.C., Jamieson, J.D., and Printz, M.P. (1974) J. Biol. Chem. 249, 3510-3518.
19. Misono, K.S., and Inagami, T. (1982) J. Biol. Chem. 257, 7536-7540.
20. Glossmann, H., Baukal, A.J., and Catt, K.J. (1974) J. Biol. Chem. 249, 825-834.
21. Munson, P.J., and Rodbard, D. (1980) Anal. Biochem. 107, 220-239.
22. Laemmli, U.K. (1970) Nature 227, 680-685.

STRUCTURAL ANALYSIS OF A 29/38kDa HEPARIN-BINDING DOMAIN OF FIBRONECTIN:

EVIDENCE THAT TWO DIFFERENT SUBUNITS OF HUMAN PLASMA FIBRONECTIN ARISE BY

ALTERNATIVE mRNA SPLICING

Hema Pande[*], Jimmy Calaycay[*], Terry D. Lee[*],
Annalisa Siri[&], Luciano Zardi[&] and John E.
Shively[*]

[*]Division of Immunology, Beckman Research Institute of
the City of Hope, Duarte, CA 91010 and [&]Laboratorio di
Biologia Cellulare Istituto Scientifico Tumori, Genova,
Italy

INTRODUCTION

Fibronectins comprise a class of high molecular weight, multi-functional glycoproteins present in soluble form in plasma, amniotic fluid and other body fluids and in insoluble form in extracellular matrices and basement membrane. Multiple functions attributed to this class of proteins include a role in malignancy, cellular adhesion, embryogenesis and development, opsonization and would healing[1-3]. These biological functions of fibronectins are based on their affinities to cell surfaces and a number of macromolecules including collagen, fibrin, heparin, DNA, actin, complement component Clq as well as to certain bacteria.

Fibronectins are composed of structurally similar, but nonidentical subunits each possessing three types of internal repeats (homology types I, II and III) and comprising a series of discrete structural domains that mediate specific interactions with the cell surfaces and other macromolecular ligands. Several distinct forms of fibronectins exist and have been studied in some detail. Plasma fibronectin which is synthe-sized and secreted by hepatocytes[4] is a soluble protein and is com-posed of two similar but non-identical polypeptide chains. A related, less soluble form found in fibrillar matrices at the surfaces of many cell types in consistently larger in the COOH-terminal region than plasma fibronectin. In addition, a third type of fibronectin has been shown to be produced by fetal tissues and a number of tumors[5]. In this form of fibronectin exists a unique domain (oncofetal domain) which is located between the COOH-terminal heparin-binding (Hep-2) and fibrin-binding (Fib-2) domains.

It has been postulated that variations in splicing of a single fibronectin gene can give rise to several different fibronectin mRNAs resulting in these diverse forms of fibronectins. At least two distinc-tive regions of differential splicing have so far been identified (6-8). This alternative splicing could result in the insertion or omission of a 90 amino acid extra-domain in one region and several sizes of IIICS (a variable region that connects the last two type III repeats) in another.

The functional significance of the insertion or deletion of these blocks of coding sequences that are predicted to result due to alternative splicing are unclear at the present time. Although a number of functions including involvement in cell adhesion, self association and fibrillo-genesis and more interestingly ontogenesis and oncogenesis could be speculated to relate to the presence or absence of these inserted segments, structure-function studies on the purified domains encoded by these highly spliced regions are necessary to answer these questions.

In the present report we desribe the isolation and characterization of the Hep-2 domains from plasma fibronetin. Two fragments (Mr29k and 38k) representing the Hep-2 domains from the heavy and light chains of plasma fibronectin were purified from the thermolysin digest. In addition a 28kDa tryptic fragment located COOH-terminal to the 29kDa Hep-2 domain was isolated. Protein structural studies including fast atom bombardment/mass spectrometry and microsequence analysis, on these purified domains reveal structural differences between the two chains of plasma fibronectin. Furthermore, these studies establish the locations of the splice-junction sites for both the heavy and light chains.

METHOD

Purification and Localization of 29/38kDa Hep-2 Domains

Purified fibronectin (1 mg/ml) in 25 mM Tris HCl, pH 7.6, 0.5 mM EDTA, 50 mM NaCl and 25 mM CaCl$_2$ was digested with 5μg/ml of thermo-lysin for 4 hr. at 22°C. After terminating the reaction by adding EDTA to 5 mM the digest was purified on a hydroxyapatite column by methods previously described[9]. The fractions containing 29/38 kDa heparin-binding domains were dialyzed against 25 mM tris HCl pH 8.6, 0.5 mM EDTA, 230 mM NaCl and loaded onto a heparin-sepharose column (1.6 x 15 cm) previously equilibrated in the same buffer. Elution with a linear gradi-ent of NaCl (from 230 mM to 1 M) resulted in the separation of 29/38 kDa heparin-binding domains from the minor contaminants (heparin-nonbinding). Further purification of the heparin-bound fraction on Biogel P-60 molecu-lar exclusion column (5 x 100 cm) using 10 mM sodium phosphate pH 7.2, 0.15 M NaCl, 1 mM EDTA yielded purified 29 kDa and 38 kDa domains (Figure 1). The identification and localization of these domains was carried out using monoclonal antibodies as previously described[10].

Fragmentation

Trypsin digestions of 29 kDa and 38 kDa domains were performed under native (0.2 M ammonium bicarbonate pH 8.0, 48 hr., 37°C) or denaturing (4 M urea, 0.1 M ammonium acetate pH 8.0, 72 hr., 37°C) conditions. In addition, in a parallel experiment, the 38 kDa domain was digested with trypsin after carboxymethylation using ^{14}C-Iodoacetic acid. The tryptic peptides were separated by reverse-phase HPLC on a Vydac C4 column (4.5 x 250 cm) using a linear gradient from solvent A (0.1% aque-ous TFA) to 70% B (0.1/9.9/90, TFA/H$_2$O/CH$_3$CN) under the conditions previously described[11]. Tryptic peptides from all the maps were used for structural analysis. Only the map produced in the presence of 4 M urea showing unique peaks for each domain is depicted here in Figure 2.

FAB/MS Analysis

Tryptic peptides (0.1-2.0 nmoles) were concentrated to dryness in polyethylene microfuge tubes on a vacuum centrifuge, redissolved in 1-2 μl of 5% aqueous acetic acid and mixed with 1μl of glycerol on a 1.5 x 5.0 mm stainless steel sample stage. Fast atom bombardment spectra were

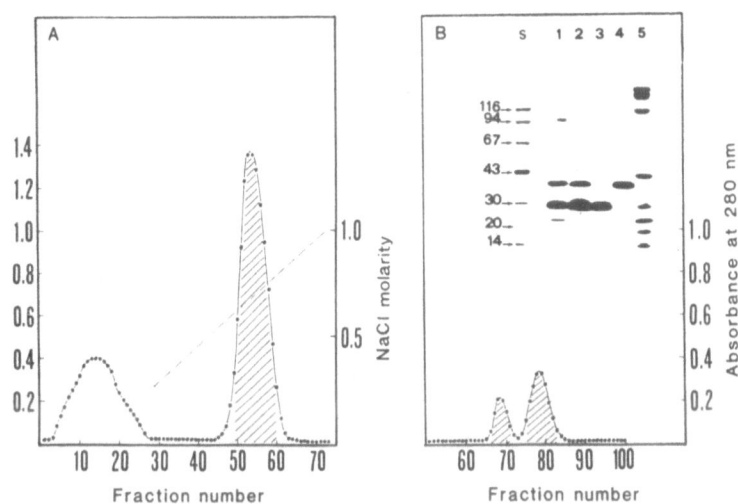

Figure 1. Purification of 29 kDa and 38 kDa Hep-2 domains.

A. A crude preparation of the Hep-2 domains was purified on a
Heparin-sepharose column (1.6 x 15 cm). The proteins were
eluted with a linear NaCl gradient (from 230 mM to 1 M in 25
mM tris HCl pH 8.6, 0.5 mM EDTA). Cross-hatched area
indicates the pooled fractions containing pure Hep-2 domain.

B. Elution profile of 29 kDa and 38 kDa domains from a
Biogel-P60 column. The column (5 x 100 cm) was equilibrated
with 10 mM sodium phosphate pH 7.2, 0.15 M NaCl, 1 mM EDTA and
0.02% NaN$_3$. A mixture of 29 kDa and 38 kDa fragments (20
ml, 1 mg/ml protein) was loaded onto the column. Elution with
the same buffer yielded pure 29 kDa and 38 kDa fragments
(pooled fractions indicated by cross-hatched areas). Inset:
analysis on a 4-18% SDS-PAGE of a crude preparation of 29 kDa
and 38 kDa domains (lane 1); 29 kDa and 38 kDa domains from
the Heparin-Sepharose column (lane 2); 29 kDa domain purified
by gel filtration (lane 3); 38 kDa domains purified by gel
filtration (lane 4); total thermolysin digested fibronectin
(lane 5). Lane s refers to molecular weight standards.

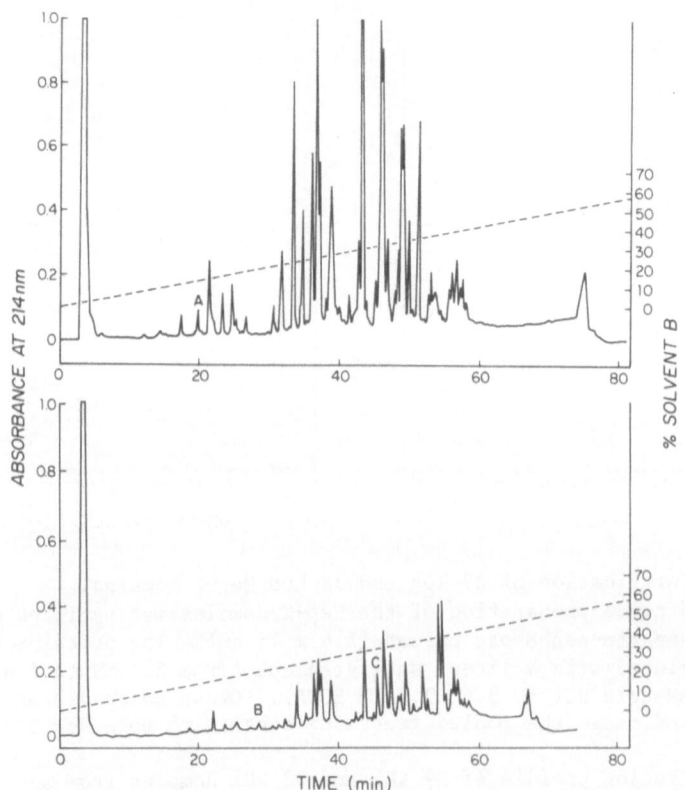

Figure 2.   Tryptic peptide map of 29 kDa and 38 kDa Hep-2 domains
            generated in the presence of 4 M urea. Approximately 5nmol of
            these domains were digested with 1% trypsin for 72 hours at
            37°C in 4 M urea, 0.1 M ammonium acetate pH 8.0. The peptides
            were fractionated on a Vydac C4 column (4.5 x 250 mm) at 40°C.
            A linear gradient was used from 100% solvent A (0.1% aqueous
            trifluoroacetic acid) to 70% solvent B (0.1:9.9:90,
            trifluoroacetic acid: H_2O:CH_3CN, v/v/v) over 90 min. at a
            flow rate of 0.8 ml/min. Two unique peptides for 38 kDa domain
            (b and c) and one for 29 kDa domain (a) were identified by
            FAB/MS and sequence analysis.

taken with a JEOL HX-100 HF mass spectrometer utilizing a 6 KV xenon atom primary beam. The instrument was calibrated to mass 5000 at 5 KV using a mixture of potassium iodide and cesium iodide. Data were collected using a JEOL DA 5000 data system. Mass assignments were accurate to within 0.2 U and values for the molecular ion clusters are reported as the nearest integer value of the monoisotopic mass.

Microsequence Analysis

Tryptic peptides that were either unique to each domain or were generated as a result of anomalous cleavage were also analyzed by microsequence analysis. Samples (0.2-0.5 nmoles) were sequenced on a City of Hope build gas phase sequenator[12]. Amino acid phenylthiohydantoin derivatives were analyzed by reverse phase HPLC on an ultrasphere (Altex) ODS column with a Waters Associates chromatograph equipped with an autosampler (WISP) and 254 and 313 nm detectors[13].

RESULTS AND DISCUSSION

Limited thermolysin digestion of human plasma fibronectin generates several functionally distinct domains[14]. Fractionation of the digest on hydroxyapatite followed by affinity chrmoatography on heparin-sepharose column results in the separation of a mixture of 29 kDa and 38 kDa domains both of which react with monoclonal antibody specific for the Hep-2 domain. These two domains can be further purified by chromatography on a molecular exclusion Biogel P-60 column. Isolation of two non-identical size Hep-2 domains suggest that the two subunits of human plasma fibronectin prossess structural differences in this region.

NH$_2$-terminal sequence analyses of 29 kDa and 38 kDa domains gave identical sequences through 16 Edman cycles revealing that the 90 amino acid extra-domain, predicted to be present in the cellular fibronectin[8] is absent in both the heavy and the light chains of plasma fibronectin. The amino acid compositions of these two domains showed that the 38 kDa domain contains one cysteine and two histidine residues both of which are absent in the 29 kDa domain.

Digestions of purified 29 kDa and 38 kDa domains under native and denaturing conditions yielded a number of fragments that were separated by reverse-phase HPLC. Unique peptides from each domain could be isolated under denaturing conditions. The cysteine containing peptide from 38 kDa domain was obtained after carboxymethylation. FAB/MS analysis was performed on each of the tryptic peptides. The M/Z values for the molecular ions of each peptide were compared with the cDNA derived peptide sequences. The peptides that were unique to each domain were also characterized by microsequence analysis.

The primary structures of 29 kDa and 38 kDa domains are shown in Figure 3. The two domains showed identical amino acid sequences through 274 residues followed by a block of sequence variations. The 38 kDa domain, after residue 274 had a deletion of the entire III CS segment (maximum size of 120 amino in rat fibronectin). This domain is located immediately adjacent to the fibrin-binding (Fib-2) domain in the light-chain of plasma fibronectin (Figure 4). It contains 359 amino acids corresponding to four repeats of type III sequences and one O-linked galactosamine residue. The 29kDa domain which is located in the heavy chain contains the same sequence for the first 274 residues corresponding to three type III repeats. However, this domain, after Thr-274 contains five amino acids from the IIICS region (not present in 38kDa domain) (Figures 3 and 4). These data revealed that the heavy

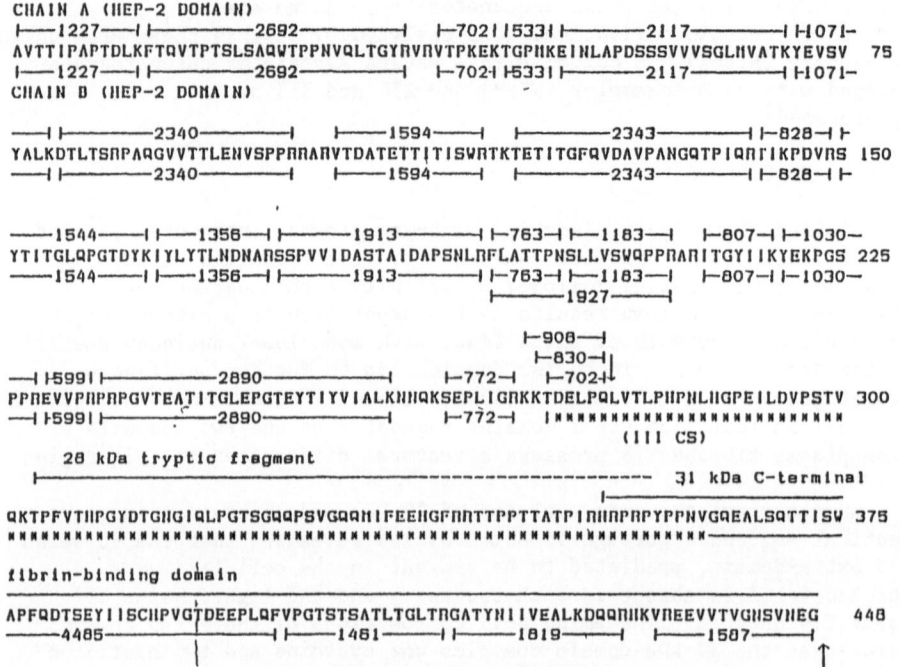

Figure 3   Amino acid sequence for the carboxy-terminal Heparin-binding
(HEP-2) domain of human plasma fibronectin. The size of the
HEP-2 domain generated by thermolysin digestion is 29 kDa for
the heavy chain (A chain) and 38 kDa for the light chain (B
chain). The thermolysin sensitive sites demarking the
C-termini of 29 kDa and 38 kDa domains are indicated with
arrows. The III CS segment (89 amino acids) is present only
in the heavy chain. The sequence of the 31 kDa C-terminal
fibrin-binding fragment [15] begins at His-354. The
sequence of the 28 kDa tryptic fragment begins at Thr-303 and
partly overlaps with the 31 kDa fibrin-binding domain.

chain contains some portion of the IIICS. The additional information regarding the size of IIICS in the heavy chain was obtained by determining NH_2-terminal sequence of a 28kDa tryptic polypeptide that was purified from the DNA unbound fraction of trypsin digested fibronectin. NH_2-terminal sequence analysis of this 28 kDa tryptic fragment which contains a major portion of IIICS corresponded to tryptic cleavage at Lys-302 (Figure 3) thus establishing that this fragment is located COOH-terminal to the 29kDa Hep-2 domain. These results suggest that the heavy chain of plasma fibronectin contains the IIICS segment of at least 89 amino acids and probably has a deletion of the COOH-terminal 31 amino acid region of IIICS (present in rat fibronection) as observed by Garcia-Pardo et al[15].

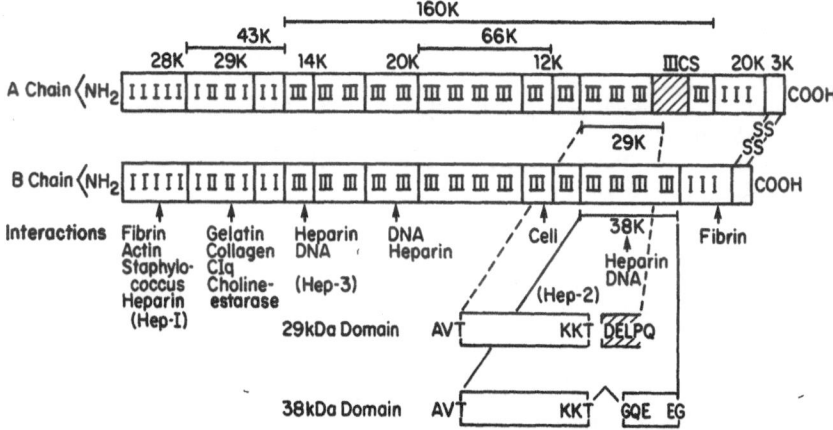

Figure 4.   Domain model for human plasma fibronectin. Two non-identical subunits are shown with the localization of internal sequence homologies (types I, II and III). The region of variability between the two subunits (IIICS, 89 amino acids) is indicated by cross-hatched area. The size of Hep-2 domain is 29 kDa in the heavy chain and 38 kDa in the light chain.

The results reported here provide the first evidence of amino acid sequence differences between the two subunits of plasma fibronectin. These results are consistent with a single gene for the two fibronectin chains and suggest that differential splicing of the initial transcript results in the elimination of an exon portion encoding a protein segment (IIICS segment) from one of the two chains of plasma fibronectin thus altering the protein sequence in the two subunits in this region.

REFERENCES

1. Pearlstein, E. Gold, L.I., and Garcia-Pardo, A. (1980) Mol. Cell. Biochem. 29, 103-128.
2. Hynes, R.O., and Yamada, K.M. (1982) J. Cell Biol. 95, 369-377.
3. Yamada, K.M. (1983) Annu. Rev. Biochem. 52, 761-799.
4. Tamkun, J.W. and Hynes, R.O. (1983) J. Biol. Chem. 258, 4641-4647.
5. Matsuura, H. and Hakomori, S. (1985) Proc. Nat. Acad. Sci. USA, 82, 6517-6521.
6. Schwarzbauer, J.E., Tamkun, J.W., Lemischka, I.R. and Hynes, R.O. (1983) Cell, 35, 421-431
7. Kornblihtt, A.R., Vibe-Pedersen, K., and Baralle, F.E. (1984) Nucleic Acid Res., 12, 5853-5868.
8. Kornblihtt, A.R., Vibe-Pedersen, K., and Baralle, F.E. (1984) The Embo J., 3, 221-226.
9. Zardi, L., Carnemolla, B., Balza, E., Borsi, L., Castellani, P., Rocco, M., and Siri, A. (1985) Eur. J. Biochem, 146, 571-579.
10. Sekiguchi, K., Siri, A. Zardi, L. and Hakomori, S., (1985) J. Biol. Chem., 260, 5105-5114.
11. Pande, H., Calaycay, J., Hawke, D., Ben-Avram, C.M., and Shively, J.E. (1985) J. Biol. Chem., 260, 2301-2306.
12. Hawke, D., Harris, D., and Shively, J.E. (1985) Anal. Biochem., 147, 315-330.
13. Hawke, D., Yuan, P.-M., and Shively, J.E. (1982) Anal. Biochem. 120, 302-311.
14. Sekiguchi, J., Fukuda, M., and Hakomori, S. (1981) J. Biol. Chem. 156, 6452-6462.
15. Garcia-Pardo, A., Pearlstein, E. and Frangione, B. (1985), J. Biol. Chem. 260, 10320-10325.

STRUCTURAL AND FUNCTIONAL STUDIES ON

THE PURIFIED INSULIN RECEPTOR

Yoko Fujita-Yamaguchi

Department of Molecular Genetics
Beckman Research Institute of the City of Hope
1450 E. Duarte Road, Duarte, CA 91010

INTRODUCTION

The insulin receptor is a membrane glycoprotein with ~ Mr=300,000, which is thought to be responsible for mediating all insulin-dependent biological actions (1). The basic structural unit of the insulin receptor is proposed to be a disulfide-linked complex composed of two 125,000 dalton (α) subunits and two 90,000 dalton (β) subunits in β-α-α-β form (2,3). Affinity-labeling experiments and cloning studies have suggested that the α subunit has insulin binding sites whereas the β subunit carries tyrosine-specific protein kinase activity (4-7).

We have purified and characterized the human placental insulin receptor with high insulin-binding and kinase activities (3,8-10). The $\alpha_2\beta_2$ form of the receptor is found as a major component in the purified receptor. In addition, we have found other minor forms, however, the complex $\alpha_2\beta_2$ is the intact form of the receptor whereas the other complexes appear to be generated from $\alpha_2\beta_2$ by proteolytic degradation and/or reduction (3). This heterogeneity is not all due to the degradation of the receptor during purification, since affinity-crosslinking and biosynthetic-labeling experiments using intact cells from different laboratories have also revealed the presence of heterogeneous forms of the receptor in plasma membranes (4,11). To date, there is no evidence that the intact insulin receptor is the physiologically active form.

We have tested whether the intact insulin receptor is fully active in terms of binding and kinase activities using purified insulin receptor preparations. Mild trypsin treatment of the purified receptor activates kinase activity (13). Removal of sialic acids from the purified receptor significantly enhances both insulin binding and kinase activities (14). These data have suggested that the intact receptor may not be the fully active form.

In this study we have examined the relationship between insulin receptor subunit structure and its kinase activity. The results suggest that one half ($\alpha\beta$) of the intact insulin receptor is more active form of the tyrosine-specific protein kinase than the intact receptor ($\alpha_2\beta_2$).

MATERIALS AND METHODS

Materials: Crystalline porcine insulin was kindly supplied by Eli Lilly. A synthetic peptide resembling the tyrosyl phosphorylation site of pp60^{src} (Arg-Arg-Leu-Ile-Glu-Asp-Ala-Glu-Tyr-Ala-Ala-Arg-Gly) was purchased from Peninsula Laboratories (15). Dithiothreitol and molecular weight markers were from Bio-Rad. [γ-^{32}P] ATP was purchased from ICN. All other chemicals used were reagent grade.

Purification of Insulin Receptor: Insulin receptor was purified 2400-fold with a yield of 40% from human placental membranes by sequential affinity chromatography on wheat germ agglutinin- and insulin-Sepharose as described previously (8).

Treatment of the Purified Insulin Receptor with Dithiothreitol: The receptor (~50 ng for kinase assay, ~500 ng for SDS-PAGE) was incubated at 4°C for 16 hr[+] with different concentrations of dithiothreitol in 50 mM Tris-HCl, pH 7.4 (20 μl for kinase assay and 40 μl for autophosphorylation) containing 0.1% Triton X-100 and subjected to either the kinase assay or the autophosphorylation reaction.

Phosphorylation of the Synthetic Peptide (Tyrosine-specific Protein Kinase Assay):   The purified receptor, treated with dithiothreitol in the presence and absence of 1 μM insulin, was incubated at 25°C for 40 min[++] in 30 μl of 50 mM Tris-HCl, pH 7.4, containing 2 mM MnCl$_2$, 15 mM MgCl$_2$, 1 mM of the src-related peptide, [γ-^{32}P] ATP (40 μM, 5-10 μCi/nmol) and 0.1% Triton X-100. The reaction was terminated by the addition of 50 μl of 5% trichloroacetic acid and 20 μl of bovine serum albumin (10 mg/ml). After incubating this solution at 0° C for 30 min, the proteins were precipitated by centrifugation for 5 min. A 35 μl aliquot of the supernatant was spotted on a piece of phosphocellulose paper (Whatman, P81). The paper was extensively washed in 75 mM phosphoric acid and placed in a vial containing Aquasol (New England Nuclear). Duplicate papers were prepared from each reaction mixture and ^{32}P incorporated into the peptide was counted in a liquid scintillation counter.

Phosphorylation of the β Subunit: After the purified receptor was treated with dithiothreitol in the presence and absence of 1 μM insulin, the phosphorylation reaction was carried out at 25° C for 1 hr in 45 μl of Tris-HCl, pH 7.4, containing [γ-^{32}P]-ATP (40 μM, 10 μCi), and 2 mM MnCl$_2$, and 0.1% Triton X-100. The reaction was terminated by adding 20 μl of 50 mM Tris-HCl, pH 6.8, containing 6% SDS, 3mM ATP, and 50% sucrose in the presence or absence of 60 mM dithiothreitol, followed by boiling for 5 min. Insulin receptor subunits were separated by SDS-PAGE (16) in the presence or absence of dithiothreitol and stained with silver. The gel was dried and autoradiographed.

Quantification of Phosphorylated α$_2$β$_2$, αβ, and β Subunit: The autoradiograms of SDS-polyacrylamide gels were monitored by an Automatic Computing Densitometer ACD-18 (Gelman Sciences). The peaks representing phosphorylated α$_2$β$_2$, αβ, and β subunit were quantified by cutting and weighing the area of the peaks.

[+] Effects of dithiothreitol can be seen after 5 min incubation, however, for the sake of convenience, we have used 16 hr incubation.

[++] Under these conditions, the phosphorylation reaction is linear so that the values obtained can be readily converted to pmol of P incorporated into the peptide/min.

RESULTS

The effect of dithiothreitol on autophosphorylation of the β subunit was first examined by incubating the purified insulin receptor with different concentrations of dithiothreitol in the presence and absence of insulin, followed by SDS-PAGE under reducing conditions. Insulin stimulated the phosphorylation of the β subunit 5-fold. Dithiothreitol enhanced the phosphorylation by a factor of 2-fold over that seen in the presence of insulin alone (Fig. 1). Phosphorylation of the β subunit was maximal at concentrations of 0.1 to 0.75 mM dithiothreitol in the presence of 1 μM insulin.

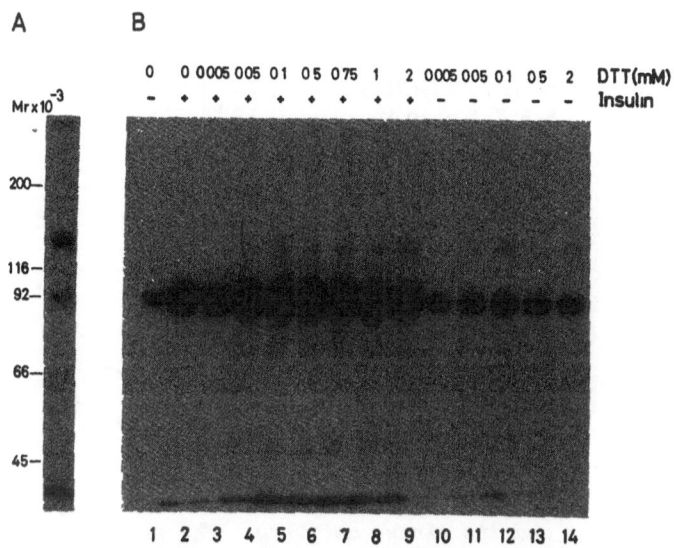

FIGURE 1. *The effect of dithiothreitol on autophosphorylation of the purified insulin receptor as analyzed by SDS-PAGE under reducing conditions.*

The purified receptor (~0.5 μg) was treated at 4°C for 16 h with different concentrations (0~2 mM) of dithiothreitol in the presence (B-2~9) and absence (B-10~14) of 1 μM insulin in 40 μl of 50 mM Tris-HCl, pH 7.4, containing 0.1% Triton X-100. The phosphorylation reaction was carried out at 25°C for 1 hr in 45 μl of the same buffer containing 2 mM MnCl$_2$, [γ-^{32}P] ATP (40 μM, 10 μCi), and 0.1% Triton X-100, and terminated by adding 20 μl of 50 mM Tris-HCl, pH 6.8, containing 6% SDS, 60 mM dithiothreitol, 3 mM ATP, and 50% sucrose, followed by boiling for 5 min. The insulin receptor subunits were separated in a 7.5% polyacrylamide gel (16), stained with silver (A). The gel was dried and autoradiographed for 45 min at -70°C using a Kodak XAR-5 film and an intensifying screen (B).

Next we analyzed the population of insulin receptor disulfide-linked complexes in dithiothreitol-treated receptor preparations by SDS-PAGE under non-reducing conditions (Fig. 2-A and B). The silver-stained SDS-polyacrylamide gel revealed three major complexes with ~ Mr=300,000, which have been previously identified as $\alpha_2\beta_2$, $\alpha_2\beta\beta_1$, and $\alpha_2(\beta_1)_2$ (3), in the absence of dithiothreitol and insulin (Fig. 2-A-1). Incubation of the receptor with insulin did not change the population of the receptor complexes (Fig. 2-A-2). These high molecular weight receptors gradually converted to αβ and $\alpha\beta_1$ forms, and eventually to free α and β subunits as the concentration of dithiothreitol increased (Fig. 2-A-2~11). There were only free subunits observed at a concentration of 2 mM dithiothreitol in the presence of 1 μM insulin.

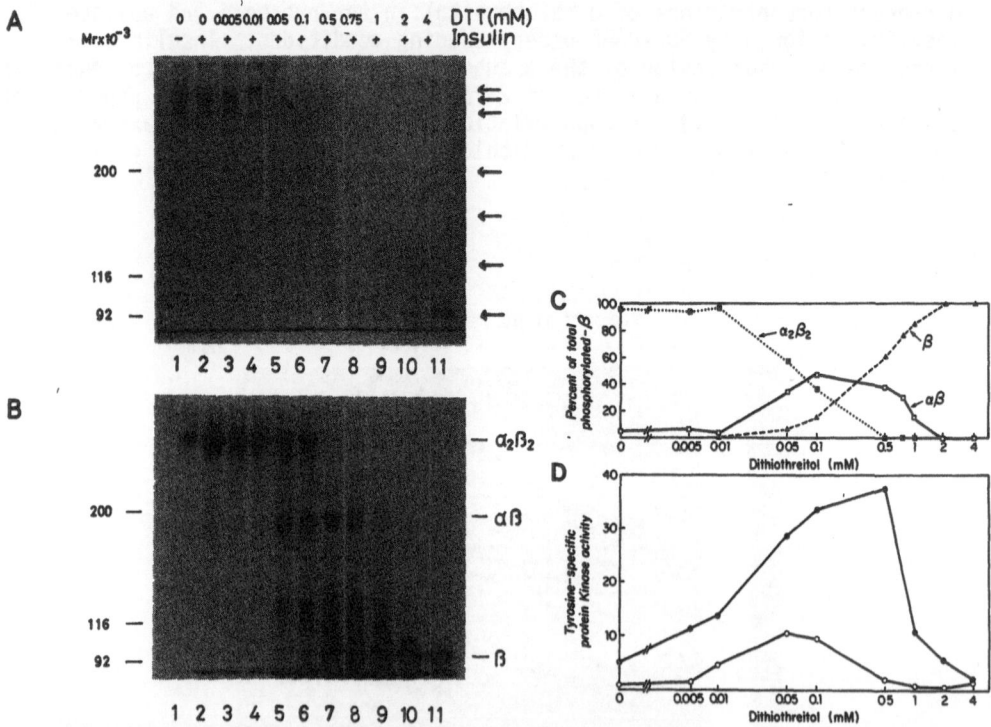

FIGURE 2. The effect of dithiothreitol on the structure of the insulin
receptor and its kinase activity: The purified receptor (~0.5
µg) was treated at 4°C for 16 h with different concentrations (0.005~4 mM)
of dithiothreitol in the presence of 1 µM insulin (A,B-2~11) and phosphorylated as described in Methods. The receptor without dithiothreitol-
treatment was phosphorylated in the presence and absence of 1 µM insulin
(A,B-1,2). The insulin receptor disulfide-linked complexes were analyzed
by a 5% SDS-polyacrylamide gel (16), under nonreducing conditions.
A: The gel was stained with silver. The arrows indicate, from the top,
$\alpha_2\beta_2$, $\alpha_2\beta\beta_1$, $\alpha_2(\beta_1)_2$, $\alpha\beta$, $\alpha\beta_1$, α and β with Mr=320,000, 300,000, 275,000,
195,000, 150,000, 125,000, and 90,000, respectively (see Ref. 3).
B: The gel was dried and autoradiographed for 1 hr at -70°C using a Kodak
XAR-5 film and an intensifying screen.
C: The autoradiogram as shown in B was used to quantitate amounts of $\alpha_2\beta_2$,
$\alpha\beta$, and β forms in dithiothreitol-treated receptor preparations. The percent of each component over total phosphorylated β subunit is shown: $\alpha_2\beta_2$
(...■...), $\alpha\beta$ (——□——) and β (---▲---).
D: Tyrosine-specific protein kinase assay was performed simultaneously
with the same receptor preparations as used for SDS-PAGE analysis (A, B,
C). The purified receptor (~50 ng) was treated at 4°C for 16 hr with different concentrations of dithiothreitol in the absence (basal activity,
—— o——) and presence (insulin dependent activity, ——●——) of 1 µM insulin in 20 µl of 50 mM Tris-HCl, pH7.4, containing 0.1% Triton X-100. The
kinase assay was carried out using the Src-related peptide as described in
Methods. The activity is expressed as a ratio of activation, as compared
to the basal activity of the receptor without dithiothreitol treatment.

The autoradiogram of the same gel clearly revealed that the receptors in $\alpha_2\beta_2$ and $\alpha\beta$ forms and β subunit are phosphorylated (Fig. 2-B). When the dried SDS-polyacrylamide gels were exposed to film for longer periods, limited incorporation of ^{32}P into an $\alpha_2\beta\beta_1$ form was observed while $\alpha_2(\beta_1)_2$ and $\alpha\beta_1$ forms were not phosphorylated at all (Data not shown). These data are consistent with our previous observation that the β_1 subunit does not exhibit the kinase activity (17), and with other data showing that autophosphorylation is an intramolecular reaction (18,19). Since this particular experiment was designed to identify the distribution of the various receptor species with increasing dithiothreitol concentrations and not to precisely determine the rate of increase in the phosphorylation, no quantitative conclusion can be drawn on the effect of dithiothreitol on the rate of the phosphorylation of each individual species. Although the extent of phosphorylation in the individual bands corresponding to $\alpha_2\beta_2$, $\alpha\beta$, and β appear to be similar to each other (Fig. 2B-6,7,8), the added phosphorylation of all forms containing β was enhanced at concentrations of 0.1~0.75 mM dithiothreitol as seen in Fig. 1. The intensity of the radioactive bands were quantified in order to calculate the ratio of each component to total phosphorylated β subunit, the sum of $\alpha_2\beta_2$, $\alpha\beta$, and β. Fig. 2-C summarizes the relative abundance of each component in the receptor preparations treated with different concentrations of dithithreitol. It appears that the enhanced phosphoryation of the β subunit observed by SDS-PAGE analysis of dithiothreitol-treated receptors under reducing conditions (Fig. 1) correlates with the amount of the receptor in the $\alpha\beta$ form.

In order to test if dithiothreitol enhances insulin receptor-kinase activity and to determine if the appearance of an $\alpha\beta$ form is correlated with activation of the insulin receptor-kinase, the rate of the kinase activity of the receptor after dithiothreitol treatment was measured using an exogenous substrate, the src-related peptide. Three independent experiments, all of which revealed results similar to Fig. 2-D except for the degree of activation (Table 1). Treatment of the receptor with dithiothreitol significantly enhanced both basal and insulin-dependent tyrosine-protein kinase activity. The maximal stimulation of the basal kinase activity achieved at a concentration of 0.1 mM dithiothreitol, was 3.3~9.4-fold. The maximal stimulation of insulin-dependent phosphorylation was achieved at concentrations of 0.5-0.75 mM, and ranged from 2.5 to 7.7 fold over controls treated with insulin alone. The concentration of dithiothreitol, at which the maximal increase in kinase activity was observed, is different for basal activity (0.1 mM) and for insulin-dependent activity (0.5-0.75 mM). This is consistent with the results of SDS-PAGE under reducing conditions (Fig. 1). The presence of insulin during dithiothreitol-treatment seems to increase the resistance of the receptor to dithiothreitol either by changing the receptor conformation or simply consuming dithiothreitol in the receptor solution. Insulin further stimulated the kinase activity of dithiothreitol-treated receptor. Overall activation of the insulin receptor-kinase in the presence of dithiothreitol and insulin was 12.4~37.5-fold.

The kinase assays were performed simultaneously with SDS-PAGE analysis. One of the results is shown in Fig. 2-A, B and C (SDS-PAGE analysis) and D (peptide assay). The results clearly suggested a correlation between the appearance of an $\alpha\beta$ form and an increase in kinase activity. Therefore, it is concluded that the $\alpha\beta$ form of the insulin receptor exhibits much higher kinase activity than the intact receptor in the $\alpha_2\beta_2$ form.

Furthermore, present studies revealed that a free β subunit can phosphorylate itself (Fig. 2-A and B-11) and an exogenous substrate in an insulin-independent manner (Fig. 2-C, 4 mM dithiothreitol) since the β subunit is free from the α subunit at this concentration of dithiothreitol.

661

TABLE 1. Activation of the purified insulin receptor kinase by dithiothreitol.

Experiment		Basal activity	Insulin-dependent activity
		%	%
1	Control	100 (0 mM)	500 (0 mM)
	Max. Activation	350 (0.1 mM)	1970 (0.75 mM)
2	Control	100 (0 mM)	560 (0 mM)
	Max. Activation	330 (0.1 mM)	1240 (0.5 mM)
3	Control	100 (0 mM)	490 (0 mM)
	Max. Activation	940 (0.1 mM)	3750 (0.5 mM)

The purified insulin receptor (~50 µg) was treated at 4°C for 16 hr with different concentrations of dithiothreitol in the absence (basal activity) and presence (insulin-dependent activity) of 1 µM insulin and assayed for its kinase activity using the src-related peptide as described in Methods.

The kinase activities are expressed as percent of control values (basal activity of the receptor without dithiothreitol-treatment). The basal and insulin-dependent activities listed correspond to the peak of maximal activation after dithiothreitol treatment (see Fig. 3-B for example). The maximal kinase activity obtained in the presence of insulin and dithiothreitol for Experiment 1, 2, and 3 was 8750, 9950, and 4055 pmol/min/mg of protein, respectively.

DISCUSSION

The present studies revealed a correlation between the appearance of an $\alpha\beta$ form and activation of tyrosine-protein kinase activity, from which it is suggested that the $\alpha\beta$ form exhibits much higher tyrosine-protein kinase activity than the intact $\alpha_2\beta_2$ form. Insulin stimulates kinase activity of the $\alpha\beta$ form of the receptor. These results are contrary to the previous report by Shia *et al* concluding that the kinase activity of the intact form is higher than that of the $\alpha\beta$ form (19). Their experimental conditions are not directly comparable to ours, however, since they used crude receptor preparations and much higher concentrations of dithiothreitol. It is not clear from their data that the peak of autophosphorylation activity correlated to the presence of $\alpha_2\beta_2$ or appearance of $\alpha\beta$ (19).

Whether the $\alpha\beta$ form exists as such, or in noncovalently associated form like $(\alpha\beta)(\alpha\beta)$ has not been determined in our system, though other investigators have observed free $\alpha\beta$ in dithiothreitol-treated solubilized receptors fractionated by sucrose density gradient sedimentation and/or gel filtration on Sepharose 6B (20-22). Our result that a half of the intact receptor shows higher insulin-dependent kinase activity may indicate that the αβ form is a functional unit for insulin signal transduction.

ACKNOWLEDGMENTS

This work was supported by Research Grants AM29770 and AM34427 from the National Institute of Arthritis, Diabetes, Digestive, and Kidney Diseases.

REFERENCES

1. Kahn, C.R., Baird, K.L., Flier, S., Grunfeld, C., Harmon, J.T., Harrison, L.C., Karlsson, F.A., Kasuga, M., King, G.L., Lang, U.C., Podskalny, J.M., and Van Obberghen, E. (1981) Recent Progress in Hormone Research 37:477-533.
2. Czech, M.P., Massague, J., and Pilch, P.F. (1981) Trends Biochem. Sci., 6:222-225.
3. Fujita-Yamaguchi, Y. (1984) J. Biol. Chem. 259:1206-1211.
4. Masague, J., Pilch, P.F., and Czech, M.P. (1980) Proc. Natl. Acad. Sci. USA 77:7137-7147.
5. Roth, R.A. and Cassell, D.J. (1983) Science 219:299-301.
6. Shia, M.A. and Pilch P.F. (1983) Biochemistry 2:717-721.
7. Ullrich, A., Bell, J.R., Chen, E.Y., Herrera, R., Petruzzelli, L.M., Dull, T.J., Gray, A., Coussens, L., Liao, Y.-C., Tsubokawa, M., Mason, A., Seeburg, P.H., Grunfeld, C., Rosen, O.M., and Ramachandran, J. (1985) Nature 313:756-761.
8. Fujita-Yamaguchi, Y., Choi, S., Sakamoto, Y., and Itakura, K. (1983) J. Biol. Chem. 258:5045-5049.
9. Kasuga, M., Fujita-Yamaguchi, Y., Blithe, D.L., and Kahn, C.R. (1983) Proc. Natl. Acad. Sci. USA 80:2137-2141.
10. Kasuga, M., Fujita-Yamaguchi, Y., Blithe, D.L., White, M.F., and Kahn, C.R. (1983) J. Biol. Chem. 258:10973-10980.
11. Crettaz. M., Jialal, I., Kasuga, M., and Kahn, R. (1984) J. Biol. Chem., 259:11543-11549.
12. Clark, S., DeLuise, M., Larkins, R.G., Melick, R.A., and Harison, L.C. (1978) Biochem. J. 174:37-43.
13. Tamura, S., Fujita-Yamaguchi, Y., and Larner, J. (1983) J. Biol. Chem. 258:14749-14752.
14. Fujita-Yamaguchi, Y., Sato, Y. and Kathuria, S. (1985) Biochem. Biophys. Res. Commun. 129:739-745.
15. Casnellie, J.E., Harrison, M.L., Pike, L.J., Hellstrom, E., and Krebs, E.G. (1982) Proc. Natl. Acad. Sci. USA 79:282-286.
16. Laemmli, U.K. (1970) Nature (London) 227:680-685.
17. Fujita-Yamaguchi, Y., and Kathuria, S. (1985) Cancer Cells Vol. 3, Growth Factors and Transformation (Feramisco, J., Ozanne, B. and Stiles, L., eds.) p. 123-129. Cold Spring Harbor Laboratory
18. White, M.F., Haring,H.U., Kasuga, M., and Kahn, R. (1984) J. Biol. Chem., 259:255-264.
19. Shia, M.A., Rubin, J.B., and Pilch, P.F. (1983) J. Biol. Chem. 258: 14450-14455.
20. Baron, M.D., Wisher, M.H., Thamm,. P.M., Saunders, D.J., Brandenburg, D., and Sonksen, P.H. (1981) Biochemistry 20:4156-4161.
21. Matuo III, J.M., Hollenberg, M.D., and Aglio, L.S. (1983) Biochemistry 22:2579-2586.
22. Velicelebi, G., and Aiyer, R.A. (1984) Proc. Natl. Acad. Sci., USA 81:7693-7697.

LOCALIZATION AND INTERACTION OF FUNCTIONAL SITES ON ANTITHROMBIN III.

USE OF AN ANTI-HAPTEN ANTIBODY AS A STRUCTURAL PROBE

Cynthia B. Peterson and Michael N. Blackburn

Louisiana State University Medical Center
Department of Biochemistry and Molecular Biology
Shreveport, LA 71130

ABSTRACT

Immunochemical and spectral techniques have been used to characterize interaction between functional sites on the protease inhibitor, antithrombin III. Site-directed modification of a single tryptophan in antithrombin with hydroxynitrobenzyl bromide (HNB) eliminates heparin binding to the protein concomitantly with loss of heparin-promoted inactivation of thrombin by antithrombin. The tryptophan involved in heparin binding is located at position 49 in the sequence, near the N-terminus of antithrombin, and an arginine present at position 393, near the C-terminus, is the site of thrombin binding and inactivation. Although these functional sites are remote in primary sequence, what is their relative orientation in the native three-dimensional structure of the protein? How do these sites communicate with each other? Spectral and haptenic properties of HNB were exploited to answer these questions. In activity measurements, antibody binding to HNB at the heparin binding site at trp-49 did not preclude thrombin inactivation by HNB-antithrombin thrombin. Also, anti-HNB antibodies were shown to bind to the HNB-antithrombin-thrombin complex more avidly than HNB-antithrombin alone, indicating structural changes to occur in the region of trp-49 upon thrombin binding. This immunochemical data was confirmed by spectral measurements on the HNB group, which served as an environmentally sensitive extrinsic probe. Upon binding of thrombin to antithrombin, a shift is observed in the visible HNB spectrum which is consistent with increased exposure of the HNB chromophore to aqueous solvent. These results indicate that: (1) the heparin-binding and thrombin-complexing sites are spatially separated on antithrombin, and (2) the functional regions of antithrombin communicate via conformational signals.

INTRODUCTION

Questions of protein structure-function relationships have traditionally been addressed using chemical modification techniques. More recently, an immunochemical approach has been used for functional "mapping" of the protein using antibodies specific for defined regions on the protein. This approach has been greatly facilitated upon the development of hybridoma technology by which monoclonal antibodies directed against unique epitopes on the surface of a protein can be produced. We have combined chemical modification with an immunochemical approach to probe functional aspects of the structure of a plasma protein, antithrombin III.

Antithrombin III, the major physiological inhibitor of the enzymes involved in blood coagulation, inactivates the plasma proteases by formation of a stoichiometric complex between enzyme and inhibitor (1). The protease-complexing region of the antithrombin molecule contains the reactive site at which thrombin interacts and is inactivated via formation of a covalent protease-inhibitor complex. Björk and co-workers have identified this reactive site as arginine-393, which lies near the carboxyl-terminal end of the inhibitor (2). The coagulation proteases are gradually inactivated by antithrombin alone, but, upon the addition of catalytic amounts of the mucopolysaccharide, heparin, the rate of this inactivation is dramatically accelerated (3). Antithrombin thus contains a heparin-binding region as well as a protease-complexing region. Blackburn and co-workers have demonstrated that heparin binding to the protein involves a single tryptophan (4) in addition to one to two lysine residues (5). The unique tryptophan has been identified as trp-49 in the primary sequence of antithrombin (6). Additionally, Koide et. al. (7) have shown that a clinical variant, antithrombin Toyama, is unable to bind heparin due to a single amino acid substitution of cysteine for arginine at position 47. Thus, the heparin-binding region of the protein is at least partially composed of residues which are located near the amino-terminus of the protein.

Chemical modification of antithrombin with the tryptophan reagent, dimethyl (2-hydroxy-5-nitrobenzyl)[1] sulfonium bromide (4), results in site-specific labeling of one of the four tryptophans contained within antithrombin (8). This single reactive residue is the critical trp-49 which is involved in heparin binding to the protein. Modification of trp-49 with HNB thus eliminates heparin binding to the protein with concomitant loss of heparin-accelerated thrombin inactivation. The HNB chromophore is a sensitive reporter group which exhibits a visible spectrum that is dependent both on pH and environmental polarity (9). Incorporation of the HNB label at trp-49 therefore provides a spectral probe which responds to protein conformational changes at a single site on antithrombin.

Antibodies to the HNB hapten were developed for use in an immunochemical approach to determination of antithrombin structure and function. Although polyclonal antibodies were developed against the HNB hapten, these antibodies react with the target protein, HNB-antithrombin, only at the HNB-label on trp-49 and thus mimic the specificity of a monoclonal antibody. Anti-HNB antibodies were used in combination with the HNB group as an extrinsic probe, sensitive to local structural changes in the heparin-binding region.

EXPERIMENTAL PROCEDURES

Materials

Antithrombin III was purified from human plasma by elution from heparin-Sepharose as previously described (10). Bovine thrombin, purchased from Parke-Davis, was purified by ion exchange chromatography on SP-Sephadex as described by Lundblad et. al. (11). Dimethyl (2-hydroxy-5-nitrobenzyl) sulfonium bromide was purchased from Calbiochem. Gly-trp-gly was purchased from Vega Biochemicals. Protein A-containing Staphylococcus aureus cells were obtained from Boehringer Mannheim. ^{3}H-NaBH$_4$ was purchased from New

---

[1]Abbreviations used are : 2-hydroxy-5-nitrobenzyl, HNB; enzyme-linked immunosorbent assay, ELISA; immunoglobulin G, IgG.

England Nuclear. Biogel P-2 and Affi-gel 401 were purchased from Biorad Laboratories, while DEAE-Sephacel, Sephadex G-50, Sepharose CL-4B, and SP-Sephadex were obtained from Pharmacia. Heparin-Sepharose was generated by coupling of unbleached heparin to cyanogen bromide-activated Sepharose CL-4B. Antithrombin concentration was determined spectrophotometrically at 280 nm using an extinction coefficient of 0.65 ml·mg^{-1}·cm^{-1} (10). Radiolabeled antithrombin was prepared according to the method of Jesty (12); following oxidation with sodium m-periodate, antithrombin was reduced with ^{3}H-NaBH$_4$. This radiolabeling procedure, which incorporates tritium into sites on the carbohydrate side-chains of the protein, yields labeled antithrombin with a specific activity of approximately 100,000 CPM per ug.

Tryptophan Modification with Dimethyl (2-Hydroxy-5-Nitrobenzyl) Sulfonium Bromide

Antithrombin was reacted with dimethyl (2-hydroxy-5-nitrobenzyl) sulfonium bromide as previously described by Blackburn and Sibley (4). Reduced and carboxymethylated lysozyme (8) was dissolved in 8 M urea, pH 2.7, and incubated 16-20 hr at 37°C. After cooling to room temperature, solid dimethyl (HNB) sulfonium bromide was added in ten-fold excess of the six tryptophan residues in lysozyme. After approximately 1 hr, the pH was adjusted to 8, and insoluble HNB alcohol was removed by centrifugation. Upon removal of excess HNB reagent by exhaustive dialysis, the extent of incorporation of the HNB hapten into lysozyme was always at least 5 moles HNB per mole protein.

Glycyl-tryptophyl-glycine was initially solubilized in 40 mM phosphate buffer, pH 7.4, containing 0.15 M sodium chloride by succinylation. Succinic anhydride (50 mg·ml^{-1} in dried acetone) was added to a suspension of 5 mg·ml^{-1} tripeptide in the pH 7.4 buffer to a final concentration of 0.2 M. After the tripeptide was dissolved, solid dimethyl (HNB) sulfonium bromide was added to a two-fold molar excess and stirred for an additional 45 min. Modified tripeptide was separated from reaction by-products by chromatography on a column of Biogel P-2 equilibrated in 10 mM ammonium acetate. Spectrophotometric determination of HNB combined with amino acid analysis of acid-hydrolyzed peptide (13), indicated stoichiometric modification of the tripeptide.

Production of Anti-HNB Antibodies

Antibodies to the HNB hapten were developed by injecting rabbits with the extensively HNB-derivatized lysozyme. The IgG fraction was purified by ammonium sulfate precipitation followed by ion exchange chromatography on a column of DEAE-Sephacel (14). HNB-specific antibodies were isolated by affinity chromatography on a column (1.5 x 1 cm) of Affi-gel 401 (a sulhydryl terminal crosslinked gel) to which HNB had been coupled. After sample application in 10 mM sodium phosphate, 0.15 M sodium chloride, pH 7.4, containing 0.02 % sodium azide, the column was washed with buffer, and then bound IgG was eluted with 4 M urea, pH 2.2. The column effluent directly flowed onto a column (1.5 x 37 cm) of Sephadex G-50 equilibrated in the same buffer.

Immunoassays

Antibody reactivity with the HNB hapten was measured using an ELISA method in which HNB-antithrombin is adsorbed onto polystyrene beads, as previously described by Peterson and Blackburn (10). A radioimmunoassay was also employed to measure antibody reactivity with the HNB-antithrombin antigen. Radiolabeled HNB-antithrombin and antibody were combined and incubated at room temperature for 1 hr, after which Protein A-containing S.

aureus was added to the incubation mixture. After centrifugation, quantitation of bound and unbound antithrombin was performed by measurement of radioactivity remaining in the supernatant as determined with an LKB Rackbeta model scintillation counter.

Spectroscopic Measurements

Spectra in the ultraviolet-visible range were obtained on a Cary Model 219 spectrophotometer interfaced with a Hewlett-Packard Model 9825T computer system. Difference spectra were calculated by subtraction of spectra of protein samples prior to mixing from spectra obtained on protein samples after mixing.

RESULTS

Antibody Specificity for HNB

Antiserum collected from rabbits immunized with HNB-lysozyme was shown to be specific for HNB in a competitive ELISA system. In this assay, HNB-antithrombin was adsorbed onto polystyrene beads, serving as the solid-phase antigen to which specific antibodies were bound. With a competing protein or ligand in solution, antibody binding is partitioned between solid-phase HNB-antithrombin and solution-phase antigen according to reactivity of the competing antigen with anti-hapten antibodies. Figure 1 shows the results of a competitive ELISA in which an HNB-labeled tripeptide (gly-(HNB)trp-gly) readily competes with solid-phase HNB for binding of antibodies. In contrast, the unlabeled tripeptide does not significantly compete for antibody binding over the wide range in antigen concentration tested. Similarly, HNB-labeled antithrombin in solution competes with solid-phase HNB-antithrombin for antibody binding in the ELISA system, whereas antithrombin alone has no effect. Nonreactivity of antibodies with unlabeled tripeptide or antithrombin in this type of analysis demonstrates specific recognition of the HNB hapten by the antibodies.

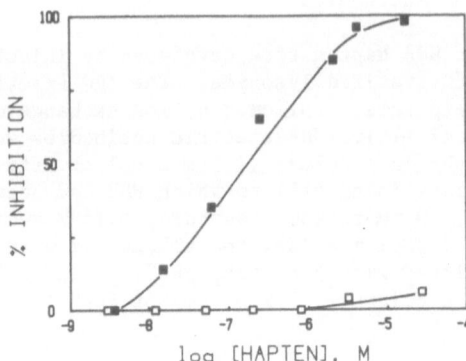

Fig. 1. *Antibodies are HNB-specific.* Antibody was incubated with solid-phase HNB in the presence of gly-(HNB)trp-gly (■) or unmodified tripeptide (□). Inhibition was calculated as 1-A/A$_0$ where A$_0$ and A are the absorbances at 405 nm for antibody alone and antibody plus competing ligand, respectively.

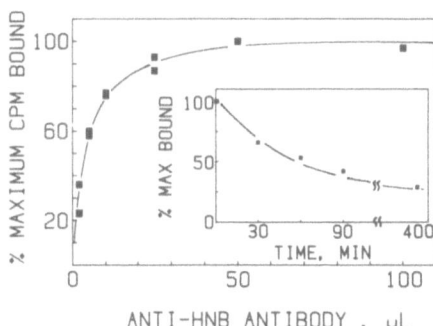

Fig. 2. *Antibodies react with HNB-antithrombin. Anti-*
body (0.45 mg/ml) and ³H-HNB-antithrombin (4 ug)
were incubated for 1 hr, and immune complexes were
precipitated with protein A-containing S. aureus.
INSERT: ³H-HNB-antithrombin is released from the
solid phase by competition of 2 x 10⁻⁵M HNB-tripep-
tide for antibody.

Anti-HNB antibody reactivity with HNB-antithrombin

Thus, the HNB-trp-49 target site in the protein was shown to be reactive with anti-HNB antibodies in the ELISA system. Interaction of the HNB-specific antibodies with HNB-antithrombin was also demonstrated in immune precipitation experiments. The antibodies were mixed with fixed levels of tritium-labeled HNB-antithrombin in antibody:antigen ratios ranging from 0.1 to 10. Protein A-containing S. aureus was added to precipitate immune complexes, and bound vs. free HNB-antithrombin was determined according to radioactivity left in the supernatant. HNB-antithrombin was bound and precipitated in a dose-dependent manner by the anti-HNB antibodies, as shown in Figure 2. At saturating levels of antibody, approximately 40 % of the HNB-antithrombin antigen was bound. The insert in Figure 2 shows release of ³H-HNB-antithrombin from the antibodies by competition of free HNB in high concentration for antibody binding. The slow off-rate for release of HNB-antithrombin, with a half-life of 60 min, is evidence for antibodies with high affinity for the hapten.

Perturbation of the HNB Chromophore upon Thrombin Binding

Although heparin binds tightly to antithrombin ($K_d \sim 10^{-7}$ M (10)), the affinity for heparin is greatly decreased upon the binding of thrombin (15). Heparin is thus released upon production of the inactive protease-inhibitor complex and then recycles to catalyze further reactions. Release of heparin from the antithrombin-thrombin complex might be effected by structural alterations in the heparin-binding domain, including the region of trp-49.

Thrombin binding to HNB-antithrombin clearly does induce structural perturbations at HNB-trp-49, as seen in the upper panel of Figure 3, which shows the difference spectrum generated upon mixing of HNB-antithrombin with thrombin. The HNB reporter group exhibits spectral maxima at 410 nm and 320 nm which correspond to the ionized and nonionized forms of the chromophore (9). Binding of thrombin to the inhibitor leads to an increased absorbance

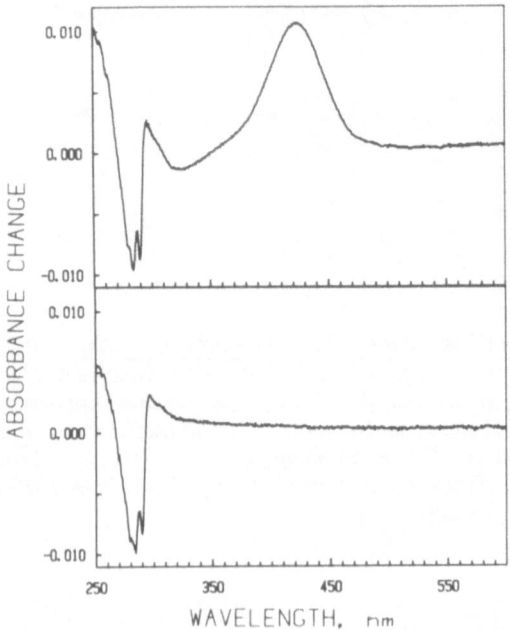

Fig. 3. Thrombin binding perturbs the HNB spectrum. Difference spectra are shown for HNB-antithrombin (top) and native antithrombin (bottom) plus thrombin.

at 410 nm accompanied by a decreased absorbance at 320 nm. This type of difference can be attributed to enhanced ionization of the hydroxyl group on HNB and would be expected upon greater exposure of the chromophore to aqueous solvent. The lower panel, which shows the difference spectrum observed upon mixing of thrombin with native antithrombin, exhibits the same spectral perturbations in the ultraviolet region as seen with HNB-antithrombin.

The local conformational change detected by the HNB reporter group was confirmed by using the anti-HNB antibodies to probe the region of HNB-trp-49. Figure 4 shows a competitive ELISA experiment which compares reactivity of HNB-antithrombin and HNB-antithrombin-thrombin complex with the anti-HNB antibodies. Both proteins are competitive with solid-phase HNB for binding of the antibodies, but the HNB-antithrombin-thrombin complex is a more effective inhibitor in the assay. Inhibition of antibody binding to the solid phase in the assay· is achieved at concentrations of the HNB-antithrombin-thrombin complex which are about one order of magnitude lower than inhibitory concentrations of HNB-antithrombin alone. This immunochemical result is consistent with the increased exposure of the HNB group to the aqueous environment, as was indicated by the spectral measurements. Thus, thrombin-induced structural alterations at a single site on antithrombin can be detected using antibodies to the extrinsic HNB probe.

DISCUSSION

Previous structure-function studies with antithrombin have delineated some essential features of the protein both for heparin binding and protease binding. Chemical modification of both tryptophan and lysine residues in antithrombin indicated the heparin binding activity of antithrombin to be

670

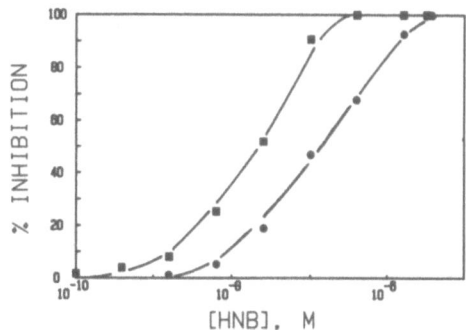

Fig. 4. Thrombin binding enhances immunoreactivity of HNB. Antibody was incubated with solid-phase HNB and competing HNB-antithrombin (●) or HNB-antithrombin-thrombin complex (■) in solution. Percent inhibition was calculated as in Fig. 1.

separable from the anti-protease activity, prompting Blackburn et. al. (6) to propose that antithrombin is organized into functional domains—one for heparin binding and one for protease complexing. Additional evidence for the two-domain structure of antithrombin comes from the guanidine denaturation studies of Villanueva and Allen (16) and Fish et. al. (17) which show a biphasic denaturation pattern.

Antibodies to HNB were developed in our laboratory as an immunological probe which would specifically react with a single locus on HNB-anti-thrombin. The antibodies bind solely to the HNB group, which is attached to trp-49, an essential residue which lies within the heparin-binding region of antithrombin. These antibodies have been used to spatially map the relative orientation of the heparin-binding and protease-complexing domains of the protein (18). Since HNB-antithrombin was shown to be capable of simultaneous binding of anti-hapten antibody and thrombin, the two functional domains appear to be spatially separated. What mechanism exists for communication between these functional regions of antithrombin?

Evidence for an Allosteric Change Induced by Protease Binding

Both immunological and spectral data indicate that protease binding to HNB-antithrombin produces a change in the immediate environment of HNB-trp-49. Villanueva and Danishefsky have also reported changes in ultra-violet and circular dichroic spectra upon complexing of thrombin and antithrombin (19), but such changes have not been localized to one or the other of the proteins. Our data indicate that the binding of thrombin at the reactive-site arginine evokes a structural change at trp-49, in the spatially distinct heparin-binding region of the protein. Communication to and from the critical regions of antithrombin must thus be transmitted via allosteric changes in the intervening protein matrix.

Usefulness of Anti-hapten Antibodies to Study Protein Conformation

We have made use of antibodies specific for a haptenic group, HNB, in order to study local changes within a defined region of structure on the protein, antithrombin III. The tryptophan reagent, HNB, can be incorporated into antithrombin at a unique residue, providing a single site for binding of the antibody probe. Use of the anti-hapten antibodies in an ELISA system indicates that thrombin binding induces structural alterations in antithrombin which perturb the local environment, and thus the immunoreactivity, of the HNB group on trp-49. This immunochemical data which indicates increased exposure of the hapten upon thrombin binding complements results obtained using HNB as a spectral reporter group. Changes in the absorbance spectrum of HNB-antithrombin which occur upon thrombin binding indicate increased ionization of the chromophore due to enhanced exposure of the hapten to its aqueous environment. In our study, the HNB-specific antibodies thus revealed an allosteric mechanism of communication which exists between the thrombin-binding and heparin-binding sites on antithrombin. In a more general sense, it should be possible to employ antibodies specific for other small haptenic groups, even those which are not themselves environmental reporter groups, to probe features of protein structure which may not be amenable by other means.

'REFERENCES

1. Rosenberg, R.D., and Damus, P.S. (1973) J. Biol. Chem. 248, 6490-6505.
2. Björk, I., Jackson, C.M., Jörnvall, H., Lavine, K.K., Nordling, K., and Salsgiver, W.J. (1982) J. Biol. Chem. 257, 2406-2411.
3. Björk, I., and Nordenman, B. (1976) Eur. J. Biochem. 68, 507-511.
4. Blackburn, M.N., and Sibley, C. C. (1980) J. Biol. Chem. 255, 824-826.
5. Pecon, J., and Blackburn, M.N. (1984) J. Biol. Chem. 259, 935-938.
6. Blackburn, M.N., Smith, R.L., Carson, J., and Sibley, C.C. (1984) J. Biol. Chem. 259, 939-941.
7. Koide, T., Odani, S., Takahashi, K., Ono, T., and Sakuragawa, N. (1984) Proc. Natl. Acad. Sci. USA 81, 289-293.
8. Petersen, T.E., Dudek-Wojciechowska, G., Sottrup-Jensen, L, and Magnusson, S. (1979) in "The Physiological Inhibitors of Coagulation and Fibrinolysis," Collen, D., Wiman, B., and Verstraete, M. (eds), New York, Elsevier, pp. 43-54.
9. Koshland, D.E., Jr., Karkhanis, Y.D., and Latham, H.G. (1964) J. Am. Chem. Soc. 86, 1448-1450.
10. Peterson, C.B., and Blackburn, M.N. (1985) J. Biol. Chem. 260, 610-615.
11. Lundblad, R.L., Kingdon, H.S., and Mann, K.G. (1976) Meth. Enzymol. 45. 156-176.
12. Jesty, J. (1979) J. Biol. Chem. 254, 10044-10050.
13. Spackman, D.H., Stein, W.H., and Moore, S. (1958) Anal. Chem. 30, 1190-1206.
14. Hurn, B.A.L., and Chantler, S.M. (1980) Meth. Enzymol. 70, 104-142.
15. Carlstrom, A., Lieden, K., and Björk, I. (1977) Thromb. Res. 11, 785-797.
16. Villanueva, G.B., and Allen, N. (1983) J. Biol. Chem. 258, 11010-11013.
17. Fish, W.W., Danielsson, A., Nordling, K., Miller, S. H., Lam, C.F., and Björk, I. (1985) Biochemistry 24, 1510-1517.
18. Peterson, C.B., and Blackburn, M.N. (1985) manuscript submitted.
19. Villanueva, G., and Danishefsky, I. (1979) Biochemistry 18, 810-817.

XIII. Characterization of Proteases

PROPERTIES OF THE HYDROLASE THAT CATALYZES REMOVAL OF THE BLOCKED

NH$_2$-TERMINAL AMINO ACID RESIDUES FROM POLYPEPTIDES

Wanda M. Jones, Lois R. Manning, and James M. Manning

The Rockefeller University

1230 York Avenue, New York, NY 10021

Acylpeptide hydrolase catalyzes the removal of the terminal blocking group together with the first amino acid residue of a peptide substrate:

$$\text{Acyl} - AA_1 - AA_2 - AA_3 \ldots AA_n \longrightarrow \text{Acyl} - AA_1 + AA_2 - AA_3 \ldots AA_n$$

The enzyme purified from human red cells has no detectable acetylase or carboxypeptidase activity. The enzyme activity is most conveniently assayed by use of blocked amino acid p-nitroanilides as substrates, or by the appearance of a new, free amino group in a peptide substrate[1].

The presence of this enzyme has been reported previously by Narita and his colleagues[2]. They found a correlation between the amounts of enzyme activity and the protein synthetic ability of a given tissue. Gade and Brown[3], Schonberger and Tschesche[4], and Witheiler and Wilson[5] reported the presence of this enzyme in various tissues. Yoshida and Lin[6] obtained a partially purified preparation of this enzyme from rabbit red cells. We have purified this hydrolase from human red cells by a variety of column chromatography procedures[1]. After the last purification step a non-denaturing gel shows one predominant band. The enzyme has a molecular weight of 300,000 and is composed of six subunits[5]. It appears that this particular enzyme has not been successfully developed into a tool for use in protein or peptide structural studies for two major reasons. One of these was the inability to remove all contaminating proteolytic enzymes and, perhaps more importantly, is the fact that the substrate specificity of this enzyme is very broad and the pH dependence varies significantly with different substrates.

A selection of blocked peptide substrates has been tested (Table 1). For purposes of comparison the activity with acetylalanine p-nitroanilide (AANA) is set at 100. The enzyme is catalytically efficient with a wide variety of di, tri, and higher peptide analogues. A chloroacetyl amino acid is also removed by the enzyme as is the formylmethionine residue and also a carbamyl amino acid at the NH$_2$-terminus. Fatty acyl groups (C$_4$–C$_6$) are also removed by the enzyme. The enzyme is not active on peptides containing the cyclized derivative pyroglutamic acid. Also shown in Table 1 are some unblocked peptides that are not substrates for the hydrolase. We have also tested a few amino acid naphthalamides with the purified hydrolase and find that these are not substrates. The hydrolase is distinct from that purified by Gade and Brown which removes the acetyl

Table 1

Relative Rates of hydrolysis for Substrates of Acylpeptide Hydrolase

Substrate	Rate of Hydrolysis (per cent)
Blocked p-Nitroanilide Substrates	
Ac-Ala-pNA (AANA)	100.0
Ac-Glu-p-NA (α) (AGNA)	0.2
Blocked Peptide Substrates	
Ac-Ala₃	35.2
Ac-Ala₂	16.0
Ac-Ser-Leu	5.9
Ac-Ala-Phe-Pro-Leu-Glu-Phe	4.0
Cl-Ac-Gly-Leu	3.4
Cbm-Met-Val	2.6[a]
f-Met-Val	2.0
Cbm-Gly-Leu	1.5[a]
Ac-Phe-Leu-Glu-Glu-Val	0.2
Ac-Gly-Ser	0.2
Ac-Lys-Glu-Gly	<0.01
Pyro-Glu-Gly-Arg-ONp	<0.01
Unblocked Peptide Substrates	
Gly-Met-Gly	2.3
Ala-Phe-Gly	<0.01
Leu-Ala-Gly-Val	<0.01
Gly-Ser-Ala	<0.01
Val-Leu-Lys-ONp	<0.01

Each substrate was incubated at a concentration of 4 mM in 0.2 M potassium phosphate, pH 7.3 with 0.2 M NaCl. [a]The Cbm-peptides were assayed in the absence of chloride and a correction factor (cf. Fig. 3) was applied to the observed values. Abbreviations: Ac = acetyl, Cl-Ac = chloroacetyl, Cbm = carbamyl, f = formyl.

group from the cleaved acetyl amino acid. It was suggested by these investigators and by others that the role of the hydrolase is for the degradation of proteins. However, whether it is present for this purpose or for the biosynthesis of proteins is not yet clear.

The purified hydrolase displays different pH optimum with different substrates (Fig. 1). For example, the blocked glutamate p-nitroanilide has a pH optimum around 6, formylated Met-Val has a pH optimum of around 7.3, and acetyl alanine p-nitroanilide has a pH optimum around 8.3. This phenomenon, i.e., different pH optima with different substrates for the same enzyme has been reported previously[7] and may represent a difference in the kinetic parameters with different intermediates in the reaction. Perhaps the descending limb of the pH optimum above 6 for AGNA might be due to the appearance of the uncharged γ-carboxyl group of the substrate. It appears that any charged amino acid side chain at the first position of the blocked peptide will not bind to the enzyme and thus these will not be substrates unless the pH is adjusted appropriately (see below). Hence, it is clear that the different pH optima for different substrates must be

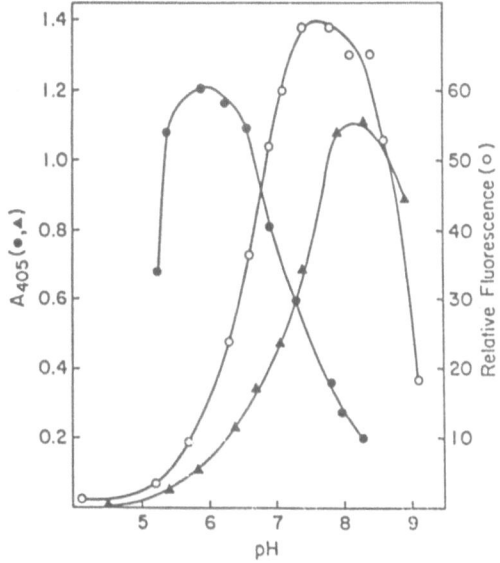

Fig. 1. pH optima of acylpeptide hydrolase for different substrates. The assays were performed in 0.2 M potassium phosphate with 0.2 M NaCl at varying pH values. The substrates tested were: acetylglutamate p-nitroanilide (-●-), formyl-Met-Val (-O-), and acetylalanine p-nitro-anilide (-▲-). The amounts of enzyme were adjusted to bring the values to the same scale.

taken into account especially if the enzyme is ever to be used in protein structural studies.

The wide spectrum of substrates for this enzyme and the fact that the pH optima are distinctly different could invite the suggestion that there are really minor contaminants in the preparation that can catalyze these other reactions. We have taken several approaches to check this possibility. One approach is to use an inhibitor of the enzyme and compare the rate of inhibition with different substrates. Thus, we studied the effect of zinc on the activity of the enzyme with two different substrates, acetylalanine p-nitroanilide (AANA) and acetylglutamate p-nitroanilide (AGNA). These two substrates show the greatest difference in catalytic efficiency and their pH optima are furthest apart. We found that the inactivation followed a very simple kinetic profile with either substrate (Fig. 2). The half-life for inactivation with ANNA was 139 minutes and the half-life with AGNA as a substrate was 132 minutes. This difference is within the experimental error of the assay. We conclude from these results that it is most likely the same enzyme that is catalyzing the reaction with either substrate.

A modulator of enzyme activity is the chloride anion (Fig. 3). Chloride is an inactivator of the enzyme with peptide substrates. However, with the p-nitroanilide substrates such as ANNA or AGNA the inactivation

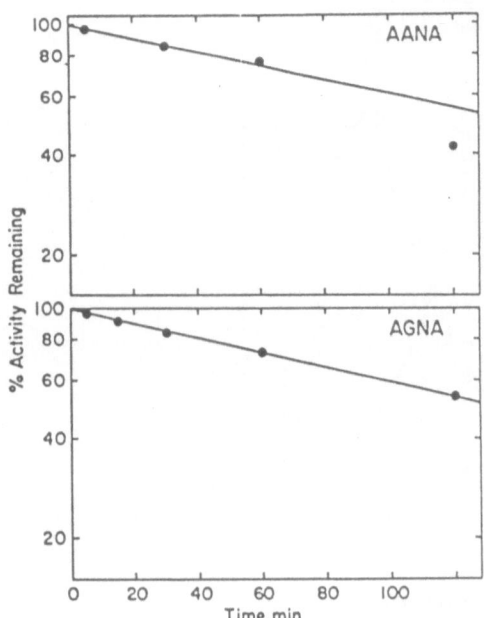

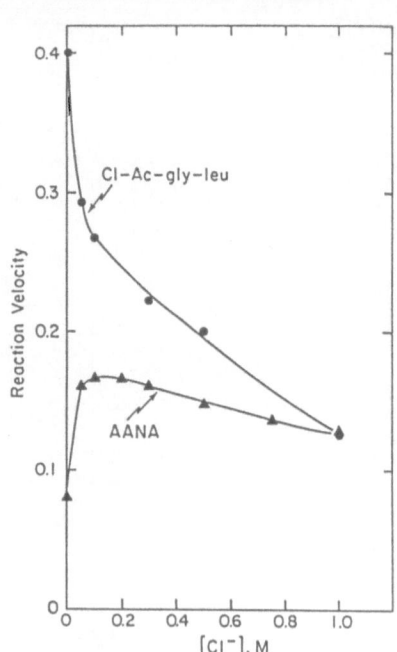

Fig. 2. Inactivation of the hydrolase by zinc ions. The enzyme was treated with Zn²⁺ at 37°C in 0.2 M Tris, pH 7.3. Aliquots were removed at varying times and assayed as described in the text.

Fig. 3. The effect of chloride on enzyme activity. The indicated concentrations of chloride in 0.2 M potassium phosphate, pH 7.3, were present during the assays.

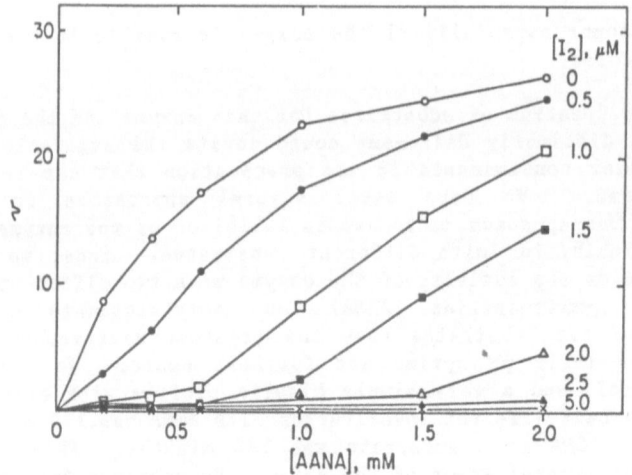

Fig. 4. Effect of iodine on enzyme activity. The indicated concentrations of iodine were incubated with the enzyme in 0.2 M Tris-acetate, pH 8.3 and the activity was measured with AANA as substrate.

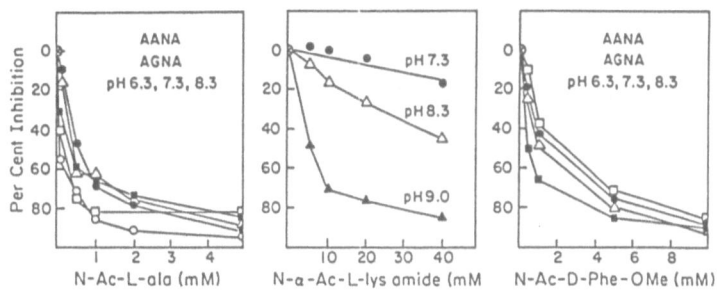

Fig. 5. Inhibition of enzyme activity by
blocked amino acids. The inhibitors were
present at the indicated concentrations in
0.2 M Tris-acetate at the appropriate pH.
The pH values are indicated by the symbols in
the middle panel. Squares represent AGNA as
substrate (2 mM). Circles and triangles
represent AANA as substrate (2 mM).

takes place only above 0.1 M. At lower concentrations the chloride anion
is an activator of the reaction. It could be that the role of chloride is
involved in the kinetic properties of the enzyme protein.

In a series of studies with other monoanionic compounds, we found
that iodide anion apparently decreased the activity of the hydrolase.
However, this inactivation by iodide was found to be actually due to the
presence of I_2 in the iodide preparation. Thus, we have found that iodine
itself is a very potent inactivator of the hydrolase (Fig. 4).

In the process of delineating the complete substrate specificity of
this enzyme, it will be necessary to synthesize many peptides. In an
approach to screen which ones should be synthesized, we have estimated the
relative efficiency of different blocked amino acids in inhibiting
competitively the activity of the hydrolase. Another reason for using this
approach with competitive inhibitors is an indirect test for the presence
of other hydrolase-like activities in the enzyme preparation. As shown in
Fig. 5, we could demonstrate that acetyl-L-alanine inhibits the hydrolase
reaction either with AGNA or with AANA as substrate nearly equally well at
pH 6.3, 7.3, or 8.3. Of particular interest is the finding that N-acetyl-
L-lysine amide is a much more effective inhibitor at pH 9 than at lower pH
values (Fig. 5). This finding reinforces the concept mentioned above that
the enzyme prefers an uncharged amino acid side chain at the terminal
position. Surprisingly, acetyl-D-phenylalanine methylester is a very good
competitive inhibitor with either the glutamate or the alanine p-
nitroanilides (Fig. 5). This finding invites further study on peptide
substrates with D-amino acids. These results add some weight to the
argument that only one enzyme protein is involved in catalyzing the
reaction.

A working model which is being developed for the possible use of this
enzyme in protein structural studies involves the derivatization of the
carboxyl groups of a peptide or a protein with glycinamide. The protein
derivative can then be hydrolyzed with trypsin or pronase if the large
proteins turn out to be poor substrates for the hydrolase. Alternatively,
we will attempt to investigate the activity of the hydrolase on denatured
proteins. The hydrolase is active in the presence of trypsin and pronase
so it can be added directly after the endopeptidase treatment. After the
second digestion, the mixture should include the blocked amino acid

corresponding to the NH$_2$-terminus of the peptide or protein and a mixture of positively-charged peptides from the interior of the protein. Application to Dowex-50 and elution with water should separate the blocked amino acid from these other peptides. Hydrolysis with HCl would provide the identification of the terminal amino acid. We have some preliminary experiments with a few peptides and a protein. The peptide acetyl-seryl-leucine, which is about 6% as active as the best substrate, can be cleaved by the hydrolase after digestion for a few hours. There is only leucine and acetylserine released and the latter compound is present in 85% yield. The larger peptide α-MSH, which is 13 amino acid residues in length and contains acetylserine at its terminus, is a substrate for the hydrolase and the blocking group is removed in 70% yield. A small fragment of ACTH was acetylated and then subjected to the treatment described above. The acetylated alanine was recovered in 83% yield. Although, these preliminary results are encouraging, more studies are needed before we know whether the hydrolase will be useful for protein structural studies.

Finally, one of the major questions about this enzyme is its biological role. Some time ago, it was postulated that the presence of an acetylated terminal residue of a protein marks that protein for degradation[8]. More recently, it has been shown that the ubiquitin system, which is related to protein degradation, requires a free amino terminal residue[9]. If the terminus is blocked, then the ubiquitin system is not functional. On the other hand, it is conceivable that the blocking group is required in some way for functional protein biosynthesis. Thus, it is well established that in prokaryotic systems the initiating amino acid is a formyl methionine residue[10]. In eukaryotic systems the picture is less clear. In some cases, a free methionine residue is present and in some other cases one can isolate it in its acetylated form at the terminus of a protein. Indeed, Bradshaw and his colleagues have recently shown with aldolase that if the protein is isolated from the tissue in the presence of a protease inhibitor then the terminal residue is blocked[11]. If the preparation of the enzyme is carried out in the absence of such an inhibitor, then the terminal blocking residue has been removed. Perhaps the hydrolase that we have described here is related to that process. If the hydrolase could be specifically inhibited without affecting the rest of the protein biosynthetic machinery, then the role of this enzyme might be more readily elucidated.

In summary, this enzyme has some interesting properties with respect to it wide substrate specificity and its varying pH optima with different substrates. Its could be useful in protein structural studies but many more studies are required. And, finally, it would be of great interest to discover the role of this hydrolase in a protein biosynthetic system.

ACKNOWLEDGEMENTS

The assistance of Judith A. Gallea was indispensable during some phases of this work. We thank Robert S. Manning for his assistance with some of the kinetic studies. This research was supported in part by NIH Grants HL-18819 and HL-29665.

REFERENCES

1. W.M. Jones, and J.M. Manning, Acylpeptide Hydrolase Activity from Erythocytes, <u>Biochem. Biophys. Res. Commun.</u> 126:933 (1985).
2. S. Tsunasawa, K. Narita, and K. Ogata, Purification and Properties of Acylamino Acid-Releasing Enzyme from Rat Liver, <u>J. Biochem.</u> 77:89 (1975).

3. W. Gade, and J.L. Brown, Purification and Partial Characterization of α-N-Acylpeptide Hydrolase from Bovine Liver, <u>J. Biol. Chem.</u> 253:5012 (1978).

4. O.L. Schonberger, and H. Tschesche, N-Acetylalanine Aminopeptidase: A New Enzyme from Human Erythrocytes, <u>Z. Physiol. Chem.</u> 362:865 (1981).

5. J. Witheiler, and D.B. Wilson, The Purification and Characterization of a Novel Peptidase from Sheep Red Cells, <u>J. Biol. Chem.</u> 247:2217 (1972).

6. A. Yoshida, and M. Lin, NH_2-Terminal Formylmethyionine and NH_2-Terminal Methionine-Cleaving Enzymes in Rabbits, <u>J. Biol. Chem.</u> 247:952 (1972).

7. V. Massey, and R.A. Alberty, Ionization Constants of Fumarase, <u>Biochim. Biophys. Acta.</u> 13:354 (1954).

8. H. Jornall, Acetylation of Protein N-Terminal Amino Groups: Structural Observations on α-Amino Acetylated Proteins, <u>J. Theoret. Biol.</u> 55:1 (1975).

9. A. Hershko, H. Heller, E. Eytan, G. Kaklij, and I.A. Rose, Role of the α-Amino Group of Protein in Ubiquitin-Mediated Protein Breakdown, <u>Proc. Natl. Acad. Sci.</u> 81:7021 (1984).

10. R.E. Webster, D.L. Engelhardt, and N.D. Zinder, In Vitro Protein Synthesis: Chain Initiation, <u>Proc. Natl. Acad. Sci.</u> 55:155 (1966).

11. H.G. Lebherz, O.J. Bates, and R.A. Bradshaw, Cellular Fructose-P_2 Aldolase has a Derivatized (Blocked) NH_2-Terminus, <u>J. Biol. Chem.</u> 259:1132 (1984).

12. F. Wold, Acetylated N-Terminals in Proteins: A Perennial Enigma, <u>Trends in Biochemical Sciences</u> 9:256 (1984).

KINETIC PRODUCT ANALYSIS OF ASPARTYL PROTEINASES UTILIZING NEW SYNTHETIC SUBSTRATES AND REVERSED PHASE HPLC

Ben M. Dunn, Melba Jimenez, Jeff Weidner, Michael Pennington, Mark Carter, and Benne Parten

Department of Biochemistry & Molecular Biology
University of Florida College of Medicine
Gainesville, FL 32610

INTRODUCTION

Kinetic characterization of the Aspartyl Proteinase family of enzymes has been hampered by the lack of convenient substrates. The enzymatic preference for cleavage between two hydrophobic residues causes most small synthetic substrates to be very insoluble in aqueous buffers. Since these enzymes do not cleave terminal residues readily, p-nitroanilides or p-nitrophenyl esters are not useful as chromophoric substrates and most previous assays required measurement of newly generated amino termini with ninhydrin or fluorescamine.

The solution to the problem of insolubility of the synthetic substrates was solved by preparing slightly larger hepta- or octapeptides and adding hydrophilic residues to overcome the hydrophobicity of the two required aromatic residues at the cleavage point[1]. The choice of the additional residues was guided by the study of Powers, et al., in which subsite preferences were deduced from a retrospective study[2] of cleavage positions during sequencing operations.

A more convenient way to monitor cleavage reactions was provided by Hofmann and Hodges[3], who found that placement of a p-nitrophenylalanine in the P1' position of a substrate resulted in a measurable change in ultraviolet absorbance upon cleavage of the peptide bond.

These ideas have been combined in the substrates synthesized to study the Aspartyl Proteinases reported here. The general structure of the substrate was:

Lys-Pro-Xaa-Yaa-Phe-Nph-Arg-Leu

P5   P4   P3   P2   P1   P1'   P2'   P3'

where Nph is p-nitrophenylalanine. (For Xaa=Val, Yaa=Glu, Asp, Val, Asn, or Ser. For Yaa=Glu, Xaa=Pro, His, Lys, Ala, Ile, Gly, or Asp). In this series, changes in the amino acid residues in the P3 and P2 positions of the substrate gave substantial changes in the measured kinetic parameters for a variety of the Aspartyl Proteinases[4,5]. It is now generally agreed that the Aspartyl Proteinase family possesses considerable sequence and conformational homology. Therefore, a relative study of the catalytic activity toward this series of substrates could reveal important features of active site binding and active site differences, especially for those enzymes whose three-dimensional structure is not known as yet. However, before such comparisons can be made, it is vital to establish that the point of attaok is identical in all cases and is limited to the desired peptide bond. Identical cleavage points will permit inferences to be made about relative subsite binding.

The purpose of this contribution is to establish, using reversed phase HPLC, that the products of digestion are identical for all Aspartyl Proteinases attacking this series of substrates. The presence of the p-nitrophenylalanine facilitates this study since all products containing the p-nitrophenylalanine have a strong absorbance and were relatively easy to detect in the HPLC.

MATERIALS AND METHODS

HPLC Conditions. All HPLC was carried out on a Waters system with a C18 microbondapak cartridge in a Waters Z-Module. Solvent A contained 10% methanol and was 0.014 M ammonium acetate, made by adjusting an acetic acid solution to pH 4.5 with ammonium hydroxide. Solvent B was HPLC grade methanol (Fisher). Typically, gradients were run from 20 to 80% Solvent B over 20 min at 1.5 ml/min. For kinetic product analysis, where aliquots are taken at time intervals as short as 15 min, isocratic conditions were used to eliminate the need for reversal and re-equilibration of the columns. In these cases, the percentage of Solvent B was chosen to elute all components within 15 min after injection. This required from 35 to 50% Solvent B depending on the sequence of the substrate peptide under study. In all cases, the order of elution was pentapeptide product (Lys-Pro-Xaa-Yaa-Phe), tripeptide product (Nph-Arg-Leu), and residual substrate last.

End Point Analyses. Samples of substrate (100-500 nmoles or 0.1 to 0.5 mg) were incubated in 0.5 to 1.0 ml of buffer of pH 4 overnight (12-16 hr) with no more than one microgram of enzyme. This is usually sufficient to allow complete hydrolysis to products. An aliquot of 5-25 microliters was injected on the HPLC to establish conditions for complete separation. Then the remaining sample was injected, normally in two batches, and all peaks were collected. The collected samples were reduced to dryness in pyrex test tubes using an Evapomix and vacuum pump. Purified HCl was then added (1 ml of 5.7 M) and the tubes were evacuated and sealed. The samples were then heated to 120 deg C for 24 hr followed by quantitative amino acid analysis using the OPA procedure[6]. For amino acid analysis, solvent A is 0.2 M sodium acetate, adjusted to pH 5.80 by addition of acetic acid. Solvent A contains 2.5% THF. Solvent B is HPLC grade Methanol. A linear gradient from 25 to 80% buffer B is run starting at 2 min up to 28 min with a flow rate of 1.5 ml per min. From injection to 2 min, a flow rate of 0.1 ml/min is used. Mixing of OPA with sample is carried out with a Waters WISP system.

Kinetic Product Analysis. To provide a continuous record of cleavage products, samples were incubated with reduced enzyme concentrations and, in some cases, at reduced temperatures to slow down the reactions. This permitted aliquots to be taken at various points during the cleavage reaction. Injection of aliquots from the kinetic solution onto the HPLC quenched the reaction by removing enzyme from the substrate rapidly and allowed separation and quantitation of the products at that point. By taking the same size aliquots repetitively, it was possible to compare the amount of substrate and all products present during the reaction. Simultaneously, the absorbance of the kinetic solution was continuously recorded in a Gilford Model 250 Kinetics Spectrophotometer at 300 or 310 nm. This permitted the comparison of the rates of product formation and substrate depletion with the rate of absorbance change that was used for enzyme kinetics.

RESULTS AND DISCUSSION

Endpoint Analysis. Reversed phase HPLC was ideal for isolation of the peptides formed on digestion of the substrates reported here. For all new substrates, digestion experiments were performed with pig pepsin and the expected pentapeptide and tripeptide products were formed in each case. A typical example is shown in Figure 1 where the HPLC profile for the substrate Lys-Pro-Val-Glu-Phe-Nph-Arg-Leu has been superimposed with the profile for the hydrolysis products following incubation with pig pepsin. In this example 270 nmoles of substrate were recovered from an injection and after incubation with pepsin an identical amount of substrate yielded 235 nmoles of the pentapeptide and 300 nmoles of the tripeptide. The recoveries from HPLC are usually 50-80% of the amount injected. These experiments, performed for all the substrates listed in the introduction, clearly establish the point of cleavage of substrate as being between the Phe and Nph as predicted.

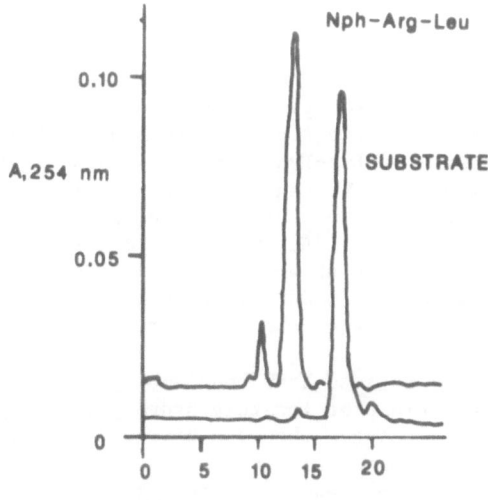

Fig. 1.   High pressure liquid chromatography of
          the substrate Lys-Pro-Val-Glu-Phe-Nph-
          Arg-Leu and the products of digestion
          of this substrate by pig pepsin. The
          largest peak in the digestion mixture
          yielded Nph, Arg, and Leu on amino acid
          analysis and is so labeled. The smaller
          peak was identified as the pentapeptide
          by its composition. A 20-80% gradient
          was run as described in the methods section.

Similar experiments have been conducted with a selection of other Aspartyl Proteinases with a variety of the new substrates. In each case, the identical peak of intermediate mobility was observed in HPLC profiles of product mixtures and amino acid analysis of the collected peak confirmed the identity of this peak as the tripeptide Nph-Arg-Leu. The pentapeptide product appears at variable elution times depending on the nature of the amino acid in the P3 and P2 position. The absorbance of this product was very small compared to the p-nitrophenylalanyl-containing tripeptide product and therefore appeared as a much smaller peak in these profiles as in the example in Figure 1.

These experiments establish that the products of digestion of the substrate series was the same for all Aspartyl Proteinases examined. Further experiments not detailed here have clearly shown that the mobility of other possible peptide products in the HPLC system used differ from the tripeptide by one to five minutes and thus can be easily distinguished.

Kinetic Product Analysis. It remains to establish that the change in absorbance observed during substrate cleavage is a faithful monitor of the rate of Phe-Nph bond cleavage and that no kinetic intermediates are accumulating during enzymatic attack.

The necessary proof can also be readily obtained through the use of reversed phase HPLC. By decreasing the amount of enzyme it is possible to slow the rate of cleavage of substrate from a time scale of minutes to a time scale of hours. In some cases, the temperature of incubation was also reduced to 10-20 deg C. This provides an additional rate reduction and has the further benefit of reducing proteinase self digestion.

After appropriate conditions were found to yield a half-life for cleavage of ca. 2 hr, incubations were prepared so that 25 microliter aliquots of the sample would contain sufficient peptide to be readily detected by HPLC. Again, the intense Nph chromophore was useful in this regard. In Figure 2 a "stacked plot" presentation is given for comparison of injections from a single kinetic solution done at the indicated time intervals in a digestion of Lys-Pro-Val-Asn-Phe-Nph-Arg-Leu by pig pepsin. It can be readily seen that the peak at about 10 min present in the bottom trace, which was due to the substrate peptide, was gradually depleted during the course of reaction. At the same time, the increase in the peak due to the tripeptide product at about 5 min is also unmistakeable. It is clear that decay of substrate leads directly to the tripeptide product without the intermediacy of any other product.

The observed peak heights have been employed to obtain an approximate pseudo-first order kinetic plot for this hydrolysis reaction. In Figure 3, plots are given for the changes observed by monitoring the absorbance at 300 nm, for the changing peak heights observed for substrate depletion in HPLC and for tripeptide appearance in HPLC. The near identity of these three lines establishes the two important relationships essential for kinetic analysis. First, the decrease in substrate concentration was correctly measured by monitoring the absorbance of kinetic solutions at 300 nm. This establishes the validity of this spectrophotometric assay for characterization of Aspartyl Proteinase kinetics. Second, the decay of octapeptide substrate leads directly to tripeptide product without the intermediacy of any other peptides containing the Nph group. This means that the kinetic measurements carried out with these substrates are uncomplicated by side reactions such as transpeptidation of either the acyl or amino group transfer type. Similar kinetic product analyses have been carried out for many of the substrates listed in the introduction. In all cases, the identity of the loss of substrate and appearance of tripeptide product has been observed.

Finally it is essential to conclude by stating that this series of experiments has provided further evidence for the predicted mode of binding of these substrates to the active site of various Aspartyl Proteinases. This binding mode places the Phe-Nph in the P1 and P1' subsites of the enzyme and results in unique cleavage of that bond alone. The design of these octapeptide substrates was intended to fill up the active site cleft in this unique fashion and decrease the likelihood of non-productive

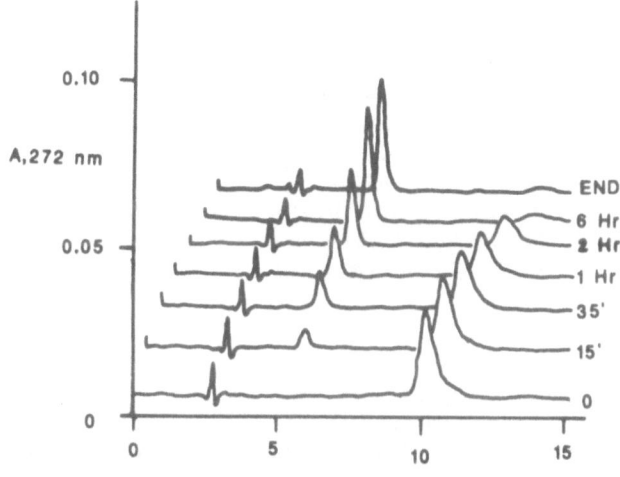

Fig. 2.   Composite plot of HPLC traces taken at the indicated times after the start of pepsin catalyzed hydrolysis of Lys-Pro-Val-Asn-Phe-Nph-Arg-Leu. These separations were done isocratically at 40% solvent B as described in the methods section.

binding or cleavage at other positions. The supporting evidence reported here is essential to allow the rational analysis of the kinetic consequences of amino acid substitution in selected positions of the substrate. All substrates in this series are cleaved at the same point, therefore the peptides must bind in the mode indicated in the introduction to place each residue in a unique subsite of the active site.

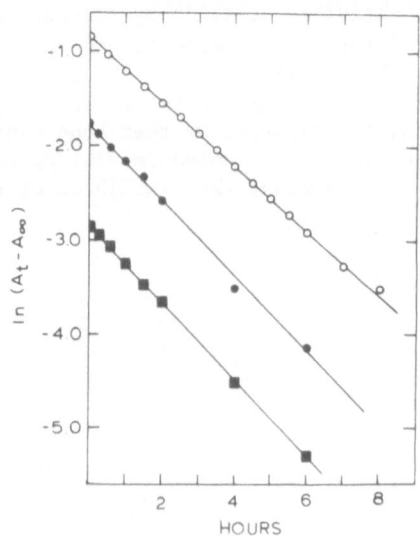

Fig. 3. *Pseudo-first order plots from the change in
absorbance recorded at 300 nm (open circles)
or from the peak heights of the substrate
peak in the HPLC shown in Fig. 2 (closed
squares). Also plotted is the change in
peak heights for the increase in the tripeptide peak from Fig. 2 (closed circles).*

Acknowledgements: We are indebted to our colleagues John Kay and
Martin Valler of University College, Cardiff, Wales for carrying out the
enzymatic digestion of various substrates with Aspartyl Proteinases other
than pig pepsin and for other inspirations. This research was supported by
NIH grant AM 18865.

REFERENCES

1. B. M. Dunn, B. Kammermann, and K. R. McCurry, The Synthesis,
    Purification, and Evaluation of a New Chromophoric Substrate for
    Pepsin and Other Aspartyl Proteases, Analyt. Biochem. 138:68
    (1984).
2. J. C. Powers, A. D. Harley, and D. V. Myers, Subsite Specificity of
    Porcine Pepsin, in: "Acid Proteinases: Structure, Function, and
    Biology", J. Tang, ed., Plenum, New York (1977).
3. T. Hofmann and R. S. Hodges, A New Chromophoric Substrate for
    Penicillopepsin and Other Fungal Aspartic Proteinases, Biochem. J.
    203:603 (1982).
4. B. M. Dunn, B. Parten, M. Jimenez, C. E. Rolph, M. Valler, and J. Kay,
    Interaction of Aspartic Proteases with a New Series of Synthetic
    Substrates and with Inhibitors Based on the Propart of Porcine
    Pepsinogen in: "Aspartic Proteinases and Their Inhibitors", V.
    Kostka, ed., Walter de Gruyter, Berlin (1985).
5. M. J. Valler, J. Kay, and B. M. Dunn, Hydrolysis of a Series of
    Synthetic Chromophoric Substrates by Aspartic Proteinases, Trans.
    of the Biochem. Soc. 13: (in press) (1985).
6. P. Lindroth, and K. Mopper, High Performance Liquid Chromatographic
    Determination of Subpicomole Amounts of Amino Acids by Precolumn
    Fluorescence Derivitization with o-Phthaldialdehyde, Analyt.
    Chem., 51:1667 (1979).

THE ESTERASE-LIKE ACTIVITY OF COVALENTLY BOUND C3[†]

Yeldur P. Venkatesh* and R.P. Levine

James S. McDonnell Department of Genetics
Washington University School of Medicine
St. Louis, Missouri 63110

In its native state, C3, the third complement protein contains an internal thiolester (1). This unique group is located in the α chain of C3 in the sequence –Gly-Cys-Gly-Glu-Glu-Asn-Met- (2). Activation by enzymic or non-enzymic means (3,4) causes a large conformational change (5), and the thiolester is either hydrolyzed, or undergoes a transacylation reaction in the presence of hydroxyl or amino groups of cell surface macromolecules (6) or small molecules (7-9); the result is either the formation of inactive C3 or the formation of a covalently-linked complex (internal acyl ester or amide bond)(7-9).

It has been previously shown that cell-surface bound C3 (acyl ester type) is released at neutral pH and 37°C, a property which is dependent on the native structure of covalently bound C3 (10). This characteristic property constitutes an esterase-like activity of covalently bound C3, and has also been demonstrated (10,11) in the fluid-phase system using C3-[^{3}H]glycerol (hereafter referred to as C3-gll). The esterase-like activity was pH-dependent and from pH-activity studies it was found that a side-chain group with a pK_a ~7.5 is crucial to this activity. Various thiol reagents enhanced the esterase-like activity of C3-gll by 3-fold, suggesting that the single sulfhydryl group, which is positioned three residues on the amino-terminal side of the internal acyl ester, is not important as a 'catalytic' group. In contrast, mercuric chloride, which is also regarded as a specific thiol reagent (12), was found to inhibit the esterase-like activity. In this paper, we describe the effect of this and other metal ions on the esterase-like activity of covalently bound C3. The mechanism by which certain metal ions bring about inhibition of the esterase-like activity is discussed. A study has also been made of the kinetic solvent deuterium isotope effect on the esterase-like activity in order to gain insight into its mechanism of action. On the basis of the results presented here, we propose a model for the mechanism of action of the esterase-like activity of covalently bound C3.

[†]Supported by Research Grant AI-16543 from the National Institutes of Health. Abbreviation used: C3b, a major fragment that is derived from C3 upon removal of the N-terminal 77 residues of the α chain by the action of a C3 convertase.
*To whom correspondence should be addressed at: Smith Kline & French Laboratories, R & D, 709 Swedeland Road, Swedeland, PA 19479

C3 was prepared from fresh human plasma (13). [2-^{3}H]glycerol (10 Ci/mmol) was obtained as an ethanol:water (1:19) solution from New England Nuclear Corp., Boston, MA. 3,6-Bis(acetoxymercuri)-o-toluidine [BAMT] was a product of Pfaltz and Bauer, Inc., Waterbury, CT. $\overline{D}_2O$ (99.84 mole% D was obtained from Bio-Rad Laboratories, Richmond, CA. Centricon 30 microconcentrators were obtained from Amicon Corporation, Danvers, MA.

Solutions of radioactive small molecules were prepared by mixing labeled with unlabeled molecules to a specific radioactivity of 50 mCi/mmol and evaporating the solvent by a stream of N_2 at room temperature; 1 ml of C3 solution (10 mg/ml) in 50 mM phosphate buffer, pH 7.0, was added. The mixture was warmed at 37°C for 5 min. 1 ml of saturated KBr solution was added and incubated at 23°C for 2 hr. At the end of the incubation period, 8 ml of 0.25 M glycerol in 0.15 M phosphate buffer, pH 7.4 were added, and unbound glycerol was removed by dialysis against 0.15 M phosphate buffer, pH 7.4, at 4°C for a total period of 20-24 hr.

The esterase-like activity of C3-g11 was followed by incubation of 3 ml of C3-g11 (protein concentration, 1 mg/ml) in 0.15 M phosphate buffer, pH 7.4, in the presence of 0.01% NaN_3 at 37°C; at various intervals, 0.5 ml aliquots were removed. 50 µl of 20% SDS (in water) were added. The samples were dialyzed immediately against 5 mM phosphate buffer, pH 7.0, at 23°C. The esterase-like activity expressed in hr^{-1} was obtained from the slope of the plot of $-\ln([\text{C3-g11}]_t/[\text{C3-g11}]_{t=o})$ against t (hr).

Metal ions were added to the protein solution during the binding reaction. Incubation was extended for another 5 h, followed by dialysis. $HgCl_2$ was added in a 1:1 molar ratio with respect to the C3 concentration. $PdCl_2$ was added as an aqueous suspension at a 4-5 fold molar excess over C3. Alkyl or aryl mercuric chloride solutions (concentration = 10 µmole/ml in dimethylformamide) was added at 4-fold molar excess over C3. BAMT was prepared as an acetone solution at 1 µmole/ml concentration, and was used at equimolar concentration with respect to C3.

For experiments involving measurement of the esterase-like activity in D_2O solutions, sodium phosphate buffer solutions (0.1 M) were made up in D_2O and adjusted to the required pD using either DCl or NaOD in the pD range 6.60-8.00. In a preliminary experiment, it was found that the initial rate was unaffected by ionic strength. For the pD range 8.01-9.20, tris buffer was employed. The concentration of tris was 50 mM, and the ionic strength was made up to 0.2 using NaCl. Trizma base and trizma hydrochloride solutions were mixed in various proportions to obtain different pH(pD) values, taking into account the temperature dependence of pH in the case of tris buffer. The pH(pD) values mentioned here refer to at 37°C. In deuterium oxide, a pD value was estimated from pD = pH-meter reading + 0.40 (14).

C3-g11 after thorough dialysis against 5 mM phosphate, pH 7.0 (or 5 mM tris, pH 8.0) was concentrated to a very small volume using Centricon 30 microconcentrators. Small aliquots (50 µl) were taken in different tubes, 1.95 ml of buffer of different pH(pD) values were added, and the solutions were mixed thoroughly. The esterase-like activity was followed as described above.

Polyacrylamide gel electrophoresis (PAGE) - denaturing and non-denaturing, protein estimation, and measurement of radioactivity were performed as described earlier (10).

RESULTS AND DISCUSSION

Effect of Metal Ions on the Esterase-like activity

Divalent mercury is known to have a high affinity for the -SH group of proteins (12). When mercuric chloride was used in stoichiometric amounts with respect to the -SH group of C3-g11, only 7% of the esterase-like activity was observed as compared to an untreated sample (Fig. 1).

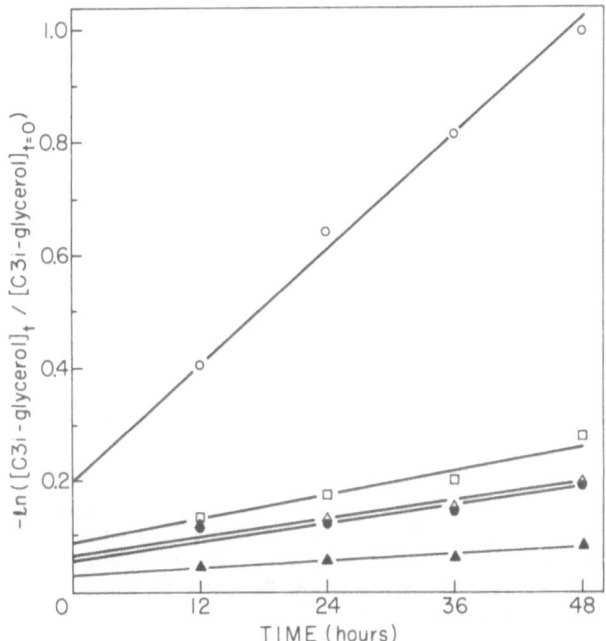

Fig. 1. The esterase-like activity of C3-g11 treated with mercuric chloride (at pH 7.4, Δ ; pH 5.6, ●), palladium chloride (□), and BAMT (▲). C3-g11 (without any metal ions) is indicated by (o). The esterase-like activity (hr^{-1}) is calculated from the slope and the values are expressed as a percentage of the value for the control sample (Table 1).

Similarly, C3-g11 treated with palladium chloride at 4-fold molar excess over the -SH group, showed only 7% of the esterase-like activity of C3-g11. In the case of C3-g11 treated with a bifunctional organic mercurial, BAMT, the esterase-like activity was found to be 5%. The observed inhibition by these reagents which are specific for the -SH group, is in apparent contrast to the effect of various thiol reagents seen earlier. Since C3-g11

Table 1. Effect of metal ions on the esterase-like activity of C3-g11

Compounds		Activity (control=1.00)
Methylmecuric chloride	$CH_3Hg^+Cl^-$	
Ethylmercuric chloride	$C_2H_5Hg^+Cl^-$	1.57
n-butylmercuric chloride	$C_4H_9Hg^+Cl^-$	
Phenylmercuric chloride	⬡—Hg^+Cl^-	0.97
p-chloromercuribenzene sulfonic acid	^-O_3S—⬡—Hg^+Cl^-	1.05
p-chloromercuribenzoic acid	^-OOC—⬡—Hg^+Cl^-	0.76
Silver nitrate	Ag^+	1.02
Mercuric chloride ($HgCl_2$)	Hg^{++}	0.07
Palladium chloride ($PdCl_2$)	Pd^{++}	0.08
3,6-Bis(acetoxymercuri)-0-toluidine (BAMT)	$CH_3COO^-Hg^+$ ⬡(CH_3)(NH_2)$Hg^+CH_3COO^-$	0.05

possesses one –SH group, the possibility of forming an intermolecular dimer through a divalent mercuric ion exists. This kind of dimer is formed by coordinate bonds between Hg^{++} and two –SH groups. However, the samples treated with $HgCl_2$, $PdCl_2$, and BAMT all appeared as a single band, their mobilities corresponding to that of untreated C3-g11 on analysis by PAGE under non-denaturing conditions (data not shown). Therefore, dimers of C3 through a Hg^{++}, Pd^{++}, or two monovalent Hg ions in the para position (BAMT) are absent in these samples.

In order to determine whether Hg^{++} inhibits the esterase-like activity by its complexation with the –SH group of C3-g11 alone or by coordinating to another side-chain functional group which is in the vicinity of the –SH group, it appeared important to test the effect of monovalent mercury compounds. Results are shown in Table 1. Among the alkyl mercuric compounds tested methyl-, ethyl, and n-butyl mercuric ions did not inhibit the esterase-like activity of C3-g11. In fact, in their presence there were slight increases in the activity. The aryl mercuric compounds, p-chloromercuribenzoic acid and p-chloromercuribenzene sulfonic acid did not affect the activity. Hence, neither alkyl nor aryl mercuric compounds which are monovalent, inhibit the esterase-like activity of C3-g11. Divalent zinc, copper, tin, nickel, and lead were found not to affect the esterase-like activity of C3-g11 (data not shown).

From the present study it appears that monovalent mercury binds to the –SH group and thus does not alter the rate of hydrolysis appreciably. The same is true of silver nitrate (Table 1), which is also monovalent. On the other hand, divalent mercury would initially form a mercaptide bond with the

-SH group and then bind to an atom with a lone pair of electrons (the oxygen of a carboxyl group or the nitrogen of an imidazole group in the unprotonated form) by virtue of the second coordination site. BAMT falls under the class of bifunctional cross-linking reagents and thus it can serve to bridge the -SH group and another amino acid side-chain of C3-gll in an intramolecular fashion and thereby bring about the inhibition of the esterase-like activity. Divalent palladium appears to inhibit the esterase-like activity in a manner analogous to that of Hg^{++}.

Kinetic Solvent Deuterium Isotope Effect

In order to determine whether the mechanism of the esterase-like activity is of the nucleophilic or general-base type, a measurement of the kinetic deuterium solvent isotope effect has been made. The esterase-like activity as a function of pH (or pD) is shown in Fig. 2. In the pH(pD) range 6.80-8.42, the $k(H_2O)/k(D_2O)$ values for the esterase-like activity were found to be 2.5-3.2. A $k(H_2O)/k(D_2O)$ value of closer to 1.0 indicates a nucleophilic catalysis, whereas values > 1.5-2.0 are indicative of a general-base mechanism (15). It is well known that in D_2O the pK_a values of side-chain functional groups in proteins shift by ~0.5 units towards the basic side (16). To arrive at a meaningful interpretation of the kinetic solvent deuterium isotope effect, it is important to obtain the $k(H_2O)/k(D_2O)$ value at a pH(pD) value in which the 'catalytic' group is in the fully unprotonated form. In the present case, this pH(pD) value would correspond to 9.00 or greater. A $k(H_2O)/k(D_2O)$ values of 1.87, 1.74, 1.54, and 1.54 were obtained at pH(pD) values of 8.62, 8.79, 9.00, and 9.20

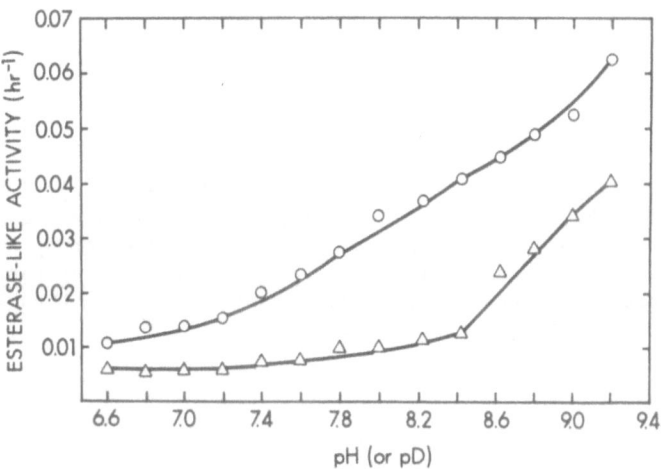

Fig. 2. The esterase-like activity of C3-gll as a function of pH(o) or pD(Δ). Sodium phosphate buffer was employed in the pH(pD) range 6.60-8.00, and tris buffer in the range 8.01-9.20. Each point represents the esterase-like activity (expressed in hr^{-1}) as calculated from the plot of $-\ln([C3-gll]_t/[C3-gll]_{t=0})$ vs. t (hr).

respectively. At pH(pD) values of 9.00, and 9.20, a k(H₂0)/k(D₂0) value of 1.54 is a borderline case, and thus the reaction mechanism could be either nucleophilic or general-base catalysis. Since the specific base catalysis of C3-g11 increases above pH 8.50 considerably (10), hydrolysis by the deuteroxide anion OD⁻, can also contribute to the observed rates of hydrolysis above pH(pD) 8.50. The contribution of OD⁻ to the observed rates of hydrolysis would result in a lowering of the k(H₂0)/k(D₂0) value, and this appears to be the case at pH(pD) values in the range 8.62-9.20.

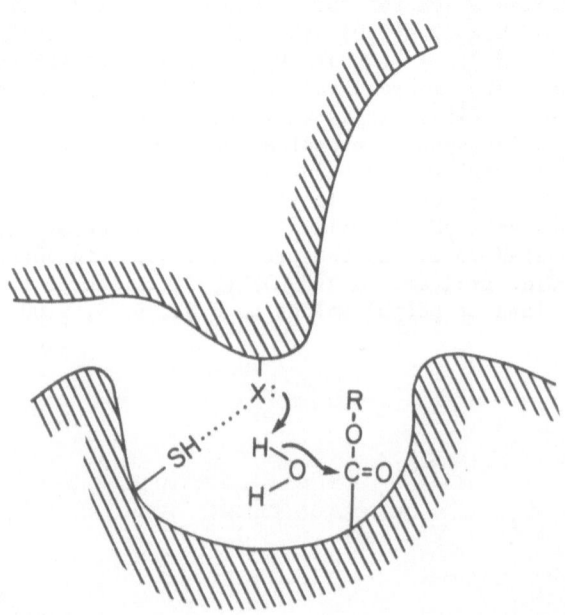

Fig. 3. Proposed mechanism for the esterase-like activity of covalently bound C3. R represents the glyceryl moiety in the case of C3-g11. In this model, X: represents the 'catalytic' group and appears to function as a general-base. The non-covalent interaction between X: and the -SH group is denoted by a dotted line.

Since the k(H₂0)/k(D₂0) values at pH(pD) values > 9.00 are not close to 1.00, a nucleophilic mechanism is less likely. The alternative mechanism of general base catalysis appears to be more likely.

Taking into account the observed results of various chemical reagents and the solvent isotope effect on the esterase-like activity, we propose a general-base mechanism for this intramolecular hydrolysis of the acyl ester bond in covalently bound C3, and it is shown in Fig. 3. It appears that the -SH group normally functions to lower the rate of hydrolysis of the acyl ester bond in covalently bound C3. This proposal requires the participation of the -SH group in some sort of non-covalent interaction with the side-chain of another amino acid residue (called the 'catalytic' group) which is situated closeby in the tertiary structure. Once the -SH group is blocked, the 'catalytic' group can attack the acyl ester bond in an enhanced manner. The side-chain group X, in its unprotonated form at physiological pH values would assist the attack of a water molecule on the carbonyl carbon of the internal acyl ester bond in covalently bound C3. It is well known that esters susceptible to general-base catalysis by imidazole group generally exhibit a marked spontaneous hydrolysis in aqueous solution (15). So, the most probable candidate for the 'catalytic' group appears to be an imidazole group of a histidyl residue.

The general-base mechanism proposed for the hydrolysis of the acyl ester bond in covalently bound C3 has led us to consider a similar mechanism for the initial formation of the acyl ester bond. According to this proposal, during activation, the 'catalytic' group X, in its unprotonated form would be positioned in such a way that it can abstract a proton from water making it nucleophilic and thus bring about the hydrolysis of the thiolester bond. The products of this reaction are a free -SH group and a carboxyl group as seen in the case of inactive C3, or fluid-phase C3b. On the other hand, if, in the fluid-phase system, there are other small molecules containing hydroxyl groups, the proton would be abstracted by X:, thus making the oxygen nucleophilic. The result is the appearance of a free -SH group and the formation of an internal acyl ester. A similar mechanism would presumably occur on cell surfaces and immune complexes to which C3b becomes covalently bound. In the case of small molecules containing amino groups (in the protonated form), the proton would be abstracted by X:; in the process the nitrogen will become nucleophilic, and attacks the carbonyl carbon of the thiolester bond resulting in the formation of an internal amide. Recently, the primary structure of human C3 has been derived from the cDNA coding sequence (17). The three-dimensional structure of human C3, which remains to be determined, should reveal the nature and the location of the 'catalytic' group which appears to be crucial to the covalent binding reaction and the esterase-like activity of covalently bound C3.

ACKNOWLEDGMENTS

We thank Ms. Dulari Shah for skillful technical assistance in the purification of C3, and Dr. Wu Liu for valuable discussion. The secretarial assistance of Ms. Kay Webb is gratefully acknowledged.

REFERENCES

1. B. F. Tack. The β-Cys-γ-Glu thiolester bond in human C3, C4, and α2-macroblogulin. Springer Semin. Immunopathol. 6:259 (1983).
2. M. L. Thomas, J. Janatova, W. R. Gray, and B. F. Tack. Third component of human complement: localization of the internal thiolester bond. Proc. Natl. Acad. Sci. USA 79:1054 (1982).
3. S. K. Law. The covalent binding reaction of C3 and C4. Ann. N.Y. Acad. Sci. 421:246 (1983).
4. J. Janatova. The third (C3) and the fourth (C4) components of complement: labile binding site and covalent bond formation. Ann. N.Y. Acad. Sci. 421:218 (1983).

5. D. E. Isenman. The role of the thioester bond in C3 and C4 in the determination of the conformational and functional states of the molecule. Ann. N.Y. Acad. Sci. 421:277 (1983).

6. S. K. Law, N. A. Lichtenberg, and R. P. Levine. Evidence for an ester linkage between the labile binding site of C3b and receptive surfaces. J. Immunol. 123:1388 (1979).

7. S. K. Law, T. M. Minich, and R. P. Levine. Binding reaction between the third human complement protein and small molecules. Biochemistry 20:7457 (1981).

8. M. K. Hostetter, M. L. Thomas, F. S. Rosen, and B. F. Tack. Binding of C3b proceeds by a transesterification reaction at the thiolester site. Nature 298:72 (1982).

9. S. K. Law. Non-enzymic activation of the covalent binding reaction of the complement protein C3. Biochem. J. 211:381 (1983).

10. Y. P. Venkatesh, T. M. Minich, S. K. Law, and R. P. Levine. Natural release of covalently bound C3b from cell surfaces and the study of this phenomenon in the fluid-phase system. J. Immunol. 132:1435 (1984).

11. Y. P. Venkatesh, T. M. Minich, and R. P. Levine. Hydrolysis of the covalent bond of C3b-/C3-small molecule complexes under physiological conditions. Ann. N.Y. Acad. Sci. 421:313 (1983).

12. T. Y. Liu. The role of sulfur in proteins in: "The proteins", H. Neurath and R. L. Hill, ed., vol. 3, p 239, Academic Press, New York (1977).

13. C. H. Hammer, G. H. Wirtz, L. Renfer, H. D. Gresham, and B. F. Tack. Large scale isolation of functionally active components of the human complement system. J. Biol. Chem. 256:3995 (1981).

14. P. K. Glasoe and F. A. Long. Use of glass electrodes to measure acidities in deuterium oxide. J. Phys. Chem. 64:188 (1960).

15. T. C. Bruice and S. J. Benkovic. Acyl transfer reactions involving carboxylic acid esters and amides in: "Bioorganic mechanisms", vol. I, p 1, W. A. Benjamin, Inc., New York (1966).

16. R. L. Schowen. Solvent isotope effects on enzymic reactions in: "Isotope effects on enzyme-catalyzed reactions", W. W. Cleland, M. H. O'Leary, and D. B. Northrop, ed., p64, University Park Press, Baltimore (1977).

17. M. H. L. de Bruijn and G. H. Fey. Human complement component C3: cDNA coding sequence and derived primary structure. Proc. Natl. Acad. Sci. USA 82:708 (1985).

XIV. Identification of Sites of Post-Translateral Modification

STRUCTURAL ANALYSIS OF CARCINOEMBRYONIC ANTIGEN (CEA) AND A RELATED TUMOR-ASSOCIATED ANTIGEN (TEX)

Raymond J. Paxton and John E. Shively

Beckman Research Institute of the City of Hope

Division of Immunology, Duarte, CA 91010

INTRODUCTION

Carinoembryonic antigen (CEA), first described by Gold and Freedman[1], is one of the most widely investigated human tumor-associated antigens. CEA was originally detected in colonic adeno-carcinoma and fetal gut, but has since been detected in other malignant and non-malignant tissues. The chemistry, biochemistry, immunology, tissue distribution and clinical aspects of CEA were recently reviewed[2]. Using anti-CEA sera, immunologically crossreacting antigens have been detected in several tissues as described in Table 1. The crossreacting antigens share at least one antigenic determinant (epitope) with CEA and for the most part share epitopes with each other.

In order to better understand the immunological crossreactivity of CEA and the other antigens we are determining the primary structures of CEA, TEX, and NCA. This report describes the techniques we are using to purify, characterize, and sequence these highly glycosylated glyco-proteins. In addition, we show that microsequence analysis can be used to directly determine sites of carbohydrate attachment glycoproteins.

Purification

CEA, TEX and NCA are extracted from liver metastases of colonic adenocarcinoma with cold 1 M perchloric acid. Subsequent purification steps include Sepharose 4B and/or Sephadex G200 chromatography, and Con-canavalin A- Sepharose chromatography. By carefully pooling the gel per-meation fractions, the three proteins can be substantially separated from one another. This method generally gives very pure CEA, however TEX and NCA often require further purification. The use of reverse-phase HPLC (RP-HPLC) has been investigated for the further purification of these antigens. Figure 1 compares Vydac C_4 RP-HPLC chromatograms of CEA and TEX. CEA elutes as a sharp single peak, however TEX elutes as broad peak and in this example contains a small amount of CEA as a contaminant. The recovery of each protein is approximately 75% and full immunological activity is retained.

A higher resolution chromatogram in which TEX is separated into three species is obtained using a Waters phenyl column (Figure 2).

Table 1. CEA Crossreacting Antigens (CEA Gene Family)

Antigen[a]	MW	Carbohydrate(%)	Tissue source
CEA	180,000	50-60%	Colon tumor[b]
			Normal colon
NCA	50,000	~30	Normal lung, spleen
			Colon tumor[b]
TEX	80,000	~45	Normal lung
			Colon tumor[b]
BGPI	83,000	~40	Bile ducts, bile
NCA-2	160,000	~50	Meconium
NFA-2	160,000	~50	Feces
NFA-1	20-30,000	~13	Feces

[a]Abbreviations are: NCA, non-specific crossreacting antigen; TEX, tumor-extracted antigen; BGPI, biliary glycoprotein I; NFA, normal fecal antigen.
[b]Isolated from liver metastases of colonic adenocarcinomas.

SDS-polyacrylamide gel electrophoresis (SDS-PAGE) shows a distinct molecular weight for each HPLC fraction (Figure 3A), and each fraction is immunologically reactive with an anti-CEA monoclonal antibody (Figure 3B). In addition, similar amino acid compositions are obtained for each fraction, however the amount of glucosamine determined by amino acid analysis is most for fraction 1 and least for fraction 3. This suggests that the separation into three species is due to differences in the amount of carbohydrate.

NCA is also separated into three species by Waters phenyl RP-HPLC (not shown). The three fractions have distinct molecular weights when analyzed by SDS-PAGE and each fraction is immunologically reactive with an anti-NCA monoclonal antibody. As with TEX, the three fractions have similar amino acid compositions, however fraction 1 contains the most and fraction 3 the least glycosamine.

Amino Acid Compositions

The amino acid and carbohydrate compositions of CEA, TEX, and NCA are presented in Table 2. The degree of relatedness is obvious. The major difference is the lack of methionine in CEA. A second difference concerns the ratio of arginine to lysine. CEA consistently analyzes with move arginine than lysine, whereas TEX and NCA contain slightly more lysine than arginine. CEA and TEX contain approximately 50% carbohydrate by weight whereas NCA contains only 30%.

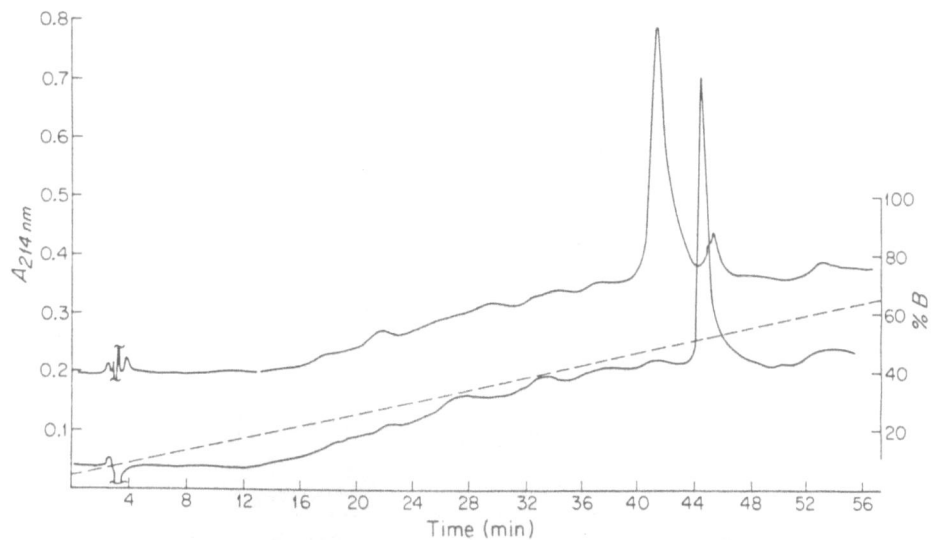

Figure 1. RP-HPLC of CEA (lower trace) and TEX (upper trace). The
samples (100-200 μg) were applied to a Vydac C_4 column and
eluted with a linear gradient, 100%A - 100%B in 90 min. A =
TFA/H_2O (0.1/99.9,v/v), B = TFA/H_2O/MeCN (0.1/9.9/90,
v/v/v).

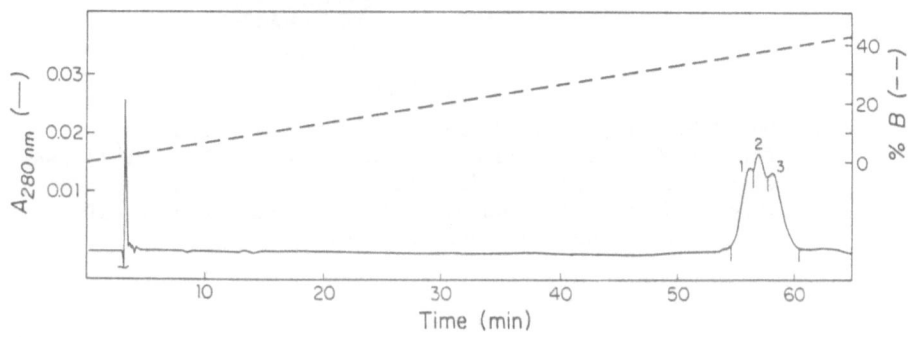

Figure 2. RP-HPLC of TEX. The sample (50 μg) was applied to a Waters
Associates μ-bondapak phenyl column and eluted with a linear
gradient, 100%A - 40%A:60%B in 90 min. The solvents are as in
Figure 1.

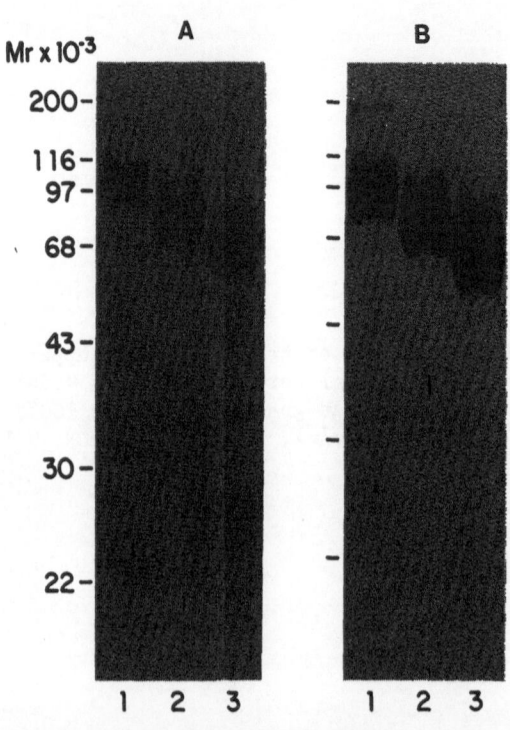

Figure 3. SDS-PAGE analysis of RP-HPLC purified TEX (Figure 2).
Duplicate samples were electrophoresed on a 10% gel[3] and
electrotransferred to nitrocellulose[4]. The proteins
were stained with amido black (A) or immunostained using
an anti-CEA monoclonal antibody (B).

Table 2. Amino Acid Compositions of CEA Gene Family Members[a]

(mole percent)

	CEA	TEX	NCA
CYS/2	• 1.7	1.6	1.4
ASX	13.4	12.7	12.2
THR	8.7	8.4	8.3
SER	10.8	9.4	9.3
GLX	10.2	11.3	11.6
PRO	8.8	7.7	7.4
GLY	5.6	6.9	8.1
ALA	5.5	6.2	6.2
VAL	8.1	7.0	7.0
MET	0	1.0	1.3
ILE	4.2	4.4	4.3
LEU	8.3	8.8	8.6
TYR	4.5	4.8	4.8
PHE	2.6	2.7	2.5
HIS	1.7	1.9	1.3
LYS	2.5	2.8	3.2
ARG	3.5	2.6	2.8
%CHO[b]	52	46	30

[a]Samples were hydrolyzed in 6N HCl containing 0.2% 2-mercaptoethanol for 48 h at 110°C. Cysteine was determined in a separate analysis after performic acid oxidation.
[b]Percent by weight. These values are estimates based on earlier studies, and on the amount of glucosamine determined during amino acid analysis.

Amino-Terminal Sequence Analysis

The amino terminal sequences of CEA and TEX were orginally determined to 25 residues by sequencing the native proteins on a modified
spinning cup sequencer. A single difference is observed at residue 21;
CEA contains valine and TEX contains alanine. By sequencing CEA and TEX
tryptic peptides (see below) the amino terminal sequences were extended
to 35 residues, as shown below:

```
              1          10          20          30
CEA    KLTIESTPF  NVAEGKEVLL  LVHNLPQHLF  GYSWYK

TEX    KLTIESTPF  NVAEGKEVLL  LAHNLPQNRI  GYSWYK
```

Additional differences are found at residues 27-29 (the differences
are underlined). The TEX sequence was confirmed by sequencing 1 nmol of
native protein on a City of Hope built gas-phase sequencer[5]. A 33%
initial yield and a 88% repetivite yield (calculated using cycles 2 and
18) was obtained. NCA (220 pmoles) was also sequenced on the gas-phase
instrument. A 55% initial yield and a 91% repetitive yield (calculated
using cycles 2 and 18) was obtained, and the sequence was identical to
that shown for TEX. In general, longer sequencer runs are obtained with
TEX and NCA. This is perhaps due to their smaller size and decreased
amount of carbohydrate compared to CEA (see Table 1).

Deglycosylation and Carboxymethylation of CEA and TEX

Proteolytic mapping of these antigens is largely unsuccessful,
probably because of their high-carbohydrate content. In order to circumvent this problem and potentially identify the glycosylation sites
directly, two methods of chemical deglycosylation are used to remove the
carbohydrate except for the ultimate N-acetylglucosamine (GlcNAc) residue
attached to asparagine (all of the carbohydrate is asparagine linked in
these antigens). The first method is deglycosylation with anhydrous HF
patterned after the original work of Mort and Lamport[6]. The sample
(1-20mg), in the Kel-F reaction vial, is dried under vacuum over P_2O_5
overnight. Anisole (1-2 ml) and a stir bar are added to the vial and it
is assembled onto a conventional HF apparatus. Freshly distilled HF is
first dried over CrF_3, and then 5-10 ml is distilled into the liquid
N_2-cooled reaction vial. The vial is sealed, warmed to 0°C, and the
reaction is stirred for 1 h. The HF is carefully removed by vacuum until
1-2 ml remain. The vial is then cooled and disconnected from the apparatus, and the remaining HF is carefully neutralized with 4-6 ml pyridine.
The sample is then transferred to a dialysis bag and dialyzed against 10%
aqueous pyridine. During the dialysis pyridinium hydrofluoride salt will
precipitate, but it will eventually redissolve. The deglycosylated antigens are generally insoluble in 10% pyridine and will precipitate from
solution. At this stage they can be collected by centrifugation or can
be resolubilized by dialysis against 10% acetic acid. However, the
deglycosylated antigens should not lyophilized or concentrated to dryness
because they are extremely difficult to redissolve.

A second deglycosylation method is patterned after Edge, et
al.[7]. The sample is dried as above in suitable vessel (polypropylene test tubes with caps or micro-centrifuge tubes work well). The
reagent is prepared by mixing 1 ml anisole with 2 ml anhydrous trifluoromethanesulfonic acid (TFMSA) at 0°C. The reagent (0.1-0.2 ml/mg of protein) is added to the protein and the solution is stirred for 5 h at 0°C.
The sample is then cooled to -70°C, diluted with 2 volumes diethylether,
and carefully neutralized by adding dropwise an equal volume of ice-cold
50% aqueous pyridine. The pyridinium TFMSA salt which forms will
dissolve when the sample is warmed to 0°C. The ether layer is removed,
the solution is re-extracted with ether, and then dialyzed against 10%
aqueous pyridine as described above. The recovery of deglycosylated
protein after this method is 70-90% with 10% retention of carbohydrate by
weight. This amount of carbohydrate is consistent with the removal of
O-glycosidically linked residues, but retention of N-glycosidically
linked GlcNAc. Figure 4 shows an SDS-PAGE analysis of DG-CEA and DG-TEX
prepared with the TFMSA reagent. The molecular weight of CEA decreases

from 180,000 to 85,000 and TEX decreases from 80,000 to 35,000. The smaller proteins are most likely generated by peptide bond cleavage, potentially at the aspartyl-proline peptide bonds known to be present in CEA and TEX. The deglycosylated antigens retain their immunological reactivity (RIA and ELISA) and can be used to produce polyclonal and monoclonal antibodies.

Based on the criteria tested to date, the two deglycosylation methods produce equivalent products. The HF method is faster but it requires specialized equipment and is potentially more hazardous. The TFMSA method is simpler, requires less equipment, and is more amenable to small amounts of samples. Less than 100 μg of TEX have been deglycosylated using 0.2 ml of reagent. Deglycosylated samples can potentially be purified by gel-permeation or reverse-phase HPLC rather than dialysis, however the recovery is poor for DG-CEA and DG-TEX.

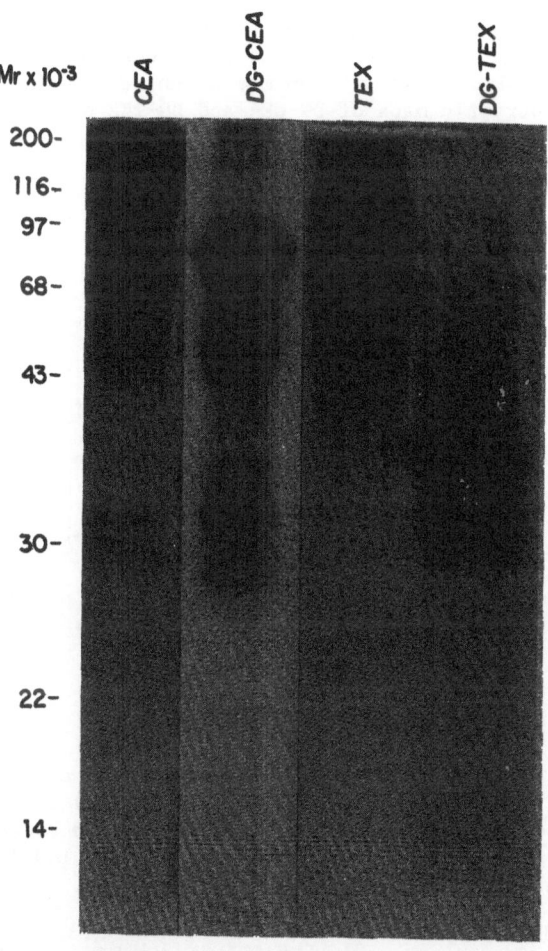

Figure 4. SDS-PAGE analysis of CEA and TEX. DG-CEA and DG-TEX were prepared using the TFMSA method described in the text. The samples were electrophoresed on a 12% gel and detected by coomassie blue staining.

DG-CEA and DG-TEX are reduced and S-alkylated using the method of Waxdal, et al.[8]. Briefly, the sample is concentrated to a small volume ($<$ 3 ml) and dialyzed against 0.5 M Tris, 0.002 M EDTA, 6 M Guanidine HCl, pH 8.1. Alternatively, the protein precipitate from the 10% pyridine dialysis (see above) is dissolved directly in this buffer. Dithiothreitol (50 moles/mole of disulfide in the protein) is added and the solution is stirred at 55°C for 4 h. After cooling to 25°C, [^{14}C]-labeled iodoacetic acid is added and the solution is stirred in the dark for 30 min. The reaction is completed by adding unlabeled iodoacetic acid (100 moles/mole of disulfide) and stirring the solution for 30 min. A large excess of dithiothreitol is then added and the protein is dialyzed exhaustively against 0.2 M $NH_4 \cdot HCO_3$, pH 7.8.

Peptide Mapping

To date, DG-CEA and DG-TEX have been digested with trypsin and chymotrypsin, and the resulting peptides have been separated by RP-HPLC on a Vydac C_{18} column. Figure 5 illustrates a tryptic map of DG-CEA. A number of well resolved peaks eluted in the early portion of thes chromatogram. These peptides were sequenced without further purification. A region of poorly resolved peptides eluted later in the chromatogram. Rechromatography of these peaks gave a few individual peptides suitable for sequencing, but in general these co-eluting peptides remain a problem. An alternative strategy is to pool these peaks and redigest with a different enzyme. This problem may be unique to the tryptic peptides because chymotryptic maps of DG-CEA and DG-TEX do not have a region of poorly resolved peptides.

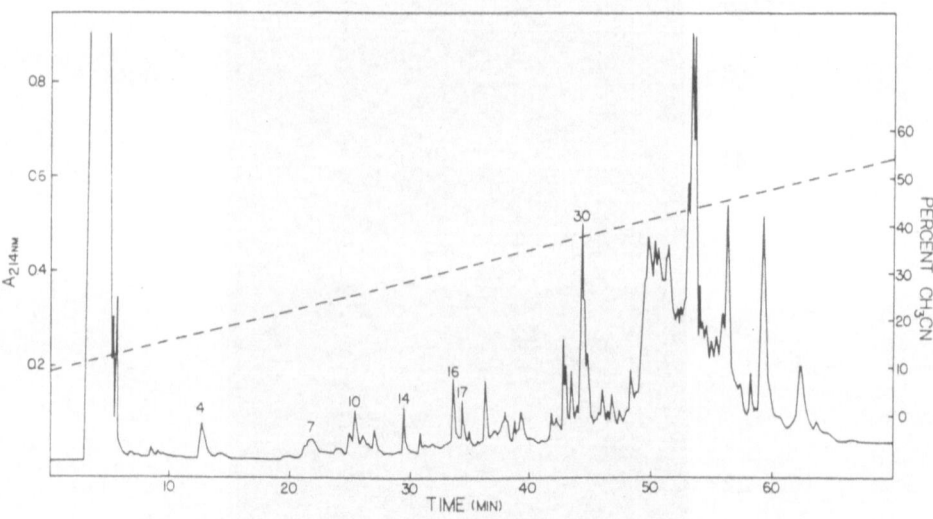

Figure 5. Tryptic map of DG-CEA. A 10nmole sample of carboxymethylated DG-CEA was incubated with 5% trypsin (wt/wt) at 37°C for 24 h. Fifty percent of the digest was applied to a Vydac C_{18} column and eluted with a linear gradient of 100%A – 30%A:70%B in 90 min. The solvents are as in Figure 1.

Cysteine-Containing Peptides

CEA and TEX cysteine-containing peptides labeled with [^{14}C]-iodoacetic acid were identified by liquid scintillation counting. The sequences of these peptides ($\sim$500 pmoles each) were determined using a gas-phase instrument[5] and are shown below:

```
TEX-1    C E T Q N P V S A R
CEA-1    C E T Q N P V S A R

TEX-2    N S G L Y T C Q A Ⓝ N S A . . .
CEA-2    N S G L Y T C Q A . . . .

TEX-3    R N D A G S Y E C I Q N P A S A Ⓝ R
CEA-3            A Y V C G I Q . . . . . . .

TEX-4    I T P N N S G T Y A C F V S N L A T G R
CEA-4    I T P N N Ⓝ G T Y A C F V S N L A T G R
```

Peptides-1 and probably peptides-2 have identical sequences, and peptides 4 have only a single difference. However, peptides-3, while apparently related, are quite different. These peptides illustrate not only the similarities that make these proteins immunologically crossreactive but also the differences that make the proteins unique. The circled asparagine residues are sites of glycosylation (see below). Suprisingly TEX-4 contains a potential glycosylation site (NNS) that is not glycosylated, however the same sequence is glycosylated in TEX-2.

The most interesting feature of the cysteine-containing peptides is their internal sequence homology, illustrated for the TEX peptides below:

```
TEX-1                         C E T Q N P V S A R
TEX-2             N S G L Y T C Q A Ⓝ N S A . . . .
TEX-3       R N D A G S Y E C I Q N P A S A Ⓝ R
TEX-4   I T P N N S G T Y A C F V S N L A T G R
```

This and other data suggests that CEA, TEX, and the other crossreacting antigens have evolved from a primitive gene sequence coding for a stretch of about 60 amino acids containing one cysteine residue. Multiple cycles of gene duplication and divergence could have given rise to a number of proteins which now comprise the CEA gene family (Table 1).

Glycosylated Peptides

As was mentioned previously, the chemical deglycosylation methods should leave a single residue of asparagine-linked N-acetylglycosamine (Asn-GlcNAc) at the sites of glycosylation. Our first experience with this derivative is illustrated by the chromatograms in Figure 6. One nanomole of a DG-CEA-tryptic peptide was sequenced on a gas-phase instrument with an initial yield of 54% (calculated at cycle 2). Cycles 2-7 are consistent with the sequence - Asp-Thr-Ala-Ser-Tyr-Lys, however the PTH-derivative for cycle 1 eluted between Glu and Asn on our PTH-HPLC system[9]. Amino acid analysis was consistent with this sequence, however one residue of Asx and one residue of glucosamine were unaccounted for in the sequence. Back hydrolysis of the cycle 1 PTH- derivative (6N HCl, 0.2% 2-mercaptoethanol, 110°C, 24 h) revealed equal amounts of Asx and glucosamine. It was concluded that the PTH-derivative was Asn-GlcNAc. The sequence of the peptide was confirmed by fast atom bombardment mass spectrometry (FAB/MS) of the intact peptide (MH^+ = 1001).

The sequences of several other CEA and TEX glycopeptides and their molecular ions as determined by FAB/MS are shown below. The glycosylated residues are circled and the peptides are alligned to illustrate their internal sequence homology.

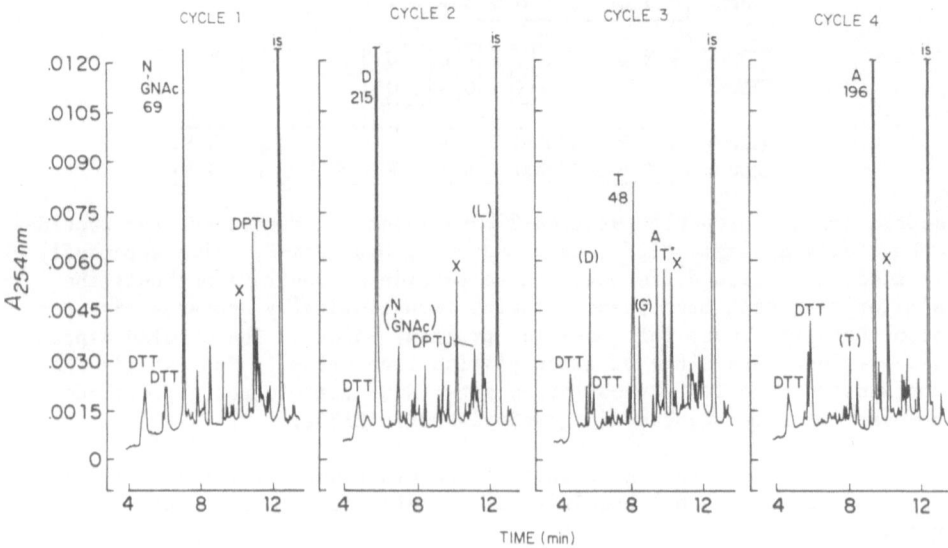

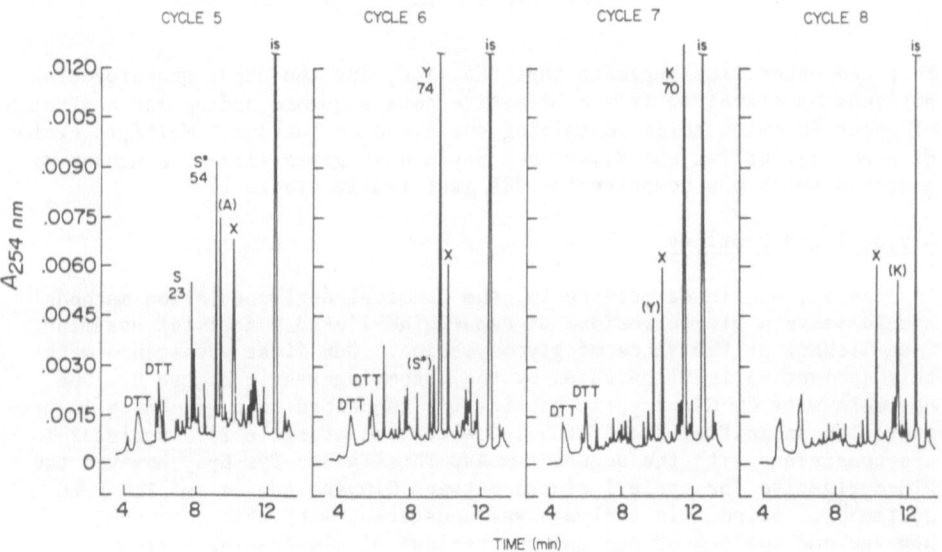

Figure 6.   Sequence analysis of a glycosylated peptide on a gas-phase
sequencer. A DG–CEA tryptic peptide (1 nmole) was sequenced
and 30% of each cycle was analyzed by RP–HPLC. The PTH–
derivative for cycle 1 was determined to be PTH-Asn-GlcNAc;
the other cycles are as labeled (see text for details). The
background peak (X) is occasionally observed on sequence
analysis and has not been chemically identified. DTPU–
diphenylthiourea, DTT–dithiothreitol, is–internal standard.

	MH+
Ⓝ D T A S Y K	1001
Ⓝ D T A S Y K	1204
T L T L F Ⓝ V T R	1268
Ⓝ V T R Ⓝ D T A S Y	1547
F V P Ⓝ I T V Ⓝ N S G T Y	1832
I P Ⓝ I T V Ⓝ N S G S Y	1685
R N D A G S Y E C E I Q N P A S A Ⓝ R	2356
T C E P E I Q Ⓝ T T Y	1560
T C E P E V Q Ⓝ T T Y	1546

The first two peptides have the same sequence yet differ by 203 mass units. This is because the second peptide contains asparagine-linked di-N-acetylchitobiose (Asn-GlcNAc-GlcNAc) due to incomplete deglycosylation at this site. A PTH-derivative for this residue was not observed. A potential explanation is that the ATZ-derivative of this residue is not extracted by butyl chloride and hence not delivered to the conversion flask of the sequencer. For the other sequences listed, a PTH-derivative for Asn-GlcNAc was observed at the indicted cycle. The presence of this amino acid derivative did not adversely affect the sequencing of these and other glycopeptides. The yield of PTH-AsnGlcNAc is often as high as 50% of the expected yield and probably varies depending on the extent of deglycosylation and the amount of ATZ-AsnGlcNAc extracted by butyl chloride.

Synthesis of PTH-AsnGlcNAc

The PTH-derivative of Asn-GlcNAc was synthesized so it could be included in the standard mix of PTH amino acids. Briefly, 2-acetamido-1-β-(L-aspartamido)-1, 2-dideoxy-D-glucose (Asn-GlcNAc, 10μmoles, Sigma) was dissolved in 0.05 ml triethylamine/pyridine/H_2O, 10/40/50 (v/v/v). Phenyl isothiocyanate (20μmoles, Pierce sequanal grade) was added and the reaction was stirred at 43°C for 30 min. The reaction was extracted 4 times with 0.05 ml of benzene and dried in a vacuum centrifuge. The yellow oil was dissolved in 0.1 ml of 25% trifluoroacetic acid/75% H_2O and stirred at 50°C for 20 min. PTH-AsnGlcNAc precipitated from solution at the end of the 20 min. The product was solubilized in 50% MeCN/50%H_2O and analyzed on our PTH-HPLC system [9]. A single peak was observed and the yield of the reaction was estimated to be 100%. FAB/MS analysis of the product gave the expected molecular ion, MH+ = 453. Amino acid hydrolysis of the product (6 N HCl, 0.2% 2-mercaptoethanol, 110°C, 24 h) gave the expected composition of Asx and glucosamine.

PTH derivatives of glutamic acid, carboxymethylcysteine (CMC), and asparagine were also synthesized and chromatographed with PTH-AsnGlcNAc on the PTH-HPLC system. The results are presented below:

Derivative	Retention Time (sec)
PTH-CMC	451
PTH-Glu	459
PTH-AsnGlcNAc	495
PTH-Asn	530

Commercially available PTH-Glu and PTH-Asn eluted at 458 and 529 sec, respectively. Clearly, PTH-AsnGlcNAc is easily distinguishable from its neighboring PTH derivatives, and its analysis does not present any problems.

SUMMARY

Highly glycosylated glycoproteins offer a unique challenge to the protein chemist. In general, these proteins are resistent to the proteolytic enzymes used to generate peptides suitable for sequence analysis. We have circumvented this problem by chemically deglycosylating CEA and TEX, leaving them susceptible to proteolytic degradation. In addition, this method leaves a single residue of asparagine-linked GlcNAc at each site of glycosylation. We have shown that Asn-ClcNAc residues behave normally during microsequence analysis and have unequivocally identified PTH-AsnGlcNAc in a number of peptide sequencing runs.

The amount of protein sequence data now available has been increased dramatically by the techniques of molecular cloning and DNA sequence analysis. Although these are powerful techniques, they can only identify potential sites of post-translational modification, including glycosylation. Using the techniques outlined in this paper, we are now able to directly determine sites of glycosylation and hence, gain a better understanding of the structures of highly glycosylated glycoproteins.

REFERENCES

1. Gold, P. and Freedman, S.O., Demonstration of Tumor-Specific Antigens in Human Colonic Carcinomata by Immunological Tolerance and Absorption Techniques, J. Exp. Med. 121:439-462 (1965).
2. Shively, J.E., and Beatty, J.D. CEA-Related Antigens:Molecular Biology and Clinical Significance CRC Critical Reviews in Oncology/Hematology 2:355-399 (1985)
3. Laemmli, U.K. Cleavage of Structural Proteins during the Assembly of the Head of Bacteriophage T4, Nature 227:680-685 (1970)
4. Towbin, H., Staehelin, T., and Gordon, J. Electrophoretic Transfer of Proteins from Polyacrylamide Gels to Nitrocellulose Sheets: Procedure and Some Applications, Proc. Natl. Acad. Sci. USA, 76:4350-4354 (1985)
5. Hawke, D.H., Harris, D.C., and Shively, J.E. Microsequence Analysis of Peptides and Proteins V. Design and Performance of a Novel Gas-Liquid-Solid Phase Instrument, Analytical Biochemistry, 147:315-330 (1985)
6. Mort, A.J. and Lamport D.T.A. Anhydrous Hydrogen Fluoride Deglycosylates Glycoproteins, Analytical Biochemistry 82:289-309 (1977)
7. Edge, A.S., Faltynek, C.R., Hof, L., Reichert, Jr., L.E., and Weber, P., Deglycosylation of Glycoproteins by Trifluoromethanesulfonic Acid, Analytical Biochemistry 118:131-137 (1981)
8. Waxdal, M.J., Konigsberg, W.H., Henley, W., and Edelman, G.M., The Covalent Structure of Hyman G-Immunoglobulin. II. Isolation and Characterization of the Cyanogen Bromide Fragments Biochemistry 7:1959-1966 (1968)
9. Hawke, D.H., Yuan, P.-M., and Shively, J.E., Microsequence Analysis of Peptides and Proteins II. Separation of Amino Acid Phenylthiohydantoin Derivatives by High-Performance Liquid Chromatography on Octadecylsilane Supports, Analytical Biochemistry 120:302-311 (1982)

DETERMINATION OF THE LOCATION OF N^G,N^G-DIMETHYLARGININE

IN A GLYCINE-RICH REGION OF NUCLEOLAR PROTEIN C23

Mark O. J. Olson, Tamba S. Dumbar, S. V. V. Rao*
and Michael O. Wallace

Department of Biochemistry, The University of Mississippi Medical
Center, 2500 North State Street, Jackson, MS 39216-4505
*Department of Biochemistry, Louisiana State University, Baton
Rouge, LA 70803

INTRODUCTION

A variety of nonribosomal, nonhistone proteins are involved in the organization of the nucleolus and in the assembly of ribosomes. One of the more extensively characterized polypeptides, designated C23 by Orrick et al.,[1] has a molecular weight of approximately 110,000 and has properties which suggest that it performs these functions; e.g., it is predominantly localized to the fibrillar regions within the nucleolus,[2,3] it has a high affinity for silver as do the chromosomal nucleolus organizer regions (NORs)[4] and has been localized at the NORs[5].

Protein C23 possesses some unusual structural features. It contains a large number of phosphoryl groups located in regions which are highly negatively charged by virtue of long polyacidic stretches[6] found in the amino terminal half of the molecule[7]. In addition, Lischwe et al.[8] recently found that protein C23 contains a substantial amount of N^G,N^G-dimethylarginine (DMA).

This investigation was initiated to determine approximately where in the molecule the methylated arginine residues are located. By use of selective chemical cleavage, HPLC purification of fragments, amino acid analyses and partial sequencing it was found that virtually all of the DMA residues are located within a glycine-rich segment near the carboxyl terminal end of the molecule. Since the details of the sequence determination are to be published elsewhere[9] this paper will concentrate on the special problems encountered and the methods used for dealing with them.

MATERIALS AND METHODS

Protein C23 was prepared from acid extracts of Novikoff hepatoma nucleoli by a method similar to that previously described[7] with minor modifications[9]. All preparations contained greater than 95% protein C23 by SDS polyacrylamide gel electrophoretic (PAGE) analyses[10]. Cleavage with N-bromosuccinimide (NBS) was done essentially as previously described[7] with a ratio of NBS to tyrosine of approximately 20. Cyanogen bromide (CNBr) cleavage was performed on samples of protein C23 (10 mg/ml) in 70% formic acid with a 100 fold excess of CNBr over methionine for 18 hr at 25°.

711

N-chlorosuccinimide cleavage was done essentially according to the method of Lischwe and Ochs[11]. Reaction mixtures were diluted with water, freeze dried and either subjected to purification by reverse phase high performance liquid chromatography (HPLC) or applied directly to SDS-PAGE gels. HPLC was done on a Perkin Elmer 3B liquid chromatograph using a Synchropak RP-P column (Synchrom, Linden, IN 47955) essentially as described previously using the trifluoroacetic acid-acetonitrile system of Mahoney and Hermodson[12]. Fragments were monitored for purity with Laemmli type SDS-PAGE and molecular weights were estimated by that system or by the gel system described by Burr and Burr[13] using low molecular weight markers from Bethesda Research Laboratories (Gaithersburg, MD 20877). Purified fragments were hydrolyzed in 5.7 N HCl at 110° for 22 hr and analyzed on a Beckman 119CL amino acid analyzer. The N^G,N^G-dimethylarginine standard was obtained from Calbiochem-Behring (La Jolla, CA 92307).

Sequencing of the fragments was done with a Beckman 890C Sequencer using the dilute Quadrol program essentially as previously described[7]. Identification of residues was done with a Varian 5020 HPLC using either the Beckman-Altex dedicated PTH amino acid analysis kit or a Waters Nova-Pak C18 column using a solvent system similar to the one recommended by Waters for PTH amino acid separations. The latter system employed solvents A (35 mM sodium acetate, pH 4.95 : acetonitrile, 5:1, v:v) and B (2-propanol:water, 3:1, v:v) using essentially a convex exponential gradient from 0 to 40% B over a 10 minute period at 39°. PTH-N^G,N^G-dimethylarginine (PTH-DMA) was prepared from DMA after first separating the DMA from its chromophoric counterion on reverse phase HPLC using the TFA-acetonitrile system used for peptide purifications (see above). The PTH derivative was prepared after first forming the anilinothiazolinone essentially according to the method of Mendez and Lai[14] and then converting to PTH DMA with aqueous acid. PTH-DMA had a retention time slightly longer than PTH-arginine using the Waters identification system.

RESULTS AND DISCUSSION

The task of locating the previously discovered DMA residues in protein C23[8] involved at least five areas of protein chemistry: 1) analyses and quantitation of DMA, 2) obtaining fragments which contained DMA, 3) determination of the sizes of the fragments, 4) locating the DMA containing fragments in the sequence and 5) sequencing the fragments. Since conventional methods have generally been used for these studies, we will focus on a number of special problems that have been encountered.

Analyses of methylated arginines. The methods for separation and quantitation of the various methylated arginine derivatives have generally been worked out[8,15] making this part of the task relatively easy. On the standard Beckman 119CL hydrolysate system the unsymmetrical (N^G,N^G_-) derivative elutes between ammonia and arginine, symmetrical (N^G,N^G_-) isomer slightly later and the monomethyl derivative emerges slightly before arginine. The use of a color factor equivalent to arginine as indicated by Paik and Kim[16] was confirmed to be generally satisfactory by later sequence studies. Protein C23 contained an average of 6.7 residues of unsymmetrical DMA, less than one residue of monomethyl arginine and no detectable symmetrical DMA[8,9]. However, there was a considerable amount of preparation-to-preparation variability in the DMA content of the protein; i.e., as few as five residues were found in some preparations while as many as nine residues were found in others.

DMA-containing fragments. To locate segments containing methylated arginine residues protein C23 was cleaved with specific chemical reagents, followed by HPLC purification and amino acid analyses. Cleavage was done initially with NBS since fragments produced by this reagent have been partially characterized and subsequently with CNBr and with N-chlorosuccinimide (NCS). The NBS and NCS digests produced a distinct, rapidly migrating fragment with an apparent molecular weight of 12,000 by Laemmli type PAGE analyses. Since the fragmentation pattern by NBS was difficult to control, subsequent digests were done with NCS (Fig. 1A) which presumably cleaves only at tryptophan. A slightly larger fragment, designated CB5, was produced by CNBr. Although the small fragments from any of the digests could be purified by HPLC most of the work was done on the CNBr digest, because of the relative ease of purification (Fig. 1B).

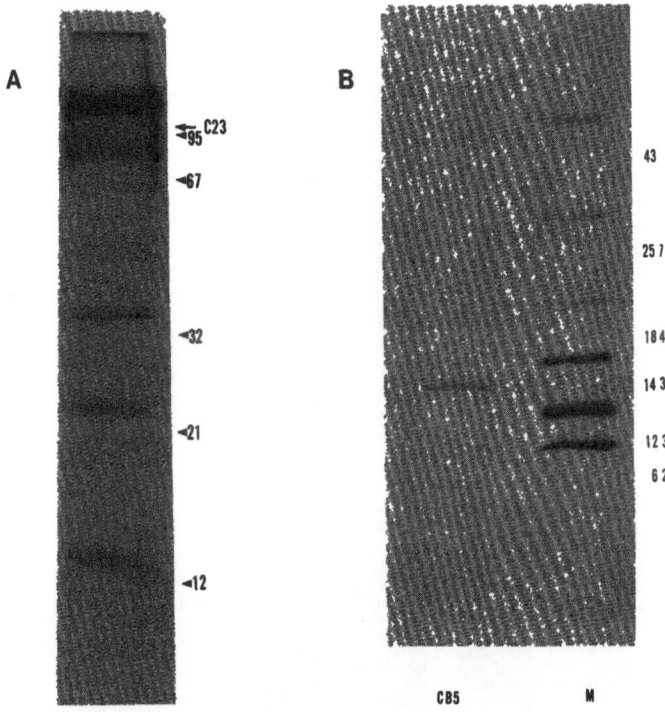

Figure 1. Cleavage of protein C23 by N-chlorosuccinimide (NCS) and cyanogen bromide (CNBr). A, protein C23 was treated with 0.05 M NCS according to the procedure of Lischwe and Ochs[11] and applied to a 15% Laemmli type gel. Approximate molecular weights of fragments in kDa are given at right. B, the purified CB5 fragment from the CNBr digest was run on the same gel system. Lane M contains molecular weight markers designated by the numbers (in kDa) on the right.

When the various HPLC fractions were subjected to amino acid analyses only the small fragments contained significant quantities of DMA. In fact, the CB5 fragment contained approximately the same number of DMA residues as did the parent molecule, suggesting that essentially all of the DMA residues were contained in this fragment. In addition to the DMA content the most striking feature was the high glycine content, accounting for nearly 50% of the residues. The composition based on the estimated molecular weight (see below) is:

$Asx_4, Thr_2, Ser, Glx_6, Pro_2 Gly_{37}, Ala_3, Val, Ile, Leu, Phe_6, Lys_7, DMA_7, Trp, Arg_2$.

Estimation of fragment sizes. By Laemmli gels the small fragments appeared to be in the order of 12-13 kDa. However, when we attempted to account for the tryptic peptides present in the CB5 fragment and cDNA information became available[9] it was clear that the earlier estimates were too high[13] We then ran the NCS and CNBr digests on the gel system of Burr and Burr[13] and found that the NCS 12 and CB5 fragments were indeed much smaller, 6.5 and 7.9 kDa, respectively (Fig. 2). Although the cause of the anomalous behavior of the small fragment on Laemmli type gels is not readily apparent, it seems likely that it is related to the high glycine content.

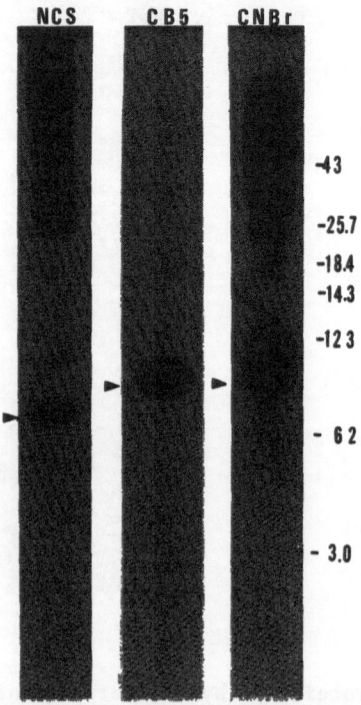

Figure 2. Estimation of fragment sizes by Burr and Burr gel system. CNBr and NCS digests were subjected to electrophoresis according to the procedure of Burr and Burr[13]. Molecular weights (in kDa) of standards are given on right. Estimated molecular weights of smallest fragments from NCS and CNBr digests are 6500 and 7900, respectively.

Location of the DMA containing fragment. Previous studies indicated that the N-terminal half of the molecule is contained in a 60 kDa fragment[7]. This polypeptide was found to contain only trace quantities of DMA indicating that essentially all of the DMA is in the C-terminal half of the molecule. By combining this information with the approximate sizes of the fragments generated by NCS it was possible to develop a map of the arrangement of fragments (Fig. 3). Because of the absence of tryptophan in the N-terminal half, under partial cleavage conditions all NCS cleavage points must be in the C-terminal half of the molecule. Therefore the 95 kDa fragment (Fig. 1A) must be at the N-terminus and the small fragment is the only one which can be placed at the C-terminus (Fig. 3). Although the molecular weights in figure 3 are from the original estimates the arrangement is not altered by the lower molecular weights of the fragments. Thus, the DMA-containing segment is at the C-terminal end of the molecule.

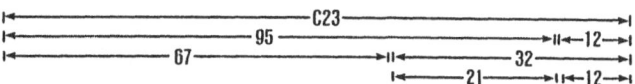

Figure 3. Map of NCS fragment alignment. Estimated sizes of fragments as shown in Fig. 1 are given in kDa. The arrangement of the fragments is based on the sizes of the fragments and the cleavage pattern by N-bromosuccinimide (NBS) in previous studies. The 12 kDa fragment is probably much smaller than indicated (see Fig. 2) although this does not change the alignment pattern.

Sequencing the DMA containing region. We had initiated sequencing studies on several fragments of protein C23 prior to the time that we knew these fragments contained DMA. In these studies, certain residues could not be identified by Edman degradation using the Beckman-Altex PTH amino acid separation kit. In this system, PTH-arginine is the last derivative to elute. When we discovered that these fragments contained DMA we re-examined the HPLC tracings and were still not able to find any evidence of unusual PTH amino acids. We then attempted to prepare a PTH-DMA standard and found that the commercially available DMA contained a counterion, namely, di(ρ-hydroxyazobenzene-ρ'-sulfonate). Since this chromophore interfered wth HPLC identification of the PTH derivative it was necessary to remove this prior to reaction with phenylisothiocyanate. This was achieved by chromatographing the derivative on the same reverse phase HPLC system used for peptide separations[12], exchanging the counterion with trifluoracetic acid. In the meantime we had begun using the Waters Nova-Pak PTH amino acid separation system in which PTH-arginine appears in the middle of the run. We found that PTH-DMA had a retention time slightly longer than PTH-arginine, but clearly separating from it. We were now able to identify the residues of DMA in the fragments during Edman degradation. The PTH-DMA was most likely missed using the earlier system because the HPLC recorder was cut off prior to the elution of PTH-DMA.

By submitting the CB5 fragment and the NBS (or NCS) small fragment and subfragments thereof to Edman degradation we have obtained a partial sequence of this region of the rat C23 molecule: Glu-Asp-Gly-Glu-Ile-Asp-Gly-Asn-Lys-Val-Thr-Leu-Asp-Trp-Ala-Lys-Pro-Lys-Gly-Glu-Gly-Gly-Phe-Gly-Gly-DMA-Gly-Gly-Gly-DMA-Gly-Gly-Phe-Gly-Gly-DMA-Gly-Gly-Gly-DMA-Gly-Gly-Gly-DMA-Gly Gly-. From these studies it was also found taht the NCS small fragment is contained in the CB5 fragment and begins at the alanine residue. It is interesting to note that all of the DMA residues placed thus far are surrounded by at least two glycine residues and there is a relatively high abundance of phenylalanine residues nearby. The complete sequence of this region from the hamster protein as deduced from the cDNA sequence is also available and contains a segment which is identical to the above sequence[9].

Protein C23 has an unusual distribution of structurally unique features (Fig. 4) with phosphorylated acidic regions on one end of the molecule and the glycine-rich, DMA-containing region on the other. Although the significance of this is not clear, the differential modification suggests that the two ends participate in different kinds of interactions with other molecules. For example, the N-terminal end may bind histones or ribosomal proteins while the C-terminal region may be a domain for interaction with RNA. It is known that heterogeneous nuclear RNA binding proteins also have a high glycine content and contain DMA[15-18]. It will be interesting to compare sequences from protein with sequences from the latter group of proteins when they become available.

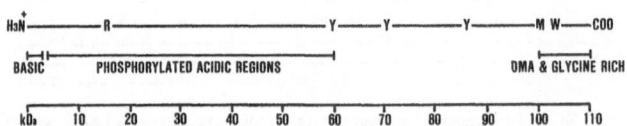

Figure 4. Structural features of protein C23. The approximate locations of various structurally unique segments were determined by a combination of fragmentation methods (NBS, NCS, CNBr, and submaxillaris protease) together with partial sequencing. Letters (single letter code) indicate cleavage points by various agents.

ACKNOWLEDGMENT: Supported by N.I.H. Grant GM28349.

REFERENCES

1. L. R. Orrick, M. O. J. Olson and H. Busch, Proc. Natl. Acad. Sci. U.S.A.. 70:1316 (1973).
2. K. Smetana, R. Ochs, M. A. Lischwe, F. Gyorkey, E. Freireich, V. Chudomel and H. Busch, Exptl. Cell. Res. 152:195 (1984).
3. M. L. Escande, N. Gas and B. J. Stevens, Biol. Cell 53:99 (1985).
4. Lischwe, M., K. Smetana, M. O. J. Olson and H. Busch, Life Sciences 25:701 (1979).
5. M. A. Lischwe, R. L. Richards, R. K. Busch and H. Busch, Exptl. Cell. Res. 136:101 (1981).
6. M. D. Mamrack, M. O. J. Olson and H. Busch (1979) Biochemistry 18:3381 (1979).
7. S. V. V. Rao, M. D. Mamrack and M. O. J. Olson, J. Biol. Chem. 257:15035 (1982).
8. M. A. Lischwe, K. D. Roberts, L. C. Yeoman and H. Busch, J. Biol. Chem. 157:14600 (1982).
9. B. Lapeyre, F. Amalric, S. H. Ghaffari, S. V. V. Rao, T. S. Dumbar and M. O J. Olson, manuscript submitted (1985).
10. U. K. Laemmli, Nature (Lond.) 227:680 (1970).
11. M. A. Lischwe and D. Ochs, Anal. Biochem. 127:453 (1982).
12. W. C. Mahoney and M. A. Hermodson, J. Biol. Chem. 225:11199 (1980).
13. F. A. Burr and B. Burr, Meth. Enzymol. 96:239 (1983).
14. E. Mendez and C. Y. Lai, Anal. Biochem. 68:47 (1975).
15. J. Karn, G. Vidali, C. L. Boffa and V. G. Allfrey, J. Biol. Chem. 252:7307 (1977).
16. T. E. Martin, P. Billings, A. Levey, S. Oroszlan, T. J. Quinlan, H. W. Swift and L. Urbas, Cold Spring Harbor Symp. Quant. Biol. 38:921 (1974).
17. D. K. Marvil, L. Nowak and W. Szer, J. Biol. Chem. 255:6466 (1980).
18. A. L. Beyer, M. E. Christensen, B. W. Walker and W. M. LeStourgeon, Cell 11:127 (1977).
19. W. K. Paik and S. Kim, "Protein Methylation," John Wiley and Sons, New York (1980).

STRUCTURAL CHARACTERIZATION OF A MURINE LYMPHOCYTE HOMING RECEPTOR

SUGGESTS A UBIQUITINATED BRANCHED-CHAIN GLYCOPROTEIN

Mark Siegelman+, Martha Bond# and Irving L. Weissman+

+Laboratory of Experimental Oncology
Department of Pathology
Stanford University School of Medicine
Stanford, CA 94305

#DNAX Research Institute
901 California Avenue
Palo Alto, CA 94306

Summary

   Partial amino acid sequence analysis of a purified lymphocyte homing receptor demonstrates the presence of two amino-termini, one of which corresponds precisely to the amino-terminus of ubiquitin. This observation extends the province of this extraordinarily conserved polypeptide to the cell surface, and leads to a proposed model of the receptor complex as a ubiquitinated branched-chain glycoprotein. Functional binding of lymphocytes to high endothelial venules (HEV) requires the accessibility of the ubiquitinated region of the receptor, suggesting a possible central role for ubiquitin in cell-cell interaction and adhesion.

Introduction

        The dynamism of the circulating lymphoid system is relieved by scattered solid collections of lymphoid elements such as thymus, lymph nodes, Peyer's patches, and spleen, which together constitute the lymphoid organs. These organs are architecturally organized so that proper inductive microenvironments insure appropriate lymphocyte differentiation and maturation of the immune response. An initial and fundamental event required for an appropriate progression of the immune response resides at the interface between a lymphocyte's mobile circulating phase and its relatively sessile phase within a particular lymphoid organ. The specific portal of entry of lymphocytes from the blood stream into peripheral lymphoid organs has been identified as specialized lymphoid organ vessels, called postcapillary high endothelial venules (HEV)(1-3). Lymphocytes specifically recognize, adhere to, and migrate through this highly specialized endothelium.

        The fundamental role of HEV-lymphocyte interaction in lymphocyte trafficking has been demonstrated (4,5). Peripheral node lymphocytes exhibit binding preference for peripheral node HEV's, while Peyer's patch lymphocytes favor binding to homologous Peyer's patch HEV (6,7). Some clonal murine T and B lymphoma lines display exquisite specificity for either peripheral node or Peyer's patch-type HEV, while others recognize neither venule type (8). Together these data are consistent with a model involving complementary cell surface recognition structures mediating organ-specific lymphocyte HEV interactions.

A crucial advance addressing more directly the existence of such cell surface recognition structures was provided by the development of the rat monoclonal antibody, MEL-14 (8). This antibody detects a cell surface determinant present only on those T and B cell lines which bind peripheral lymph node HEV is absent on those which either bind Peyer's patch HEV only or have no HEV binding activity at all (8). Furthermore, presaturation with MEL-14 of either normal or neoplastic lymphocyte populations which normally bind peripheral node HEV completely ablates specific binding to these HEV's. MEL-14 specifically precipitates from the cell surface of lymph node HEV binding lymphocytes and lymphoma cell lines a protein band of 85-95,000 daltons molecular weight, the size varying slightly with the source of antigen. This property has made feasible the examination of this sytem in fine detail at the molecular and genetic levels.

In this and additional work, we demonstrate that MEL-14 binds a ubiquitin dependent-determinant on the homing receptor protein. Primary structural analysis of various ubiquitins and their encoding genes has revealed remarkable evolutionary conservation, with identical amino acid sequence existing between organisms as distant evolutionarily as insects and man (9-12). Such rigid conservation bespeaks a fundamental role for ubiquitin in cellular function. Independent investigations of the protein structure of a nucleosomal protein A24 revealed ubiquitin to be a component of a branched chain complex with two amino-termini and one carboxy-terminus (13). The linkage is a covalent bond between the carboxyl group of the carboxy-terminal glycine of ubiquitin in isopeptide linkage to the epsilon-amino group of lysine 119 of histone 2A. Subsequently, covalent conjugation of ubiquitin to cytoplasmic proteins was demonstrated, further generalizing the principle (14,15).

Our findings have indicated that MEL-14 screening of gt11 libraries results in the isolation exclusively of independent expressed cDNA clones encoding ubiquitin (16). This suggested that ubiquitin might in fact be present as all or a portion of this cell surface interactive molecule, and would establish a precedent for the presence of ubiquitin or ubiquitinated proteins at the cell surface.

In this manuscript we have isolated by antibody affinity and SDS gel purification the cell surface species recognized by MEL-14, which we designate gp90 MEL-14, for structural analysis. By partial amino-terminal amino acid sequence analysis using instrinsic labeling techniques, we demonstrate the presence of two amino-termini, one of which corresponds precisely with the amino-terminus of ubiquitin. These findings extend the locus of this highly conserved polypeptide to the third major cellular compartment, the cell surface, and invite investigations into its role at this site.

Results

Isolation and purification of the MEL-14 antibody-defined cell-surface glycoprotein, gp90MEL-14

The cell line utilized for these studies was EL-4/MEL-14hi, a variant of the continuous T cell lymphoma cell line, EL-4, selected by fluorescence flow cytometry for high level expression of the MEL-14 antigen, a property which cosegregated with the capacity to bind peripheral node venules (17).

The molecular species specifically recognized by MEL-14 antibody was isolated by immune complex formation between MEL-14, an IgG2a rat monoclonal antibody (8), and affinity purified rabbit or goat anti rat IgG antibody. Figure 1A shows the profile of an immunoprecipitation from cell surface ^{125}I-iodinated cell lysates of EL-4/MEL-14hi, similar to that shown in a previous report (8).

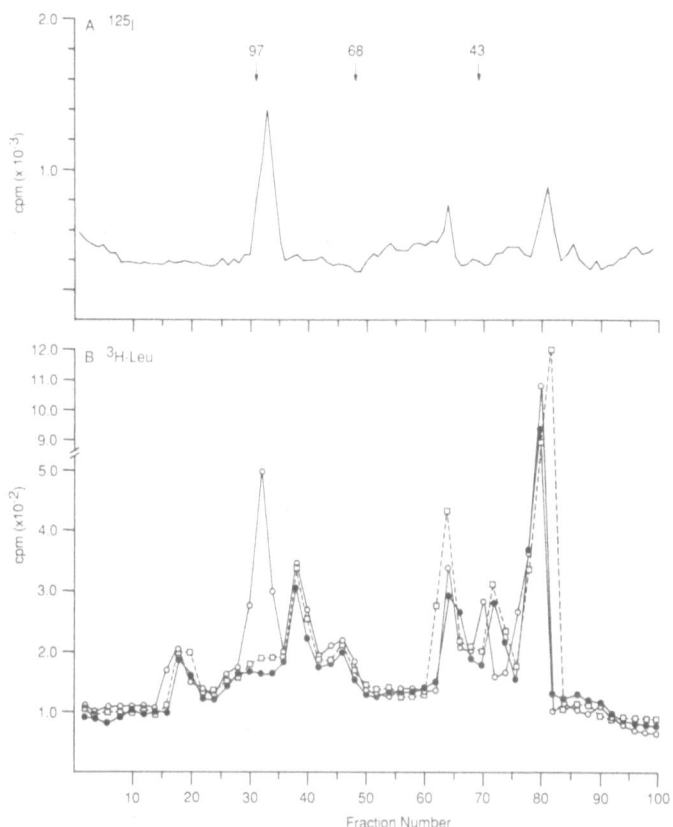

Figure 1. Immunoprecipitation and SDS-PAGE analysis of the putative lymphocyte homing receptor, gp90MEL-14. A) Immunoprecipitation of cell surface 125I-iodinated EL4/MEL-14hi by MEL-14 antibody. 2×10^7 cells were surface radioiodinated by lactoperoxidase catalysis (37), then solubilized in 2 mls PBS containing 1% Triton X-100, 0.5% deoxycholate, 0.1% SDS, 0.1M NaCl, 0.01 M Na phosphate, pH 7.5, and 5 mM PMSF according to the method of Witte, et al. (38) and clarified by ultracentrifugation (30 minutes at 30.000 rpm). The lysate was incubated with a 20X concentrated MEL-14 hybidoma supernatant (8), equivalent to 10-20 micrograms of monoclonal antibody, for 3-4 hours at 4oC, followed by the addition of a four-fold excess over first stage of affinity purified goat anti-rat IgG, incubated overnight at 4oC to effect formation of a solid precipitate. The precipitate was centrifuged at 3,000 rpm, and washed three times in 0.01 Tris-HCL, pH 7.4, 0.15 M NaCl, 0.2% Nonidet P40. Remaining complexes were solubilized by heating to 90o for 3 minutes in Laemmli sample buffer, and analyzed on 10% SDS polyacrylamide tube gels in the Laemmli discontinuous gel system (39), as modified by Cullen (40). The profile was obtained by gel fractionation at 1mm intervals followed by counting. Gel fraction number is plotted against counts per minute in each fraction. B) Immunoprecipitation by MEL-14 antibody of EL4/MEL-14hi cells metabolically labeled with 3H-Leucine. 2×10^8 cells were labeled with 10 mCi of 3H-Leucine as previously described (41). Briefly, cells were harvested in rapid growth phase and placed in culture at 10^7 cells/ml for 4-6 hours in Spinner balanced salt solution (Gibco), 10% fetal calf serum, supplemented with all amino acids except leucine, which was added only as isotope at 200 Ci/ml. Cells were washed and solubilized in 0.5% Nonidet P40, 20mM PMSF, for one hour at 4oC. Nuclei and debris were removed by centrifugation 15

minutes at 3,000 rpm. The lysate was applied to a column of Lens culinaris lectin conjugated to Sepharose 4B equilibrated in 0.01 M Tris-HCl, pH 7.4, 0.15M NaCl, 0.25% Nonidet P40. The glycoprotein enriched pool was eluted with 0.3 M methyl,-D-mannopryranoside. Precipitation of 5 X 106 cell equivalents and SDS-PAGE analysis was as described above in Fig. 1A. Gel fractions were incubated in 0.1% SDS overnight at 4oC to elute radioactivity. Radioactivity was counted in Biofluor scintillation fluid (New England Nuclear) in a Beckman LS counter (Model LS-230). Symbol o, precipitation with MEL-14; and precipitation with IgG2a isotype-matched rat monoclonal antibody controls ● 30G12 (anti-T200), and □ 9B5 (produced against a human lymphocyte surface marker (18), gift of E. Butcher) Molecular weight markers: phosphorylase b, 97,400; bovine serum albumin, 68,000; ovalbumin, 43,000.

To isolate the MEL-14 reactive antigen directly from whole cell lysates labeled internally with 3H-leucine, 35S-methionine, or 35S-cysteine, we interposed a lentil-lectin enrichment step before immunoprecipitation, thereby removing 90-95% of total lysate cpm. Figure 1B shows typical gel profiles of a MEL-14 antibody immunoprecipitation from 1.5 x 106 cpm of a lectin adherent pool of an EL-4/MEL-14hi lysate labeled metabolically with 3H-leucine. While a number of nonspecific bands are present, a single specific species of molecular weight 94,000, identical in size to that precipitated from cell surface iodinated lysates is found. This species, gp90MEL-14, was routinely eluted from appropriate gel fractions of immunoprecipitates made from approximately 2-3 x 108 cells metabolically labeled singly with a variety of tritiated amino acids, and 35S-methionine.

An aliquot of the gp90MEL-14 thus isolated was subjected to two dimensional gel analysis (data not shown). This internally labeled material migrates with the identical isoelectric mobility as cell surface iodinated gp90MEL-14 (18), in the pH range 4.0-4.5, establishing further identity between the MEL-14 antigen isolated from cell surface iodinated an internally labeled preparations. The microheterogeneity in the internally labeled material appeared as extensive as that in the iodinated material, with a pattern consistent with glycosylation differences. This indicates that the material isolated from internally labeled lysates for amino acid sequence analysis is unlikely to represent a selected subset of the antigen isolated from cell surface iodinated material. Nor did it appear, by these criteria, to be detectably more complex than the iodinated material.

Partial amino-terminal amino acid sequence analysis of the gp90MEL-14 lymphocyte homing receptor

Cells were labeled as described in Figure 1. A number of different essential amino acids were used. In a typical experiment, 20-50,000 cpm of affinity and gel purified material were subjected to automated amino-terminal Edman degradation (19, See Figure 2). Positions containing radioactivity significantly above background indicated the presence of the particular tritiated amino acid residue at that position within a polypeptide chain. Amino acid sequence determination of metabolically labeled gp90MEL-14 indicated the presence of two types of amino termini, easily distinguishable by their relative initial yields. A demonstration of this is given in Figure 2 for the purified gp 90MEL-14 labeled with 3H-lysine. The predominant sequence identified is reflected by radioactive peaks at positions 8, 17, 20, and 32. A reproducible secondary sequence is also present at a lower specific activity at positions 6, 11, 27, and 29. This secondary sequence corresponds precisely with the amino-terminal sequence of ubiquitin (9). The pattern of a major and minor amino-terminal sequence in absolute yields and relative proportions similar to that illustrated in Fig. 2 has been obtained with all amino acid labels. A summary of the data for this sequence is given in Figure 3 in comparison to the known amino-terminal sequence of ubiquitin. It is of note that in searches of known protein and nucleic acid data files, the presence of leucines at positions 8 and 15 and the valine at position 5 were sufficient to select exclusively

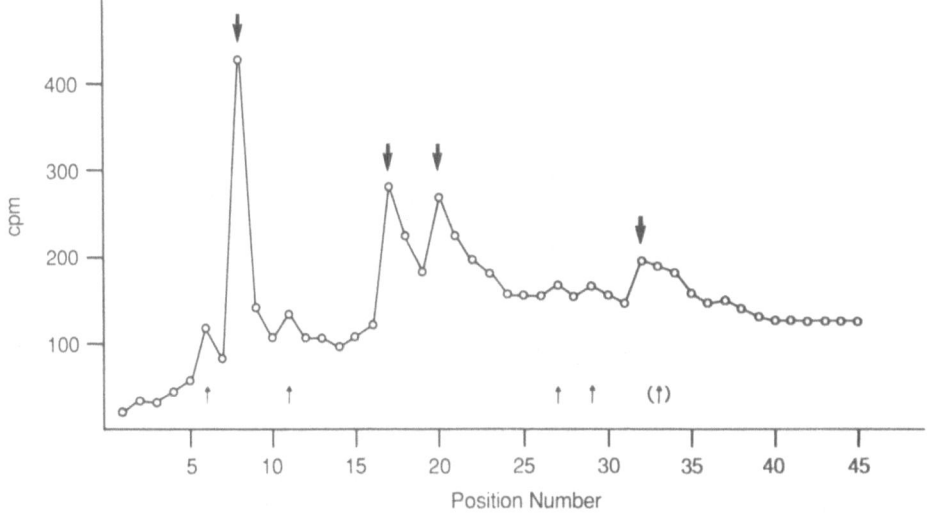

Figure 2: Amino terminal automated Edman degradation of gp90MEL-14 intrinsically labeled with 3H-lysine. Automated sequence analysis was performed on an Applied Biosystems Model 470A gas-liquid phase protein sequinator, modified to bypass the flask for conversion of thiazolinone derivatives. Entire butyl chloride extracts containing the 2-anilino-thiazolinone derivatives at each step were transferred into vials directly for scintillation counting in toluene/PPO. Each sample was counted in duplicate for 10 minutes on a Beckman LS counter (Model LS-7500). Positions containing radioactivity above background indicate the presence of 3H-lysine at that position. A plot of position number against total cpm at that position is given. Downward arrows indicate the major high specific activity assignments. Upward arrows indicate the lower specific activity assignments. The upward arrow in parenthesis indicates a tentative determination.

ubiquitin as a sequence match, an indication of the specifity of this sequence.

The selection of cDNA clones in the gt11 expression system showed that MEL-14 recognizes a ubiquitin associated determinant (16). It was therefore important to demonstrate that the relatively minor ubiquitin sequence was covalently associated with a specific core polypeptide, and not representative of a background of ubiquitin or of ubiquitinated proteins precipitated by MEL-14 and distributed over the entire SDS gel from which the specific MEL-14 species was isolated. To rule out this possibility the remainder of the gel fractions in preparative runs from which gp90MEL-14 had been isolated were similarly eluted and collected in four separate pools. The pools were subjected to amino-terminal amino acid analysis (data not shown). There were no detectable sequences corresponding to a ubiquitin amino-terminus in any other portions of the gels, and the relatively minor sequence of ubiquitin in the MEL-14 specific species could not be explained by contamination with a neighboring molecule containing ubiquitin as a major sequence. In addition, a light chain biosynthetically labeled with 3H-leucine was isolated from a B cell lymphoma line, 38C13, which also bears gp90MEL-14 (8). Isolation was performed by immunoprecipitation with MEL-14 and goat anti-mouse IgG in strictly analogous fashion to that routinely used for obtaining the MEl-14 reactive species. Amino terminal sequence analysis revealed the expected mouse kappa chain leucine positions at 11 and 15 with no evidence of a minor sequence at position 8 to suggest the presence of a ubiquitin amino-terminus. Thus, by two such divergent approaches as amino acid sequence analysis

```
        5         10        15        20        25        30        35
M Q I  F V K T L T G K T I  T L E V E P S D T I  E N V K A K I Q D K E G
M - I  F V K T L T - K T I  T L - V - - - -(T) I  - - - K - K - - -(K)- -
```

Figure 3-   Compilation of amino-terminal amino acid sequence
assignments for the low specific activity component in gp90 MEL-14.
The sequence is compared to the previously reported sequence of the
amino-terminus of ubiquitin (9) given in the top line. The second line
represents positions determined from independent single amino acid
labels. A dash indicates that no assignment at that position has been
made. Parentheses indicate tentative assignments.

```
        5          10        15        20        25        30
- T - H - - - K - M - - - - - - K F - K - - - - - - V (V) I - L K - -
```

Figure 4.   Compilation of amino-terminal amino acid sequence
assignment for the major high yield sequence in gp90 MEL-14.
Determinations ere made from independent single amino acid labels.
Dashes indicate that no assignment at that position has been made. The
parentheses at position 28 indicate tentative assignment.

and filter paper immunoselection to select cDNA clones (16), evidence indicates
that ubiquitin is indeed a component of this lymphocyte homing receptor.

A summary of the second (major) sequence obtained is given in Figure
4. The repetitive yield of lysine calculated from the two peak heights at positions
8 and 17 (92%) compares well with the repetitive yield calculated, minus
background, for the lysines in the ubiquitin chain at positions 6 and 11. This
observation is consistent with the interpretation that the major sequence, like the
ubiquitin sequence, is derived from a single polypeptide chain. This interpretation
is reinforced by our failing to encounter any evidence for more than a single amino
acid determination at any position for this major sequence in all single amino acid
labels examined.   In thorough computer searches of data files, the sequence
appears to be unique, although limited information to date prevents adequate
searches for homologies. It should be noted that the relatively lysine rich amino-
terminus should be regarded as a minimal estimate of the number of lysine
positions. Ubiquitin is known to conjugate to proteins in an isopeptide bond betwen
its carboxy-terminal glycine and epsilon-amino groups of lysines (13). Should such
a conjugated lysine position exist in the amino-terminal 35 residues of the core
polypeptide, it is likely that after Edman degradation a 77 residue ubiquitin
polypeptide (or ubiquitin multimer) attached to this lysine would not be extracted
from the filter and would remain undetected by this analysis.

The finding that ubiquitin may be synthesized as a poly-ubiquitin
transcriptional unit in head-to-tail array (16,20-23) raises the issue whether
ubiquitin exists as a concatamer in association with mature proteins as well. This
issue was addressed by taking advantage of the single methionine at position one in
ubiquitin and performing cyanogen bromide (CNBr) cleavage of filter bound
gp90 MEL-14, as is illustrated in Fig. 5.   Glass fiber filters to which protein
samples are applied for amino acid sequence analysis in gas phase sequencers can
be subjected to CNBr cleavage reactions even after amino-terminal sequence
analysis of the protein on the filter has been performed (24). This allows one to
assess the presence of internal methionine residues by the generation of new
amino-termini. After CNBr cleavage, Edman degradation is again performed. This
CNBr maneuver was conducted on purified MEL-14 antigen labeled with single
tritiated amino acids following 30 Edman degradation cycles. The profiles for the
initial ten residues for 3H-valine before and after CNBr digestion are given in
Figure 6.   A shift by one in the valine from position five in the native form to

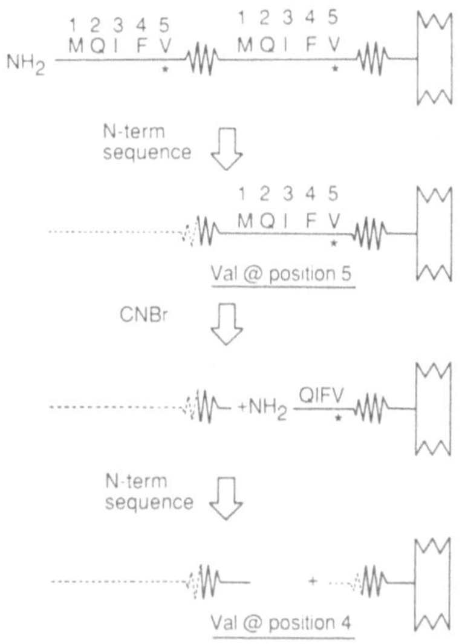

Figure 5. Diagramatic representation of an interpretation of cyanogen bromide experiments, exemplified by an hypothetical 3H-valine preparation of gp90MEL-14. The diagram depicts two ubiquitin monomers connected in linear arrangement. The first five amino-terminal amino acids of each subunit are given. The carboxy-terminus of the carboxy-terminal ubiquitin unit is shown attached to a truncated core polypeptide. Automated amino terminal sequencing is performed as described in Fig. 3. CNBr digestion was performed basically as described (24). Briefly, the glass fiber filter containing the sample is removed from the sequinator, acylated with trifloroacetic anhydride to block remaining free amino groups, and digested by wetting with 25 ul of CNBr solution (100mg/ml in 60% trifloroacetic acid) in a closed container at room temperature for 20 hours. The filter is then returned to the gas phase sequencer for resumption of sequencing analysis. See text for further discussion.

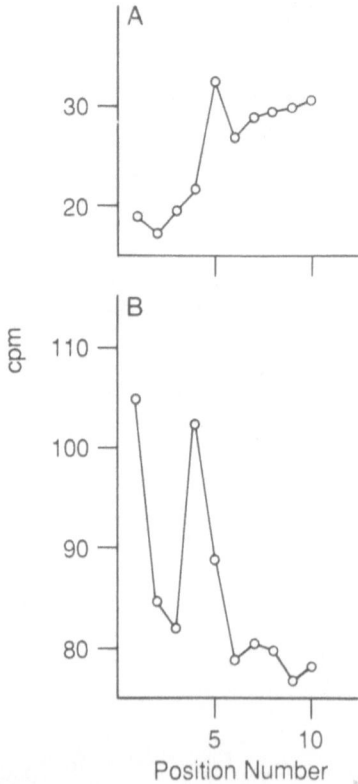

Figure 6. Automated Edman degradation of 3H-valine labeled gp90MEL-14 A) before, and B) after CNBr digestion. Amino acid sequence analysis and digestion was performed as described in Fig. 6 and in the text. Position number is plotted against total cpm at that position.

position four in digested material is noted. Amino-terminal analysis of CNBr digested polyubiquitin would result in such a shift in sequence by one position, while single ubiquitin subunits would generate no additional amino-termini in ubiquitin. A similar shift by one after CNBr digestion was seen for the first ubiquitin lysine and threonine positions. Background counts are relatively higher after CNBr digestions and ubiquitin positions later in the sequence tend to be somewhat obscured by background and apparent high yield determinations, likely deriving from internal cleavages in the second (major) chain. Since the amino-terminal ubiquitin sequence through the first 30 positions is presumably absent, a second, shifted sequence after CNBr digestion suggests that a ubiquitin sequence exists internally within this complex, consistent with a head-to tail arrangement.

Alternatively, it is possible that one or more ubiquitin chains is present exclusively as a monomer and that a proportion of the ubiquitin amino-termini, initially blocked to Edman degradation, are available for Edman degradation after CNBr digestion. Isolation of purified CNBr-derived fragments will likely be required to resolve this question.

Discussion

Lymphocyte migration occurs in a regulated fashion and with a high degree of specificity. Populations of lymphocytes exhibit preferential homing patterns in vivo, and preferential adherence to sites of entry from blood into secondary lymphoid organs, the high endothelial venules of peripheral lymph node and Peyer's patches, in vitro (27-30). This adherence is mediated via cell surface molecules in a presumed ligand-receptor interaction, the lymphocyte component of which has been designated the "lymphocyte homing receptor." A monoclonal antibody, MEL-14, specifically and completely inhibits binding of lymphocytes to the HEV of peripheral lymph nodes. This monoclonal identifies a single predominant cell surface species, designated gp90 MEL-14, which we infer to represent the lymphocyte homing receptor for peripheral lymph nodes. In the present studies we present initial amino-terminal amino acid sequence analysis and structural characterization of this lymphocyte homing receptor, which we conclude undergoes a complex series of posttranslational modifications. Our evidence suggests the receptor exists as a branched chain polypeptide consisting of a core polypeptide most likely in isopeptide linkage to ubiquitin and highly glycosylated in at least N-linked form as well (manuscript submitted). This represents to our knowledge, the first description of a ubiquitinated cell surface protein, and only the second instance of characterization of any isolated naturally occurring ubiquitinated protein.

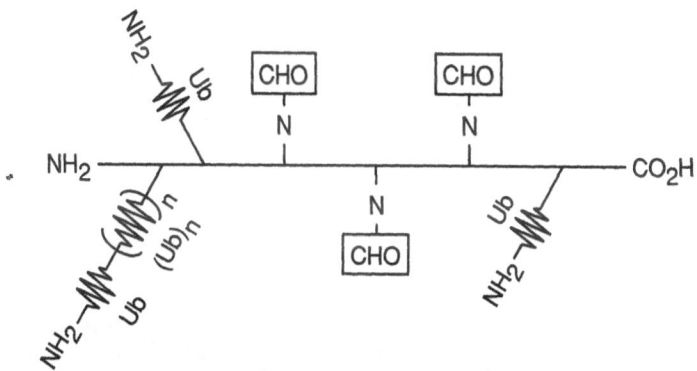

Fig. 7. Hypothetical model of gp90 MEL-14, a putative lymphocyte homing receptor for peripheral lymph node high endothelial venules. CHO, sites of N-linked glycosylation; Ub, ubiquitin subunits. The latter are represented in both monomeric (UB), and in head-to-tail arrangement (Ub)n. The disposition and number of ubiquitin moieties is arbitrary as is the placement of carbohydrates. Both carbohydrates and ubiquitin chains are attached to a central core polypeptide. See text for discussion.

A schematic model for the overall structure of the murine lymphocyte homing receptor for peripheral lymph nodes is given in Figure 7. A core protein of maximum size 46.5 KD is modified by a minimum of three N-linked glycosylations (manuscript submitted), constituting approximately 45% of the mass of this molecule. This is a lower estimate of the extent of glycosylation, since the presence of O-linked oligosaccharides was not addressed by these studies. The second major and rather unexpected structural feature is the covalent association of ubiquitin with the core polypeptide, inferred from the identification of a clear ubiquitin amino-terminal amino acid sequence present in addition to that of a putative core polypeptide. A precedent for ubiquitin-protein conjugates was set for the nuclear protein A24, which was sequenced in its entirety, including a detailed analysis of the ubiquitin-H-2A junction, deduced to be a novel linkage between the carboxyl group of the carboxy-terminal glycine residue of ubiquitin to the epsilon-amino group of a lysine at position 119 of histone H-2A (13). Since the isolation of gp90MEL-14 is performed after complete reduction and since ubiquitin contains no cysteine residues, a disulfide protein linkage between the chains is not possible. We therefore expect that a similar isopeptide bond exists linking ubiquitin to the core polypeptide of this murine lymphocyte homing receptor. Alternatively, it might be considered that the two amino-terminal sequences derive from completely separate and unassociated chains, a ubiquitin sequence from a ubiquitin protein polymer of approximately 10-12 sequential ubiquitin subunits, giving the appropriate molecular weight species. The other sequence would derive from a completely unrelated glycoprotein, possibly representing the "true" homing receptor, which fortuitously bears a determinant crossreactive with but unrelated to ubiquitin sequences. Both chains would be recognized and precipitated by the MEL-14 antibody. This alternative can be regarded as highly unlikely for several reasons: 1) although it is apparent that the potential for large plyubiquitin protein polymers exists (16,20-23), these in fact have not as yet been found; 2) the intrinsically labeled receptor isolated for sequence analysis migrates as a single major species on two dimensional gel analysis, which is identical to the cell surface iodinated molecule; 3) since ubiquitin is generally known not to be glycosylated and contains no canonical sequences that would serve as an acceptor site for N-linked glycosylation, free or unmodified polyubiquitin should be selected against in the lectin adherent step used to enrich for glycoproteins; 4) ubiquitin contains a single methionine at position one, so that complete CNBr digestion of a hypothetical 10-12 unit concatamer should result in a 10-12 fold increase in a ubiquitin associated sequence, and though such a sequence was identified (see results, Fig. 6), this extent of possible concatamerization was not seen in our experiments; 5) the gp90MEL-14 migrates in nonreducing SDS-PAGE more rapidly than under reducing conditions (31), a property of proteins rich in internal disulfide bonds; ubiquitin contains no cysteines. Therefore, we favor the ubiquitin-core polypeptide relationship indicated in Figure 7. Detailed peptide analysis will of course be required to test this model.

While ubiquitination as a postranslational modification poses no special problems for nuclear and cytoplasmic proteins, ubiquitination of cell surface proteins is more difficult to envision. If ubiquitination occurs contranslationaly on the cytoplasmic face of the rough endoplasmic reticulum (E.R.), transport of a branched chain polypeptide to the lumen of the E.R. may require special transport vehicles. It is more difficult to imagine the ubiquitination event to occur within the lumen of the E.R. or Golgi, as ubiquitin and so far all ubiquitin transcripts lack a classical signal sequence at the amino-terminus. Perhaps a new form of transport in a unique vesicle is involved. Why only a subset of cell surface molecules is ubiquitinated is also unclear. It shall be of interest to determine whether isopeptide bond formation occurs only with lysines surrounded by a "canonical" sequence, analogous to glycosylation sites, and, if so, with what frequency such sequences are ubiquitinated. Resolution of this should reveal new insights into sites and functions of ubiquitination and protein targeting mechanisms.

Other details regarding the relationship of ubiquitin to the core polypeptide remain to be ascertained. We cannot, from our available data, determine the number of ubiquitin units associated with the receptor, since the stoichiometry in the yield in counts per minute in the two polypeptide chains is unlikely to reflect their molar ratios (see below). Experiments using puified isopeptidase, an activity in whole cell lysates shown to remove ubiquitin specifically from proteins (32,33), when available, should allow us to determine the molecular weight of the core peptide and the number of ubiquitin subunits present. Another unresolved issue is whether ubiquitin exists in linear arrangements of more than one ubiquitin unit in association with this receptor (Figure 7). Evidence has been presented from sequence analysis of CNBr digests of the receptor molecule on filters which have already been subjected to 30 cycles of amino-terminal amino acid degradation (Figure 6), suggesting that such head-to-tail organization might exist in this system. More direct evidence for this type of structure requires isolation of CNBr fragments of the receptor complex for sequence analysis. A head-to-tail structure predicts that an 8,500 dalton molecular weight peptide, consisting of a precise ubiquitin polypeptide sequence beginning with its second position and ending with a carboxyterminal methionine from the succeeding ubiquitin unit can thus be isolated.

It has been noted above in several contexts that by amino acid sequence analysis the yield in counts per minute for the non-ubiquitin chain are quite different from the counts in the ubiquitin sequence, which generally represents 10-20% of the yield found in the putative core polypeptide. Since we have reason to believe, from gt11 expressed cDNA clones (16) that the MEL-14 determinant is ubiquitin dependent, we expect that MEL-14 may only recognize a ubiquitinated form of the receptor. Therefore, one expects at least one ubiquitin molecule per receptor complex. As outlined in the Results section, we have ruled out the trivial explanation that MEL-14 immunoprecipitates contain a general background of ubiquitin or ubiquitinated proteins, or that the ubiquitin sequence associated with a MEL-14 specific peak represents a contaminant from a neighboring nonspecific band containing ubiquitin in higher yield. By the same analysis, we exclude the alternative that after immunoprecipitation ubiquitin-core polypeptide conjugates spontaneously dissociate. This follows from amino acid sequence analysis of pools containing the smallest species on SDS polyacrylamide gels, which would be expected to contain free ubiquitin. No trace of such a ubiquitin sequence is found.

Another explanation for the relatively low yield of the ubiquitin sequences is that the amino-terminus of ubiquitin chains are preferentially blocked to Edman degradation, either as a consequence of cell culture or during protein isolation, although it would be unexpected for an amino-terminal methionine residue in a eukaryotic system to be blocked in such a fashion. We regard as much more likely an alternative which takes into consideration the intracellular synthethic properties which can be surmised about ubiquitin. Ubiquitin is likely to comprise a large intracellular pool for conjugation to a large number of proteins destined for various cellular compartments. Ubiquitin is also almost certainly to have a long half-life, as evidenced by its refractoriness to a battery or proteolytic digestions in its native state (9), and requirement for maneuvers such as malealation before yielding to enzymatic digestions for peptide isolation. Strong evidence exists from NMR studies (34) and recent X-ray crystallography (35) that ubiquitin assumes a highly globular compact conformation, consistent with its resistance to degradation. It follows then, that by intrinsic labeling it would be particularly difficult to pulse efficiently a given tritiated amino acid through a large pool of long-lived ubiquitin over the 4-5 hour period of labeling, compared to a molecule of presumed normal or even fast turnover, as we would suggest the core polypeptide may be. Therefore, the differential yields in the two polypeptide amino-termini may simply reflect differential turnover rates.

The convergence of observations from two such disparate approaches as amino acid sequence analysis of a cell surface receptor purified with a monoclonal antibody to a lymphocyte homing receptor, and the isolation of antibody reactive expressed cDNA clones (16), provides compelling evidence that this antibody recognizes a ubiquitin-dependent determinant. However, the specificity of this antibody by cell and tissue section staining and by functional assays (27-30) raises a possible paradox. How can a determinant on a polypeptide as prevalent as ubiquitin be so selectively expressed, and what might that determinant be? Crystallographic studies of ubiquitin reveal a tight globular conformation (35), and native ubiquitin monomers may therefore present relatively few epitopes at the solvent surface which would serve as immunogens. Furthermore, since the amino acid sequences of ubiquitin is very conserved, unless immunologic tolerance is overcome, we would expect only a very restricted number of ubiquitin associated-determinants to be immunogenic. The most logical site for a relatively unique ubiquitin determinant to be generated would be at its carboxy-terminal end, where conjugation to other proteins is effected. Additional evidence attesting to the relative availability of the carboxy-terminus derives from X-ray crystallography of ubiquitin, demonstrating that the carboxy-terminal amino acids extend away from the otherwise globular structure, and that this portion of the molecule appears particularly free of constraints on its motion (35). MEL-14 may therefore recognize a ubiquitin-receptor conjugate specific determinant, or less likely, the conformational alteration in the ubiquitin moiety proper induced by its conjugation to the receptor. We can postulate that the particular ubiquitin containing cDNA clones selected by MEL-14 produce proteins with amino acid sequences carboxy-terminal to the ubiquitin sequence which mimic the ubiquitin-receptor conjugate determinant; or that these sequences induce similar conformational changes in ubiquitin as exists in the receptor complex. In any case, these observations are provocative and suggest that this system may prove informative at a fundamental level about the nature of what constitutes an antigenic determinant.

We have only begun to characterize the polypeptide which we suggest is the core polypeptide of this lymphocyte homing receptor. By Endoglycosidase F treatment we have deduced the presence of at least three N-linked glycosylation sites, leaving a 55 KD protein species (manuscript submitted). If the equivalent of a single ubiquitin moiety is subtracted, to be conservative, we are left with a maximal estimate of the receptor's core polypeptide size as 46.5 KD, less than one half of the apparent molecular weight of the native receptor complex. We again emphasize that this represents an upper limit in size of the core polypeptide; if a single di-ubiquitin is present the estimate is lowered to 38 KD.

A striking feature about the initial available amino-terminal sequence of this chain is its richness in lysine, with four residues in the first 32 amino acids. This is of interest for several reasons. Isopeptide bonds between ubiquitin and other proteins occurs at the epsilon - amino group of lysine, so it is curious that abundant opportunity for such bonds exists at the amino-terminus of the presumed core polypeptide, where interaction with the cell surface of HEV is likely to happen. Moreover, it has been shown in the rat (3) and more recently in the mouse (18), that HEV binding lymphocytes treated with mild trypsinization lose their capacity to bind to HEV and are believed to lose an amino-terminal fragment of the homing receptor molecule. In the murine system, in addition to HEV bindin, MEL-14 reactivity also is lost (18). Since MEL-14 completely inhibits peripheral lymph node HEV binding and appears to recognize a ubiquitin-depenent cell surface determinant, the evidence together suggests that a functional HEV binding may require an amino-terminal trypsin-sensitive ubiquitinated peptide. Finally, a relatively positively charged amino-terminus would be expected to help stabilize a required interaction with a generally negatively charged cell surface of an HEV. Complete sequence information of the core polypeptide will likely require full-length cDNA clones whose sequence encodes the known amino acid positions.

Evidence, presented herein, establishes a precedent for the existence of a cell surface ubiquitinated protein. Assignments of a particular role for ubiquitin on cell surfaces can only be speculated upon at this point. The remarkable degree of conservation of ubiquitin throughout evolution clearly suggests a fundamental cellular function. One such fundamental role, for which a body of evidence already exists, proposes that it is required for intracellular protein degradation (36). It is conceivable that ubiquitination of these cell surface proteins is incidental and simply serves as a tag for surface proteins which are destined for degradation. Indeed it may be that a surface molecule such as a lymphocyte homing receptor has a special requirement for rapid internalization and degradation by either high endothelial cells, or the lymphocytes bearing it so that entry into a secondary lymphoid organ is directional and not easily reversible. Ubiquitination may then ensure rapid turnover. This could be a general property of many developmental cell surface molecules which target cells to particular sites.

Consideration should also be given to a possible independent role for ubiquitin at the cell surface. It can be recalled that the MEL-14 antibody specifically and completely inhibits binding of lymphocytes to peripheral lymph node HEV, and that, from our evidence, it appears to recognize a ubiquitin-dependent determinant. It is likely that this class of cell surface molecules can be bound by antibody without interfering with functional HEV attachment, and this implies that ubiquitin may indeed be critial for binding. We may extrapolate this principle hypothetically to suggest that ubiquitin may subserve a vital role in cell-cell interaction and adhesion. It is possible that the degree of motion allowed the carboxy-terminal portion of ubiquitin (35) in combination with the potential for conjugation to a number of cell surface proteins provides a sufficiently varied conformational repertoire at the site of conjugation to largely determine receptor specifity. If so, a strong evolutionary pressure for the conservation of ubiquitin primary structure may be its requirement to interact with a large number of proteins around such sites of conjugation. Alternatively, the homing receptor backbone could provide specificity for a particular cell type, in this instance high endothelium, while ubiquitin might provide additional stability to the adhesive interaction. X-ray crystallography indicates that beta-pleated sheet comprises 37% of the ubiquitin chain, and appears concentrated along one outer aspect of the molecule (35). External beta-pleated regions may present a broad hydrophilic surface with which a complementary surface can interact. Such avid association between two cell surfaces could be important for a lymphocyte to migrate through a vessel wall into a lymphoid organ. This mechanism could be a general requirement in many biological systems in which transmigration through vessel walls or other tissues is essential.

Acknowledgements: The authors thank Pila Estess for careful, critical review of the manuscript and Graciela Rodriguez for preparation of the manuscript. This work was supported by USPHS grant AI 19512. M.S. was supported by the Tumor Biology Training Grant CA 09151.

REFERENCES

(1) J.L. Gowans, E.J. Knight, Proc. R. Soc. B159, 257 (1964)
(2) V.T. Marchesi, J.L. Gowans, Proc. R. Soc. B159, 283 (1964)
(3) H.B. Stamper, J.J. Woodruff, J. Exp. Med. 144, 828 (1976)
(4) G.A. Gutman, I.L. Weissman, Transplantation. 16, 621 (1973)
(5) J.C. Howard, S.V. Hunt, J.L. Gowans, J. Exp. Med. 135, 200 (1972)
(6) S.K. Stevens, I.L. Weissman, E.C. Butcher, J. Immun. 128, 844 (1982)
(7) E.C. Butcher, I.L. Weissman, Ciba Fdn. Symp. 71, 265 (1979)
(8) W.M. Gallatin, I.L. Weissman, E.C. Butcher, Nature. 303, 30 (1983)
(9) D.H. Schlessinger, G. Goldstein, H.D. Niall, Biochemistry. 14, 2214 (1975)
(10) D.H. Schlessinger, G.Goldstein, H.D. Niall, Nature 255, 423 (1975)
(11) D.C. Watson, W.B. Levy, G.H. Dixon, Nature 276, 196 (1978)
(12) J.G. Gavilanes, et al., J. Biol. Chem. 257, 10267 (1982)
(13) I.L. Goldknopf, H. Busch, Proc. Natl. Acad. Sci. U.S.A. 74, 864 (1977)
(14) A. Hershko, A. Ciechanover, Annu. Rev. Biochem. 51, 335 (1982)
(15) A. Ciechanover, C. Finley, A.Varshavsky, J. Cell. Biochem. 24, 27 (1984)
(16) T. St. John, et al., submitted (1985)
(17) W.M. Gallatin, manuscript in preparation
(18) S.T. Jalkanen, et al., unpublished observations
(19) R.M. Hewick, et al., J. Biol. Chem. 256, 7990 (1981)
(20) E. Dworkin-Rastl, A. Shrutkowski, M.B. Dworkin, Cell 39, 321
(21) E. Ozkaynak, D. Finley, A. Varshavsky, Nature 312, 663 (1984)
(22) U. Bond, M.J. Schlessinger, Mol. Cell. Bio., 5, 949 (1985)
(23) O. Wiborg, M.S., et al., EMBO J. 4, 75 (1985)
(24) D.L. Urdal, et al., J. Chomatog. 296, 171 (1984)
(25) J.H. Elder, S. Alexander, Proc. Natl. Acad. Sci. U.S.A. 79, 4540 (1982)
(26) H.T. Smith, V.A. Fried, manuscript in preparation
(27) E.C. Butcher, R.G. Scollay, I.L. Weissman, J. Immun. 123, 1996 (1979)
(28) S.K. Stevens, I.L. Weissman, E.C. Butcher, J. Immun. 128, 844 (1982)

(28) S.K. Stevens, I.L. Weissman, E.C. Butcher, J. Immun. **128**, 844 (1982)

(29) R.A. Reichert, et al., J. Exp. Med. **157**, 813 (1983)

(30) M.O. Dailey, et al., J. Immun. **128** 2134 (1982)

(31) B.M. Gallatin, et al., unpublished observations

(32) S. Matsui, et al., Proc. Natl. Acad. Sci. U.S.A. **79**, 1535 (1982)

(33) A. Hershko, et al., J. Biol. Chem. **258**, 8206 (1983)

(34) R.E. Lenkinski, et al., Biochem. Biophys. Acta **494**, 126 (1977)

(35) S. Vijay-Kumar, et al., Proc. Natl. Acad. Sci. U.S.A. **82**, 3582 (1985)

(36) A. Hershko, Cell **34**, 11 (1983)

(37) M.F. Loken, L.A. Herzenberg, Ann. N. Y. Acad. Sci. **254**, 163 (1975)

(38) O.N. Witte, et al., Proc. Natl. Acad. Sci. U.S.A. **75**, 2488 (1978)

(39) U.K. Laemmli, Nature **227**, 680 (1970)

(40) S.E. Cullen, T.N. Freed, S.G. Nathenson, Transplant. Rev. **30**, 236 (1976)

(41) E.S. Vitetta, et al., Proc. Natl. Acad. Sci. U.S.A. **73**, 905 (1976)

XV. New Sequences

HUMAN C3b/C4b RECEPTOR (CR1): ISOLATION, PROTEIN SEQUENCE ANALYSIS, AND CLONING OF A PARTIAL cDNA FROM HUMAN TONSIL

John A. Smith, Winnie W. Wong, Lloyd B. Klickstein, John Weis, and Douglas T. Fearon

Departments of Molecular Biology and Pathology, Massachusetts General Hospital; Department of Rheumatology and Immunology, Brigham and Women's Hospital, Boston, MA 02114

The human receptor for the C3b/C4b fragments of complement, designated CR1, is a large glycoprotein comprised of a single polypeptide chain that exhibits genetically regulated structural and quantitative polymorphisms. The structural polymorphisms are manifested in an autosomal codominant expression of different allotypic proteins; all allotypic forms are capable of binding C3b[1,2,3,4]. The most common proteins have MW of 250,000 (F allotype) and 260,000 (S allotype) by NaDodSO$_4$/PAGE, although rarer forms differ by as much as 90,000 daltons[4]. Removal of the N-linked oligosaccharides does not eliminate this polymorphism, suggesting that the primary structures of the various allotypic forms are different[4]. CR1 is an integral membrane glycoprotein residing on erythrocytes, polymorphonuclear leukocytes, monocytes/macrophages, all B and some T lymphocytes, and renal glomerular epithelial cells[5,6]. Quantitative polymorphism refers to the number of receptor sites on erythrocytes, which varies by 10-fold among different individuals. Population and family studies have shown that there are two autosomal codominant alleles that determine the quantitative polymorphism.

To gain a molecular understanding of these polymorphisms, we have begun to study the molecular biology of the human CR1.

MATERIALS AND METHODS

CR1 Protein Analysis

CR1 was purified from human erythrocytes (0.6 x 10^{13}) by Matrex Red A (Amicon) and monoclonal affinity chromatography[7] (Table 1). The N-terminus of the protein was blocked and refractory to Edman degradation. Six nmoles of CR1 were fully reduced and alkylated, precipitated by 4 volumes of chloroform /methanol (1:1, v/v), and digested with 120 pmol trypsin (TPCK-treated) in 0.1 M NH$_4$HCO$_3$/0.2% Zwittergent 3-14 (Calbiochem-Behring) for 22 hr at 37 C. Tryptic peptides were isolated by reverse-phase HPLC. Fractions were rechromatographed isocratically using the formula: P=0.9E-2, where P is the percentage of CH$_3$CN in the mobile phase and E is the percentage of CH$_3$CN during the linear gradient separation. The tryptic fragments were sequenced by automated Edman degradation using an Applied Biosystems 470A sequencer as described[7].

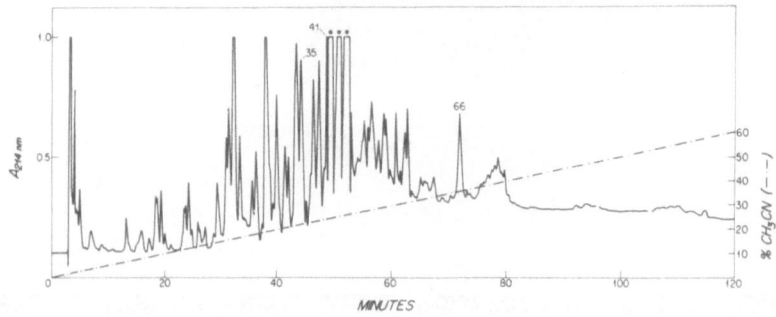

Fig. 1. HPLC separation of CR1 tryptic peptides. Six nanomoles of purified CR1 was reduced, alkylated, digested with trypsin, and chromatographed on a 0.46 x 25 cm Vydac phenyl column with 0.1% CF_3COOH with a linear gradient of 0-60% CH_3CN (120 min). Numbers 35, 41, and 66 refer to the tryptic peptides, and the sequences are shown in Table 2. Because of a sudden baseline shift during the chromatography, the absorbance of the asterisked peaks is uncertain.

Characterization a CR1 cDNA Clone in the Human Tonsilar Library

Only one of 400,000 recombinant cDNA clones hybridized with CR20.1. This clone (lambda T8.3)(Fig. 2) also hybridized to CR20.2 and CR15.1. Its cDNA insert was characterized by restriction enzyme digestion, Southern blot analyses, and nucleotide sequencing. As shown in Fig. 2, the CR20.1 and CR20.2 probes hybridized to EcoRI fragments, CR1-1 and CR1-2. In order to characterize this nucleotide sequence homology, the degree of cross-hybridization of the CR1-1 and -2 fragments was determined by Southern blot analysis (data not shown). The results suggested that these were repetitive sequences and that the disparate restriction maps of the EcoRI fragments of the cDNA insert (Fig. 2) indicate that the repetition is not the result of duplication. Sequencing of the total cDNA insert showed that the these two fragments were >90% homologous (Klickstein et al, unpublished).

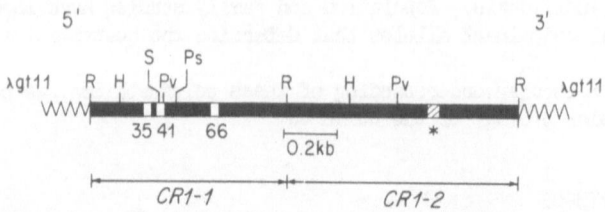

Fig. 2. Partial restriction map of the cDNA insert of lambda T8.3. The black bar denotes human cDNA insert, while CR1-1 and -2 are two of its EcoRI fragments. An additional small EcoRI fragment located between CR1-1 and -2 is known to be present and is currently being sequenced. The white areas denote the locations of the coding sequences for the tryptic peptides 35, 41, and 66 (Table 2). Oligonucleotides CR15.1 and CR20.1/20.2 were derived from tryptic peptides (Table 2) and hybridized to CR1-1 during Southern blot analysis. The hatched area designates the hybridization site for CR20.1/20.2 on the fragment CR1-2. The enzymes used for mapping were EcoRI(R), HindIII(H), SalI(S), PvuII(Pv), and PstI(Ps). Multiple Pv sites were present in the insert, but only the site used for M13 subcloning is shown.

Construction and Screening of Human Tonsillar cDNA Library

The construction and screening of this lambda gt11 library, as well as the synthesis of the redundant oligonucleotide probes was previously described[8].

Blot Analysis and DNA Sequencing

Restriction fragments of the lambda phage insert, hybridizing to the oligonucleotide probes, were subcloned into phage vectors M13 mp 18 or 19, and both strands were sequenced by the dideoxynucleotide chain termination method[9]. Northern blot hybridization was carried out with human tonsilar poly(A)+ RNA[10,11], and Southern blot analysis was done with peripheral blood leukocyte genomic DNA[12].

RESULTS

Sequence Analysis of Tryptic Peptides and Oligonucleotide Probe Synthesis

Tryptic peptides were separated by HPLC in 0.1% CF_3COOH using a linear gradient of 0-60% CH_3CN (0.5%/min)[13] (Fig. 1) Sequence data was determined for 14 of the tryptic fragments, accounting for approximately 10% of total CR1's primary structure (data not shown). Table 2 shows the partial sequence data used to construct three oligonucleotide probes: two overlapping 20 base probes (CR20.1 and CR20.2 (128-fold and 256-fold degenerate)), corresponding to the 10 amino acids of the peptide in peak 66 and a 15 base probe (CR15.1(32-fold degenerate)), based on 5 amino acids in the peptide in peak 41.

Table 1. Purification of CR1

	Total Protein (mg)	Total CR1 (mg)
Detergent solubilized ghosts	37,000	10.2
Matrex Red A eluate	880	7.4
Concentrated and dialyzed Matrex Red A pool	800	4.7
Eluate of Sepharose-YZ-1	1.47	3.7

[1]Lowry's protein assay; [2]Radioimmunoassay for CR1

Table 2. Partial nucleotide and amino acid sequences of the CR1-1 fragment of lambda T8.3

Tryptic Peptide	Sequence Data
35	(Lys)*-Cys-Gln-Ala-Leu-Asn-Lys-Trp-Glu-Pro-Glu-Leu
	AAG -TGC-CAG-GCC-CTG-AAC-AAA-TGG-GAG-CCG-GAG-CTA-
	Pro-Ser-Cys-Ser-Arg
	CCA-AGC-TGC-TCC-AGG

CR15.1
———————————

41	(Arg)*-Asp-Lys-Asp-Asn-Phe-Ser-Pro-Gly-Gln-Glu
	AGG -GAC-AAG-GAC-AAC-TTT-TCA-CCC-GGG-CAG-GAA
	Val-Phe-Tyr-Ser-Cys-Glu-Pro-Gly-Tyr-Asp-Leu-Arg
	GTG-TTC-TAC-AGC-TGT-GAG-CCC-GGC-TAT-GAC-CTC-AGA

CR20.2
———————————
CR20.1
———————————

66	(Lys)*-Val-Asp-Phe-Val-Cys-Asp-Glu-Gly-Phe-Gln-Leu
	AAA -GTG-GAT-TTT-GTT-TGT-GAT-GAA-GGA-TTC-CAA-TTA
	Lys-Gly-Ser
	AAA-GGC-AGT

* These amino acids were derived from the nucleotide sequences and were not a part of the original tryptic peptide sequence.

<u>Northern Blot Analysis</u>

CR1-1 and -2 were subcloned into pBR327, and these plasmids, pCR1-1 and pCR1--2, were used as Northern probes. Fig. 3 shows that pCR1-2 hybridized to two mRNA's of approx. 9 and 11 kb. Since the donor was homozygous for the F structural allotype of CR1 (data not shown), this presumably reflects a heterogeneity of message for CR1 proteins of an identical size.

<u>Correlation of CR1 Genetic and Structural Polymorphism</u>

Genomic DNA was isolated from family members of a FF x FS cross (F and S phenotypes defined in introduction), and the restriction fragments from a BamHI digest were probed with $[^{32}P]$-labelled CR1-2 (Fig. 4). The F and S phenotypes were determined by immunoprecipitation with rabbit anti-CR1 and $NaDodSO_4$/PAGE (Fig. 4). The expression of the S allotype in one parent and two offspring correlated with an extra BamHI fragment of 14.5 kb. Hence, there is a cosegregation between a restriction fragment and protein structural polymorphism.

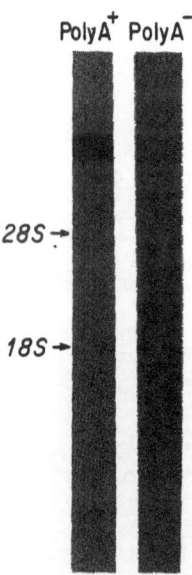

Fig. 3. Autoradiogram of Northern blot-hybridization analysis of tonsilar mRNA. Poly(A)$^+$ RNA and poly (A)$^-$ RNA (2.5 ug) from a single donor was denatured by 50% (vol/vol) formamide at 55 C, electrophoresed in 1.2% (wt/vol) agarose in Mops/acetate buffer and 0.74% formaldehyde, transferred to nylon membranes, hybridized with $[^{32}P]$CR1-2 for 24 hr in 50% formamide containing 5X NaCl/Citrate (1X=0.15 M NaCl/0.015 M sodium citrate, pH 7), and washed with 0.2X NaCl/Citrate at 65 C. Migration of the 28S and 18S ribosomal RNA subunits is indicated. The molecular weights of these subunits are 2 and 5 kb, and these markers were used to estimate the size of CR1 mRNA's.

DISCUSSION

The cloning strategy relied on purification of CR1 receptor, protein sequence analysis of reverse-phase HPLC-separated tryptic peptides, and synthesis of oligonucleotide probes. Such random tryptic fragments, rather than N-terminal sequence alone, circumvented the requirement for a full-length cDNA clone and increased the probability of identifying a specific cDNA clone. A positive cDNA clone was identified by its hybridization with three oligonucleotide probes derived from two peptide sequences. Sequence analysis of this clone, lambda T8.3, confirmed the presence of the nucleotide sequences of the probes and the protein sequences of flanking amino acids in the tryptic fragments (Table 2).

Analysis of poly(A)$^+$ RNA from human tonsils demonstrated mRNA that is large and heterogeneous (9 and 11 kb). Both of these messages are twice the expected size for a mRNA encoding a protein of MW 200,000-250,000. It is not known whether this large message encodes a precursor of CR1 or if a large proportion of the message does not encode CR1.

This is the first evidence for the presence of repetitive, homologous domains within a partial cDNA insert for CR1. Such repetitive coding sequences may extent to the entire CR1 gene, and duplication or deletion of these via homologous recombination may generate the various CR1 allotypic proteins. Additionally, the cosegregation of a restriction fragment polymorphism with the expression of the S allotype of CR1 (Fig. 4) results from changes at the level of the CR1 gene. Further studies are aimmed at sequencing the total cDNA and determination of the structure of the genomic DNA.

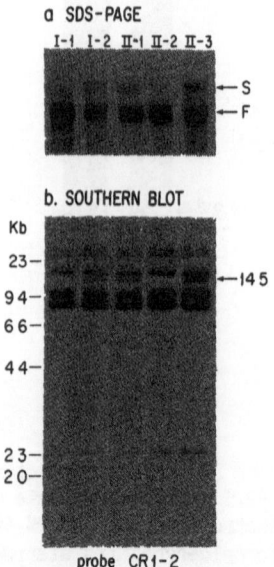

Fig. 4. Correlation of restriction fragment length polymorphism with CR1 structural phenotypes in a human family. (a) CR1 was isolated from erythrocyte membranes by immunoprecipitation with rabbit F(ab)$_2$ anti-CR1-Sepharose and analyzed by NaDodSO$_4$/PAGE. CR1 protein was identified with silver staining. The S and F allotypes are indicated. (b) DNA was prepared from peripheral blood leukocytes of the same individuals, and 15 ug of each was digested with BamHI. The restriction fragments were electrophoresed in 0.8% agarose in Tris-acetate buffer, transferred, hybridized with [^{32}P]CR1-2 for 36 hr, and washed as described in Figure 3. Lanes: I-1 and I-2, parents; II-1, II-2, and II-3, offspring. Sizes are shown in kb.

ACKNOWLEDGEMENTS

J.A.S. was supported by a grant from Hoechst Aktiengesellschaft (W. Germany). L.B.K. was supported in part by NIH Training Grant GM07753-06. This work was supported by NIH grants AI 22833, AI 10356, and AM 35907.

REFERENCES

1.   W.W. Wong, J.G. Wilson, and D.T. Fearon, Genetic regulation of a structural polymorphism of human C3b receptor, J. Clin. Invest. 72: 685 (1983).

2.   T.R. Dykman, J.L. Cole, K. Iida, and J.P. Atkinson, Polymorphisms of human erythrocyte C3b/C4b receptor, Proc. Natl. Acad. Sci. USA 80: 1698 (1983).

3.   T.R. Dykman, J. Hatch, and J.P. Atkinson, Polymorphisms of the human C3b/C4b receptor, J. Exp. Med., 159:691 (1984).

4.   T.R. Dykman, J.A. Hatch, M.S., and J.P. Atkinson, Polymorphism of the C3b/C4b receptor (CR1): characterization of a fourth allele, J. Immunol. 134: 1787 (1985).

5.   D.T. Fearon and W.W. Wong, Ligand-receptor interactions that mediate biological responses, Ann. Rev. Immunol. 1:243 (1983).

6.   S.H. Yoon and D.T. Fearon, Characterization of a soluble form of the C3b/C4b receptor (CR1) in human plasma, J. Immunol. 134:3332 (1985).

7.   W.W. Wong, R.M. Jack, J.A. Smith, C.A. Kennedy, and D.T. Fearon, Rapid purification of the human C3b/C4b receptor (CR1) by monoclonal antibody affinity chromatography, J. Immunol. Methods 22:in press (1985).

8.   W.W. Wong, L.B. Klickstein, J.A. Smith, J.H. Weis, and D.T. Fearon, Identification of a partial cDNA clone for the human receptor for complement fragments C3b/C4b, Proc. Natl. Acad. Sci. USA 82:in press 1985.

9.   F. Sanger and A.R. Coulson, A rapid method for determining sequences in DNA by primed synthesis with DNA polymerase, J. Mol. Biol. 94:441 (1975).

10. H. Lehrach, D. Diamond, J.M. Wozney, and H. Boedtker, RNA molecular weight determinations by gel electrophoresis under denaturing conditions, a critical reexamination, Biochemistry 16:4743 (1977).

11. P.S. Thomas, Hybridization of denatured RNA and small DNA fragments transferred to nitrocellulose, Proc. Natl. Acad. Sci. USA 77:5201 (1980).

12. E. Southern, Detection of specific sequences among DNA fragments separated by gel electrophoresis, J. Mol. Biol. 98:503 (1975).

13. H.P.J. Bennett, A.M. Hudson, C. McMartin, and G.E. Purdon, Use of octadecasilyl-silica from the extraction and purification of peptides in biological samples, Biochem. J. 168:9 (1977).

HTLV-III/LAV PARTICLE-ASSOCIATED PROTEINS: I. VIRUS PURIFICATION,

INACTIVATION AND BASIC CHARACTERIZATION

Michael Phelan, J. Willard Hall, Martha Wells, Michele
Kowalski, Luba Vujcic, Gerald Quinnan, Jr., and Jay Epstein

Division of Virology, Office of Biologics Research and
Review, Center for Drugs and Biologics, Food and Drug
Administration, 8800 Rockville Pike, Bethesda, MD 20205

INTRODUCTION

Human T-lymphotropic virus type III/lymphadenopathy virus (HTLV-
III/LAV) is the etiologic agent of acquired immunodeficiency syndrome
(AIDS),[1-4] although cofactors may play a significant role in disease
expression.[5] Changes in social behavior may reduce the rate of AIDS
transmission, but complete curtailment of the epidemic seems unlikely
without immunological or pharmacological intervention. Much effort has
been invested in the direct examination of the virus and its infection
processes to elucidate promising strategies of intervention, one of which
may be vaccination. Detailed information on HTLV-III/LAV components,
especially proteins, should aid the rational design of vaccines by
focusing attention on those viral constituents likely to induce protective
immunity.

In this report, a foundation is laid for the characterization of
HTLV-III/LAV particle-associated proteins by describing procedures for
bulk purification of the virus and its inactivation without any
appreciable modification of the viral polypeptides. By several electrophoretic techniques, these polypeptides are assessed for apparent
molecular weights, isoelectric patterns and glycosyl prosthetic groups to
yield a particle "fingerprint" of some complexity. Finally, certain
virus-associated proteins, for example actin, are tentatively identified
based on their biochemical characteristics.

MATERIALS AND METHODS

Virus Purification

HTLV-III/LAV (Electronucleonics Inc., Silver Spring, MD) was obtained
by zonal centrifugation of supernatant fluids from HTLV-III infected H-9
cells (H9/HTLV-III$_B$) through ribonuclease-free sucrose gradients.[7] Virus
harvested from the 1.13-1.18 g/cm^2 density zone was dialyzed against
0.2M NaCl, 0.01M Tris and 1mM EDTA, pH 7.0 (NTE buffer), diluted to about
2 mg/ml and stored at -70°C prior to secondary purification using equilibrium isopycnic centrifugation through a 20-60% sucrose gradient.[8]

Fractions (1 ml) were taken from the top to the bottom of the gradient and assayed for reverse transcriptase (RT) activity[9] and polypeptide composition as described below.

Virus Inactivation

Purified HTLV-III/LAV was inactivated by adding 8 volumes of phosphate-buffered saline, pH 7.2 (PBS), and 1 volume of a solution containing 100 ug/ml of 4'-amino-methyl trioxalen (Calbiochem, La Jolla, CA) and 10% dimethyl sulfoxide (DMSO) in PBS.[10] The mixture containing 200 ug/ml viral protein was exposed in plastic Petri dishes without lids to long-wave ultraviolet light (LWUV; 365 nm) with a minimum radiation flux of 6×10^3 ergs/sec/cm^2 at the sample surface. Aliquots of irradiated and untreated mixtures were taken at various times and tested for infectivity by culturing with H-9 cells[11] for 45 days followed by indirect immunofluorescence assessment of acetone-fixed cells using normal or AIDS sera developed with fluorescein conjugated anti-human IgG. Polypeptide integrity of treated and untreated virus was evaluated by polyacrylamide gel electrophoresis and ELISA essentially as described[12] except that 4-methyl umbelliferyl phosphate (Sigma Chemical Co., St. Louis, MO) was used as substrate for the alkaline-phosphatase conjugate. The product of conjugate enzymatic activity, methyl umbelliferone, was measured by fluorimetry using a microtiter plate reader (MicroFluor; Dynatech Laboratories, Alexandria, VA).

Analyses of Viral Proteins

The polypeptide composition of purified virus particles and the apparent molecular weights of HTLV-III/LAV particle-associated proteins were determined by one-dimensional sodium dodecyl sulfate polyacrylamide gel electrophoresis (SDS-PAGE)[13] using discontinuous 12.5% acrylamide slab gels. For certain experiments, samples were not heated or reduced in the presence of SDS-containing lysis buffers. Isoelectric patterns of HTLV-III/LAV polypeptides were analyzed by two-dimensional PAGE (2D-PAGE)[8] using thin-layer, broad-range, polyacrylamide isoelectric focusing slab gels (LKB, Bromma, Sweden) instead of separate tube gels in the first dimension. These slab gels were soaked for 1 hr in a volume of 10M urea containing 2% Nonidet-P40 (NP-40) and 1% 2-mercaptoethanol (2-ME) equal to twice the gel mass. The samples were applied and focused for 1.5 hr at 1400V and 1mA maximum current/gram of gel. The influenza virus, A/USSR/90/77, was coelectrophoresed in some gels to calibrate the focusing dimension.[8]

Western blots were prepared by electroblot transfer of HTLV-III or uninfected H9 cell proteins resolved by SDS-PAGE to 0.45 um cellulose nitrate membranes (Schleicher and Schuell, Keene, NH).[14] After transfer, excess reactive groups on the membranes were blocked with 2.5% highly-purified bovine serum albumin (BSA; #A-7030 Sigma) in PBS containing 0.05% NP-40 (BSA buffer). HTLV-III/LAV particle-associated glycoproteins on Western blots[15] were identified by exposure for 1 hr to Concanavalin-A (20 ug/ml) in PBS/0.05% NP-40 buffer containing 1mM MgCl and 1mM CaCl, washing 5 times for 1 min in this buffer and then exposure to horseradish peroxidase (10 ug/ml) in PBS/NP-40 buffer, again for 1 hr. After 5 more washes, the glycoblots were developed using the chromogenic substrate, 4-chloro-1-napthol, which had been dissolved in absolute methanol to a final concentration of 0.3% and then diluted 1:6 with NTE buffer containing 6mM H_2O_2. HTLV-III/LAV particle-associated proteins were tested for reactivity with rabbit antisera by immunoblotting as described[16] using ^{125}I-labelled goat anti-rabbit IgG to develop bound immunoglobulin followed by autoradiography.

RESULTS AND DISCUSSION

Zonal centrifugation of supernatants harvested from H9/HTLV-III$_B$ cells yielded a broad density fraction enriched for live virus. This material contained viral particles of variable protein constitution separable by secondary equilibrium isopycnic centrifugation as shown in Figure 1. Fractions 8 and 9 of the secondary gradient contained all of the virus-associated proteins (Figure 1, Panel A), glycoproteins (Panel B), and the greatest reverse transcriptase activity. We, therefore, inferred that these fractions contained the bulk of intact viral particles. Other fractions such as 3 and 4 contained traces of some but not all of the viral polypeptides and some RT activity. These may have been aggregates of proteins from fragmented rather than complete viral particles. Alternatively, fractions 3 and 4 may have contained that portion of the total intact viral harvest which exhibits nonideal sedimentation properties as previously reported for other retroviruses.[7]

The fractions collected from the dense region of the sucrose gradient (11 and 12) appeared to contain aberrant assemblies. Such fractions, nonetheless, proved useful in purification efforts, not reported here, since they were partially enriched for certain proteins with molecular weights around 45 kD (p45) and 66 kD (p66). The determination that these two proteins were of cellular origin, given later in the text, lent

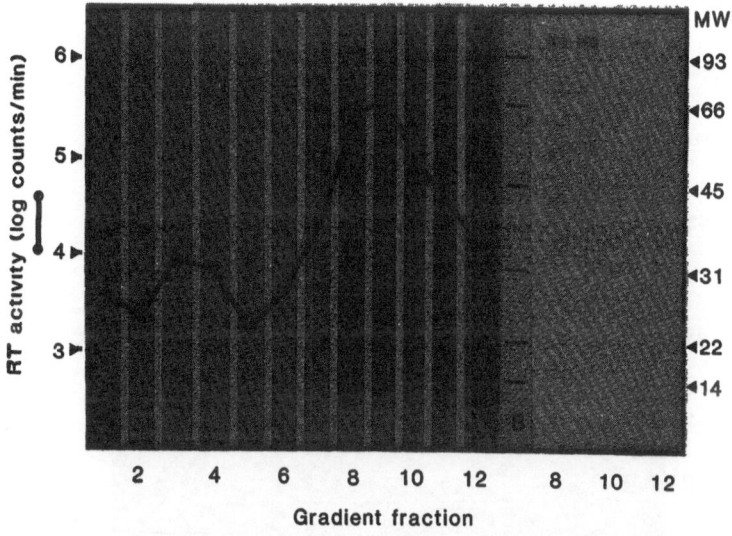

Figure 1: Secondary sucrose density gradient fractionation of HTLV-III/LAV particles. The polypeptides associated with the viral particles in each fraction have been resolved by SDS-PAGE and stained for total protein (Panel A; Fxs 1-12) or for glycoproteins (Panel B; Fxs 8-12) following electroblot transfer. Reverse transcriptase (RT) activities found in the gradient fractions have been superimposed on Panel A. Kilodaltan molecular weights (MW) of coelectrophoresed markers are indicated on the right margin in the appropriate positions.

justification for excluding fractions enriched for p45 and p66 from pools of those gradient fractions (8 and 9) inferred to contain complete virus.

To ensure safety during experimentation with HTLV-III/LAV, we adapted a simple, rapid method for virus inactivation based on the successful trials with RNA and DNA viruses already reported.[10],[11] The psoralen analogue, 4'-amino-methyl trioxalen, was added to virus suspensions and the mixture exposed to LWUV for up to 2 hr during which time samples were taken and tested for infectivity. Exposure periods longer than 30 min completely ablated any infectivity measurable by up to 45 days of cultivation with H-9 cells. This effective inactivation method was chosen since the psoralen photoreaction appears to be specific for nucleic acid polymers by forming covalent adducts which block replication and transcription.[17] Although some studies suggest that oxidative photoreactions between psoralen and proteins may occur, we observed no differences in the SDS-PAGE polypeptide patterns of the HTLV-III/LAV before or after treatment. Furthermore, the reactivity in ELISA of whole virus antigens with AIDS serum was not altered as a result of psoralen treatment.

The polypeptide constituents of HTLV-III/LAV were resolved by 2-D PAGE to produce molecular weight and isoelectric point fingerprints for general biochemical characterization. A complex pattern of single or multiply-charged species, generally with acidic or neutral isoelectric points dominated the profile shown in Figure 2. The major protein near the center of the figure with an apparent molecular weight (M_r) of 28 kD

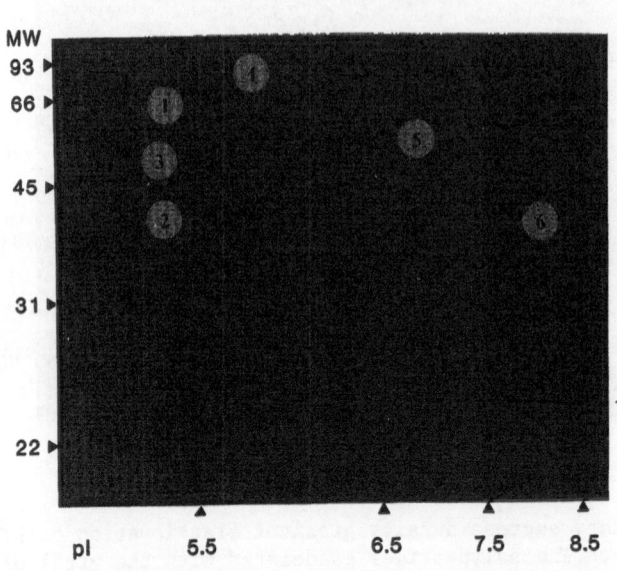

Figure 2: Two-dimensional fingerprint of HTLV-III/LAV associated polypeptides. The isoelectric point (pI) and kilodalton molecular weight (MW) scales indicated on the margins were determined by coelectrophoresis of markers in an identical procedure. Certain virus-associated polypeptides indicated by numbers or arrows placed above and to the right of the appropriate spot are discussed in the text.

and an isoelectric point of 6·3 has been variously named core or gag protein, p24 or p25.[16,18,19] It has here been labelled "core" to avoid confusion considering the disparities among reported apparent molecular weights and the results found in this study. The minor charge species, pI=5·9, toward the acidic (left) side of "core" protein may have resulted by deamidation of the 28 kD protein during handling or electrophoresis.

Another major protein with M_r=55 kD and pI<5·5 (labelled 1 in Figure 2) exhibited the same isoelectric point as the major protein below it with M_r=38 kD (labelled 2). The staining characteristics and diffuse banding patterns indicated that both species were glycoproteins. As shown in Figure 1 (Panel B), glycoproteins of these molecular weights were detected among virus-associated proteins resolved by one-dimensional SDS-PAGE. Interestingly, the identical pIs in 2D-PAGE suggested that the 55 and 38 kD glycoproteins were associated throughout the isoelectric focusing procedure to be separated later in the second dimension. These molecules could have been associated as subunits of a higher molecular weight species of about 93 kD. Alternatively, the 38 kD glycoprotein could have been a subunit of the parent 55 kD glycoprotein. In either case, the observed pIs of the 38 and 55 kD moeities would probably not reflect the actual isoelectric points.

The polypeptide with M_r=45 kD and pI<5·6 (labelled 3 in Figure 2) exhibited multiple charge species possibly due to post-translational modification. Based upon the observed M_r and pI, this protein was tentatively identified as actin, a common virus particle constituent.[8] Other common cytoskeletal proteins such as alpha & beta-tubulin[20] may also have been present focusing in the acidic region (pI=5·2) around 53-55 kD just below the 55 kD glycoprotein.

The polypeptide labelled 4 in Figure 2 also displayed some charge heterogeneity, a pI<5·6 and M_r=66 kD. This protein, when tested by immunoblotting, bound IgG from rabbit antisera made against the proteins of uninfected H9 cells indicating that p66 was of cellular, not viral, origin. This protein was easily removed from virus suspended in NTE buffer by diluting 1:5 in water or alkaline Tris (pH 9.0) followed by rapid sedimentation (40,000 rpm x 10 min). The p66 remained in the supernatant after this relatively gentle procedure suggesting that it was peripherally associated with the virus perhaps by ionic interactions. This protein did not react with anti-BSA in immunoblots.

Two other major polypeptides, marked 5 and 6 in Figure 2, focused in the neutral region of the 2D-PAGE. The protein with M_r=50 and pI=6·6 appeared to be related to "core" protein based on the similarity of isoelectric points and staining characteristics. Our preliminary one-dimensional immunoblot studies using a monoclonal antibody to p24 (a gift from Dr. Robert Gallo, NCI, Bethesda, MD) showed specific binding to "core" protein and another protein corresponding in apparent molecular weight to this 50 kD protein strongly suggesting an inter-relationship. A gag gene product in HTLV-III/LAV with an M_r=55 kD has already been described as a precursor protein yielding upon processing, "core" protein and another product of M_r=34 kD.[21]

The second major polypeptide (labelled 6 in Figure 2) displayed a M_r=40 kD and a pI=8·0. By inference drawn from other work,[21-23] this protein was tentatively identified as a cleavage product of the HTLV-III/LAV env gene precursor molecule. Since the results in Figure 1 (Panel B) showed that a glycoprotein banded around the 40-43 kD region, the tentative identification was strengthened. However, the nearby species focusing between pI 7·4 and 7·8 at M_r=42 kD (indicated by the arrow in Figure 2) bore the multiple charge characteristics typical of some viral glycoproteins.[8] Therefore, certain designation was not possible. Similarly, many of the other minor polypeptides associated with HTLV-III/LAV and resolved by the 2D-PAGE remain to be identified.

ACKNOWLEDGEMENTS

We are grateful to Mrs. Cathy Hobbs for preparing this manuscript.

REFERENCES

1. M. Essex, M. F. McLane, T. H. Lee, L. Falk, C. W. S. Howe, J. I. Mullins, C. Cabradilla, and D. P. Francis, Antibodies to cell membrane antigens associated with human T-cell leukemia virus in patients with AIDS, Science 220:859 (1983).

2. F. Barre-Sinoussi, J. C. Chermann, F. Rey, M. T. Nugeyre, S. Chamaret, J. Gruest, C. Dauguet, F. Axler-Bain, F. Vezinet-Brun, C. Rouzioux, W. Rozenboum, and L. Montagnier, Isolation of a T-lymphotropic retrovirus from a patient at risk of acquired immunodeficiency syndrome (AIDS), Science 220:868 (1983).

3. R. C. Gallo, S. Z. Salahuddin, M. Popovic, G. M. Shearer, M. Kaplan, B. F. Haynes, T. J. Palker, R. Redfield, J. Oleske, B. Safai, G. White, P. Foster, and P. D. Markham, Frequent detection and isolation of cytopathic retroviruses (HTLV-III) from patients with AIDS and at risk for AIDS, Science 224:500 (1984).

4. J. A. Levy, A. D. Hoffman, S. M. Kramer, J. A. Landis, J. M. Shimabukuro, and L. S. Oshiro, Isolation of lymphocytopathic retroviruses from San Francisco patients with AIDS, Science 225:840 (1984).

5. R. J. Albin, AIDS: A noncommunicable cofactor, JAMA 253(23):3398 (1985).

6. H. W. Haverkos and R. Edelman, Female-to-male transmission of AIDS, JAMA 254(8):1035 (1985).

7. I. Toplin and P. Sottong, Large-volume purification of tumor viruses by use of zonal centrifuges, App. Microbiol. 23:1010 (1972).

8. J. C. Leavitt, M. A. Phelan, A. H. Leavitt, R. E. Mayner, and F. A. Ennis, Human influenza A virus: Comparative analysis of the structural polypeptides by two-dimensional polyacrylamide gel electrophoresis, Virology 99:340 (1979).

9. B. J. Poiesz, F. W. Ruscetti, A. F. Gazdan, P. A. Bunn, J. D. Minna, and R. C. Gallo, Detection and isolation of type C retrovirus particles from fresh and cultured lymphocytes of a patient with cutaneous T-cell lymphoma, Proc. Nat'l. Acad. Sci. USA 77:7415 (1980).

10. D. C. Redfield, D. D. Richman, M. N. Oxman, and L. H. Kronenberg, Psoralen inactivation of influenza and herpes simplex viruses and of virus-infected cells, Infect. Immun. 32:1216 (1981).

11. M. Popovic, M. G. Sarngadharan, E. Read, and R. C. Gallo, Detection, isolation, and continuous production of cytopathic retroviruses (HTLV-III) from patients with AIDS and Pre-AIDS, Science 224:497 (1984).

12. B. R. Murphy, M. A. Phelan, D. L. Nelson, R. Yarchoan, E. L. Tierney, D. A. Alling, and R. M. Chanock, Hemagglutinin-specific enzyme-linked immunosorbent assay for antibodies to influenza A and B viruses, J. Clin. Micro. 13(3):554 (1981).

13. U. K. Laemmli, Cleavage of structural proteins during assembly of the head of bacteriophage T4, Nature (London) 277:680 (1970).

14. H. Towbin, T. Staehelin, and J. Grodon, Electrophoretic transfer of proteins from polyacrylamide gels to nitrocellulose sheets: Procedure and some applications, Proc. Nat'l. Acad. Sci. USA 76:4350 (1979).

15. J. C. S. Clegg, Glycoprotein detection in nitrocellulose transfers of electrophoretically separated protein mixtures using Concanavalin A and peroxidase: Application to Arenavirus and Flavivirus proteins, Anal. Biochem. 127:389 (1982).

16. M. G. Sarngadharan, L. Bruch, M. Popovic, and R. C. Gallo, Immunological properties of the Gag protein p24 of the acquired immunodeficiency syndrome retrovirus (human T-cell leukemia virus type-III), Proc. Nat'l. Acad. Sci. USA 82:3481 (1985).

17. C. V. Hanson, Inactivation of viruses for use as vaccines and immunodiagnostic reagents, in: "Medical Virology 2," L. M. de la Maza and E. M. Peterson, eds., Elsevier Science Publishing Co., Inc., New York (1983).

18. L. Montagnier, F. Clavel, B. Krust, S. Chamaret, F. Rey, F. Barre-Sinoussi, and J. C. Chermann, Identification and antigenicity of the major envelope glycoprotein of lymphadenopathy-associated virus, Virology 144:283 (1985).

19. V. S. Kalyanaraman, C. D. Cabradilla, J. P. Getchell, R. Narayanan, E. H. Braff, J.-C. Chermann, F. Barre-Sinoussi, L. Montagnier, T. J. Spira, J. Kaplan, D. Fishbein, H. W. Jaffe, J. W. Curran, and D. P. Francis, Antibodies to the core protein of lymphadenopathy-associated virus (LAV) in patients with AIDS, Science 225:321 (1984).

20. J. L. Saborio, F. Diaz-Barriga, G. Duran, V. Tsutsumi, and E. Palmer, Purification and characterization of GP-55, a protein associated with actin-based cytoplasmic gels derived from brain tissue, J. Biol. Chem. 260:8627 (1985).

21. W. G. Robey, B. Safai, S. Oroszlan, L. O. Arthur, M. A. Gonda, R. C. Gallo, and P. J. Fischinger, Characterization of envelope and core structural gene products of HTLV-III with sera from AIDS patients, Science 228:593 (1985).

22. J. Schneider, H. Bayer, U. Bienzle, and G. Hunsmann, A glycopoly-peptide (gp 100) is the main antigen detected by HTLV-III antisera, Med. Microbiol. Immunol. 174:35 (1985).

23. J. S. Allan, J. E. Coligan, F. Barin, M. F. McLane, J. G. Sodroski, C. A. Rosen, W. A. Haseltine, T. H. Lee, and M. Essex, Major glycoprotein antigens that induce antibodies in AIDS patients are encoded by HTLV-III, Science 228:1091 (1985).

STRUCTURE AND FUNCTION OF FISH ANTIFREEZE POLYPEPTIDES

Choy L. Hew and Peter L. Davies

Research Institute, Hospital for Sick Children
Toronto and Depart. of Biochemistry, Queen's University
Kingston, Ontario, Canada

INTRODUCTION

The ability of many organisms to survive in cold environments is of considerable scientific interest and importance. During the past two decades, two entirely different animal groups have been investigated for their ability to survive at subzero temperatures. These are the marine fishes inhabiting the ice-laden sea water of the polar regions (1,2,3) and the overwintering terrestrial arthropods (4). Generally, the members of these groups that do not undertake a seasonal migration to warmer environments produce serum antifreezes. These may be the macromolecular antifreezes (glycoproteins and polypeptides)found in fishes and some insects, or the colligative antifreezes such as glucose, glycerol, sorbitol and other small molecules that are common in the latter group. Some organisms also appear to survive in a super-cooled state (5). In the present communication, only the structural and functional aspects of antifreeze proteins from the marine fishes will be discussed.

Depending on latitude and season, the sea water temperatures in the polar and sub-polar regions can be as low as -1.8°C. However, the serum of most fishes freezes at -0.6°C which results in the death of the organism. Scholander et al. (5) were the first to report that the sera of Arctic fish had a lower freezing temperature than did the sera of fish not adapted to the cold. The lowering of the serum freezing temperature was due to the presence of macromolecular materials. DeVries, Feeney and co-workers (6) isolated and characterized these materials from Antarctic fish and found them to be glycoproteins of unusual chemical structure. The glycoprotein antifreeze (AFGP) from the two species of Antarctic fish studied, T. borchgrevinki and D. mawsoni are identical and contain eight polypeptides made up of a repeating tripeptide unit of alanine-alanine-threonine linked to galactose-N-acetylgalactosamine. This glycopeptide unit is repeated up to 50 times to give eight components of 2,600 to 33,000 daltons. Since then, AFGP have been isolated from the Northern cods, B. saida, E. gracilis, G. ogac and more recently the Atlantic cod, G. morhua and frostfish, M. tomcod. All of these AFGP have similar, if not identical structures. Some of the smaller glycopeptides contain proline or arginine which replace alanine or threonine at certain positions.

Presence of antifreeze proteins in Newfoundland fishes

Because of the Labrador cold current, the temperature in Newfoundland coastal waters fluctuates around -1.0°C to -1.3°C in the winter. For the past several years, our laboratories have investigated the occurence, distribution and structure of antifreeze proteins in various species inhabiting these waters. Our studies have been facilitated by the use of two instruments, namely the freezing point osmometer (and more recently the nanoliter osmometer) and reverse phase HPLC. The osmometer provides a simple, fast and non-destructive detection method for antifreeze activity in crude serum while the reverse phase HPLC has been used to resolve multiple antifreeze components. As a result we have identified antifreeze proteins in at least six species of fish (Table 1). Their sera show thermal hysteresis (commonly referred to as antifreeze activity).

Whereas the antifreeze glycoproteins are present all year round in Antarctic fish, the occurence of antifreeze proteins in Newfoundland fish is seasonal. The best documented example is the winter flounder. Its antifreeze is present at a concentration of 8-10 mg/ml in the winter, but is not detectable in the summer (14).

Table 1. Occurence of antifreeze proteins in Newfoundland marine fishes. Thermal hysteresis is defined as the difference between the freezing and melting temperatures.

Species	Freezing Temperature[a] in °C	Thermal Hysteresis[a] in °C	References
Atlantic cod Gadus morhua	-1.01	0.27	7
Frostfish Microgadus tomcod	-0.92	0.20	8
Winter flounder Pseudopleuronectus americanus	-1.20	0.50	9,10
Shorthorn sculpin Myoxocephalus scorpius	-1.20	0.40-0.50	11
Sea raven Hemitripterus americanus	-1.20	0.40-0.50	12
Ocean pout Macrozoarces americanus	-1.62	0.87	13

[a] The freezing temperature and thermal hysteresis were measured using a freezing point osmometer (Model 3R, Advanced Instruments, Needham Height M A) or a nanoliter osmometer (Clifton Technical Physics, Hartford, NY).

Structure of antifreeze polypeptides

In addition to the direct measurement of thermal hysteretic
activity in crude serum, the presence of antifreeze proteins, their
relative contribution to the freezing activity and molecular size can be
simultaneously determined by Sephadex G75 chromatography. Freshly prepared and concentrated serum was applied directly to the column. As seen
in Fig. 1, both the sera of shorthorn sculpin and sea raven contained
antifreeze proteins of molecular weights 10,000 and 17,000, respectively.
Although the antifreeze proteins from winter flounder, shorthorn sculpin,
sea raven and ocean pout showed only one single molecular component on
Sephadex G75 chromatography, their microheterogeneity was revealed by
reverse phase HPLC (Fig. 2). Ocean pout antifreeze is most ·complex
having at least 12 active components of similar size and electrophoretic
charge. The microheterogeneity of these polypeptides is largely due to
their being encoded in multigene families as well as some post-translational modifications of the nascent chains.

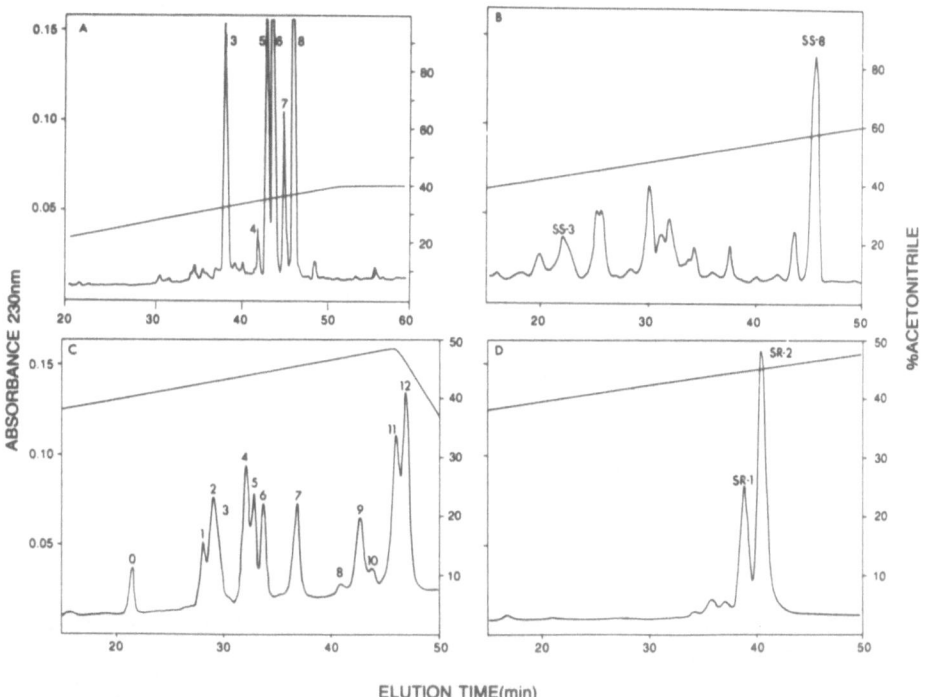

*Figure 1. The identification of antifreeze polypeptides on Sephadex G75
chromatography. Freshly prepared, concentrated sera (2.3 ml) was applied
on the column (1.6 x 86 cm) in 0.2 M NH_4HCO_3. After lyophilization the
fractions were redissolved in 0.5 ml of 0.01 M NH_4HCO_3 and their antifreeze activity were monitored with a freezing point osmometer (Model
3R, Advanced Instruments). (A) Shorthorn sculpin (B) Sea raven. The
elution position of cytochrome C (M_r=12,000) was indicated.*

The structures of most of these antifreezes have been characterized through a combination of automated Edman degradation and cDNA sequencing (Table 2). Antifreeze(s) from Atlantic cod(7) and frostfish(8) were found to be similar if not identical to the Antarctic AFGP. However the antifreezes from winter flounder (15-17), shorthorn sculpin (18), ocean pout (19) and sea raven (11) are not glycoproteins. Based on their amino acid compositions, primary and secondary structures, as well as immuno-logical properties, these antifreeze proteins (or more appropriately anti-freeze polypeptides (AFP))can be further divided into at least 3 distinct groups, namely the alanine-rich AFP (winter flounder and shorthorn sculpin), cystine-rich AFP (sea raven) and a third type of AFP which is neither alanine nor cystine-rich (ocean pout). Some of the structural characteristics of various antifreezes are shown in Table 3.

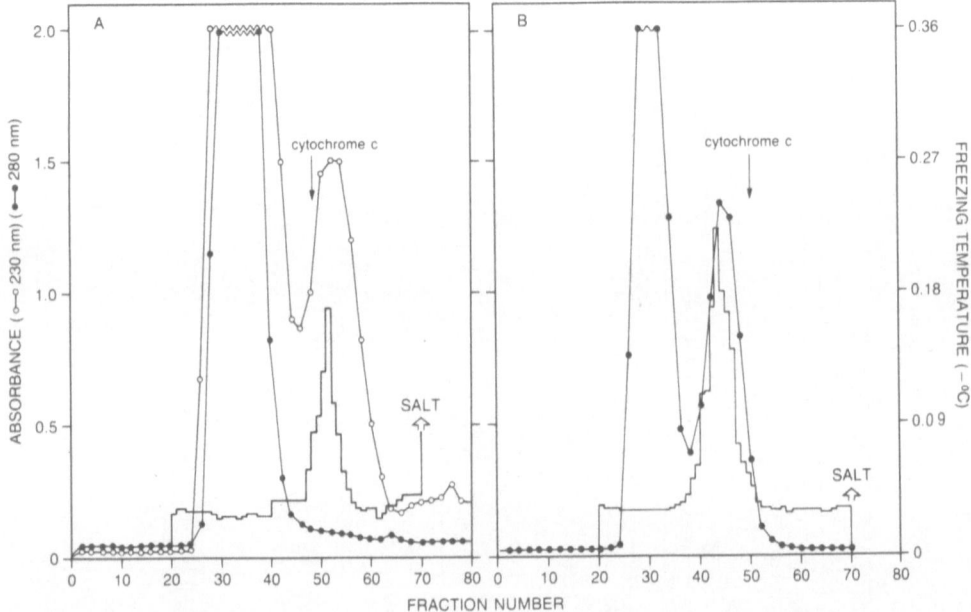

Figure 2. Analysis of antifreeze polypeptides by reverse phase HPLC. The AFP's were intitially purified by Sephadex G75 chromatography as in Figure 1. (a) Winter flounder, Altex Ultrasphere ODS column (4.6 mm x 25 cm), flow rate 1 ml/min with 0.01 M triethylamine phosphate buffer, pH 3.0-acetonitrile gradient. Antifreezes are designated from 3 to 8. The conditions for (b) shorthorn sculpin (c) ocean pout (d) sea raven are Waters μ Bondapak C 18 column (7.1 mm x 30 cm), flow rate 1 ml/min with 0.05% trifluoroacetic acid acetonitrile gradient. All chromatro-graphy were done at room temperature. In (c) component 0 is not an AFP-related peptide.

Although these proteins are diverse in their structures, they nonetheless exhibit similar biological activity. All of them lower the freezing temperature via a non-colligative mechanism, i.e. there is a difference between the freezing and melting temperatures. On a molal basis, these antifreeze proteins are 200-300 fold better antifreeze than NaCl. Studies of the freezing behaviour of AFGP and AFP indicated that they interacted with and absorb to ice during freezing (1,3). Electron microscopic studies showed that in the presence of AFGP or AFP, ice formation was in the form of long thin spicules with the direction of growth parallel to the c-axis (20). A mechanism involving inhibition of crystal growth at the ice water interface has been proposed (3).

In the case of AFGP, chemical modification of the hydroxyl side chain of the carbohydrate moiety correlated with the loss of activity suggesting that hydrogen bonding was important. For the antifreeze to be tightly absorbed to ice, potential hydrogen-bonding residues would need to occupy positions in the antifreeze molecule to allow them to align with the opposite oxygens and hydrogens in the ice lattice in a regular pattern. Many of the hydroxls of the carbohydrate side chains are separated by a distance of 4.5 $\overset{\circ}{A}$, a distance that also separates the oxygens parallel to the a-axis in the ice lattice (3).

Table 2. Amino acid sequences of some antifreeze polypeptides

Spec.[a]	Fract.[b]	Sequences
W F.	#6	DTASDAAAAAALTAANAKAAAELTANNAAAAAAATAR
	#8	DTASDAAAAAALTAANAVAAAKLTADNAAAAAAATAR
S S.	SS-8	MNGETPAQKAARLNVVVVLAAKTAADAVVVAKAAIAAAAASA
	SS-3	MN----------AFARAAAKTAADALAVAKKTAADAAAAAAA
O P.	HPLC-6	-Q-SVVAQLIPINTALTPAMMQGKV-INFIGIPFAE-MSQIVGK
	HPLC-12	NQASVVANQLIPINTALTLVMMRSEVVII-PVGIP-AEDIPRLVSM
		QVNTPVAKGDTLMPNMVKLYVAG
		QVNRAVPLGTTLMPDMVKGYPPA

[a] The abbreviations are W.F., winter flounder; S.S., shorthorn sculpin; O.P., ocean pout. The dashed lines in S.S. and O.P. are used to maximize structural homology.

[b] The fraction numbers correspond to the numbers designated in Figure 2.

The presence of a triplicated eleven amino acid repeating structure of Thr-$(X)_2$-polar amino acid-$(X)_7$ where X is a non-polar amino acid (mainly as alanine) in the winter flounder AFP has provided an additional insight on the structure and function of these AFP. The distance between the threonyl and the polar amino acid is apparently 4.5 Å. DeVries (3) has proposed a model indicating the flounder AFP hydrogen bonded to the prism face of hexagonal ice, parallel to one of the a-axes. This model is consistent with loss of activity upon chemical modification of the carboxyl side chains (10). With the recent elucidation of the amino acid sequence of shorthorn sculpin AFP, it is apparent that besides the presence of these repeating structures, both flounder and sculpin AFP exist as amphiphilic helical structure (18). The flounder AFP has now been crystallized in our laboratories (21). The crystal has a space group of $P2_1$ with unit cell parameters, a=38.14, b=37.19, c=21.82 and d=101.5°. There are two molecules in the asymmetric unit. The subsequent x-ray diffraction studies should provide the information needed to delineate the structural and conformational requirements for antifreeze activity.

Table 3. Classification of proteinaceous antifreezes

TYPE	PRIMARY STRUCTURE		SECONDARY STRUCTURE	MOLECULAR WEIGHT	BIOSYNTHETIC PRECURSOR
Antifreeze glycoproteins (AFGP) 2 Antarctic fish Atlantic cod Polar cod Frostfish	disaccharide (AlaAlaThr	)n	expanded	2,600 to 33,000	Not known
Antifreeze polypeptides (AFP)					
(I) White flounder Shorthorn sculpin Alaskan plaice[a]	Alanine-rich 11-amino acid repeat		α helical amphiphilic	3,300 to 4,500	preproAFP
(II) Sea raven	Cystine-rich		β structure	17,000	preAFP
(III) Ocean pout	Neither alanine nor cystine-rich		Compact non α helical non β structure	6,000	preAFP

[a]Based on amino acid composition and partial sequence only (3).

The documentation of new types of AFP from sea raven and ocean pout has added additional complexity to the mechanism of action of the antifreezes. The lack of obvious repeating structures with 4.5 A spacing between two hydrophilic amino acids in the ocean pout AFP has made it difficult to generalize about a common mechanism for antifreeze action. More work is required to understand the mode of action of these molecules.

Evolution of Antifreezes

The presence of at least four different types of fish antifreezes (1 AFGP and 3 types of AFP) is puzzling. Furthermore some closely related fishes have very different antifreeze protein types while some distantly related fishes have almost identical antifreezes. For example, both the sea raven and the shorthorn sculpin are morphologically and phylogenetically related as members of the same cottid family but they each produce different types of AFP. On the other hand, the winter flounder and shorthorn sculpin which belong to different orders synthesize the same type of alanine-rich AFP.

Scott et al. have attempted to rationalize the distribution of antifreeze protein types within marine teleosts(22). They postulate that antifreeze protein genes have been independently elaborated from several genetic elements during the teleost radiation in response to the Cenozoic ice ages which began in the southern hemisphere approximately 3×10^7 years ago and in the northern hemisphere 2.5×10^6 years ago. If, as it seems likely, the present day cottids obtained their antifreeze protein gene identitiy in the Northern hemisphere, then in terms of a time frame they could have done so after the cottid genera were distinct. Instead of receiving the same antifreeze type from a common ancestor the sea raven and shorthorn sculpin predecessors were already evolving along distinct lines and apparently utilized different genetic elements to formulate their antifreezes. Antifreeze types that evolved in the Southern hemisphere will have appeared much earlier on and therefore be common to whole families or even suborders of fishes. The time which has elapsed since the earlier antarctic glaciation events began is probably sufficient for the bipolar distribution of some of these antifreeze types.

CONCLUSION

Antifreeze glycoproteins and antifreeze polypeptides present an intriguing and important area for the study of protein structure, function and evolution. The structural information accumulated during the past few years has provided a basis for further experimentation using such techniues as chemical modifications and site specific mutagenesis to prove the mode of action of AFP's in further detail. X-ray difraction studies of winter flounder and other AFP's will make a significant contribution to our understanding of the interaction of these macromolecules with ice crytals.

ACKNOWLEGEMENT

This investigation was supported by the MRC and NSERC, Canada.

REFERENCES

1. Feeney, R.C. and Yeh, Y., (1978), Adv. Protein Chem., 32: 191-282.
2. Ananthanarayanan, V.S. and Hew, C.L., (1978), In Srinavasan, B. (ed.) Biomolecular structure, conformation and evolution. Pergamon Press, New York, 39: 191-198.
3. DeVries, A., (1984), Phil. Trans. R. Soc. London, B., 304: 575-588.
4. Duman, J.G.,(1977), Cryobiol., 19: 613-627.
5. Scholander, P.G., vanDane,L., Kanwisher, J.W., Hammel, H.T. and Gordon, M.S.,(1957), J. Cell. Comp. Physiol., 49: 5-24.
6. DeVries, A., Komatsu, S.K. and Feeney, R.E. (1970), J. Biol. Chem., 245: 2901-2913.
7. Hew, C.L., Slaughter, D.,Fletcher, G.L. and Joshi, S.B., (1981), Cdn. J. Zool., 59: 2186-2192.
8. Fletcher, G.L., Hew, C.L. and Joshi, S.B., (1982), Can.J. Zool., 60: 348-355.
9. Hew, C.L. and YIP, C., (1976), Biochem. Biophys. Res. Commun., 71: 845-850.
10. Duman, J.G. and DeVries, A.L., (1976), Comp. Biochem. Physiol.,53B: 375-380.
11. Hew, C.L., Fletcher, G.L. and Ananthanarayanan, V.S.,(1980), Can. j. Biochem., 58:377.
12. Slaughter, D., Fletcher, G.L., Ananthanarayanan, V.S. and Hew, C.L., (1981), J. Biol. Chem., 256:2022.
13. Hew, C.L., Slaughter, D., Joshi, S.B.,Fletcher, G.L. and Ananthanarayanan, V.S., (1984), J. Comp. Physiol, 155: 81-88.
14. Slaughter, D. and Hew, C.L., (1982), Can. J. Biochem., 60: 824.
15. DeVries, A.L. and Lin, Y., (1977), Biochem. Biophys. Acta., 495: 380-392.
16. Davies, P.L., Roach, A.H. and Hew, C.L., (1982). Proc. Natl. Acad. Sci. USA, 79: 335-339.
17. Pickett, M.H., Scott, G., Davies, P., Wang, N., Joshi, S.B. and Hew, C.L., (1984), Eur. J. Biochem., 143: 35-38.
18. Hew, C.L., Joshi, S.B., Wang, N.C., Kao, M.H. and Ananthanarayanan, V.S., (1985), Eur. J. Biochem., 151: 167-172.
19. Li, X.M., Trinh, K.Y., Hew, C.L., Buettner, B., Baenziger, J., Davies, P. L.,(1985), J. Biol. Chem., 260: 12,904-12,909.
20. Raymond, J.A. and DeVries, A.L., (1977), Proc. Natl. Acad. S 74: 2589-2593.
21. Yang, D. and Hew, C.L., (1985), Am. Crystallography Assoc., August Standord, CA. PBI.
22. Scott, G.K., Fletcher, G.L. and Davies, P.L., (1986), Can. J. of Fisheries and Aquatic Sce., 43: 1028-1034.

SEQUENCE OF A GLYCINE-RICH PROTEIN FROM LIZARD CLAW:

UNUSUAL DILUTE ACID AND HEPTAFLUOROBUTYRIC ACID CLEAVAGES

Adam S. Inglis, J. Morton Gillespie, Charles M.
Roxburgh, Lois A. Whittaker and Franca Casagranda

Division of Protein Chemistry, CSIRO, Parkville
Melbourne, Vic

INTRODUCTION

The hard keratins of mammals and birds have been extensively studied over recent years and a considerable body of knowledge is now available on the structure and composition of the constituent proteins and of their arrangement within the keratin structure[1]. Mammalian keratins have a very characteristic structure in which filaments of about 70A diameter are embedded in a non-filamentous matrix composed of sulfur-rich and glycine-tryosine-rich proteins. The filaments are responsible for the α-type X-ray pattern. In contrast avian keratins contain only one family of proteins arranged as small filaments of about 30A diameter which are responsible for a characteristic X-ray pattern, often referred to as the feather type. No sequence homology has been found between avian and mammalian keratin proteins and it is generally considered that they represent separate evolutionary developments.

There is a third group of hard keratins produced by reptiles (lizard claw and turtle scute) which have received scant attention. It is known from X-ray and electron microscope studies that a feather-type structure is present and the proteins are ordered into 30A filaments. The major constituent proteins (SDS components 1-3) are similar to those of avian keratins in having molecular weights around 15,000 and high contents of glycine (30%) but they differ by being rich in half cystine (13%) and in this respect are like some mammalian high-sulfur proteins which are comparable in size[2,3]. Since only the determination of a complete amino acid sequence of a lizard claw protein will enable the inter-relations between these keratin proteins to be assessed, we chose to determine the amino acid sequence of SDS 1 which is homogeneous and accounts for 50% of the claw, and for 70% if the homologous SDS 2 is included.

The sequence of SDS 1 was determined almost entirely by automatic sequencing of fragments obtained by high performance liquid chromatography. Difficulties were encountered because the N-terminus is blocked and the N-terminal region is very rich in glycine and half-cystine (S-carboxymethylcysteine in the soluble protein) residues. Novel cleavage sites were found during attempts to open up this region. Contrary to earlier work[4], under the conditions used here heptafluorobutyric acid (HFBA) cleaved glycylserine bonds rather than aspartyl bonds, and this was advantageous in completing the determination. Dilute formic acid[5] cleaved

an asparaginylproline bond and partially cleaved glycylglycine bonds which
provided essential overlapping peptides. Both acidic reagents
substantially cleaved after the blocked N-terminal alanine.

METHODS

Varanus gouldii claws were dissected to obtain the outer sheath of
translucent keratin, this was ground, washed and the proteins solubilised
with 8M urea-0.2M mercaptoethanol at pH 11. The proteins were alkylated
with iodoacetate, dialysed and SDS 1 obtained by a fractional precipitation
with ammonium sulfate followed by chromatography on DEAE-cellulose[2].

Enzymic cleavages with *A. mellea* enzyme, trypsin (TPCK treated),
chymotrypsin and thermolysin, and chemical cleavages with cyanogen bromide
and dilute formic acid were carried out by conventional methods. HFBA
cleavage was made with a chromic acid treated preparation[6] by adding 200μl
to 0.2 mg of protein in a tapered glass test tube (pretreated in a muffle
oven at 500°C), flushing briefly with nitrogen, stoppering and incubating
at 55°C for 60 minutes. After drying down, the residue was redissolved in
98% formic acid and redried before loading onto the HPLC or gas phase
sequencer.

For HPLC of peptides, dried digests were dissolved usually in 0.1%
trifluoroacetic acid (TFA) containing 20% acetonitrile. Where solubility
was poor, less dilute TFA was used. Reversed phase separations were done at
50° or 60° (HT 101 column oven, Nucleus instruments, Melbourne) on a wide
pore C_{18} column, 250 x 4.6mm (Vydac 218 TP54 300A°), or for larger
preparations on the corresponding 10mm column (Vydac 218 TP 510), with an
appropriate gradient and flow rate between 1 and 2ml/min. Solvent A was
normally 0.1% TFA in water, and solvent B 0.1% TFA in 90% acetonitrile.
Where a different pH was required for solubility or separation, an ammonium
formate (0.01M, pH 6.8)/acetonitrile system was used. Detection was at
220nm and 280nm. The chromatograph was an LDC assembly with a CCM
controller. The modules included were two Constametric III pumps, a dynamic
mixer, two Spectromonitor III variable wavelength detectors, a FRAC-100
fraction collector (Pharmacia). Initially peaks were collected in silanised
test tubes but use of these was discontinued because of large losses for
some preparations after removing the solvent. They were replaced with 1.5ml
polypropylene micro centrifuge tubes (Labconinc, Cal.).

Hydrolysis of peptides was carried out for 22 hours at 110°C in 10 x
75 mm muffled tubes enclosed and evacuated in vials attached to the Picotag
Work Station. The acid, 200μl of 5.8M HCL (0.1% phenol) was added to the
bottom of the vial. Amino acid analyses were made on a computer controlled
amino acid analyser (Waters), usually with the ninhydrin reaction module
and a 490 programmable detector.

Automatic sequencing was made either in an Edman and Begg type
sequenator[7] or a gas phase sequencer[8] (Applied Biosystems). Identification
was made on a Hewlett Packard 1084 HPLC with a Zorbax Gold ODS column (8 x
6.2 mm) with an 8 min gradient comprising buffer A (dilute ammonium acetate
pH 4) and buffer B (acetonitrile).

RESULTS

Table 1 and Figures 1 and 2 illustrate some of the analytical data
obtained during the course of the investigation, and Figure 3 shows how the
sequence was constructed from the peptide sequence data.

Table 1. Amino acid compositions[a] of lizard claw protein (SDS 1), the N-terminal tryptic (T_1) and chymotryptic (C_1) peptides

Amino Acid	SDS 1		T_1		C_1	
	Fd.	Seq.	Fd.	Seq.	Fd.	Seq.
CMC	20	22	6.6	7	6.4	6
Asp	6.0	6	3.1	3	1.3	1
Thr	5.3	5	4.0	4	-	-
Ser	11	13	4.0	4	1.1	1
Glu	3.3	3	1.3	1	-	-
Pro	12	11	3.7	4	1.0	1
Gly	38	41	24	24	9.7	10
Ala	6.8	7	4.2	4	1.2	1
Val	8.7	9	3.9	4	-	-
Met	0.7	1	-	-	-	-
Ile	3.6	4	1.0	1	-	-
Leu	6.0	6	2.0	2	-	-
Tyr	3.2	3	0.9	1	0.9	1
Phe	2.8	3	1.0	1	-	-
Lys	1.0	1	-	-	-	-
His	4.0	4	0.8	1	-	-
Arg	3.3	3	1.0	1	-	-
Total		142		62		21

a) No corrections made for losses on hydrolysis (110°, 22h)

Table 1 demonstrates the good correlation between the amino acid analyses and the amino acid sequence data for the intact protein and the blocked N-terminal tryptic (T_1) and chymotryptic (C_1) fragments. Analysis of C_1 highlights the unusual composition (48% glycine, 29% half-cystine) of this peptide.

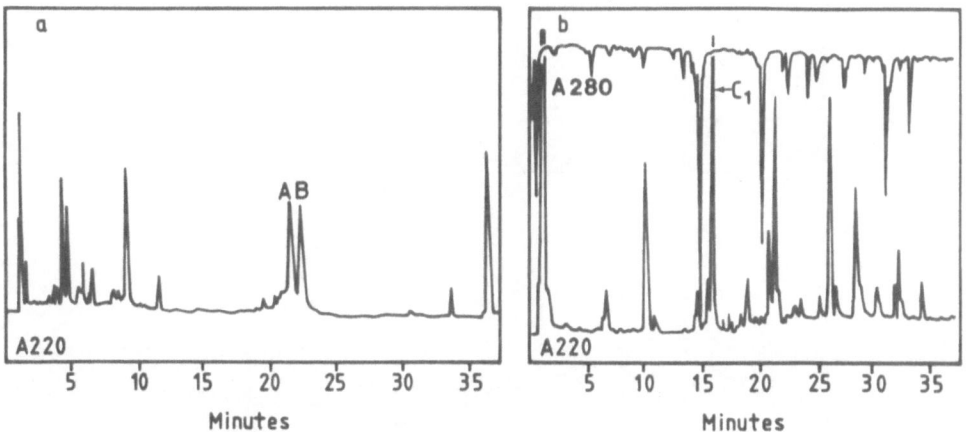

Figure 1. *HPLC traces of (a) tryptic digest of SDS 1 and (b) the chymotryptic digest of the B peak (containing the blocked N-terminal tryptic peptide). C_1 is the blocked N-terminal peptide.*

Examples of the HPLC gradient elution of tryptic (a) and chymotryptic (b) digests of lizard claw SDS 1 are given in Figure 1. The upper trace shows the absorption at 280nm, the lower (offset, and at a lower sensitivity) at 220nm. Peaks A and B (Figure 1a) both analyse for the N-terminal tryptic peptide. The sequence data presented here is for peptides from the B peak. Peak C, (Figure 1b) is the N-terminal chymotryptic peptide. The gradients vary of course with the digest but often begin at 5% buffer B with an initial linear increase of 1% buffer B per minute.

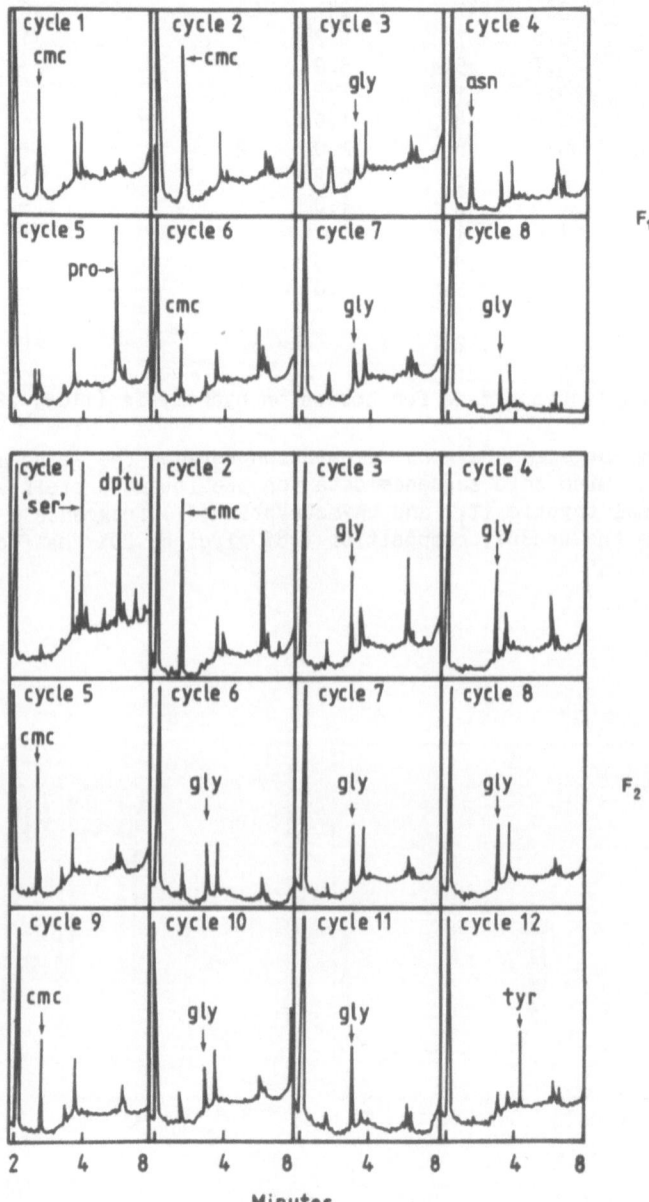

Figure 2. HPLC traces of PTH-amino acids from successive cycles of the gas phase sequencer for the products F_1 and F_2 of HFBA cleavage of the blocked chymotryptic peptide (C_1). They correspond to the residues 2-9 and 10-21 respectively.

Figure 2 shows the sequence data obtained for peptides F_1 and F_2 resulting from HFBA cleavage of C_1. The rapid HPLC run (8min, 11 min turn around) was developed to meet the huge demand for manual sequence operations in the Division. Small adjustments to the gradient and an increase in the run time to 12 minutes are required to separate PTH-tryptophan and diphenylthiourea. F_1 had a minor sequence present which actually sequenced to the tyrosine with a possible Cmc substitution for a Gly at residue 10. F_2 was clean and gave unequivocal data despite the potential problems associated with the multiple Gly and Cmc residues.

The primary structure found for SDS 1 is given in Figure 3 which includes the peptide sequence data used to piece it together. The products of CNBr and <u>A. mellea</u> proteinase digests were not chromatographed but loaded directly onto the sequenator since there is only one residue each of methionine and lysine in the protein. The whole protein was digested with trypsin and with chymotrypsin to generate the peptides required to establish the order of 80 residues at the C-terminal end of the protein. Chymotryptic, thermolysin and dilute acid treatments of the blocked N-terminal tryptic fragment (B in Fig. 1) provided a further 42 residues towards the N-terminus, then dilute acid and HFBA cleavages of the N-terminal chymotryptic peptide (C_1 in Figure 1) gave fragments that enabled the last 20 residues to be arranged unequivocally.

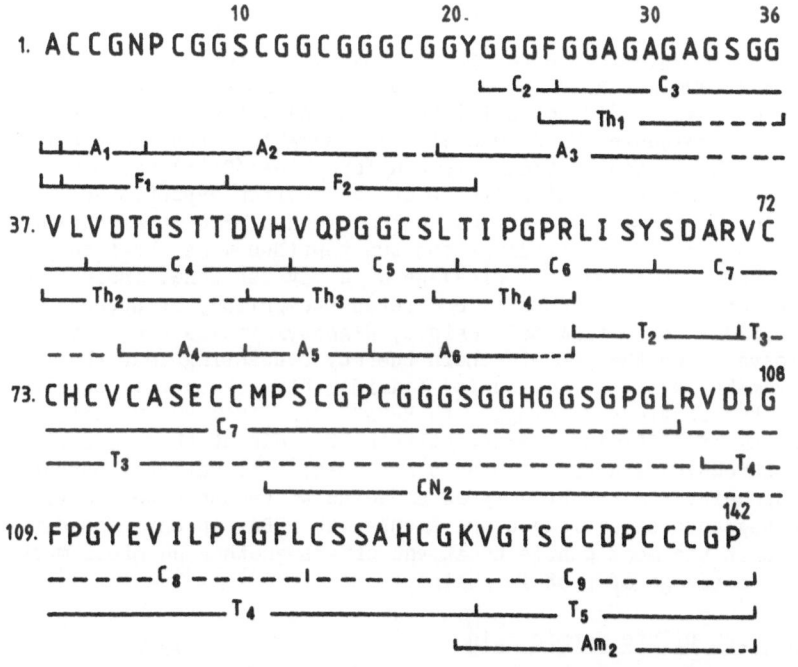

Figure 3. Amino acid sequence of lizard claw SDS 1 protein with the sequence data (solid lines) and peptides (solid lines plus dotted lines) obtained to establish the structure. Cleavages were effected by: dilute acid (A), anhydrous HFBA (F), thermolysin (Th), chymotrypsin (C), trypsin (T), cyanogen bromide (CN) and A. mellea enzyme (Am).

DISCUSSION

    Lizard claw component SDS 1 is a small protein molecule of 142
residues with a molecular weight in the native form of 13,095 and in the SCM
form of 14,392. It is rich in glycine (29%), half-cystine (16%), serine
(9%) and proline (8%). Glycine and cystine are not uniformly distributed
through the molecule. All except two of the half-cystine residues are found
in three cystine-rich segments, namely residues 2-18 (6), 72-89 (8), and
127-140 (6); also most of the glycine is found in two glycine-rich segments,
residues 1-36 (20) and 87-108 (11).

    Although amino acid analyses and some minor amino acid sequence data
imply an alanine residue at the N-terminus, the composition of the
N-terminal amino acid is still to be established. It is highly likely that
it is N-acetylalanine in accord with the N-termini of the similar avian[9]
and mammalian keratins[10,11]. Similarly, the C-terminal residue, proline,
has not been determined per se, but amino acid analyses and sequence
analyses of C-terminal peptides, plus the failure of carboxypeptidases Y
and P to liberate a proline residue, all support this assignment. Also,
there remains the possibility of heterogeneity in the N-terminal region.
Analyses of two well resolved peaks that are obtained by HPLC of tryptic
digests (A and B, Figure 1a) indicate that they are very similar, with
perhaps small differences in the glycine and carboxymethylcysteine
analyses. The sequence of the 'B' peak is given here.

Cleavage with HFBA

    Anhydrous HFBA has been used to remove N-acetyl groups from proteins
with N-acetylated serine residues at their N-termini[12]. Studies by Brandt
et al.[4] showed that HFBA also cleaved at preferred sites along the chain -
after aspartic acid residues, to a lesser extent before serine and
threonine, and between Gly-Gly bonds - after 30 hours at 55°C. In the
present work in which lizard claw keratin was digested for only 1 hour,
N-terminal analysis in the gas phase sequencer revealed that serine was
the major N-terminal amino acid in the digest. However, subsequent cycles
established that it did not originate from the N-terminus but from Gly-Ser
bonds in the molecule. There was also evidence for cleavage before a
threonine (Thr^{57}) and an S-carboxymethylcysteine (CMC^2) residue. On the
other hand there was no significant cleavage after aspartic acid residues.
Application of the same treatment to the blocked N-terminal chymotryptic
peptide confirmed these findings and two peptides were obtained in good
yields (approx. 40-50%) from cleavage of a Gly-Ser bond. Their sequences
(see Figures 2 and 3) virtually completed the primary structure
determination. The N-terminal residue, N-acetylalanine presumably, was
also cleaved from the protein chain thereby suggesting that the adjacent
S-carboxymethylcysteine residue(s) considerably weaken this peptide bond.
Possibly the previous findings with HFBA[4] arose from an averaging out of
the effects of differing susceptibilities to acid of the bonds linking
serine, threonine and aspartic acid residues; alternatively perhaps there
is a propensity for the hydroxy amino acids to become either acylated or
further degraded on prolonged treatment with HFBA. In any case our results
suggest that the more gentle treatment offers another chemical method for
selective cleavage of protein chains.

Cleavage with Dilute Formic Acid

    Dilute acid hydrolysis[5] of the N-terminal chymotryptic and tryptic
peptides also effected unusual cleavages, besides those expected after
aspartic acid residues, namely between Ala^1-CMC^2 and Asn^5-Pro^6, and to a
lesser extent Gly^{15}-Gly^{16}, Gly^{19}-Gly^{20} and Gly^{52}-Gly^{53}. A minor sequence
beginning with alanine was also generated. The peptide sequences obtained

for the products provided the overlapping data for the alignment of HFBA and chymotryptic peptides. These data highlighted the virtues of a powerful separation technique such as HPLC and a high sensitivity sequencing technique for analysis of the minor cleavage products of reactions. This strength was further exemplified when it was found that chymotrypsin cleaved predominantly at Arg^{62}; although the overlapping peptide (Thr^{57-66}) for the abutting Arg^{62} Leu^{63} was therefore only one of the minor peaks, sufficient material was obtained for confirmation of the sequence.

Relationships with Avian and Mammalian Keratins

Just as the sensitivity and quantitative nature of the modern analytical techniques accelerate the primary structure determination, protein sequence data bases provide a valuable adjunct for further analysis of the structure. Computer search of the data of the Protein Identification Resource (using segments of 36 for comparison with the test piece and overlaps of 5 residues for each test piece) gave a very interesting analysis of this reptilian structure as compared with the avian and mammalian structures obtained previously. It contains regions homologous with both mammalian and avian keratins; residues 1-36 show homologies with human epidermal keratin[13], residues 1-36, 63-98 and 132-142 are homologous with high-sulfur wool keratins[10,11], whereas residues 32-67 resemble sequences found in feather keratins[9]. Residues 27-34 also comprise the repeating sequence, Gly Ala Gly Ala Gly Ala Gly Ser, of the fibrous protein silk.

The homology with the mammalian keratins raises the question of whether the lizard claw proteins are in fact related more closely to the mammalian Type II high-tyrosine proteins than to feather, and this examination is proceeding.

REFERENCES

1. R.D.B. Fraser and T.P. MacRae, Current views on the keratin complex, in: The Skin of Vertebrates, R.I.C. Spearman and P.A. Riley, eds., Academic Press, London (1980).
2. J.M. Gillespie, R.C. Marshall, and E.F. Woods, A comparison of lizard claw keratin proteins with those of avian beak and claw, J. Mol. Evol. 18: 121 (1982).
3. R.C. Marshall and J.M. Gillespie, The tryptophan-rich keratin protein fraction of claw of the lizard: Varanus gouldii, Comp. Biochem. Physiol. 71B:623 (1982).
4. W.F. Brandt, A. Henschen, and C. von Holt, Nature and extent of peptide bond cleavage by anhydrous heptafluorobutyric acid during Edman degradation, Hoppe-Seyler's Z. Physiol. Chem. 361:943 (1980).
5. A.S. Inglis, Cleavage at aspartic acid, in: Methods in Enzymology, 91, C.H.W. Hirs and S.N. Timasheff, eds., Academic Press, New York (1983).
6. P.Edman, and G. Begg, A protein sequenator, Eur. J. Biochem.1:80 (1967).
7. A.S. Inglis, P.M. Strike, W.C. Osborne and R.W. Burley, Sequenator determination of the amino acid sequence of apovitellenin I from turkey's egg yolk, FEBS Lett. 97:179 (1979).
8. M.W. Hunkapiller, R.M. Hewick, W.J. Dreyer, and L.E. Hood, High sensitivity sequencing with a gas-phase sequenator, in: Methods in Enzymology, 91, C.H.W. Hirs and S.N. Timasheff, eds., Academic Press, New York (1983).

763

9. I.J. O'Donnell, and A.S. Inglis, Amino acid sequence of a feather
    keratin from silver gull (Larus novae-hollandiae) and comparison
    with one from emu (Dromaius novae-hollandiae), Aust. J. Biol. Sci.
    27:369 (1974).
10. T. Haylett, and L.S. Swart, Studies on the high-sulfur proteins of
    reduced Merino wool, Textile Res. J., 39:917 (1969).
11. J.M. Gillespie, T. Haylett, and H. Lindley, Evidence of homology in
    a high-sulfur protein fraction (SCMK-B2) of wool and hair
    α-keratins, Biochem. J. 110:198 (1968).
12. W.N. Strickland, M.S. Strickland, P.C. de Groot and C. von Holt,
    The primary structure of histone H2A from the sperm cell of the sea
    urchin Parechinus angulosus, Eur. J. Biochem., 109:151 (1980).
13. P.M. Steinert, A.C. Steven, and D. R. Roof, Structural features of
    epidermal keratin filaments reassembled in vitro, J. Invest,
    Dermatology 81:86 (1983).

PRIMARY STRUCTURE STUDIES ON SERINE HYDROXYMETHYLTRANSFERASE

Donatella Barra, Filippo Martini, Sebastiana Angelaccio,
Stefano Pascarella, Francesco Bossa and LaVerne Schirch*

Department of Biochemistry and CNR Center of Molecular
Biology, University of Rome, 00185, Italy and *Virginia
Commonwealth University, Richmond, Virginia

INTRODUCTION

Serine hydroxymethyltransferase (EC 2.1.2.1) is a pyridoxal-phosphate containing enzyme which, among other reactions, catalyzes the tetrahydrofolate-dependent interconversion of serine and glycine, thus representing the major source of one-carbon groups required for the biosynthesis of several cell components[1]. Both cytosolic and mitochondrial forms exist in eukaryotic cells and preliminary sequence studies have demonstrated that these two isoenzymes are homologous proteins[2]. Cytosolic serine hydroxymethyltransferase from rabbit liver is the enzyme form by far the best characterized from the functional and spectroscopic point of view[1]. As far as the protein moiety is concerned, this enzyme is a tetramer of identical subunits with a molecular weight of about 215,000. We initiated the determination of its primary structure by purifying and analysing two active-site peptides[3,4]. Additional sequence information has been accumulated after analysis of tryptic, chymotryptic and, now, CNBr peptides.

Recently, the nucleotide sequence of the Escherichia coli glyA gene has been determined and the deduced protein sequence was supposed to be that of serine hydroxymethyltransferase[5]. Comparison of the structure of the available peptides from the cytosolic enzyme with the sequence of the putative enzyme from E. coli , confirmed the identification of the latter and was useful in reconstructing the sequence of the former.

MATERIALS AND METHODS

A sample (80 mg) of carboxymethylated serine hydroxymethyl-transferase was treated with an equal weight of CNBr in 70% formic acid for 20 h at room temperature. The peptide mixture was first fractionated by gel-filtration through a column (2.5 cm x 135 cm) of Sephadex G-50 superfine eluted with 10% acetic acid; the fractions containing the largest peptides were subdigested with an appropriate protease (trypsin, chymotrypsin or S. aureus protease) either directly or after further purification by hplc on preparative macroporous columns (Brownlee Labs, Aquapore RP-300, 10 um, 7 mm x 25 cm) with gradients of 0.2% trifluoroacetic acid in acetonitrile. The subdigestion mixtures and the medium- and low-size fractions from the gel-filtration step were also purified by hplc and the resulting pure peptides analyzed. Details on the carboxymethylation procedure and on the analytical techniques adopted for peptide structure determination are reported in recent publications on sequence work from our laboratory [6].

For the selective isolation of a C-terminal fragment of the protein, 25 mg of carboxymethylated serine hydroxymethyltransferase were subjected to a modified version of the procedure reported by Hargrave and Wold [7], outlined in the following scheme:

1) *The protein is reduced and carboxymethylated.*
2) *All carboxyl groups are blocked with glycinamide, after activation with carbodiimide.*
3) *The modified protein is digested with trypsin. All the resulting peptides except the C-terminal one have a free carboxyl group.*
4) *The peptide mixture is treated with carboxypeptidase B to remove the carboxyl terminal arginine or lysine from the tryptic peptides.*
5) *The trypsin-carboxypeptidase B digest is subjected to chromatography in alkaline conditions on an anionic resin (QAE-Sephadex A-25). All peptides containing a free carboxyl group are retained, the C-terminal peptide, together with free arginine, is not retained.*
6) *Final purification of the C-terminal peptide is achieved by reverse-phase hplc.*

For the isolation of the N-terminal blocked peptide of the protein the classical procedure modified from Yoshida was applied [8,9]; also in this case final purification of the blocked peptide and identification of the blocking group was achieved by hplc.

The secondary structure of the E. coli protein and of the fragments from cytosolic serine hydroxymethyltransferase were predicted according to the method of Chou and Fasman by using a Basic microcomputer program [10]. The hydrophobicity and hydrophilicity profiles were calculated according to Cid et al. [11] and to Kyte and Doolittle [12], respectively, by means of a simple program written in Applesoft Basic on an Apple IIe.

RESULTS AND DISCUSSION

The tentative identification of the protein coded by the glyA gene of E. coli with serine hydroxymethyltransferase was substantiated by our preliminary sequence data on both cytosolic and mitochondrial isoenzymes from rabbit liver and by the recent isolation of the bacterial enzyme and study of some of its functional and structural properties [13]. Thus, we could confidently attempt the alignment of the structure of peptides obtained from the cytosolic isoenzyme from rabbit liver with the sequence of the protein from E. coli. Data reported in Figure 1 show that a continuous tract of 365 residues could easily be aligned from the N-terminus with the sequence of the E. coli protein, because of the presence of extended regions of high homology and on the basis of analysis of some peptides, which proved to be crucial in establishing the overlaps. Particularly useful to this purpose were the CNBr peptides: all the expected fragments could be obtained by extensive application of hplc purification procedures on macroporous columns. Their sequence was either directly determined or almost completely reconstructed after purification and analysis of the peptides derived from one or more subdigestions. Of great help was also the purification and analysis of the N-terminal blocked peptide obtained by a selective method.

By an analogous procedure it was possible to reconstruct the sequence of a tract of 85 residues which represents the C-terminal portion of the enzyme, since it ends with a structure identical to that of the fragment obtained with the specific procedure outlined in the Materials and Methods section, although not homologous with that of the corresponding portion of the E. coli protein. Moreover, the proposed structure is in accord with the results obtained after carboxypeptidase digestion of the whole protein[2].

The results of the final hplc step for the purification of the C-terminal fragment from the cytosolic enzyme are shown in Figure 2.

In Figure 1 is also reported the sequence of a third unoverlapped fragment of 37 residues from the cytosolic enzyme which does not show a significant degree of homology with any tract of the sequence of the E. coli protein.

Figure 1. The partial amino acid sequence of cytosolic rabbit liver serine hydroxymethyltransferase is aligned with that of the E. coli protein. —— indicate gaps introduced in the bacterial sequence to obtain maximal homology; .... indicate undetermined portions of sequence of the rabbit enzymes. Boxes: conserved residues.

On the whole, the sequence data contained in the three reported fragments should account for at least 97% of the entire sequence. The extent of homology among the two proteins is about 41%. The distribution of conserved and substituted regions is clearly not uniform throughout the two sequences. In particular, homology is very high in central regions of the two proteins, such as that adjacent to the lysine which binds pyridoxal-phosphate (Lys 229 in the E. coli protein) and is very low in the C-terminal portion.

A number of segments can also been observed which represent insertions in the mammalian protein; on the whole these account for the expected larger size (about 70 residues) of this protein in respect with the bacterial one.

Finally, we carried out predictions of the secondary structures and calculated the hydrophobicity and hydrophilicity profiles of both the E. coli and the cytosolic rabbit enzyme. The results obtained after prediction of the secondary structure of E. coli serine hydroxymethyltransferase according to Chou and Fasman are reported, as an example, in Figure 3. The amount of residues in predicted ordered structures is: 31% α-helix and 23% β-sheet. More interesting would be a comparison of these predictions with those carried out for the eukaryotic isoenzyme, but this will be more significant when made with the reconstruction of the complete sequence of the latter enzyme. However, the following observations can already be made: 1) the homology of predicted secondary structures between the two proteins are more extended than that of the primary structure; 2) the predicted structural features of the stretches of sequence of the eukaryotic enzyme which

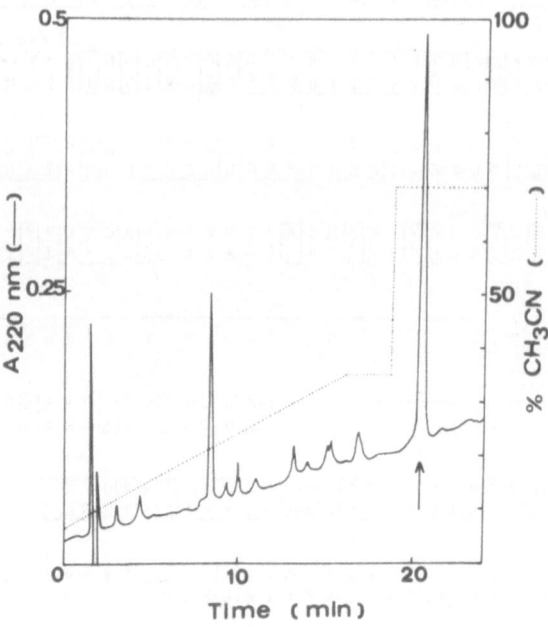

Figure 2. Hplc final purification of the C-terminal peptide from cytosolic serine hydroxymethyltransferase. The peak corresponding to the C-terminal peptide is indicated by the arrow.

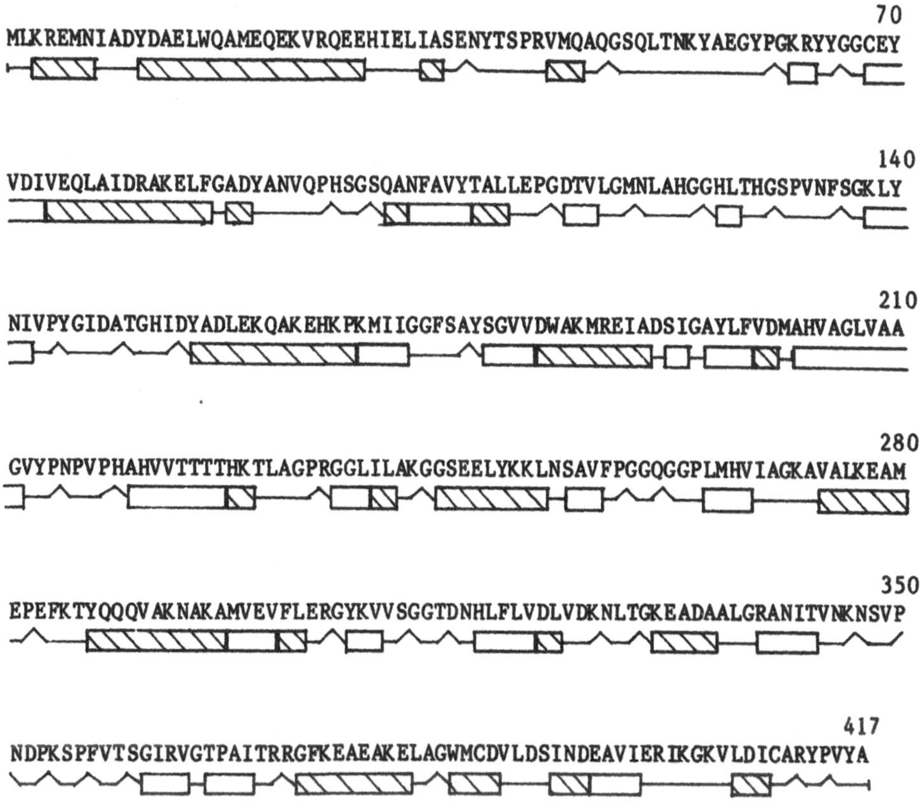

MLKREMNIADYDAELWQAMEQEKVRQEEHIELIASENYTSPRVMQAQGSQLTNKYAEGYPGKRYYGGCEY

VDIVEQLAIDRAKELFGADYANVQPHSGSQANFAVYTALLEPGDTVLGMNLAHGGHLTHGSPVNFSGKLY

NIVPYGIDATGHIDYADLEKQAKEHKPKMIIGGFSAYSGVVDWAKMREIADSIGAYLFVDMAHVAGLVAA

GVYPNPVPHAHVVTTTTHKTLAGPRGGLILAKGGSEELYKKLNSAVFPGGQGGPLMHVIAGKAVALKEAM

EPEFKTYQQQVAKNAKAMVEVFLERGYKVVSGGTDNHLFLVDLVDKNLTGKEADAALGRANITVNKNSVP

NDPKSPFVTSGIRVGTPAITRRGFKEAEAKELAGWMCDVLDSINDEAVIERIKGKVLDICARYPVYA

Figure 3. Prediction of the secondary structures of E. coli serine hydroxymethyltransferase. ▨▨: α –helix; ▭: β –sheet; ⋀ : β –turn.

represent insertions with respect to the sequence of the E. coli protein are compatible with their location at the surface of the protein molecule.

REFERENCES

1. L. Schirch, Serine hydroxymethyltransferase. Adv. Enzymol. 53:83 (1982).
2. D. Barra, F. Martini, S. Angelaccio, F. Bossa, F. Gavilanes, D. Peterson, B. Bullis and L. Schirch, Sequence homology between prokaryotic and eukaryotic forms of serine hydroxymethyltransferase. Biochem. Biophys. Res. Commun. 116:1007 (1983).
3. F. Bossa, D. Barra, F. Martini, L. Schirch and P. Fasella, Serine transhydroxymethylase from rabbit liver. Sequence of a nonapeptide at the pyridoxal-5´-phosphate-binding site. Eur. J. Biochem. 70:397 (1976).
4. L. Schirch, S. Slagel, D. Barra, F. Martini and F. Bossa, Evidence for a sulfhydryl group at the active site of serine transhydroxymethylase. J. Biol. Chem. 255:2986 (1980).
5. M.D. Plamann, L. T. Stauffer, M.L. Urbanowski and G. V. Stauffer, Complete nucleotide sequence of the E. coli glyA gene. Nucleic Acids Res. 11:2065 (1983).
6. D. Barra, M.E. Schininà, M. Simmaco, J. V. Bannister, W.H. Bannister, G. Rotilio and F. Bossa, The primary structure of human liver manganese superoxide dismutase. J. Biol. Chem. 259:12595 (1984).

7. P.A. Hargrave and F. Wold, A. preparative method for the isolation of carboxyl terminal tryptic peptides from proteins. Int. J. Peptide Protein Res. 5:85 (1973).

8. A. Yoshida, Micro method for determination of blocked NH -terminal amino acids of protein. Anal. Biochem. 49:320 (1972).

9. F. Martini, S. Angelaccio, D. Barra, S. Doonan and F. Bossa, Partial amino-acid sequence and cysteine reactivities of cytoplasmic aspartate aminotransferase from horse heart. Biochim. Biophys. Acta 789:51 (1984).

10. A. J. Corrigan and P.C. Huang, A BASIC microcomputer program for plotting the secondary structure of proteins. Computer Programs in Biomed. 15:163 (1982).

11. H. Cid, M. Bunster, E. Arriagada and M. Campos, Prediction of secondary structure of proteins by means of hydrophobicity profiles. FEBS Lett. 150:247 (1982).

12. J. Kyte and R. F. Doolittle, A simple method for displaying the hydropathic character of a protein. J. Mol. Biol. 157:105 (1982).

13. L. Schirch, S. Hopkins, E. Villar and S. Angelaccio, Serine hydroxymethyltransferase from Escherichia coli : purification and properties. J. Bacteriol. 163:1 (1985).

THE PROTEINS IN AVIAN EGGS:

A TEST OF PROTEIN SEPARATION METHODS BASED ON HPLC

R.W. Burley and J.F. Back

CSIRO Division of Food Research

P.O. Box 52, North Ryde, NSW, 2113, Australia

ABSTRACT

A study has been made of the separation of egg proteins by high-pressure-liquid chromatography (HPLC). Proteins in four parts of the egg were studied: .

(i)   The vitelline membrane. A Cl8 column was suitable for separating six proteins from the outer layer of this membrane. Three of these were isolated and partial N-terminal sequences determined on a gas-phase sequencer. One proved to be egg lysozyme; another VMOI (vitelline membrane outer I) and the third has been named VMOII.

(ii) Egg yolk low density lipoprotein. The apoproteins from this lipoprotein were separated on a phenyl hydrophobic column using a gradient of acidic urea. The results were similar to those obtained with phenyl Sepharose.

(iii) Egg yolk livetins. A column of DEAE – 5 PW was used to separate the main proteins using an ionic-strength gradient. A comparison was made with blood serum.

(iv) Albumen (i.e. egg white). The same column was used with the same eluant, which separated the main egg-white proteins.

For each of these analyses the samples required a minimum of preparation.

INTRODUCTION

The proteins of eggs are of interest for subjects as diverse as food technology and chemical embryology. They have also been favourites with protein chemists for more than a century, and most analytical methods have been applied to their separation; see, for example the review of Vadehra and Nath (1). We have been applying HPLC methods to the separation of these proteins with the aim of improving the speed with which known egg proteins can be isolated and of separating and isolating proteins that remain to be idéntified. A summary of our recent work is presented here. Some aspects of it have been reported in abstract form (2).

EXPERIMENTAL

Materials

Unless stated otherwise, eggs were from Australorp hens and were freshly laid. After the shell had been broken, the yolk and albumen were separated on a separator. The yolk was rolled on filter paper and then pierced. Yolk granules were removed by centrifugation (1 h, 50,000 g 5°C) after 1:1 dilution with 0.16 M sodium chloride (3). The rest of the yolk was added to an equal volume of 4 M sodium chloride and recentrifuged for 20 h at 100,000 g. The low-density yolk lipoprotein floated as a yellow waxy solid, and the yolk soluble proteins, the livetins, remained in the subnatant solution (4, 5). Lipid was removed from the lipoprotein with chloroform-methanol (2:1) and the apoproteins dissolved in 6 M urea acidified with either hydrochloric acid (to 0.02 M) (6) or trifluoroacetic acid (to 0.025 M). For chromatography the apoprotein mixture was dialyzed into 1.2 M urea, 0.025 M trifluoroacetic acid. The livetin fraction was recentrifuged to remove all of the lipoprotein and dialysed into 0.2 M ammonium acetate.

The vitelline membrane was obtained as described previously (7) and the proteins of the outer layer dissolved in 0.5 M sodium chloride, 0.1 M sodium acetate pH 5. Before chromatography the solution was dialysed into 0.05% trifluoroacetic acid (TFA).

The albumen was diluted with four volumes of water and centrifuged (20 h, 50,000 g, 2°C) to sediment most of the ovomucin, which was discarded. The solution was applied directly to the columns. A crocodile's egg was the gift of Dr G.C. Grigg of the Zoology Department, University of Sydney.

Solvents and buffers were of HPLC grade.

METHODS

Chromatography. The HPLC apparatus used was from ETP-Kortec Pty Ltd Ermington, NSW, Australia, and consisted of two dual piston K35M pumps, a K45 gradient computer, and a K95 variable wavelength ultraviolet detector. The columns used are mentioned under the chromatograms.

To isolate the individual proteins, solvent from the appropriate chromatographic peak was collected in silanized bottles and removed by freeze drying. Non-volatile solvents were diluted with water and concentrated repeatedly on a Centricon Microcentrator (Amicon Corp. Danvers MA) before drying.

Electrophoresis. Proteins were identified when necessary by comparison with standards on polyacrylamide gel electrophoresis in the presence of SDS (sodium dodecyl sulfate). Minor modifications of the procedure of Weber and Osborn (8) were used (9).

Protein sequencing. A model 470A gas-phase sequencer from Applied Biosystems Inc., Foster City, CA, was used to determine N-terminal sequences. Phenylthiohydantoins of amino acids were identified by HPLC

Table 1. N-terminal sequences of proteins from the outer layer of the vitelline membrane (see Fig. 1).

Protein	Sequence
Lysozyme	Arg-Val-Phe-Gly-Arg-Cys-Asp-Leu-Ala-Ala-Ala-Met-Lys-Arg-His-Gly-Leu-Asp-Asn-Tyr-Arg-Gly-Tyr-Ser-Leu-Gly-Asn-Trp-Val-
VMOI	Arg-Thr-Arg-Glu-Tyr-Thr-Ser-Val-Ile-Thr-Val-Pro-Asn-Gly-Gly-His-Trp-Gly-Lys-Trp-Gly-Ile-Arg-Gln-Phe-
VMOII	Leu-Pro-Arg-Asp-Gln-

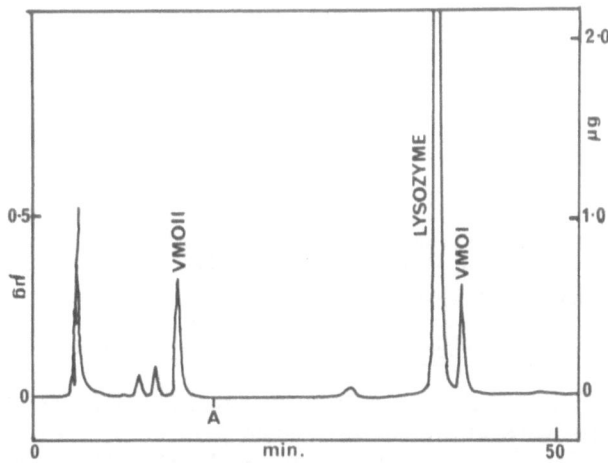

Fig. 1. Proteins of the outer layer of the vitelline membrane. A Vydac C-18 column (4.6 mm x 25 cm), 5 μm particles, was used eluting with a gradient of (A) 0.05% TFA, 5% B; (B) acetonitrile:isopropanol:water 2:2:1 according to the program 10%-30% B in 30 min, then 30%-80% B in 10 min, finally 80% B. The volume applied (25 μl) contained 50 μg of the total protein. The flow rate was 1.5 ml/min at 20°C. Proteins were detected at 280 nm. The microgram scales were standardized using egg-white lysozyme and may not be strictly applicable to the other peaks. At point "A" the sensitivity of the detector was decreased from the left-hand scale to the right-hand scale.

using a du Pont Gold C-18 column. Solvent A was 82% 12.5 mM ammonium
acetate pH 3.91, 18% acetonitrile and solvent B was 20% 12.5 mM ammonium
acetate, 80% acetonitrile. A gradient of 7% to 65% B in 9 minutes was
used.

For VMOI (Fig. 1) the first 20 residues were confirmed by isolating
the N-terminal peptide (Arg to Lys) produced by digestion with
lysylendopeptidase (10). This peptide was isolated by HPLC on a C18
column.

Amino-acid sequences were compared with those of known proteins by
means of the computer-based Molecular Biological Information Service of
CSIRO Division of Molecular Biology, North Ryde, NSW.

RESULTS AND DISCUSSION

The vitelline membrane

The vitelline membrane surrounds the yolk and provides the last
barrier to micro-organisms. It has recently been shown by conventional
chromatographic methods (7) that the outer layer of this membrane is
rich in the anti-microbial proteins, ovomucin and lysozyme, and also
includes a new protein "VMOI". Fig. 1 shows the separation of the
proteins of the outer layer by HPLC. Six well-separated peaks were
obtained. Three of these were isolated and their N-terminal sequences
determined (Table 1). One of them was identical to hen's egg lysozyme
(11). Evidently the difference between lysozyme in albumen and the
insoluble form in the vitelline membrane is not related to this part of
the sequence. VMOI has not previously been sequenced. The N-terminal
sequence is not related to any known protein sequence.

VMOII has not previously been reported. On gel-electrophoresis in
SDS it had the same mobility as lysozyme so its molecular weight is
approximately 14,000. As for VMOI the N-terminal sequence is not
related to any known protein sequence.

Egg yolk low-density lipoprotein

The apoproteins of this lipoprotein are notable because they are
soluble only in disaggregating solvents such as acidic urea (6, 9).
They have previously been fractionated by gel-filtration chromatography
in 6 M urea (6) and more recently (9) by using a urea gradient on a
column of phenyl Sepharose or other hydrophobic material. Fig. 2 shows
the adaptation of this technique to HPLC. Better separations were
achieved than with large columns, but there were differences in the
order of elution and there are still problems with the elution of the
very hydrophobic proteins, apovitellenins III and V.

Egg yolk livetins

These proteins, which are closely related to avian blood proteins,
have previously been fractionated by a combination of gel-filtration and
ion-exchange chromatography (5). Fig. 3 shows a comparison of hens' egg
livetins with blood serum proteins from a laying hen. The serum protein
patterns of different hens showed considerable variations whereas that
for the livetins was relatively constant. Nevertheless it is clear that
there is unlikely to be an unselected transfer of blood proteins to egg
yolk.

Albumen (i.e. egg white)

The albumen proteins have received more study than those of any other part of an egg and many methods of separation have been described (1). The HPLC separation on a DEAE column (Fig. 4) is not as good as that achieved by electrophoresis (eg. 1), but can more easily be used for isolations. The high concentration of ovalbumin has not interfered with the separation. Fig. 4 also gives an example of the comparison of albumen proteins from different species.

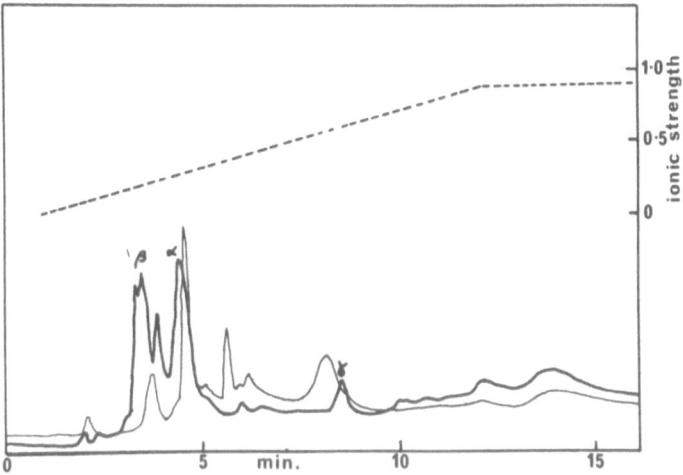

Fig. 2. Hydrophobic chromatography of the apoproteins (i.e. apovitellenins) from the low-density lipoprotein of hens' egg yolk. A TSK Phenyl-5 PW column was used with a gradient of urea plus 0.025 M trifluoroacetic acid, pH 3.5. The flow rate was 1.6 ml/min, 20°C. Proteins were detected at 280 nm and identified by gel electrophoresis in SDS. The Roman numbers refer to apovitellenins I-VI. A 100 μl sample containing 0.4 mg of total protein was applied. The dashed line refers to the urea gradient.

CONCLUSION

We have described how HPLC can be applied to the separation of the main proteins in avian eggs. Other eggs can also be used. In our tests the protein mixtures were subjected to a minimum of prior treatment. In particular, the only protein excluded from the analysis was ovomucin from egg white. These HPLC methods should be applicable to problems such as the comparison of different species and variations between individuals. Of particular interest is the possibility of isolating and identifying minor proteins of which eggs contain a large number. Some of these may have uses as metabolic regulators.

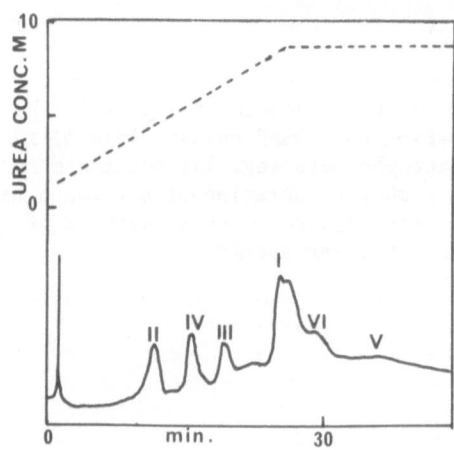

Fig. 3. *Proteins of the livetin fractions of hens egg yolk compared with blood serum proteins from a laying hen. An Ultropac DEAE-5PW, 10 μm column (7.5 mm x 7.5 cm) was used eluting with an ionic strength gradient made from 0.05 M and 1.0 M ammonium acetate. The flow rate was 0.7 ml/min. The dashed line refers to ionic strength. Proteins were detected at 280 nm. Thick curve (A): livetins (25 μl containing 0.12 mg of proteins) in 0.16 M ammonium acetate. The three main livetins, α, β and γ were identified by gel-electrophoresis in SDS. Thin curve (B): laying hens' blood serum, after removal of lipoproteins by prolonged centrifuging and dialysis into 0.16 M ammonium acetate.*

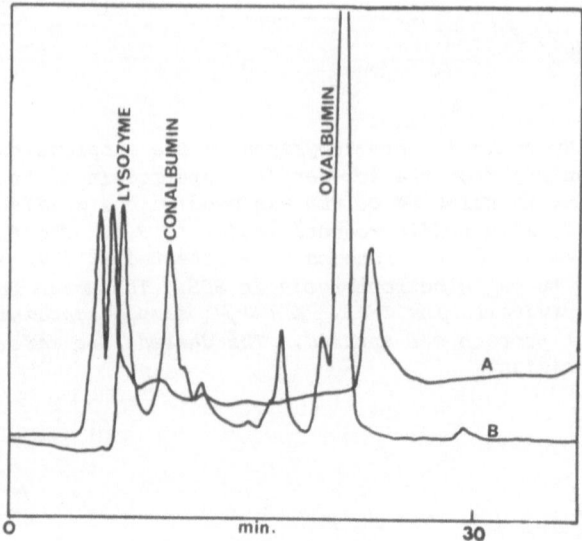

Fig. 4 *Proteins of hens' egg albumen and a comparison with albumen from an egg of a crocodile (Crocodylus porosus). The column and conditions used were the same as those described for Fig. 3. Curve A: Albumen proteins from hens' egg. 100 μl containing 250 μg of total protein was applied. Curve B: Albumen proteins from crocodiles egg. The same volume was applied, but the amount of protein was less. Before application each albumen sample was diluted with four volumes of water and centrifuged to remove ovomucin and similar proteins. Incomplete removal may have been responsible for the sloping baseline of Curve B.*

ACKNOWLEDGEMENTS

We thank Dr A. Reisner, CSIRO Division of Molecular Biology for doing the computer search of sequence data.

REFERENCES

1. D.V. Vadehra and K.R. Nath (1973) CRC Crit. Rev. Food Technol. 4, 193-309.
2. J.F. Back and R.W. Burley (1985) Proc. Aust. Biochem. Soc. 17, 38.
3. R.W. Burley and W.H. Cook (1961) Can. J. Biochem. Physiol. 39, 1295-1307
4. R.W. Burley (1978) Aust. J. Biol. Sci. 2 31, 587-592.
5. R.W. Burley and D.V. Vadehra (1979) Anal. Biochem. 94, 53-59.
6. R.W. Burley (1975) Aust. J. Biol. Sci. 28, 121-132.
7. J.F. Back, J.M. Bain, D.V. Vadehra, and R.W. Burley (1982) Biochim. Biophys Acta 705, 12-19.
8. K. Weber and M. Osborn (1969) J. Biol. Chem. 244, 4406-4412.
9. R.W. Burley and R.W. Sleigh (1983) Biochem. J. 209, 143-150.
10. K. Maeda, S. Kakabayashi and H. Matsubara (1985) Biochim. Biophys. Acta 828, 213-221.

AMINO TERMINAL SEQUENCE OF 7B2: A NOVEL, HUMAN PITUITARY/BRAIN

POLYPEPTIDE PRESENT IN GONADOTROPHS AND RELEASED BY LHRH

R. Leduc, N.G. Seidah, S. Benjannet, J.S.D. Chan,
M. Marcinkiewicz, C. Lazure and M. Chrétien

Clinical Research Institute of Montreal
110 Pine Avenue West
Montreal, Quebec, Canada   H2W 1R7

INTRODUCTION

The pituitary gland is a rich source of biologically active polypeptides and current work on the isolation of other proteins hint that additional active agents remain to be found. This paper deals with a highly
conserved protein (Mr 21,000) recently isolated in our laboratory (1,2),
which has been given the name "7B2". The amino terminal sequence of 7B2
has been found to vary only in one position between the human and porcine
homologs, within the first 78 residues (2). The availability of a specific antiserum (3) allowed the RP-HPLC purification of sufficient
amounts of human 7B2 (1-6 µg/pituitary) to permit the extension of its
sequence to 128 residues. An exhaustive computer search confirmed the
novelty of such a protein sequence, and its classification as belonging
to a new protein superfamily. Furthermore, immunocytochemical studies
revealed (i) the wide distribution of 7B2 within brain and pituitary,
(ii) its presence in secretory granules, and (iii) its co-localization
with both LH and FSH in anterior pituitary cells. This led to studies
which demonstrated its co-release with both gonadotrophic hormones following an LHRH challenge in vitro.

RESULTS

A. RP-HPLC purification of human 7B2

Frozen human pituitaries were extracted with acetone/HCl at 4°C
(2). The resultant extract was then partially purified on a preparative
Sephadex G-75 superfine column. The fractions containing human growth
hormone were pooled and repurified on a preparative µ-Bondapak C-18
column (2.5 x 30 cm, Waters) and eluted with a 30-90% aqueous 0.1%
TFA/CH3CN linear gradient (Fig. 1A). The elution position of 7B2 was
monitored by a specific radioimmunoassay directed against the N-terminal
23-39 sequence (3). The immunoreactive (IR-) material (eluting at 48%
CH3CN) was then further purified on a semipreparative µ-Bondapak C-18
(0.78 x 30 cm) using a 20-60% aqueous TFA/CH3CN linear gradient. The
fractions containing IR-7B2 (eluting at 46% CH3CN; Fig. 1B) were then
pooled and repurified on an RP-300 C-8 column (0.7 x 25 cm, Brownlee).
The elution was made with 20-60% aqueous TFA/2-propanol linear gradient.

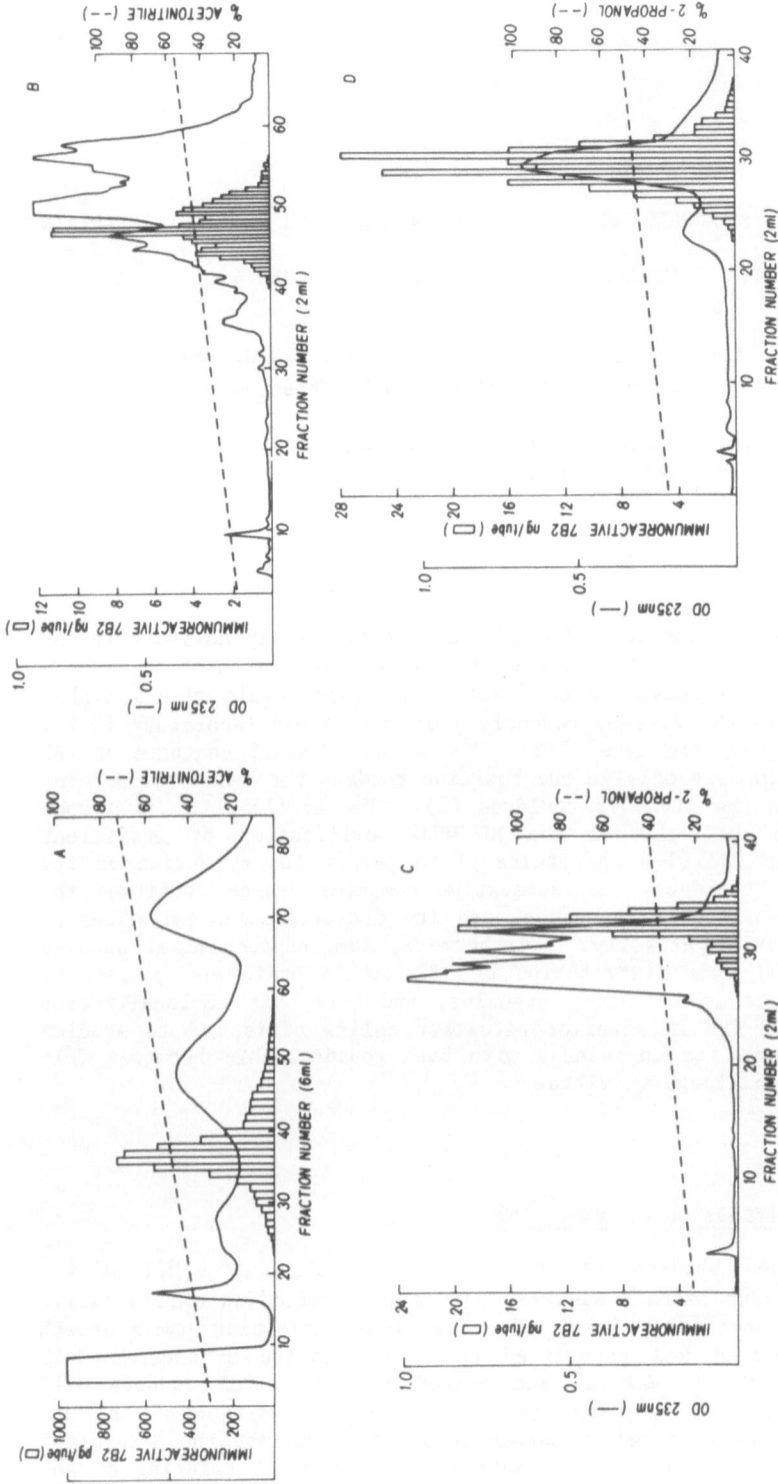

Fig. 1 Reverse-phase (RP-) HPLC purification of human 7B2. In all cases the elution position of 7B2 was monitored by a specific radioimmunoassay (3) in which 1% of each fraction was taken for assay. The flow rate used was 6 ml/min. in Fig. 1A, and 2 ml/min. for Figs. 1B, C and D. In all cases the linear gradient shown (dotted line), is 0.5% organic solvent/min. This procedure allows the purification of about 1.5 mg of 7B2 from 250 human pituitary glands.

The IR-7B2 elutes at 36% 2-propanol (Fig. 1C). The final repurification was made on the same column using a 30-60% aqueous 0.13% HFBA/2-propanol linear gradient, where IR-7B2 elutes at 44.7% 2-propanol (Fig. 1D).

B. Sequence analysis

The material so obtained was then subjected to automated amino acid sequence analysis on a Beckman 890 M sequencer using a 0.3 M Quadrol program (1,2). Phenylthiohydantoin (PTH) amino acids were identified by RP-HPLC as described (4). This allowed the unambiguous determination of the first 78 amino acids of human 7B2 (2). In order to extend the sequence, the reduced and carboxymethylated (CM-) 7B2 was digested with Lysobacter-C protease and the resultant fragments purified by RP-HPLC (Fig. 2). The N-terminal sequence of two of these fragments (L-1 and L-2) allowed the identification of the first 127 amino acids of human 7B2 (Fig. 3). The N-terminal Asp-Phe- sequence of L-1, which should be preceded by Lys-, thus forming the triad Lys-Asp-Phe-, is identical to the sequence of the residues 79-81 of the porcine homolog (2). Based on the high degree of conservation of the 7B2 sequence between these two species, the overlap between residues 78-81 was established. The overlap between residues 97-100 was determined by homology of the L-2 primary structure to that of the segment 90-118 obtained from porcine 7B2 following acid cleavage at the Asp 89-Pro 90 bond. Tryptic digestion of the carboxymethylated 7B2 followed by RP-HPLC purification (not shown) permitted the isolation of a peptide which upon sequencing, confirmed the C-terminal sequence of L-2 up to its C-terminal Lys-128 (residues 117-128, see Fig. 3).

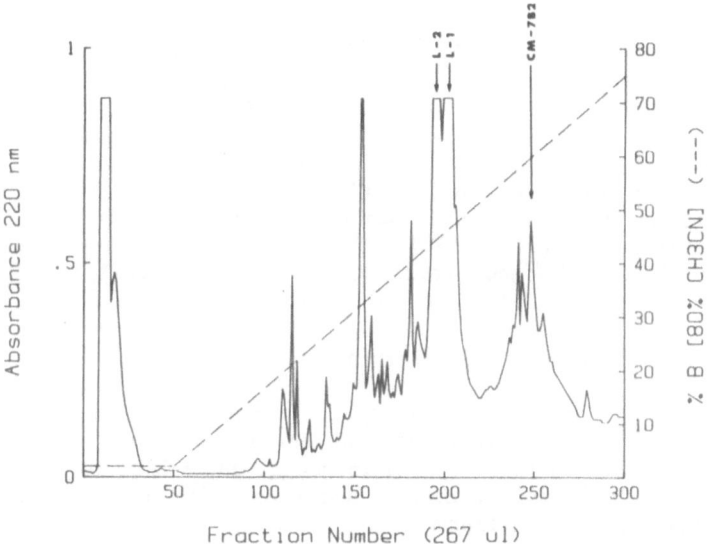

Fig. 2 RP-HPLC of fragments derived from Lysobacter-C (Boehringer Mannheim) digest of CM-7B2, on μ-Bondapak C-18 using TFA/CH3CN at 1 ml/min. The protein (200 μg) was dissolved in 0.1 M NH4HCO3, pH 9.0, and digested with the enzyme at a 1/30 enzyme to substrate ratio (w/w) for 19 hrs at 37°C.

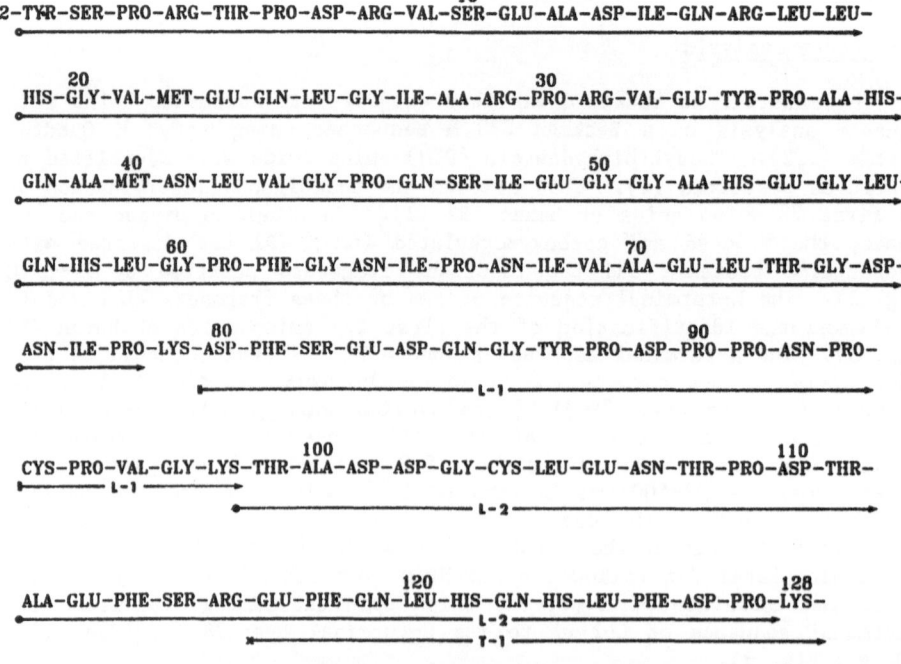

Fig. 3 The N-terminal sequence of human 7B2 up to residue 128 is depicted. The Lysobacter–C fragments L–1 and L–2 and the tryptic
fragment T–1 allowed the extension of the N–terminal sequence
previously reported (2) up to 128 residues.

C. Immunocytochemical distribution of 7B2 in pituitary and brain

The unlabeled antibody (PAP) technique of Sternberger (5) was used
for light and electron microscopy. Co-localization studies were performed on serial 5 µm sections of female rat anterior pituitary which were
stained with antibodies to rat β–LH (Fig. 4A) and β–FSH (Fig. 4C) (a gift
from Dr. Parlow, NIH, USA), and compared to complementary slices stained
with 7B2 antiserum (Figs. 4B, 4D). Thus, LH and FSH co-localized consistently with 7B2, while some cells were stained uniquely with 7B2 antiserum. Preadsorption of 7B2 antiserum with LH and FSH did not affect 7B2
staining. Each antiserum, preadsorbed with an excess of respective
antigen, failed to produce any staining.

Preembedding immunocytochemistry for electron microscopy, performed
as in (6) (Fig. 4E), has revealed that 7B2 is localized within FSH–like
cells. These cells contain both LH in small, and FSH in big, vesicles as
shown in (7). Only small secretory vesicles, 180–250 µm, were specifically labeled with 7B2 antiserum.

Neurons and fibers containing immunoreactive 7B2 were localized in
rat brain. The regions of the hypothalamus containing a higher concentration of cell bodies are shown in panel 5A; these include the paraventricular (PVN) and supraoptic (SON) nuclei. The fibers were highly concentrated in whole hypothalamus, but only a few were visible in the
thalamic (Th) nuclei.

Finally, Figure 5B shows that in the median eminence (ME), some highly immunoreactive fibers and varicosities cross both inner (MEi) zone and the external (MEe) portal capillary-containing zone.

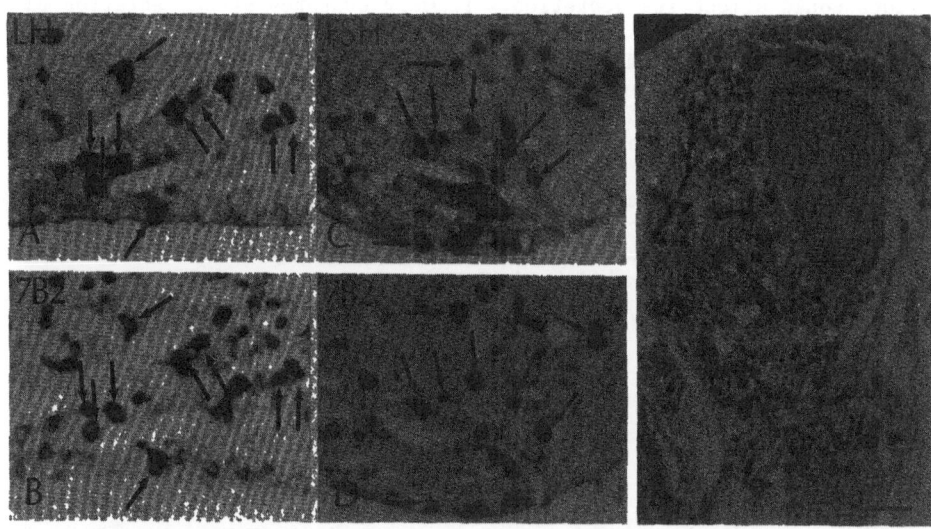

Fig. 4 Co-localization of LH (A), FSH (C), with 7B2 (B,D), in rat anterior pituitary lobe. Arrows indicate cells which label with both LH and 7B2 antibodies (A,B) or FSH and 7B2 antibodies (A,B) or FSH and 7B2 antibodies (C,D). Magnification x250. In (E), ultrastructural localization of 7B2. Immunoreaction is confined to small 180 to 250 nm vesicles (small arrow) while big 400 to 700 nm vesicles (big arrow) where unstained. N = nucleus. Magnification x6600. Bar = 1 μm.

Fig. 5 (A) Highly labeled hypothalamic regions (Hy). PVN, paraventricular nucleus; SON, supraoptic nucleus. Only slight staining was observed in thalamus (Th). No immunoreaction was found in fornix (F) and chiasma optica (ChO). 3v, third ventricle. Magnification x20. (B) Portion of median eminence. Fibers and varicosities are seen (arrows) both in inner (MEi) and external (MEe) zones. Magnification x250.

D. Specific release of 7B2 by LHRH

Based on the above mentioned co-localization of 7B2 with both LH and FSH in anterior pituitary cells, it was then decided to investigate whether LHRH would also release 7B2 and whether other hypothalamic factors would affect its secretion. Primary cultures of rat anterior pituitary cells (8,9) were incubated for 3 or 5 hrs with various concentrations of either luteinizing hormone releasing hormone LHRH, ovine corticotropin releasing factor oCRF, or human pancreatic growth hormone releasing factor, hpGRF (obtained from Dr. N. Ling, Salk Institute, La Jolla, California). Subsequently, the concentrations of immunoreactive 7B2 (IR-7B2), adrenocorticotropin (IR-ACTH) and growth hormone (IR-GH) released into the medium were quantitated by specific radioimmunoassays (3,8,9). As shown in Fig. 6A, LHRH effectively stimulates the release of 7B2, LH and FSH from normal rat anterior pituitary cell cultures. In contrast, oCRF or hpGRF, while stimulating the release of ACTH or GH respectively (see Figs. 6B, C), do not alter the basal release of 7B2 in vitro.

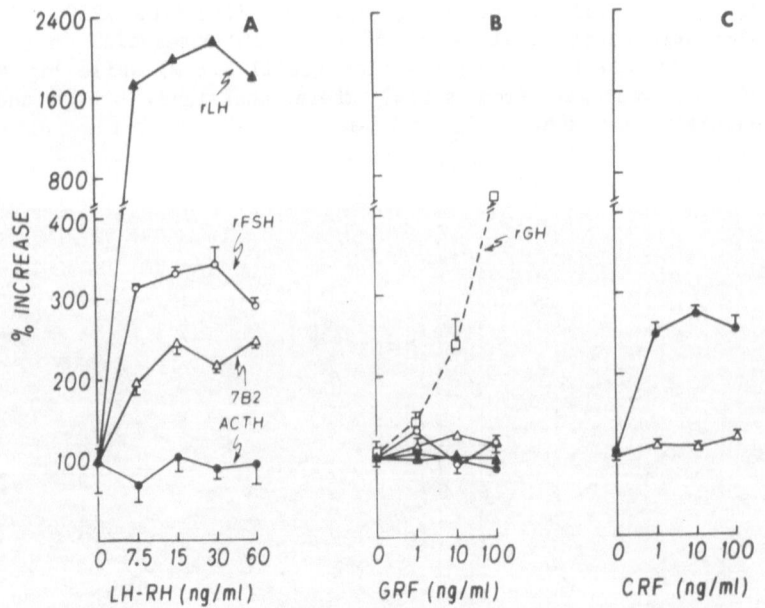

Fig. 6 Dose-dependent response of rat (r) 7B2, LH, FSH and GH to LHRH (A), hpGRF (B) and oCRF (C) in rat anterior pituitary cell cultures. Vertical lines represent the standard errors of the means of the response (n = 3). The basal levels of IR-7B2, IR-FSH, IR-LH and IR-ACTH were 153 ± 22 pg/ml, 3,130 ± 120 pg/ml, 4,820 ± 590 pg/ml and 3,628 ± 717 pg/ml respectively in Fig. 6A. In Fig. 6B, the basal levels were 115 ± 6 pg/ml, 1,290 ± 230 pg/ml, 3,240 ± 298 pg/ml and 4,905 ± 1,057 pg/ml respectively. The basal levels of IR-7B2 and IR-ACTH were 159 ± 40 pg/ml respectively (Fig. 6C).

DISCUSSION

The purification of human 7B2 proved to very difficult using conventional chromatographic procedures. This is partly due to the very minute quantities of this protein within pituitary tissues. Therefore, the development of a specific antiserum against a synthetic region of 7B2 (segment 23-39) permitted the establishment of efficient reverse-phase HPLC procedures for the purification of sufficient quantities for sequence purposes. This methodology initially involved the use of large C-18 columns (2.5 x 30 cm) for the handling of kilogram quantities of human pituitary extracts. Monitoring column efluents by RIA allowed the detection of IR-7B2 and the repurification of the resulting immuno-positive fractions using a succession of reverse-phase columns eluted under different conditions (Fig. 1). This represents one of the proteins purified in our laboratory solely and completely by RP-HPLC.

The proposed amino terminal 128 residue sequence for human 7B2 was extensively compared to that of any known sequence in the National Biomedical Foundation data base (Georgetown University, Washington, D.C., USA). Out of a total of 3,309 protein sequences, representing a total of 738,997 residues searched, and included in 1,077 protein superfamilies, it was found that this sequence is unique. This confirmed earlier work classifying 7B2 as belonging to a new protein superfamily (2). Considering the previously reported Mr of 21,000 for 7B2, the total number of remaining amino acids is expected to be in the order of 40 to 60. Based on the previously reported amino acid composition of human 7B2 (2), and that calculated from the 128 amino acid sequence presented, the positions of 6 other Lys residues remain to be determined. This might explain the difficulty we found in isolating either Lysobacter-C or Trypsin derived fragments, long enough to complete the sequence.

So far the physiological role of 7B2 remains obscure. The tissue and cellular distribution of this molecule has been studied in detail in order to define its sites of synthesis and its projections either in endocrine or neural tissues. This work has demonstrated that 7B2 is widely distributed in various tissues. Quantitative radioimmunoassays demonstrated that the pituitary represents the organ in which 7B2 is present in highest amounts (3). However, the cellular localization of 7B2 within the pituitary permitted the unambiguous demonstration of the presence of 7B2 in gonadotrophs of the anterior lobe. Furthermore, electron microscopic evidence reveals that it is a component of secretory granules in these same gonadotrophs. Its presence in the posterior pituitary and hypothalamus (2), led us to study in more details its hypothalamic cellular localization. It was thus shown to co-localize with vasopressin in the supraoptic and paraventricular nuclei (10). The fact that 7B2 co-localizes with LH and FSH in the anterior pituitary, led us to study their possible co-release from anterior pituitary cells in culture following incubation with the specific gonadotroph releasing factor, LHRH. Our results show that indeed 7B2 was release in a dose dependent manner in response to an LHRH challenge. However, no significant change in the basal release of IR-7B2 was observed when these same cells were stimulated by oCRF or hpGRF at a dose as high as 100 ng/ml. This specific response indicates that the secretion of this protein from the anterior pituitary may be mediated by LHRH in vivo, as are LH and FSH.

Recent data demonstrate the presence of 7B2 in pancreas and in spinal cord (11,12). Ontogeny studies show that in pancreas (11) and in pituitary 7B2 levels are elevated at parturition and just before the onset of puberty. These results should stimulate interest in the physiology of this novel highly conserved protein, in order to illuminate its possible role in reproductive biology. The wide distribution of 7B2 in various brain cells should also help in delienating its function(s) in this organ.

REFERENCES

1. K.L. Hsi, N.G. Seidah, G. De Serres, and M. Chrétien, Isolation and NH2-terminal sequence of a novel porcine anterior pituitary polypeptide: Homology to proinsulin, secretin and Rous sarcoma virus transforming protein TVFV60, FEBS Lett. 147:261 (1982).

2. N.G. Seidah, K.L. Hsi, G. De Serres, J. Rochemont, J. Hamelin, T. Antakly, M. Cantin, and M. Chrétien, Isolation and NH2-terminal sequence of a highly conserved human and porcine pituitary protein belonging to a new superfamily: Immunocytochemical localization in pars distalis and pars nervosa of the pituitary and in the supraoptic nucleus of the hypothalamus, Arch. Biochem. Biophys. 225:525 (1983).

3. H. Iguchi, J.S.D. Chan, N.G. Seidah, and M. Chrétien, Tissue distribution and molecular forms of a novel pituitary protein in the rat, Neuroendocrinology 39:453 (1984).

4. C. Lazure, N.G. Seidah, M. Chrétien, R. Lallier, and S. St-Pierre, Primary structure determination of Escherichia coli heat-stable enterotoxin of porcine origin, Can. J. Biochem. Cell Biol. 61:287 (1983).

5. L.A. Sternberger, in: "Immunocytochemistry", L.A. Sternberger, ed., Wiley, New York (1979).

6. M. Marcinkiewicz and C. Bouchaud, Formation and maturation of axo-glandular synapses and concomitant changes in the target cells, Biol. Cell (in press).

7. K. Inoue and K. Kurosumi, Ultrastructural immunocytochemical localization of LH and FSH in the pituitary of the untreated male rat, Cell Tissue Res. 235:77 (1984).

8. J.S.D. Chan, C.L. Lu, N.G. Seidah, and M. Chrétien, Corticotropin releasing factor (CRF): Effects of the release of pro-opiomelanocortin (POMC)-related peptides by human anterior pituitary cells in vitro, Endocrinology 111:1388 (1982).

9. J.S.D. Chan, L. Gaspar, H. Iguchi, N.G. Seidah, N. Ling, and M. Chrétien, Synergistical effects of ovine corticotropin-releasing factor (CRF) and arginine-vasopressin (AVP) on the release of pro-opiomelanocortin (POMC) related peptides by pituitary adenoma of a patient with Nelson's syndrome in vitro, Clin. Invest. Med. 7:205 (1984).

10. M. Marcinkiewicz, S. Benjannet, N.G. Seidah, M. Cantin, and M. Chrétien, Immunocytochemical localization of a novel pituitary protein "7B2" in rat brain and pituitary gland, Fifth International Washington Spring Symposium, Washington, D.C., May 28-31, Abstract #153 (1985).

11. H. Suzuki, N.D. Christofides, M. Chrétien, N.G. Seidah, J.M. Polak, and S.R. Bloom, Developmental changes in the immunoreactive content of a novel pituitary protein (7B2) in the human pancreas and its identification in pancreatic tumours, Diabetes (in press).

12. H. Suzuki, N.D. Christofides, P. Anand, M. Chrétien, N.G. Seidah, J.M. Polak, and S.R. Bloom, Regional distribution of a novel pituitary protein (7B2) in the rat spinal cord: Effect of neonatal capsaicin treatment and thoracic cord transection, Neurosci. Lett. 55:151 (1985).

C. SUMMARY OF SYMPOSIUM WORKSHOP

THE SYMPOSIUM TEST PEPTIDE (STP): SYNTHESIS AND CHARACTERIZATION OF A MODEL PEPTIDE

J. Rivier, R. Galyean, W. Woo, D. Karr, T. Richmond and
J. Spiess

Clayton Foundation Laboratories for Peptide Biology
The Salk Institute
10010 N. Torrey Pines Road
La Jolla, California 92037

ABSTRACT

A 19-residue polypeptide with the sequence: H-Trp-Asn-His-Ala-Ala-Ala-Thr-Ser-Gln-Ala-Pro-Arg-Val-Ile-Ile-Glu-Glu-Asn-Asp-OH has been synthesized using Merrifield's solid phase approach, purified by preparative reverse-phase high pressure liquid chromatography (HPLC), tested for purity in several HPLC systems and characterized by analysis of its amino acid composition and sequence.

INTRODUCTION

The following 19-peptide (called Symposium Test Peptide:STP) was assembled on a Merrifield type resin (1) using N$^{\alpha}$ Boc protection and the side chain protecting groups shown in parenthesis: H-Trp-Asn-His(Tos)-Ala-Ala-Ala-Thr(Bzl)-Ser(Bzl)-Gln-Ala-Pro-Arg(Tos)-Val-Ile-Ile-Glu(Obzl)-Glu(Obzl)-Asn-Asp(OBzl)-OCH$_2$-resin, cleaved and fully deprotected in HF (2) and purified using preparative HPLC (3). This peptide was to be used as a model for evaluation of the sensitivity of the presently available characterization techniques including column chromatography, amino acid analysis, sequence analysis, and mass spectrometry. Thirteen of the naturally occurring amino acids are present, all in the L-configuration. No D amino acid was introduced since their natural occurrence is relatively limited. For long term stability of the preparation, methionine or cysteine were not included. Since this peptide was to be used for direct characterization without prior derivatization or enzymic treatment, cysteine or pyroglutamyl and acetylated N-termini (all naturally occurring substitutions) were not considered. This peptide is relatively hydrophilic with three basic functional groups including the imidazole and guanidino side chains and four acidic groups, the latter assembled toward the C-terminus to favor retention in the sequencer reaction chambers. As a challenging feature for amino acid analysis, we introduced the sequence -Val-Ile-Ile- which will not be fully hydrolysed under the usual (110°C, 24 h) acidic hydrolysis conditions and thus requires kinetic investigation. To allow for the determination of carryover, we introduced a series of alanine, isoleucine and glutamic acid residues, respectively. Threonine-7 and serine-8 were added because these two amino acids are usually derivatized to multiple forms during Edman degradation and thus often limit the overall sensitivity of sequence analysis. An alanine was introduced in position 10 to allow repetitive yield calculations on the basis of the yields of alanine-4 and alanine-10. The sequence of proline-11 and arginine-12 was designed to permit testing of the efficiency of the cleavage conditions during Edman degradation. The C-terminal sequence (-Glu-Asn-Asp-OH or END) should have alerted the investigator that the C-terminus was reached.

During synthesis, this 19-peptide presented coupling difficulties associated with some putative steric hindrance at the level of the Arg-Val-Ile-Ile sequence or some recognized chemical rearrangements associated with the glutamic residues during HF cleavage (4). Other recognized problems are associated with racemisation of histidine during coupling (5) or the introduction of asparagine and glutamine which we coupled unprotected in the presence of two equivalents of HOBt (6).

EXPERIMENTAL

Synthesis. Boc-Asp(OBzl) was attached to a chloromethyl resin from Lab Systems (0.9 mM Cl/g resin) using the method of Horiki et al. (7). Substitution was approximately 0.4 meq/g resin. The peptide was assembled manually on a starting resin (10 g) utilizing standard procedures (8) which included TFA (60% in CH_2Cl_2) for deblocking of the BOC protecting group, neutralization with Et_3N, washes with isopropanol, MeOH and CH_2Cl_2. Coupling was achieved with dicyclohexyl carbodiimide and HOBt in most cases. Couplings were carried out for an average 2 h. Completion was monitored by the ninhydrin test of Kaiser et al. (9). Isoleucine-15, valine-13, arginine-12, alanine-10, serine-8, alanine-4, and trypotophan-2 were recoupled once, histidine-3 and asparagine-2 were recoupled twice. The chain was not terminated by acetylation at any point. Protecting groups for side chains are those shown in the introduction. All protected amino acids were purchased from Bachem (Torrance, California). Boc-His(Tos) was liberated from its dicyclohexylammonium salt and crystallized just prior to usage. Final weight of the dried resin peptide was 18 g, indicating a weight gain of approximately 8 grams.

The cleavage and deprotection of the resin-peptide (9 g) was accomplished by treatment with approximately 130 ml HF for 1 h at 0°C in the presence of cresol (ca 2 g/g resin) as scavenger. After removal of HF under vacuum, the peptide/resin mixture was treated with anhydrous diethyl ether, filtered and taken up in dilute acetic acid (AcOH) and lyophilized. Crude peptide (6g), an HPLC analysis of which showed one main peak as well as a significant number of impurities, was purified using preparative HPLC techniques as previously described (3); gradient of acetonitrile (ACN) applied to the preparative cartridge in TEAP 2.25 was 12 to 27% in 1 h; analysis of the fractions was achieved using isocratic conditions: 19.8% ACN in .1% TFA. Desalting was carried out using an ACN gradient from 6 to 36% ACN in 30 min. Purified fractions were lyophilized yielding STP as the trifluoroacetate (550/mg.

Characterization. STP was characterized using the bulk preparation.

a) Molecular weight was calculated to be 2121.9. Molecular ion, was 2122 when using fast atom bombardment mass spectrometry with a JEOL HX-100HF mass spectrometer utilizing a 3 KV xenon atom primary beam. Sample (10 nmole) was dissolved in 1 µl glycerol on a 1.5 x 6 mm stainless steel sample stage. Fragment ions were consistent with the reported sequence (personal communication from Dr. T. Lee, City of Hope).

b) Optical rotation was measured in 1% AcOH $[\alpha]_D^{20}$ = -97.9 (C = 1).

c) Reverse phase HPLC in several systems (Table I and Fig I) indicated that the preparation consisted of one major component (>95%) and several small impurities (<1.5% each).

TABLE 1. HPLC ANALYSIS OF STP

Solvent Systems A	B	Gradient Conditions	Retention Time (min)
TEAP 2.25	60% ACN in A	20%B 30' 35% B	16.45
TEAP 7.0	"	"	20.60
.01% HCl,.1% KCl	60% ACN/H_2O,.01% HCl,.1% KCl	"	20.29
.1% TFA	60% ACN/H_2O,.1% TFA	25%B 20' 35% B	15.04

Liquid chromatograph was from Waters Associates with Kratos model 757 variable UV detector set at 210 nm (0.2 AUFS). Load STP were 5,20,20 and 10 µg resp. Flow rate was 2 ml/min. Vydac C_{18} (5µ) column 25 x 0.46 cm was used at room temperature.

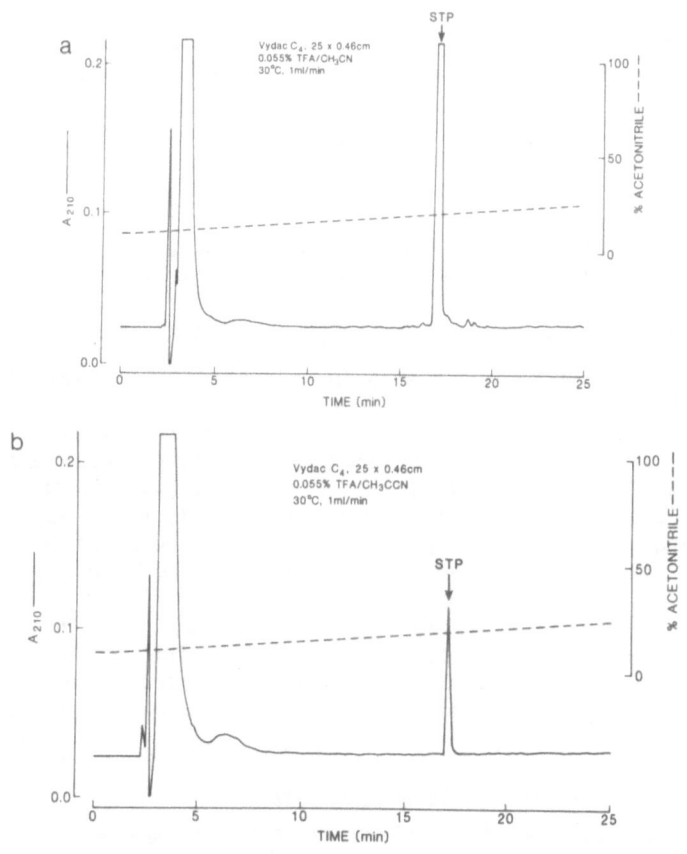

Fig. 1 (a: top; b: bottom). Liquid chromatograph was a
         Hewlett Packard. Load was 10 µg, 4.7 nmol top
         panel and 1 µg, 0.47 nmol bottom panel. % ACN
         was 9 to 24.7 in 25 min.

TABLE 2.   AMINO ACID ANALYSIS OF STP

| Amino | MSA Hydrolysis | | | Edman | |
Acid	24 hr	48 hr	72 hr	Degradation	Expected
Asx	2.98(3)			3	3
Thr	0.97(1)			1	1
Ser	0.93(1)			1	1
Glx	3.11(3)			3	3
Pro	1.14(1)			1	1
Gly	0.10(0)			0	0
Ala	4.01(4)			4	4
Val	0.48(0)	0.77(1)	0.79(1)	1	1
Ile	1.13(1)	1.58(2)	1.69(2)	2	2
Lys	0.16(0)			0	0
His	1.01(1)			1	1
Trp	0.66(1)			1	1
Arg	1.08(1)			$\underline{1}$	$\underline{1}$
				19	19

d)   The amino acid composition of STP was obtained after methane
sulfonic acid hydrolysis using a Beckman 121MB Amino Acid Analyzer.
Norleucine was used as an internal standard to calculate concentrations
with the Beckman Model 126 Data System (for details see 10).   Hydrolysis
experiments were performed for 24, 48 and 72 hours at 110°C under vacuum
in sealed glass tubes to determine the changes in concentration.   Complete
hydrolysis of the valine and isoleucine containing partial peptide was not
reached even after 72 hours (Table 2).

e)   Sequence analysis was performed using 300 pmoles and 10 nmoles STP
in order to confirm the sequence and purity of the preparation:

i.   sequence determination with 300 pmoles of STP

STP (300 pmol) was degraded in an Applied Biosystems 470A Protein
Sequencer. Degradation was carried out in the presence of 2 mg Biobrene.
Each cycle sample (33% of total) was subjected to Phenylthiohydantoin
(PTH) amino acid analysis.   PTH amino acid standards were eluted with
reverse phase HPLC on a Dupont Zorbax PTH column (Woo, W. and Spiess, J.
in preparation).   The Edman degradation yielded 19 amino acids (Fig. 2)
with only one amino acid per cycle. No preview was observed and percentage
of carryover was normal except for alanine in cycle 5 and cycle 6.   The
butyl chloride (S3) extraction of the anilinothiazolinone derivative of
basic amino acids, such as arginine and histidine from the filter,
resulted in low yields.   Low yields for threonine and serine were
attributed to the formation of the respective dehydro forms which were not
well detected at 266 nm.   The sequence data agreed with the amino acid
analysis of STP.

ii. Determination of purity (10 nmol of STP)

STP (10 nmol) was then subjected to Edman Degradation to determine
purity of the peptide.   All degradation conditions were identical to that
obtained from the 300 pmol STP degradation except for a TFA treated filter
and the presence of 0.27 mg sodium chloride in the Biobrene (as suggested
by ABI bulletin 12).   PTH-amino acid analysis was performed on 10% of each
cycle sample.   The recovery of histidine was improved to 60% while the
arginine yield remained low.   A 2-4% preview may have been a result of

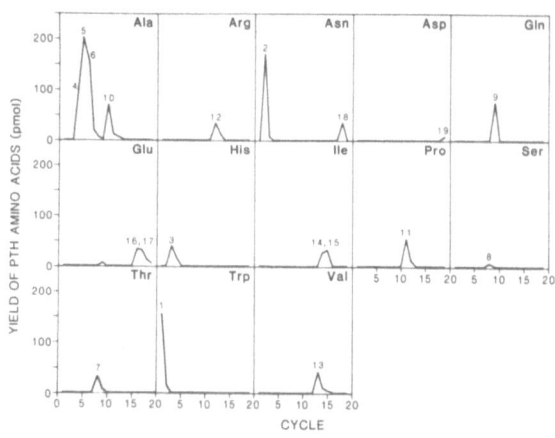

Fig. 2. Automatic Edman degradation of 300 pmol STP
in the ABI 470A Protein Sequencer. Total
yields of PTH-amino acids per Edman cycle
are presented.

excessive TFA cleavage or a small amount of tryptophan deleted peptide.
On the basis of these sequence data and in agreement with the chromatographic results (Fig. 1a, 1b), the investigated preparation contained
the 19 residue STP with a purity of at least 95%.

f)   Aliquots of STP for distribution were obtained as follows:
microfuge 1.5 ml tubes (Eppendorf, catalogue No. 2236 411-1) were washed
with aqueous 10% nitric acid, subsequently bidistilled water, methanol
(HPLC grade, Burdick & Jackson), again with bidistilled water and finally
with methanol (see above). The tubes were then evaporated in a Savant
Speed Vac Concentrator. Subsequently, 40 µl of an STP stock solution,
which contained 10 pmol of STP per µl according to amino acid analysis,
was added to each tube. For pipetting, an oxford sampler (10-50 µl range)
was used. Each tube was then evaporated in a Savant Speed Vac Concentrator before distribution. This last step of drying the sample was only
undertaken for economic reasons. Normally, we send small amounts of
peptide (especially if it represents a valuable natural product) as frozen
solution.

g)   Peptide recovery from sample tubes was examined. Aliquot samples
from the batch of distributed tubes stored in the laboratory at room
temperature were extracted twice with aqueous 1 M acetic acid (100 µl).
The extracts of one tube were immediately transferred to a hydrolysis tube
containing norleucine as internal standard, dried in a Savant Speed Vac
Concentrator and subjected to amino acid analysis as described above.
Whereas initially approximately 400 pmol of peptide was extracted, the
extractable amount of peptide decreased rapidly within two weeks after
sample preparation and then more slowly over the following 6 weeks (Fig.
3). The amount of extractable peptide was found to be always above the
250 pmol level. This was also true for samples which were sent back to
our laboratories from both New Zealand and Chicago at least two months
after aliquoting.

The average recovery for these samples was above 250 pmol. The lowest
amount of extractable peptide was found to be 220 pmol. As many as 41
samples were tested in these recovery experiments. None of these experiments revealed any significant contamination with foreign amino acids as

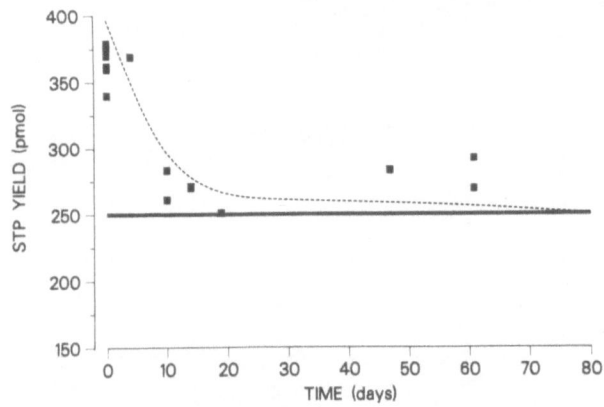

Fig. 3.   STP recovery from randomly selected sample
tubes.   Total yields were determined by
amino acid analysis as described in the text.

demonstrated by the observation that all analyses satisfied the expected
amino acid composition of STP with the exception of three analyses which
showed a too high relative content of glycine.

It is concluded from the results of these experiments that at least
220 pmol of STP was available for every Workshop participant who received
a peptide sample.   This amount was thus within the desired range for
microanalytical experiments to be presented at the Workshop.

h)   The stability of the peptide preparation was tested with reverse-
phase HPLC.   Two months after aliquoting, a sample was recovered from an
Eppendorf cap as described above, and applied to reverse-phase HPLC on a
Vydac C_{18} column using a gradient (15% to 30.67% in 24.5 min) of acetonitrile in 0.055% TFA.   Two peaks were detected at 210 nm.   The first
eluting peak (23% of total), had the same retention time as STP; the
second peak may be a degradation product, the structure of which was not
investigated.

DISCUSSION

We started with the objective to design and synthesize a polypeptide
named Symposium Test Peptide (STP) which would be unknown to all Workshop
participants, could serve as model peptide for microanalytical studies and
thus could help to facilitate discussion about the quality of microanalytical procedures such as manual and automatic Edman degradation,
amino acid analysis and mass spectrometry.   Every structural feature which
we introduced into STP represented a problem which we have had to face in
the characterization of various natural products.

STP was thoroughly analyzed in our laboratories by reverse-phase HPLC,
amino acid analysis and automatic Edman degradation on low and high levels
of peptide applied to the sequencer.   On the basis of these experiments,
we assume a purity of approximately 95% for this peptide.   That STP was
found to be partially converted to another product upon storage at room
temperature was not foreseen.   However, this property does not reduce its
value as a model peptide because similar conversions may also occur with
natural products.

The extractable amount of STP which was expected to be available to every Workshop participant was in the range of 200-300 pmol of peptide as requested by the Organizing Committee of the Symposium and the organizers of the Workshop. This amount was thought to be significantly larger than that required by currently available analytical instrumentation. Nevertheless, it was expected that handling of this amount for preparation of microanalytical experiments would be a challenging task as is the case in the handling of small amounts of natural product.

Synthesis of a peptide specifically designed for an international protein chemical symposium was a novel undertaking. It represented an experiment to promote communication between protein chemists. The results of this experiment are summarized elsewhere in this book by J. L'Italien and J. Spiess.

ACKNOWLEDGEMENTS

Research was supported by NIH Grant AM 26741. Research was conducted in part by the Clayton Foundation for Research, California Division. Joachim Spiess is a Clayton Foundation Investigator. We thank Dr. T. Lee for analyzing the stock preparation of STP by mass spectrometry after the symposium. We are indebted to J. Dykert for technical assistance and R. Hensley for manuscript preparation.

REFERENCES

1. Merrifield, R.B. (1963) Solid phase peptide synthesis. I. The synthesis of a tetrapeptide. J. Am. Chem. Soc. 85, 2149-2154
2. Sakakibara, S., Shimonishi, Y., Kishida, Y., Okada, M., and Sugihara, H. (1976) Use of anhydrous hydrogen fluoride in peptide synthesis. I. Behavior of various protective groups in anhydrous hydrogen fluoride. Bull. Chem. Soc. Jpn. 40, 2164-2167
3. Rivier, J., McClintock, R., Galyean, R. and Anderson, H. (1984) Reversed phase HPLC: Preparative purification of synthetic peptides. J. of Chromatography 288, 303-328
4. Feinberg, R.S. and Merrifield, R.B. (1975) Modfication of Peptides containing glutamic acid by hydrogen fluoride-anisole mixtures. γ-acylation of anisole or the glutamyl nitrogen. J. Am. Chem. Soc. 902, 3485-3491
5. The Peptides: Analysis, Synthesis, Biology, Gross & Meienhofer, Vol. 2, p.179, (1980)
6. Koenig, W., and Geiger, R. (1970) New method for the synthesis of peptides: activation of the carboxyl group with dicyclohexyl-carbodiimide by using 1-hydroxybenzotriazoles as additives. Chem. Ber. 103, 788-798
7. Horiki, K., Igano, K., and Inouye, K. (1978) Amino acids and peptides. Part 6. Synthesis of the Merrifield resin esters of N-protected amino acids with the aid of hydrogen bonding. Chem. Lett., pp. 165-168
8. Rivier, J., Kaiser, R., and Galyean R. (1978) Solid phase synthesis of somatostatin and glucagon selective analogs. Biopolymers 17, 1927-1938
9. Kaiser, E., Colescott, R.L, Bossinger, C.D. and Cook, P.I. (1970) Color test for detection of free terminal amino groups in the solid-phase synthesis of peptides. Anal. Biochem. 34, 595-598
10. Spiess, J., Villarreal, J. and Vale, W. (1981) Isolation and sequence analysis of a somatostatin-like polypeptide from ovine hypothalamus. Biochemistry 20, 1982-1988

WORKSHOP IN MICROANALYTICAL METHODS - EXPERIENCES WITH AN UNKNWON SYNTHETIC PEPTIDE

James L'Italien* and Joachim Spiess+

*Molecular Genetics, Inc.   +The Clayton Foundation
 Minnetonka, MN 55343   Laboratories for Peptide
                                Biology
                                The Salk Institute
                                La Jolla, CA 92037

ABSTRACT

This report summarizes the results obtained by investigators who took part in the structural characterization of an unknown peptide sample as a part of a workshop on micromethods of polypeptide characterization at the First Symposium of American Protein Chemists, San Diego, September 30 through October 2, 1985.

INTRODUCTION

One of the aims of the First Symposium of American Protein Chemists was to provide a forum for discussion of modern methods of structural analysis of polypeptides and of common problems encountered therein. To facilitate this goal, attendees at the symposium were invited to characterize an unknown polypeptide (made available through the symposium) and then to bring their results to the workshop discussion on micromethods of polypeptide oharacterization. This paper will summarize the data that was presented at the symposium workshop and collected in a subsequent written survey.

RESULTS

Of the 150 peptide samples which were sent to conference attendees at their request, only 26 investigators reported their results at the symposium workshop or responded to our written survey. The methods of polypeptide characterization which were used by those who reported their results are summarized in Table I. This table shows that the Edman degradation was the most frequently used method of characterization used by 71% of those reporting) and, as will be described later in Table III, gave the most complete results. It is interesting to note that only 42% of those reporting employed multiple methods of characterization. Seventy percent of those using multiple methods employed amino acid analysis and Edman degradation.

Table I   Summary of Methods Used to Characterize the Symposium Test Peptide (STP)

Methods of Characterization	Number of Reports	Usage of Specific Methods				Multiple Methods
		HPLC	FAB	A	SA	
RP-HPLC only	3	3				
RP-HPLC/AAA/Edman Degrad.	1	1		1	1	1
FAB only	1		1			
RP-HPLC/AAA/FAB	1	1	1	1		1
RP-HPLC/Edman Degrad.	3	2			2	2
AAA only	2			2		
AAA/Edman Degrad.	6			6	6	6
Edman Degrad. only	9				8	
Totals	26	7	2	10	17	10

List of abbreviations used: RP-HPLC or HPLC - reverse phase high performance liquid chromatography; AAA or A - amino acid analysis; Edman Degradation or SA - amino terminal polypeptide-sequence analysis using the Edman Degradation (Phenylisothiocyanate) Chemistry; FAB - fast atom bombardment mass spectrometry.

Table II Summary of Results Reported for Characterization of Symposium
Test Peptide by Amino Acid Analysis

	Hydrolysis Method	Analysis Method		Sample		Recovery pmol	Results
		Column	Identification	%	pmol		
1	HC1	RP–HPLC	PTC	10	14	140	Not Reported
2	HCL	RP–HPLC	PTC	25	–		Injected only 10% of Hydrolysate which was below integration level
3	HC1	RP–HPLC	PTC	10	37.5	375	2B, T, 2S, 3Z, P, 2G, 4A,
	MES	RP–HPLC	PTC	10	37.5		V, 2I, K, H, W, R
4	HC1	ION EX.	Post Col. Fluorescamin	10	35	350	3B, T, 1.45, 3.4Z, G, 3.8A, 0.9V, 1.6I, 1.1H, 1R; W + P Not Det.
5	HC1	ION EX.	PC-OPA	2X5	<2	<40	No Results
6	HC1	ION EX.	PC-Nin	25	51	204	3.2B, 1.6T, 2.25S, 3.9Z, 1.9P, 4A, 1.2V, 0.6I, 0.5R, +H
7	HC1	ION EX.	PC-Nin	40	120	300	2B, T, 2S, 4–5Z, P, 3G, 4A, 0–1V, 1–2I, 0–1L, H, 0–1R
							3B, T, S, 3Z, P, 4A, V, 2I, H, R, W
8	HC1 24hr.	ION EX.	PC-Nin	2X12.5			Nanomole Quantities of Amino Acid
	72hr.	ION EX.	PC-Nin	2X12.5			Nanomole Quantities of Amino Acid
9	HC1	ION EX.	PC-Nin	60% of each of 2 pks			Not Reported
10	HC1	ION EX.	PC Fluorescamine				Not Reported
11	HC1	ION EX.	PC-Nin	10			Not Reported

List of abbreviations: HCL–hydrochloric acid; MES–methane
sulfonic acid; RP-HPLC–reverse phase high performance liquid
chromatography; ion ex.-ion exchange chromatography; PTC-
precolumn phenylthiocarbamyl derivatization of amino acids;
Post Col. Fluorescamine or P.C. Fluorescamine-post column
fluorescamine derivatization of amino acids; PC-OPA – post
column o – phthaldialdehyde derivatization of amino acids;
PC-Nin – post column ninhydrin derivatization of amino acids;
pmol – picomole; the standard single letter code for the
amino acids.

Table III Summary of Results Reported for Characterization of the
Symposium Test Peptide by Automated Edman Degradation

	Seq. Method	% Sample	PTH Id.	Yield Cycle 1 (pmol)	Residues Identi-fication	Results
1	GP	10%	online HPLC	K7.5	19/19	WNHAAATSQAPRVIIEEND
2	GP	75%	HPLC	NR	11/19	WNHAAATSQAP...
3	GP	80%	HPLC	39	12/19	WNHAAATXEAPR(V)...
4	GP	50%	HPLC	NR	19/19	WNHAAATSQAPRVIIEEND
5	GP	80%	HPLC			No Results
6	GP	50%	HPLC	18	11/19	WNHAAATSQAP...
7	GP	37.5%	online HPLC	21	18/19	WNHAAATSQAPRVIIEEN(N/D)
8	GP	100% each of 2 HPLC Peaks	online HPLC	6.5 11.2	18/19 18/19	WNHAAATSQAPRVIIEENX WNHAAATSQAPRVIIEEN(D/N)
9	GP	25%	HPLC	30	15/19	(W)NHAAAT(S)QAPXVII(E/I)E(N)
10	GP	100%	HPLC	<10	17/19	WNHAAXTSQAPRVIIEEN
11	GP	100%	online	141	19/19	WNHAAATSQAPRVIIEEN(D)
12	GP	100%	HPLC	88	19/19	WNHAAATSQAPRVIIEEND
13	GP LP	50% 50%	HPLC HPLC			No Results No Results
14	LP	100%	HPLC			No Results
15	LP	100%	HPLC	NR	6/19	WXHAAXTXQ...
16	SP	100%	HPLC			No Results
17	GP	100%	online HPLC	149	19/19	WNHAAATSQAPRVIIEEND
18	GP	100%	HPLC	2	19/19	WNHAAATSQAPRVIIEEND

List of abbreviations: Seq. Method – sequencing method used to characterize the polypeptide samples; GP – gas-phase Edman degradation; LP – liquid phase Edman degradation; SP – solid-phase Edman degradation; % sample – percentage of sample used to characterize polypeptide by Edman degradation; PTH Id. – method used to identify the PTH's resulting from the automated sequencer; HPLC – high performance liquid chromatography; online HPLC – the HLPC used for PTH identification was directly attached to the polypeptide sequencer such that no operator interventions was required between the Edman degradation and PTH identification steps.

As described in the accompanying paper "The Symposium Test Peptide: Synthesis and Characterization of a Model Peptide" by Rivier et al., a random survey of the samples prepared by the organizers demonstrated by amino acid analysis that >200 pmol of symposium test peptide (STP) should have been present in each sample sent to the participants. Of 11 investigators who performed amino acid analysis, four recovered >200 pmol of STP, one found 140 pmol, two observed <40 pmol, one obtained only qualitative results due to operator error and two did not report results. In contrast with these findings, one investigator observed nanomole amounts of STP. Those investigators who reported results employed >30 pmol of STP to pre- or post column techniques (Table II). None of the amino acid analyses performed provided the true amino acid composition of STP. Actually, the results (Table II) were only poorly correlated with the true composition.

There were two reports of peptide characterization by FAB-MS. One of these employed FAB-MS directly on 30% of the sample, to obtain two species of molecular weights of 2121.99 and 2134.02. It was suggested that both species were approximately equally abundant. In a second experiment, 100% of the peptide sample was applied to reverse phase HPLC and two major species were collected. One third of each of the collected species was subjected to FAB-MS and the remaining 2/3 was used for amino acid analysis. These species corresponded to the molecular weights previously determined. Unfortunately, the results from amino acid analysis were inconclusive and could shed not light on the nature of the mass difference.

Three investigators performed RP-HPLC only. Two of these reports were only verbal but the third for which we received documentation also showed two peaks. In fact, two peaks were described in each of five RP-HPLC reports for which extensive data was given. Only two of these performed Edman degradation on each HPLC purified peak. These results, which are summarized as samples 8 and 17 in Table III show that the only discernable difference between the species was in the amount of preview present.

Of the 18 investigators who subjected STP to Edman degradation, 14 used gas phase sequencers, two employed spinning cup sequencers, one applied a gas phase and a spinning cup sequencer and one worked with a solid phase sequencer. The most complete results were obtained by those investigators who used gas phase sequencers with on-line RP-HPLC for PTH-amino acid analysis (Table III). Those investigators (nine of 15) who worked with gas phase sequencers without on-line RP-HPLC (Table III), as a group obtained the second best results. Significant results from spinning cup sequencers and solid-phase sequences were not presented.

DISCUSSION

This was the first time that an unknown synthetic peptide (STP) was offered to the participants of an international meeting to perform microanalytical studies on it and compare data with other participants. Since STP was synthesized on the basis of methodological requirements, communication of structure analysis could take place without the fear of disclosure of valuable biologic information.

The decision to dry the samples for shipment was exclusively based on cost considerations and certainly contrasted with the general experience that optimal recoveries of small amounts of peptide are accomplished when peptides are kept in a frozen solution during their transport. However, as mentioned above, it was established that at least 200 pmol of STP should have been available in each sample sent to the participants. On this background, the most striking observation at the workshop was the poor recovery of peptide, the contamination with foreign amino acids and the significant deviation of all presented amino acid analyses from the true composition of STP. In our opinion, these results could indicate significant problems in handling minute amounts of peptide and demonstrate at the current time, the lack of powerful methods of amino acid analysis operating in the range of 10-50 pmol of total peptide sample. Thus, the value of amino acid analysis on polypeptide samples in the low picomole range is questionable at this time. Perhaps information from this survey could be used to influence editors who look for compositional proof of each sequence analysis.

In contrast to the results of amino acid analysis, Edman degradation and mass-spectormetry (as FAB) produced significant information about STP. Edman degradation was the most frequently used and most successful method of characterization. Its superiority over amino acid analysis in the characterization of STP may at least in part be explained by the generally accepted observation that single amino acid contaminations are deleterious for amino acid analysis but relatively unimportant for (stepwise) Edman degradation beyond the first cycle. The rationale for this observation is that during Edman degradation, free-amino acids are usually either removed before the extraction of the anilino-thiazolinone derivative of the cleaved N-terminal amino acid or they are derivatized by PITC and appear in the first sequencer cycle. Thus, the majority of single amino acid contaminates are removed from the sample at the end of the first Edman degradation cycle. Such purification does not take place in the process of amino acid analysis. Even if longer con-taminating peptides are present, the careful qualitative and quan-titative analysis of the RP-HPLC of the PTH-amino acids of each cycle allows usually unambiguous identification of the residues of the main peptide by background subtraction.

As mentioned above, investigators using gas phase sequencers and on line HPLC were the most successful in obtaining substantial or complete sequence information. In our opinion, these results are indicative for the high sensitivity of gas phase sequencers. The results with online HPLC demonstrate that PTH-amino acid analysis represents another handling problem which can be circumvented by analysis with on-line-HPLC.

There were only two reports which employed FAB-MS to characterise the test peptides and thus general conclusions have to be drawn with caution. In both studies, the true molecular weight of STP was found, and thus it may be tentatively concluded that at least molecular weight determinations of peptides of the size of STP or smaller can be accomplished on a picomole level.

If one would have to design a general microanalytical approach of sequence analysis of small polypeptides (containing up to 30-40 residues) on the basis of the data provided at this workshop one would have to recommend to subject the peptide to FAB-MS and to automatic Edman degradation with a gas phase sequencer equipped with on line RP-HPLC.

CONTRIBUTORS

Tetrahydrofuran, 370
Thermolysin, 653, 758, 761
Thermostability, enzymic
  by engineering, genetic, 529–
      537
Thioglycerol, 256–258
Thiolester, 689
Thymidine and DNA, 76
Threonine, 217–219
Thrombin, 335, 471, 487
  -binding, 669–671
Thromospondin, 471–477
T-lymphotrophic virus, type III
      741–747
  and 4'-aminomethyl trioxalen
      744
  characterization, 742–744
  culture in H-*9* cell, 741–742
  inactivation, 742, 744
  polypeptide
    composition, 742, 744–746
    fractionation, 743–746
  proteins, 742–743
  purification, 741–742
Tomato glycoprotein, 16
Tooth pulp peptides, human,
      275–276, 278
Torpedo californica acetyl-
      cholinesterase, 599–605
Torpedo marmorata organ, elecric,
      607
Trace enrichment technique, 9,
      22, 39
Transferrin, 18, 19, 286
  receptor, 10, 22
Tridecapeptide
  synthetic, 388
  of yeast, 286
Triethylamine, 350
Trifluoracetic anhydride, 337,
      339, 340
Trifluoroacetic acid, 317, 318,
      323, 632, 775
Trinitrobenzene sulfonate, 553–
      568
  and ribulose bisphosphate
      carboxylase, 558
Triosephosphate isomerase, 430
      529–535
Troponin-C, 289
Trypsin, 47–49, 346, 348, 758
      -761, 766
  soybean inhibitor of, 110, 111,
      114, 115, 117, 124
  purity of commercial samples, 47
Tryptophan, 210, 217–219, 367, 379
Tyrosine kinase, 445

Ubiquitin, 445, 718, 721, 722
      726–729
Urea, 344, 346

Urea (continued)
  gel gradient, 503–507
    electrophoresis, 503–504
Urogastrone, 29–30
  and growth factor, epidermal, 30

Vaccinia virus polypeptide
  amino acid composition, 43
Valylhistidine, 156–159
Vapor phase hydrolysis, 188
Varanus gouldii, see Lizard
Vasopressin, 286
Venule, endothelial, 717
Vesicular stomatitis virus, 479
4-Vinylpyridine, 380

Weight, molecular, determination
  of peptide, 242–248, 253, 255,
      256, 258
Western blot technique, 121, 742

Yeast
  mating factor, 286
  tridecapeptide, 286
Yellow fever virus, 117
Yolk lipoprotein, low-density,
      772, 774, 776